GLENCOE

Automotive Excellence

 Glencoe
McGraw-Hill

New York, New York   Columbus, Ohio   Woodland Hills, California   Peoria, Illinois

NOTICE

Publisher and Author do not make any representation, warranty, guarantee or endorsement of any of the products, their use or the methods or techniques described in this book. Publisher and Author have not, and expressly disclaim any obligation to independently test any products or to verify facts, information, methods or techniques described in this book.

The reader is expressly advised to consider and use all safety precautions described in this book or that might also be indicated by undertaking the activities described herein. In addition, common sense should be exercised to help avoid all potential hazards.

Publisher and Author assume no responsibility for the activities of the reader or for the subject matter experts who prepared this book. Publisher and Author make no representation or warranties of any kind, including but not limited to, the warranties of fitness for particular purpose or merchantability, nor for any implied warranties related thereto, or otherwise. Publisher and Author will not be liable for damages of any type, including any consequential, special or exemplary damages resulting, in whole or in part, from reader's use or reliance upon the information, instructions, warnings or other matter contained in this book.

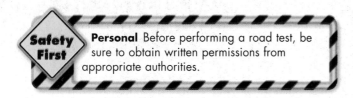

Safety First

Personal Before performing a road test, be sure to obtain written permissions from appropriate authorities.

Glencoe/McGraw-Hill

A Division of The **McGraw·Hill** *Companies*

Copyright © 2000 by Glencoe/McGraw-Hill. All rights reserved. Except as permitted under the United States Copyright Act, no part of this publication may be reproduced or distributed in any form or by any means, or stored in a database or retrieval system, without prior written permission from the publisher.

Send all inquiries to:
Glencoe/McGraw-Hill
3008 W. Willow Knolls Drive
Peoria, IL 61614

ISBN: 0-02-831363-1 (Student Text)
ISBN: 0-02-831365-8 (Technical Applications)
ISBN: 0-02-831366-6 (Academic Applications)
ISBN: 0-02-831368-2 (Instructor's Resource Binder)
ISBN: 0-07-821215-4 (Transparency Package)
ISBN: 0-02-831371-2 (Instructor's Productivity CD-ROM)

Printed in the United States of America

3 4 5 6 7 8 9 10 **071** 04 03 02 01 00

GLENCOE'S AUTOMOTIVE EXCELLENCE TEAM

Leroy Frazier
Henry County High School
Paris, Tennessee

Ron Chappell
Santa Fe Community College
Gainesville, Florida

Robert Porter
Delaware Co. Technical High School
Folcroft, Pennsylvania

Al Blethen
Bessemer State Technical College
Toyota T-TEN
Bessemer, Alabama

Terry Wicker
Franklin County High School
Carnesville, Georgia

John R. Gahrs
Ferris State University
Big Rapids, Michigan

Betty Tibbitts
Valley Forge High School
Parma Heights, Ohio

Debbie Massari
Cuyahoga Community College
Parma, Ohio

Jan Adams
Bowling Green State University
Firelands Campus
Sandusky, Ohio

Darrell L. Parks
NATEF Educational Consultant
Columbus, Ohio

Jessica Levy
Levy, Powell, & Associates Consulting
Rochester, New York

Bob Weber
Write Stuff Communications
Purcellville, Virginia

Mike Dale
Walbro Corporation, Electronics Div.
Cass City, Michigan

Gary E. Goms
Buena Vista Auto Clinic
Buena Vista, Colorado

Mike Mavrigian
Birchwood Automotive Group
Creston, Ohio

Contents in Brief

Automotive Technician's Handbook

Brakes

Electrical & Electronic Systems

Contents in Brief

Engine Performance

Suspension & Steering

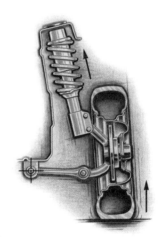

Table of Contents

Automotive Technician's Handbook

DANGER
SAFETY GLASSES
REQUIRED IN
THIS AREA

3 Automotive Tools & Equipment HB-34

Brakes

CHAPTER 1 Brake System Operation...................................... BR-2

CHAPTER 2 Diagnosing & Repairing the Hydraulic System BR-14

Diagnosing & Repairing Drum Brakes BR-32

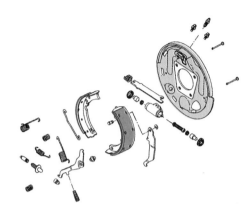

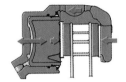

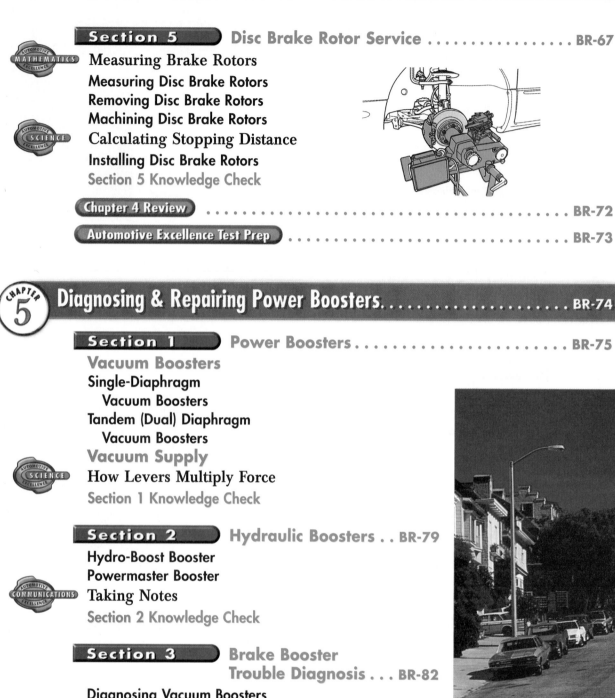

Diagnosing & Repairing Power Boosters BR-74

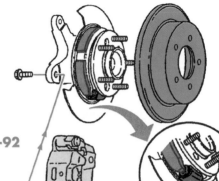

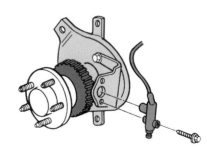

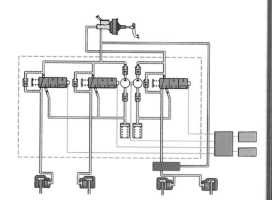

Electrical & Electronic Systems

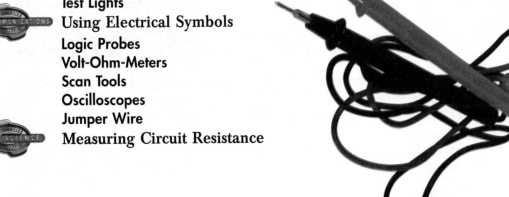

Engine Performance

Diagnosing & Repairing Emission Control Systems EP-136

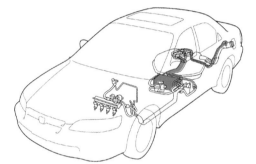

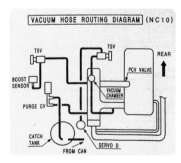

Suspension & Steering

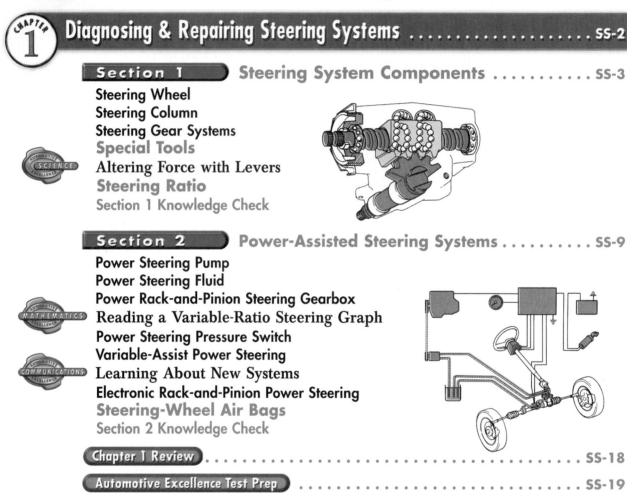

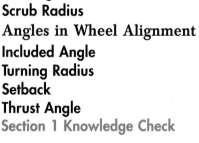

CHAPTER 3   Diagnosing, Adjusting, & Repairing Wheel Alignment SS-44

Your Road Map to Excellence

Safety First The contents of this eye-catching feature stress **Personal** and **Material** safety in the lab/shop and on the job.

TECH TIP These timesaving bits of technical advice will help you perform diagnostic and repair procedures.

Career Focus Activity

This feature appears at the beginning of each part of the text (e.g., Brakes). Three activities appear beneath a group of want ads. Each activity will give you an opportunity to explore a career within the automotive industry.

AUTOMOTIVE EXCELLENCE — COMMUNICATIONS

Each feature correlates a NATEF Communications Standard to automotive content. It provides you with background information or a situation and then prompts you to **Apply it!** by responding to a set of questions.

AUTOMOTIVE EXCELLENCE — MATHEMATICS

Each feature correlates a NATEF Mathematics Standard to automotive content. It provides you with information and then prompts you to **Apply it!** by working through a set of problems.

AUTOMOTIVE EXCELLENCE — SCIENCE

Each feature correlates a NATEF Science Standard to automotive content. Placed at "point of use," it provides you with critical information and then prompts you to **Apply it!** by performing a hands-on experiment.

A Closer Look at Volume 1

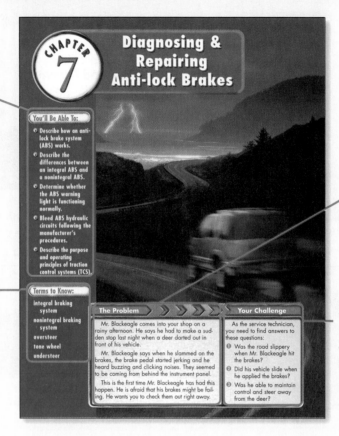

You'll Be Able To: Major learning objectives are identified at the beginning of each chapter and help students focus their reading.

Terms to Know: Key terms important to an understanding of the subject area. Printed in boldface and defined within the chapter, these terms are also included in the Glossary.

The Problem Opens the chapter with a customer complaint that motivates students to use the chapter information to find solutions.

Your Challenge Poses questions that will help students find answers to The Problem.

From the chapter page:

CHAPTER 7

Diagnosing & Repairing Anti-lock Brakes

You'll Be Able To:
- Describe how an anti-lock brake system (ABS) works.
- Describe the differences between an integral ABS and a nonintegral ABS.
- Determine whether the ABS warning light is functioning normally.
- Bleed ABS hydraulic circuits following the manufacturer's procedures.
- Describe the purpose and operating principles of traction control systems (TCS).

Terms to Know:
- integral braking system
- nonintegral braking system
- oversteer
- tone wheel
- understeer

The Problem

Mr. Blackeagle comes into your shop on a rainy afternoon. He says he had to make a sudden stop last night when a deer darted out in front of his vehicle.

Mr. Blackeagle says when he slammed on the brakes, the brake pedal started jerking and he heard buzzing and clicking noises. They seemed to be coming from behind the instrument panel.

This is the first time Mr. Blackeagle has had this happen. He is afraid that his brakes might be failing. He wants you to check them out right away.

Your Challenge

As the service technician, you need to find answers to these questions:
1. Was the road slippery when Mr. Blackeagle hit the brakes?
2. Did his vehicle slide when he applied the brakes?
3. Was he able to maintain control and steer away from the deer?

Safety First The contents of this eye-catching feature stress **Personal** and **Material** safety in the lab/shop and on-the-job.

Tech Tip These timesaving bits of technical advice will help you perform diagnostic and repair procedures.

Knowledge Check These short reviews appear throughout each chapter. They will help you evaluate what you have learned in the section you have just read. This will make it easier to complete the Chapter Review at the end of the chapter.

From the second sample page:

Safety First

Personal Carbon monoxide is a poisonous gas. It is invisible, it has no odor, and it has no taste. Avoid working on vehicles in enclosed, unventilated spaces while a vehicle's engine is running. Make sure all engine exhaust is properly and completely vented from the workspace. To further reduce the possibility of carbon monoxide poisoning, be sure that your workspace is properly ventilated.

Symptoms of carbon monoxide poisoning include drowsiness, dizziness, headache, and nausea.

Figure 5-7(b) also shows a form of abnormal combustion known as *detonation* or *spark knock*. The flame starts across the combustion chamber. However, before the flame reaches the far side, a portion of the mixture explodes rather than burns. The two flame fronts meet, producing a very rapid high-pressure rise. The result is a high-pitched metallic pinging noise called detonation. A low-octane fuel or a high compression ratio can cause detonation.

The sudden power shocks of detonation can damage an engine. Damaged pistons, rings, and spark plugs are the most frequent result of prolonged (or frequent) detonation.

Composition of Exhaust Gases

When complete combustion occurs, all the hydrogen and carbon in the fuel combines with oxygen. The resulting exhaust contains only water

(H_2O), carbon dioxide (CO_2), and nitrogen (N). These are basically harmless compounds. But fuel combustion is not always complete, so products of incomplete combustion occur.

Fuel that does not burn completely produces unburned hydrocarbons (HC) and carbon monoxide (CO). In addition, the high temperature of combustion (over 3,000°F [1,650°C]), forms nitrogen oxides (NO_X). **Fig. 5-8.**

Carbon monoxide is a poisonous gas. Unburned hydrocarbons and nitrogen oxides combine chemically in the presence of sunlight to form smog. Various federal, state, and local agencies regulate the emission of all three of these gases.

Tech Tip **Exhaust Fumes.** Make sure all engine exhaust is removed from any enclosed workspace. Attach a length of heat-resistant flexible tubing over the opening of the exhaust pipe(s) of the vehicle. Route the tubing to the outside through a window or suitable opening. Make sure the tubing is attached whenever the engine is operating.

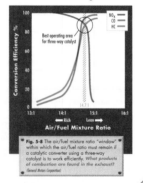

Fig. 5-8 The air/fuel mixture ratio "window" within which the air/fuel ratio must remain if a catalytic converter using a three-way catalyst is to work efficiently. *What products of combustion are found in the exhaust?* (General Motors Corporation)

SECTION 2 KNOWLEDGE CHECK
1. What element in the air does fuel combine with to complete the combustion process?
2. What is a normally aspirated engine?
3. What can cause fuel detonation?
4. Complete combustion produces what three harmless compounds?
5. What three compounds are produced by fuel that does not burn completely?

SCIENCE

Levers in Braking Systems

Cables play an important part in parking brake systems. They "pull" on the integral or auxiliary pads or shoes to set the parking brakes. Can cables also "push"? At first glance, the question may seem strange. How can a flexible cable push?

In this activity you will make a working cable brake system. Using a bicycle caliper brake, you will show that the answer to the above questions depends on the path of the cable.

Apply it!

Testing a Brake Cable

Meets NATEF Science Standards for understanding energy, force, and levers.

Materials and Equipment
- Flat board, about 8" wide and 24" long [20 cm x 60 cm]
- Bicycle handlebar brake lever assembly
- Bicycle front brake cable
- Bicycle side-pull brake caliper assembly
- 6" [15 cm] piece of wooden dowel or broomstick, the same diameter as a bicycle handlebar
- Nails or screws to fasten the dowel to the board

1. Clamp the brake lever to the dowel. Fasten the dowel to the board with screws or nails. **Fig. A.**
2. Drill two holes in the board at positions X and Y. The center shaft of the caliper assembly must fit tightly into each hole (see illustrations).
3. Put the center shaft of the brake caliper assembly into hole X. **Fig. A.**
4. Cut the cable sheath and cable so that the cable will run in a straight line between the lever and the caliper. Thread the cable. Install the cable so that the caliper is slightly closed. **Fig. A.**
5. Push the brake lever down. Notice which ...

6. Move the caliper to position Y. **Fig. B.** Press the lever down. What parts of the caliper move? Try holding one part of the caliper still. Observe what happens. Then hold the other part of the caliper. What happens?

Results and Analysis

This experiment demonstrates that a cable system can sometimes both pull on one part and push on another part. Look at **Figs. A** and **B.** What changed to allow the cable system to "push?" What special conditions are needed for a cable to act this way?

Notice the similarity between the bicycle brake mechanism and the disc brakes on a vehicle. Which do you think came first? Research the history of bicycle technology to find when caliper-rim brakes were first used.

Fig. B

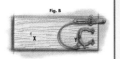

Each feature correlates NATEF Academic Standards to automotive content. It provides you with information or a situation and then prompts you to **Apply it!** by responding to a set of questions, working through a set of problems, or by performing a hands-on experiment.

Key Points   This summary of important chapter points refers to the NATEF Automotive Standards addressed by the chapter.

SS-42 Suspension & Steering

CHAPTER 2 REVIEW

Key Points

Meets the following NATEF Standards for Suspension and Steering: diagnosing and repairing leaf, coil, torsion bar, and strut suspension systems.

- The basic parts of a suspension system are springs, shock absorbers, anti-sway bars, control arms, ball joints, axles, and wheel bearings.
- Compression or jounce occurs when the suspension moves closer to the body. Rebound occurs when the suspension moves away from the body.
- Front or rear suspensions can use a variety of spring types.
- The two most common types of shock absorbers are "traditional" and struts. Shock absorbers can also be spring-assisted, adjustable, and air-assisted.
- The most common type of strut is the MacPherson strut. It combines many of the standard suspension components.
- An active suspension system uses an on-board computer and sensors to maintain full control of the suspension during all suspension activity.
- There are a variety of front suspension designs including leaf, coil, torsion bar, and strut.
- There are a variety of rear suspension designs including leaf, coil, and strut.

Review Questions

1. Describe the types of springs used in front and rear suspension systems.
2. What factors determine the spring rate of a coil spring?
3. In an air-spring suspension, how do the springs adjust for road and load conditions?
4. What are the different front suspension designs?
5. Once a MacPherson strut is removed from a vehicle, what steps do you take to service the strut?
6. Describe the three types of rear suspension designs.
7. What is the advantage of a leaf spring rear suspension?
8. **Critical Thinking** Based on their construction and effect, which rear suspension is safer—one that uses coil springs or one that uses leaf springs? Why?
9. **Critical Thinking** Based on th... and effect, which front suspens... that uses coil springs, one that u... or one that uses leaf springs? W...

Review Questions   These questions reinforce the chapter objectives and provide an opportunity to use critical thinking skills.

TECHNOLOGY FORECAST

Electronic Suspension Systems

Automakers want to give their cars a quiet, comfortable ride and stable handling. New electronic suspension systems should make their job easier. Ride-motion and steering wheel sensors will be able to tell if the pavement dips, curves, or has a bump. They will also sense how the driver is steering the car or truck.

Using information from the sensors, a control module will signal the suspension to be firmer or softer as needed. Electronically controlled dampers in the shock absorbers or struts will make

these changes. The system can a... the vehicle's body flat while turn... This control is possible because... nected to the stabilizer bars.

Also in the future, automake... forming. This metal-shaping pr... pressure to shape suspension a... nents. Benefits include parts that... precisely and that are very stro... Their strength will improve ri... Squeaks and rattles will be redu...

Technology Forecast   This quick glimpse of emerging technologies in the automotive industry will lead you to think about the future.

AUTOMOTIVE EXCELLENCE TEST PREP

Answering the following practice questions will help you prepare for the ASE certification tests.

1. Technician A says the high current for the starter comes from the battery. Technician B says high current comes from the ignition switch. Who is correct?
   - Technician A.
   - Technician B.
   - Both Technician A and Technician B.
   - Neither Technician A nor Technician B.

2. A truck engine with a manual transmission does not crank when the key is turned to START. The battery is good. The clutch is depressed. Which of the following is the most likely cause?
   - Low resistance in the high-current circuit.
   - The clutch pedal position switch is bad.
   - The truck is in high gear.
   - All of the above.

3. A starter current draw test shows a reading of 100 amps. The manufacturer's specifications indicate the reading should be 190 amps. Which of the following can cause the problem?
   - Corroded battery terminals.
   - A bad ground connection to the engine block.
   - Worn brushes in the starter.
   - All of the above.

4. Sometimes an engine starts right up. Sometimes it cranks slowly. Technician A says it may be high resistance in the high-current circuit. Technician B says it may be low resistance in the low-current circuit. Who is correct?
   - Technician A.
   - Technician B.
   - Both Technician A and Technician B.
   - Neither Technician A nor Technician B.

5. Technician A says a gear-reduction starter gives better cranking performance. Technician B says it provides greater torque by using a higher gear ratio. Who is correct?
   - Technician A.
   - Technician B.
   - Both Technician A and Technician B.
   - Neither Technician A nor Technician B.

6. Higher than normal cranking current draw could result from:
   - A bad battery.
   - A starter or engine mechanical problem.
   - Poor cable connections.
   - All of the above.

7. A car has a grinding noise during cranking. Technician A says the starter drive mechanism may be damaged. Technician B says the starter shims may be missing. Who is correct?
   - Technician A.
   - Technician B.
   - Both Technician A and Technician B.
   - Neither Technician A nor Technician B.

8. A cranking current draw test shows a reading twice as high as specifications, and the engine does not crank. Technician A says the cause could be a defective starter. Technician B says it could be a seized engine. Who is correct?
   - Technician A.
   - Technician B.
   - Both Technician A and Technician B.
   - Neither Technician A nor Technician B.

9. Technician A says the overrunning clutch prevents the pinion gear from being driven by the ring gear once the engine starts. Technician B says the clutch is used to remove the pinion gear from the ring gear once the engine starts. Who is correct?
   - Technician A.
   - Technician B.
   - Both Technician A and Technician B.
   - Neither Technician A nor Technician B.

10. Technician A says worn brushes in the starter can cause high current flow. Technician B says worn bearings or bushings in the motor can cause high current flow. Who is correct?
    - Technician A.
    - Technician B.
    - Both Technician A and Technician B.
    - Neither Technician A nor Technician B.

Automotive Excellence Test Prep   These ten practice questions will help you begin to prepare for the ASE certification exams.

Attention Students!

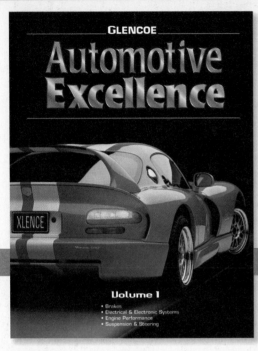

You are using the only textbook specifically designed to address both the technical and applied academic standards as defined by the National Automotive Technicians Education Foundation (NATEF), the organization that oversees ASE certification. *Automotive Excellence* can help you become a successful automotive technician who has:

- a thorough knowledge of automotive systems and components
- good computer skills
- excellent communication skills
- above average mechanical aptitude
- good reasoning ability
- ability to read and follow instructions
- manual dexterity

Automotive Excellence, Volume 1

Automotive Excellence will focus your education and training on the necessary technical skills *and* the correlated communications, mathematics, and science knowledge needed to enter the changing workforce. In *Volume 1* of *Automotive Excellence* you will study:

- Brakes
- Electrical & Electronic Systems
- Engine Performance
- Suspension & Steering

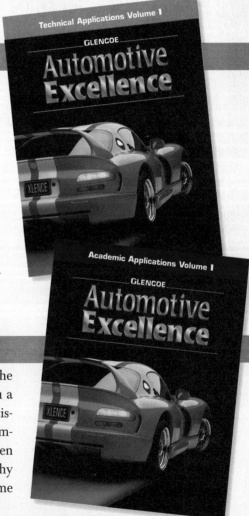

Technical Applications

In addition to your textbook, a spiral-bound set of materials called *Technical Applications* provide four different ways to further your skills.

- *Career Profiles* encourage you to use print and computerized resources to find out where 21 different "pathways" within the automotive industry can lead.
- *Diagnostic Sheets* provide you with opportunities to sharpen your knowledge of how automotive systems and components work.
- *Job Sheets* focus your mechanical aptitude and manual dexterity on reading and following instructions as you diagnose and repair vehicles.
- *Challenge Sheets* help you develop reasoning ability and fine-tune communication skills by responding to customer complaints about their vehicles.

Academic Applications

The worksheets inside *Academic Applications* will help you learn the applied communications, mathematics, and science necessary to give you a thorough knowledge of automotive systems and components. These exercises will help improve your ability to read and follow instructions, basic communication skills, and reasoning ability. Each presents information and then asks you to answer questions or perform a hands-on experiment to see why the academic standard is important. Then, you will be asked to answer some follow-up questions to reinforce learning.

Automotive Technician's Handbook

CHAPTER 1
The Automotive Industry

CHAPTER 2
Automotive Safety Practices

CHAPTER 3
Automotive Tools & Equipment

CHAPTER 1

The Automotive Industry

You'll Be Able To:

⊗ Identify the major systems of an automotive vehicle and briefly explain the purpose of each system.

⊗ Identify the purposes and explain the importance of ASE, NATEF, and STS to the automotive service industry.

⊗ Identify career opportunities in the automotive industry.

⊗ Explain how to become a certified automotive technician.

Terms to Know:

Automotive Service Excellence (ASE)

automotive system

master technician

National Automotive Technicians Education Foundation (NATEF)

Service Technicians Society (STS)

The Problem 〉 〉〉〉〉〉〉 Your Challenge

Your goal is to prepare for a technical career after finishing your education. You are looking for solid information on a promising career.

One of your main interests has always been cars. Because of this, you're considering a career as an automotive technician. But you're not really sure what an automotive technician does on the job. You wonder if there are any jobs available. You would also like to know if you could make a good income as an automotive technician.

Search for the answers to these questions:

❶ What do automotive technicians do on the job?

❷ What are the employment opportunities for automotive technicians?

❸ What is the expected annual income for an automotive technician?

Section 1

The Automotive Profession

In the United States, the automotive industry and its 350,000 related industries employ about 7 million workers. More than 30 percent of all passenger cars are in the United States. Ninety percent of Americans own at least one car, and 55 percent own two or more. Americans drive 2.25 trillion miles a year. Their vehicles consume 140 billion gallons of fuel per year. Annually, motorists in the United States spend $100 billion on vehicle insurance. They also use more than $300 billion of credit in the purchase and maintenance of vehicles. The automotive industry affects many areas of the economy.

A Century on the Road

Starting with the first patented gasoline-powered *Benz Motorwagen* in 1886, the automotive industry has seen continuous and exciting changes for over a century. **Fig. 1-1.** The early gasoline-powered automobile generated 1.5 horsepower. It reached top speeds of 3–5 mph [5–8 kph]. Today's vehicles have 110–450 horsepower engines and cruise at 65–75 mph [106–121 kph], depending on the legal speed limit. **Fig. 1-2.** For many people around the world, automobiles are the most important means of personal transportation.

Fig. 1-2 A modern automotive engine. *What is the horsepower range of today's automotive engines?* (Ford Motor Company)

Automotive Systems

Automobiles are highly complex vehicles with multiple computer-controlled systems. A car today has more computers in it than the first spaceship. A new car today may have as many as 15 computers operating everything from the engine to the radio.

Automotive vehicles are available in a wide variety of models, sizes, and body styles. Vehicles range from compacts to full-size cars and from minivans to sport utility vehicles (SUVs). There are also sedans, convertibles, hatchbacks, station wagons, and light-duty trucks. Luxury models are also available.

Fig. 1-1 An 1886 Daimler motor carriage, the forerunner of today's modern automobile. *What was the top speed of this vehicle?* (Mercedes-Benz of North America, Inc.)

Technology has improved automotive vehicles over the past century. Vehicles are designed according to a variety of factors, including the number of engine cylinders, the type of drive-train system, and vehicle application.

About 15,000 separate parts are assembled to make an automotive vehicle. These parts are grouped into several systems, known as automotive vehicle systems. An **automotive system** is a system made up of two or more parts that work together to perform a specific task. **Fig. 1-3.**

As automotive vehicles become more complex, vehicle service, maintenance, and repair must keep up with changing technology. Many people who serviced and repaired their own vehicles have turned to highly skilled professional service technicians. Specialized equipment is used to troubleshoot and diagnose modern-day automotive vehicle problems to determine service and repair options. **Fig. 1-4.**

Fig. 1-4 An automotive service technician using a scan tool to check engine performance. *Why do people who serviced their own vehicles now rely on professional service technicians?* (Jack Holtel)

Anyone whose vehicle has failed to start knows the importance of an automotive technician's job. A competent automotive technician can diagnose the source of the problem quickly and accurately. The technician can then perform the necessary service or repair.

Automotive Service Excellence (ASE) Certification

As the need for skilled technicians has grown, a number of automotive service-related organizations have been created. These organizations support automotive service technicians as they grow in their careers. These organizations also help ensure the quality of training received by today's automotive technicians.

The National Institute for Automotive Service Excellence In 1972 the automotive service industry created the independent, nonprofit National Institute for Automotive Service Excellence. The purpose of **Automotive Service Excellence (ASE)** is to improve the quality of vehicle service and repair. ASE does this through the testing and certification of service and repair professionals. **Fig. 1-5.** ASE is governed by a member board of directors.

We employ technicians certified by the
National Institute for
AUTOMOTIVE SERVICE EXCELLENCE
Let us show you their credentials

Fig. 1-5 The industry-recognized ASE Blue Seal of Excellence. *Why do consumers look for the ASE seal?* (National Institute for Automotive Service Excellence)

Fig. 1-3 The basic automotive systems. *What task does each system perform?* (General Motors Corporation)

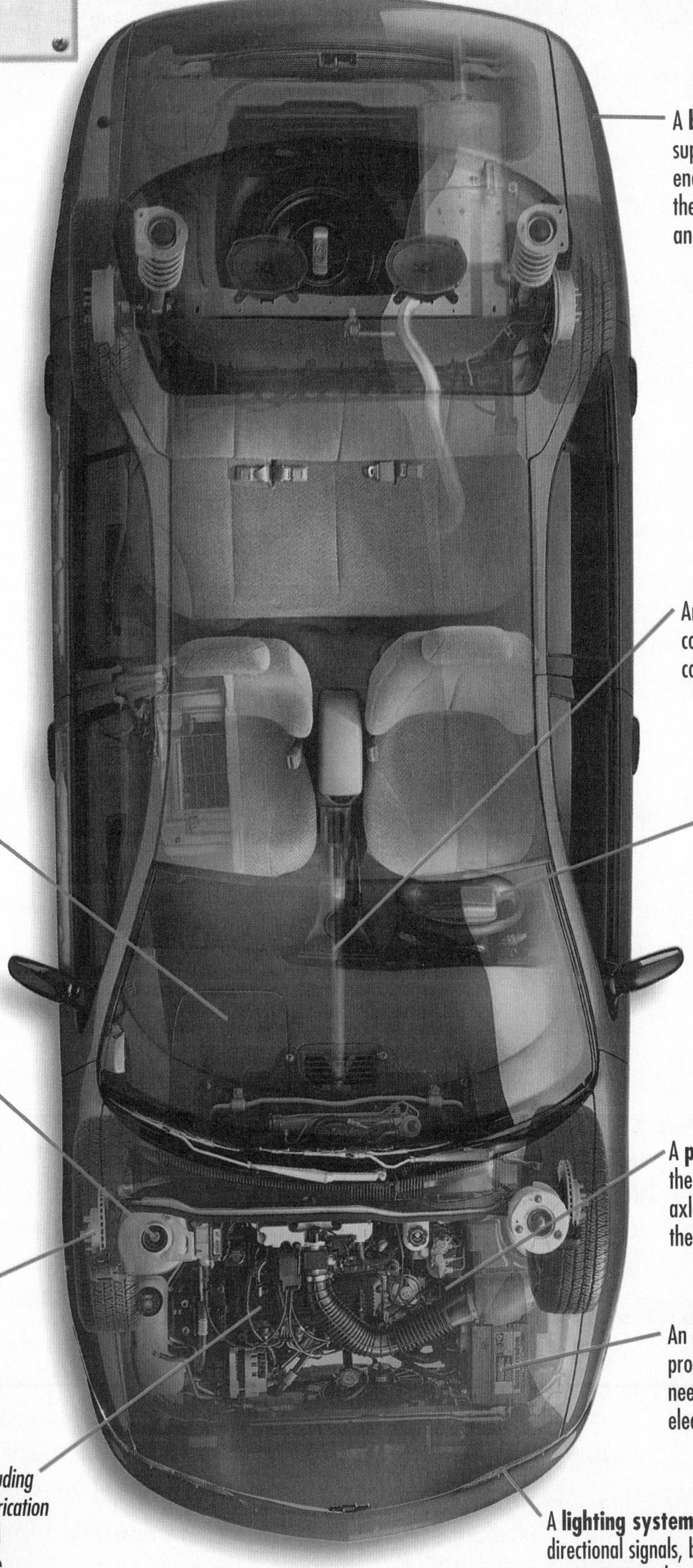

A **body and frame system** supports the vehicle and provides enclosures or compartments for the engine, passengers, luggage, and cargo.

An **accessory system** includes comfort, safety, security, and convenience devices.

A **safety system** provides added passenger safety and includes devices such as air bags, anti-lock brakes, side-door panel steel rails, and shock-absorbing bumpers.

A **steering system** controls the direction of vehicle travel.

A **suspension system** absorbs the shocks of the tires and wheels from bumps and holes in the road.

A **power train system** includes the transmission, drive train, and axles that carry the power from the engine to the drive wheels.

A **braking system** slows and stops the vehicle.

An **electrical/electronic system** produces and directs electrical power needed to operate the vehicle's electrical and electronic components.

An **engine** or **power plant** *(including the fuel, exhaust, cooling, and lubrication systems)* produces dependable and efficient power to move the vehicle.

A **lighting system** includes headlights and taillights, directional signals, brake warning lights, interior convenience and courtesy lights, and instrument control-panel information and warning lights.

ASE administers a series of certification exams. Individuals can become certified as automobile and light truck technicians, collision repair technicians, medium- and heavy-duty truck technicians, alternate fuels technicians, engine machinists, and parts specialists. There are a total of 25 areas in which one or more individual certification exams are given. **Table 1-A.**

A technician can achieve ASE certification by doing the following two things. First, the technician must pass at least one certification exam. Second, the technician must provide proof of two years of relevant work experience. A technician who is certified in all eight auto/light truck areas is a **master technician**. These areas are:
• Engine Repair.
• Automotive Transmission/Transaxle.
• Manual Drive Train and Axles.
• Suspension/Steering.
• Brakes.
• Electrical/Electronic Systems.
• Heating and Air Conditioning.
• Engine Performance.

ASE Certification Exams The exams stress knowledge and job-related skills. They are not easy to pass. About one of every three test-takers fails. ASE certification exams are offered twice a year at more than 750 locations in the United States.

Nearly 430,000 professional technicians have been certified under this voluntary certification program. To reduce hiring risks, many employers now make ASE certification a condition for employment. Many employers pay higher wages to ASE-certified technicians than to noncertified technicians. Certification, however, is not for life. To remain certified, a technician must pass a recertification test every five years.

ASE-certified technicians work in every area of the automotive service industry. They work in car and truck dealerships, independent garages, fleets, and service stations. ASE-certified technicians usually wear the blue-and-white ASE logo on their shirts. They also carry credentials listing their area(s) of expertise, such as brakes or engine performance. Employers often display their technicians' credentials in the customer waiting area.

Fig. 1-6 The National Automotive Technicians Education Foundation (NATEF) logo. *Why is the work of NATEF important?* (National Institute for Automotive Service Excellence)

NATEF's Role

In 1978 the **National Automotive Technicians Education Foundation (NATEF)** was created as an organization that certifies automotive training programs. **Fig. 1-6.** NATEF was formed through the work of the Industry Planning Council (IPC) of the then American Vocational Association (AVA). The AVA is now known as the Association for Career and Technical Education (ACTE). The IPC was composed of representatives from the automotive service industry and vocational education.

The IPC was concerned with the quality of automotive education. They directed a multiyear study that resulted in recommended task lists, tools and equipment, and program standards.

The IPC, working with ASE, developed a process for certifying automotive training programs. NATEF was formed to carry out this process. NATEF evaluates individual automotive training programs. It also recommends programs for ASE certification and certifies those programs.

Service Technicians Society

Formed by the Society of Automotive Engineers (SAE), the **Service Technicians Society (STS)** is an association for automotive and transportation professionals. It provides a forum to exchange technical information and share industry trends.

Many automotive service and repair technicians are STS members. Students enrolled in automotive technician training programs are eligible to join STS and are encouraged to do so.

Table 1-A ASE CERTIFICATION AREAS

Auto/Light Truck

- Engine Repair
- Automatic Transmission/Transaxle
- Manual Drive Train and Axles
- Suspension/Steering
- Brakes
- Electrical/Electronic Systems
- Heating and Air Conditioning
- Engine Performance

Medium/Heavy Truck

- Gasoline Engines
- Diesel Engines
- Drive Train
- Brakes
- Suspension/Steering
- Electrical/Electronic Systems
- Heating, Ventilation, and Air Conditioning
- Preventive Maintenance Inspection

School Bus

- Body Systems and Special Equipment
- Diesel Engines
- Drive Train
- Brakes
- Suspension/Steering
- Electrical/Electronic Systems
- Air Conditioning

Collision Repair

- Painting and Refinishing
- Nonstructural Analysis and Damage Repair
- Structural Analysis and Damage Repair
- Mechanical and Electrical Components
- Damage Analysis and Estimating

Engine Machinist

- Assembly Specialist
- Cylinder Block Specialist
- Cylinder Head Specialist

Alternate Fuel

- Light Vehicles—Compressed Natural Gas

Parts Specialist

- Automobile Parts Specialist
- Medium/Heavy Truck Parts Specialist

Advanced Series

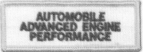

- Automobile Advanced Engine Performance Specialist
- Truck Electrical Diesel Engine Diagnosis

SECTION 1 KNOWLEDGE CHECK

❶ Explain how automotive engines have changed in the past 100 years. What are the similarities between today's automobile and the very early ones?

❷ Identify the major automotive systems and relate the major purpose of each system.

❸ What does ASE stand for? What is its mission, and how does one achieve ASE certification?

❹ What is NATEF and what is its mission? How does it differ from ASE?

❺ What does STS stand for? Why are many automotive technicians members of this organization?

Section 2

Automotive Career Opportunities

Each year thousands of jobs in automotive technology are available in the United States. The demand for automotive technicians has never been greater. **Fig. 1-7.**

According to the *Occupational Outlook Handbook,* 60,000 additional automotive technicians will be needed in the near future. Jobs will be plentiful for those who finish training programs in high school, vocational or technical school, or two- or four-year colleges. Most technicians who enter the field can expect steady work. Changes in economic conditions have little effect on the automotive service and repair business.

Automotive Employers

A variety of businesses employ trained automotive technicians. **Fig. 1-8.** These include:

- **Dealerships** that sell and service specific brands of vehicles.
- **Independent garages** that service all types of automotive vehicles.
- **Service stations** where fuel, oil, and related products and automotive services are sold.
- **Large retail facilities** owned by mass merchandisers.
- **Specialty centers** that handle brakes, tune-ups, transmission repair, and wheel alignment.
- **Fleet garages** operated by the government or private companies to service their own vehicles.

Dealerships There are about 22,600 new-car dealers in the United States. There are also several hundred truck dealers. The dealer must prepare and service each new vehicle before delivery to the customer. This is the *predelivery service.* Sometimes vehicle problems develop after the sale. The dealer must fix vehicles under warranty. *Warranty work* is repair work paid for by the vehicle manufacturer. The warranty covers repairs only for a specified period of time or mileage amount after the date the vehicle was sold.

BRAKE STANDARDS ENGINEER

Immediate opening for a brake standards engineer. Specific tasks include collecting and collating existing brake requirements and

AUTOMOTIVE ANALYSIS ENGINEER

Opening for an analysis engineer with an MS in mechanical engineering and 3-5 years of experience in the field. Duties include developing mathematic models of vehicle dynamics a control systems; simulating ? validating math models aga test results for ride and hand tire dynamics, ABS, and tra control systems; and deve and modifying algorithms.

AUTOMOTIVE CUSTOMER RELATIONS

Immediate opening for a "GREETER" in Service Department. Consists of greeting customers, filing, follow-

Automotive Technicians

☆ **$2000 HIRING BONUS** ☆

If you are an experienced technician looking to call home, give at our shop. We nefits, a busy g bonus.

Automotive Service Advisor

Extremely busy Service Department is seeking career-Minded person to join our management team. Individual Must be sensitive to customer needs and capable of over delivering on customer expectations. Earning potential is in the 40K-50K range. If you feel you're the best and would like to join a winning team, **call us.**

Fig. 1-7 Classified ads for positions in the automotive field. *Why is the demand for automotive technicians so great?*

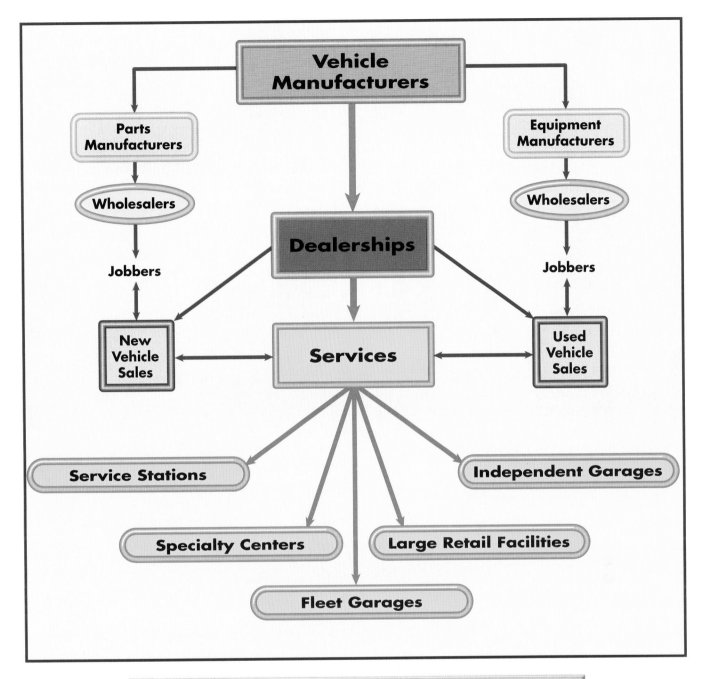

Fig. 1-8 Key automotive employers include people in the manufacturing, dealership, and service areas of the automotive industry. *What is the difference between a dealership and an independent garage?*

A manufacturer may recall a vehicle. This means the manufacturer asks the owner to return the vehicle to the dealer for inspection and possible repair. The manufacturer pays for any required parts or service. Recalls are frequently safety related. Technicians should never allow the vehicle owner to ignore a recall.

New-car dealers also sell the used vehicles traded in when customers buy new vehicles. After inspecting and reconditioning the vehicle, the dealer sells the trade-in, often with a limited warranty. This warranty protects the buyer if a defect shows up shortly after the sale. Dealers also sell used vehicles "as is." In this case, the buyer does not receive any warranty.

Fig. 1-9 Today's technicians must have a variety of skills including the ability to use computer diagnostic equipment. *(Gary D. Landsman/The Stock Market)*

Fig. 1-10 A service department in an automotive dealership. *Why is a warranty important to the consumer?* *(Terry Wild Studio)*

The dealership has a service department to handle the automotive service work. This is a properly equipped shop staffed with trained technicians who do warranty work and work for which the customer pays. Diagnostic equipment, special tools, and parts must be available. Some service departments are small, employing only one to five technicians. Larger dealerships may employ 25 or more technicians.

Most dealerships have personal computers in the service department so they can be online with the vehicle manufacturer. **Fig. 1-9.** This means the computers at the dealership are connected to the manufacturer's computer. A telephone line or satellite link connects the two. If a vehicle has a difficult problem, the technician can connect to the manufacturer's computer, which shows how to locate and repair the problem. The screen may also display the parts needed, along with their prices and availability.

Independent Garages There are about 129,000 non-dealership automotive repair shops in the United States. These are independent garages. Some are one-person specialty shops. Others are general repair shops that employ many automotive technicians. The larger shops often operate like dealership service departments. **Fig. 1-10.**

Service Stations There are over 58,000 service stations in the United States. **Fig. 1-11.** At one time nearly all service stations performed some automotive service and repair work. However, today about 25,000 service stations are self-serve. They sell only fuel, oil, and other automotive fluids. They do not perform service work.

Large Retail Facilities Service facilities owned by mass merchandisers operate much like independent garages. However, these centers usually specialize in engine tune-up, electrical/electronic systems, suspension and steering, lighting systems, and tire replacement and repair.

Specialty Centers Specialty centers provide various "trade services" for the automotive service

Fig. 1-11 A full-service station. *Why do you think fewer service stations today are full service?* (Bel-Air Shell)

industry. These services usually include brake and wheel alignments, tune-ups, transmission repair, lube and oil changes, exhaust and muffler replacement, and radiator service. Many large and small dealership service departments now take their machine work to specialized automotive machine shops. Such shops refinish brake drums, rotors, and engine flywheels. The machine shop may also repair cracks in cylinder blocks and heads, as well as bore and sleeve engine cylinders.

Fleet Garages A fleet is a group of five or more vehicles owned and maintained by a single company. Automotive dealers and independent garages maintain some fleet vehicles. However, there are about 39,000 fleet garages in the United States. Companies that do their own service and repair work operate these shops. They include bus and trucking companies as well as taxicab and delivery fleets.

The technician in a fleet garage usually works for the company that owns the fleet. Often, the work is done on a preset schedule. This scheduled periodic service, or preventive maintenance, helps prevent unexpected vehicle breakdown and costly repairs.

The fleet driver checks some items daily before driving the vehicle. Then at scheduled intervals, the vehicle is brought into the shop. The technician checks the fluid levels, changing the fluids and filters if necessary. A visual inspection determines the condition of the belts, tires, and other parts. Meters and gauges show the condition of the battery and electrical system. A road test may be conducted to check vehicle performance. Any problems found during these checks are corrected.

Jobbers A jobber is a person who buys goods for resale. An automotive parts store may be both a wholesaler and a retailer. As a wholesaler, the store sells parts at a discount to other automotive service businesses. As a retailer, the store sells parts to those who service their own vehicles. **Fig. 1-12.**

The primary job of the parts counterperson is to identify the needed part in a paper or microfiche parts catalog. Parts catalogs may contain hundreds of pages that include illustrations, part numbers, and related information. The price is then found on a price list.

Fig. 1-12 An automotive parts store. *What is the primary job of a parts counterperson?* (Pep Boys)

Electronic parts catalogs are rapidly replacing paper and microfiche parts catalogs. **Fig. 1-13.** A computer terminal displays the illustrations, part numbers, prices, and other information. The computer also checks the inventory and indicates if the part is in stock. Some service and repair shops have

Fig. 1-13 An electronic parts catalog. *What information is found in an electronic parts catalog?* (Reynolds & Reynolds)

a computer terminal in or near the service bay. The technician can use it to check a part's availability or to order parts.

The Automotive Work Environment

Automotive technicians enjoy varied and interesting day-to-day encounters. Many consider diagnosing hard-to-find problems one of their most challenging and satisfying duties. While the use of hand tools is still an important part of the technicians' responsibilities, they also use electronic analyzers and computerized diagnostic equipment. This technology helps technicians find problems and make precision adjustments. Modern automotive service and repair facilities are clean and well ventilated. They are pleasant and safe places to work.

Technicians generally work a standard 40-hour week. Opportunities for overtime work are usually available. To better satisfy customer needs, some dealerships and independent repair garages provide evening and weekend hours.

There are career opportunities in the automotive service industry other than that of a technician. The following are some of these careers.

Service Writer The service writer greets the customer and listens carefully as the customer describes the vehicle's problem. The service writer writes the work order in enough detail to assist the technician in diagnosing and repairing the problem.

Service Dispatcher The service dispatcher schedules each vehicle service and repair job. He or she also assigns the technician to perform the job. The service dispatcher tracks all repair jobs. This helps ensure that the jobs are completed correctly and that billings for the technicians' time are recorded.

Automotive Salesperson The automotive salesperson works in the dealer's showroom or used-car department. He or she is responsible for selling vehicles to interested buyers. The salesperson must have good sales skills. He or she must also know about each vehicle's features, options, and price.

Shop Supervisor The shop supervisor is responsible for the overall supervision of the service department. The shop supervisor helps other technicians troubleshoot problems in all automotive areas. He or she provides a communication link between the shop floor, the service manager, and the parts department. The shop supervisor is also responsible for effectively communicating with technicians and resolving any personnel disputes. The shop supervisor usually has been a successful automotive technician before being promoted.

Service Manager The service manager coordinates the service and repair activities of the service and parts departments. The service manager also responds to customer questions and complaints and ensures quality service.

Parts Department Manager The parts department manager knows the components and systems of an automotive vehicle. He or she can quickly locate, price, and dispense requested parts.

Additional formal education will present other career options in the automotive industry. These include engineers, designers, technical writers, and automotive technician trainers and instructors.

Preparing for an Automotive Career

For a successful technician, learning is never finished. While all vehicles still have the same basic parts, those built more than 20 years ago are antiques compared to today's high-tech vehicles. Technological improvements continue with each new generation of cars and trucks. Automotive technology is becoming more complex. State and federal regulations affecting the automotive industry are also increasing. For these reasons, technicians must stay up-to-date if they are to be successful. **Fig. 1-14.**

Employability Skills

Technicians must have technical skills as well as academic skills in mathematics, science, and language arts. Technicians must also be punctual and

Fig. 1-14 A student learning from a professional automotive repair technician. *Why is it important for students to visit automotive businesses?* (Jack Holtel)

Table 1-B	EMPLOYABILITY SKILLS NEEDED FOR SUCCESS	
Skill/Trait	**Purpose**	**Example**
1. Mathematical skills	Analyze and solve problems	Measure wear in an engine; calculate resistance in electrical circuits
2. Science knowledge (especially physics)	Understand science concepts like force, friction, hydraulics, and electrical circuits	Diagnose brake and electrical problems
3. Communication skills	Locate, read, and interpret technical information from service manuals, flat rate manuals, service bulletins, parts manuals, and computer databases	Effectively deal with customers and coworkers; write work orders, process warranty claims, and prepare reports
4. Steady, reliable, and punctual	Improve ability to succeed in workplace; ensure that employer has employees for workload	Arrive at work on time and on a regular basis
5. Pride in personal appearance and job performance	Influence motivation, job satisfaction, and sense of achievement	Practice good personal hygiene and dress appropriately; take responsibility for work performed by ensuring quality and accuracy
6. Willingness to follow directions, manage time, and get along with coworkers	Enhance job performance, including ability to perform quality work in a reasonable time frame	Listen to supervisor's directions; stay on task when performing work; show sensitivity and understanding to coworkers

dependable. They should be skilled in teamwork, problem solving, decision making, and time management. **Table 1-B.**

The Automotive Technician Today

There are three major requirements for getting started as a technician in the automotive service industry. The first requirement is to have a mechanical aptitude and an interest in automotive vehicles and how they work. Visit businesses that do automotive service and repair. Observe the automotive service and repair technicians' work environment. Discuss job requirements and career opportunities with working technicians.

The second requirement is to complete an automotive technician training program in high school, a vocational or technical institution, or a two- or four-year college. If available, the training program of choice should be ASE certified. Such certified programs have met rigid industry standards for training automotive technicians. These programs should display a NATEF/ASE logo. **Fig. 1-15.**

To be ASE certified in a certain area, a technician must pass a written exam in that program area. He or she must also have at least two years of relevant work experience. Completion of an ASE-certified automotive technology training program may be substituted for one year of work experience toward ASE technician certification. This could lead to finding employment more quickly as well as a higher-paying job.

A third requirement is to gain work experience in a modern automotive service and repair facility. This could be at a dealership or an independent garage. Ideally, some of the work experience should occur while going to school. This experience could take place through cooperative education or part-time employment. Work experience adds relevance and value to an in-school program. It also provides an excellent opportunity to develop technical and applied academic skills. These skills are important to the technician's success.

A Well-Paid Profession

Today's automotive technician is a respected, well-paid professional. The average annual wage for an experienced technician ranges from $30,000 to $60,000. Most technicians work on what the industry refers to as a *flat rate*. The industry has predetermined the amount of time it should take to perform a specific service or repair job. If it takes the technician an hour to complete a job listed in the flat rate manual as a two-hour job, he or she is paid for two hours of labor.

However, if it takes the technician three hours to do that same job, the technician is paid only the two-hour flat rate.

Many experienced technicians employed by automotive dealers and independent repair shops receive a guaranteed annual salary plus a commission. This *commission* is a percentage of the labor costs charged to customers. Under this method, weekly earnings depend on the work completed by the technician. A master technician may earn from $70,000 to $100,000 a year. This salary range indicates the demand for experienced and certified automotive technicians.

Fig. 1-15 An ASE-certified automotive training program. *How can ASE certification expand career opportunities?* (Jack Holtel)

SECTION 2 KNOWLEDGE CHECK

❶ List five types of businesses that employ automotive service technicians.

❷ Explain the difference between a dealership service department and an independent garage. How do their services and the vehicles they work on vary?

❸ Describe an automotive service technician's work environment in a modern automotive service and repair establishment.

❹ List the three major requirements for getting started as a technician in the automotive service industry.

❺ Why is an automotive service technician's learning never finished?

CHAPTER 1 REVIEW

Key Points

Addresses NATEF program guidelines and career opportunities in the automotive industry.

- The automobile has evolved from a very simple to a very complex vehicle.
- Each major system has a specific purpose in the vehicle's operation.
- Three organizations that serve the automotive service industry are the National Institute for Automotive Service Excellence (ASE), the National Automotive Technicians Education Foundation (NATEF), and the Service Technicians Society (STS).
- The employment outlook for automotive technicians is very positive over the next several years. Sixty thousand new technicians will be needed in the near future.
- Many automotive technicians receive a guaranteed annual salary plus a commission. A commission is a percentage of the labor costs charged to customers.

Review Questions

1. How have automobiles changed over the past century? How have they remained the same?

2. Identify the major automotive systems and explain the purpose of each.

3. Distinguish between an automotive part, system, and task.

4. Identify the purposes of these three organizations that represent the automotive service industry: ASE, NATEF, and STS.

5. What is the employment outlook for automotive technicians?

6. What are the advantages of being an ASE-certified master technician?

7. How does one get started on the pathway to a successful career as an automotive technician?

8. (Critical Thinking) Explain how strong skills in each of the following academic areas can help you in the automotive services field: science, mathematics, and communications.

TECHNOLOGY FORECAST FOR AUTOMOTIVE EXCELLENCE

Automotive Industry Changes

It should come as no surprise that the automotive industry has a global future. Cars and trucks are built in countries all over the world. And tens of millions of them are produced and sold each year. Increasingly, automakers are looking for ways to take advantage of these international markets.

The "global market" is leading automakers to design cars and trucks built on what are called "global platforms." These vehicles will likely share many of the same components and basic styling. They will differ to suit laws and driving styles that can vary from one country to the next. For example, safety equipment requirements differ from one country to the next. Some countries have stricter exhaust emissions regulations or higher speed limits. For example, on Germany's autobahn, the recommended speed is about 130 kph [about 80 mph]. The cost of fuel–and its supply–can also differ dramatically from country to country.

Another development will be the growth of modularization. Increasingly, automakers are studying ways to manufacture vehicles by using preassembled systems, or modules. These modules replace–yet include–the dozens of components used separately today.

Perhaps the biggest change of all could be a move away from the internal combustion engine. As automakers work to meet clean air laws, they are looking to vehicles that produce little or no emissions. Electric and hybrid-powered vehicles, some already on the road, may be the solution.

AUTOMOTIVE EXCELLENCE
TEST PREP

1. Which of the following statements about today's new automobiles is correct?
   - a There are more computers in a car today than aboard the first spaceship.
   - b A new car today may have as many as 15 computers onboard.
   - c Both a and b.
   - d Neither a nor b.

2. Which of the following is <u>not</u> considered an automotive system?
   - a Power train.
   - b Computer.
   - c Electrical/electronic.
   - d Engine/power plant.

3. The purpose of Automotive Service Excellence (ASE) is to:
   - a Improve the quality of vehicle service and repair.
   - b Test and certify automotive service and repair professionals.
   - c Both a and b.
   - d Neither a nor b.

4. To be ASE certified in a certain area, a technician must:
   - a Pass a written exam.
   - b Have at least two years relevant work experience.
   - c Complete an ASE-certified automotive training program.
   - d Both a and b.

5. Which of the following statements about the automotive industry outlook is true?
   - a Changes in economic conditions have little effect on the automotive service and repair business.
   - b The demand for automotive technicians has never been lower.
   - c Both a and b.
   - d Neither a nor b.

6. The National Automotive Technicians Education Foundation (NATEF) does which of the following?
   - a Provides a forum for technical information and exchange.
   - b Tests automotive technicians on their knowledge and skills.
   - c Evaluates automotive training programs and recommends programs for ASE certification.
   - d Recommends automotive students to prospective employers.

7. Predelivery, the preparation and service of a new vehicle before delivery to a customer, is performed in a(n):
   - a Independent garage.
   - b Service station.
   - c Dealership.
   - d Fleet garage.

8. An automotive parts catalog comes in which of the following forms?
   - a Paper.
   - b Microfiche.
   - c Online.
   - d All of the above.

9. Which of the following would be considered a "nontechnical" automotive career?
   - a Service technician.
   - b Brake specialist.
   - c Automotive salesperson.
   - d Engine machinist.

10. To get started in a career as an automotive service technician, you should:
   - a Have a mechanical aptitude and interest in automotive vehicles and how they work.
   - b Complete an automotive training program.
   - c Gain work experience in a modern automotive service facility.
   - d All of the above.

CHAPTER 2

Automotive Safety Practices

You'll Be Able To:

- ⊗ Recognize hazardous materials and wastes and proper methods for their disposal.
- ⊗ Identify the types of safety information posted in an automotive service center.
- ⊗ Identify personal protective equipment for use in the automotive repair facility.
- ⊗ Use hand and power tools in a safe manner.
- ⊗ Demonstrate a knowledge of fire protection and safety techniques.

Terms to Know:

hazardous materials

material safety data sheet (MSDS)

National Institute for Occupational Safety and Health (NIOSH)

Occupational Safety and Health Administration (OSHA)

personal protective equipment (PPE)

spontaneous combustion

The Problem ⟩ ⟩ ⟩ ⟩ ⟩

You are reinstalling an idle air control valve in Mr. Layman's car. Terry, a new technician in your facility, is assisting you in this procedure. Terry offers to clean the part, using solvent.

You continue to work on Mr. Layman's car. Suddenly you hear a loud commotion from the rear of the shop. It is Terry asking for help. She was not wearing safety glasses when she cleaned the part. Terry is now complaining of pain in her eyes and is rubbing them. Some of the solvent splashed in her eyes.

Your Challenge

As the service technician, you need to find answers to these questions:

1. What should you do in this emergency?
2. Where is the first aid information that explains what Terry needs to do to care for her eyes?
3. How could her accident have been avoided?

Section 1

General Workplace Safety

Safety means protecting yourself and others from danger and possible injury. Just as important, technicians must look out for the safety of others around them. Safety must be the first priority of every employee in the automotive workplace.

OSHA

In 1970, more than 14,000 workers in the United States died due to job-related accidents. Nearly 2.5 million workers were disabled. The estimated number of cases of occupational disease was 300,000. The **Occupational Safety and Health Administration** (OSHA) was created that year to deal with safety hazards. OSHA's purpose is:

> "To assure safe and healthful working conditions for working men and women . . . by authorizing enforcement of the standards developed under the [Occupational Safety and Health] Act . . . by providing for research, information, education and training in the field of occupational safety and health . . . "

All employees must comply with all OSHA regulations. Observance of OSHA regulations and standards has helped make the workplace safer. **Fig. 2-1.**

Workplace Precautions

Every year 6,000 Americans die from workplace injuries. Another 50,000 die from illness caused by chemical exposures in the workplace. Six million suffer non-fatal workplace injuries.

Most hazards are due to careless work habits or unsafe working conditions. Seventy to ninety percent of all accidents are caused by unsafe acts rather than unsafe work conditions. Not wearing proper protective equipment, not practicing safe work procedures, and abusing chemicals and other substances are unsafe acts.

Major hazards in the automotive workplace include slippery floors, blocked doors, unsafe ladders, and poor ventilation.

Floors Slips and falls are a common cause of injuries. Prevent these by keeping floors clean and slip resistant. Contain and clean up all spills and

Fig. 2-1 Everyone working in an automotive repair facility has a responsibility to make safety the first priority. *What government agency is responsible for assuring a safe and healthful workplace?* (Terry Wild Studio)

leaks immediately. Cover or guard all pits and floor openings. **Fig. 2-2.** Provide aisle space and keep it clear, especially around machinery. Keep all workspaces clean, orderly, and well lit. This will make it easier to complete repairs safely.

> **Safety First** **Personal** Diagnostic and repair tasks must be accomplished in accordance with the manufacturer's recommended procedures. All workplace practices must be in accordance with federal, state, and local regulations.

Exit Doors and Aisles Keep areas around exit doors and aisles leading to exits free of obstructions. Clearly identify aisleways with safety tape or other approved material. If there is a need to evacuate the area, a blocked exit could result in serious injury or death.

Ladders Ladders are sometimes used in the repair facility and parts room. Because they are not used frequently, they may not be maintained properly. To prevent accidents, inspect ladders before each use. Maintain them on a regular basis.

Fig. 2-2 All automotive bay pits and floor openings must be covered or guarded when not in use. *(Terry Wild Studio)*

Always use the right ladder for the job. Never use a metal ladder near electric wires. Shock and electrocution can occur if the ladder contacts the wires.

Ventilation Repair facilities must have an exhaust evacuation system for use on running vehicles. **Fig. 2-3.** It is a hazard if a vehicle is running without being connected to the evacuation system. Exhaust gas contains carbon monoxide (CO) and other poisonous materials. CO is a colorless, odorless, tasteless, and poisonous gas. In just three minutes, an engine running in a closed one-car garage can produce enough CO to cause death. A small amount of CO can cause nausea and headaches.

Exhaust System

Fig. 2-3 Automotive repair facilities must have a working exhaust evacuation system for use on running vehicles. *(Terry Wild Studio)*

Hazardous Materials and Wastes

Many products used on vehicles and workplace equipment contain hazardous materials. The waste from these products might also be hazardous waste. **Hazardous materials** and wastes are materials that pose a danger to human health and the environment.

Many laws have been written to protect people from hazardous materials. The *Environmental Protection Agency (EPA)* requires facilities to keep track of, handle, and dispose of hazardous materials properly. The "Right to Know" law, enforced by OSHA, requires that employers tell employees about the dangers of hazardous materials in their facilities.

Hazardous materials and wastes can cause illness or death. They affect humans, animals, and plants. They may cause long-term damage if released into the environment. Common automotive hazardous materials and wastes include:

- Batteries.
- Battery acid.
- Used oil.
- Some used transmission fluid.
- Used antifreeze.
- Brake fluid.
- Solvent.
- Carburetor cleaner.

It is in the best interest of every repair facility to prevent pollution and the buildup of hazardous wastes. Methods include:

- Preventing spills on the floor by using different drain pans for different fluids. Use an oil drain pan for oil and an antifreeze drain pan for antifreeze.
- Using floor soaps and solvents that are biodegradable or easily decomposed by bacteria.
- Recycling and/or disposing of all hazardous wastes including fluids, greases, and lubricants. Dispose of all wastes according to state or federal regulations.

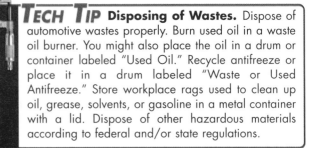

TECH TIP Disposing of Wastes. Dispose of automotive wastes properly. Burn used oil in a waste oil burner. You might also place the oil in a drum or container labeled "Used Oil." Recycle antifreeze or place it in a drum labeled "Waste or Used Antifreeze." Store workplace rags used to clean up oil, grease, solvents, or gasoline in a metal container with a lid. Dispose of other hazardous materials according to federal and/or state regulations.

Table 2-A	USING FIRE EXTINGUISHERS			
	A	**B**	**C**	**D**
Class of Fire				No icon designated for Class D fires
Material Burned	Combustibles: Wood, paper, cloth, etc.	Flammable liquids: Grease, gasoline, oil, etc.	Electrical equipment: Motors, switches, etc.	Combustible metals: Iron and magnesium
Carbon Dioxide (Carbon Dioxide Gas Under Pressure)		●	●	
Pump Tank (Plain Water)	●			
Multi-Purpose Dry Chemical	●	●	●	●
Ordinary Dry Chemical		●	●	
Halon		●	●	
Dry Powder				●

Type of Extinguisher

Information reprinted with permission from *Portable Fire Extinguishers for the Home*, Copyright © 1998, National Fire Protection Association, Quincy, MA 02269.

Fire Safety

One of the most serious hazards in automotive workplaces is fire. Flammable and combustible substances, such as gasoline and solvents, can easily catch fire or cause an explosion. It is actually the vapors that catch fire. If materials are handled properly, fires and explosions can be avoided. Every automotive employee needs to be aware of fire safety issues, preventive measures, and how to respond in emergency situations.

Fire Extinguishers A fire extinguisher is a portable container filled with a fire retardant material. It can be discharged to put out small fires. There are different extinguishers for different types of fires. **Table 2-A** shows various fires and the type of extinguisher to use for each. You should be able to locate the fire extinguishers in the repair facility. Be sure to have the right type of extinguisher on hand. Learn how to use it correctly. Once used, fire extinguishers must be recharged.

Fire Emergency Plan Every workplace should have a fire emergency plan. This plan should include the location of fire exits, where employees should meet outside, and what is expected of each person. Technicians should know what is expected of them in case of a fire.

Fire Prevention Follow these simple rules to prevent fires:

- Do not smoke around gasoline or other flammable liquids.
- Immediately wipe up any spills, especially flammable and combustible liquids.
- Leaking or spilled gasoline quickly vaporizes. Keep workplace doors open and the ventilation system running. Follow workplace rules for disposal, as gasoline is a hazardous material.
- If working on a vehicle that is leaking gasoline, exercise caution. Follow workplace rules for catching and disposing of the gasoline. Avoid tasks that could cause sparks, such as connecting a test light to the battery.
- Store no more than five gallons [19 L] of gasoline in an approved safety container. Never store gasoline in a glass container. The glass could break. An explosion and fire could result.
- When removing or replacing flammable or combustible liquids, use only an OSHA-approved container or portable holding tank. The tank stores the liquid safely. Ground wires prevent sparks that might jump between the tank or hose and the vehicle.
- When pumping a flammable liquid from one container to another, make sure that both

containers are electrically grounded. Connect both containers using a grounding wire. This prevents static electricity from causing a spark.

Fig. 2-4 Store flammable and combustible liquids in fire-resistant cabinets. *What materials in the automotive workplace might be stored in a fire-resistant cabinet?* (Akron Safety)

• Store all flammables and combustibles in a fire-resistant cabinet. **Fig. 2-4.**
• Oily rags can catch fire without a spark or flame. This is called spontaneous combustion. **Spontaneous combustion** is fire caused by chemical reactions with no spark. To prevent this, always store rags containing oil, grease, paint, or solvent in a fireproof safety container with a tightly fitted lid. The lid will prevent air from reaching the rags. **Fig. 2-5.**
• Use caution when using a grinding wheel. Flying sparks can catch clothes and other materials on fire.

Fig. 2-5 Store used rags with oil, grease, or solvents in an OSHA-approved waste can. *Why is a lid required for this can?* (Sears, Roebuck & Co.)

Safety Notices

Accidents are often caused by carelessness and inattention. They can also be caused by using damaged tools or the incorrect tool. Chemicals and substances can impair a person's judgment. It is important that you are aware of the layout of your facility. This will help you react well in an emergency. The term *facility layout* means the location of work benches, equipment, special tools, vehicle lifts, storage areas, and sources of compressed air and water.

OSHA requires automotive service facilities to post certain information for their employees. **Fig. 2-6.** This information includes:
• An OSHA Job Safety and Health Protection poster.
• Emergency telephone numbers.
• Signs indicating exits and fire extinguishers.
• The ingredients of containers holding hazardous materials.
• OSHA also requires that an accident form be completed after each accident. Employers must advise employees of the location of accident forms.

Fig. 2-6 Examples of safety signs posted in an automotive repair facility.

Employers are also required to inform employees of the location of all material safety data sheets (MSDS) covering all of the hazardous chemicals and substances in the building.

A **material safety data sheet** (**MSDS**) is an information sheet that identifies chemicals and their components. The sheet also lists possible health and safety problems and describes safe use of the chemical.

Posted safety signs are designed to maintain a safe and healthy workplace. Other federal, state, and local agencies may require additional postings or employee notifications. Pay attention to all posted signs. Standard colors are used to identify various physical hazards. Red is used for "danger" signs, and to identify fire protection equipment. Yellow means caution. Orange indicates warning. It identifies hazardous equipment or hazardous parts of a machine. Be aware of the location of exits, where chemicals and flammables are stored, and when personal protective equipment should be worn. Also note the recommended evacuation routes for fires, tornadoes, and hurricanes.

Auditory alerts can also help maintain a safe workplace. *Auditory alerts* are sounds that are used to caution people in the area. For example, vehicles such as forklifts emit a distinct sound when driven in reverse. A warning sound alerts others in the area that the forklift is in operation. Flashing lights sometimes accompany these auditory alerts.

Evacuation Routes

All evacuation routes, including aisles and exits, must be clearly marked. The routes should be used when evacuation of an area is required. Emergencies or natural events such as tornadoes, hurricanes, or earthquakes may cause an evacuation.

Lockout/Tagout

Lockout/tagout is an OSHA procedure required in all workplaces. This procedure is designed to prevent electrical equipment from being started while being repaired or maintained. With lockout/tagout, all of the necessary switches must be opened, locked out, and tagged. **Fig. 2-7.** Any person that may be involved with a piece of equipment must be trained on this procedure.

Fig. 2-7 A lockout/tagout station. *When might a technician use a lockout/tagout station?* (Snap-on Tools)

SECTION 1 KNOWLEDGE CHECK

❶ What are the most common types of injuries encountered in the workplace?

❷ Why must hazardous materials be used, stored, and disposed of properly?

❸ What are the four classes of fire?

❹ What is spontaneous combustion?

❺ What is a lockout/tagout procedure?

Section 2
Personal Safety Practices

Automotive technicians can follow a variety of personal safety practices to help ensure a safe workplace. Following these basic rules of safety will help to make the automotive service center a safe and enjoyable place to work.

Personal Protective Equipment

Vehicles and automotive equipment, tools, and chemicals can create hazardous situations for a technician. Everyone should be protected from workplace hazards, hazardous work procedures, and substances that can cause injury or illness. Technicians must use personal protective equipment. **Personal protective equipment (PPE)** is equipment worn by workers to

protect against hazards in the environment. Typical automotive PPE includes:
- Safety glasses, goggles, and face shields.
- Steel-toed boots and shoes with skid-resistant soles.
- Gloves.
- Respirators.
- Protective sleeves.
- Ear plugs and earmuffs.
- Bump hats.

Automotive employers must provide the correct PPE for technicians. Employers must also provide training on the proper use, fit, and care of most required PPE. It is the technician's responsibility to wear PPE.

Eye Protection Approximately 1,000 eye injuries occur every day in the workplace. Three out of five workers injured in this way are not wearing eye

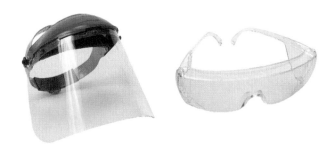

Fig. 2-8 Some types of eye protection available. *What protection should you use when grinding?*

protection at the time of the accident. Seventy percent of eye injuries result from flying or falling objects or sparks striking the eye. Another 20 percent of the eye injuries are caused by contact with chemicals. These accidents can occur in an automotive workplace. Wear protective eye devices to prevent injury. **Fig. 2-8.** Wear a face shield with glasses or goggles when operating machinery. Technicians' eyes can be harmed by:

• Dust thrown off by grinding.
• Welding sparks.
• Small particles produced when chiseling and hammering.
• Radiation from welding.
• Air-conditioning refrigerant.
• Paints, thinners, and solvents.

Foot Protection Accidents account for thousands of disabling foot injuries each year. Automobile engines and automotive parts and equipment are heavy. Service area floors are sometimes slippery. Steel-toed shoes will protect feet from heavy objects. **Fig 2-9.** Footwear must meet OSHA's standards (ANSI Z 41.1) for impact resistance. This standard should be marked on the shoe label. Slip-resistant soles will offer protection from slips and falls.

Fig. 2-9 Steel-toed shoes. *What other type of PPE should you wear when working under a vehicle?* (David Hwang)

Hand and Arm Protection Some of the most common injuries in the automotive workplace are cuts to the hands. Such cuts can be dangerous. Many chemicals and hazardous wastes can cause injury and illness when absorbed through the skin. Some of these chemicals and materials are:

• Used oil, carburetor cleaner, some used antifreeze, and many solvents, paints, and thinners that contain carcinogens and substances that irritate or injure skin.
• Welding produces sparks and flying metal particles that can injure skin.
• Hot coolant that can burn skin.
• Battery electrolyte contains acid that can irritate or burn skin.

Wear gloves to prevent injury to your hands. There are different gloves for different situations. When gloves cannot be worn, such as around moving machinery, use *barrier creams.* Barrier creams can help protect hands from harsh chemicals and other substances.

Technicians routinely work near very hot objects. Repairing exhaust systems or changing oil can expose bare arms to very hot conditions. Technicians should wear protective sleeves to prevent burns. **Fig. 2-10.**

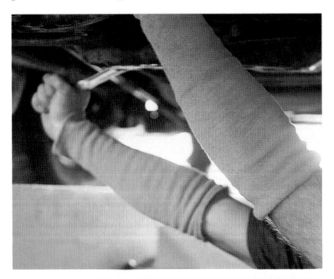

Fig. 2-10 Protective sleeves help prevent burns to bare arms. *What are some situations in which a protective sleeve might be beneficial?* (Terry Wild Studio)

Lung Protection Some brake linings contain asbestos. Exposure to airborne brake dust or asbestos fibers is extremely hazardous. Exposure to asbestos fibers can cause several disabling diseases.

These diseases include asbestosis and gastrointestinal cancer. Technicians should exercise caution when working around asbestos.

- When handling containers of asbestos dust or other asbestos waste, wear a NIOSH-approved respirator. **Fig. 2-11. NIOSH** is the **National Institute for Occupational Safety and Health.** NIOSH tests and certifies respirators and other items of safety equipment. It also recommends filters rated at 100.
- Use a barrier cream on hands along with the standard PPE.
- Never use an air hose to blow dust from a brake assembly. Remove brake dust by one of the two approved methods. Use a wet wash-recycle system to capture the dust. You might also use a *high-efficiency particulate air (HEPA)* filter vacuum system.
- Wash hands after brake work to prevent the transfer of asbestos fibers to food.

Fig. 2-11 A NIOSH-approved respirator. *What safety concerns are associated with asbestos?* (Jack Holtel)

Ear Protection Some automotive tools and equipment, such as air-operated tools and high-speed drills, are very noisy. At certain levels and over a period of time, such noise can cause permanent hearing loss.

Pay attention to noise levels. There are decibel meters available to check the noise level. Wear ear plugs or muffs when appropriate. **Fig. 2-12.**

Fig. 2-12 Examples of ear protection. *In what automotive repair situations should ear protection equipment be worn?* (Jack Holtel)

Head Protection Bump hats will protect a technician's head from hot exhaust systems and sharp protruding objects. These hats also protect against grease, water, dirt, and sparks.

Ergonomics

Ergonomics is the study of workplace design. It studies the tools used, the lighting, and the type of movements required by the employee on the job. Repetitive motions, forceful exertions, vibration, and sustained or awkward postures can cause arm and hand injuries. Common automotive injuries caused by ergonomic stresses are carpal tunnel syndrome, tendonitis, and back injuries.

Preventing Arm and Hand Injuries Whenever possible, use tools that do the twisting for you, such as automatic screwdrivers or electric drills. If using a tool that produces excessive vibrations, wear gloves or put rubber sleeves on the tool. Make sure that the tool fits your hand properly. **Fig. 2-13.**

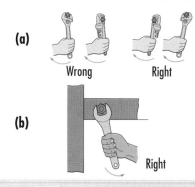

(a) Wrong   Right

(b) Right

Fig. 2-13 Holding tools correctly can prevent serious injury. **(a)** Never apply pressure that would force or spread the moveable jaws of a wrench. **(b)** Never push a wrench. Always pull it. *What safety precaution can be taken when using tools that vibrate excessively?*

Preventing Back Injuries Automotive repair work requires twisting, bending, and heavy lifting. These actions frequently cause back problems. Lift with your legs and keep your back straight. Always ask for help if an object is too heavy. **Fig. 2-14.**

Fig. 2-14 Technicians should lift heavy objects in the workplace using the correct lifting procedures. *What is the correct lifting procedure?* (Jack Holtel)

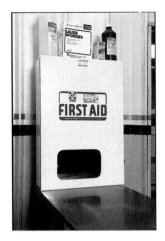

Fig. 2-15 First aid kit. *What precautions must be taken when responding to an emergency?* (Terry Wild Studio)

Responding to Emergencies

The only way to prevent accidents and injuries is to follow safety rules and regulations. Every year there are thousands of accidents and injuries in automotive repair workplaces. There are also numerous deaths. The most common injuries are lacerations, cuts, slips, falls, eye injuries, strains, hernias, and back injuries.

If someone is injured, notify your manager at once. Every repair facility must have a set procedure for emergencies. Every person should know this procedure. If possible, a trained person should be available to give first aid. Everyone in the facility should be able to locate the posted emergency phone numbers, evacuation routes, first aid kits, eyewash stations, and safety showers. **Fig. 2-15.**

Emergencies call for fast action. Remember the following when responding to an emergency:
- If chemicals enter the eyes, flush immediately. See the MSDS for the substance.
- Think twice before giving any first aid. If the injured person is bleeding, special precautions must be taken. Improperly moving a person with a serious injury could worsen the injury.
- CPR requires special training. Technicians who are not trained in the proper procedure should find someone who is trained.

Avoiding Bloodborne Pathogens A *pathogen* is something that causes a disease. Bloodborne pathogens are microorganisms such as the hepatitis B virus and the HIV virus, which causes AIDS. Pathogens are found in the blood of people infected with the virus. The chance of being exposed to these diseases on the job is small. However, it is important to think before reacting when someone is bleeding.

It is easy to prevent exposure. Avoid contact with blood or other bodily fluids. First aid kits contain gloves for protection. Do not give first aid to a person unless you have been trained on how to avoid bloodborne pathogens.

SECTION 2 KNOWLEDGE CHECK

① What eye protection should be worn when operating machinery?

② Where should the OSHA standard for impact resistance be found on a shoe?

③ What precautions should be taken if using a tool that produces excessive vibrations?

④ List two pieces of PPE and describe the use of each one.

⑤ What are some common injuries caused by ergonomic stresses in the automotive workplace?

Section 3

Tool and Equipment Safety

Automotive technicians use many different types of equipment and tools. It is important to learn how to safely use every tool and piece of equipment in the workplace.

Hand Tools

Misuse and improper maintenance pose the greatest hazard in using hand tools. Observe the following when using hand tools:

- Keep hand tools clean and in good condition. Greasy and oily tools are difficult to hold and use.
- Wipe tools clean before and after each use.
- Use a hand tool only for the task for which it was designed. For example, use a screwdriver only for turning screws.
- Always wear safety glasses or goggles when using hand and power tools. This prevents chips and particles from flying into your eyes.
- Use hand tools correctly. For example, you should pull a wrench rather than push it. **Fig. 2-16.**
- Do not use a hardened hammer or punch on a hardened surface. Hardened steel is brittle and can shatter from heavy blows. Slivers may fly out and cut a hand or enter an eye.
- Never use a tool that is in poor condition, such as a hammer with a broken or cracked handle. Do not use chisels and punches with mushroomed heads and broken or bent wrenches.
- Do not use a screwdriver on a part that is being handheld. The screwdriver can slip and hit the hand. This is a common cause of injury.

Fig. 2-16 The correct use of tools. *What might affect what tools are easier and safer for you to use?*
(Jack Holtel)

Fig. 2-17 An electric drill with a three-pronged plug. *What is the function of the third prong?*

Electric Power Tools

There are a number of hazards when using power tools. Observe the following safety practices when using power tools:

- Use only power tools that have been properly grounded or double insulated.
- Make sure that all extension cords are the three-wire grounded type.
- Make sure the three-pronged plug is used in a grounded receptacle. **Fig. 2-17.**
- Make sure that all power tools are equipped with a constant pressure switch that cannot be locked in the ON position. This will prevent a dropped tool from continuing to run, possibly causing injury.
- Avoid accidental startups by keeping fingers off the START switch when carrying a tool.
- Secure work with clamps or a vise. This will free both hands to operate the tool.
- Keep electrical cords away from oil and sharp edges.
- Make sure that all electrical cords are free of frays or breaks in the insulation.
- Disconnect power tools when not in use, before servicing, and when changing accessories such as bits and cutters. This will prevent a sudden, hazardous start-up.
- Keep tools as sharp and clean as possible for best performance.
- Wear the appropriate PPE, such as safety glasses and a face shield.
- Do not wear loose clothing, ties, or jewelry that can become caught in moving machinery.
- Do not wear gloves while operating power tools or equipment that rotates. Gloves can get caught in the moving parts.
- Do not stand in water while working on equipment. Electrical shock could occur.

Fig. 2-18 When in use, the machine guard must be down to prevent injury to the technician. Here the machine guard is shown directly behind the wheel. *(Hunter Engineering Company)*

Machine Guards Machine guards are important safety devices. Always make sure the machine guard is in place when working on grinders and wire and abrasive wheels. Always use the hood guard with a dynamic (spin) tire balancer. **Fig. 2-18.**

Power presses can create enormous pressure on parts in addition to producing sparks, dust, and flying chips. Under such pressure, bearings can burst open and shoot off at high speed. Be sure machine guards are in place. In addition, wear PPE such as face shields with goggles and gloves when using machines with machine guards.

Pneumatic Tools

Pneumatic tools are powered by compressed air. These tools include:
• Impact wrenches for removing and tightening nuts and bolts.
• Grinders for removing sharp edges and burrs, or finishing rough surfaces.
• Cutting tools for removing rusted parts like exhaust systems.
• Blowguns for cleaning parts and removing debris.

Air pressure for these tools is usually regulated to less than 30 psi [207 kPa]. However, uncontrolled or unregulated compressed air can move loose particles and debris at high speed. These particles can cause serious personal injury, especially to the eyes and exposed skin.

The proper use of compressed air is extremely important. When using pneumatic tools, always follow these suggested safety procedures.
• When using a blowgun, always wear the correct PPE, such as safety glasses or goggles, and a face shield. **Fig. 2-19.** Always direct the airflow away from you.
• Always carry a pneumatic tool by its frame or handle. Do not carry the tool by the attached compressed air hose.
• Make sure all pneumatic tools are securely attached to the compressed air line.
• Never use a pneumatic tool, such as a blowgun, to remove debris from your clothing or your body.
• When using compressed air to clean and dry a part such as a ball bearing, avoid spinning the part. A spinning part can eject debris or material remaining from the cleaning process. Use a brush or vacuum system to remove as much debris from the part as possible.
• Securely position a pneumatic tool before operating it. Many of these tools operate at high speed or under high pressure. Incorrect placement may result in component damage or personal injury.

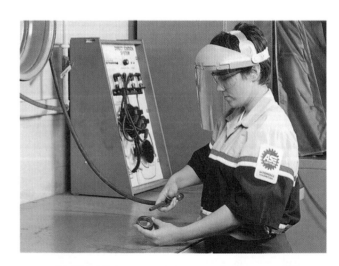

Fig. 2-19 A blowgun is used to blow parts dry or clean. Always wear safety glasses or goggles and a face shield when operating a blowgun. *Why are safety glasses and a face shield important?* *(Jack Holtel)*

Properly maintain compressors to keep contaminants out of the system and hoses. Proper maintenance prevents air hoses from breaking. A broken air hose can cause air and particles to blow into the workplace at high speed.

Routinely check air hoses for leaks and damage. Repair immediately. Lockout and bleed the system before any repair. This will prevent accidental start-up and the escape of pressurized air from a broken hose.

Drive Belts and Pulleys Air compressors have pulleys and drive belts. Pay close attention when using these machines. It is easy to get your fingers caught in moving drive belts and pulleys. Be sure that guards have been placed over any pulleys and drive belts.

Fans can also injure fingers and hands. Keep your hands away from fans. Some fans can come on with the engine and ignition turned off. Always electrically disconnect a fan before beginning any repair job.

Lifts, Jacks, and Safety Stands

The safe use of lifts, floor jacks, and safety stands can prevent many unnecessary accidents.

Many of these accidents are caused by:
- Vehicles placed incorrectly on a lift or floor jack.
- Failure of a floor jack or lift to support the vehicle's weight.
- Vehicles slipping from a floor jack due to vehicle movement or imbalance.

- Incorrect placement of, or failure to use, safety stands.
- Placing a floor jack or safety stand at a lift point other than the designated vehicle lift points.

Vehicle Lifts Always use the manufacturer's recommended lift points when lifting a vehicle. Use of any other point can cause damage to the vehicle or the vehicle could slip off the lift. **Figure 2-20** shows the lift points for one type of vehicle. Check each vehicle's service manual for the correct points.

When using a lift, observe these safety tips. Also, always refer to the lift manufacturer's manual for specific information.
- Only trained technicians should operate a lift.
- Keep the lift area free of oil, grease, and other debris.
- Never overload a lift. Check each lift's maximum weight capacity.
- Never operate a lift with anyone inside the vehicle.
- Never lift a vehicle while someone is working under it.
- When driving a vehicle onto a lift, never hit or run over lift arms, adapters, or axle supports. This could damage the lift and the vehicle.
- If working under a vehicle, make sure the lift has been raised high enough to engage the locking device.
- Before lowering the lift, clear away all tools, trays, and other equipment from under the vehicle.

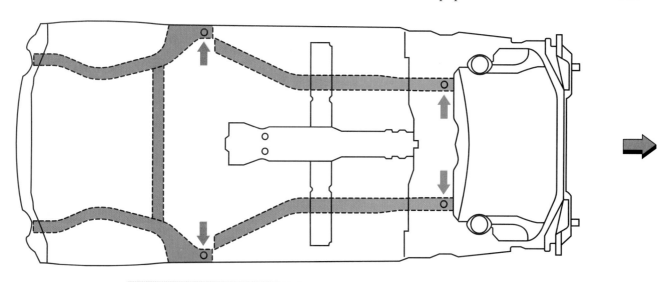

Fig. 2-20 When using a lift, make sure the vehicle is positioned at the lift points recommended by the manufacturer. The lift points here are shown in red. *Why is it important to use the lift points noted by the manufacturer?* (Ford Motor Company)

Fig. 2-21 A floor jack properly lifting a vehicle. *What safety precaution do you see in this photo?* (Terry Wild Studio)

Floor Jacks and Safety Stands   When using a floor jack, position it properly under the vehicle. **Fig. 2-21.** Do not allow the jack to slip out! Always put safety stands in place before going under a vehicle. **Fig. 2-22.** When using safety stands, place the stands and the vehicle on a flat area of the floor where the vehicle weight cannot make the stands sink.

Fig. 2-22 Never go under a vehicle supported by a floor jack unless safety stands are in place. Place safety stands at the lift points recommended by the manufacturer. *Why is it important to place the stands and the vehicle on a flat, hard surface?*

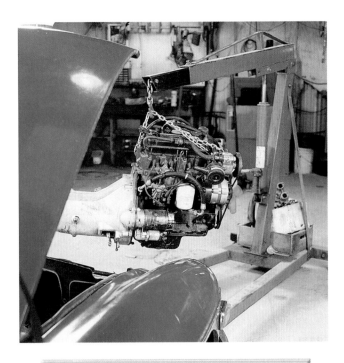

Fig. 2-23 When using a hoist, make sure the load rating of the hoist can support the weight of the load. (Terry Wild Studio)

Transmission Jacks   A transmission jack is used to support a transmission when it is serviced or replaced. Removing a transmission changes a vehicle's center of gravity. The vehicle will be easier to control if it is on a lift or hoist supporting the wheels.

Hoists   A hoist is used to pick up and move heavy objects. For example, technicians use a hoist to remove an engine from a vehicle. **Fig. 2-23.**

Verify that the hoist has the correct load rating for the load. If the load is too heavy or held too far from the body of the hoist, the hoist can fall over. To help prevent this, use an appropriate sling or chain. The bolts attaching the chain to the object must be strong, tight, and threaded well into the object. This will allow lifting without tearing out the threads. When moving a load with a hoist, keep the load close to the body of the hoist to keep the hoist stable.

Electrical Safety

While most vehicles use 12 volts, automotive workplaces use 110 and 220 volts. These voltages and the amperages involved are dangerous. They can cause burns and death. Practice these electrical safety habits when working in the workplace:

- Never use your fingers or bare hands to determine if a circuit is live.
- Always replace a circuit breaker or fuse with a circuit breaker or fuse of the same capacity.
- Place "Out of Order" signs on equipment that is defective.
- Consider every electrical wire to be live until it is positively known it is dead.
- To prevent sparks from broken bulbs, use a fluorescent tube drop light. If a bulb cage light is used, always use a shatterproof bulb. **Fig. 2-24.**

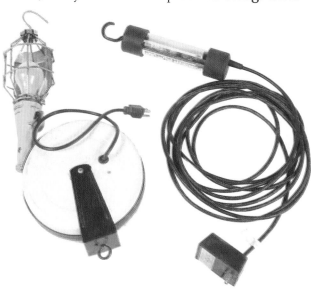

Fig. 2-24 The two types of drop lights most commonly used. *Why is the fluorescent drop light safer?* (Terry Wild Studio)

Welding Safety

Only trained workers should weld or cut with a torch. Head, eye, face, and body protection must be worn when welding or cutting.

Welding uses compressed gases. This presents other safety issues. OSHA has strict regulations for the storage of compressed gases. Technicians must be properly trained before working with compressed gases.

Compressed Gases

Compressed gases include oxygen and acetylene. Cylinders containing these gases must be stored upright, capped, and chained in approved racks. **Fig. 2-25.** There must be at least 20 feet [6 m] or a solid barrier wall between stored oxygen cylinders and stored acetylene cylinders.

To prevent fire or explosion, store cylinders away from room heaters or other heat sources. Never store cylinders in unventilated lockers or closets. Never use cylinders as supports or rollers to move an object. The cylinder could explode.

Fig. 2-25 Compressed gas cylinders must be stored properly. *Why must gas cylinders be stored away from heat sources?* (David Hwang)

SECTION 3 KNOWLEDGE CHECK

❶ What PPE should be worn when using all hand and power tools?

❷ Why should all power tools be equipped with a constant pressure switch?

❸ Name four types of pneumatic tools and their uses.

❹ Name three pieces of equipment that require machine guards when operating.

❺ What safety tips need to be followed when using a lift?

CHAPTER 2 REVIEW

Key Points

Addresses NATEF program guidelines for automotive safety practices, including the work environment and the storage, handling, and use of hazardous materials.

- OSHA was created to establish standards and procedures that help decrease the number of accidents, injuries, and deaths in the workplace.
- Hazardous wastes in the automotive repair facility should be properly disposed of to avoid danger to human health.
- Every employer is required to post certain safety information for its employees.
- Personal protective equipment can prevent exposure to hazards in the workplace.
- It is important to learn how to safely use every tool and piece of equipment.
- It is important for all employees to know emergency procedures.
- Safe work habits can prevent many accidents, injuries, and even deaths.

Review Questions

1. Who is responsible for complying with OSHA?
2. Identify several automotive hazardous materials and wastes and how to properly dispose of them.
3. Name three types of safety information employers are required to post for their employees.
4. Name three items of personal protective equipment and their uses in the automotive repair facility.
5. List three safety concerns when using hand tools.
6. What are several hazards technicians must be aware of when using power tools?
7. List three items that should be included in a fire emergency plan.
8. **Critical Thinking** Why do you think it might be important that CPR training be given to all employees?
9. **Critical Thinking** What substances used in the automotive workplace must have MSDS available?

TECHNOLOGY FORECAST
FOR AUTOMOTIVE EXCELLENCE

Safety Is Everyone's Responsibility

Common sense will continue to be the most important safeguard a technician can use in the workplace. By being alert, using the proper tools, and wearing safety equipment, technicians can easily avoid injury.

Service facility owners have taken on the responsibility of workplace safety, knowing that good conditions make for happy and healthy employees. Increasingly, technicians work in properly ventilated workplaces. Special equipment captures dust, dirt, and fumes, as well as vehicle exhaust, fluids, and lubricants.

Water-based cleaning systems are becoming more common. Looking like large dishwashers, these machines clean vehicle components for the technician. They free the technician from working

with harmful chemicals and make time available for other tasks. Safer citrus-based and non-chlorinated cleansers are also becoming more widely used.

Electrical hazards continue to be a concern. Among the solutions are fluorescent lights instead of the traditional incandescent. These bulbs are cooler and less likely to burn skin. They are also safer around gasoline and other flammable liquids.

Increasingly, building codes require that shock-proof and explosion-proof outlets be installed in service facilities. These sealed outlets lessen the risk of accidentally igniting flammable liquids. Many employers have one or more employees trained in CPR and basic first aid.

AUTOMOTIVE EXCELLENCE
TEST PREP

1. OSHA was created to:
   a Assure safe and healthful working conditions for working men and women.
   b Authorize enforcement of the standards developed under the Act.
   c Provide research, information, education, and training in the field of occupational safety and health.
   d All of the above.

2. Technician A says that carbon monoxide poisoning can occur in a workplace with poor ventilation. Technician B says opening windows will always prevent carbon monoxide poisoning. Who is correct?
   a Technician A.
   b Technician B.
   c Both Technician A and Technician B.
   d Neither Technician A nor Technician B.

3. Technician A says the "Right to Know" law only applies to large automotive workplaces. Technician B says having access to a material safety data sheet is required by the "Right to Know" law. Who is correct?
   a Technician A.
   b Technician B.
   c Both Technician A and Technician B.
   d Neither Technician A nor Technician B.

4. Examples of PPE include:
   a Respirators.
   b Steel-toed shoes with skid-resistant soles.
   c Safety glasses, goggles, and face shields.
   d All of the above.

5. Technician A says MSDS list possible health and safety problems of a chemical. Technician B says MSDS identify chemicals and their components. Who is correct?
   a Technician A.
   b Technician B.
   c Both Technician A and Technician B.
   d Neither Technician A nor Technician B.

6. Which of the following is a good safety practice for power tool use?
   a Keep fingers on the start switch when carrying a tool.
   b Make sure the three-pronged plug is used in a grounded receptacle.
   c Use power tools that have a constant pressure switch that can be locked in the "on" position.
   d All of the above.

7. Technician A says safety stands must always be used with a floor jack. Technician B says the floor jack can be used alone. Who is correct?
   a Technician A.
   b Technician B.
   c Both Technician A and Technician B.
   d Neither Technician A nor Technician B.

8. Technician A says that used rags with gasoline on them should be immediately thrown away in the trash. Technician B says the rags should be stored in a fireproof container with a lid. Who is correct?
   a Technician A.
   b Technician B.
   c Both Technician A and Technician B.
   d Neither Technician A nor Technician B.

9. Technician A says that lockout/tagout is used only on construction sites. Technician B says that any person involved with a machine has to be trained on the lockout/tagout procedures used in the facility. Who is correct?
   a Technician A.
   b Technician B.
   c Both Technician A and Technician B.
   d Neither Technician A nor Technician B.

10. Which of the following are automotive service facilities not required to post for employees?
   a MSDS covering all of the hazardous chemicals and substances in the building.
   b Signs indicating exits and hazardous substances.
   c Emergency telephone numbers.
   d The location of the nearest fire hydrant.

CHAPTER 3

Automotive Tools & Equipment

You'll Be Able To:

- Identify and properly use tools and equipment.
- Explain the relationship among weight, volume, and linear measurements in the metric system.
- Convert measurements from USC to SI and SI to USC.
- Describe the functions of gaskets and where they are used.
- List the three purposes of thread dressings.
- Describe typical uses of aerobic sealant and anaerobic sealant.

Terms to Know:

antiseize compound

pitch

pneumatic motors

System of International Units (SI)

United States Customary (USC) System

The Problem

Buying and maintaining a good set of tools is a costly investment. Be careful to buy proper tools and shop around to get the best prices. When purchasing tools, consider quality as well as price. It doesn't make much sense to buy the cheapest tool you find if it breaks the first time you use it.

You can't buy all the tools you need immediately. Start by buying those basic tools that will be used more frequently in your work as an automotive technician.

Your Challenge

When looking for tools to buy, you need to find answers to these questions:

1. What tools should I buy as a starter set?
2. How much can I afford to spend on my initial tool investment?
3. What warranty does the manufacturer of each tool offer?

Section 1

Hand Tools

Automotive repair technicians must be familiar with a wide variety of tools. The correct tools must be used in the correct manner. As automobiles have become more complex, many of the required tools have become more specialized.

Specialized tools for working on brakes and suspension and steering will be covered in the sections of this book in which they have application. This chapter of the handbook will identify and illustrate the tools and equipment used in automotive maintenance and repair. These are the basic tools of the automotive trade.

Tool Safety

Practice good safety habits when using tools. Chapter 2 of this handbook, "Automotive Safety Practices" provides specific recommendations for tool safety.

Many workplace hazards are created by the careless use of hand tools. Some hazards are obvious and easy to avoid. For example, if you hammer a chisel against a metal part, it is likely that chips of metal could fly out and strike a hand or eye. Always wear eye protection.

Not all tool hazards are obvious. For example, using a pedestal grinder with a wheel out of balance can result in serious injury. A grinding wheel that has turned glassy from heat can break up. Pieces of the wheel can fly off. This can cause injury or death.

There are many possible safety hazards in an automotive repair facility. To avoid these hazards, technicians must always be aware of them. Keep informed of the safety issues involved with the use of tools by following the manufacturer's instructions. All of your questions about tool safety should be answered before you use the tool.

Paying attention to the task at hand and to the potential hazards are the bottom line in tool safety. In most situations there will be some warning of a hazard before it causes an accident.

There is no "acceptable risk level" when using tools. There should be no dangerous situation in auto repair. Every operation and procedure is safe if performed

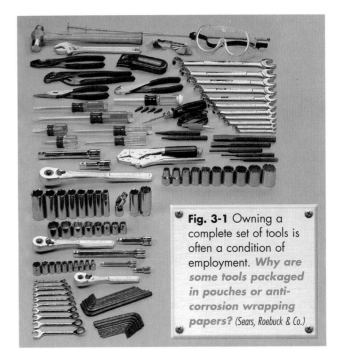

Fig. 3-1 Owning a complete set of tools is often a condition of employment. *Why are some tools packaged in pouches or anti-corrosion wrapping papers?* (Sears, Roebuck & Co.)

properly. If a situation seems unsafe, something may be wrong. Correct the problem immediately.

Using Tools Properly

Most shops will require the technician to have a complete set of hand tools as a condition of employment. A starter set of hand tools can cost $1,000 to $2,000. A basic set of hand tools could easily cost more than $5,000. **Fig. 3-1.** Technicians must upgrade their tools as needed. Because tools are costly, it makes sense to take care of them.

Tool manufacturers usually include care and storage instructions with their tools. Some tools come with custom packages for storage. For example, measuring tools often come in pouches or in anti-corrosion wrapping papers. These protect the tools' delicate precision surfaces. Carefully read the manufacturer's instructions for tool use and care.

To take good care of tools, technicians must know the capabilities and limits of the tools. They must know when and how the tools should be used. A technician must also know how tools should not be used. For example, using a screwdriver in place of a chisel or punch might get the job done. But it might also damage the screwdriver and the part on which it is being misused.

A good technician has respect for his or her tools. Such a technician uses the correct tool for the job. For example, pliers are grasping tools useful for holding nonprecision surfaces, such as a thick

Fig. 3-2 Always use the right tool for the job. *What problems can occur from misusing a tool?* (Jack Holtel)

Fig. 3-4 Power tools are best suited for certain tasks. *What are two types of motors that drive power tools?* (Jack Holtel)

pipe. But pliers are not a good choice for holding bolt heads or the threaded surface of a bolt. Pliers can easily damage a threaded surface or the corners of a nut or bolt head. Using a hand socket on a powerful impact wrench, instead of using a special impact socket, defies common sense. Such misuse of tools also ignores manufacturer's instructions and can void the warranty.

It can be tempting to use the tool-in-hand rather than make the effort to find the right tool. It is important to know which tool is best for a job and to use that tool. **Fig. 3-2.**

Types of Tools

In some cases different types of tools can be used to do the same work. Different tools can provide different efficiencies in doing the same job.

Hand Tools The most basic tools are hand tools such as wrenches, screwdrivers, and hammers. Hand tools are used to tighten and loosen fasteners and move tightly fitted parts. Your hand supplies the energy to use hand tools. **Fig. 3-3.**

Fig. 3-3 Hand tools are the most basic tools. *What supplies the energy for hand tools?* (Jack Holtel)

Power Tools Power tools are driven by electric or pneumatic motors. **Pneumatic motors** are motors that are powered by air pressure. This pressure is created by an air compressor. Tools driven by pneumatic motors should be lubricated daily with oil. The air compressor should be drained of air at the end of each day.

You can often work more quickly with power tools than with hand tools. However, power tools are usually not as useful for performing sensitive operations or for making delicate adjustments. In many cases, a power tool is the only practical tool to use. For example, power tools are best suited for quickly removing nuts and bolts or for unthreading long-threaded fasteners. **Fig. 3-4.** Power tools sometimes give better access to a vehicle part.

Storing and Maintaining Tools

Keep your tools clean, organized, and in good working condition.
- Clean hand tools after every use.
- Remove oil and grease from tools before storing.
- Store tools properly.

Tools left lying on the shop floor are a safety hazard, causing someone to slip or trip. Technicians who store their tools carelessly will waste time looking for the right tool. Properly storing tools in tool chests and other appropriate containers will protect them from impact, abrasion, and corrosion.

Table 3-A identifies and describes the basic hand tools usually required for employment as an automotive technician. Before applying for a job, a technician should have all the tools in this table. He or she should know how to properly use each tool.

Table 3-A INDIVIDUAL HAND TOOLS

Tool	Description	The Job
Wrenches	• Long-handled tools with fixed or adjustable jaws. • A 1/4″–1″ USC set and a 7 mm–19 mm metric set will handle most jobs. The measurement given is the width of the jaw opening. Purchase in sets of open-end, box, and combination wrenches.	• Use to turn bolts, nuts, and screws.
❶ **Adjustable Wrench**	• Has a movable jaw that can be adjusted to fit nuts and bolt heads of various sizes. • The two most common sizes are 6″ and 12″.	• Use when other types of wrenches will not fit the nut or bolt head. • Because of its flexibility, it is a good tool to carry for emergencies. • Use only when applying relatively light torque.
❷ **Allen Wrench**	• Small L-shaped or T-shaped six-sided bar. Sometimes the bar has a six-pointed star shape. • A complete set should include standard 0.05″–3/8″, as well as metric 2 mm–7 mm. • Allen wrenches are available in sizes ranging from 3/32″–5/8″ and from 2 mm to 19 mm.	• Use to turn cap screws that have an internal hex turning configuration.
❸ **Combination Wrench**	• Has a box on one end and is open on the other. • The two ends are usually the same size. • The most common box-wrench has 12 notches or "points" in the head. A six-point box wrench holds better on a nut or bolt but needs a greater swing. • Provides the flexibility of a box on one end and an open wrench on the other. • A complete set should include standard 1/4″–1″ and metric 7 mm–19 mm.	• Designed for a wide variety of work. • A 15° angle on the open end allows use when a minimum of wrench-swing space is available. • Because of its weak jaws, the open end should not be used on extremely tight nuts or bolts. • Use the box end for loosening extremely tight nuts or bolts.

Table 3-A	INDIVIDUAL HAND TOOLS (continued)	
Tool	**Description**	**The Job**
❹ Crowfoot Wrench Set	• Consists of a set of crowfoot drives and holders or extension bars. • Specialized open- and box-end wrenches that are turned with a socket handle. • The drives have slots to accept the holders or extension bars. • Available in standard and metric sizes.	• Use with an extension bar to turn nuts and bolts in hard-to-reach locations. • Use with torque wrenches when there is no other way to access a fastener head.
❺ Flare-Nut (Tubing) Wrench	• Special type of combination or box-end wrench. • The ends of the flare-nut wrench are thicker than those on other wrenches. This helps prevent slipping and rounding off the points on soft-metal tube fittings. • A complete set should include 3/8"–3/4" standard and 10 mm–17 mm metric sizes.	• Use to attach or remove a flare nut or tubing nut from brake lines. • Use to grip the jamb nuts on fittings. • The flare nut pattern, with its hex box opening, is especially useful for air conditioning work where tubing terminates on flare nuts.
❻ Ignition Wrench Set	• Open-ended wrench with a 15° offset on one end and a 60° offset on the other end. • Sizes range from 1/8" to 3/4" and from 3.2 mm to 11 mm.	• Use for turning small nuts for ignition system adjustments. Used on contact point distributor ignitions. Distributorless systems do not have these components.
❼ Torque Wrench	• Includes a gauge that measures force expressed in units of distance and weight. • A complete set should include a 3/8" drive (10–250 lb-in), a 3/8" drive (5–75 lb-ft), and a 1/2" drive (50–250 lb-ft). • Torque wrenches are available in 1/4", 3/8", and 1/2" drives.	• Use when fastening pressure is critical to sealing or operation.

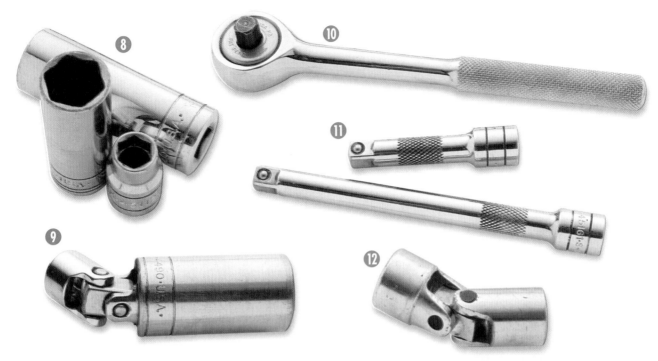

Table 3-A	INDIVIDUAL HAND TOOLS (continued)	
Tool	**Description**	**The Job**
Socket Sets	• These are probably the most widely used tools in a service center. • A socket wrench has two basic parts—a detachable socket and a ratchet handle.	• Use to increase the speed for removing or installing bolts or nuts.
⑧ **Socket**	• Cylinder-shaped, extended box-end tool that is placed on the ratchet handle. • Sockets are available in 6-point, 8-point, or the standard 12-point socket. Both regular-depth sockets and deep sockets are available. • The drive end of the sockets come in four sizes: 1/4", 3/8", 1/2" and 3/4". • Sizes range from 1/4"–1-1/8" and from 6 mm–25 mm.	• Use 6-point socket if a bolt head or nut has rounded corners or excessive resistance to turning. • Use 8-point socket for turning square heads such as drain plugs, fill plugs, and pipe plugs. • Use 12-point socket to turn a bolt or nut in tight spots.
⑨ **Spark Plug Socket**	• A special deep socket that has a rubber insert.	• Use for removing and installing spark plugs.
⑩ **Ratchet**	• Connects to a socket and provides the leverage and ratchet mechanism for turning nuts and bolts. • Has a mechanism that permits free motion in one direction but lockup in the other. Select the direction for lockup by moving the reversing lever on the ratchet handle.	• Use to provide leverage and to allow quick ratchet removal or tightening.
⑪ **Extension**	• Fits into the ratchet handle at one end and receives a socket at the other end. • A complete set should include 3", 6", 12", and 18" extensions.	• Allows you to reach with a ratchet otherwise unreachable nuts or bolts.
⑫ **Universal Joint**	• Connecting piece with a swivel. • Connects to the driver at one end and the socket at the other end.	• Allows you to turn a nut or bolt while holding the driver at an angle.

Table 3-A	INDIVIDUAL HAND TOOLS (continued)	
Tool	**Description**	**The Job**
Pliers	• Hand tools with a pair of adjustable pivoted jaws.	• Use for cutting, crimping, bending, or gripping.
⑬ 6″ Combination Pliers	• Standard pliers that can be adjusted to hold various sizes of items.	• Provide great versatility.
⑭ 6″ Needle-Nose Pliers	• Pliers that have long narrow gripping area.	• Use for holding or positioning small parts and when there is minimum space for maneuvering the plier head.
⑮ Hose-Clamp Pliers	• Have special teeth for holding hoses. • There are two types—locking and nonlocking.	• Use to remove hose clamps.
⑯ Locking-Jaw Pliers	• Have locking jaws that are adjusted by turning a screw at the end of one handle. • Have a release lever on one of the handles.	• Use when you need a vise-like grip on the part being held or turned.
⑰ Side-Cutting Pliers	• Often called "side cutters" or "diagonals."	• Use for cutting thin sections of rod or wire or removing or installing cotter pins.
⑱ Slip-Joint (Water-Pump) Pliers	• Tongue and groove pliers.	• Use for various gripping jobs.

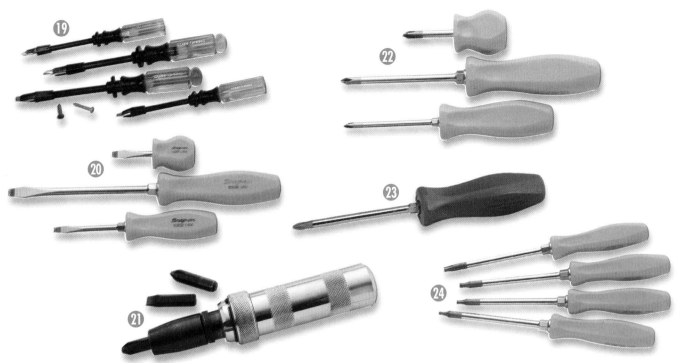

Table 3-A	INDIVIDUAL HAND TOOLS (continued)	
Tool	**Description**	**The Job**
Screwdrivers	• Made in a variety of sizes, shapes, and special-purpose designs. • Hand-held turning devices that can be equipped with a wide variety of head or drive configurations.	• Use for driving or turning screws.
⑲ **Screw Starter**	• Tool for holding either a Phillips or standard screw. • Spring-loaded steel bits grip screw-slot wall.	• Use to grasp a screw to more easily start the screw in a threaded hole.
⑳ **Single-Blade Screwdriver**	• Has a single blade that fits into the slot of a screw head. • A complete set should include a Stubby, a 6", a 9", a 12", and offset screwdrivers.	• Use for turning slotted-head screws.
㉑ **Impact-Driver Screwdriver**	• A hand-held hex-shaped screwdriver that is approximately 6" long. • Uses the impact of a hammer blow to seat a driver blade in a screw and turn it with great momentary force.	• Use for removing a screw that is "frozen" in its thread.
㉒ **Phillips-Head Screwdriver**	• Has two crossing blades that fit into a similarly shaped screw slot. • A complete set should include a Stubby #1 and #2; a 6" #1 and #2; a 12" #3, and an Offset #2.	• Use for turning Phillips-head screws.
㉓ **POZIDRIV® Screwdriver**	• The walls on POZIDRIV® tips are not as tapered as on Phillips tips and have wedges for a tight fit. • A complete set should include a #1, #2, #3, and #4.	• Use for turning POZIDRIV® screws.
㉔ **Torx® Head Screwdriver**	• A complete set should include the following: T-8, T-10, T-15, T-20, T-25, T-27, T-30, T-40, T-50, T-55.	• Use for turning Torx® head screws.

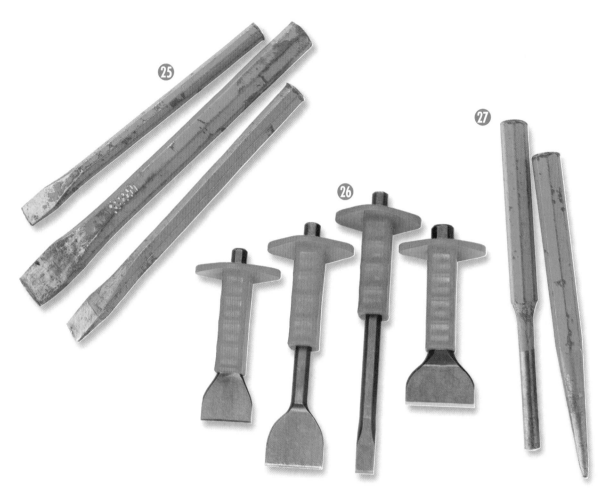

Table 3-A	INDIVIDUAL HAND TOOLS (continued)	
Tool	**Description**	**The Job**
Chisels and Punches	• A chisel is a cutting tool with a single cutting edge. • Chisels are available in a variety of shapes and sizes. • Chisels are made with tough steel alloys.	• Use chisels for cutting and shearing. • Use punches to drive round objects, such as pins that are tightly fitted in precision holes.
㉕ Chisel	• A set should include a 5/16" cape chisel and a 3/8" and a 3/4" cold chisel.	• Use to cut off damaged or rusted nuts, bolts, and rivet heads.
㉖ Chisel Holder	• Has two basic parts. One is a doughnut-shaped part that is placed around the chisel. The other is a plastic handle that is attached to the doughnut. The handle is turned to tighten the doughnut around the chisel.	• Use to protect your hand and to get a strong grip on the chisel.
㉗ Punch	• A set should include a center punch; a brass drift punch; 1/8", 3/16", 1/4" and 5/16" pin punches; and 3/8", 1/2", and 5/8" taper punches.	• Use a center punch to mark the spot where a hole is to be drilled. • Use a pin punch after marking the spot with a center punch. It will provide a more stable starting point for the drill bit. • Use a brass punch to avoid damaging the surface of the part being driven. • Use punches to knock out rivets and pins or to align parts for assembly.

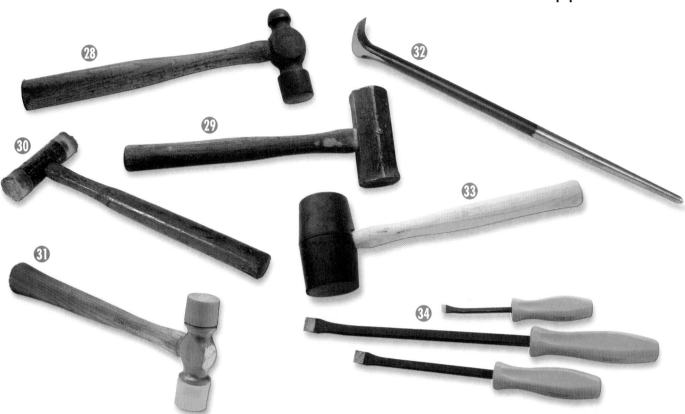

Table 3-A	INDIVIDUAL HAND TOOLS (continued)	
Tool	**Description**	**The Job**
Hammers and Prying Tools	• A hammer consists of a weighted striking head and a wooden, composite, or plastic handle.	• Use hammers for striking. • Use prying tools for leveraging or separating parts.
㉘ **16-oz Ball Peen Hammer**	• Most commonly used hammer in an automotive shop. • Has a flat face on one end of the head and a round end on the other.	• Use the flat end for general striking of objects. • Use the round end for shaping metal parts or rounding rivets.
㉙ **Brass Hammer**	• Hammer head is made of brass to prevent steel parts from chipping when struck. • The relatively soft head is designed to accept the energy of the blow. This protects the surface being struck.	• Use for striking parts you wish to protect from being marred.
㉚ **Dead-Blow Plastic Mallet**	• Designed to deliver impact with a minimum of rebound. • Plastic coated, metal face. • Has a hollow head partially filled with small metal shot.	• Use to reduce bounce-back of hammer head.
㉛ **Plastic-Tip Hammer**	• Plastic surface prevents damage to item being struck.	• Use when only light blows are required and when protecting the striking surface from being marred.
㉜ **Rolling-Head Pry Bar**	• Has a pry bar with a 90° bent head. • Also known as a lady foot.	• Use when leverage is needed for jobs such as tightening pulleys or separating tightly fitted assemblies.
㉝ **Rubber Mallet**	• Hammer head is made of solid rubber.	• Use when striking easily marred surfaces.
㉞ **Straight Pry Bar**	• A long shank made of hardened alloys with a plastic handle. • Range in size from 8" to 24".	• Use for removing and aligning parts.

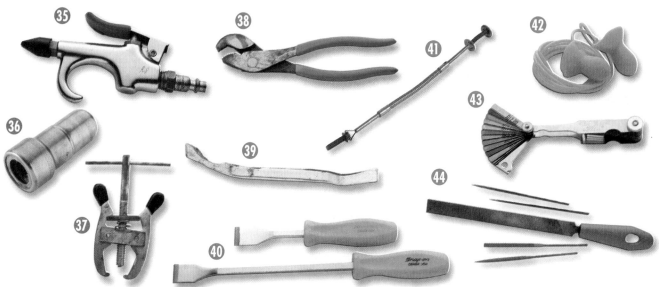

Table 3-A	INDIVIDUAL HAND TOOLS (continued)	
Tool	**Description**	**The Job**
Miscellaneous Tools	• All other hand tools that are used in the automotive repair facility.	• Uses include vehicle support, vehicle charging, grinding, and protection.
㉟ **Compressed-Air Blowgun**	• Connects to an air hose. • Brings air to small tip for high pressure. • Has a shut-off valve to stop flow of air.	• Use to direct compressed air at parts to dry or clean them.
㊱ **Battery Post Cleaner**	• Fine wire or scraper. • Reamer encased in a holder.	• Use for cleaning battery posts and terminals.
㊲ **Battery Terminal Puller**	• Jaws engage below battery terminal with a center screw to lift the battery terminal.	• Use to remove battery terminals.
㊳ **Battery Terminal Pliers**	• Compact parrot-nose pliers with especially strong gripping power.	• Use to remove battery terminals.
㊴ **Brake Spoon**	• A wide-blade tool similar to a flat pry bar. • Approximately 12" long.	• Use to adjust drum brakes.
㊵ **Carbon Scraper (1")**	• Carbon blade has special sharpened angles that won't gouge metal surfaces.	• Use to remove carbon and other debris from metal parts.
㊶ **Claw-Type Pickup Tool**	• A mechanical finger on a flexible section that is operated by a button that opens and closes the jaws.	• Use for picking up small objects in tight spaces.
㊷ **Ear Protection**	• Something to cover or insert in ears.	• Use to protect ears from loud or piercing shop noises.
㊸ **Feeler Gauge (Blade Type) Set**	• A set of precision-thickness metal blades. • Complete set should include a set of standard feeler gauges, 0.002"–0.040", and a set of metric feeler gauges, 0.006 mm–0.070 mm.	• Use for adjusting spark plug gap to prescribed measurement.
㊹ **File**	• On a coarse file, there are only a few cutting edges per inch. On a fine file, the cutting edges are close together. • Set of files should include a coarse 6" and 12," a fine 6" and 12," a half-round 12", and a round 6" and 12".	• Use for removing burrs and cuts, smoothing sharp edges, and reducing and shaping metal.

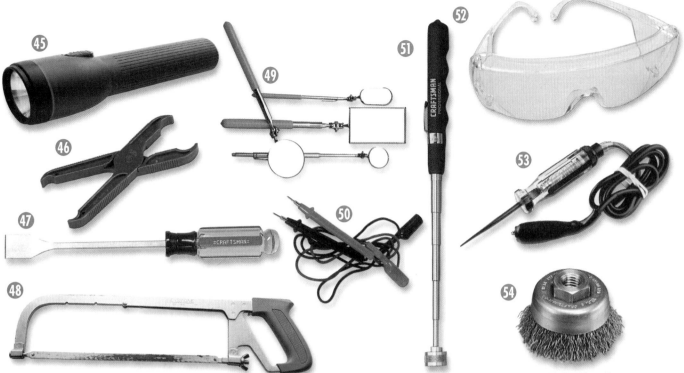

Table 3-A	INDIVIDUAL HAND TOOLS (continued)	
Tool	**Description**	**The Job**
㊺ Flashlight	• Standard battery-operated, hand-held flashlight.	• Use instead of a utility light.
㊻ Fuse Puller	• Tool for gripping and removing blade- and glass-type fuses.	• Use to remove fuses.
㊼ Gasket Scraper (1")	• A 1"-wide blade connected to a handle.	• Use to remove gaskets or carbon and other debris from parts.
㊽ Hacksaw	• An adjustable metal frame that holds a steel saw blade. • Blade is replaceable and has from 14 to 32 teeth per inch.	• Use for making a smooth cut in metal or hard plastics.
㊾ Inspection Mirror	• Small mirror with a telescoping handle.	• Use to inspect areas that would not otherwise be visible.
㊿ Jumper-Wire Set	• A wire lead with a terminal or probe on each end.	• Use to bypass and test circuits.
51 Magnetic Pickup Tool	• Consists of a magnet hinged to the end of a rod. • Most magnetic pickup tools can be adjusted for length and swiveled to reach into otherwise inaccessible areas.	• Use for retrieving dropped metal parts from hard-to-reach locations.
52 Safety Glasses (meeting OSHA requirements)	• Made with top and side protection and shatterproof lenses.	• Use to protect eyes from flying debris.
53 Test Light (12 V)	• Has a probe on one end and an encased lightbulb on the other.	• Use for testing circuits and locating shorts, grounds, and open circuits up to 12 volts.
54 Wire Brush	• Wire brush either connected to a handle or incorporated into a wheel for use on power-driven equipment.	• Use to clean surfaces.

SECTION 1 KNOWLEDGE CHECK

❶ Why must automotive technicians know how to use a wide variety of tools?

❷ Explain the phrase "no risk is acceptable" as it refers to automotive repair.

❸ Why must a technician know a tool's capabilities and limits?

❹ List the general difference in the way hand tools and power tools work.

❺ What are torque wrenches and when are they used?

Section 2

General Workplace Equipment

Automotive technicians usually supply their own tools. The employer usually supplies the stationary equipment. This equipment is shared by many technicians.

General workplace equipment is used for a wide variety of tasks. These tasks include hoisting cars off a shop floor, cleaning and surfacing parts, and analyzing performance defects. Some general equipment, such as holding fixtures and calibrating devices, is highly specialized for certain operations or for use on specific cars. **Fig. 3-5.** Other equipment, like cleaning tanks and hydraulic presses, is more general. It is used for many different tasks and on many types of vehicles.

Using General Equipment General equipment can be more difficult to learn to use. Some equipment, like electronic diagnostic devices, is routinely upgraded. As a result, technicians need training to operate the equipment effectively.

There is one aspect of automotive work that new technicians can find confusing. There are often many different ways to do the same job. Different kinds of equipment will sometimes provide similar results. For example, special equipment can surface brake drums either by grinding them or by cutting them. Each method has advantages and disadvantages.

A manufacturer may recommend or require one specific procedure or piece of equipment when several others might also seem appropriate. As a result, you need to know more than just what a piece of equipment can do. You also need to know how it works, as well as its strengths and weaknesses. It is also important to know the applications for which the equipment is recommended or required.

Table 3-B identifies and describes the general tools and equipment required in a well-equipped automotive workplace. These tools and equipment are generally not considered to be individual hand tools.

Fig. 3-5 Many automotive service stations require using stationary workplace equipment, such as a hoist. *(Jack Holtel)*

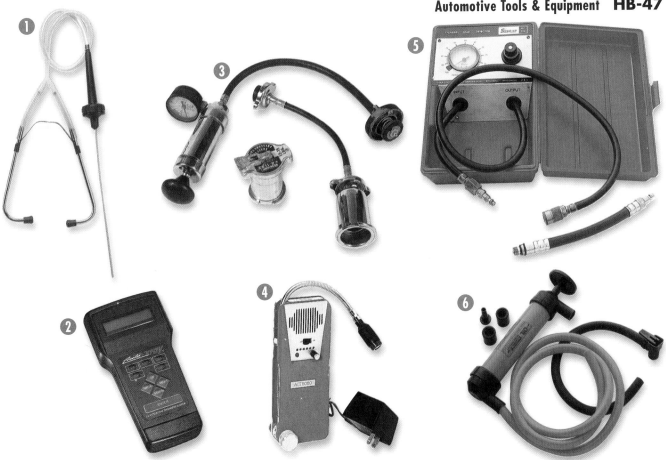

Table 3-B	GENERAL EQUIPMENT	
Tool	**Description**	**The Job**
Diagnostic Tools	• Tools that diagnose various vehicle conditions.	• Uses include locating abnormal noises, checking leaks, and gathering data from the vehicle.
❶ **Automotive Stethoscope**	• Consists of a headset connected to a hollow flexible tube that is attached to a sound-amplification probe. • An electronic stethoscope is preferred.	• Use to listen to very specific areas in cars to find abnormal noises.
❷ **Computer Scan Tool**	• Either a handheld computer scan tool or a personal computer with interface capability for on-board diagnostics.	• Use to gather data from the various electronic functions of a car and sensors.
❸ **Cooling System Pressure Tester and Adapters**	• A handheld pump to pressurize radiator and cap.	• Use to determine the ability of a cooling system to hold pressure. • Use to check for antifreeze leaks.
❹ **Cooling-Combustion Gas Detector**	• Electronic device that employs a chemical reaction to identify combustion gases in a vehicle's cooling system.	• Use to analyze cooling systems.
❺ **Cylinder Leakage Tester**	• Kit used with air compressor to test leakage on gasoline and diesel engines. • Consists of a neoprene hose, air gauge, and an adapter on each end of the hose. One end connects to the air supply. The other end threads into the cylinder chamber.	• Use as a test to measure ability to hold air. • Use to locate worn rings, defective valves, a cracked cylinder, or a leaky head gasket.
❻ **Hand-Held Vacuum Pump**	• A hand-held pump, with a gauge, used to create a vacuum, usually by pumping action.	• Use to simulate engine vacuum to check various vacuum-operated and calibrated devices on automobiles.

Table 3-B	GENERAL EQUIPMENT (continued)	
Tool	**Description**	**The Job**
Electrical Tools	• Tools that require electricity for operation.	• Uses include drilling, jump starting, and soldering.
⑦ **Drill**	• 3/8" variable speed, reversible and 1/2" variable speed, reversible.	• Use for heavy-duty drilling.
⑧ **Extension Cord**	• 12- or 16- gauge 3-conductor electrical cord with ground plug.	• Use for electrical equipment that has a cord too short to reach an electrical outlet.
⑨ **Hot Plate (or equivalent)**	• A heating element that can be heated to a desired temperature.	• Use to test thermostats or to heat parts for precise fitting.
⑩ **Jumper Cables**	• Cables with clamps on both ends.	• Use to jump start a vehicle with a dead battery, from either the battery of another vehicle or from a stationary battery.
⑪ **Remote Starter Switch**	• Two wires with clamps on each end, connected to a switch.	• Use to crank an engine or start it without turning the ignition key.
⑫ **Soldering Gun**	• A pistol-looking tool with a heating element.	• Use to melt solder-to-solder electrical connections.
⑬ **Soldering Iron (25-Watt Pencil Tip)**	• A pencil-looking tool with a heating element.	• Use to melt solder-to-solder electrical connections.
⑭ **Twist-Drill Set 1/64"–1/2"**	• A hand-held air or electric motor or a drill press drives the drill bit. • The three major parts of a drill bit are the shank, body, and cutting head. The grooves along the body carry the removed chips out of the hole.	• Use for drilling holes in metal or other materials.

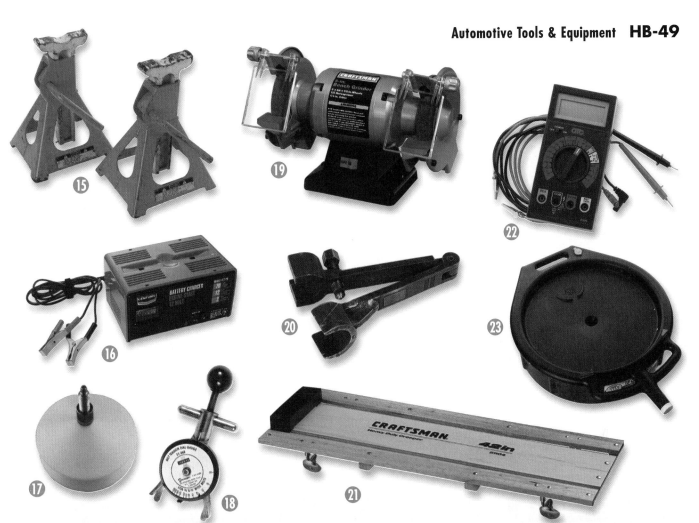

Table 3-B	GENERAL EQUIPMENT (continued)	
Tool	**Description**	**The Job**
Miscellaneous Tools	• All other tools that are used in the automotive repair facility.	• Uses include vehicle support, vehicle charging, grinding, and protection.
⑮ **Axle Stands (Safety Stands)**	• Strong steel stands that have a center post that can be adjusted for height.	• Use to safely support a vehicle so you can work under it.
⑯ **Battery Charger**	• A converter box that changes 120-volt current to 6-volt, 12-volt, or 24-volt current.	• Use for charging vehicle batteries.
⑰ **Bearing Packer (hand operated)**	• Cup, usually transparent, filled with grease between two plastic cones.	• Lubricates a bearing with new grease.
⑱ **Belt-Tension Gauge**	• Spring-loaded gauge that shows pounds per square inch and fits over the belt.	• Use to measure the tightness of a belt.
⑲ **Bench or Pedestal Grinder**	• A grinding wheel attached to an electric motor.	• Use for grinding steel.
⑳ **Constant Velocity (CV) Universal Joint Service Tools**	• Requires a boot-installation tool and boot-clamp pliers or crimping ring.	• Use to remove, repair, or install CV joints.
㉑ **Creeper**	• Four swivel wheels attached to a flat board with a padded headrest.	• Use to move around under a vehicle.
㉒ **Digital Multimeter with Various Lead Sets**	• Digital continuity meter reading volts and amps.	• Use to check electrical voltage.
㉓ **Drain Pans**	• Plastic or steel containers.	• Use to catch liquids such as antifreeze or oil.

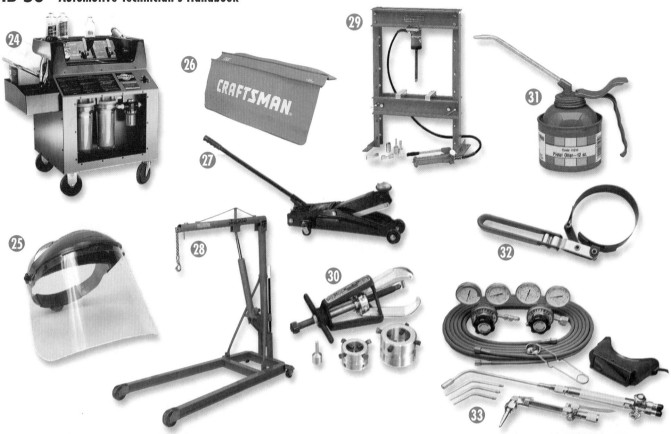

Table 3-B	GENERAL EQUIPMENT (continued)	
Tool	**Description**	**The Job**
㉔ **Engine Coolant Recovery Equipment (Recycler or Coolant Disposal Contract Service)**	• A suction pump connected to a storage tank.	• Use for removing and storing antifreeze.
㉕ **Face Shield**	• Transparent plastic shield to cover face. • Held on head by adjustable strap.	• Use to protect eyes from splashing chemicals or flying debris.
㉖ **Fender Cover**	• Plastic material or cloth that fits over fenders.	• Use to protect body of vehicle from scratches and other damage while working on it.
㉗ **Floor Jack (1-1/2 Ton Minimum)**	• A steel arm raised by a hydraulic cylinder.	• Use to jack up a vehicle.
㉘ **Hoist**	• Large piston-driven device used to lift heavy items.	• Use to remove engines.
㉙ **Hydraulic Press with Adapters (25 Ton)**	• Vertically oriented press that accepts items to be pressed together by hydraulic pressure.	• Use to press together parts that require more than normal pressure to join; used, for example, to place bearings.
㉚ **Master Puller Set**	• Assortment of pulling tools for a wide range of tasks.	• Use to remove hubs, bearings, pulleys, and so forth.
㉛ **Oil Can—Pump Type**	• Stores small amount of oil and has a mechanism for pumping the oil from the can.	• Use to oil hinges, latches, and other parts that require lubrication.
㉜ **Oil Filter Wrench**	• A strap or cup that fits and tightens around a filter.	• Use for removing or installing oil filters.
㉝ **Oxy Acetylene Torch**	• Tool that mixes oxygen and acetylene to produce a strong flame.	• Use to cut or braze metal.

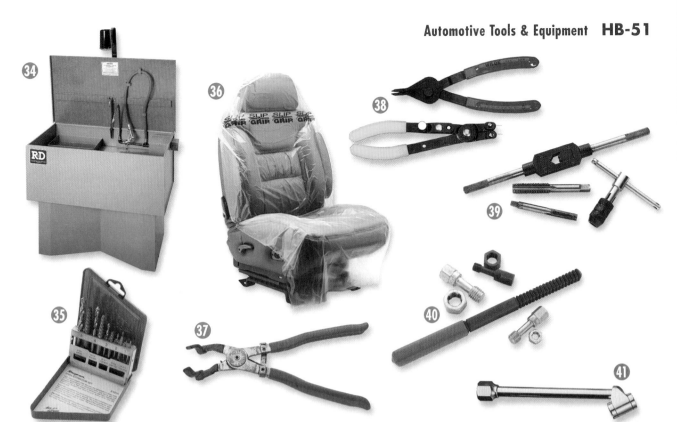

Table 3-B	**GENERAL EQUIPMENT (continued)**	
Tool	**Description**	**The Job**
㉞ Parts Cleaning Tank (nonsolvent-based cleanser suggested)	• Sink-like tank used for cleaning various parts in a cold liquid solvent.	• Use gloves to protect your hands from harsh chemicals. • Use to remove dirt, grease, and adhesives from gaskets or other parts.
㉟ Screw Extractor Set	• Drill-bit set used to extract broken screws and bolts.	• Use to remove broken screws and bolts.
㊱ Seat Cover	• Piece of plastic, cloth, or paper designed to fit over seats.	• Use to protect vehicle seats from grease, tears, and other damage while working on vehicle.
㊲ Spark Plug Boot Puller	• Special type of pliers with a unique grabbing end that is designed to fit over spark plug wire boot.	• Use to remove stubborn spark plug wires.
㊳ Snap-Ring Pliers	• Have sharp, pointed tips. • Internal and external available.	• Use external for installing and removing special clips called snap rings. • Use internal to hold a bearing in a housing.
㊴ Tap and Die Set	• Made of very hard steel alloy that is strong enough to rethread nuts or bolts. • Available in standard and metric.	• Use tap to cut internal threads in holes. • Use die to cut external threads on bolts, studs, or rods.
㊵ Thread Repair Kit	• Includes tough drill bits for drilling out damaged threads, and inserts that fit in the drilled-out hole, providing new threads. • The inserts are designed to provide the same size thread as the original.	• Use to replace threads that cannot be repaired.
㊶ Tire Inflator Chuck	• Adapter that connects to air-compressor hose allowing air to be added to tires.	• Use to add air to tires.

Table 3-B	GENERAL EQUIPMENT (continued)	
Tool	**Description**	**The Job**
㊷ **Trouble/Work Lights (Fluorescent Preferred)**	• Electrical light.	• Use to light a wide area. • Because they produce less heat, fluorescent lights should always be used when working near gas tanks or with substances that could become dangerous with the addition of heat.
㊸ **Tube Quick-Disconnect Tool Set**	• Plastic cups that fit over various supply lines.	• Use to disconnect air-conditioning, radiator, and fuel lines.
㊹ **Tubing Bender**	• Special tool used for bending copper, brass, or steel tubing. • There are various kinds of tubing benders.	• Use for bending tubing.
㊺ **Tubing Cutter/Flaring Set (Double-Lap and ISO)**	• A cutting wheel and feed mechanism that is adjustable to fit over tubing.	• Use to fit over tubing and cut the tubing.
㊻ **Valve-Core Removing Tool**	• Convenient guide that slips into the tire-valve stem.	• Use to remove or install valve-stem core.
㊼ **V-Blocks**	• A steel block with a V-shape machined into it.	• Use to hold parts for precision measuring.
㊽ **Waste-Oil Receptacle with Extension Neck and Funnel**	• A receptacle with an adjustable telescopic tube that has a funnel at its top.	• Use to collect oil.
㊾ **Wheel Chocks**	• A block of wood or rubber.	• Use to keep vehicle from rolling.
㊿ **Workbench**	• Stable metal or wooden tables on which work can be performed.	• Use for working on components that require assembly or disassembly.
51 **Vise**	• A steel or cast-iron tool used for holding parts. • The five basic parts of a vise are the handle, jaws, anvil, base, and position-adjustment bar.	• Use to free both hands for working on parts that require cutting, drilling, hammering, or gluing.

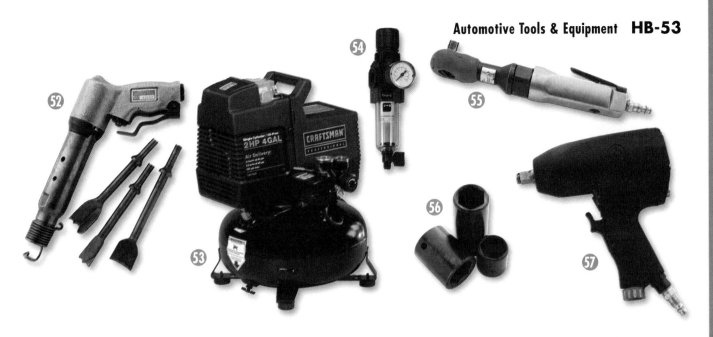

Table 3-B	GENERAL EQUIPMENT (continued)	
Tool	**Description**	**The Job**
Pneumatic Tools	• Tools that are powered by compressed air.	• Use for a variety of tasks that would take much longer using hand tools.
⑤ Air Chisel Set with Various Bits	• Series of special chisel bits designed to be used with pneumatic tools. • Chisels include the following: flat chisel, sitting chisel, bent-end chisel, curved flat chisel, scraper, double-blade panel cutter, and muffler cutters.	• Use to speed up chiseling projects such as chiseling off rivets, chiseling off mufflers, or disconnecting ball joints.
⑤ Air Compressor and Hoses	• Supplies compressed air for many shop uses.	• Use for running air tools.
⑤ Air Pressure Regulator	• A moisture trap that supplies clean dry air at constant pressure.	• Use to regulate air pressure.
⑤ Air Ratchet (3/8" Drive)	• A ratchet set powered by compressed air.	• Use to quickly remove bolts.
⑤ Impact Socket	• Made of exceptionally strong metal to absorb extra torque. • A basic set includes a 3/8" drive and a 1/2" drive. Also available in metric.	• Use to remove stubborn bolts.
⑤ Impact Wrench	• Made of exceptionally strong metal to absorb extra torque. • A basic set includes both a 1/2" drive and a 3/8" drive.	• Designed for quick loosening or tightening of hard-to-turn nuts and bolts.

SECTION 2 KNOWLEDGE CHECK

❶ Why is it usually more difficult to learn to use general workplace equipment than it is to learn to use hand tools?

❷ Why is it important for technicians to know the specific uses for each item of general equipment?

❸ What is a cooling-combustion gas detector?

❹ What tool is used to check electrical voltage?

❺ Name two items that can help protect a vehicle while it is being serviced.

Section 3

Measuring Systems and Measuring Tools

There is a need to take a wide variety of measurements in the automotive shop. Sometimes, such as in the measuring of the inside or outside dimension of a cylinder, the measurement must be accurate to one thousandth of an inch. Other times, such as for measuring a length of tubing, the measurement need not be so precise.

Measuring Systems

It is necessary to make many kinds of measurements in the automotive shop. These measurements detect if parts are worn or damaged. They also identify parts that are out of adjustment or out of spec, and by how much. Sometimes technicians measure engine vacuum or power, generator output, or battery voltage. Alignment specialists measure angles in the front-suspension system. But for most service work, technicians measure length, diameter, or clearance. They might, for example, measure the bore, or diameter, of the engine's cylinders.

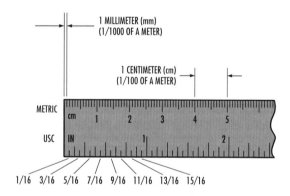

Fig. 3-6 Ruler marked in both the metric and USC system. *Which system is most common throughout the world?*

Measurements can be made in the metric system or in the United States Customary (USC) system. The *metric system* is also known as the System of International Units (SI). The **System of International Units (SI)** is a system of measurement that uses meters, liters, and grams. **Figure 3-6** compares the inch of the USC system with the millimeter of the SI system.

The United States is the only major country that does not use the SI system. The SI system is used for the measurements in all imported cars and most cars made in the United States. In this book, USC measurements are given first. The SI equivalent follows in brackets. For example, 1 inch [25.4 mm].

The **United States Customary (USC) system** is a system of measurement that uses inches, feet, miles, pints, gallons, and pounds. The USC system is often referred to as standard.

The SI System The SI system is based on multiples of ten, just like the monetary system in the United States. In the United States, ten pennies equal one dime, and ten dimes equal one dollar. Similarly, in the SI system, 10 mm equal 1 centimeter (cm), 10 cm equal 1 decimeter (dm), and 10 dm equal 1 meter (m). One thousand meters equal 1 kilometer (km), which is 0.62 mile.

Each type of measurement (such as length, volume, weight) is represented by a specific metric unit. The three basic metric measurements are:
• Meter (m) for length.
• Liter (L) for volume.
• Gram (g) for weight.

Each of these units is called a stem unit. A stem unit is made larger or smaller by the addition of a prefix. The only stem unit that does not use prefixes is Celsius, which measures temperature. Prefixes such as milli, centi, deci, and kilo have special meanings. For example:
• Kilo means 1000 (one thousand).
• Deci means 0.10 (one-tenth).
• Centi means 0.01 (one-hundredth).
• Milli means 0.001 (one-thousandth).

Table 3-C	SI METRIC AND CUSTOMARY UNITS	
Type of Measurement	SI Metric Unit	Approximate Size of Customary Unit
1. Length and distance	a. meter [m] b. centimeter [cm] c. millimeter [mm] d. kilometer [km]	a. 1.1 yard b. 0.4 inch c. 0.04 inch d. 0.6 mile
2. Mass or weight	a. gram [g] b. kilogram [kg]	a. 1/28 ounce b. 2.2 pounds
3. Volume (liquid)	a. milliliter [mL or ml] b. liter [L]	a. 1/5 teaspoon b. 1.06 quart
4. Temperature	a. degrees Celsius [°C]	a. $1.8 \times °C + 32 = °F$
5. Pressure	a. 1 kilopascal [kPa]	a. 0.145 psi (pounds per square inch)
6. Energy	a. 1 kilojoule [kJ]	a. 0.239 Calories

Within the ruler figure:
1 MILLIMETER (mm) (1/1000 OF A METER)
1 CENTIMETER (cm) (1/100 OF A METER)
METRIC cm 1 2 3 4 5
USC IN 1 2
1/16 3/16 5/16 7/16 9/16 11/16 13/16 15/16

METRIC UNITS

Table 3-D

Prefix	Stem	Results	Abbreviation	Size in USC units
milli	meter	millimeter (1/1000 of a meter)	mm	tiny fraction of an inch
centi	meter	centimeter (1/100 of a meter)	cm	about 0.4 inch
kilo	meter	kilometer (1000 meters)	km	about 0.6 mile
mega	meter	megameter (a million meters)	Mm	more than 1 million yards
milli	gram	milligram (1/1000 of a gram)	mg	very tiny fraction of an ounce
centi	gram	centigram (1/100 of a gram)	cg	a tiny fraction of an ounce
kilo	gram	kilogram (1000 grams)	kg	about 2.2 pounds
mega	gram	megagram (a million grams, or a metric ton)	Mg	about 2200 pounds
milli	liter	milliliter (1/1000 of a liter)	ml	very tiny fraction of an ounce

Table 3-C shows the types of measurements in the metric system. It shows the common terms and abbreviations in use, along with the size of each measurement. There is also an indication of the size of each measurement relative to familiar USC measures. **Table 3-D** shows how a prefix added to a stem unit, a base unit, changes the value.

Weight and Volume The metric unit of weight is the gram. It is the weight of 1 cubic centimeter (cc) of water. A cubic centimeter is a cube that measures 1 cm [1/100 m] on a side. **Fig. 3-7.** One thousand grams is 1 kilogram (kg). This is equal to 1000 cc.

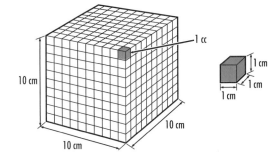

Fig. 3-7 A liter is 1000 cubic centimeters. A gram equals the weight of how many cubic centimeters of water?

CONVERTING FRACTIONS

Table 3-E

Inches

Fraction	Decimal	mm	Fraction	Decimal	mm
1/64	0.0156	0.3969	31/64	0.4844	12.3031
1/32	0.0312	0.7938	1/2	0.5000	12.7000
3/64	0.0469	1.1906	33/64	0.5156	13.0969
1/16	0.0625	1.5875	17/32	0.5312	13.4938
5/64	0.0781	1.9844	35/64	0.5469	13.8906
3/32	0.0938	2.3812	9/16	0.5625	14.2875
7/64	0.1094	2.7781	37/64	0.5781	14.6844
1/8	0.1250	3.1750	19/32	0.5938	15.0812
9/64	0.1406	3.5719	11/16	0.6875	17.4625
5/32	0.1562	3.9688	45/64	0.7031	17.8594
11/64	0.1719	4.3656	23/32	0.7188	18.2562
3/16	0.1875	4.7625	47/64	0.7344	18.6531
13/64	0.2031	5.1594	3/4	0.7500	19.0500
7/32	0.2188	5.5562	49/64	0.7656	19.4469
15/64	0.2344	5.9531	25/32	0.7812	19.8438
1/4	0.2500	6.3500	51/64	0.7969	20.2406
23/64	0.3594	9.1281	13/16	0.8125	20.6375
3/8	0.3750	9.5250	53/64	0.8281	21.0344
25/64	0.3906	9.9219	27/32	0.8438	21.4312
13/32	0.4062	10.3188	55/64	0.8594	21.8281
27/64	0.4219	10.7156	7/8	0.8750	22.2250
7/16	0.4375	11.1125	57/64	0.8906	22.6219
29/64	0.4531	11.5094	29/32	0.9062	23.0188
15/32	0.4688	11.9062	59/64	0.9219	23.4156

Table 3-F CONVERTING MEASUREMENTS

To Change Customary	To Metric	Multiply By
Inches	Centimeters	2.540
Feet	Meters	0.305
Yards	Meters	0.914
Miles	Kilometers	1.609
Square inches	Square centimeters	6.451
Ounces (U.S. liquid)	Milliliters	29.573
Pints	Liters	0.473
Quarts (U.S. liquid)	Liters	0.946
Gallons (U.S. liquid)	Liters	3.785
Pounds	Kilograms	0.454
Short tons (2,000 lbs)	Metric tons	0.900
Miles per hour	Kilometers per hour	1.609
Pounds per square inch	Kilopascals	6.895
Miles per gallon	Kilometers per liter	0.425
Degrees Fahrenheit	Degrees Celsius	(°F − 32) ÷ 1.8

To Change Metric	To Customary	Multiply By
Centimeters	Inches	0.394
Meters	Feet	3.280
Meters	Yards	1.094
Kilometers	Miles	0.621
Square centimeters	Square inches	0.155
Milliliters	Fluid ounces	0.034
Liters	Pints	2.113
Liters	Quarts	1.057
Liters	Gallons (U.S. liquid)	0.264
Kilograms	Pounds	2.205
Metric tons	Short tons (2,000 lbs)	1.102
Kilometers per hour	Miles per hour	0.621
Kilopascals	Pounds per square inch	0.145
Kilometers per liter	Miles per gallon	2.354
Degrees Celsius	Degrees Fahrenheit	(°C × 1.8) + 32

The metric unit of volume for fluid, or liquid, measurements is the liter (L). It is slightly larger than a quart. The liter is the volume of a cube that measures 10 cm on a side (or 1000 cc). This is the same measurement for weight. One liter of water weighs 1 kg [1000 g]. One kg equals 2.2 pounds.

The USC System Making small measurements in the USC system often requires dealing with fractions of an inch such as 1/4, 1/8, 1/16, 1/32, and 1/64. Sometimes these may not be small enough. Many automotive measurements are in thousandths

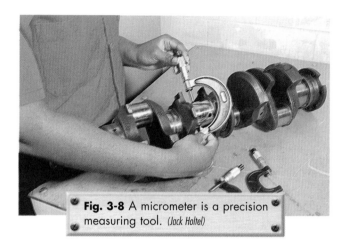

Fig. 3-8 A micrometer is a precision measuring tool. *(Jack Holtel)*

and sometimes ten-thousandths of an inch. For example, 1/64 inch is 0.0156 inch. A bearing clearance may be 0.002 inch. To convert fractions of an inch into decimal fractions, technicians may need a table of decimal equivalents. **Table 3-E.**

Unlike the SI system the USC system has no direct relationship among linear measurements, weight, and volume. For example:
- 12 inches = 1 foot
- 3 feet = 1 yard
- 1760 yards = 1 mile, or 5280 feet, or 63,360 inches

In measuring liquid volume in the USC system:
- 16 fluid ounces (fl oz) = 1 pint (pt)
- 2 pints = 1 quart (qt)
- 4 quarts = 1 gallon (gal)

In measuring weight in the USC system:
- 16 ounces (oz) = 1 pound (lb)
- 2000 pounds = 1 ton

Converting Measurements Sometimes technicians must convert USC measurements to SI measurements. At other times they need to convert SI measurements to USC measurements. **Table 3-F** will help in these conversions.

Measurement Tools

A wide variety of tools are available for measuring and calibrating. To calibrate means to adjust precisely for a specific function. Some measuring tools, such as a tape measure, are used to make approximate measurements. Other measuring tools, such as a micrometer, can precisely make measurements as small as 0.0001" (one ten-thousandth of an inch). **Fig. 3-8.**

Table 3-G describes some of the basic measurement tools used in the automotive workplace.

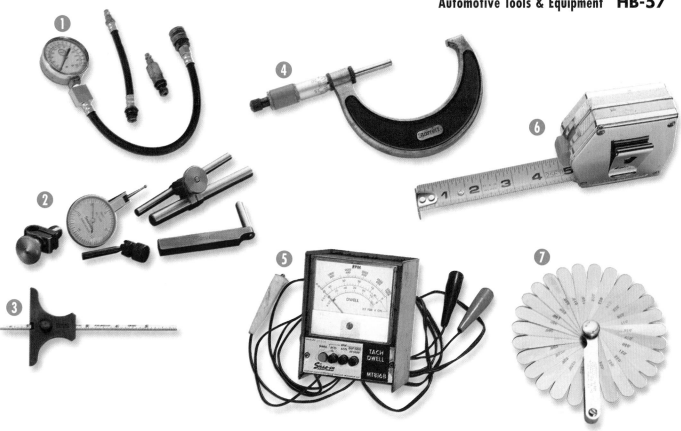

Table 3-G	MEASUREMENT TOOLS	
Tool	**Description**	**The Job**
❶ **Compression Tester**	• Pressure gauge with hose that can be connected to cylinder spark plug hole.	• Use to measure the pressure in the engine cylinder at top dead center of the compression stroke.
❷ **Dial Indicator with Flex Arm and Clamp Bar**	• Has a dial face and a needle to register measurement. • The needle moves in relation to movement of a movable arm or plunger. • As the plunger moves, the needle shows the distance or variation.	• Readings may be in thousandths of an inch or hundredths of a millimeter. • Measures end-play in shafts or gears by pushing the plunger against the part to be measured until the needle moves.
❸ **Depth Gauge**	• A type of micrometer. • Consists of a ratchet stop, thimble, barrel, base, and measuring rod.	• Use to measure the depth of grooves or holes.
❹ **Outside Micrometer**	• Hand-held precision measuring instrument. • Has a frame and a movable spindle with precision screw threads. • Available to measure in USC or SI. • A complete USC set includes the following dimensions in outside type: 0.1", 1–2", 2"–3", 4"–5".	• Use to accurately make linear measurements as small as one-thousandth or one ten-thousandth of an inch. • Outside micrometer is designed to measure the outside diameter of cylindrical forms.
❺ **Tach/Dwell Meter**	• Hand-held electric meter.	• Use to measure engine rpm.
❻ **Tape Measure**	• Available to measure in USC and SI measurements.	• Use to make approximate measurements.
❼ **Thickness Gauges**	• Strips or blades of metal of various thicknesses. • Most are made of steel; however, some are made of nonmagnetic metals such as brass.	• Also called feeler gauges. • Use to measure small gaps or distances such as the clearance between two parts.

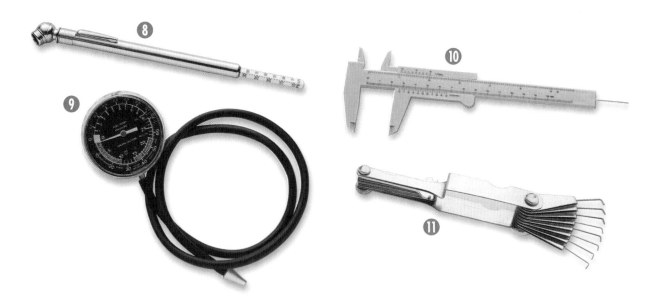

Table 3-G	MEASUREMENT TOOLS (Continued)	
Tool	**Description**	**The Job**
⑧ **Tire Pressure Gauge**	• There are various types of tire-pressure gauges. They all measure the amount of air in a tire.	• Use to measure air pressure in tires.
⑨ **Vacuum Gauge**	• Measures vacuum in inches of mercury.	• Use to measure vacuum and small pressures, such as engine vacuum.
⑩ **Vernier Calipers**	• Has a movable scale that runs parallel to a fixed scale. • Available to measure in either one-thousandths of an inch or one-hundredths of a millimeter. • One vernier caliper measures standard distances of 0" to 6". Another measures metric distances of 0 mm to 125 mm. • Use to make quick and reasonably accurate measurements of small distances between surfaces.	• Can be used to measure the diameter of a rod. • Can be used to take inside or outside measurements.
⑪ **Wire Gauges**	• Precisely sized pieces of round wire. • The diameter is usually marked on the handle or holder.	• Use to measure spark plug gaps and other openings. • The specified gauge should fit into the gap snugly, without bending.

SECTION 3 KNOWLEDGE CHECK

❶ Give examples of two types of measurements a technician might make while working on a vehicle.

❷ What do SI and USC stand for?

❸ Which measurement system is used for all imported cars and most cars made in the United States?

❹ List the basic metric measurement units for length, volume, and weight.

❺ Why is it a good idea to have SI and USC conversion charts in an automotive workplace?

Section 4

Fasteners, Gaskets, and Sealants

There are many different ways to hold parts together. Fasteners, such as nuts and bolts, and sealants, such as epoxy and RTV, are used to hold parts together. Sometimes gaskets are placed between two flat metal surfaces to make up for small irregularities in the metals. Otherwise, the result might be a leakage of air or liquid.

Characteristics of Fasteners

Fasteners hold automotive parts together. Screws, nuts, bolts (screws with hex heads), and studs (bolt-like fasteners with no head on either end) are examples of fasteners. Most fasteners are removable so the assembly can be taken apart. There are also more permanent ways of fastening parts together. This is usually done through soldering and welding.

Fastener Strength Groups   Both nuts and bolts are available in different strength classifications. These are identified as SAE grades. Fastener strength is most important for the fasteners that secure critical parts.

In the standard and metric measurement systems, bolts are marked on their heads. Nuts bear similar markings as indentations on one of their end faces. For example, three radial marks on a bolt head indicate a medium carbon steel, quenched and tempered (SAE Grade 5). Six radial marks indicate the strongest common threaded fasteners (SAE Grade 8). **Fig. 3-9.** Beyond that there are designations like S for "super." However, that is beyond the needs of most automotive applications. Bolts with unmarked heads have low strength. They should not be used in any critical application.

The strength of metric bolts is identified by a number on the head of the bolt. Strength can range from 4.6 (low strength) to 10.9 (very high strength). The strength noted is tensile strength. This is the amount of stress a bolt can take without breaking.

Screw Threads   A screw thread is a fastener that has a spiral ridge, or screw thread, on its surface. Threads can be internal (as on a nut) or external (as on a bolt or stud). Screws are manufactured with either USC or SI screw threads. USC and SI threads are not interchangeable. A USC screw will not fit an SI nut or vice versa. Some vehicles have metric fasteners. Others have USC fasteners. Some have both.

Complete Thread Designation   Bolts are identified by all of the factors noted above. A common bolt designation might be *1/4-20×2, Group 5*. This means that the bolt has a 1/4" diameter shank, has 20 threads per inch, is 2" long, with Group 5 strength. A metric bolt might be designated M12×1.75×25. This means that the bolt has a 12-mm shank diameter, 1.75 mm between threads, and is 25 mm long.

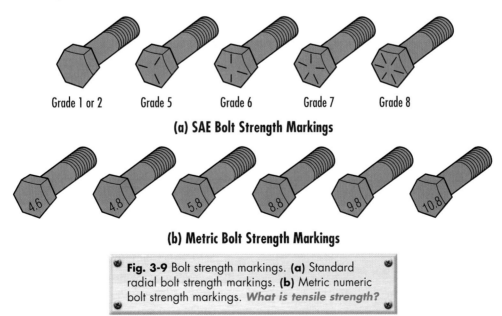

Grade 1 or 2   Grade 5   Grade 6   Grade 7   Grade 8

(a) SAE Bolt Strength Markings

4.6   4.8   5.8   8.8   9.8   10.8

(b) Metric Bolt Strength Markings

Fig. 3-9 Bolt strength markings. **(a)** Standard radial bolt strength markings. **(b)** Metric numeric bolt strength markings. *What is tensile strength?*

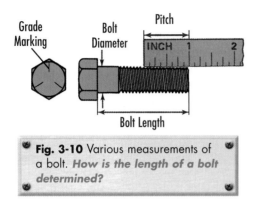

Fig. 3-10 Various measurements of a bolt. *How is the length of a bolt determined?*

Fastener Size Threaded fasteners are classified by length, pitch, series, and class. Fasteners are available in a wide variety of lengths and diameters. The length of a bolt or screw is the distance from the bottom of its head to the end of its threads. **Fig. 3-10.**

Fastener Pitch To determine the size of a fastener, you must measure the fastener's pitch. **Pitch** is the length from a point on a fastener thread to a corresponding point on the next thread. Thread pitch is calculated by dividing one inch by the number of threads per inch. Threaded fasteners can have several common pitches. These are UNC (coarse thread), UNF (fine thread), and UNEF (extra fine thread). **Fig. 3-11**

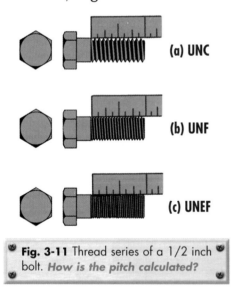

(a) UNC

(b) UNF

(c) UNEF

Fig. 3-11 Thread series of a 1/2 inch bolt. *How is the pitch calculated?*

It is important to know exactly what thread pitch you need. Attempting to mate fasteners of different pitches will strip the threads of one or both parts. Different pitches in threaded holes, nuts, studs, or bolts may be difficult to determine by eye.

The same wrench will fit the turning heads of different fasteners with different thread pitches.

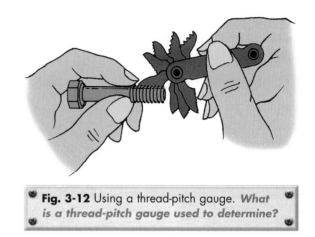

Fig. 3-12 Using a thread-pitch gauge. *What is a thread-pitch gauge used to determine?*

Wrenches are designated by head size, not by bolt shank or nut bore diameter. Studs often have different thread pitches at either end. A thread-pitch gauge will help determine which fasteners will work in different situations. **Fig 3-12.**

Nuts and Lock Washers

Figure 3-13 shows several nuts, including the following:
- The hex nut which is the most common nut in the automotive service center.
- The slotted hex and the castle nut, which are used with a cotter pin.
- The acorn nut which covers the end of a screw or bolt. This gives the assembly a neat appearance.
- The speed nut, which has little holding power. However, it can be quickly installed by pushing it over the threads. The speed nut provides a light clamping force when fast assembly is needed.

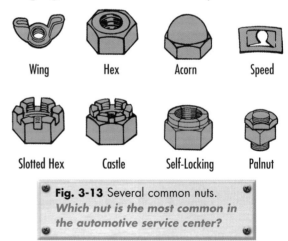

Wing   Hex   Acorn   Speed

Slotted Hex   Castle   Self-Locking   Palnut

Fig. 3-13 Several common nuts. *Which nut is the most common in the automotive service center?*

A lock washer placed under a nut or bolt head helps lock the fastener in place. The sharp edges of the lock washer bite into the metal. This helps prevent the nut or bolt from turning.

Other Fasteners

Many other types of non-threaded fasteners are used in automobiles. Some are used for special purposes.

Snap Rings   Snap rings are used to secure or locate the ends of shafts. There are two types of snap rings (external and internal). **Fig 3-14.** External snap rings fit on shafts to prevent gears or collars from sliding on the shaft. Internal snap rings fit in housings to keep shafts or other parts in position.

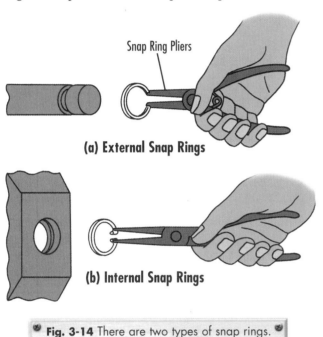

Snap Ring Pliers

(a) External Snap Rings

(b) Internal Snap Rings

Fig. 3-14 There are two types of snap rings. *How do their functions differ?*

Thread Inserts   Damaged or worn threads can sometimes be replaced by installing a thread insert in a threaded hole. To install one type of thread insert:
• Drill out the old threads.
• Rethread the hole with the special thread-cutting tool or the tap from a thread repair kit.
• Install the thread insert in the tapped hole. **Fig. 3-15.**

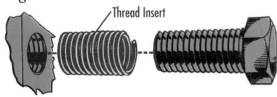

Thread Insert

Fig. 3-15 A thread insert is used when the original threads on a fastener are damaged or worn. *How is a thread insert installed?*

Setscrews   A setscrew secures a collar or gear on a shaft. Setscrews are usually loosened or tightened using an allen wrench. Tightening the screw "sets" the collar or gear into place.

Self-Tapping Screws   Self-tapping screws are screws that cut their own threads when turned into drilled holes.

Rivets   Rivets are metal pins used to fasten parts together. One end has a head. After placing the rivet, use a driver (or hammer and rivet set) to form a head on the other end. To remove a rivet, cut off the rivet head with a chisel and hammer. Then use a punch and hammer to drive the rivet out of the hole.

Rivets are used when there is little likelihood that they will ever have to be removed for routine maintenance and repair. A blind hole is a hole where you cannot reach the end to form a head. Blind, or "pop," rivets are used in blind holes. **Fig 3-16.**

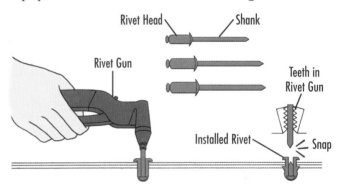

Rivet Head   Shank
Rivet Gun   Teeth in Rivet Gun
Installed Rivet   Snap

Fig. 3-16 Installing a blind rivet. The teeth inside the rivet gun pull the shank through the rivet head. The shank snaps when the rivet is tight.

There are many other specialized fasteners, and new ones are being invented. It is important for automotive technicians to know the characteristics of each type of fastener. Technicians should also master any special procedures or techniques needed to work with these fasteners.

Thread Dressings

Many threaded-fastener applications require specific thread dressings. These dressings lubricate, lock, or seal threads. Often, some of these purposes are combined, as in compounds that both seal and lock threads.

Thread Lubrication Sometimes it is necessary to install a steel bolt in an aluminum part such as a cylinder head or engine block. Before doing so, coat the bolt threads with an antiseize compound. **Fig. 3-17.** An **antiseize compound** is a lubricant that prevents bolt threads from locking or seizing. Removing seized bolts may damage or pull out the aluminum threads. Coating the bolt threads with antiseize compound helps prevent this.

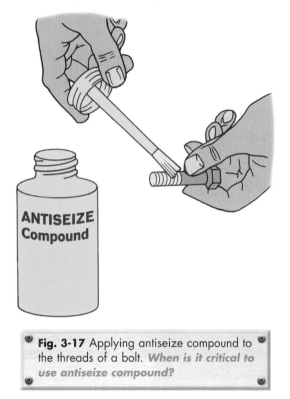

Fig. 3-17 Applying antiseize compound to the threads of a bolt. *When is it critical to use antiseize compound?*

Lubricating fastener threads is critical when the fasteners are to be torqued. *Torque* is a measurement of force expressed in units of distance and weight, such as "foot/pounds" or "Newton/meters." Where fastening pressure is critical to sealing or operation, bolt torque will be specified. For example, engine head bolts are torqued to a certain specification. This guarantees even and correct clamping pressure on an engine's head gasket between the head and block.

Torque value specifications assume that threads have been treated with an antiseize compound. To work effectively, an antiseize compound should be brushed onto the threads of at least one member of a threaded fastener pair. It should also be brushed onto the underside of the head of a bolt or nut that is being torqued. This is because there is as much friction on the underside as on the threads.

Thread-locking Compounds Thread-locking compounds are used to prevent threaded fasteners from loosening. They are specified in many critical automotive applications. They are available as a liquid or a paste.

Thread-locking compounds are applied by dripping or rubbing the compound onto one member of a threaded pair. Most of these compounds are of the anaerobic type. **Fig. 3-18.** Anaerobic compounds can harden in the absence of air. They can harden, or "cure," in the spaces between tightened threads. Once cured, the compounds prevent unthreading by conforming to the space between the threads. Thus, they prevent motion.

Thread-lockers work only in clean, grease-free environments. Some thread-lockers will not cure without a "primer." The primer must be applied to the threads before the thread-locker is applied to the threads. Primers also speed the cure times for anaerobic sealants. They are sometimes applied for that reason.

Thread-lockers come in different grades. Always use the thread-locker grade specified for the strength of the locking action that is required. Milder grades of thread-locking compounds can be undone by heating the parts that they lock to. Heat the parts at a relatively low temperature, such as 400°F [205°C].

Simple thread sealants are sometimes used where threads are used to seal liquids and gases. They come in various form (pastes, liquids, caulks, tapes, and sprays). Sealants sometimes require a curing period before parts can be put back into service.

Fig. 3-18 Applying an anaerobic thread-locking compound to the outside of a bushing. *When must a primer be used prior to applying a thread-locking compound?* (Terry Wild Studio)

Gaskets and Sealants

Many types of gaskets and sealants are used in automobiles. When using any type of sealant, follow safety rules for handling. These materials may cause injury or pollute the environment if not handled properly. One of the most important gaskets is the head gasket, which seals between the cylinder head and the cylinder block. **Fig. 3-19.**

Preformed Gaskets A gasket is a thin layer of soft material such as paper, cork, rubber, copper, synthetic material, or a combination of these. The gasket is preformed or precut to the desired shape and thickness. Clamping a gasket between two flat surfaces makes a tight seal.

The clamping force that results from the tightening of the fasteners squeezes the gasket. The soft material then fills any small irregularities in the mating surfaces. This prevents leakage of fluid, vacuum, or pressure from the joint. Holes through the gasket allow it to seal in fuel, oil, or coolant. The gasket material keeps dirt, water, and air out of the passages. Sometimes the gasket serves as a shim to take up space.

Formed-in-Place Gaskets Some gaskets are formed in place. This is done by squeezing a bead of plastic gasket material or sealant from a tube onto one of the mating surfaces. Typical surfaces include valve covers, thermostat housings, water pumps, and differential covers. When using sealants always follow the manufacturer's instructions.

There are two kinds of plastic gasket material. *Aerobic sealant* hardens in the presence of air. It is sometimes referred to as room-temperature vulcanizing (RTV) sealant. This is a silicone-rubber sealant. It vulcanizes, or cures, at room temperature when exposed to air.

RTV sealant can be used with or without a preformed gasket. RTV sealant can be used on a surface

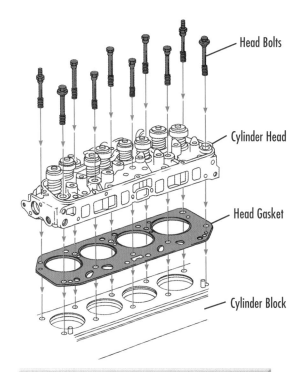

Fig. 3-19 A head gasket is placed between the cylinder head and the cylinder block to seal the joint. *(General Motors Corporation)*

that flexes or vibrates slightly, such as on an engine valve cover. Never use RTV sealant around parts with high temperatures and pressures, such as head gaskets. Clean the mating surfaces thoroughly before applying the sealant. It will not seal to dirty, greasy, or oily surfaces.

Anaerobic sealant material hardens in the absence of air. One way to remove the air is to squeeze the material between two surfaces. Such material can be used as an adhesive, a sealer, and a locking cement. It also serves as a chemical thread-locker on bolts, nuts, screws, and bushings.

Unlike aerobic sealant, anaerobic material should not be used on parts that flex. Anaerobic and aerobic sealants each have their own uses. The two should not be used interchangeably.

SECTION 4 KNOWLEDGE CHECK

❶ Name three types of automotive fasteners.

❷ Why is it important to classify nuts and bolts by strength?

❸ What is the purpose of an antiseize compound?

❹ What is the difference between a preformed gasket and a formed-in-place gasket?

❺ Why is it important to thoroughly clean an area before applying a thread-locker or a sealant?

CHAPTER 3 REVIEW

Key Points

Addresses NATEF program guidelines for individual hand tools and general shop tools and equipment.

- Today's automotive technician must be familiar with a wide variety of tools and equipment.
- Ownership of a basic tool kit is often required for employment.
- The metric system is based on multiples of ten.
- Automotive technicians must convert measurements between standard, or USC, (United States Customary) and metric, or SI (System of International Units, and vice versa.
- Many threaded-fastener applications require specific thread dressings.
- Gaskets are used to create a seal between two flat surfaces. One example is the head gasket, which seals the cylinder head and the cylinder block.
- An aerobic sealant hardens in the presence of air. An anaerobic sealant hardens in the absence of air.

Review Questions

1. What tool is best suited for turning nut or bolt heads—pliers or a wrench? Why?

2. Name several tools that are used for calibrating.

3. Explain what technicians should know about proper use of tools and equipment.

4. Explain the relationship among linear measurements, weights, and volume in the metric system.

5. Connecting-rod-bearing clearance for an imported car engine is given as 0.51 mm. What is this in the USC system?

6. What is the function of a gasket? Name four surfaces that use formed-in-place gaskets.

7. What are the three purposes of thread dressings?

8. Define anaerobic sealant and aerobic sealant and name their typical uses.

9. **Critical Thinking** What type of sealant should be used on a brake hose?

TECHNOLOGY FORECAST
FOR AUTOMOTIVE EXCELLENCE

Toolbox of Tomorrow

Technical knowledge is perhaps the most important tool the automotive technician can have. Equally important is the ability—and willingness—to learn new information and skills as technology develops. Tomorrow's technicians will use laptop computers to stay current with technology. Laptops will be an important part of the well-stocked toolbox of tomorrow, providing diagnostic as well as educational services.

Laptop computers can be used in place of today's hand-held scan tools and larger diagnostic machines. These small computers can be programmed to process input to diagnose problems on many different vehicles. They can be updated easily to keep pace with technology and used to update the vehicle's computer.

The educational benefits of computers are numerous. Laptops can be used to search the Internet for technical help. Automotive experts can be contacted by e-mail for advice. Laptops can run interactive educational CD-ROMs, and can be used for word processing and bookkeeping purposes.

Elsewhere in the toolbox, tomorrow's technicians will have more battery-powered tools. Lighter and quieter than air-impact equipment, these tools can be used for routine service. Tomorrow's technicians will have wrenches that grab the flat area of a fastener to avoid rounding the corners. Another improvement will be the use of multiple-jointed tools that help get to hard-to-reach nuts and bolts.

AUTOMOTIVE EXCELLENCE
TEST PREP

1. A starter set of hand tools costs about:
   - ⓐ $100.
   - ⓑ $1,000.
   - ⓒ $5,000.
   - ⓓ $10,000.

2. Pneumatic motors are powered by:
   - ⓐ Electricity.
   - ⓑ Solar power.
   - ⓒ Air pressure.
   - ⓓ Water pressure.

3. Technician A has decided to use a wrench to turn a bolt head. Technician B has decided to use a pair of pliers. Who is correct?
   - ⓐ Technician A.
   - ⓑ Technician B.
   - ⓒ Both Technician A and Technician B.
   - ⓓ Neither Technician A nor Technician B.

4. Technician A says a soldering gun should be used to solder electrical connections. Technician B says a soldering gun should not be used to solder electrical connections. Who is correct?
   - ⓐ Technician A.
   - ⓑ Technician B.
   - ⓒ Both Technician A and Technician B.
   - ⓓ Neither Technician A nor Technician B.

5. Technician A and Technician B are discussing what they need to do to install a steel bolt into an aluminum cylinder head. Technician A says they need to coat the bolt threads with an anti-seize compound. Technician B says that would prevent the bolt threads from locking. Who is correct?
   - ⓐ Technician A.
   - ⓑ Technician B.
   - ⓒ Both Technician A and Technician B.
   - ⓓ Neither Technician A nor Technician B.

6. Two service technicians are discussing the best way to thread a rod. Technician A says they should use a tap and die set. Technician B says they should use a thread repair insert kit. Who is correct?
   - ⓐ Technician A.
   - ⓑ Technician B.
   - ⓒ Both Technician A and Technician B.
   - ⓓ Neither Technician A nor Technician B.

7. In the metric system the prefix kilo means:
   - ⓐ One-tenth.
   - ⓑ One-thousandth.
   - ⓒ Ten.
   - ⓓ One thousand.

8. The metric unit of measurement for volume is:
   - ⓐ Meter.
   - ⓑ Liter.
   - ⓒ Gram.
   - ⓓ Celsius.

9. Two service technicians are discussing the best way to make a very accurate measurement of a cylinder. Technician A says they should use a depth gauge. Technician B says they should use a dial indicator. Who is correct?
   - ⓐ Technician A.
   - ⓑ Technician B.
   - ⓒ Both Technician A and Technician B.
   - ⓓ Neither Technician A nor Technician B.

10. Two service technicians are discussing the best way to remove an engine from an automobile. Technician A says they should use a jack. Technician B says they should use a hoist. Who is correct?
    - ⓐ Technician A.
    - ⓑ Technician B.
    - ⓒ Both Technician A and Technician B.
    - ⓓ Neither Technician A nor Technician B.

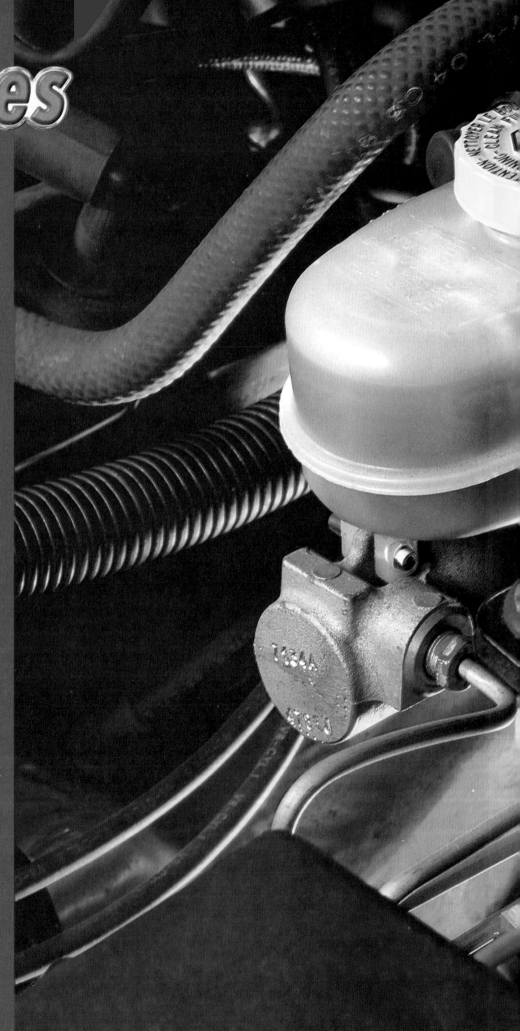

Brakes

AUTOMOTIVE CUSTOMER RELATIONS

Immediate opening for a GREETER in our busy automotive service department. Responsibilities consist of greeting customers, maintaining customer service files, making follow-up phone calls, handling customer complaints, and cashiering. We offer competitive wages, a generous benefits package, advancement opportunities.

Automotive Technician

Responsibilities include:

- Alignment
- Engine analysis/drivability problems
- Air conditioning/cooling/ heating systems
- Computer control systems
- Fuel injection systems
- Brake systems
- Steering & suspension systems
- Exhaust systems
- Struts & shocks

Applicants should have current ASE certifications in above areas, valid driver's license, and your own tools. Requires 2 years of related work ... oodyear is an EOE.

BRAKES TECHNICIAN

A technician with an Associate's degree will be trained from the ground up in brake work. Duties include relining brake shoes, resurfacing drums, servicing master and wheel cylinders, and checking the hydraulic brake system asembly and mechanic...

Automotive Technicians
☆ $2000 HIRING BONUS ☆

If you are an experienced technician looking for a great dealership to call home, give us a call or stop in to look at our shop. We offer good pay, good benefits, a busy workload, and our $2000 hiring bonus!

Career Focus Activity

After reading the above job ads, do the following:

- Imagine that you are applying for one of these positions. Write a one-page résumé that shows your qualifications.
- Research the amount of time it would take you to obtain ASE certification in Brake Systems.
- Where could you obtain the education, training, and work experience that would lead to continuing career advancement within the automotive industry?

CHAPTER 1

Brake System Operation

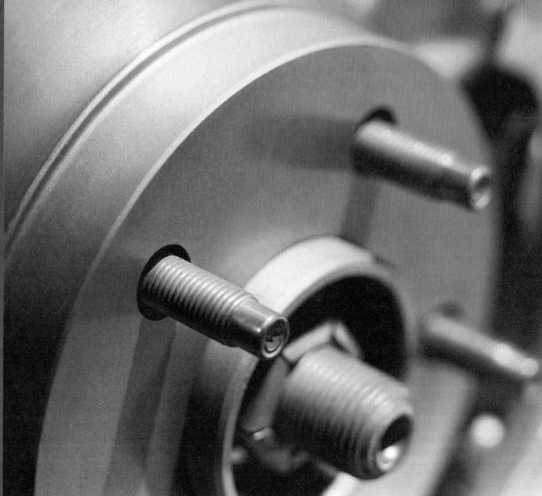

You'll Be Able To:

- Identify the two basic types of wheel brakes.
- Explain how friction provides braking action in a motor vehicle.
- Explain the advantage of a dual-braking system.
- Activate a vehicle's service braking system.
- Select and install brake fluid.
- Safely support a vehicle for brake servicing.

Terms to Know:

friction
kinetic friction
parking brakes
service brakes
static friction

The Problem ⟩⟩⟩⟩⟩ Your Challenge

A vehicle's braking system is extremely important to the safe operation of the motor vehicle. Without it, the vehicle could not be slowed or stopped.

Understanding the principles used in braking systems helps technicians in diagnosing brake problems. What important role does friction play in braking action? What are the major components of a braking system? Does the vehicle have drum brakes, disc brakes, or a combination? Does it have an anti-lock brake system (ABS)?

As the service technician, you need to find answers to these questions:

1. How is the braking system activated?
2. What is a master cylinder?
3. What is special about brake fluid? Why are there different types of brake fluid?

Section 1

Types of Brakes

All vehicles must have brakes. Without brakes, there is no way to safely slow or stop a moving vehicle. Most vehicles have two types of brakes–service brakes and parking brakes.

Service Brakes

Service brakes are the primary braking system. Force applied to the service brake pedal is converted to hydraulic pressure by the master cylinder. This pressure is transferred through the service braking system until it reaches the wheel brakes. Here, brake force is applied to the vehicle's wheels, causing the vehicle to slow or stop.

Figure 1-1 shows the major components (parts) of a vehicle's service brake system.

- The master cylinder serves as the brake fluid reservoir for the service brakes and converts mechanical force to hydraulic pressure for the braking system.
- The brake hoses/lines carry brake fluid under pressure from the master cylinder to the wheel brakes.
- The disc brakes include rotors and disc pads, which transfer brake force to the vehicle's wheels.
- The drum brakes include drums and brake shoes, which transfer brake force to the vehicle's wheels.
- The *power brake booster* supplies the increased forces needed by the brakes. The booster does this without requiring increased brake pedal pressure.
- The brake pedal activates the master cylinder.

Parking Brakes

Although sometimes mistakenly called emergency brakes, **parking brakes** are brakes that are used to keep a parked vehicle from moving. They are usually on the rear wheels and are mechanically operated.

Friction and Braking

Friction and braking are closely related in vehicles. Without friction, there would be no braking.

Friction

Friction is the resistance to motion between two objects or surfaces that touch. Friction varies by the amount of force (often referred to as the load) between the surfaces. Friction also varies by the roughness of the surfaces and the materials from which the objects are made. For instance, there is more friction between a piece of sandpaper and a block of wood than between an ice cube and a countertop.

A lubricant between the objects can reduce friction. Water between the melting ice cube and the countertop allows the ice cube to move with less friction. Similarly, wet brakes will not stop a vehicle as effectively as dry brakes.

Without friction, no moving object would slow down or stop. A toy top, once set spinning, would continue to spin. It is the friction between the top, the surface it spins on, and the air around it that slows it down.

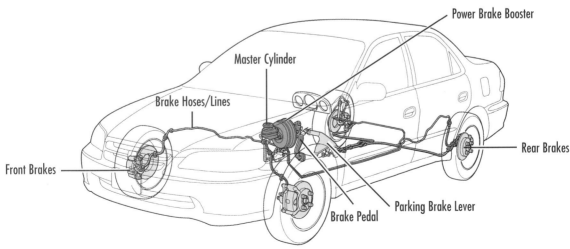

Fig. 1-1 The major components in a typical braking system. *What are the differences between service brakes and parking brakes?* (American Honda Motor Company)

Another example is the space shuttle as it reenters the earth's atmosphere. Friction between the shuttle's surface and the atmosphere reduces the shuttle's speed. The reduction in speed creates massive amounts of heat. Friction slows the space shuttle from an orbital speed of about 18,000 mph [28,800 kph] to a landing speed of about 500 mph [800 kph].

There are two types of friction—static friction and kinetic friction. **Fig. 1-2.**

Friction Stops!

A good driver's braking skills allow the brakes to stop the vehicle. Unfortunately, if a driver locks his wheels and skids, the brakes cannot do their job. Then the friction between the tires and the road must stop the vehicle. The stopping distance is much greater in a skid. The driver also has little control of steering with locked wheels.

Let's examine an unfortunate situation. Two older pickup trucks are traveling down the freeway at 55 mph [88 kph]. The trucks are identical except that one truck is fully loaded and the other is empty. Suddenly several deer step onto the roadway, right into the path of the trucks! The drivers see the deer immediately and overreact. Each slams on his brakes at exactly the same instant and locks the wheels on his truck.

Under normal braking conditions more energy will be required to stop the fully loaded truck than the empty truck. The loaded truck would normally require a greater stopping distance. The two trucks with the locked brakes in this example have the same tire surface area in contact with the road, but they differ in weight. Under these conditions will the lighter truck still be able to stop in a shorter distance? Does the weight of the truck make a difference if the truck is skidding?

Apply it!

Experimenting with Friction

Meets NATEF Science Standards for friction and deceleration.

Materials and Equipment
• Three large metal flat washers
• Small piece of double-faced tape
• Ruler or other straightedge

Fig. A.

① Tape two of the washers together with the double-faced tape. No tape should touch the bottom surface of the stack. This stack represents the heavy, loaded truck. The single washer represents the lighter, empty truck.

② Place the ruler on a smooth, level surface.

③ Put the single washer and the double washer side-by-side, touching the straight edge of the ruler. **Fig. A.**

④ Quickly shove and retract the ruler to set the washers in motion. Which one stops first?

Results and Analysis As you should have observed, the stopping distances of the objects are not dependent on their weight. Why?

Explanation The force of friction stops the trucks and washers. The amount of friction is proportional to the weight of the object. Each truck's kinetic energy of motion is also proportional to its mass. You might expect that the double washer would take longer to stop because of its greater kinetic energy. However, twice the mass results in twice the frictional stopping force. Finally, doubling the amount of friction causes the double washers to stop in exactly the same distance as the single washer. The single washer has half the kinetic energy but also half the frictional stopping force.

Static Friction   *Static* means "at rest." **Static friction** is the resistance between objects that are in contact but at rest. Parking brakes are an example of static friction at work. Friction between the brake lining and the brake drum or rotor keeps the wheels from moving. Friction between the vehicle's tires and the pavement keeps the vehicle from sliding.

Kinetic Friction   *Kinetic* means "in motion." The resistance between objects that are in contact and in relative motion is called **kinetic friction.**

Like the space shuttle, vehicles on earth use kinetic friction to change their motion into heat energy. Friction between the moving and nonmoving parts of the service brakes creates heat.

As the vehicle slows and stops, kinetic energy is changed to heat energy. The brakes pass this heat to the air and other wheel parts.

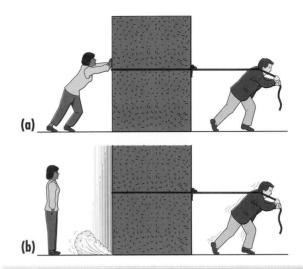

(a)

(b)

Fig. 1-2 There is more friction between objects at rest **(a)** than between objects in relative motion **(b)**. More force is needed to move an object at rest than to keep an object moving. *Objects at rest have what type of friction?*

SECTION 1 KNOWLEDGE CHECK

❶ What are service brakes?

❷ What is friction?

❸ Name the two types of friction.

❹ Give an example of static friction.

❺ Describe how kinetic friction is used to slow or stop a vehicle.

Section 2

Service Brake System

The service brake system is one of the most important systems on a vehicle. The engine may get the vehicle moving, but it is the brakes that slow or stop it. The service brakes are designed to be more powerful than the engine. When applied, brakes can even stall the engine.

Brake Action

Service brakes, those operated by the vehicle's brake pedal, have two basic parts. The first is the master cylinder. The *master cylinder* is the part that applies hydraulic (fluid) pressure through the *brake lines*. The second is the *wheel brake mechanisms*. These are located at each of the vehicle's wheels. They are activated by hydraulic pressure.

Two types of service brakes are used on vehicles. *Drum brakes* use hydraulic pressure to press *brake*

shoes against the inside of a rotating brake drum. **Fig. 1-3(a).** *Disc brakes* use hydraulic pressure to clamp *brake pads* against a rotating disc called a *brake rotor.* **Fig. 1-3(b).**

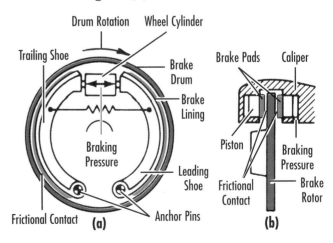

Drum Rotation   Wheel Cylinder

Trailing Shoe

Brake Drum

Brake Lining

Braking Pressure

Leading Shoe

Frictional Contact   **(a)**   Anchor Pins

Brake Pads   Caliper

Piston

Braking Pressure

Frictional Contact

Brake Rotor

(b)

Fig. 1-3 There are two types of service brakes: drum brakes **(a)** and disc brakes **(b)**. *Why do they have these names?* (Robert Bosch GmbH)

Drum brakes have shoes that press against the inside of the rotating brake drum. **Fig. 1-4.** Early bicycle coaster brakes worked this way.

Disc brakes have pads that clamp against a rotating rotor. **Fig. 1-5.** Hand-operated bicycle brakes work in a similar way.

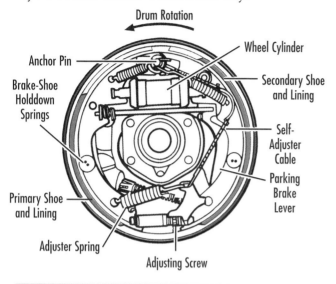

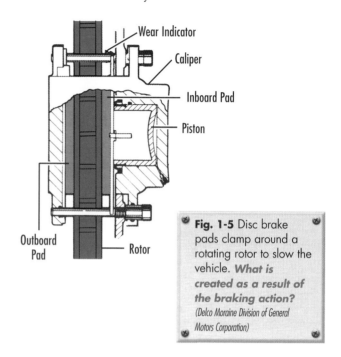

Fig. 1-4 Drum brakes have shoes that push against the inside of a rotating drum. *What type of friction is at work here?* (Ford Motor Company)

Fig. 1-5 Disc brake pads clamp around a rotating rotor to slow the vehicle. *What is created as a result of the braking action?* (Delco Moraine Division of General Motors Corporation)

Remember–Safety First!

Safety is important to both the customer and the technician. If you, as the technician, are not careful, you may harm yourself or your customer. And you may do serious damage to your customer's vehicle.

An action as simple as using old brake fluid that has been contaminated with water can cause brakes to fail. Using the wrong fluid or not checking the fluid level properly can also lead to problems. To assure personal safety and proper service of your customer's car–read!

Technicians must always look for visual clues. Sometimes a manual will warn you about specific service products. Sometimes, the label on a can or package will tell all. When you read the *Safety First* sections in this chapter, you will be given specific warnings about brake fluid. Stay on the alert for safety information about all products and procedures that you use.

Apply it!

Meets NATEF Communications Standards for collecting, evaluating, and using information.

Read the *Safety First* features later in this chapter.

❶ Where could you find this information and other similar warnings?

❷ Make a list of sources that warn or advise about the products used in servicing a vehicle.

❸ Keep a list of specific warnings that you find as you service cars. Use this list as an on-the-job resource.

Pushing the vehicle's service brake pedal forces fluid through the braking system lines and hoses. The pressure of this fluid activates the brakes. The brake shoes and brake pads are stationary. The drums and rotors are moving. Kinetic friction between the brake shoes and drums, or brake pads and rotors, slows and then stops wheel rotation. Friction between the tires and the road stops the vehicle.

The service brakes can be applied hard enough to lock the wheels (stop them from rotating). When brakes are locked, they no longer convert the vehicle's motion into heat. Success in stopping a vehicle then depends entirely on the kinetic friction between the tires and the road. If the kinetic friction is not great enough, the tires lose their grip on the road and the vehicle begins to skid.

The best braking performance occurs just before the tires begin to lose traction. This is the principle of *anti-lock brakes.*

Dual-Braking Systems

Before 1967 most vehicles had a single-piston master cylinder. This meant that brakes on all four wheels were activated by a single hydraulic system. If any part of the hydraulic system failed, the whole braking system failed. In 1967, the U.S. Department of Transportation (DOT) passed legislation requiring vehicles to have dual-braking systems.

A *dual-braking system* has a dual-piston master cylinder, two fluid reservoirs, and two separate hydraulic systems. One hydraulic system controls the brakes of two wheels. The other hydraulic system controls the remaining two wheels. This arrangement provides additional safety. If one system fails, the other system continues to work.

The first dual-braking systems separated the front brakes from the rear brakes. **Fig. 1-6(a).** More recent dual-braking systems are separated diagonally. Each system on a diagonally split system controls one front brake and one rear brake. **Fig. 1-6 (b).**

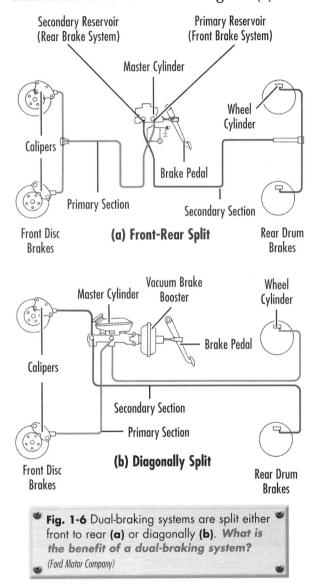

Fig. 1-6 Dual-braking systems are split either front to rear **(a)** or diagonally **(b)**. *What is the benefit of a dual-braking system?* (Ford Motor Company)

SECTION 2 KNOWLEDGE CHECK

❶ What are the two basic parts of the service brakes?

❷ What are the two types of service brakes used on vehicles?

❸ What is the effect of kinetic friction between the brake shoes and drums (or brake pads and rotors)?

❹ What kind of pressure do drum brakes use to press the brake shoes against the inside of a rotating brake drum?

❺ What are the main components of a dual-braking system?

Section 3

Brake Fluid

Brake fluid is used to transfer hydraulic pressure from the service brake's master cylinder to the wheel brake mechanisms.

Brake Fluid Properties

Brake fluid must remain stable through a wide range of temperatures and operating conditions.

Brake fluid must:
• Be compatible with the metals in the brake system.
• Lubricate the moving parts of the braking system.

Brake fluid must not:
• Become too thin or too thick as the temperature changes.
• Evaporate easily.
• Soften or damage rubber parts of the braking system.
• Boil at high temperatures.

Converting Temperatures

Vapor lock, overheated brakes, boiling radiators, and overheated engines all involve heat and temperature changes.

Most service manuals show temperatures in both degrees Celsius and degrees Fahrenheit. However, you will sometimes need to convert temperature readings from one system to the other. You could use thermometers that give temperatures both in degrees Celsius (°C) and degrees Fahrenheit (°F). **Fig. A**. But it's often easier to convert a temperature mathematically.

Refer to **Fig. A.** Look at the temperature difference between the boiling point and the freezing point of water. The change in Fahrenheit temperature from 212°F to 32°F corresponds to a change in Celsius from 100°C to 0°C.

The rate of change in Fahrenheit to Celsius is:

$$\frac{212-32}{100-0} = \frac{180}{100} = 1.8$$

For each increase of 1.8° Fahrenheit (1.8°F), there is an increase of 1° Celsius (1°C). Note that 0°C equals 32°F.

To convert Celsius temperatures to Fahrenheit temperatures, use this formula:

$$F = 1.8\ C + 32$$

The rate of change in Celsius to Fahrenheit is:

$$\frac{100-0}{212-32} = \frac{100}{180} = \frac{5}{9} = 0.556$$

For each increase of 5°C there is an increase of 9°F. Note that 32°F corresponds to 0°C.

To convert Fahrenheit temperatures to Celsius temperatures, you can use this formula:

$$C = \frac{5}{9}(F-32) \text{ or } C = 0.556(F-32)$$

Apply it!

Meets NATEF Mathematics Standards for using formulas to convert measurements between English and metric systems.

❶ What is 98.6°F in °C?

❷ DOT 3 brake fluid boils at 401°F. What is the equivalent temperature on the Celsius scale?

❸ What is the Fahrenheit equivalent temperature for 40°C?

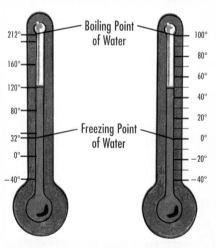

Fig. A

When brake fluid boils, it becomes a vapor (gas), much like boiling water turns to steam. A vapor can be compressed; a liquid cannot be compressed. Pressing on the service brake pedal will compress the vapor in the hydraulic lines instead of transferring the pressure through the fluid. This can lead to partial or complete braking system failure, sometimes called *brake pedal fade*. In fact, vapor in the hydraulic braking system is the primary reason for brake pedal fade.

Brake Fluid Types   The Society of Automotive Engineers (SAE) and the DOT have standards for brake fluid. The DOT specification is typically the one referred to. The three currently approved types are DOT 3, DOT 4, and DOT 5. The higher the number, the more strict the specifications, especially for the boiling point.

> **TECH TIP** **Proper Brake Fluids.** Always follow the manufacturer's specifications. A vehicle's warranty may be voided if incorrect or incompatible brake fluids are used.

DOT 3 and DOT 4 brake fluid types are polyglycol-based. These are the most commonly used. They are inexpensive and compatible with most service brake systems.

DOT 5 brake fluid is a silicone-based product. Silicone-based brake fluid is more expensive than polyglycol-based brake fluid. It does not readily blend with DOT 3 and DOT 4 types, so it must not be mixed with them.

DOT 5 brake fluid offers some advantages. It has a higher boiling point than DOT 3 or DOT 4 types. It does not damage paint and does not absorb moisture. However, because of its higher cost, it is usually used only in heavy-duty applications. There are only a few applications where silicone-based brake fluid is in common use. They are:
- Military vehicles.
- Postal vehicles.
- Race cars.
- Motorcycles.

Water in Brake Fluid   The major disadvantage of polyglycol-based brake fluid is that it absorbs moisture. Because water boils at 212°F [100°C] and DOT 3 brake fluid boils at 401°F [205°C], any moisture in the brake fluid lowers its boiling point. Lowering the boiling point of brake fluid increases the chance of having vapor in the braking system.

Safety First

Materials DOT 3 and DOT 4 brake fluids are also strong paint solvents that can damage a vehicle's finish. Take care to avoid spilling these fluids on painted surfaces. Clean a spill immediately, using nonabrasive (nonscratching) soap and water.

Moisture gets into brake fluid through damaged seals and loose or faulty connections on the master cylinder. Moisture also enters through damaged brake hoses and seals on wheel cylinders and calipers.

After containers of brake fluid have been opened, they must be kept tightly capped when not in use. Brake fluid should not be stored for a long period of time. The longer it is stored, the more moisture it can absorb.

Pour brake fluid directly into the master cylinder reservoir from a sealed and clean container. This will decrease the chance of brake fluid contamination. It is important to keep dust and dirt out of the master cylinder reservoir.

Installing Brake Fluid

As the disc brakes wear, the fluid level in the reservoir will drop. If the level gets too low, there is a possibility that air can get into the hydraulic system. To prevent this, the fluid level must be kept above the minimum level at all times. **Fig. 1-7.**

Additionally, whenever the hydraulic system is serviced, the brake fluid will have to be replenished.

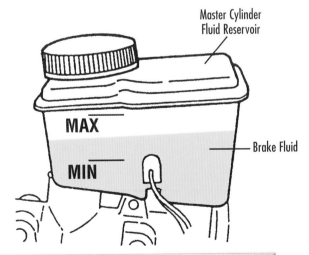

Master Cylinder Fluid Reservoir

MAX

MIN

Brake Fluid

Fig. 1-7 Always maintain the brake fluid level above the MIN mark on the reservoir. Do not overfill. *What might cause the fluid level to become critically low?* (Chilton)

To install brake fluid in the master cylinder reservoir, do the following:

1. Clean all dirt and grease from the cap (or cover) with a clean shop rag. **Fig. 1-8(a)**.

2. Remove the cover by unscrewing the cap or prying off the wire bale with a screwdriver.

3. With the cover off, carefully pour fresh brake fluid into the reservoir until the level reaches the full or maximum mark. **Fig. 1-8(b).**

4. If there is a screen in the reservoir, inspect it for any debris. If needed, clean the screen.

5. Replace the cap or cover. Secure it snugly, but do not over-tighten.

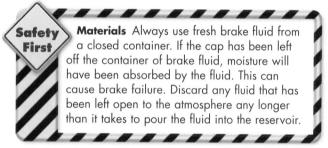

Safety First

Materials Always use fresh brake fluid from a closed container. If the cap has been left off the container of brake fluid, moisture will have been absorbed by the fluid. This can cause brake failure. Discard any fluid that has been left open to the atmosphere any longer than it takes to pour the fluid into the reservoir.

Brake Service Safety

Proper service procedures are vital to a technician's safety. Safety procedures, both personal and material, must always be observed.

Vehicle Support

Servicing wheel brake components requires lifting the vehicle. Portable floor jacks are designed only for lifting a vehicle. They cannot safely support a vehicle. Lift a vehicle with a floor jack. Then support the vehicle with safety stands.

When lifting a vehicle with a hydraulic or electric hoist, be sure to properly position the lift blocks under the vehicle's intended lift points. If you are unsure about the location of the vehicle's lift points, refer to its service manual. **Fig. 1-9.** Make sure the hoist's safety support mechanism is engaged before working under the vehicle.

Brake Dust and Chemical Safety

Brake linings were once made of asbestos. Asbestos is a health hazard. It is no longer used for brakes on cars and light trucks. However, asbestos-lined brake shoes may still be in use on older vehicles.

Fig. 1-8(a) Clean the master cylinder reservoir cap before unscrewing. *Why is this important?* (Jack Holtel)

Fig. 1-8(b) Always use fresh brake fluid from an unopened container. *What could happen if dirt or debris was in the brake fluid?* (Jack Holtel)

Therefore, do not use compressed air to blow brake dust from a brake assembly. As an alternative, use a brake vacuum equipped with a high efficiency particulate air (HEPA) filter. **Fig. 1-10.**

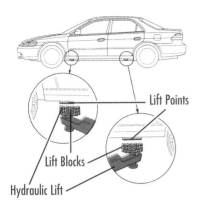

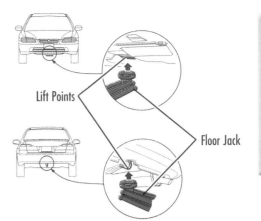

Lift Points

Lift Points

Lift Blocks

Hydraulic Lift

Floor Jack

Fig. 1-9 Always raise and support a vehicle with the proper equipment positioned in the correct locations. Refer to the vehicle's service manual. *What are some possible consequences of not lifting a vehicle correctly?* (American Honda Motor Company)

Safety First

Personal When servicing wheel brakes, avoid creating potentially harmful brake lining dust.

When cleaning brake components, use an approved solvent, preferably in a brake washer. Spray solvents are also acceptable. As a technician you must service brakes in a safe manner. Do not:

• Grind or scrape brake shoes or pads.
• Sand or file brake shoes or pads.
• Remove brake dust with compressed air or a brush.

When using cleaning solvents:

• Always wear goggles, especially when using spray solvents.
• If you get any brake fluid or other chemicals in your eyes, rinse them immediately and thoroughly with clean water.
• Always make sure there is adequate ventilation.
• Avoid breathing chemical vapors.
• Wear protective clothing such as aprons and gloves.
• Wash immediately after exposure to chemicals, including brake fluid.
• Change any clothes that get chemicals on them.

Fig. 1-10 When cleaning brake mechanisms, use a brake vacuum with a HEPA filter. This will help prevent brake dust from escaping into the environment. *What is the potential danger of brake dust?* (Nilfisk–Advance America, Inc.)

SECTION 3 KNOWLEDGE CHECK

❶ What is used to transfer hydraulic pressure from the service brake's master cylinder to the wheel brakes?

❷ What is the primary reason for brake pedal fade?

❸ Why should you avoid spilling DOT 3 and DOT 4 brake fluids on a vehicle's finish?

❹ What is the major disadvantage of polyglycol-based brake fluid?

❺ When lifting a vehicle with a hydraulic or electric hoist, what should you position under the vehicle's intended lift points?

CHAPTER 1 REVIEW

Key Points

Meets the following NATEF Standards for Brakes: describing brake functions and principles; selecting, handling, storing, and installing brake fluids.

- The service braking system is activated hydraulically. The parking brake is activated mechanically.
- Friction provides braking action in a vehicle.
- Service brakes have two main components: the master cylinder and the wheel brake mechanisms.
- The two basic types of service brakes are drum and disc.
- Dual-braking systems can be identified by the number of separate fluid reservoirs in the master cylinder.
- Brake fluids are chosen on the basis of intended use.
- A vehicle must be safely supported for brake servicing.

Review Questions

1. What are the two types of braking systems on most motor vehicles?
2. Explain how friction provides braking action in a motor vehicle.
3. What is the advantage of a dual-braking system?
4. How many separate fluid reservoirs does a dual-braking system master cylinder have?
5. How is the service braking system activated?
6. Explain the process for selecting and installing brake fluid.
7. Describe the safe way to support a motor vehicle for brake servicing.
8. **Critical Thinking** Explain why you should avoid "locking up" your service brakes.
9. **Critical Thinking** Is it safe to use brake fluid from a container that has been opened, capped, and stored for six months? Why or why not?

TECHNOLOGY FORECAST
FOR AUTOMOTIVE EXCELLENCE

Future Braking Systems

Electronics will play a large role in the future of braking systems. This will help increase safety and efficiency, while reducing weight. The use of hydraulic fluid will likely remain the same, due to its proven stopping power.

These new systems will work much like today's brakes, but with a "brake-by-wire" design. The driver puts pressure on the brake pedal. At that instant, hydraulic fluid causes calipers and wheel cylinders at each wheel to apply braking force. This action stops the vehicle.

There is one key difference in this new brake design. Hydraulic fluid will no longer be stored in the master cylinder. Instead, it will be held in containers, called reservoirs, at each wheel. The hydraulic fluid applies stopping force with the help of small motors. These units are turned on by an electronic signal from a computer when the brake pedal is depressed.

This setup should simplify routine maintenance and repair. It will also allow the use of software-driven safety features on even the lowest priced models. These features might include anti-lock brakes (ABS), traction control, and stability enhancement systems.

Meanwhile, engineers are exploring other brake options. One design uses electric power only, and no hydraulic fluid. In this design, an electric motor gearbox controls braking.

Is the public ready for electric-only braking systems? Only time will tell.

AUTOMOTIVE EXCELLENCE
TEST PREP

Answering the following practice questions will help you prepare for the ASE certification tests.

1. Two service technicians are discussing a car's primary braking system. They are talking about its:
   - a Parking brake system.
   - b Air brake system.
   - c Service brake system.
   - d Emergency brake system.

2. Which of the following types of wheel brakes are found in a car's service braking system?
   - a Drum and coaster brakes.
   - b Disc and coaster brakes.
   - c Coaster and hand brakes.
   - d Drum and disc brakes.

3. Technician A and Technician B are talking about which components in a moving vehicle's wheel brake system are in relative motion. Technician A says they are the drums and rotors. Technician B says they are the brake shoes and drums. Who is correct?
   - a Technician A.
   - b Technician B.
   - c Both Technician A and Technician B.
   - d Neither Technician A nor Technician B.

4. When a vehicle's wheel brakes are locked up where would you expect to find kinetic friction at work?
   - a In the wheel brake system.
   - b In the parking brake system.
   - c Where the tires meet the road.
   - d In the entire service braking system.

5. Technician A and Technician B are discussing at what instant a vehicle's braking system is most effective. Technician A says it is just before the wheels lock up. Technician B says it is just after the brakes have been applied. Who is correct?
   - a Technician A.
   - b Technician B.
   - c Both Technician A and Technician B.
   - d Neither Technician A nor Technician B.

6. You would expect your car's tires to have better traction with the road when the road is:
   - a Just wet.
   - b Wet and icy.
   - c Ice covered.
   - d Dry.

7. Two technicians are working on a car's service brakes. The technicians determine that the car has a dual-braking system. This means that the car has:
   - a Two hydraulic braking systems.
   - b Two types of brakes.
   - c Two sets of brakes.
   - d Two master cylinders.

8. Technician A and Technician B are discussing the dual-braking system of a brand new front wheel drive car. Technician A says it is more likely that the brakes will be separated from front to rear. Technician B says it is more likely that they will be separated diagonally. Who is correct?
   - a Technician A.
   - b Technician B.
   - c Both Technician A and Technician B.
   - d Neither Technician A nor Technician B.

9. The most commonly used brake fluids are:
   - a DOT 3 and DOT 5.
   - b DOT 3 and DOT 4.
   - c DOT 4 and DOT 5.
   - d DOT 5 and DOT 6.

10. DOT 3 and DOT 4 brake fluids are:
    - a Water-based.
    - b Silicone-based.
    - c Polyglycol-based.
    - d Water- and polyglycol-based.

CHAPTER 2

Diagnosing & Repairing the Hydraulic System

You'll Be Able To:

- Describe how hydraulics is used to transfer motion and force.
- Explain the function of a master cylinder.
- Remove, bench bleed, and replace the master cylinder.
- Describe the function of double flare and ISO brake lines.
- Identify the control valves used in hydraulic braking systems.
- Demonstrate bleeding a hydraulic brake circuit.

Terms to Know:

brake bleeding
hydraulics
metering valve
pressure differential valve
proportioning valve
residual pressure check valve

The Problem

Mr. Ortiz says that while suddenly braking for another car, he had to repeatedly pump the brake pedal. He also says that later that day the brake pedal slowly sank to the floor. This happened as he was waiting for a traffic light to turn green.

You get into the driver's seat and push the brake pedal. As Mr. Ortiz said, the pedal slowly goes down. You open the hood and check the master cylinder fluid level. You note that it is almost full.

Your Challenge

As the service technician, you need to find answers to these questions:

1. What could be causing the brake pedal to drop to the floor?

2. Could the problem lie with the brake shoes, pads, or hydraulic system?

3. How would you isolate the problem?

Section 1

Hydraulics

Hydraulics is the process of applying pressure to a liquid to transfer force or motion. The pressure applied to a liquid to create the force is called *hydraulic pressure*. In an automotive brake system, hydraulic pressure creates the force to move the brake shoes or pads into contact with the brake drums or rotors.

Properties of Hydraulics

A gas that is compressed has a smaller volume. Gas can be compressed by applying pressure to it. For example, a small scuba tank with a volume of a few cubic feet can hold nearly 80 cubic feet [2.24 m³] of compressed air. If gases were used in an automotive brake system, part of the brake pedal force would be used to compress the gas.

Liquids cannot be compressed. **Fig. 2-1.** That is why hydraulic pressure is used to transmit motion and force in an automotive brake system. All the pressure applied to the liquid is transferred as motion and force.

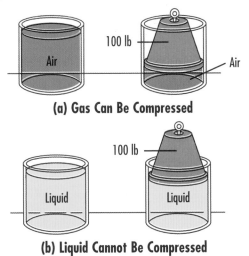

(a) Gas Can Be Compressed

(b) Liquid Cannot Be Compressed

Fig. 2-1 (a) Gases can be compressed when pressure is applied. **(b)** Liquids cannot be compressed. *Why are liquids used to transfer motion and force?*

The Transfer of Motion   An *apply* or *input piston* and an *output piston* are shown in the same cylinder. **Fig. 2-2.** There is liquid between the pistons. When enough pressure is applied to move the apply piston 8 inches [203 mm], the output piston moves the same distance and in the same direction.

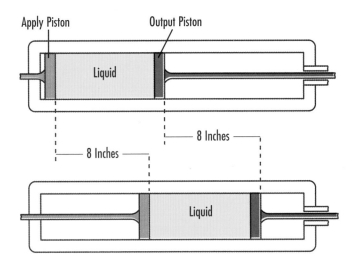

Fig. 2-2 A liquid can transmit both force and motion. When the apply piston is moved, the output piston moves the same distance. *What is the advantage of using a fluid instead of a solid connection?*

A metal rod or solid connection between the two pistons would serve the same function as the liquid. But a solid connection could not transfer motion and force through the hoses and lines of the hydraulic system.

The two pistons need not be in the same cylinder. A tube can be used to transfer hydraulic pressure from one cylinder to another. **Fig. 2-3.** In automotive brake systems, the input cylinder is called the *master cylinder*. The output, or remote, cylinders are the vehicle's *wheel cylinders* or *calipers*.

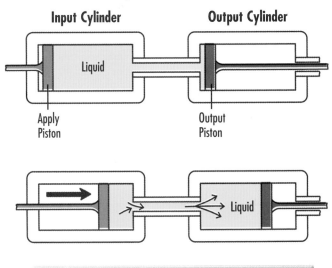

Fig. 2-3 A liquid can transmit motion and force through a tube from one cylinder to another. *Does the length of the connecting tube change the amount of movement of the output piston?*

Calculating Hydraulic Pressure

You can calculate the pressure in a hydraulic brake system using the formula:

$$P = \frac{F}{A}$$

P is the pressure in pounds per square inch (psi). F is the mechanical force applied to a piston in pounds (lb). A is the surface area of the piston in square inches (in²).

Here's an example. A mechanical force of 100 pounds is applied to the brake pedal pushrod. This force is transferred onto a master cylinder piston with a surface area of 2 square inches. What is the pressure in the braking system?

$$P = \frac{F}{A} = \frac{100\ \text{lb}}{2\ \text{in}^2} = 50\ \text{psi}$$

What is the pressure if the same force is applied to a master cylinder piston with twice the surface area?

$$P = \frac{F}{A} = \frac{100\ \text{lb}}{4\ \text{in}^2} = 25\ \text{psi}$$

Some manufacturers' specifications show pressure measured in the metric unit kPa, or kilopascals. You can convert pressure measured in psi to the metric form by remembering that:

$$1\ \text{psi} = 6.9\ \text{kPa}$$

To convert the pressure in the first example, use the formula:

$$50\ \text{psi} \times 6.9 = 345\ \text{kPa}$$

Apply it!

Meets NATEF Mathematics Standards for manipulating algebraic equations, using symbols, equivalent forms, and converting between the English and metric systems.

❶ You need to calculate the pressure in a hydraulic brake system. What is the pressure in the system if a force of 100 pounds is applied to a piston with an area of 0.5 square inches? Give your answer in psi and kPa.

❷ How would you calculate the force needed to produce a pressure when you know the surface area of the piston?

Rearrange the equation $P = \frac{F}{A}$ to find F.

❸ How much force must be applied to a piston with a surface area of 0.25 square inches to create a pressure of 160 psi?

Applying force to the apply piston transfers hydraulic pressure through the connecting tube to the remote cylinder. Pressure in the remote cylinder forces the output piston to move. If the master cylinder and a wheel cylinder are the same diameter, the amount of force and movement is the same in both cylinders. Neither the length nor the diameter of the tube connecting the two cylinders affects the amount of movement of the output piston.

The Transfer of Force   When force is applied to a liquid, the pressure created by the force is transmitted equally in all directions. **Fig. 2-4.**

In **Fig. 2-4,** the piston has a surface area of 1 square inch [6.45 cm²]. A force of 100 pounds is applied to the piston. The piston therefore is applying 100 pounds per square inch (psi) [690 kPa] pres-

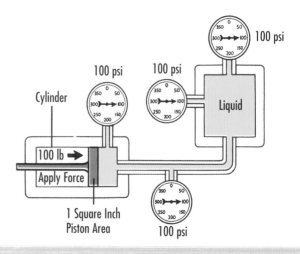

Fig. 2-4 The pressure of a liquid is the same throughout the system. *What do you think the gauges would read if there were two tubes leading from the cylinder?*

(Pontiac Division of General Motors Corporation)

sure to the liquid. Notice that all gauges show the same pressure.

The size of the piston is a factor in the amount of pressure applied to the fluid. If the surface area of the piston is doubled and the force applied is the same, half the pressure results. For example, consider that we apply a force of 100 pounds [45 kg] to a piston surface area of 2 square inches [12.9 cm²]. The resulting pressure would be 50 psi [345 kPa]. **Fig. 2-5.**

In a system with more than one remote cylinder, the applied force is equally distributed to each cylinder. The output force of the remote cylinders will differ from each other, however, if the remote cylinder pistons are of varying sizes.

A remote cylinder with a piston half the size of the input piston delivers only half the force. A remote cylinder with the same size piston delivers the same amount of force. A remote cylinder with a piston twice the size of the input piston delivers twice the force. In other words, the larger the output piston, the greater the output force. **Fig. 2-6.** This difference in applied force and output force is why light brake pedal pressure can stop a heavy car.

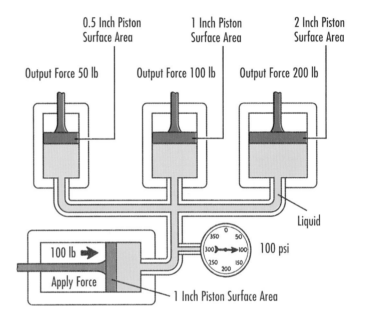

Fig. 2-6 The force created by the apply piston is called system pressure. The output force of each remote cylinder depends upon the output piston's surface area. *What would the output force be if the output piston had a surface area of 4 square inches [25.8 cm²]?* (Pontiac Division of General Motors Corporation)

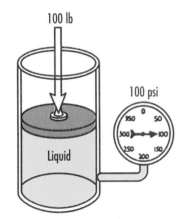

(a) Piston Surface Area of 1 Square Inch

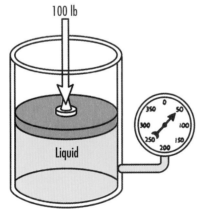

(b) Piston Surface Area of 2 Square Inches

Fig. 2-5 The pressure in a hydraulic system can be calculated by dividing the applied force by the surface area of the piston. *What would the fluid pressure be if the force were 100 pounds [45 kg] and the surface area of the piston were 0.5 square inch [3.23 cm²]?* (Pontiac Division of General Motors Corporation)

Force and motion are related in an automotive braking system. As the designed output force is increased, the amount of motion (distance of travel) is decreased. The reverse is also true. As designed required input force is reduced at the brake pedal, greater movement must be provided.

In a disc brake system, the master cylinder piston is much smaller in diameter than the piston in the wheel brake caliper. This means the brake pedal will have to move the master cylinder piston a relatively long distance. Being larger, the caliper piston will move only a short distance but will have increased force.

SECTION 1 KNOWLEDGE CHECK

❶ What is the process of applying pressure to a liquid to transfer force or motion?

❷ Which materials cannot be compressed by applying pressure–liquids or gases?

❸ What other names are given to the input cylinder and the output cylinder when they are used in a braking system?

❹ What does the apply piston do when force is applied to it?

❺ What happens to output force if the output piston is larger or smaller than the input piston?

Section 2

Brake Master Cylinders

The brake master cylinder provides the pressurized fluid to the remote cylinders. In an automobile the remote cylinders that activate the drum brakes are called the wheel cylinders. The cylinders that activate the disc brakes are called *calipers*.

Integral Master Cylinders

An *integral master cylinder* is a one-piece, cast-iron component with a dual reservoir. **Fig 2-7.** It is called an integral master cylinder because the fluid reservoirs are integrated into the master cylinder assembly. This type of master cylinder is still found on some older large cars and light trucks.

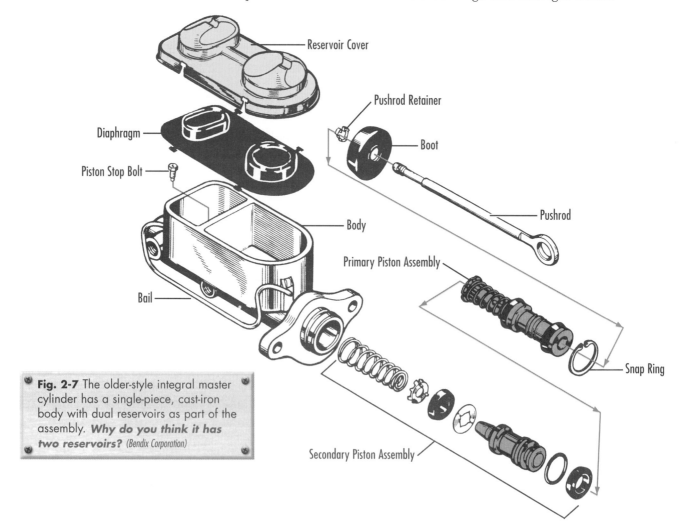

Reservoir Cover
Diaphragm
Piston Stop Bolt
Pushrod Retainer
Boot
Pushrod
Body
Bail
Primary Piston Assembly
Snap Ring
Secondary Piston Assembly

Fig. 2-7 The older-style integral master cylinder has a single-piece, cast-iron body with dual reservoirs as part of the assembly. **Why do you think it has two reservoirs?** (Bendix Corporation)

Composite Master Cylinders

Most vehicles now have *composite master cylinder* assemblies. The master cylinder is made of cast aluminum. A plastic reservoir, holding the brake fluid, is attached directly to the master cylinder with rubber grommets that maintain a tight seal. **Fig. 2-8.**

Master-Cylinder Operation

A master cylinder converts mechanical brake pedal force into hydraulic pressure. Both integral and composite master cylinders work in the same way. The cylinder bore is filled with brake fluid. When the driver presses the brake pedal, the piston inside the bore moves forward. The force of the piston acting on the fluid transfers pressure to

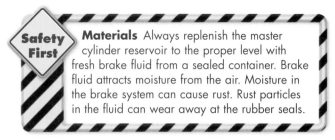

Safety First

Materials Always replenish the master cylinder reservoir to the proper level with fresh brake fluid from a sealed container. Brake fluid attracts moisture from the air. Moisture in the brake system can cause rust. Rust particles in the fluid can wear away at the rubber seals.

the wheel cylinders. When the driver releases the brake pedal, a spring moves the piston back to its original position. As the piston moves back, piston force and resulting fluid pressure are removed.

In a dual master cylinder, two pistons move back and forth in the same bore. **Fig. 2-8.** The space in front of each piston is the fluid chamber. It is kept full by the reservoir above it.

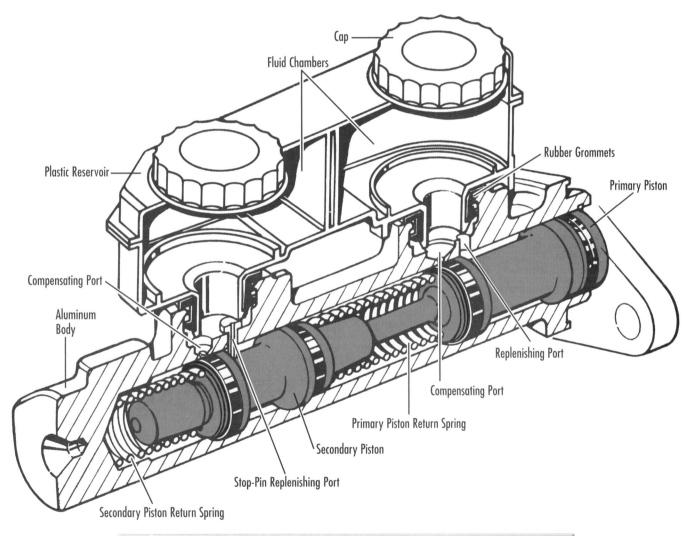

Cap

Fluid Chambers

Plastic Reservoir

Rubber Grommets

Primary Piston

Compensating Port

Aluminum Body

Replenishing Port

Compensating Port

Primary Piston Return Spring

Secondary Piston

Stop-Pin Replenishing Port

Secondary Piston Return Spring

Fig. 2-8 The composite master cylinder has a lightweight, cast-aluminum body with a separate plastic reservoir attached. The two pistons travel back and forth in the same bore. *What is the effect of the piston on the brake fluid pressure?*
(DaimlerChrysler)

There are two openings at the bottom of each reservoir. The openings toward the front are the compensating ports. These ports are open to the reservoir so fluid can enter and leave the high-pressure chambers. The openings toward the rear are the replenishing ports. These ports are open to the reservoir so fluid can keep the low-pressure chambers filled.

When the brakes are applied, the piston cup (seal) moves past the compensating port and seals it off from the bore. The cup seals the chamber and piston so fluid cannot flow into the low-pressure chamber. **Fig. 2-9(a).** As the piston continues its forward travel, fluid pressure builds. Pressure is transferred through the brake lines to the wheel cylinders.

The replenishing port lets fluid fill the low-pressure chamber behind the piston. If this chamber were left empty, a vacuum would be created. The vacuum would try to pull the piston back.

When the brake pedal is released, return springs move the pistons to their original position. The pistons move faster than the fluid can return through

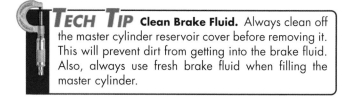

TECH TIP **Clean Brake Fluid.** Always clean off the master cylinder reservoir cover before removing it. This will prevent dirt from getting into the brake fluid. Also, always use fresh brake fluid when filling the master cylinder.

the lines. This motion creates a partial vacuum in the high-pressure chamber. The piston cup allows fluid to flow from the low-pressure chamber into the high-pressure chamber. The fluid flows around the piston cup or through small holes in the piston. **Fig. 2-9(b).**

Quick-Takeup Master Cylinders This master cylinder design is used on disc brake systems with low-drag calipers. The calipers used in low-drag disc brake systems have special seals. These seals quickly pull the piston back into the bore.

In a low-drag disc brake system, there is a relatively large gap between the brake pads and the rotor. As a result, it takes a lot of fluid to move the caliper pistons the greater distance. The caliper pistons must travel the distance quickly but not require any excessive brake pedal travel.

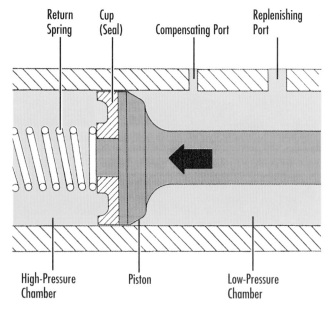

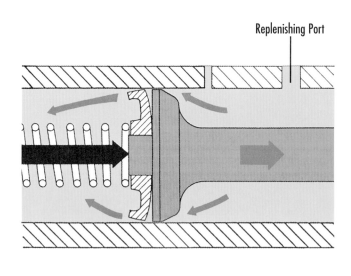

(a) Brakes Applied

(b) Brakes Released

Fig. 2-9 (a) With the brakes applied, the piston is sealed in the bore and pushes pressurized fluid ahead of it. **(b)** When the brakes are released, a spring pushes the piston back. The seal relaxes, and fluid from the low-pressure chamber flows into the high-pressure chamber. *What happens to the pressurized fluid in the brake lines when the brakes are released?*

In the quick-takeup master cylinder, the primary piston operates in a *step-bore cylinder*. This is a cylinder with two different diameters. **Fig. 2-10.**

When the brakes are first applied, fluid from the large-diameter low-pressure bore flows past the cup (seal) into the primary high-pressure chamber. This large amount of fluid quickly applies the brakes connected to the primary chamber. The fluid action also forces the secondary piston to quickly apply the brakes connected to the secondary chamber.

The quick-takeup design supplies a large amount of fluid with relatively short brake pedal travel. Both chambers are supplied with more fluid than is needed to apply the brakes. A quick-takeup valve opens as the pressure builds and allows the excess fluid to flow back into the reservoir.

Some master cylinders have a fluid-level sensor that illuminates a warning light on the instrument panel when the fluid level drops too low.

TECH TIP Removing the Master Cylinder. Whenever the master cylinder is removed, the entire hydraulic system may need to be bled. Some master cylinders can be bench bled before installation. This can save time in the procedure.

Master Cylinder Diagnosis

If the brake fluid reservoir is low, inspect the master cylinder for leaks. Check the master cylinder booster mount, vacuum check valve, vacuum supply hose, or the rear master cylinder seal. It may be necessary to unbolt the master cylinder from the power booster. Refer to the manufacturer's manual for specific procedures.

If the front or rear brakes lock up during normal brake use, the master cylinder pushrod may need adjustment. Inspect the entire vehicle brake system for other causes before adjusting the master cylinder pushrod and brake pedal height.

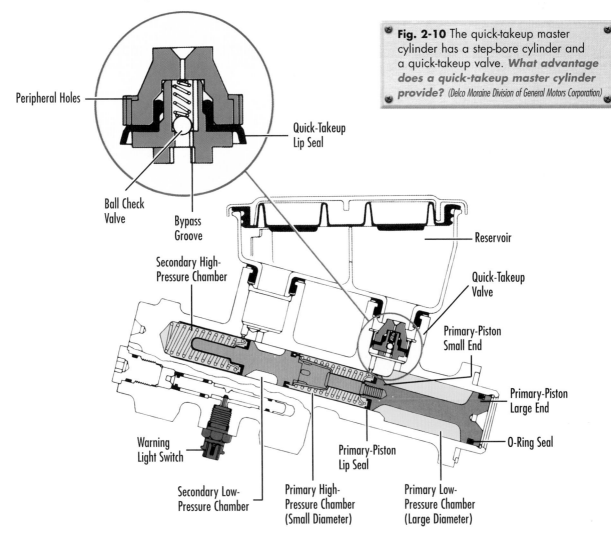

Fig. 2-10 The quick-takeup master cylinder has a step-bore cylinder and a quick-takeup valve. *What advantage does a quick-takeup master cylinder provide?* (Delco Moraine Division of General Motors Corporation)

Peripheral Holes

Quick-Takeup Lip Seal

Ball Check Valve

Bypass Groove

Secondary High-Pressure Chamber

Reservoir

Quick-Takeup Valve

Primary-Piston Small End

Primary-Piston Large End

O-Ring Seal

Warning Light Switch

Primary-Piston Lip Seal

Secondary Low-Pressure Chamber

Primary High-Pressure Chamber (Small Diameter)

Primary Low-Pressure Chamber (Large Diameter)

Recording Information

Note taking is an informal method used to record information. You can write in a form you can easily understand. You don't need to use complete sentences. However, you should write neatly and spell words correctly so you can share your notes with others if necessary. Taking notes as a form of communication can be especially helpful in diagnosing a vehicle's problem.

You should take notes during a discussion with a customer about a vehicle's problem. Listen carefully. Write down what you believe are the customer's important observations. Include observations of your own.

Using these notes and your own knowledge, you can start to eliminate unlikely causes. You can also jot down ideas for testing to diagnose the problem.

Apply it!

Meets NATEF Communications Standards for taking notes and organizing and editing information.

Reread *The Problem* at the beginning of this chapter. Assume you are the technician.

❶ On a sheet of paper, take notes on Mr. Ortiz's observations about brake failure in his vehicle.

❷ Note what you have observed.

❸ Next to Mr. Ortiz's observations and your own, write down what you think may be causing the brake failure. Also note the causes you think you can eliminate and why.

Master Cylinder Bench Bleed

Bench bleed the master cylinder as follows:

1. Remove the master cylinder using the manufacturer's service manual procedures. Some master cylinders can be "bench bled" prior to installation.

2. Mount the master cylinder in a vise, with the reservoir attached.

3. Attach the bleed lines to outlet ports in the master cylinder.

4. Run the bleed lines to a clean catch container that contains clean brake fluid.

5. Place the catch container below the level of the master cylinder outlet ports.

6. Fill the master cylinder reservoir with the clean brake fluid.

7. Using a wooden dowel, push the master cylinder primary and secondary pistons slowly. This compresses and drives air from the master cylinder.

8. Be sure fluid in the reservoir is maintained at the full level.

9. Repeat Step 7 several times until there are no more air bubbles in the fluid that moved to the catch container.

10. Remove the bleed tubes and cap the outlet ports.

11. Replace the reservoir filler cap.

12. Remove the master cylinder from the vise and reinstall per the manufacturer's recommended procedures.

Brake Lines

Brake lines are the tubes and hoses that carry the brake fluid from the master cylinder to the wheel cylinders and calipers. There are two types of brake lines, rigid and flexible. Rigid lines run from the master cylinder to a connection point near the wheel brake. Flexible lines run from the connection point to the wheel cylinder or caliper.

Rigid brake lines are made of steel. Copper or brass tubing is too soft to safely handle the fluid pressure.

Replacement rigid brake lines are available as straight tubing or preformed (with all the required bends) to fit a particular vehicle model. In all cases the ends are flared to provide leakproof connections. A *double flare* looks like the bell of a trumpet bent back over itself. The *ISO (International Standards Organization) flare* resembles a bubble. **Fig. 2-11.** Connections are made with fittings called flare nuts.

Special tools are used to make the flares. Technicians must be careful when forming the flares. Any damaged or cracked flares must be cut off and new ones formed. For this reason it is better to install tubing that has been "factory" (previously) flared. It comes in many lengths, and new flare nuts are included. When possible, use preformed brake lines. Replace brake lines, fittings, or supports as needed.

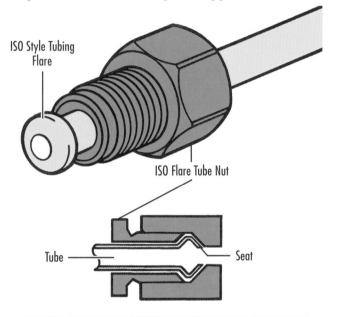

Fig. 2-11 The International Standards Organization (ISO) flare resembles a bubble. This flare is becoming the worldwide standard. *Why are the ends of brake lines flared?* (DaimlerChrysler)

A vehicle's chassis and suspension move in relative motion to each other. To allow for this relative motion, there must be a flexible brake line between the chassis and the wheel brakes. Brake hoses made of rubber allow suspension movement. To prevent these hoses from bursting under high pressure, they are made of three layers:
• A flexible inner tubing.
• Reinforcement of braided fabric.
• Outer tubing for protection.

The fitting on one end of the brake hose fits the flared fitting and flare nut of a steel brake line. The other end of the brake hose has a threaded fitting or

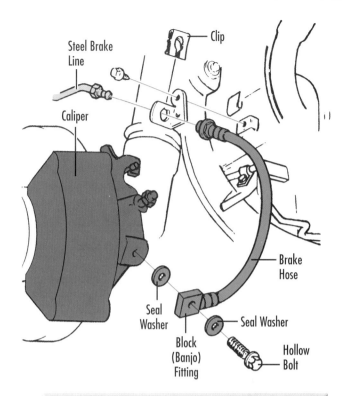

Fig. 2-12 A flexible brake hose connects the stationary, rigid brake line and the movable parts of the vehicle such as the caliper. *Why can rigid lines not be used throughout?* (General Motors Corporation)

a banjo fitting sealed with copper sealing washers. **Fig. 2-12.** Inspect rubber brake hoses at the brake line connection. Hoses should be free of cracks, dry rot, bulging, or other damage. Replace hoses, fittings, or supports as needed.

Hydraulic Circuits

Early brake systems had a single hydraulic circuit with a single-piston master cylinder. In 1967 the U.S. Department of Transportation mandated dual master cylinders for all vehicles. The use of dual master cylinders in dual hydraulic circuit brake systems decreases the number of brake system failures. If one circuit fails, partial braking is still available. Only very rarely do both brake circuits fail at the same time. Dual-circuit brake systems are either front-rear split or diagonally split.

Front-Rear Split Systems

Front-rear split systems are found mostly on rear-wheel-drive (RWD) vehicles with front disc brakes and rear drum brakes. One circuit serves the front brakes; the other circuit serves the rear brakes. **Fig. 2-13(a).**

Front-wheel-drive (FWD) vehicles have much more weight in the front than in the rear. Therefore, the front brakes must provide up to 85 percent of the braking. If a FWD vehicle has a front-rear split system and the front half fails, the rear brakes may not be able to stop the car effectively.

Diagonally Split Systems

In a diagonally split system, the left front and right rear wheels share one hydraulic circuit and the right front and left rear wheels share the other. **Fig. 2-13(b).** Should there be a failure in one circuit, the other circuit can still provide stopping power.

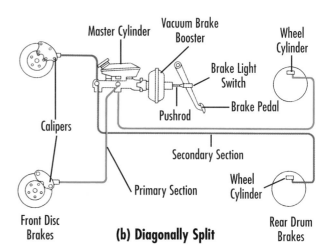

Fig. 2-13 The two basic types of split, or dual, braking systems are **(a)** front-rear split and **(b)** diagonally split. **Why are dual systems safer than single systems?** (Ford Motor Company)

SECTION 2 KNOWLEDGE CHECK

① Explain how a master cylinder transmits force to the wheel brakes.

② Explain the difference between compensating and replenishing ports in the master cylinder.

③ Explain the need for a quick-takeup master cylinder in some brake systems.

④ Describe the difference between a double flare and an ISO flare for brake lines. Are they interchangeable?

⑤ What is the difference between a front-rear split and a diagonally split system?

Section 3

Brake System Valves

There are typically five types of control valves in automotive braking systems:

- Metering valves.
- Pressure differential valves.
- Proportioning valves.
- Residual pressure check valves.
- Combination valves.

Metering Valve

Most vehicles that have front disc and rear drum brakes are equipped with metering valves. **Fig. 2-14.** The **metering valve** is the valve that controls, or delays, the flow of brake fluid to the front brakes.

The rear drum brakes take longer to respond than the front disc brakes. The metering valve ensures that front disc brakes do not act before the rear drum brakes. The delay is long enough to allow the drum brakes time to react. This delay is particularly necessary during light braking and on slick road surfaces.

Metering valves are not usually found on FWD cars that have diagonally split brake systems. Vehicles with similar type brakes (front and rear) may not need a metering valve. Brakes of the same type front and rear generally require the same amount of time to apply.

Pressure Differential Valve

In a dual system, a **pressure differential valve** is the valve that senses the pressure in each circuit. **Fig. 2-14.** The valve piston stays centered as long as the hydraulic pressure in both circuits is balanced. If the pressure drops in either circuit, the piston moves away from center. This movement triggers a switch that turns on the brake warning light. The warning light indicates a partial failure of the braking system. Refer to the manufacturer's manual for specific diagnosis and test procedures.

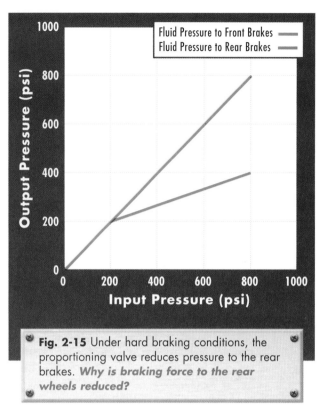

Fig. 2-15 Under hard braking conditions, the proportioning valve reduces pressure to the rear brakes. *Why is braking force to the rear wheels reduced?*

Proportioning Valve

A **proportioning valve** is a valve that reduces the amount of braking force at the rear wheels on front disc and rear drum brakes systems. During hard braking forward momentum shifts the vehicle weight toward the front. The weight shift requires less braking force at the rear wheels. Too much braking force at the rear could cause the rear wheels to lock up.

The proportioning valve has no effect on braking force during normal braking. However, during hard braking the proportioning valve reduces the amount of pressure to the rear brakes once the pressure reaches a certain point. This is called the *split point.* **Fig. 2-15.**

A proportioning valve should not be confused with a metering valve. A proportioning valve regulates braking force to the rear wheels. A metering valve delays fluid flow to the front brakes.

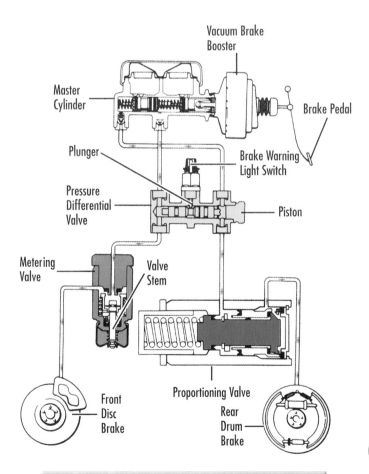

Fig. 2-14 Control valves used on a typical front disc brake and rear drum brake system. *What type of valve is not used on systems with similar type brakes?* (Ford Motor Company)

TECH TIP Servicing Hydraulic Components. When servicing or replacing components in the brake hydraulic system, it is important to keep dirt and moisture out of the hydraulic lines. Whenever you disconnect a line, cover the open end with a piece of tape, or a rubber or plastic cap.

Proportioning valves are most commonly found in the master cylinder outlets. In some vehicles they are found in the brake lines near the rear brakes. If the vehicle has a problem with either front or rear brake lockup under normal braking, inspect the proportioning valve. Diagnose and test the valve using the manufacturer's recommended procedures. Clean, adjust, or replace as needed.

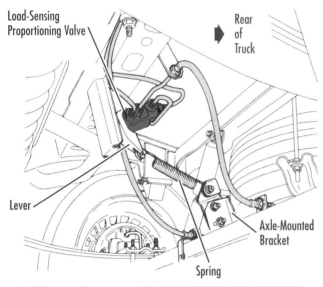

Fig. 2-16 Depending on the distance between the truck bed and the axle, the load-sensing proportioning valve controls the hydraulic pressure to the rear brakes. *Why does a vehicle with a heavier load require more braking force at the rear wheels?* (DaimlerChrysler)

Some light-duty trucks have *load-sensing proportioning valves.* **Fig. 2-16.** The load-sensing proportioning valves adjust the pressure to the rear brakes according to changes in the load. The valves sense vehicle load by measuring the distance between the truck bed and the axle. As the load increases, the distance decreases. The heavier the load, the greater the pressure supplied to the rear brakes. Inspect the load-sensing proportioning valves. Diagnose and test per manufacturer's recommended procedures. Clean, adjust, or replace as needed. Inspect the linkage or spring actuator of the valve for free movement. Binding or restriction could keep it from operating properly.

Vehicles with anti-lock brake systems (ABS) usually do not have load-sensing proportioning valves. The ABS prevents wheel lockup.

Residual Pressure Check Valve

Some master cylinders, used mainly with drum brakes, have a residual pressure check valve in the outlet port. The **residual pressure check valve** is a valve that maintains a residual pressure of about 6–18 psi [41–124 kPa] in the brake lines. The residual pressure keeps the seals in the wheel cylinders expanded. The expanded seals prevent air from leaking past them due to the sudden drop in pressure when the brakes are released.

Check valves are not used in disc brake circuits. Disc brakes do not usually use return springs. A check valve would allow pressure buildup and cause brake drag.

Generally, residual pressure check valves have not been used in brake systems since the 1970s. Instead, the wheel cylinders have cup expanders behind the seals to prevent fluid from leaking past them.

Combination Valve

Many vehicles with front disc and rear drum brakes have a three-function combination valve. **Fig. 2-17.** This valve combines the pressure differential valve, metering valve, and proportioning valve into one unit. Two-function combination valves combine either the proportioning valve or the metering valve with the pressure differential valve.

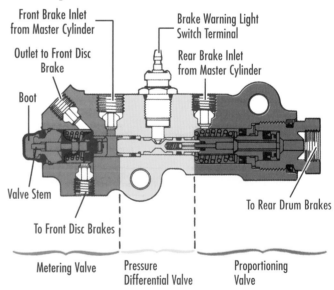

Fig. 2-17 The combination valve combines the metering valve, pressure differential valve, and proportioning valve in one assembly. *Do all combination valves combine three valves into one assembly?* (General Motors Corporation)

Using Hydraulics to Multiply Force

For centuries, people have used fluids to do work. But it was not until Blaise Pascal (1623–1662) studied fluids under pressure that the relationship between fluids and pressure was understood. The *Pa* in the metric system pressure unit kPa (kilopascal) derives from Pascal's name.

In this activity you will make a simple hydraulic system and show how it multiplies forces.

Apply it!

Make a Hydraulic Lift

Meets NATEF Science Standards for transferring force in a hydraulic system.

Materials and Equipment
- 1 strong balloon
- Large rubber bands
- 6-foot length of plastic tubing with about ½-inch inside diameter
- Measuring cup or baby bottle marked in fluid ounces
- Funnel to fit the tubing
- Thin, lightweight board about one square foot
- Four small blocks of wood about 1-inch thick
- Some objects of known weight (for example, a 12-ounce canned soft drink, a quart of milk [2 pounds], a 15-ounce can of soup)

❶ Fill the balloon with as little water as possible. Do not inflate or stretch the balloon. Add just enough water to remove any trapped air.

❷ Using a rubber band, tightly seal the mouth of the balloon over one end of the tubing.

❸ Hang the other end of the tubing so the balloon lies flat on the floor.

❹ Gently balance the board on the balloon with your hands. Don't press down. Notice that the weight will force some water up into the tube.

❺ Use the blocks to support the four corners of the board.

❻ Add water to the tube until the board just begins to rise off the blocks. Your hydraulic lift system is now ready for testing.

❼ Put a 12-ounce weight at the center of the board.

❽ Fill the measuring cup or baby bottle with water.

❾ Using the funnel, slowly pour water into the tube until the board just begins to lift.

❿ Write down how many ounces of water you poured into the tube. It wasn't nearly 12 ounces, was it?

Here's what is happening. The weight of the water you added is the apply force, or input force. A small amount of water at the bottom of the tube is acting as the apply piston. The balloon which is in contact with the board, is the output piston. The difference in the areas of these two "pistons" is multiplying the force.

⓫ Repeat the test using the other objects.

⓬ Write down the total number of ounces of water you added to the tube to lift each weight.

⓭ Divide each weight by the number of added ounces to find the approximate ratio of the piston areas.

⓮ Compare the ratios you calculated for the various weights.

The ratios are not exactly the same in each test, are they? Try to think of the reasons this simple hydraulic lift does not follow the mathematical formula exactly.

More things changed than just the weight of the objects. The size of the footprint of the balloon varied as the weights increased. This meant that the area of the "output piston" varied.

Bleeding a Hydraulic Braking System

Servicing or replacing components in a hydraulic braking system usually allows air to enter the brake system. Air, like any gas, is compressible. In the system the trapped air acts like a cushion, creating a soft brake pedal feel and an unsafe condition. All trapped air must be bled out.

Water may also enter the hydraulic system through small pores in the flexible brake lines. Contaminated fluid should be flushed out periodically. **Brake bleeding** is the process of flushing air or contaminated fluid from the braking system. There are four basic methods. They are manual bleeding, gravity bleeding, pressure bleeding, and vacuum bleeding.

Manual Bleeding

Manual bleeding requires an assistant to operate the brake pedal from inside the vehicle. When the brakes are pumped, pressure is created by the master cylinder. The pressure pumps the fluid and air out of the system through the bleeder valves on each wheel.

To manually bleed hydraulic brakes (non-anti-lock brake system), do the following:

1. With the engine and ignition OFF, pump the brake pedal several times. This will expel any remaining vacuum in the power brake booster.

2. Remove the master cylinder reservoir cover. Fill the reservoir to the full level. Place the cover loosely over the opening.

3. Pour a little brake fluid into a catch container. The bleeder hose used to bleed the system must be submerged in the fluid during the bleeding process. The bleeder hose must be clear tubing. This will allow you to see more easily the air bubbles being expelled from the brake system.

4. The bleeding process sequence depends on the braking system. For a front-rear split, the sequence is right rear wheel, left rear wheel, right front wheel, and left front wheel. For a diagonally split, the sequence is right rear, left front, left rear, and right front.

5. Always check and, if needed, refill the master cylinder reservoir with fluid after bleeding each caliper or wheel cylinder. Refill more often if needed.

6. Loosen the bleeder valve at the wheel to be bled. Then, tighten it just enough to allow it to be easily loosened and tightened.

7. Install one end of the clear tubing over the bleeder valve. Leave room for a wrench to fit over the flats of the valve. Place the other end of the tubing into the catch container. Make sure the end is near the bottom and submerged. **Fig. 2-18.**

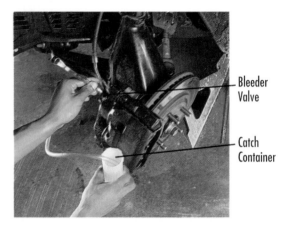

Bleeder Valve

Catch Container

Fig. 2-18 When bleeding brakes, a tube runs from the bleeder valve to the catch container. *What is the purpose of the catch container?* (Jack Holtel)

8. Have an assistant slowly press the brake pedal down as far as possible. Ask the assistant to hold the pedal in the down position.

9. While the assistant holds the pedal down, open the bleeder valve enough to allow fluid to flow through the tubing and into the catch container. Watch for air bubbles to come out of the tube. The fluid flow will slow down in a few seconds. When the air bubbles stop, tighten the bleeder valve. Have your assistant remove their foot from the brake pedal.

10. Repeat the prior two steps until no air bubbles can be seen coming from the tube. Then, securely tighten the bleeder valve. Remove the tube and container. Move on to the next caliper or wheel cylinder.

11. Check and, if needed, add brake fluid to the master cylinder reservoir.

12. Repeat the prior six steps for each caliper or wheel cylinder. Do this in the sequence required by the type of system.

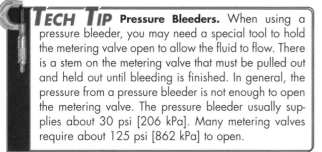

TECH TIP **Pressure Bleeders.** When using a pressure bleeder, you may need a special tool to hold the metering valve open to allow the fluid to flow. There is a stem on the metering valve that must be pulled out and held out until bleeding is finished. In general, the pressure from a pressure bleeder is not enough to open the metering valve. The pressure bleeder usually supplies about 30 psi [206 kPa]. Many metering valves require about 125 psi [862 kPa] to open.

13. After bleeding all four calipers and/or wheel cylinders, check and, if needed, add brake fluid to the master cylinder reservoir. Secure the master cylinder reservoir cover.

14. Check the operation of the brakes. The brake pedal should stop firmly when pressed down. If the brakes feel spongy, repeat the bleeding procedure.

Gravity Bleeding

Gravity bleeding uses gravity and atmospheric pressure to bleed the brake system. Gravity and atmospheric pressure provide the pressure to force the air and fluid from the system. This method is very slow, but it can be effective. It may be the only alternative if no power bleeder or helpers are available. It does not work, however, if the system has a residual pressure check valve.

To gravity bleed, remove the master cylinder cover and add fluid if necessary. Open a bleeder valve and let drip until all air is purged from the brake line and wheel cylinder and caliper. Close the bleeder valve. Do this for all wheels. Top off the master cylinder and replace cover.

Pressure Bleeding

Pressure bleeding, also called power bleeding, is the most popular method. The pressure bleeder is filled with brake fluid, attached to the master cylinder, and pressurized with shop air. The pressure bleeder provides the pressure normally provided by the master cylinder. **Fig. 2-19.** Every pressure bleeder works differently. Follow the procedures in the manufacturer's manual.

Vacuum Bleeding

A single technician using a hand-held vacuum pump and appropriate adapter can do *vacuum bleeding.* The vacuum pump is used to draw fluid and trapped air from the system at the bleeder valves. Vacuum bleeding may not work on systems that do not have cup expanders. This method will not work on rear calipers with stroking seals. Refer to the vehicle's manual to determine if this method is appropriate and for specific directions.

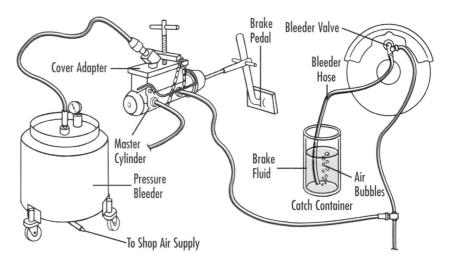

Fig. 2-19 Of the four brake-bleeding methods, pressure bleeding is the most common. *What are some of the disadvantages of the other methods?* (General Motors Corporation)

SECTION 3 KNOWLEDGE CHECK

❶ What is the function of the metering valve in a disc/drum braking system?

❷ Describe the purpose and function of a proportioning valve.

❸ What is the purpose of the residual pressure check valve? Why don't all vehicles have one?

❹ Explain the concept of a combination valve and explain what functions are combined.

❺ What are the pros and cons of the various bleeding techniques?

CHAPTER 2 REVIEW

Key Points

Meets the following NATEF Standard for Brake Systems: bleeding brake systems.

- Hydraulic pressure creates the force needed to operate an automotive braking system.
- The master cylinder contains pistons that act on the brake fluid to transfer pressure to the wheel cylinders.
- Braking system components are connected by rigid and flexible brake lines.
- Technicians must use special tools to make brake line flares. Any cracked or damaged flares must be cut off and new ones formed.
- Dual-circuit brake systems are either front-rear split or diagonally split.
- Some braking system control valves delay the arrival of brake pressure at the wheel brakes. Others limit the amount of fluid pressure.
- Brake bleeding flushes air and contaminated brake fluid out of a braking system.
- The most popular method of bleeding brakes is pressure bleeding, or power bleeding.

Review Questions

1. How is hydraulics used to transfer motion and force?
2. What is the function of a master cylinder?
3. Why is it important to install factory-made brake lines?
4. Describe the process for removing, bench bleeding, and replacing the master cylinder.
5. What is the function of double flare and ISO brake lines?
6. Identify the control valves used in hydraulic braking systems.
7. Describe the process for bleeding a hydraulic brake circuit.
8. **Critical Thinking** Why is it critical to have a sealed hydraulic system?
9. **Critical Thinking** Why should a hydraulic brake system use two circuits? Which type of split system is the best? Why?

TECHNOLOGY FORECAST
FOR AUTOMOTIVE EXCELLENCE

Electrohydraulic Brake Systems

The proper repair of automotive brake systems is of vital importance. Passenger safety is dependent on brakes that are in good working order. There is no room for error. Finding and servicing automotive brake problems should get easier in the future, thanks to electrohydraulic brake systems.

Electrohydraulic brake systems will use brakes with both electric motors and hydraulic fluid. With this new system, the fluid won't be stored in the master cylinder. Hydraulic fluid will instead be kept in small containers, called reservoirs, at each wheel.

A common problem with today's brake systems is that air sometimes gets trapped in the hydraulic lines. Air can enter through leaks or when brake parts are repaired or replaced. Hydraulic fluid and air don't mix, so the brakes may not stop the vehicle safely.

To repair this problem, the technician removes the unwanted air by bleeding it out of the system. This repair isn't difficult, but it usually requires two people to do it correctly. This increases the cost of the repair.

With electrohydraulic brakes, a single technician can fix the problem. By simply turning a valve, the technician can release the unwanted air. The vehicle can then stop normally.

AUTOMOTIVE EXCELLENCE TEST PREP

Answering the following questions will help you prepare for the ASE certification tests.

1. Technician A says brake fluid transmits motion. Technician B says brake fluid transmits force. Who is correct?
   a Technician A.
   b Technician B.
   c Both Technician A and Technician B.
   d Neither Technician A nor Technician B.

2. All the following are true about integral master cylinders except:
   a The primary and secondary pistons share the same bore.
   b The reservoir is cast in one piece with the cylinder housing.
   c It is made of cast iron.
   d The reservoir attaches to the master cylinder with grommets.

3. In brake master cylinders, the replenishing ports are used to:
   a Keep the fluid reservoirs filled.
   b Keep low-pressure chambers filled.
   c Keep high-pressure chambers filled.
   d None of the above.

4. Technician A says a quick-takeup master cylinder allows the fluid to return quickly when the brakes are released. Technician B says quick-takeup master cylinders are used on low-drag brake systems. Who is correct?
   a Technician A.
   b Technician B.
   c Both Technician A and Technician B.
   d Neither Technician A nor Technician B.

5. Technician A says brake lines can be made from copper tubing. Technician B says compression couplings may be used in brake systems. Who is correct?
   a Technician A.
   b Technician B.
   c Both Technician A and Technician B.
   d Neither Technician A nor Technician B.

6. In diagonal braking circuits, the wheel brakes are split:
   a Front to rear.
   b Side to side.
   c Left front to right rear.
   d Individually.

7. Metering valves on front disc and rear drum brake systems are used to:
   a Reduce pressure at rear drum brakes.
   b Increase pressure at rear drum brakes.
   c Delay fluid pressure to rear brakes.
   d Delay fluid pressure to front disc brakes.

8. Technician A says the proportioning valve has no effect during normal braking. Technician B says proportioning valves are found only on cars with four-wheel disc brakes. Who is correct?
   a Technician A.
   b Technician B.
   c Both Technician A and Technician B.
   d Neither Technician A nor Technician B.

9. Technician A says a two-function combination valve combines the proportioning valve and the metering valve into one component. Technician B says a two-function combination valve combines either the proportioning valve or the metering valve with the pressure-differential valve. Who is correct?
   a Technician A.
   b Technician B.
   c Both Technician A and Technician B.
   d Neither Technician A nor Technician B.

10. The most commonly used method of bleeding brake systems is:
    a Gravity bleeding.
    b Pressure bleeding.
    c Manual bleeding.
    d Vacuum bleeding.

CHAPTER 3

Diagnosing & Repairing Drum Brakes

You'll Be Able To:

- Describe drum brake construction and operation.
- Diagnose drum brake problems.
- Explain how to manually adjust drum brake self-adjusters.
- Describe the process used to service or replace a wheel cylinder.
- Describe the process used to service brake drums.

Terms to Know:

backing plate

duo-servo drum brakes

leading-trailing drum brakes

self-adjuster

The Problem >>>>>>

Cathy Drummond says she notices a metallic grinding noise when she applies the brakes. At first, the noise was only occasional. It has been getting louder over the past few days. Cathy now hears the noise every time she applies the brakes.

Cathy's car is more than five years old. She tells you that no brake work has been done on it since she bought it new.

Cathy is concerned about the safety of her vehicle. She wants to correct the problem as soon as possible. Cathy brings her car to your service center.

Your Challenge

As the service technician, you need to find answers to these questions:

1. How often does Cathy use her parking brake?
2. When she pushes on the brake pedal, does it go down farther than normal?
3. Does the noise seem to be coming from the front or rear of the car?

Section 1

Drum Brake Basics

Most current vehicles have drum brakes only on their rear wheels. They provide less than 40 percent of the overall braking effort for front engine, rear-wheel drive cars and only about 20 percent for front-wheel drive cars. Older cars had drum brakes on all four wheels.

Hydraulic pistons activate drum brakes. The drum brakes are retracted by spring tension. They can also be engaged mechanically. For this reason, they are widely used as the mechanisms for parking brakes.

Drum Brake Construction

The brake assembly attaches through the backing plate to an axle housing or a strut-spindle assembly. **Fig. 3-1.** The **backing plate** is the metal plate on which many of the drum brake components are mounted. A metal brake drum encloses these components and provides a friction surface for the brake shoes. Two curved brake shoes are pushed outward against the inside of the brake drum. Friction between the shoes and the drum acts to slow or stop drum and wheel rotation.

A wheel cylinder provides the hydraulic force required to push the brake

shoes against the drum. Brake hardware consists of return springs, holddown springs, pins, and a self-adjuster. This hardware attaches the various components to the backing plate or to each other.

When working properly, a self-adjuster assembly keeps the brakes correctly adjusted at all times.

Backing Plate The backing plate bolts to the axle housing or spindle. The rest of the brake components are attached to the backing plate. **Fig. 3-2.**

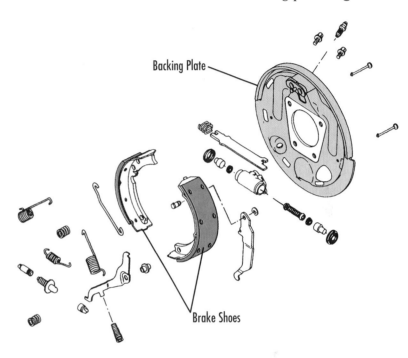

Backing Plate

Brake Shoes

Fig. 3-2 The backing plate serves as a mounting point for the brake hardware and brake shoes. *What would happen if the backing plate were damaged?* (Ford Motor Company)

The backing plate has raised metal pads. The brake shoes ride on these pads. These pads cannot have any grooves in them; they must remain smooth.

There are several holes in the backing plate. These are for mounting the wheel cylinder and hold-down pins. Some vehicle models have a hole through which the brakes can be adjusted.

The backing plate helps to keep road debris from entering the drum brake assembly.

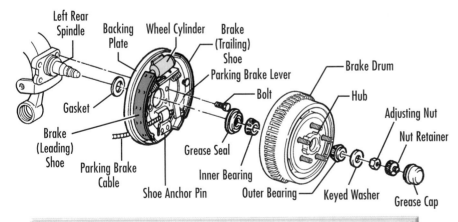

Left Rear Spindle — Backing Plate — Wheel Cylinder — Brake (Trailing) Shoe — Parking Brake Lever — Bolt — Brake Drum — Hub — Adjusting Nut — Nut Retainer — Gasket — Grease Seal — Inner Bearing — Keyed Washer — Grease Cap — Brake (Leading) Shoe — Parking Brake Cable — Shoe Anchor Pin — Outer Bearing

Fig. 3-1 Exploded view of a leading-trailing drum brake commonly found on front-wheel drive cars. Only about 20 percent of the braking is done by the rear drum brakes. *Why are self-adjusters necessary on drum brakes?* (DaimlerChrysler)

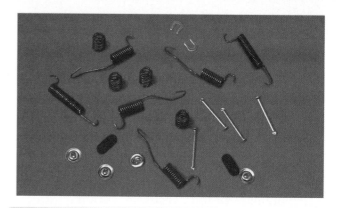

Fig. 3-4 The hardware consists of the springs, pins, cups, and other small pieces that hold the shoes to the backing plate and retract the brakes when they are released. *Why should hardware be replaced with every job?* (Wagner Brakes)

The backing plate must be firmly attached to the axle housing and must not be bent or otherwise damaged. A damaged backing plate can cause the brakes to malfunction.

Brake Shoes Brake shoes are made of metal, usually steel. **Fig. 3-3.** The friction material, called the brake lining, is attached to the shoe by rivets or adhesives. Cemented linings are often called bonded linings. Together, the lining and shoe assembly are referred to as the *brake shoe.*

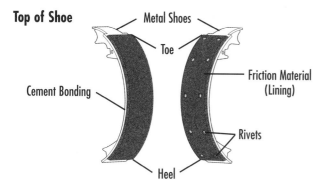

Top of Shoe — Metal Shoes — Toe — Friction Material (Lining) — Cement Bonding — Rivets — Heel

Fig. 3-3 Rivets or adhesives are used to attach the lining to the metal shoe. *What might happen if the lining wears down to expose the rivets?* (Bendix Corporation)

Drum brakes use friction between the drum and brake shoe to slow the motion of the vehicle. This friction creates enormous amounts of heat. The resulting high temperatures limit the materials suitable for brake shoe linings.

On older vehicles brake shoes were lined with asbestos fiber. Less hazardous materials are now used for linings. The linings are usually made of a semimetallic material or fiberglass. The friction material must be able to withstand the heat produced during braking.

Brake Hardware The assorted small parts that attach to the shoes are known as *brake hardware.* **Fig. 3-4.** This hardware includes springs, pins, cups, and clips that hold the drum brake assembly together.

Pin and cup assemblies hold the brake shoes against the backing plate. The pins pass through the backing plate, through the brake shoe and hold-down spring, and lock to the cup. The cups can be unlocked from the pins by rotating one-quarter turn in either direction.

Return springs return the shoe to the anchor pin when the brakes are released.

Wheel Cylinders When the driver pushes the brake pedal, brake fluid is forced out of the master cylinder through the brake lines to the wheel cylinder. **Fig. 3-5.** The wheel cylinder converts the hydraulic pressure from the master cylinder into mechanical movement.

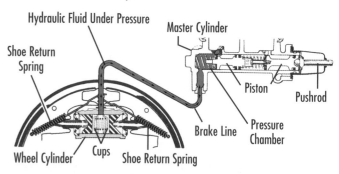

Hydraulic Fluid Under Pressure — Master Cylinder — Shoe Return Spring — Piston — Pushrod — Brake Line — Pressure Chamber — Wheel Cylinder — Cups — Shoe Return Spring

Fig. 3-5 When the brakes are applied, hydraulic fluid moves under pressure from the master cylinder to the wheel cylinders. When the brakes are released, spring pressure retracts the shoes and forces the fluid back to the master cylinder. *Why are the front wheel cylinder pistons larger on some vehicles with four-wheel drum brakes?* (General Motors Corporation)

"Reading" Your Customer

Listening to a customer's comments about a vehicle's problem can provide essential clues to help you solve that problem. Your customer drives the car daily. A person who knows the car can quickly respond to a technician's questions.

Be aware that there are many types of drivers. These include people who play their radios or music systems loudly enough to cover the noises made by their cars. Others pay no attention to how their cars are running.

As a technician, you need to be able to "read" your customer. Often, you can learn more from customers' actions than from what they say. Take, for instance, the driver who plays his radio loudly as he drives in. He might say that he just noticed the noise. You might guess that the noise has been there for a while. The music may have been too loud for him to notice the noise.

Assuming that this driver has not been as attentive as he could have been may help you as you begin to diagnose the problem.

If a vehicle is clean, you may guess that the customer also pays attention to mechanical problems. If the vehicle is not well cared for, you may draw different conclusions.

However, you need to be cautious when making such guesses. You are working with clues, not hard facts.

Apply it!

Meets NATEF Communication Standards for using and evaluating verbal and nonverbal cues.

Reread the "The Problem" at the beginning of the chapter. Look for clues that tell you something about the customer.

① What does the condition of the customer's car suggest to you about her?

② What does her comment that she has "not had any brake work done" suggest to you about her?

③ What can you learn about her from the complaint she makes?

④ What do you think the problem could be? Base your answer on what she has said and what you seem to know about her.

A wheel cylinder has two pistons. There are seals or cups on each piston. There is also a spring between the two pistons. As the pressure increases, the pistons overcome the brake shoe return springs. This causes the pistons to push the shoes outward into contact with the drum.

Early-model vehicles with four-wheel drum brakes usually had larger pistons in the front wheel cylinders than in the rear ones. This was necessary due to the greater braking force required on the front wheels. Braking transfers more of the vehicle's weight to the front wheels.

SECTION 1 KNOWLEDGE CHECK

① Why are drum brakes frequently used as the mechanism for parking brakes?

② What is the primary function of the backing plate?

③ How are brake linings attached to the shoes?

④ Name three pieces of brake hardware.

⑤ What is the purpose of a wheel cylinder?

Section 2

Drum Brake Operation

There are two types of drum brakes: leading-trailing or non-servo, and duo-servo. **Fig. 3-6. Leading-trailing drum brakes** are brakes in which the action of one shoe does not affect the other shoe. **Duo-servo drum brakes** are brakes in which the action of one shoe reinforces the action of the other shoe.

Leading-trailing drum brakes are used on the rear wheels of some front-wheel drive vehicles. Duo-servo drum brakes are usually found on the rear wheels of rear-wheel drive vehicles.

Leading-Trailing Drum Brakes

In the leading-trailing (non-servo) drum brake system, the tops of the brake shoes rest against the wheel cylinder. **Fig. 3-6(a).** The bottoms of the brake shoes rest against a fixed anchor. Depressing the brake pedal overcomes the return spring tension and causes the wheel cylinder pistons to move the tops of the brake shoes outward against the drums. Friction between the forward, or leading, brake shoe and the drum causes the leading shoe to try to rotate with the drum. This self-energizing action of the leading shoe forces the bottom of the brake shoe against the anchor pin. As a result, the leading shoe does most of the braking.

TECH TIP **Brake Shoe Linings.** The brake shoe linings may be the same length on both shoes, but there may be a difference in their lining materials. If the brake shoes are marked, do not install them in the wrong locations. Brake function will be reduced.

When the rear, or trailing, brake shoe contacts the drum, the rotation tries to force the brake shoe away from the drum. There is no self-energizing action. Therefore, the trailing shoe usually wears less than the leading shoe. When the vehicle is moving backwards, the brake shoes switch jobs and the rear brake shoe does most of the braking. This is because the rear brake shoe is now the leading shoe. The front brake shoe is now the trailing shoe.

The leading-trailing brake is less self-energizing and more dependent on the force supplied by the wheel cylinder.

Duo-Servo Drum Brakes

In duo-servo drum brakes, the tops of the brake shoes rest against a single anchor pin. **Fig. 3-6(b).** The bottoms of the brake shoes are linked by a floating adjuster assembly. When the brakes are applied, the wheel cylinders force the shoes out against the drum.

(a) Leading-Trailing Drum Brake

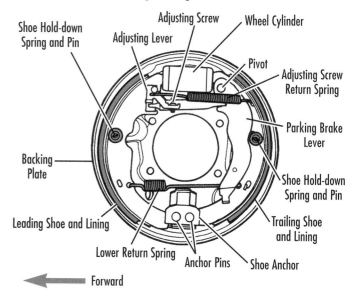

(b) Duo-Servo Drum Brake

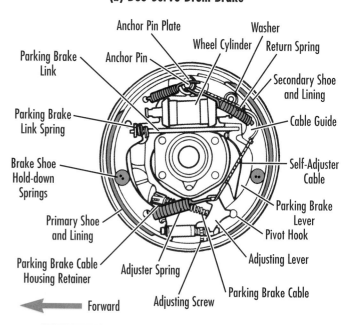

Fig. 3-6 (a) The leading-trailing drum brake system and **(b)** the duo-servo drum brake system. *What kind of brakes would you expect to find on the rear wheels of a rear-wheel drive vehicle?* (Ford Motor Company)

The primary (forward) shoe contacts the drum and tries to rotate with the drum, pulling the top of the shoe away from the anchor pin. This causes the bottom of the shoe to transfer force through the adjuster assembly to the rear (secondary) brake shoe. This, in turn, forces the secondary shoe harder against the drum.

In duo-servo brakes, the self-energizing action of both shoes makes total braking force greater than the amount supplied by the wheel cylinder.

> **TECH TIP** **Length of Lining.** You can usually identify the brake shoes in a duo-servo brake system. The primary (forward) shoes will usually have a shorter lining than the secondary (rearward) shoes. The lining is longer on the secondary shoe because it does more of the braking than the primary shoe.

Drum Brake Self-Adjusters

A **self-adjuster** is a device on drum brakes that compensates for lining wear by automatically adjusting the shoe-to-drum clearance. This adjustment prevents the brake pedal from getting lower and lower during normal use. The two types of self-adjusters are the one-shot and the incremental.

One-Shot Adjusters The *one-shot adjuster* makes a single adjustment once the clearance between the lining and drum reaches a predetermined gap. At this point, no additional adjustments can be made. The shoes must be replaced and the self-adjuster reset.

Incremental Adjusters The *incremental adjuster* moves the shoes outward whenever the gap is large enough to turn the adjusting screw. **Fig. 3-7.** On some vehicles, adjustment occurs when the vehicle is braked while moving either forward or rearward. On other vehicles, adjustment occurs only when the vehicle is braked while moving rearward. And on yet other vehicles, particularly Asian imports, the brakes are adjusted whenever the parking brake is applied.

An incremental self-adjuster is attached to the secondary shoe of a duo-servo brake. The adjusting

> **TECH TIP** **Adjusting Screws.** The adjusting screws on cable-type adjusters are right hand and left hand. If they are reversed, the screws retract the shoes instead of expanding them. If unsure about which side an adjuster goes on, pull the end cap from the adjuster. It will be marked RIGHT or LEFT on the end of the adjusting screw.

lever attaches to the self-adjuster cable that passes around a cable guide and fastens to the anchor pin. The brakes are adjusted each time they are applied while the vehicle is moving backward. Friction forces the upper end of the primary shoe against the anchor pin. The wheel cylinder forces the upper end of the secondary shoe downward and away from the anchor pin. This causes the cable to pull the adjusting lever upward.

If the brake linings have worn enough, the lever passes over and engages the end of a new tooth on the adjusting wheel. This wheel is commonly called the *star wheel*. The incremental-adjuster assembly is a one-way, ratcheting mechanism.

> **Safety First** **Personal** Remind drivers to use their parking brakes regularly if that is the method used to keep their brakes adjusted correctly. Parking brakes that are not used regularly could freeze up.

When the brakes are released, the adjuster spring pulls the adjusting lever downward. **Fig. 3-7.** This turns the tooth and slightly lengthens the adjusting screw. The brake shoes move closer to the drum.

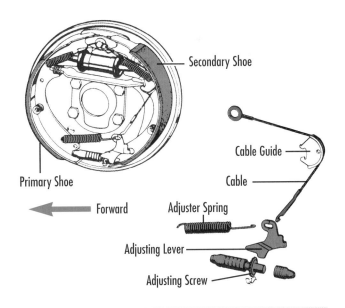

Fig. 3-7 Some incremental adjusters have a cable and some have levers. The cable is attached to the secondary shoe, and the lever moves the star wheel in a ratcheting action. *What might happen if the self-adjuster does not work correctly?* (Bendix Corporation)

SECTION 2 KNOWLEDGE CHECK

❶ Name the two types of drum brakes.

❷ Are both the leading and trailing shoes self-energizing in leading-trailing (non-servo) drum brakes?

❸ Are both shoes self-energizing in duo-servo drum brakes?

❹ Name the two types of self-adjusters.

❺ Are all self-adjusters activated by action of the brake pedal?

Section 3

Diagnosing Drum Brakes

When a vehicle with a brake problem comes in, talk to the owner. Ask him or her questions about the problem. Listen carefully to what the owner says. It is important to fully understand the problem before you begin your diagnosis. Once you have the facts and understand the complaint, you can begin to consider possible causes. **Table 3-A.**

After identifying the possible causes, test each of them, beginning with the most likely ones. Continue to eliminate possible causes until the problem is found.

To correct the problem, you may need to repair, adjust, or replace a component. Or, you may only need to bleed the braking system.

Safety First

Personal Use caution when working around brake dust. When working on brakes, you must:

• Wear safety glasses at all times.
• Use a vacuum cleaner with a high efficiency particulate air (HEPA) filter, approved brake washer, or aerosol cleaner to remove brake dust. Never use an air hose to blow out the dust.
• Wear a respirator to avoid breathing airborne brake dust. Treat all brakes as though they have asbestos linings.
• Dispose of brake dust in a sealed container in accordance with hazardous waste laws.
• Wash your hands after performing brake service and do not wear your work clothes home.

One Brake Drags

When one brake drags, it indicates that the shoes are not pulling back from the drum. The following are some causes for this.

• Wheel cylinder piston stuck in the applied position.
• Weak or broken brake return springs.
• Loose wheel bearings that cause the drum to wobble.
• Backing plate is bent or damaged, and the shoes are hanging up.
• Brake adjuster is overextended.
• Damaged brake hose.
• Seized parking brake cable.

All Brakes Drag

When all brakes are dragging, it is because the fluid is still under pressure. There are several reasons fluid pressure will remain in the system.

• Master cylinder piston not returning far enough to uncover the compensating port.
• Mineral oil in the system caused by topping off the master cylinder with automatic transmission fluid, engine oil, or power steering fluid, which will swell the rubber seals.
• Improperly adjusted master cylinder pushrod.

Brake Pedal Free Play

Without the correct amount of free play, the brake pedal will not return fully and the brakes may not fully release. There should be at least one-quarter to one-half inch of free play at the brake pedal. **Fig. 3-8.** The master cylinder pushrod may need adjusting or something may be interfering with pedal travel.

Table 3-A	DRUM-BRAKE DIAGNOSIS	
Complaint	**Possible Cause**	**Check or Correction**
1. Pedal goes to floor, loss of pedal reserve	a. Linkage or shoes out of adjustment or damaged b. Brake linings worn c. Lack of brake fluid d. Air in hydraulic system e. Defective master cylinder	a. Check and repair linkage or adjusters b. Replace c. Add fluid, bleed system d. Add fluid, bleed system e. Repair or replace
2. One brake drags	a. Shoes out of adjustment b. Clogged brake line c. Wheel cylinder defective d. Weak or broken return spring e. Loose wheel bearing	a. Adjust b. Clear or replace c. Repair or replace d. Replace e. Adjust or replace
3. All brakes drag	a. Incorrect linkage adjustment b. Defective master cylinder c. Mineral oil in system	a. Adjust b. Repair or replace c. Replace damaged rubber parts; flush, fill, and bleed system
4. Pulls to one side when braking	a. Oil on brake linings b. Brake fluid on brake linings c. Brake shoes out of adjustment d. Tires not uniformly inflated e. Brake line clogged f. Defective wheel cylinder g. Backing plate loose h. Mismatched linings	a. Replace linings and oil seals; avoid overlubrication b. Replace linings; repair or replace wheel cylinder c. Adjust d. Adjust tire pressure e. Clear or replace line f. Repair or replace g. Tighten h. Install matched linings
5. Soft or spongy pedal	a. Air in hydraulic system b. Brake shoes out of adjustment c. Defective master cylinder d. Loose connections or damaged brake line e. Loss of brake fluid	a. Add fluid, bleed system b. Adjust c. Repair or replace d. Tighten connections, replace line e. See item 9, below
6. Poor braking requiring excessive pedal force	a. Brake linings wet with water b. Shoes out of adjustment c. Brake linings hot d. Brake linings burned e. Brake drum glazed f. Power brake inoperative g. Wheel cylinder pistons stuck	a. Allow to dry b. Adjust c. Allow to cool d. Replace e. Refinish or replace f. Repair or replace g. Repair or replace
7. Brakes grab	a. Shoes out of adjustment b. Wrong linings c. Brake fluid, oil, or grease on lining d. Drums scored e. Backing plate loose f. Power brake booster defective	a. Adjust b. Install correct linings c. Replace shoes and repair leaks d. Refinish or replace drums e. Tighten f. Repair or replace
8. Noisy brakes	a. Linings worn, shoes warped, or rivets loose b. Drums worn or rough c. Loose parts	a. Replace shoes b. Refinish or replace c. Tighten
9. Loss of brake fluid NOTE: After repair, add brake fluid and bleed system	a. Master cylinder or wheel cylinder leaks b. Loose connectors, damaged brake line	a. Repair or replace b. Tighten connections, replace line
10. Brakes do not self-adjust	a. Adjusting screw stuck b. Adjusting lever does not engage adjusting wheel c. Adjuster incorrectly installed	a. Free and clean b. Repair or replace adjuster c. Install correctly
11. Brake warning light comes on while braking	a. One section of hydraulic system has failed b. Pressure differential valve defective	a. Inspect and repair b. Replace
12. Pedal pulsation	a. Loose wheel bearings b. Wheel or tire problems c. Broken or weak brake return springs d. Debris in or on brake drums or shoes e. Out of round drums	a. Adjust b. Inspect wheel for damage; test wheel and tire balance c. Replace springs d. Clean and machine brake drum; replace brake shoes e. Machine or replace

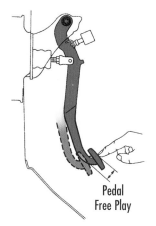

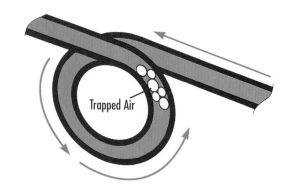

Fig. 3-8 With the engine off, pump the brakes a couple of times to remove the vacuum from the booster. Measure the brake pedal free play. It should be one-quarter to one-half inch. *What might happen if there is no free play?*

Fig. 3-9 Air trapped in the hydraulic system is a major cause of soft or spongy pedal. *Why does trapped air cause this problem?* (DaimlerChrysler)

Rear Brakes Drag

The parking brakes may not release completely if they are over adjusted or if the cables are corroded. Also, the brakes can drag if the parking brake pedal or lever does not return to the full release position.

Pulls to One Side When Braking

When a vehicle pulls to one side when braking, it can be because the brakes on that side of the vehicle are defective. However, it is also possible that the brakes on the opposite side of the vehicle are not working. If one side is not working, the vehicle will pull to the side that is working. Verify normal brake operation on the opposite side of the vehicle to help isolate the problem. A pull to one side could be caused by:
• Linings that are contaminated with oil or grease.
• Brakes that are out of adjustment.
• A defective automatic brake adjuster.
• A faulty or leaking wheel cylinder.
• A restricted hydraulic brake line.

Soft or Spongy Pedal

A soft or spongy pedal is usually the result of a problem in the hydraulic system. Air in the hydraulic system will cause a soft or spongy pedal. **Fig. 3-9.** This is because air is compressible and liquids are not. Conditions that lead to air in the braking system include:
• Low fluid level in the master cylinder.
• Plugged air vents in the master cylinder cover or cap.
• Residual pressure check valve failure.

Low Fluid Level in Master Cylinder Air, instead of fluid, may enter the system when the pedal is stroked if the master cylinder is low on fluid. Check for fluid leaks and add fluid. Then bleed the hydraulic system. **Fig. 3-10.**

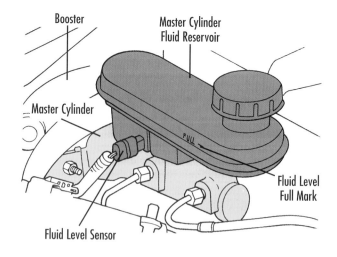

Fig. 3-10 Clean off any dirt or other foreign matter before removing the cover to the brake fluid reservoir. *What damage could dirt do in the hydraulic system?* (DaimlerChrysler)

Measuring Static Friction!

A driver sees an obstacle ahead and hits the brakes in a panic. Will he stop more quickly if he locks the wheels of his older-model car? A car requires almost twice the stopping distance in a skid with the wheels locked as it does in a controlled stop that does not lock the wheels. This happens because the force of kinetic (or sliding) friction is much less than the force of static (or stationary) friction.

With this easy experiment, you can measure the coefficient of static friction for any pair of surfaces. You can use it for rubber and wood or metal and wood.

Apply it!

Determining the Coefficient of Friction

Meets NATEF Science Standards for friction and linear motion.

Materials and Equipment
- Eraser with one flat side
- Coin
- Smooth wooden board

❶ Place the wooden board on top of a table or any smooth, level surface.

❷ Place the rubber eraser and the coin beside each other near one end of the board. **Fig. A.**

❸ Slowly lift the end of the board that has the eraser and the coin on it.

❹ Carefully observe which object begins to slide first. This object has overcome the force of static friction.

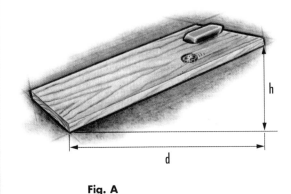

h

d

Fig. A

Results and Analysis

You should have observed that the coin begins to slide first. This is because the force of static friction is less for metal on wood than it is for soft rubber on wood.

The *coefficient of friction* is defined as the ratio of the force of friction to the normal force that presses the two surfaces together. Measure the height (h) of the object when it began to slide. Measure the horizontal distance (d) that the object slid. **Fig. A.** The coefficient of friction is h/d.

The coefficient of friction is greater for rubber on wood than it is for metal on wood. You needed to raise the end of the board at a steeper angle for the soft rubber eraser before it began to slide. This demonstrates one reason vehicle tires are made of rubber instead of metal. The coefficient of friction for rubber on concrete is even larger than it is for rubber on wood.

Plugged Air Vents in Master Cylinder Cover or Cap Plugged air vents in the cover or cap can cause a partial vacuum in the master cylinder. If this occurs, air may be allowed to bypass the primary cup. Eliminate this condition by clearing the vent and bleeding the hydraulic system of any trapped air.

Residual Pressure Check Valve Failure If the residual pressure check valve does not hold pressure, air can be sucked past the wheel cylinder cups when the brakes are released. Correct this condition by replacing the residual pressure check valve and bleeding the hydraulic system.

Poor Braking Requiring Excessive Pedal Force

Poor braking requiring excessive pedal force can be a temporary or a persistent condition. Some of its causes are:

- Obstruction of the brake pedal.
- Pinched brake lines.
- Defective power booster.
- Defective vacuum check valve for power booster.
- Low engine vacuum.
- Glazed linings or brake drums.
- Wet brakes.
- Overheated brakes.

Failure of the power booster will cause a very hard pedal and reduced braking power. Power booster failure can result from insufficient vacuum from the engine. Failure will also occur if the vacuum check valve is not working correctly.

 TECH TIP **Inspecting Brakes Regularly.** Many vehicle manufacturers recommend regular brake inspections. Technicians should check brake pedal action and fluid level in the master cylinder. They should inspect the condition of the brake lines, hoses, and wheel brake assemblies.

Excessively long intervals of braking create high temperatures that may burn or char the brake lining. Overheating can also glaze the brake drum. When this happens, the low friction surface between the lining and the drum will lead to poor braking. To correct this condition, replace the brake shoes, wheel cylinders (if necessary), and brake hardware. Machine or replace the brake drum. Going through deep water can cause temporary brake fade until the brakes dry off.

Brakes Grab

Brake grabbing is a condition in which the brakes apply more quickly than expected compared to the amount of pedal effort applied. Some causes include:

- Linings contaminated with grease or oil.
- An inoperative proportioning valve.
- Shoes out of adjustment.
- A leaking wheel cylinder.
- A loose backing plate.

Linings contaminated with grease, oil, or brake fluid will grab on light pedal force. If contaminated linings are found, they must be replaced. Always check grease seals to assure they are not leaking.

A loose backing plate, where it bolts to the axle housing, allows the shoes to twist in the drum. Backing plates should be inspected any time brake work is being performed.

Noisy Brakes

Brake system noise results from the vibration of brake system components. Some sources of brake system noise are:

- Excessively worn linings.
- Warped shoes.
- Threaded drums.
- Loose hardware.
- Loose backing plate.

If the linings are worn down to the rivets or the steel backing and are touching the drum, the metal-to-metal contact will make a grinding or scraping noise. Any loose parts in the drum brake system can make noise when the brakes are applied. A bent backing plate will give a high-pitched noise as the drum rotates.

A drum may have spiral grooves from improper machining. These grooves will act as threads. The linings may ride up the grooves until they reach the end. Then they will make a clunk as they return to their original position.

 Safety First **Materials** Never add anything but brake fluid to the reservoir. Petroleum products are particularly dangerous. They cause rubber parts to soften, swell, and fail. This could result in complete brake failure.

Loss of Brake Fluid

A leak will cause loss of brake fluid. A leak can occur anywhere in the hydraulic system. When checking for leaks, look for wet spots around connected components or brake lines. Pay particular attention to the following:

- Brake lines and component connections.
- Brake hoses and their connections.
- Master cylinder.
- Wheel cylinders.
- Calipers.

Leaks at most components, connections, and lines are obvious, but don't forget to check the vacuum supply to the booster. **Fig. 3-11.** Although there may be

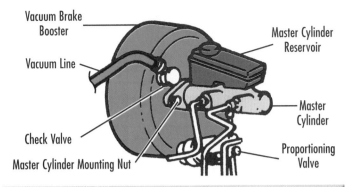

Vacuum Brake Booster

Vacuum Line

Check Valve

Master Cylinder Mounting Nut

Master Cylinder Reservoir

Master Cylinder

Proportioning Valve

Fig. 3-11 A leak will sometimes be apparent where the master cylinder bolts to the booster. Frequently, however, the fluid enters the booster itself. *Where else might you find brake fluid leaking from a master cylinder?* (Ford Motor Company)

moisture where the master cylinder meets the vacuum booster, the fluid will often leak into the booster itself. It may also leak into the vacuum line leading from it.

Brakes Do Not Self-Adjust

If the self-adjusters do not work, pedal travel will be excessive. Reasons self-adjusters may not work include:
- Driver seldom applies brakes in reverse.
- Driver seldom applies parking brake.
- Self-adjuster screws (star wheels) are on wrong sides of vehicle or frozen in place.
- Teeth on star wheel are rounded off.
- Adjuster cable is worn or stretched.

Always check the brake hardware and replace any worn items when doing a brake job.

If the adjuster assembly is installed backwards, it will not function. All adjuster parts should be checked.

Brake Warning Light Comes On While Braking

In split brake systems, the pressure differential valve turns on the brake warning light, if fluid pressure is lost in one-half of the system. Locate the problem, which is likely a leak, and repair it. Refer to the

vehicle's service manual for instructions on resetting the pressure differential valve if necessary.

Pedal Pulsation

Pedal pulsation is when the brake pedal feels as if it is pushing back against the driver's foot when the brakes are applied. This is normal on vehicles that have anti-lock brakes. When the brakes are applied suddenly, such as in a panic situation, the brake pedal will lightly pulsate as the anti-lock feature is engaged. This is not normal with any other brake system. In a brake system with drum brakes, pulsation can be caused by:
- Loose wheel bearings.
- Wheel or tire problems.
- Broken or weak brake return springs.
- Debris in or on the brake drum or shoes.
- Warped or out-of-round drums.

Loose Wheel Bearings If the wheel bearings are not adjusted to the manufacturer's specifications, pulsation can result at the brake pedal. Refer to the vehicle service manual for the specific adjustments.

Wheel and Tire Problems Use a wheel balancer to check the wheel runout for bent, cracked, or loose wheels. Check tires for tread separation or sidewall damage. Check that the wheel-and-tire assembly is balanced.

Broken or Weak Brake Return Springs Inspect the brake springs and shoes. Springs not retracting the brake shoes properly may cause pulsation. The shoes may have glazed areas. Replace the brake springs and brake hardware. Clean or replace brake shoes as needed.

Debris in or on the Brake Drum or Shoes Inspect the brake shoes and drum. Warped or out-of-round drums, embedded material, or other foreign substances can cause high spots, resulting in brake pedal pulsation. Clean and machine the brake drum if within the manufacturer's specifications. Replace the brake shoes if damaged or contaminated.

SECTION 3 KNOWLEDGE CHECK

1. What are two causes for one brake dragging?
2. List three causes for all brakes dragging.
3. What could cause the brakes to pull to one side?
4. What is the major cause of a spongy pedal?
5. Name three causes of poor braking requiring excessive pedal force.

Section 4

Servicing Drum Brakes

Sometimes the only brake service needed will be brake adjustment. There will be many more cases, though, in which you will need to remove the brake drum to access the brake shoes and brake hardware.

Drum Brake Self-Adjusters

Few vehicles come without self-adjusters. These vehicles require periodic adjustment to keep the shoes in close proximity to the drums.

Brakes with self-adjusters should need adjustment only after servicing of brake shoes or drums. However, if the vehicle needs only an adjustment of the drum brake system, perform the adjustment as follows:

Brake self-adjusters can be manually adjusted by turning the adjuster's star wheel. Most duo-servo brakes have a slot, or knockout, in the backing plate or drum through which you can access the star wheel. Turn the star wheel using a brake adjusting tool called a *spoon*. Turning the star wheel in one direction will move the shoes closer to the drum. Turning the star wheel in the opposite direction will move the shoes away from the drum. The correct direction for shortening or lengthening the adjuster varies from vehicle to vehicle. Refer to the vehicle's service manual for additional information.

Complete the adjustment by making several alternating forward and reverse stops with the vehicle.

If adjustment requires use of the parking brake, apply and release it several times–sometimes up to 50 cycles may be required.

Removing the Brake Drum

To remove brake drums refer to the vehicle's service manual. In general, do the following:
- Raise the car and install safety equipment if using a jack stand.
- Loosen and remove the wheel cover, if there is one.
- Loosen and remove the wheel.
- Remove the brake drum from the drum brake system assembly.

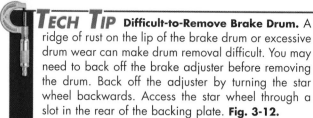

TECH TIP **Difficult-to-Remove Brake Drum.** A ridge of rust on the lip of the brake drum or excessive drum wear can make drum removal difficult. You may need to back off the brake adjuster before removing the drum. Back off the adjuster by turning the star wheel backwards. Access the star wheel through a slot in the rear of the backing plate. **Fig. 3-12.**

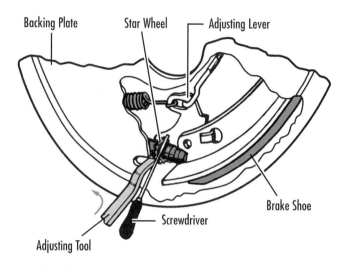

Backing Plate   Star Wheel   Adjusting Lever

Brake Shoe

Screwdriver

Adjusting Tool

Fig. 3-12 Use a screwdriver to push the adjuster lever away from the star wheel and turn it in the direction opposite to its normal rotation. *Why are drums occasionally difficult to remove?* (Bendix Corporation)

Inspecting Brake Drums

Clean the brake drum with a clean shop towel using cleaning solvent or denatured alcohol. Inspect the cleaned brake drum for the following defects: **Fig. 3-13.**
- Scoring–scratches or grooves in the drum's machined brake surface.
- Hard spots–areas of the machined brake surface that have not worn down with the rest of the machined surface.
- Bell mouthing–drum brake surface shows evidence of excessive wear toward the drum's rim.
- Barreling–machined brake surface is no longer flat and shows evidence of having greater wear toward the center of the brake drum's machined surface.
- Threads–brake drum's machined surface has the appearance of being threaded; results from improper machining.

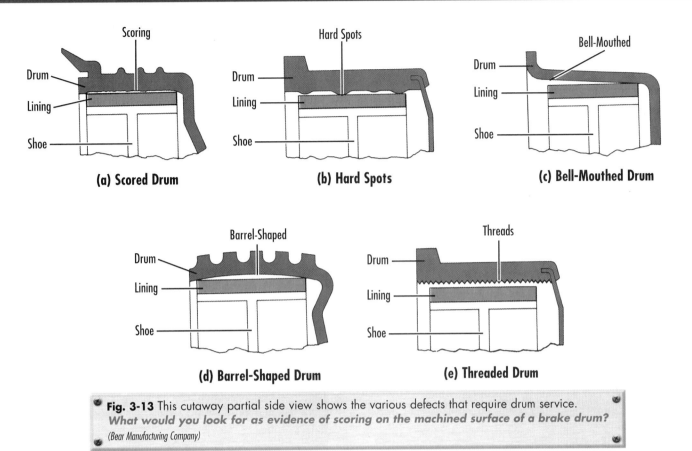

Fig. 3-13 This cutaway partial side view shows the various defects that require drum service. *What would you look for as evidence of scoring on the machined surface of a brake drum?* (Bear Manufacturing Company)

Measuring Brake Drums

Measure the diameter of the drum using a brake drum micrometer. If the damage is not extensive, the drums can be machined. In machining brake drums, a brake lathe removes metal to restore the drum's braking surface.

If the drum is worn or scored too deeply, it must be replaced. The discard diameter is cast into the drum. This is the maximum allowable diameter.

Machining Brake Drums

To machine the brake drum, perform the following steps: (Follow the brake lathe manufacturer's instructions when using the lathe.)

1. Determine the maximum diameter for the machined surface of the brake drum. (The value can be found cast into the body of the brake drum.)
2. Measure the inside diameter of the brake drum using a brake drum micrometer or measuring calipers.
3. Calculate the amount of material that can be safely removed from the drum's machined surface.
4. Mount the brake drum on the brake lathe.
5. Machine the brake drum, being careful not to remove too much material.
6. Remove the brake drum from the lathe.
7. Clean the brake drum by wiping the machined surface with cleaning solvent and a clean rag. This will remove metal dust left in the drum. If not removed, it will be forced into the lining on the first brake application.

Inspecting Brake Shoes

Brake shoes must be replaced when their lining wears down below minimum thickness. The shoes may be replaced with new shoes or relined shoes. Relined brake shoes are recycled shoes with new friction material attached.

Inspect the entire wheel brake mechanism, paying particular attention to brake lining wear.

Before removing the brake hardware, make a note of where all the parts belong. Be sure to install brake shoes in the forward or rearward position as designed. There are likely to be differences between the shoes, even if they look the same.

Fig. 3-14 Replacing brake shoes on a rear drum-brake system. *Does it matter which replacement shoe is installed in the leading shoe position?* (Terry Wild Studio)

Follow the manufacturer's instructions for replacement of shoes. **Fig. 3-14.**

Removing Brake Shoes

1. Remove the return springs from the backing plate. Unhook them from the brake shoes.
2. Remove the retainer pins and cups holding the shoes to the backing plate.
3. Remove the brake shoes from the backing plate.
4. Remove the self-adjuster. Be sure to note its mounting orientation.

TECH TIP **Hardware.** Check brake hardware with every brake job. Replace worn items. The return springs, for instance, can become weak if the brakes are overheated. If you intend to reuse a spring, it should be tested. To test a spring's condition, drop it on the floor. If it makes a ringing sound it is still strong. A spring that has been overheated will make a dull sound when dropped.

Inspecting Brake Hardware

If possible, replace all brake hardware when servicing drum brakes. If this is not possible:

1. Clean all pins, clips, springs, levers, and retaining hardware with brake cleaning solvent.
2. Carefully inspect the brake hardware for damage, including rust damage.

3. Replace any damaged components.
4. Disassemble the self-adjuster and clean it thoroughly in brake cleaning solvent.
5. Lubricate the threads on the self-adjusters with high-temperature grease.
6. Reassemble the self-adjuster; wipe off any excess lubricating grease.
7. The brake hardware is ready for reassembly.

Removing a Wheel Cylinder

Before servicing the wheel cylinder, try to loosen the bleeder screw on the back of the wheel cylinder. Bleeder screws often become seized and break off very easily. If the bleeder screw breaks off, replace the wheel cylinder.

TECH TIP **Special Tools.** Special tools are available to loosen stubborn bleeder screws. They fit over the bleeder and are struck with a hammer while being turned by a wrench.

A leaking wheel cylinder with deep scores should be replaced. Leaking wheel cylinders with no serious damage may often be rebuilt using a repair kit. The repair kit contains all the parts except the cylinder housing, bleeder, and pistons. Although wheel cylinders can usually be serviced on the vehicle, some manufacturers suggest removing them. The wheel cylinder must be assembled and installed correctly or it will not function properly. **Fig. 3-15.**

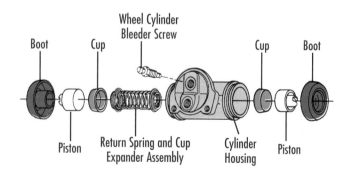

Fig. 3-15 Wheel cylinders can be rebuilt by replacing the boots, cups, and spring. *Should a wheel cylinder with deep scores be rebuilt or replaced?* (Ford Motor Company)

Measuring Out-of-Round Brake Drums

Courtney Leruchi says the brakes on her car are pulsating. You suspect the brake drums may be out of round.

The brake specifications give an 11.000″ drum diameter with a maximum allowance of 11.090″ diameter. You will resurface, or machine, the drum to a maximum of 11.060″, leaving 0.030″ for further wear.

You use a brake drum micrometer to measure the inside diameter at points 90° apart. What action would you recommend if the smallest diameter measurement is 11.018″ and the largest diameter measurement is 11.028″?

Write a sentence stating how much the drum is out of round. Explain your recommendation to the customer. Support your answer mathematically.

You find the difference between the largest and smallest measurements of diameter:

11.028″ − 11.018″ = 0.010″.

The drum is 0.010″ out of round.

The maximum brake drum diameter desired is 11.060″. So the diameter available to work with is

11.060″ − 11.018″ = 0.042″.

Because there is enough metal to work with, you can use a brake drum lathe to remove up to 0.021″ of metal. Note that removing 0.021″ around the entire drum would increase the diameter by 0.042″.

In practice, a technician would probably remove just enough metal to make the drum round. Then he or she would check to make sure the inside diameter was acceptable.

Apply it!

Meets NATEF Mathematics Standards for using multiplication and division and working with tolerance specifications.

The customer was so pleased with your explanation that she recommended you to her brother. His vehicle has an 8.640″ drum diameter with a maximum allowance of 8.730″ diameter.

1. What action would you recommend if two measurements 90° apart are 8.648″ and 8.674″, respectively?
2. Write a sentence stating how much the drum is out of round.
3. Explain your recommendation to the customer. Support your answer mathematically.

Remove the wheel cylinder by following this procedure:

1. Remove the wheel and brake drum.
2. Remove brake shoes and brake hardware.
3. Disconnect the brake line from the wheel cylinder.
4. Plug the end of the brake line with a rubber vacuum line plug to keep out dirt.
5. Loosen and remove the attaching bolts or retainer clip.
6. Remove the wheel cylinder for servicing.

Rebuilding a Wheel Cylinder

Whether on or off the vehicle, the wheel cylinder is serviced as follows:

1. Remove the boots.
2. Push out the pistons, cups, and spring with your finger.
3. Clean all the parts, which will be reused, in clean brake fluid.
4. Dry each part using compressed air.
5. Inspect the wheel cylinder bore for wear or scoring.
6. Remove surface corrosion with crocus cloth.
7. Remove slight imperfections with a brake hone. **Fig. 3-16.** Do not remove more than 0.003″ [0.08 mm] beyond the original inside diameter. Overboring the wheel cylinder can cause it to leak. (Note: Do not use crocus cloth or a wheel cylinder hone on aluminum cylinders.)

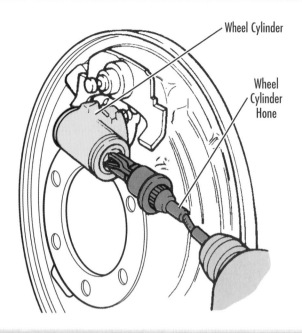

Wheel Cylinder

Wheel Cylinder Hone

Fig. 3-16 Minor scratches and pits can be removed with a hone. *What could happen if you increase the diameter of the bore too much?* (Raybestos Division, Brake Systems Inc.)

Fig. 3-17 After cleaning the self-adjuster, lubricate the threads. Use a high-temperature grease. (Terry Wild Studio)

8. Clean the bore with a lint-free cloth.

9. Dry with compressed air.

10. Lubricate all the parts with clean brake fluid.

11. Reassemble the cylinder using an appropriate parts kit.

Installing a Wheel Cylinder

Install a new or serviced wheel cylinder by following the removal instructions in reverse order.

The hydraulic system must be bled following any servicing or replacement of a wheel cylinder.

Inspecting Drum Brake System Hardware

If possible, replace all brake hardware when servicing drum brakes. If this is not possible, then:

1. Carefully inspect the brake hardware for damage.

2. Replace any damaged components.

3. Disassemble the self-adjuster and clean it thoroughly.

4. Lubricate the threads on the self-adjuster with high-temperature grease. **Fig. 3-17.**

5. Reassemble the self-adjuster and wipe off any excess lubricating grease.

6. The brake hardware is ready for reassembly.

Installing Brake Shoes

1. Reattach the brake shoes to the backing plate using the pin, spring, and cup assemblies.

2. Install the parking brake hardware.

3. Install the self-adjuster.

4. Attach the brake shoe return and hold-down springs. **Fig. 3-18.**

Fig. 3-18 Attach the brake shoe return and hold-down springs between the shoes and the anchor pin. *What step needs to occur to all brake hardware before reattaching?* (Terry Wild Studio)

Adjusting Brake Shoes

If a brake shoe adjusting gauge is not available, manually adjust the brake shoes.

Brake Shoe Adjusting Gauge   Use a brake shoe adjusting gauge to adjust the brake shoes. This gauge can be used only before the brake drum is reinstalled.

 TECH TIP **Star Wheel Adjustment.** You may not need to adjust the star wheel if a brake shoe adjusting gauge is used.

When using a brake shoe adjusting gauge, measure the drum and adjust the shoes to match its diameter. **Fig. 3-19.** Any additional adjustment, if required, can be made after the wheels are installed by accessing the star wheel through the backing plate.

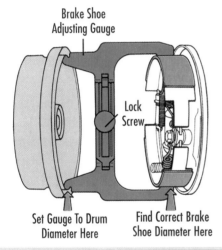

Brake Shoe
Adjusting Gauge

Lock
Screw

Set Gauge To Drum
Diameter Here

Find Correct Brake
Shoe Diameter Here

Fig. 3-19 Use a brake shoe adjusting gauge to set the preliminary clearance or adjustment. *Can you access the star wheel without removing the brake drum?* (Ford Motor Company)

Manual Brake Shoe Adjustments Manually fit the brake shoes to the drum when no gauge is available as follows:

1. Extend the self-adjuster a small amount and then place the brake drum over the shoes.

2. Continue to extend the self-adjuster and fit the drum until the shoes lightly brush the drum during installation.

3. You are now ready to install the brake drum.

Installing Brake Drums

Be sure you have repaired or replaced and correctly reinstalled all wheel brake hardware before proceeding with brake drum installation.

To install a new or machined brake drum, perform the following steps:

1. Install the drum over the wheel brake assembly.

2. Be sure that the drum is free to move.

3. Install the wheel. Tighten and torque the wheel nuts.

4. Reinstall the wheel cover, if there is one.

5. Test the brakes. If they are too tight, back off the star wheel on the self-adjuster.

6. Remove all tools from beneath the vehicle.

7. Remove the jack stands or other safety hardware from beneath the vehicle.

8. Lower the car to the ground.

9. Use the correct method to activate the vehicle's self-adjuster. Repeat activation of the self-adjuster until you are satisfied that the brakes are properly adjusted.

10. Test drive the vehicle to assure the brakes are working properly.

 Safety First **Personal** Before performing a road test, be sure to obtain written permissions from appropriate authorities.

SECTION 4 KNOWLEDGE CHECK

❶ How would you adjust the brakes on a vehicle without removing the drums?

❷ What instrument is used to check the diameter of the brake drum?

❸ What are relined brake shoes?

❹ Why should you try to loosen the bleeder screw on the wheel cylinder before attempting to service it?

❺ What is the purpose of a brake shoe adjusting gauge?

CHAPTER 3 REVIEW

Key Points

Meets the following NATEF Standards for Brakes: diagnosing and servicing drum brakes and master cylinders.

- Most current vehicles have drum brakes only on their rear wheels.
- Drum brakes have two metal shoes lined with friction material. The shoes are pushed against the inside of the brake drum during braking.
- There are two types of drum brakes: leading trailing or non-servo, and duo-servo.
- Leading-trailing drum brakes are used on the rear wheels of front-wheel drive vehicles.
- Duo-servo drum brakes are usually found on the rear wheels of rear-wheel drive vehicles.
- A drum brake diagnosis table can be used to identify potential problems with brake systems.
- Self-adjusters can be manually adjusted by using a brake "spoon" tool to turn the star wheel.
- Wheel cylinders can be serviced on or off the vehicle. They can also be replaced.
- Brake drums can be machined to remove minor defects.

Review Questions

1. Describe drum brake construction and operation.
2. Describe the process used to diagnose drum brake problems.
3. Explain the process for manually adjusting drum brake self-adjusters.
4. Describe the process used to service or replace a wheel cylinder.
5. How does a technician know when a brake drum needs to be replaced?
6. Describe the process used to service brake drums.
7. **Critical Thinking** Describe the components of a drum brake system. Explain how they operate together to provide braking for the vehicle. What are the consequences of neglecting noises or pedal problems?
8. **Critical Thinking** What are the advantages of replacing brake hardware with each brake job? What are the possible consequences if the hardware is not replaced?

TECHNOLOGY FORECAST FOR AUTOMOTIVE EXCELLENCE

Improving Drum Brake Efficiency

Many consumers prefer disc brakes over drum brakes. But drum brakes are still in wide use. They provide effective stopping power. They are also less costly to make.

To further reduce costs, some automakers are installing drum brakes on the rear wheels instead of disc brakes on all four wheels. Consumers don't notice a change in the vehicle's operation. Safety isn't affected. On front-wheel drive cars, only 20 percent of the braking is done by the back wheels.

Automakers are looking for ways to improve the efficiency of drum brakes. The use of aluminum

drums with cast iron inserts on the friction surface will provide strength and durability at less weight. An added benefit of aluminum is that it quickly transfers heat from the brake to the surrounding air.

Another improvement is the use of "leading trailing." With this setup, the front and rear brake shoes are the same length. Used with an anti-lock brake system (ABS), they provide more linear (natural) braking than the older dual-servo design.

One automaker has developed an "advanced design" drum brake. A single, horseshoe-shaped spring takes the place of two retracting springs.

AUTOMOTIVE EXCELLENCE
TEST PREP

Answering the following practice questions will help you prepare for the ASE certification tests.

1. Technician A says the return springs should be replaced with every brake job. Technician B says the return springs will become weak if they are overheated. Who is correct?
   - ⓐ Technician A.
   - ⓑ Technician B.
   - ⓒ Both Technician A and Technician B.
   - ⓓ Neither Technician A nor Technician B.

2. Technician A says brake hardware is attached to the backing plate. Technician B says the backing plate keeps road debris from entering the drum-brake assembly. Who is correct?
   - ⓐ Technician A.
   - ⓑ Technician B.
   - ⓒ Both Technician A and Technician B.
   - ⓓ Neither Technician A nor Technician B.

3. Brake shoe linings are typically attached to the shoes using:
   - ⓐ Adhesives.
   - ⓑ Screws.
   - ⓒ Rivets.
   - ⓓ Either a or c.

4. Relative to the secondary shoe on a duo-servo drum brake system, you would expect the lining on the primary shoe to be:
   - ⓐ Longer.
   - ⓑ Shorter.
   - ⓒ Thicker.
   - ⓓ Thinner.

5. Technician A says braking while moving in reverse activates self-adjusters. Technician B says using the parking brake activates self-adjusters. Who is correct?
   - ⓐ Technician A.
   - ⓑ Technician B.
   - ⓒ Both Technician A and Technician B.
   - ⓓ Neither Technician A nor Technician B.

6. Glazing of brake drum friction surfaces is caused by:
   - ⓐ Excessive use.
   - ⓑ Overheating.
   - ⓒ Defective brake hardware.
   - ⓓ Improper machining.

7. Technician A says oil on the brake linings can cause the brakes to grab. Technician B says water on the linings can cause the brakes to grab. Who is correct?
   - ⓐ Technician A.
   - ⓑ Technician B.
   - ⓒ Both Technician A and Technician B.
   - ⓓ Neither Technician A nor Technician B.

8. Technician A says a leak in the master cylinder can leak brake fluid into the power booster. Technician B says the brake fluid can leak into the power booster vacuum lines. Who is correct?
   - ⓐ Technician A.
   - ⓑ Technician B.
   - ⓒ Both Technician A and Technician B.
   - ⓓ Neither Technician A nor Technician B.

9. If a cable type self-adjuster is installed backwards, it will:
   - ⓐ Function normally.
   - ⓑ Not function.
   - ⓒ Operate backwards.
   - ⓓ Fall out.

10. The brake warning light can be triggered by:
   - ⓐ A faulty self-adjuster.
   - ⓑ Badly worn brake shoe linings.
   - ⓒ A loose parking brake cable.
   - ⓓ The pressure differential valve.

CHAPTER 4

Diagnosing & Repairing Disc Brakes

You'll Be Able To:

- ⊗ Describe how disc brakes work.
- ⊗ Diagnose disc brake problems.
- ⊗ Inspect brake pads and hardware.
- ⊗ Inspect, disassemble, and clean the caliper assembly.
- ⊗ Measure a rotor with a micrometer.

Terms to Know:

brake pads

calipers

fixed caliper disc brakes

floating caliper disc brakes

on-car brake lathe

rotors

scores

sliding caliper disc brakes

The Problem ⟩ ⟩⟩ ⟩⟩ ⟩⟩ ⟩⟩ Your Challenge

Beth Richmond stops in at your service center. She hears noises from her car as she drives it. She is concerned that rocks or debris may have gotten into her wheels. Beth would like you to test-drive her car to determine the problem.

Beth says her car makes a high-pitched scraping sound when she is driving. When she pushes the brake pedal, the noise goes away. As far as she can tell, her car seems to be stopping properly. Beth is planning to leave on a business trip soon and she is concerned about her car's safety.

As the service technician, you need to find answers to these questions:

❶ How long has Beth been hearing this noise?

❷ Is the noise coming from the front or the rear of the car?

❸ When was the last time Beth had her vehicle's brakes serviced?

Section 1

Disc Brake Construction

Most vehicles have front-wheel disc brakes. Rear-wheel brakes can be either disc or drum. Disc brakes are engaged by hydraulic pressure. Parking brakes on vehicles with four-wheel disc brakes are activated by a mechanical system. Some systems mechanically activate the calipers. Other systems use a small drum brake inside the rotor to serve as the parking brake.

Disc brakes use friction between the rotor and brake pads to convert vehicle motion into heat. As with drum brakes, great amounts of heat are produced during braking. This high heat limits the types of materials that can be used for brake pads. Unlike drum brakes, however, disc brakes release heat quickly. Some rotor designs are better suited than others for releasing heat.

Disc brake systems have four major parts. These are the:

- *Mounting brackets.* The mounting brackets hold the calipers in place.
- *Rotors.* The rotors are metal discs supported by the suspension. They are clamped by the calipers, slowing the wheel rotation. This slows and stops the vehicle.
- *Calipers.* The calipers are housings that contain the pistons and the brake pads. Connected to the hydraulic system, the calipers hold the brake pads so they can straddle the rotor.
- *Brake pads.* The brake pads are friction surfaces that convert motion into heat.

Disc brakes are constructed so the friction surfaces are external and in contact with the air flowing under the vehicle. This makes them more effective than drum brakes. On drum brakes, friction surfaces are enclosed in the drum and covered by the backing plate. This makes it harder to dispel the heat generated during braking. This is one reason disc brakes are more effective in making repeated stops. Disc brakes are easier to service, due to their simpler design.

Rotors

Rotors are disc-shaped devices made from cast iron or sintered iron. They have flat friction surfaces machined onto both sides. They are attached to the spindles or hubs and rotate on the wheel bearings.

There are two type of rotors:

- *Vented rotors* have fins in the space between their machined surfaces. These fins allow air to pass through the rotor. This airflow helps carry away excess heat built up during braking. Vented rotors are used on heavier vehicles that produce large amounts of heat during braking. **Fig. 4-1(a).**
- *Nonvented rotors* are solid metal with no air vents to aid in heat removal. Nonvented rotors are used mainly on smaller vehicles where brake overheating is not a serious problem. **Fig. 4-1(b).**

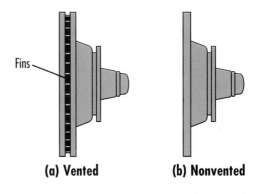

Fins

(a) Vented   **(b) Nonvented**

Fig. 4-1 Of the two types of rotors, vented rotors are usually found on larger, heavier cars and light trucks. *Why don't smaller, lighter vehicles require vented rotors?*

Calipers

Calipers are housings that contain the pistons and the brake pads. They convert hydraulic pressure from the master cylinder into linear motion. Each caliper includes a housing with one, two, or four cylinder bores. Each cylinder bore has a single piston.

When the driver presses the brake pedal, brake fluid is forced out of the master cylinder through the brake lines and into the calipers. Fixed calipers have two sides, an inboard side and an outboard side. There are pistons in each half. Fluid passes between the caliper halves—either through outside crossover lines or through inside passages. Hydraulic pressure from the master cylinder forces pistons in both halves of the caliper to push the pads against the rotor. This provides the desired braking action. Floating or sliding calipers use only one piston. They allow caliper movement to equalize force to both sides of the rotors.

Brake Pads

Brake pads are the parts of disc brake systems that convert vehicle motion into heat. They are flat steel plates with friction material attached by rivets or adhesive. The pads are in the caliper housing. They are pushed against each side of the rotor by the caliper piston(s). Linings are made from non-asbestos or semimetallic materials on late-model vehicles. *Caution:* Some replacement pads for older vehicles may have some asbestos content.

When the brakes are applied, the pads generate high braking forces on both sides of the brake rotor. Because both sides of the rotor are exposed to the cooling air, heat can be rapidly removed from the rotor. This allows repeated stops without brake fade.

When the brakes are applied, the vehicle weight transfers toward the front of the vehicle. Because of this weight transfer, the front brakes do more work than the rear brakes.

Brakes must be able to handle the vehicle's weight. The way a vehicle is built decides how much weight is on the front wheels. Due to weight distribution in rear-wheel drive automobiles, about 60 percent of braking is handled by the front disc brakes and 40 percent by the rear brakes.

Front-wheel-drive vehicles have their drivetrains mounted over the front wheels. This results in about 80 percent of the braking effort being handled by the front disc brakes. The rear brakes provide only 20 percent of the braking effort.

SECTION 1 KNOWLEDGE CHECK

① On a vehicle with both disc and drum brakes, would you expect to find the disc brakes on the front or the rear?

② What rotor design would you expect to find on a large, heavy vehicle? Why?

③ How many pistons might you find on current vehicle calipers?

④ On fixed caliper disc brakes, how does the pressurized fluid pass between the two halves?

⑤ What friction materials are used on vehicle brake pads?

Section 2

Disc Brake Operation

The operation of a disc brake depends on its caliper arrangement. The three types of disc brakes are:
- Fixed caliper disc brakes.
- Floating caliper disc brakes.
- Sliding caliper disc brakes.

Fixed Caliper Disc Brakes

Fixed caliper disc brakes are disc brakes that use a caliper which is fixed in position and cannot move. The caliper has pistons on both sides of the rotor, in both the inboard and outboard caliper halves. There may be two or four pistons in each caliper. **Fig. 4-2.**

Two major drawbacks of fixed caliper designs are their high costs to manufacture and to repair.

These drawbacks result from the complexity of fixed caliper designs.

Floating Caliper Disc Brakes

Floating caliper disc brakes are disc brakes that use a caliper that is free to move sideways on bushings and guide pins. A support bracket is mounted to the spindle or steering knuckle. The caliper housing is mounted to the support bracket. Floating caliper disc brakes have only one piston. This piston is on the inboard side of the caliper. **Fig. 4-3.**

The piston applies direct pressure to the inboard pad. As the inboard pad contacts the rotor, it forces the caliper to slide along the pins. This pulls the outboard pad against the rotor.

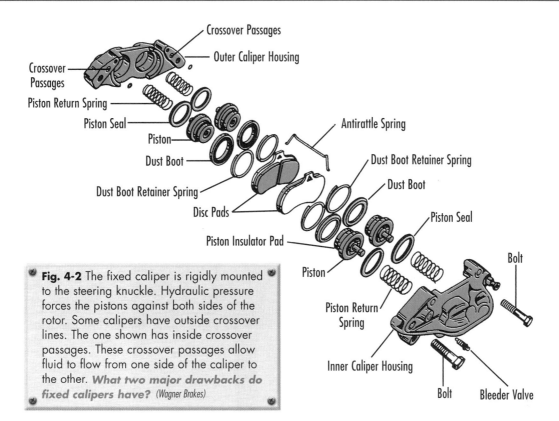

Fig. 4-2 The fixed caliper is rigidly mounted to the steering knuckle. Hydraulic pressure forces the pistons against both sides of the rotor. Some calipers have outside crossover lines. The one shown has inside crossover passages. These crossover passages allow fluid to flow from one side of the caliper to the other. *What two major drawbacks do fixed calipers have?* (Wagner Brakes)

The actual movement of floating calipers on the guide pins when brakes are applied is very small. When the brakes are applied, the caliper grips the rotor. It relaxes upon release of the brake pedal. The brake pads stay at zero clearance to the rotor. **Fig. 4-4.**

Sliding Caliper Disc Brakes

In operation, sliding calipers are similar to floating calipers. The difference is in the way the caliper is mounted and moved. **Sliding caliper disc brakes**

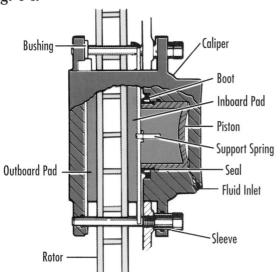

Fig. 4-3 Construction of a floating caliper disc brake, which has only one piston. When force is applied, the outboard pad is drawn against the rotor. *Is the piston in a floating caliper on the inboard side or on the outboard side of the caliper?* (General Motors Corporation)

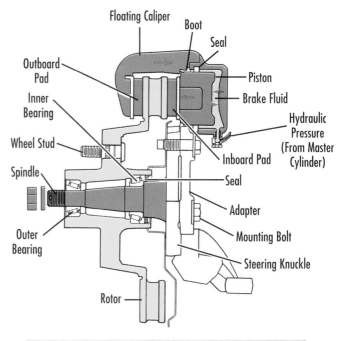

Fig. 4-4 The red arrows show the movement of floating calipers. *What is the normal clearance between the rotor and the brake pads when the brakes are at rest?* (DaimlerChrysler)

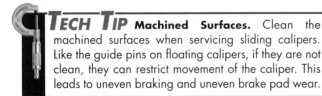

TECH TIP Machined Surfaces. Clean the machined surfaces when servicing sliding calipers. Like the guide pins on floating calipers, if they are not clean, they can restrict movement of the caliper. This leads to uneven braking and uneven brake pad wear.

use a caliper that is held in place by a retainer, a spring, and a bolt. There are no guide pins. Sliding calipers slide on mating surfaces. These surfaces are machined onto both the caliper and the mounting bracket. **Fig. 4-5.**

Self-Adjusting Disc Brakes

Disc brakes will self-adjust. They automatically adjust for brake pad wear. The caliper's cylinder bore has a rectangular groove with a square-cut seal. This seal fits tightly around the piston. The seal provides a barrier between the piston and the caliper bore. This barrier keeps brake fluid from leaking. A dust boot on the open end of the caliper bore keeps out dirt and water.

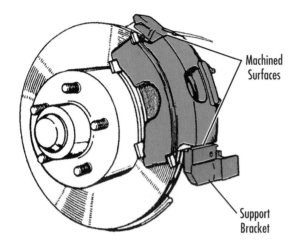

Machined Surfaces

Support Bracket

Fig. 4-5 The sliding caliper's machined surfaces move along the machined surfaces of the support bracket. *What must be kept clean to allow free movement of the caliper?*

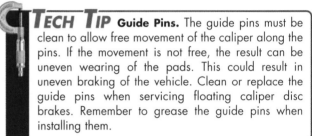

TECH TIP Guide Pins. The guide pins must be clean to allow free movement of the caliper along the pins. If the movement is not free, the result can be uneven wearing of the pads. This could result in uneven braking of the vehicle. Clean or replace the guide pins when servicing floating caliper disc brakes. Remember to grease the guide pins when installing them.

COMMUNICATIONS

Reading Diagrams

Diagrams are visual representations of written words. The purpose of the "pictures" is to make clear how parts fit together or function. Many complex components or systems cannot be easily described in words alone.

Sometimes people who do not like to read will try to understand a subject using only visual information. However, they may have problems because the picture may not tell the whole story. A written explanation may be necessary to understand how a system functions.

You need to be able to "read" a diagram as well as the text that applies to it. The written words tell you much of what you need to know. The diagram lets you "see" what the words are describing. Using these two tools together, you may more easily understand what you need to

know about the function or construction of the part or system.

The text plus the diagram equals the whole picture.

Apply it!

Meets NATEF Communications Standards for reading strategies.

❶ Refer to **Fig. 4-3** and **Fig. 4-4.** Read the section on calipers and the text that applies to these figures.

❷ Referring only to the above figures, explain how the floating caliper disc brake is assembled and how it operates.

❸ Did you read information in the text that helped you understand the diagrams? Explain.

When the brakes are applied, the piston moves toward the rotor. This causes the square-cut seal to deflect slightly. The seal continues to grip the piston. When the brakes are released, the square-cut seal acts to pull the piston back into the bore and away from the rotor. This small movement keeps the pads at zero clearance with the rotors.

As the linings wear, the piston travel when the brakes are applied becomes greater than the amount the seal can deflect. When this happens, the piston slides outward through the seal and assumes a new position. **Fig. 4-6.** This moves the pads closer to the rotor and adjusts for brake pad wear.

Brake fluid from the master cylinder fills the additional space behind the piston in the caliper bore. This lowers the fluid level in the master cylinder. This is why a disc brake reservoir is larger than a drum brake reservoir.

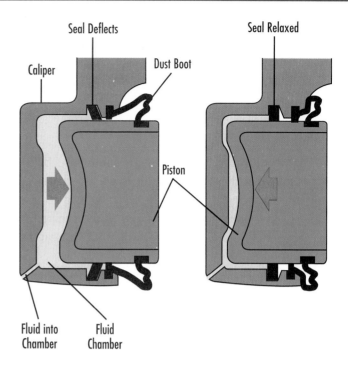

(a) Brakes Applied   (b) Brakes Released

Fig. 4-7 The seal groove in the low drag caliper is cut at a slight angle to allow the piston to move farther out and back. *Why is it important that the pads not drag against the rotor?*

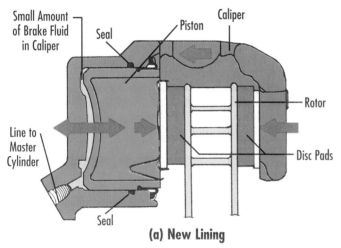

(a) New Lining

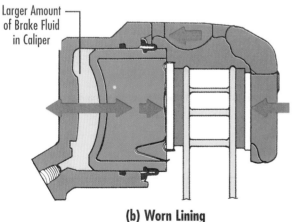

(b) Worn Lining

Fig. 4-6 As the linings wear, the piston moves farther out of the caliper bore. *What happens to the fluid level in the master cylinder reservoir?* (General Motors Corporation)

In the 1980s car manufacturers searched for a way to increase fuel economy. That is when the low drag caliper was invented. In design and operation, this caliper is similar to other calipers, with the exception of the square-cut seal and groove.

The shape of the seal groove in the caliper determines how far the seal travels with the piston. The amount of seal deflection also determines how far the piston retracts into the caliper bore. On a low drag caliper, the seal groove in the caliper's bore is cut at a small angle. **Fig. 4-7.** This design allows the seal more movement.

Greater movement allows the seal to deflect farther when the brakes are applied. **Fig. 4-7(a).** It also permits them to retract more when they are released. **Fig. 4-7(b).** This increased piston movement allows the brake pads to be pulled farther from the rotors. As a result, when the brakes are released, the pads do not drag against the rotor. Dragging pads would waste energy and increase fuel consumption.

The greater piston movement of low drag calipers requires more brake fluid movement into and out of the master cylinder. As a result, systems with low drag calipers require the use of quick take-up master cylinders.

Disc Brake Wear Indicators

Wear indicators are found on some disc brake pads. Wear indicators alert the driver to the need for brake service. Ignoring the warning and continuing to use worn pads may cause costly damage to the rotors. There are two types of wear indicators: mechanical and electrical.

Mechanical Wear Indicators Mechanical wear indicators are attached to the metal part of the pad. As the lining wears down, the wear indicator rubs against the rotor, making a high-pitched sound. This tells the driver to have the brakes serviced. **Fig. 4-8.**

Electrical Wear Indicators Disc pads with electrical wear indicators have buried sensor wires. When the pads wear down far enough, these sensor wires contact the metal rotor. The metal rotor conducts the current across the sensor wires. This turns on the warning light on the dash.

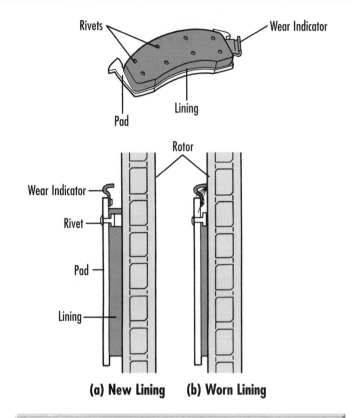

(a) New Lining   (b) Worn Lining

Fig. 4-8 When the lining gets so thin that it should be replaced, the wear indicator scrapes the rotor, making an audible warning. *What can happen if the driver ignores the wear indicator's warning?* (General Motors Corporation)

SECTION 2 KNOWLEDGE CHECK

❶ What caliper design has pistons on both sides of the rotor?

❷ What two types of calipers allow their housings to move on guide pins or machined surfaces?

❸ Why must guide pins be kept clean? What can happen if their movement is not free?

❹ How do disc brake calipers adjust for lining wear?

❺ How do electrical wear indicators alert the driver that the brakes need servicing?

Section 3

Disc Brake Diagnosis

A disc brake diagnosis table can help identify possible causes of disc brake complaints. After identifying the possible causes, check each of them.

Begin with the most likely. Repairing some problems may require only an adjustment to the system. Repairing other problems may require complete disassembly of the entire wheel brake. **Table 4-A.**

Table 4-A	DIAGNOSING DISC BRAKES	
Complaint	**Possible Cause**	**Check or Correction**
1. Excessive pedal travel	a. Excessive rotor runout b. Air in hydraulic system c. Low brake fluid d. Bent brake pad and lining or loose insulators e. Loose wheel bearing f. Damaged piston seal in caliper g. Power brake inoperative h. Failure of one section of hydraulic system i. Caliper parking brake out of adjustment j. Loose calipers k. Floating or sliding caliper not moving freely	a. Refinish or replace rotor b. Bleed brakes c. Fill, inspect, repair, and bleed system d. Inspect and replace e. Adjust or replace f. Replace g. Repair or replace h. Inspect and repair; check brake warning light i. Inspect rear calipers; repair j. Tighten retaining bolts k. Inspect, clean, repair, or replace guide pins or slides
2. Pedal pulsations	a. Excessive rotor runout (wobble) b. Rotor out of parallel (uneven rotor thickness) c. Loose wheel bearings d. Tight caliper slides e. Wheel and tire vibration	a. Refinish or replace rotor b. Refinish or replace rotor c. Adjust or replace d. Free and repair e. Diagnose vibration
3. Excessive pedal force, grabbing, uneven braking	a. Power brake booster defective b. Brake fluid, oil, or grease on lining c. Linings worn or glazed d. Wrong lining e. Piston stuck in caliper f. Failure of one section of hydraulic system	a. Repair or replace b. Repair leakage; replace pads c. Replace pads d. Install correct lining e. Service or replace caliper f. Inspect and repair; check brake warning light
4. Pulls to one side while braking	a. Brake fluid, oil, or grease on linings b. Seized caliper c. Piston stuck in caliper d. Incorrect tire pressure e. Bent or warped brake pads f. Incorrect wheel alignment g. Broken rear spring h. Restricted line or hose i. Unmatched linings j. Loose caliper k. Loose suspension parts l. Broken shocks	a. Repair leakage, replace pads b. Service or replace caliper c. Service or replace caliper d. Adjust e. Replace f. Align g. Replace h. Repair or replace i. Replace j. Tighten k. Tighten l. Replace shocks
5. Brake noise	a. Typical of some disc brakes	a. Not detrimental, no action required. Often eliminated when driver increases or decreases brake-pedal force slightly
• Frequent or continuous squeal	a. Brake pad loose on caliper b. Worn or missing parts c. Improper clearance on rear disc brakes d. Rotor scored or improperly machined e. Antisqueal compound not applied to pads f. Glazed linings g. Rivets loose in brake pad and lining	a. Clinch brake pad tabs to caliper b. Check for damaged or missing antirattle springs, antisqueal shims or insulators, or worn caliper guide pins and bushings c. Check clearance between rotor and lining d. Remachine or replace rotor e. Remove and apply antisqueal compound to back of pads f. Inspect rotor; replace pads g. Replace brake pads
• Occasional squeal	a. Typical of many disc brakes	a. Not detrimental, no action required. Explain to driver the occasional squeal cannot always be eliminated
• Scraping	a. Disc rubbing caliper b. Loose wheel bearing c. Worn lining; wear sensor scraping on disc d. Improper assembly	a. Remove rust or mud from caliper; tighten caliper bolts b. Adjust or replace c. Inspect disc; replace pads d. Assemble properly
• Rattle at low speed	a. Pads loose on caliper b. Loose antirattle clips c. Loose attaching hardware	a. Clinch pad tabs to caliper; install new pads b. Adjust or tighten clips c. Tighten attaching hardware
• Groan when slowly releasing brakes (creep groan)	a. Brake pads loose on caliper	a. Clinch brake pad tabs to caliper; install new brake pads

Table 4-A	DIAGNOSING DISC BRAKES (continued)	
Complaint	**Possible Cause**	**Check or Correction**
6. Brakes fail to release	a. Power brake booster defective b. Brake pedal binding c. Master cylinder push rod improperly adjusted d. Driver rides pedal e. Incorrect stoplight switch adjustment f. Caliper piston not retracting g. Speed control switch improperly adjusted h. Incorrect parking brake adjustment i. Restricted pipes, hoses, or banjo bolts	a. Repair or replace b. Free and repair c. Adjust d. Notify driver e. Adjust f. Service or replace g. Adjust h. Adjust i. Repair
7. Fluid leaking from caliper	a. Damaged or worn piston seal b. Scores or corrosion on piston or in caliper bore c. Defective caliper seal d. Loose brake line fitting e. Damaged copper sealing washer on banjo bolt	a. Replace b. Service caliper c. Rebuild or replace caliper d. Tighten fitting or replace brake line e. Replace copper sealing washer
8. Front disc brakes grab (rear drum brakes okay)	a. Defective metering valve b. Incorrect brake pad linings c. Improper surface finish on rotor d. Contaminated brake pad linings	a. Replace b. Replace c. Refinish rotor d. Replace
9. No braking with pedal fully depressed	a. Piston pushed back in caliper b. Leak in hydraulic system c. Damaged piston seal d. Air in hydraulic system e. Leak past primary cup in master cylinder f. Low fluid in reservoir	a. Pump brake pedal; check brake pad position b. Repair c. Replace d. Add fluid; bleed system e. Service or replace master cylinder f. Add fluid
10. Fluid level low in master cylinder	a. Leaks b. Worn linings	a. Repair; add fluid; bleed system b. Replace
11. Warning light comes on while braking	a. One section of hydraulic system has failed b. Pressure differential valve defective	a. Check both sections; replace b. Replace
12. Caliper parking brake will not hold vehicle	a. Improper parking brake cable adjustment b. Defective rear actuators c. Ineffective rear lining d. Defective parking brake pedal assembly	a. Adjust b. Adjust, repair c. Inspect and replace d. Repair
13. Caliper parking brake will not release	a. Improper cable adjustment b. Vacuum release system inoperative	a. Adjust b. Repair

Excessive Pedal Travel

Excessive pedal travel exists when the piston must travel farther than normal. Excessive pedal travel indicates the need for a greater volume of brake fluid to compensate. Causes include:
• Excessive rotor runout knocking the pistons back into their bores. **Fig. 4-9.**
• Loose calipers.
• A damaged piston seal in the caliper.
• Rear caliper parking brake out of adjustment.
• Floating or sliding caliper not moving freely.

Pedal Pulsations

Pedal pulsations are often due to:
• A rotor with excessive runout.
• A rotor that is out of parallel (uneven rotor thickness).

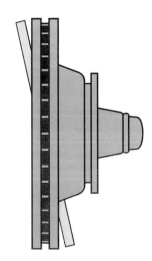

Fig. 4-9 Runout describes the wobble of the rotor. This can knock the piston back into the bore. *Why does this cause excessive pedal travel?*

- Loose wheel bearings.
- Tire and wheel vibration.

Uneven rotor thickness knocks the pistons back into their bores. This motion is carried through the brake fluid, back to the master cylinder, and then to the pedal. Often these problems can be corrected by machining or replacing the rotor.

Pedal pulsations are normal on some ABS brakes. Pulsations are noticeable only when ABS brakes are activated in emergency or panic stops.

Excessive Pedal Force, Grabbing, Uneven Braking

Excessive pedal force, grabbing, and uneven braking are usually caused by a problem with the power brake booster. They could also result from:
- Contaminated linings (brake fluid, grease, and oil).
- Worn or glazed linings.
- The wrong linings.
- Piston stuck in the caliper.
- Glazed rotor. **Fig. 4-10.**

If the rotor is glazed, check rotor thickness to see if it can be machined on a lathe. If the rotor is not thick enough to machine, replace it. Replace glazed pads at the same time. Always service both sides of the vehicle, even if the condition is only on one wheel. This will help eliminate uneven braking. Check for problems in the hydraulic system.

Fig. 4-10 When there is glazing on the pads or rotor, it is difficult to produce the friction needed for good braking. *How might this be corrected?* (Terry Wild Studio)

Pulling to One Side While Braking

When brakes pull to one side during braking, it is usually due to uneven brake operation. First, verify proper brake operation on the opposite brake. The vehicle will pull to the side with the working brake. Causes of brake pull include:
- Linings ruined by brake fluid, oil, or grease.
- A seized caliper.
- Piston stuck in the caliper.
- Loose caliper.
- Unmatched linings.
- Bent or warped brake pad(s).

Non-brake-related causes include:
- Low tire pressure.
- Bad wheel alignment.
- Broken rear spring.
- Broken shocks.
- Loose suspension parts.

TECH TIP **Antisqueal Compound.** Antisqueal compound can be placed on the metal backing of brake pads to eliminate some squealing noises. Do not apply coating to the friction surface of the pads.

Noise

Brake noise indicates vibration of brake parts. Some occasional noises are normal. High humidity or using a very light touch on the brake pedal when applying the brakes can cause brake squeal. Another cause of brake squeal is use of a hard brake pad compound. Other noises may indicate brake problems.

Constant Squealing A constant brake squeal may be the result of:
- Pads loose in the caliper.
- Worn or missing antirattle springs, guide pins, or bushings.
- Loose or missing antisqueal shims. **Fig. 4-11.**
- Loose lining attachment rivets.
- Scored or improperly machined rotor.
- Antisqueal compound not on backs of pads.
- Glazed linings.

Scraping Sounds   Scraping sounds may be caused by:
• The caliper contacting the rotor.
• Brake pad worn down to the metal backing.
• Mechanical wear indicators rubbing the rotor.

Rattling Noises   Rattling noises may be caused by:
• Loose antirattle clips.
• Loose attaching hardware.
• Loose pads.

Some non-disc brake related causes include loose wheel bearings and debris or rust between the rotors and splash shields.

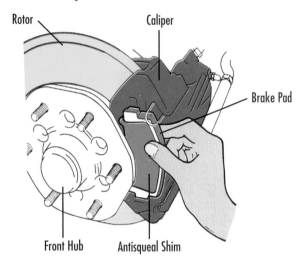

Fig. 4-11 An antisqueal shim or compound can sometimes control brake squeal. *Name two non-disc brake related causes of brake noise.* (DaimlerChrysler)

Failing to Release

Disc brakes might fail to release due to:
• Brake pedal binding.
• Master cylinder pushrod out of adjustment.
• Incorrect stoplight switch adjustment.
• Caliper piston not retracting properly.
• Parking brake out of adjustment.
• Hydraulic lines (pipes, hoses, banjo bolts) restricted.

Fluid Leaking from Caliper

Visible brake fluid on the caliper indicates a leak. This may be caused by:
• A defective piston seal.
• A scored or pitted piston.
• A loose brake line fitting.
• A damaged copper sealing washer on a banjo bolt.

Front Disc Brakes Grabbing

If the front brakes grab (and the rear drum brakes are in good condition), the causes may be:
• Incorrect brake pad lining.
• Defective surface finish on rotor.
• Contaminated brake pad lining.

On vehicles with rear drum brakes, a defective metering valve can allow the front brakes to grab. The metering valve is designed to hold off hydraulic pressure to the front disc brakes until the rear brakes have overcome the return spring tension.

No Braking with Pedal Fully Depressed

No braking with the pedal fully depressed may be caused by the piston (and thus the pads) being too far from the rotor after servicing. It may take more than several pumps to extend the pistons. It may also be the result of low brake fluid in the master cylinder reservoir.

Low Fluid Level in Master Cylinder

Low fluid level in the master cylinder is usually due to worn brake pads. As the pistons move farther out of the caliper bore, the fluid level drops. Check the hydraulic system for leaks, as well.

Brake Warning Light On

If the brake warning light comes on while braking, there may be unequal pressure between the two hydraulic braking circuits. This could mean that one side of the hydraulic system has failed. It may also indicate that the pressure differential valve is not working correctly.

TECH TIP **Low Fluid Level in Master Cylinder.** When the fluid level is low in the master cylinder and there are no visible leaks, remove the vacuum hose from the booster. Check for visible signs of brake fluid in the hose. The leak may be in the rear seal of the master cylinder. This could allow fluid to be drawn into the booster and through the vacuum hose into the engine. Loosen the bolts holding the master cylinder to the power booster. Look for wet areas between the master cylinder and the booster. If fluid is escaping from the rear of the master cylinder, it must be rebuilt or replaced. Most master cylinder cores are returned for remanufactured or new units.

Caliper Parking Brake Not Holding

If the parking brake is not holding the vehicle, inspect the parking brake cable or assembly for proper operation. Parking brake adjustment is important, especially on vehicles with rear disc brakes. If the vehicle has four-wheel disc brakes, refer to the vehicle's service manual for the correct procedure to adjust the parking brakes.

Caliper Parking Brake Not Releasing

When the parking brake does not release prop-

Safety First

Materials Never put oil, transmission fluid, or power steering fluid in the brake system. It will damage the rubber parts and can lead to brake fluid leaks and eventual brake failure.

erly, check the parking brake cable adjustment. Inspect the cable for rust, which may cause binding. Some vehicles have a parking brake vacuum release system. Check to see that it is working correctly.

SECTION 3 KNOWLEDGE CHECK

❶ List three reasons the car may pull to one side.

❷ What can cause brake squealing?

❸ What might cause failure of brakes to release?

❹ What does a low master cylinder reservoir fluid level indicate?

❺ If the brake warning light comes on, what could be the problem?

Section 4

Disc Brake Pad and Caliper Service

When preparing to service disc brakes, check the brake pads first. They may be the only problem. Also, always check the vehicle service manual before doing any repair work. Procedures vary from vehicle to vehicle. The procedures outlined here represent only one possible variation.

You can usually replace the pads in fixed calipers without removing the calipers from the vehicle. Only the wheel will need to be removed. On floating or sliding calipers, however, you must remove the wheel and caliper to replace the pads.

Inspecting Disc Brake Pads

Disc brake pad inspection requires removing the vehicle's wheel. Inspect the surface of the pads for signs of glazing, cracking, or other problems. Replace pads if they are worn to minimum thickness or if there is uneven wear between the pads on the same wheel. If there is uneven wear, inspect the slides, guides, and pistons of the caliper for binding.

Inspect the brake pads as follows:

1. Use a hoist or other device to safely raise the vehicle. Place safety stands under the vehicle.

2. Remove the lug nuts holding the wheel to the axle or hubs. Remove the wheel.

3. Visually inspect the brake pads. Check the pads for even wear. Uneven wear indicates a problem with the caliper. Refer to the vehicle service manual for minimum thickness. If the manual does not specify minimum thickness, compare the thickness to that of a new pad.

4. Inspect brake pad retaining hardware. Repair or replace as necessary.

5. If pads appear to be more than about 75 percent worn, replace them and reinstall the wheel. If pads do not need to be replaced, reinstall the wheel.

6. Remove safety stands and lower the vehicle to the ground.

7. Verify that the brakes work normally. Then remove the vehicle from the hoist.

8. Road test the vehicle.

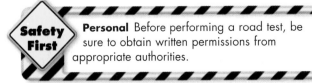

Safety First

Personal Before performing a road test, be sure to obtain written permissions from appropriate authorities.

Replacing Disc Brake Pads

Before removing and installing disc brake pads, remove and discard one-half to two-thirds of the fluid from the master cylinder reservoir. This prevents the fluid from overflowing when the caliper pistons are pushed back into their bores. **Fig. 4-12.**

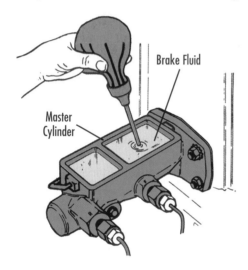

Brake Fluid

Master Cylinder

Fig. 4-12 Remove and discard one-half to two-thirds of the fluid from the reservoir. Do this before pushing the pistons back into their bores. *What could happen if there is too much fluid in the reservoir?*

Replacing Disc Brake Pads (Fixed Calipers)
Before installing new disc brake pads, coat the back of the pads with the antisqueal compound. Do this 15–20 minutes before you expect to install them to allow the antisqueal compound to dry. Remove the old brake pads as follows:

1. Remove and discard one-half to two-thirds of the fluid from the master cylinder.

2. Use a hoist or other device to safely raise the vehicle. Place safety stands under the vehicle.

3. Remove the lug nuts holding the wheel to the axle or hub. Remove the wheel.

4. Remove the brake pad retaining bolts or pins from the caliper housing.

5. Use a piston retraction tool or pry bar to push the caliper pistons back into the cylinder bores. This is necessary to make room for the thicker replacement brake pads.

6. Remove the brake pads.

7. Coat the metal sides of the disc brake pads with antisqueal compound or install new shims.

8. Install the new pads with the linings toward the friction surface of the rotor.

9. Reinstall the brake pad retaining bolts or pins in the caliper housing.

10. Reinstall the wheel.

11. Remove the safety stands and lower the vehicle to the ground.

12. Pump the brake pedal several times to displace the caliper pistons and verify that the brakes operate normally.

13. Top off the fluid in the master cylinder.

14. Remove the vehicle from the hoist.

15. Test drive the vehicle.

Removing and Replacing Disc Brake Pads and Calipers (Floating or Sliding Calipers)

Before removing the disc brake calipers, remove and discard one-half to two-thirds of the fluid from the master cylinder reservoir. This prevents the fluid from overflowing when the caliper pistons are pushed back into their bores.

Remove a floating disc brake caliper as follows:

1. Use a hoist or other device to safely raise the vehicle. Install safety stands under the vehicle.

2. Remove the lug nuts holding the wheel to the axle or hub. Remove the wheel.

3. Use a piston retraction tool or a C-clamp to force the pistons back into the bore of the caliper. Back the pistons and disc brake pads away from the rotor to make caliper removal easier. **Fig. 4-13.**

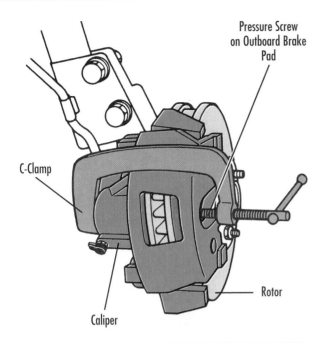

Pressure Screw
on Outboard Brake
Pad

C-Clamp

Rotor

Caliper

Fig. 4-13 Use a retraction tool or a C-clamp to force the piston back into the caliper bore. *Why is it necessary to push the caliper piston(s) back into the bore?* (General Motors Corporation)

4. Remove the two caliper guide pins or mounting bolts and positioners. These attach the caliper to the steering knuckle adapter. On sliding-type calipers, remove the bolt, retainer key, and flat spring.

5. Remove the caliper. Support or hang the caliper to the vehicle frame with a piece of wire. Never allow the caliper to hang by the hose as this may do internal damage to the hose.

6. Remove the brake pads, support spring, and outer bushings from the caliper. Discard the bushings. Slide the inner bushings off the guide pin and discard them.

7. Inspect the caliper for wear and damage.

8. Inspect the caliper mounting bracket, slides, and guide pins. Slides and guide pins must be rust-free, smooth, and undamaged. Repair or replace as necessary.

9. Inspect the rotor for damage.

10. Clean all disc brake components. Lubricate caliper slide and guide parts with special brake caliper lubricant. Do not get brake caliper lubricant on the rotor friction surface or on the brake pads.

TECH TIP Supporting the Caliper. Support the caliper with wire. Without stretching the brake hose, attach one end of a piece of wire to a strong suspension member. Do not let the caliper hang from the brake hose. The hose might be damaged.

11. Install the new disc brake pad with the spring clip on the inboard pad. Install the outboard pad.

12. Install the caliper containing the new pads over the rotor. Install new bushings on the guide pins. Secure the caliper using the appropriate hardware. If necessary, bend the ears (tabs) of the outboard pad to remove any play or rattle.

13. Add fresh brake fluid. Gently pump the brakes until the pedal comes up. Recheck the brake fluid in the reservoir. Add fluid, if necessary.

14. Reinstall the wheel.

15. Remove the safety stands and lower the vehicle to the ground.

16. Test drive the vehicle.

17. Instruct the vehicle owner to avoid panic or emergency stops and hard brake use for the next 100–200 miles. This gives the pads and rotors time to be burnished.

Safety First **Personal** Before performing a road test, be sure to obtain written permissions from appropriate authorities.

Repairing Disc Brake Calipers

Perform caliper service if the caliper shows signs of damage or fluid leakage. If the caliper is damaged beyond repair, replace it. The damage that can occur in caliper assemblies includes:

• Scratches, pits, and scoring on the piston or cylinder bore surfaces.
• Damaged square-cut seals.
• Damaged dust boots.
• Damaged cylinder grooves.
• Damaged or leaking bleeder valves.

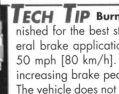**TECH TIP Burnishing.** New linings must be burnished for the best stopping performance. Make several brake applications (not panic stops) from about 50 mph [80 km/h]. Gently apply the brakes at first, increasing brake pedal pressure as the vehicle slows. The vehicle does not need to come to a complete stop. Resume driving, leaving a minute or two between brake applications.

Scoring is serious damage to the piston surface or the inside surface of the cylinder bore. If you cannot remove damage to these surfaces by honing the cylinder bore or by polishing the piston, replace the caliper. Otherwise, perform service on the caliper as follows:

1. After removing the pads, remove the brake hose. Do not damage or lose the copper washers. To prevent brake fluid from leaking, cap or plug the brake hose.

2. Loosen and retighten the bleeder valve on the caliper to verify that the threads are not damaged or seized. If the bleeder valve breaks off, replace the caliper.

Safety First **Personal** Keep your fingers away from the front of the piston when applying air pressure to caliper. Wear goggles to prevent brake fluid from getting in your eyes.

3. Take the caliper to the workbench to remove the piston. Place several shop towels against the outboard side of the caliper to cushion it. Carefully use air pressure at the hose inlet to force the piston out of the caliper.

4. Do not drop or damage the piston. Damage to the piston could cause brake fluid leaks after reassembly. **Fig. 4-14.**

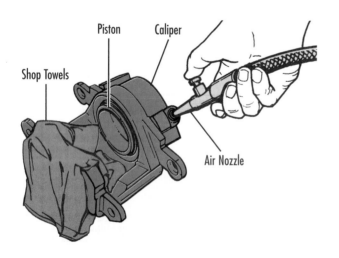

Shop Towels — Piston — Caliper — Air Nozzle

Fig. 4-14 Place shop towels in front of the piston to cushion it. *What could happen if the piston were damaged?* (General Motors Corporation)

5. After removing the piston, remove the dust boot. Using a plastic or wooden stick, take the square-cut piston seal out of the cylinder bore. Be very careful not to damage the seal groove. Damage to the seal groove can lead to brake fluid leaks and damage to the new square-cut seal. **Fig. 4-15.**

6. Clean the caliper housing and pistons in brake cleaning fluid. Make sure the square-cut seal groove and cylinder bore are clean.

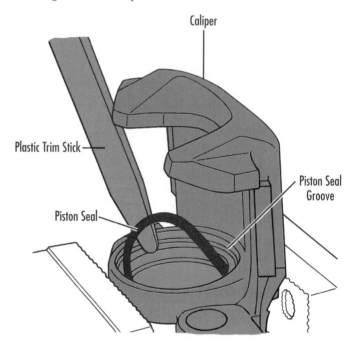

Caliper — Plastic Trim Stick — Piston Seal — Piston Seal Groove

Fig. 4-15 To avoid damage to the seal groove, remove the piston seal with a plastic or wooden stick. *What might happen if the seal groove is damaged?* (DaimlerChrysler)

7. Examine the piston for scratches, pits, or scoring. **Scores** are grooves or deep scratches on the piston's surface. Scoring is caused by debris or metal-to-metal contact. The piston surface rides on the square-cut seal. If the seal is damaged, it can cause a leak.

8. Inspect the caliper bore for pitting, corrosion, or roughness. Small imperfections can be honed out.

9. Hone the caliper's cylinder bore to remove imperfections. **Fig. 4-16.** If honing cannot repair the bore, replace the caliper.

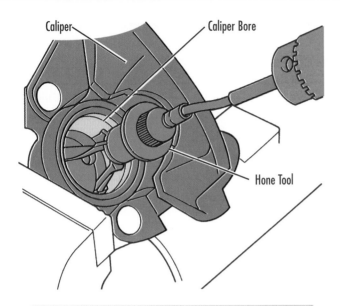

Caliper — Caliper Bore — Hone Tool

Fig. 4-16 Minor imperfections can be removed with light honing. *What solvent should be used to clean the bore after honing?* (DaimlerChrysler)

10. Clean the caliper, including the cylinder bore. Install the square-cut seal in the seal groove in the cylinder bore. Verify that the seal is not twisted or rolled. An improperly installed square-cut seal can cause the caliper to leak brake fluid.

11. Coat the piston with clean brake fluid. Assemble the new boot onto the piston.

12. Push the piston into the bore until it bottoms out in the bore.

TECH TIP Cleaning Parts. Never use an oil-based solvent to clean disc brake parts. Oil-based solvents will weaken rubber parts. Use only brake cleaning solvent, clean brake fluid, or denatured alcohol.

13. Position the boot in the caliper counterbore and install the boot in the counterbore using the appropriate tool (see vehicle's service manual).

14. Slide the new inboard pad into place. **Fig. 4-17.** Be sure to use the proper brake pad. Some brake pads are specific to the inboard or outboard side. Install the outboard pad.

15. Install the caliper on the vehicle. Follow the instructions in "Removing and Replacing Disc Brake Pads and Calipers (Floating or Sliding Calipers)."

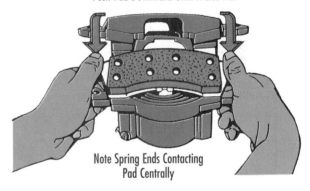

Push Pad Downward Until It Lies Flat
Note Spring Ends Contacting Pad Centrally

Fig. 4-17 Inboard pad installation. *Can either brake pad be used on the inboard side?* (General Motors Corporation)

SECTION 4 KNOWLEDGE CHECK

1. Before removing the caliper, what should you do at the master cylinder?
2. Why must you support the caliper with a piece of wire?
3. Why should you loosen the bleeder valves before servicing the caliper?
4. What tool should you use to remove the square-cut seal?
5. How do you remove minor imperfections from the caliper bore?

Section 5

Disc Brake Rotor Service

Rotor service follows removal of the caliper and pads. First you need to measure the rotor's thickness.

This determines if there is enough material to allow machining and still have the rotor within the vehicle manufacturer's specifications.

Measuring Brake Rotors

The disc brakes on a 1998 Pontiac Grand Prix need to be serviced. You suspect the rotors are the problem. You will measure the thickness of the rotor at six points. Then you can decide if the rotor is still thick enough to be machined. If it is thick enough, you will check the lateral runout. You will also use the measurements to decide if the thickness variation (parallelism) is within tolerance limits.

You find the following brake rotor information in the 1998 Grand Prix service manual:

Original Thickness	1.27"	32.200 mm
Minimum Wear Thickness	1.25"	31.700 mm
Discard Thickness	1.21"	30.700 mm
Maximum Lateral Runout	0.003"	0.080 mm
Parallelism Tolerance	0.0005"	0.013 mm
(Allowable Thickness Variation)		

You start with the left front brake. You use metric measurements for your calculations.

Lateral runout is 0.075 mm, and the six measurements for parallelism are 32.000 mm, 31.850 mm, 31.882 mm, 32.000 mm, 31.916 mm, and 31.997 mm.

You first make a quick mental estimate to see if the rotor should be discarded. Because the smallest measurement of the rotor's thickness is greater than the discard thickness (30.700 mm), you do not have to discard the rotor.

The lateral runout of 0.075 mm is less than 0.080 mm and is within tolerance limits. You now check the measurements for parallelism. You find that:

$$32.000 \text{ mm} - 31.850 \text{ mm} = 0.150 \text{ mm}$$

and $0.150 \text{ mm} > 0.013 \text{ mm}$

Because the variation is greater than allowed, you decide to machine, or refinish, the rotor.

Apply it!

Meets NATEF Mathematics Standards for mentally comparing tolerances and understanding conditional solutions.

You look at the right front brake and make the following measurements: 32.140 mm, 32.135 mm, 32.140 mm, 32.138 mm, 32.139 mm and 32.140 mm. Lateral runout is 0.015 mm.

❶ Make a quick mental estimate to see if the rotor will have to be discarded because it is too thin.

❷ If the rotor can be reused, determine if the lateral runout measurement is within tolerance limits.

❸ Determine if the rotor needs to be refinished.

Measuring Disc Brake Rotors

A rotor should be evaluated while still on the vehicle.

1. Use a micrometer to measure the thickness of the rotor in several places around the rotor.

2. Compare the smallest measurement to the minimum thickness allowed for the rotor. The minimum thickness for the rotor is stamped into the rotor. **Fig. 4-18.**

3. Compare the thickness measurements around the rotor. If the variation is more than 0.0005, the friction surfaces of the rotor are not parallel with each other. This is a test for parallelism.

Fig. 4-18 The rotor must be discarded if it is beyond the minimum thickness after machining. *What is the minimum thickness of this rotor?* (Terry Wild Studio)

4. Measure runout by first adjusting the wheel bearing to zero end play. Then mount a dial indicator. Rotate the rotor and measure the runout as it turns. Maximum allowable lateral runout is usually about 0.004 to 0.005 inch [0.10–0.13 mm]. If runout is excessive, replace the rotor.

5. If a rotor has smooth friction surfaces (not scored or damaged) and is within parallelism, runout, and minimum thickness specifications, it can remain in service. Readjust wheel bearing to specifications

6. Minor scoring, runout, or parallelism can be corrected by machining. Proceed with rotor machining only if the measurements indicate enough material thickness remains to safely allow machining.

7. If the rotor does not have enough thickness to allow machining, it must be replaced. A rotor that is machined below its minimum thickness can fail during use.

Removing Disc Brake Rotors

With the caliper removed, remove the rotor as follows:

1. Remove the dust cap, cotter pin, hex nut, and thrust washer holding the rotor to the spindle assembly. (On front-wheel-drive and rear disc brakes, there may be a retaining screw holding the rotor to the hub assembly. It may be possible to remove the screw and the rotor without removing the hub and bearings from the vehicle.)

2. Remove the rotor and wheel bearings (inner and outer) from the spindle. **Fig. 4-19.**

3. Remove the inner and outer wheel bearings from the rotor. Then clean the rotor with brake solvent. The rotor must be clean before you can inspect it for damage.

Machining Disc Brake Rotors

If the rotors have only surface scoring, they may not require additional service. However, if the scoring is deep, or if the rotor has lateral runout, lack of parallelism, or a taper, it must be machined.

You can perform machining using a standard brake lathe or an on-car brake lathe. An **on-car brake lathe** is a device used to perform brake rotor machining on the vehicle.

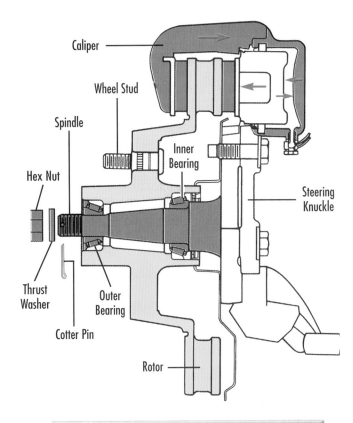

Fig. 4-19 Wheel bearings are positioned on each side of the rotor to support smooth rotation of the rotor. *Why do you need to clean the rotor after removal?* (DaimlerChrysler)

Machined rotors must have a nondirectional finish. If not applied, the brakes may "walk" during use and cause brake noise.

Most car manufacturers suggest using an on-car brake lathe. **Fig. 4-20.**

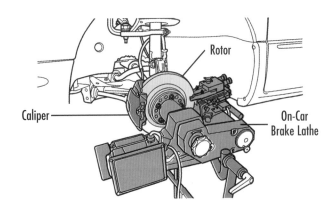

Fig. 4-20 Use an on-car brake lathe to restore the rotor to its original condition. *What is the advantage of using an on-car brake lathe?* (DaimlerChrysler)

AUTOMOTIVE SCIENCE EXCELLENCE

Calculating Stopping Distance

One day on a clean dry street, you make an abrupt stop for a traffic light. The anti-lock brake system does its job and the car stops quickly and safely. You wonder, "What is the relationship between speed and stopping distance?"

When you stop a vehicle, its kinetic energy of motion is converted to heat. Friction stops the car. The brake rotors become warm. The equation for finding the stopping distance, *x*, at a certain speed, *v*, is:

$$x = \frac{v^2}{(2g\mu_s)}$$

In the denominator, g is the acceleration of gravity (32 ft per second per second) and μ_s is the coefficient of static friction.

The coefficient of static friction is used because the stopping distance is the shortest when the tires do not skid. The coefficient of static friction between rubber and dry concrete is approximately 0.60.

The table shows the minimum stopping distance for various initial speeds. The stopping distance of a vehicle moving 20 mph [32 kph] is four times the distance for a vehicle moving 10 mph [16 kph]. Doubling the speed usually quadruples the stopping distance. You can understand how quickly stopping distance increases with speed by plotting a graph.

Apply it!

Meets NATEF Science Standards for using graphs and describing the role of friction in deceleration.

❶ Use a sheet of standard graph paper. Graph the information in the table. Label the horizontal axis "Speed in mph." Mark speed values of 0, 10, 20, 30, 40, 50 and 60 on the horizontal axis.

❷ Near the left edge of the page, label the vertical axis "Stopping Distance in Feet". Mark distance values of 0, 20, 40, 60, 80, 100, 120, and 140 on the vertical axis.

❸ Plot the values in the table on your graph.

STOPPING DISTANCE AND VEHICLE SPEED	
Speed (mph)	**Stopping Distance (ft)**
0	0.0
10	5.6
20	22.4
30	50.4
40	89.6
50	140.0
60	201.6

❹ Design a safe experiment to verify the values in the table.

On-Car Brake Lathe Perform disc-brake rotor machining with an on-car brake lathe as follows:

1. Attach the on-car brake lathe to the wheel brake assembly. Follow the instructions given in the on-car brake lathe manual.

2. Machine the rotor according to the manual.

3. After machining, the rotor must have a non-directional finish put on it before it is returned to service. Many brake lathes have the ability to apply a nondirectional finish. **Fig. 4-21.** Remove the on-car lathe.

TECH TIP Rotor Surface Finish. Carefully inspect the rotor's surface finish. If machining grooves are left in the surface, the pads will follow those grooves and "walk" until they can go no further. They will then make a noise as they snap back to their original position. The pads will also wear prematurely.

4. Thoroughly clean the wheel brake assembly using brake solvent or denatured alcohol.

5. Check the rotor for thickness. Use a micrometer to measure rotor thickness to determine that the rotor has not been machined beyond minimum thickness.

Fig. 4-21 After machining and before reinstallation, the rotor must receive a nondirectional finish. *What might happen if the finish is not correct?* (Wagner Brakes)

6. Replace any components that were removed in order to install the on-car lathe.

Standard Brake Lathe Perform brake rotor machining with a standard brake lathe as follows:

1. Mount the rotor on the brake lathe. Follow the instructions given in the brake lathe manual.

2. Machine the rotor according to the brake lathe manual.

3. Check the rotor for thickness. Use a micrometer to measure rotor thickness to determine that the rotor has not been machined beyond minimum thickness.

TECH TIP **Bearing Lubrication.** Clean the bearings and inspect them for damage. They should be free of pits and color. If a slight blue color is evident, install new bearings. If they are not damaged, lubricate them with wheel-bearing grease and install them in the rotor.

Installing Disc Brake Rotors

After you have machined and cleaned the rotor, reinstall it as follows:

1. Pack the bearings (if appropriate) with wheel-bearing grease. Install the inner bearing and a new seal in the rotor hub.

2. Place the rotor on the vehicle. (On front-wheel-drive vehicles or those with rear disc brakes, replace the retaining screw when installing the rotor on the hub.)

3. Install the outer bearing. Then install the thrust washer and spindle retaining nut. Snug the nut down.

4. Tighten and torque the spindle retaining nut and adjust bearing preload to the value given by the vehicle service manual.

5. Install the cotter pin and bend the tabs to hold it in place. Install the dust cap.

SECTION 5 KNOWLEDGE CHECK

❶ How can you determine the minimum allowable thickness for a rotor?

❷ What must you do if the rotor does not have enough thickness to allow machining?

❸ What device can you use to machine a rotor while it is still on the vehicle?

❹ Why must you put a nondirectional finish on the machined brake rotor?

❺ How would you determine the correct torque value for the spindle retaining nut?

CHAPTER 4 REVIEW

Key Points

Meets the following NATEF Standards for Brakes: diagnosis, service, and repair of disc brake calipers, brake pads, rotors, and master cylinder.

- Disc brakes use friction between the rotor and the brake pads to reduce the motion of the vehicle.
- Disc brake systems have four major parts: mounting brackets, calipers, brake pads, and rotors.
- Vented rotors are used on heavier vehicles. Nonvented rotors are used mainly on smaller vehicles.
- Disc brake problems can include noisy brakes, pulling, grabbing, and pedal pulsation.
- Brake pad inspection should include inspecting for uneven wear and signs of glazing, cracking, or other problems.
- Caliper service is performed if an inspection of the caliper indicates damage or signs of fluid leakage.
- A rotor's thickness should be measured while the rotor is on the vehicle.
- Disc brake rotor defects, such as grooves, can sometimes be removed by machining.

Review Questions

❶ How are disc brakes engaged?

❷ Describe the procedure(s) for diagnosing disc brake problems.

❸ What is the procedure for inspecting brake pads and hardware?

❹ What should you look for when inspecting calipers? What parts should be cleaned and lubricated?

❺ Describe the procedure for measuring a rotor with a micrometer.

❻ What rotor measurement should you compare with the manufacturer's specified minimum thickness allowed?

❼ Why are disc brake rotors machined?

❽ (Critical Thinking) Explain the purpose of the square-cut seal.

❾ (Critical Thinking) Explain why disc brakes are used on the front wheels of vehicles.

TECHNOLOGY FORECAST
FOR AUTOMOTIVE EXCELLENCE

Disc Brake Developments

Disc brakes are very popular with consumers and automakers. They are likely to remain popular in the future. Disc brakes are efficient and provide proven stopping power.

But there are refinements that can be made. One recent development is the aluminum metal matrix composite disc rotor. The material in this rotor is so hard that it requires a special diamond-tipped tool for machining!

This aluminum composite weighs half as much as traditional cast iron and sheds heat faster. It is also non-magnetic. Its hardness comes from the use of a 20 percent silicon-carbide material. As a result, these rotors cannot be machined with conventional equipment. Instead, the technician must use the diamond-tipped tool when machining.

Another new technology is called the low-pressure floating rotor full-contact foundation brake system. This design allows the brake's friction material to be applied to all 360 degrees of the rotor's contact surface.

Engineers working on the full-contact foundation brake system say it causes less noise, vibration and harshness (NVH) when the brakes are used. Other benefits include a lower rotor temperature and less rotor runout for more efficient braking.

Also, these brakes have seals on the driving hub that position the rotor in the ideal place. These seals help to reduce drag and NVH when the vehicle's brakes are not being applied.

AUTOMOTIVE EXCELLENCE
TEST PREP

Answering the following practice questions will help you prepare for the ASE certification tests.

1. Technician A says rear disc brakes require a mechanical parking brake. Technician B says parking brakes are not adjustable. Who is correct?
   - **a** Technician A.
   - **b** Technician B.
   - **c** Both Technician A and Technician B.
   - **d** Neither Technician A nor Technician B.

2. Technician A says sliding calipers are mounted with guide pins. Technician B says the floating calipers slide on machine surfaces. Who is correct?
   - **a** Technician A.
   - **b** Technician B.
   - **c** Both Technician A and Technician B.
   - **d** Neither Technician A nor Technician B.

3. Technician A says the square-cut seal provides a barrier to the brake fluid. Technician B says the square-cut seals keep grease from getting on the brakes. Who is correct?
   - **a** Technician A.
   - **b** Technician B.
   - **c** Both Technician A and Technician B.
   - **d** Neither Technician A nor Technician B.

4. What does a dust boot do in a caliper?
   - **a** Keeps pressurized brake fluid away from the piston.
   - **b** Keeps dirt and water out of the caliper's cylinder bore(s).
   - **c** Serves as a backup for the square-cut seal.
   - **d** None of the above.

5. What should be checked first when servicing disc brakes?
   - **a** The brake pad linings.
   - **b** The dust boot.
   - **c** The rotor.
   - **d** The square-cut seal.

6. Why are shims or brake compound placed on the back of disc brake pads?
   - **a** To attach the brake pads to the caliper piston.
   - **b** To help reduce disc brake noise.
   - **c** To hold the brake pads in the proper position.
   - **d** To stop brake fluid leaks at the caliper piston.

7. Why shouldn't petroleum products be used to clean brake parts and lube the piston?
   - **a** Oil will cause the rubber parts to swell.
   - **b** Petroleum products will contaminate the brake fluid.
   - **c** Both a and b.
   - **d** Neither a nor b.

8. Why should caliper pistons be inspected?
   - **a** To see if the bleeder valve will loosen easily.
   - **b** To see if the pressure differential valve is working properly.
   - **c** Both a and b.
   - **d** Neither a nor b.

9. Rotors must be checked for:
   - **a** Parallelism.
   - **b** Scoring.
   - **c** Glazing.
   - **d** All of the above.

10. Technician A says rotors must have a nondirectional finish on the machined surface. Technician B says the rotor must be discarded if it is beyond the minimum thickness. Who is correct?
   - **a** Technician A.
   - **b** Technician B.
   - **c** Both Technician A and Technician B.
   - **d** Neither Technician A nor Technician B.

CHAPTER 5

Diagnosing & Repairing Power Boosters

You'll Be Able To:

- Identify the two basic types of brake power boosters.
- Explain the operation of a vacuum assisted brake booster and a hydraulic assisted brake booster.
- Test the vacuum storage system.
- Perform the procedure for testing hydro-boost booster operation.
- Test the hydraulic pressure storage system.

Terms to Know:

accumulator
control valve
power piston
spool valve

The Problem

Ted telephones your service center saying his car is hard to stop. He says it requires an unusually hard effort on the brake pedal. It seems that every time he brakes, he really has to hit the brake pedal hard!

Ted says he didn't have this problem until he brought his car to your center less than two months ago. At that time, Ted's car received an oil change, lube, and filter, along with a brake inspection. You check your service records. They show that you serviced the car's brakes six weeks ago.

Your Challenge

As the service technician, you need to find answers to these questions:

1. When did this problem start? Did it start right after the brake service or more recently?

2. Can he bring the car safely to a stop using extra effort on the pedal?

3. Is steering effort normal?

Section 1

Power Boosters

Power brakes were introduced in the 1950s. They became very popular in the 1960s as vehicles became larger and heavier.

Disc brakes were also becoming more popular. They required more braking force than drum brakes. The added force is needed because drum brakes are self-energizing while disc brakes are not. *Power boosters* were designed to provide the additional force. Power boosters are also called power-assist units.

Power boosters are mounted on the engine side of the vehicle's bulkhead, between the brake pedal lever and the master cylinder. Power boosters supply the increased force needed by the brakes without requiring increased brake pedal pressure. There are two basic types of power boosters: vacuum assisted and hydraulic assisted.

Vacuum assisted designs can be single diaphragm or tandem (dual) diaphragm. Hydraulic assisted units can be mechanical-hydraulic (hydro-boost) or electrohydraulic (Powermaster™).

Vacuum Boosters

Vacuum boosters use vacuum to supply additional energy to boost brake application pressures. **Fig. 5-1.** This allows the driver to apply the brakes with less force on the brake pedal. The vacuum can be supplied by the engine's *intake manifold* or by a separate *vacuum pump,* or both.

Vacuum boosters rely on a pressure differential across a diaphragm to multiply the force coming from the brake pedal. The diaphragm is suspended between two chambers of the booster housing. When the system is at rest, the vacuum on each side of the diaphragm is the same. When the driver presses the brake pedal, atmospheric pressure enters the side nearest the pedal. This raises the pressure on this side of the diaphragm.

Because there is vacuum on the master cylinder side of the diaphragm, the diaphragm moves in the direction of the vacuum. It applies boosted pressure to the master cylinder's primary piston.

There are two types of vacuum boosters, the single diaphragm and the tandem (dual) diaphragm.

Single-Diaphragm Vacuum Boosters

The vacuum booster housing is a steel canister. It is separated into two chambers by a flexible rubber diaphragm. The diaphragm is positioned between two pushrods. One pushrod extends from the rear of the booster through the bulkhead and attaches to the brake pedal lever. A second pushrod extends out the front of the power booster and makes contact

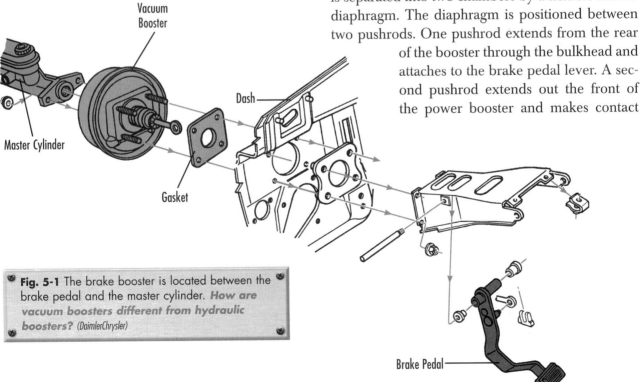

Fig. 5-1 The brake booster is located between the brake pedal and the master cylinder. *How are vacuum boosters different from hydraulic boosters?* (DaimlerChrysler)

Vacuum Booster

Master Cylinder

Gasket

Dash

Brake Pedal

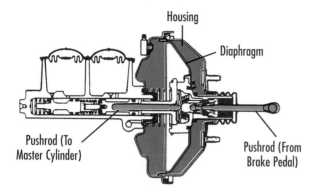

with the primary piston located inside the master cylinder. **Fig. 5-2.**

The pushrod connected to the brake pedal is attached to the power piston. The **power piston** is the interface between the front and rear pushrods. The power piston is attached to, or suspended from, the diaphragm. **Fig. 5-3.**

Inside the power piston is a control valve. The **control valve** is a valve that regulates the flow of atmospheric pressure and vacuum to the two separate chambers in the booster.

The control valve has two ports. One port (the *vacuum valve*) is normally open, allowing vacuum to enter the rear chamber. The second port (the *air valve*) is normally closed. It opens when the driver pushes the brake pedal. It allows air, at atmospheric pressure, to enter the rear chamber. At the same time, the vacuum valve to the rear chamber closes. This creates a pressure differential between the two chambers. The high-pressure side forces the diaphragm and power piston to move the pushrod toward the master cylinder. When the driver

stops pushing the pedal, both valves close. This is the "hold" position.

When the driver releases the pedal, the air valve closes and the vacuum valve opens. The vacuum equalizes on both sides of the diaphragm. A spring on the front of the diaphragm helps push the brake pedal back to its rest position.

When brakes are applied, air rushing into the rear chamber would normally make an undesirable hissing noise. Filters installed on the inlet side of the booster prevent this. The filters also keep dirt out of the booster.

A rubber reaction disc and a reaction piston are located at the end of the brake pedal pushrod. Together, the reaction piston and the reaction disc provide brake "feel" feedback to the driver. **Fig. 5-4(a).**

As the driver presses on the brake pedal, the reaction disc compresses slightly. At the same time, the master-cylinder pushrod presses on the reaction piston in the opposite direction. Because there is an equal and opposite reaction for every action, the driver gets a good pedal feel.

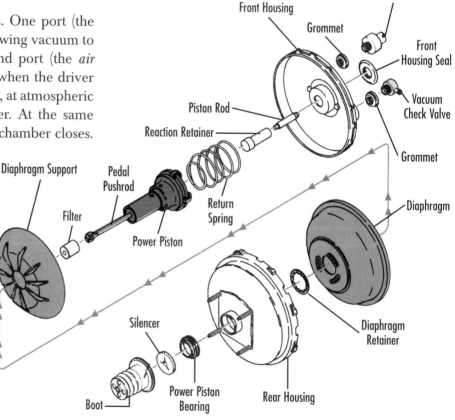

Tandem (Dual) Diaphragm Vacuum Boosters

Some vehicles are equipped with tandem diaphragm vacuum boosters. This type of booster has two diaphragms instead of one. The tandem diaphragm has the following advantages:

- It doubles the surface area to provide more boost. This is important for some heavier vehicles.
- The same amount of diaphragm area (and boost) can be achieved with a smaller diameter booster housing. Or, more boost can be achieved with the same size booster. This is important for small cars and those with low hood lines. **Fig. 5-4(b)**.

Vacuum Supply

A gasoline engine is basically a large air pump. It draws in air through the carburetor or throttle body. There the engine compresses the air, mixes it with fuel, burns it in the cylinders, and pumps it out the exhaust pipes.

The restriction to air drawn in through the throttle body (or carburetor) creates a vacuum in the intake manifold. This vacuum can be used for numerous purposes. One is to supply vacuum to the power brake booster.

Some vehicles, notably those with diesel engines, need a separate vacuum pump. This is because diesel engines run unthrottled and do not create a vacuum in the manifold. The vacuum pump can be driven by a fan belt or mechanically from the engine. The camshaft usually drives them.

Manifold vacuum is highest at idle and lowest at wide-open throttle. To prevent these changes

from affecting the amount of vacuum available to the booster, vacuum is often stored in a reservoir. **Fig. 5-5**.

A check valve is installed between the manifold and reservoir (or in the reservoir). It keeps the vacuum from flowing in the opposite direction under low manifold vacuum conditions.

The vacuum check valve at the booster maintains vacuum inside the booster housing. Vacuum storage is important because, should the engine stall, power boost remains available for at least one brake application. It may be enough for two or three applications.

The brakes will still work even after all vacuum has been used. The difference is that the driver will have to push harder on the brake pedal.

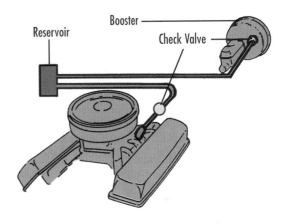

Fig. 5-5 The engine supplies vacuum for the booster. A reservoir holds a small supply of vacuum for emergencies. *When is manifold vacuum highest—when the engine is at idle or at wide-open throttle?* (Ford Motor Company)

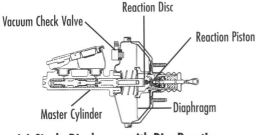

(a) Single Diaphragm with Disc Reaction

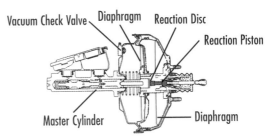

(b) Tandem Diaphragm with Disc Reaction

Fig. 5-4 With a tandem diaphragm, more boost can be achieved with the same size booster assembly or the same boost can be achieved with a smaller booster assembly. *What is the advantage of using tandem vacuum boosters on small cars with low hood lines?* (General Motors Corporation)

How Levers Multiply Force

On a sunny day, sixteen-year-old Sarah is traveling down the highway at cruising speed. She puts her foot on the brake pedal to stop her car. She is barely 5 ft [1.5 m] tall and weighs 105 lb [48 kg]. The car weighs thousands of pounds. What happens? The car slows to a controlled stop. How can such a small person apply enough force to stop a heavy vehicle?

The driver receives a mechanical assist and a power assist from the brake system. The brake pedal attaches to a lever that provides the mechanical assist. Let's investigate the mechanical advantage of a lever.

Apply it!

Demonstrating Mechanical Advantage

Meets NATEF Science
Standards for levers and force.

Materials and Equipment
- Yardstick
- 1.0 or 1.5 liter unopened bottle of water
- Two short lengths of string or fishing line
- Spring scale
- C-clamp

❶ Tie one end of a string tightly around the top of the water bottle. Tie the other end into a loop. Hang the loop on the scale. Record the weight of the bottle as F_1.

❷ Set up the experiment as shown below. Clamp one end of the yardstick to a table. Slide the loop of the string attached to the water bottle along the yardstick to the middle of the stick. Keep the bottle supported from below until Step 5.

❸ Record the distance in inches from the edge of the table to the bottle's attachment point, as L_1.

❹ Tie the other string into a loop. Use the loop to attach the scale near the end of the yardstick as shown.

❺ Pull up on the scale until the yardstick just barely supports the hanging bottle. Then measure on the scale the force needed to just support the water bottle. Record this force as F_2.

❻ Measure and record the distance from the end of the table to the point where the scale is attached. Record this as L_2.

❼ Calculate the ratio of $F_1:F_2$. This is the mechanical advantage of the lever.

❽ Calculate the ratio of $L_2:L_1$. How does this ratio compare to the mechanical advantage of the previous step?

Results and Analysis You should have discovered that the mechanical advantage of a lever is the same as the ratio of the distances, or $F_1:F_2 = L_2:L_1$.

Check this by sliding the water bottle to several other points along the yardstick and repeating Steps 5–8. If you slide the bottle toward the table, the mechanical advantage increases. Half the force is required. Slide the bottle toward the spring scale and the mechanical advantage decreases. The scale will show that the force, F_2, increases.

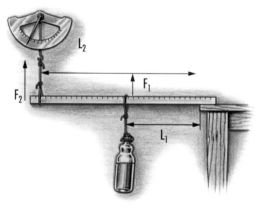

SECTION 1 KNOWLEDGE CHECK

① Why are power boosters necessary for disc brake systems?

② What is the source of vacuum for the booster?

③ What does the reaction piston do?

④ What causes the diaphragm to move when the pedal is depressed?

⑤ Why is there a check valve in the booster?

Section 2

Hydraulic Boosters

In the early days of emission controls, engines were designed with *camshafts* that had long overlaps. Coupled with the retarded ignition timing specifications, these engines developed little useful vacuum. Because there was concern that vacuum boosters may be inadequate, another source of brake boost was developed—the *hydraulic booster.*

Unlike vacuum boost, hydraulic boost is constant even if the engine is under load. This is because the boost pressure is obtained from the power steering pump. The power steering pump is driven by an engine belt and delivers a constant source of pressure. These booster systems are known as mechanical-hydraulic, or hydro-boost. They use mechanical and hydraulic systems in combination. More recent hydraulic booster systems are powered by electric pumps. These are known as electrohydraulic, or Powermaster, systems.

The Bendix hydro-boost system was first used on GM cars. The booster is located between the master cylinder and bulkhead (similar to a vacuum booster). The booster receives high pressure fluid from the power steering pump by way of a high pressure hose. Two more lines lead from the booster. One high pressure line runs to the steering gear. One low pressure line runs back to the power steering reservoir. **Fig. 5-6.**

Hydro-Boost Booster

The main components of the hydro-boost system are the lever, spool valve, master cylinder pushrod, input rod, and power piston.

The **spool valve** is a valve that shuttles back and forth to open and close ports for the pressurized

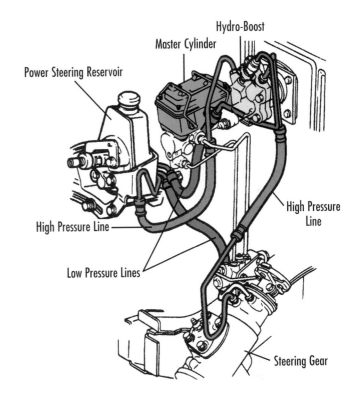

Fig. 5-6 The hydro-boost system uses hydraulic pressure from the power-steering pump to boost brake pedal application force. *Where does the low pressure line from the hydro-boost unit lead?* (General Motors Corporation)

power steering fluid. It resembles a spool that carries sewing thread.

When the driver applies the brakes, the input rod moves a lever forward against the power piston. It also moves the spool valve forward, closing the return port to the power steering pump.

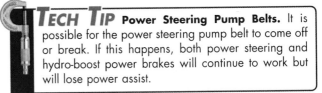

TECH TIP Power Steering Pump Belts. It is possible for the power steering pump belt to come off or break. If this happens, both power steering and hydro-boost power brakes will continue to work but will lose power assist.

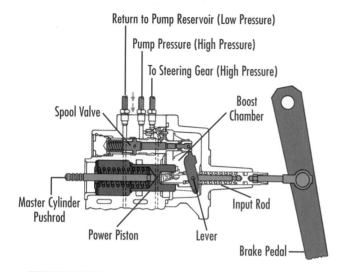

Fig. 5-7. As the pressure increases, the spool valve is pushed forward. As it moves, high pressure fluid flows through the spool valve into the boost chamber. The power piston pulls the lever forward with it. This increases the brake application force.

When the driver releases the brakes, a spring opens the spool valve. The fluid goes from the power steering pump to the steering gear. **Fig. 5-8.**

The hydro-boost system includes an accumulator. This provides pressure to the booster in case the engine stalls or the power steering belt breaks.

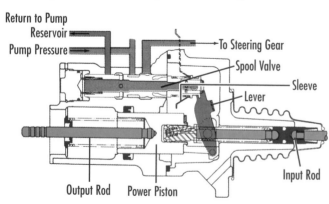

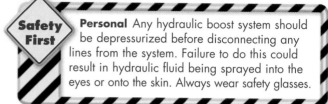

An **accumulator** is a device that stores fluid under pressure. A strong spring or pressurized nitrogen gas provides the pressure. The accumulator stores enough pressure for one or two brake applications. After that the brakes will behave like brakes without power assist. Power steering fluid should be changed routinely on vehicles with hydro-boost systems.

Powermaster Booster

The Powermaster booster operates very much like the hydro-boost system, with one major exception. Instead of using the power steering pump to supply hydraulic pressure, it uses a separate hydraulic pump driven by an electric motor. **Fig. 5-9.**

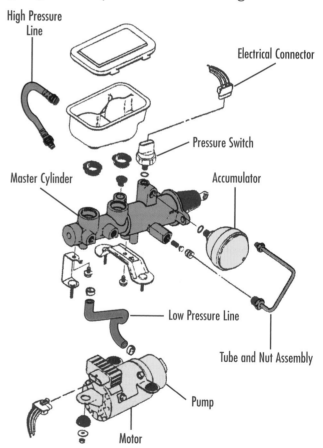

Taking Notes

Taking notes is a life skill. The ability to take good notes can be useful to you in many situations. You can use this skill when listening to a person describe how to do a procedure. You can take notes when you want to buy an item. For example, you can organize information from magazines, catalogs, and advertisements. Your notes will help you compare prices and features and make a wise decision.

A successful technician learns how to take notes to identify new parts or systems. Good notes will help you learn a procedure you are doing for the first time. There is often a blank page for notes in service manuals. In this text, you will read about new components and systems. Take notes on the material so you have a reference for study.

Often, important terms are printed in *italics* or **bold-faced** type. These print types signal you to take note of them. Using the titles of chapters and sections is a good way to organize your notes.

Apply it!

Meets NATEF Communications Standards for study habits and note-taking.

❶ Read through this chapter.

❷ As you read, look for *italicized* or **bold-faced** words and phrases.

❸ Write down and define or explain each term.

❹ Make these terms part of your notes for this chapter.

Powermaster boosters use brake fluid from the master cylinder reservoir rather than power steering fluid. There are several advantages to Powermaster boosters. They are less complex, smaller, and the pump needs to run only when there is a demand. Powermaster boosters are also well suited for use on anti-lock brake systems.

Like hydro-boost systems, Powermaster systems have an accumulator to store fluid under pressure. Inside the accumulator, a diaphragm separates the brake fluid from the section filled with high pressure nitrogen.

Safety First

Personal Power steering fluid should be added only to the power steering pump. Hydraulic brake fluid should be added to the master cylinder as approved by the manufacturer. Adding power steering fluid to the master cylinder is a very dangerous mistake that could lead to total brake failure.

Materials The Powermaster booster derives pressure to provide boost from the accumulator. A pressure switch turns the electric pump on and off at proper pressure levels. This keeps the accumulator charged.

SECTION 2 KNOWLEDGE CHECK

❶ From what device does the hydro-boost booster get its hydraulic pressure?

❷ What valve controls where hydraulic pressure is distributed in a hydro-boost booster?

❸ What component provides hydraulic pressure in a hydro-boost booster if the engine stalls?

❹ What device provides hydraulic pressure in a Powermaster booster?

❺ What keeps the accumulator pressurized in a Powermaster booster?

Section 3

Brake Booster Trouble Diagnosis

Brake boosters are generally not serviced. When they stop working, they are replaced as a unit. No field-serviceable parts are available, except for the vacuum check valve.

Diagnosing Vacuum Boosters

Generally, the only symptom of a vacuum booster problem is a hard brake pedal. This is when the driver must use extra effort to stop the car.

Testing Vacuum Booster Operation To test vacuum booster operation, do the following:

1. Start the engine and let it idle. Apply and release the brakes several times.

2. Turn the engine off. Apply and release the brake pedal until the pedal gets hard to push. Press down on the brake pedal and start the engine. The pedal should move down a small amount. If it does not, the booster is not working.

There are several possible causes. To isolate the failure, inspect the:
- Vacuum check valve(s) for restrictions or leaks.
- Vacuum supply lines and reservoirs (if present) for leaks.
- Vacuum supply hose for soft spots that could collapse during use and restrict flow.

If the vacuum check valves, reservoir (if present), or the vacuum supply hose are defective, replace them. Otherwise, you must replace the entire vacuum booster.

Testing Vacuum Storage System Operation To test the vacuum storage system, start the engine and run it at a medium idle. Turn off the engine and wait for about five minutes. Press down on the brake pedal. It should provide one or two soft applications before it gets hard to push. If it does, the vacuum storage system is operating normally. **Fig. 5-10.**

If the pedal is hard to push during the first and second applications of the test, vacuum is not being stored in the vacuum system. If this is the case the check valve(s), hose(s), reservoir if used, or booster is leaking.

Diagnosing Hydro-Boost Boosters

As with vacuum boosters, the only symptom of a hydro-boost booster problem is a hard brake pedal.

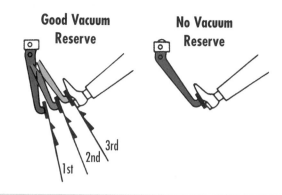

Good Vacuum Reserve   No Vacuum Reserve

1st 2nd 3rd

Fig. 5-10 Testing the vacuum storage system. *How many soft applications should you get after pressing the brake pedal?* (General Motors Corporation)

Testing Hydro-Boost Booster Operation To check hydro-boost booster operation, do the following:

1. Start the engine. With the engine running, apply the brakes several times.

2. Turn the engine off. Apply the brakes several more times until the pedal gets hard to push. Keep your foot pressed down on the pedal and start the engine. If the booster is working properly, the pedal will fall a small amount and then push back against your foot. If the pedal does not fall, the booster is not working.

If a hydro-boost booster is not working, check the power steering pump belt and fluid level. Inspect all hoses and connections for fluid leaks. If everything checks out and power steering has normal power assist, the booster is defective. You must replace it as a unit.

Testing Hydraulic Pressure Storage System Operation To test the hydraulic pressure storage system, do the following:

1. Start the engine and let it idle.

2. Turn the steering wheel from lock to lock several times. Apply the brakes a few times. This allows the accumulator to charge with fluid.

3. Shut off the engine and allow the car to sit. After about 30 minutes, apply the brakes. If the system is working normally and the accumulator is holding a charge, you should get one or two assisted brake applications before the pedal gets hard to push. If the pedal is hard to push on the first application, the accumulator is leaking.

Brake Pedal Ratios

You have learned that the brake hydraulic system increases both pressure and braking force. The amount of increase depends on the ratio of the areas of the master cylinder piston and the output piston in the brakes. This chapter covers power boosters, which also increase the pressure and braking forces.

There is another force multiplier–the brake pedal ratio. In the figures below, the brake pedal ratios are B to A or B/A. The brake pedal ratio is the ratio of the lengths of the lever arms on the pedal assembly. The brake pedal ratio for **Fig. A** is 3 to 1 or 9″ divided by 3″. The brake pedal ratio for Fig. B is 5 to 1 or 10″ divided by 2″.

The mechanical advantage supplied by the brake pedal assembly is determined by the brake pedal ratio. The ratio chosen depends on whether the brake system is equipped with a power booster.

If a booster is present, the pedal assembly does not need to increase the force as much.

A driver applies a force of 50 lb [23 kg] to the unboosted brake pedal. **Fig. B.** The force on the master cylinder pushrod will be:

$$50 \text{ lb} \times \frac{10''}{2''} = 250 \text{ lb}$$

Apply it!

Meets NATEF Mathematics and Science Standards for levers and ratios.

❶ The driver applies the same 50 lb. of force to the pedal of the power brake system. **Fig. A.** What is the force on the master cylinder pushrod?

❷ How much additional force must be provided by the booster to create the same brake input force as in the unboosted system?

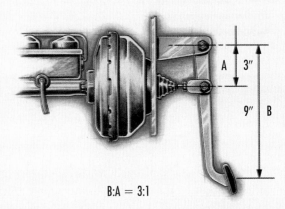

B:A = 3:1

Fig. A Brake pedal ratio for a power assisted brake.

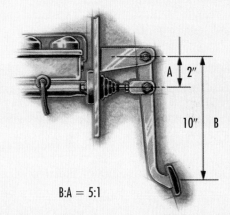

B:A = 5:1

Fig. B Brake pedal ratio for a brake without power assist.

SECTION 3 KNOWLEDGE CHECK

❶ What part on a vacuum assisted power booster is serviceable?

❷ How can you tell if you have a power booster problem?

❸ What should you check for if the vacuum storage system fails?

❹ Should the pedal fall when you test a hydro-boost system?

❺ What should you check if the hydro-booster is not working?

CHAPTER 5 REVIEW

Key Points

Meets the following NATEF Standards for Brakes: diagnosis and repair of power boosters.

- Power assist systems supply the increased forces needed by brakes without requiring increased brake pedal pressure.
- Vacuum boosters use a pressure differential across a diaphragm to boost brake pedal force.
- Hydraulic boosters control the flow of pressurized fluid to boost brake pedal force.
- The main components of the hydro-boost system are the lever, spool valve, master cylinder pushrod, input rod, and piston.
- The vacuum storage system should store enough vacuum for one or two brake applications with the engine turned off.
- A hard brake pedal when the vehicle's engine is running may indicate hydro-boost booster failure.
- Applying the brakes and turning the wheel lock to lock during the hydraulic pressure storage test allows the accumulator to charge with fluid.

Review Questions

1. Identify the two most common types of power boosters.
2. Explain the operation of a vacuum assisted brake booster.
3. Why do vacuum boosters rely on a pressure differential across the diaphragm?
4. Explain the operation of a hydraulic assisted brake booster.
5. To test the vacuum storage system, a technician should start the engine, run it at medium idle, and then turn off the engine and wait about five minutes. What happens after this?
6. What is the procedure for testing hydro-boost booster operation?
7. Explain the procedure for testing the hydraulic pressure storage system.
8. **Critical Thinking** Why do automobiles need power boosters?
9. **Critical Thinking** Why do diesel vehicles need a separate vacuum pump?

TECHNOLOGY FORECAST
FOR AUTOMOTIVE EXCELLENCE

Electronic Braking Systems

With the development of electrohydraulic and electronic braking systems, power assist brakes will eventually become a thing of the past. However, power assist brakes will not go away overnight. Engineers must prove that the new braking systems are safe. Automakers must then convince buyers of their safety.

With electrohydraulic brakes, the hydraulic fluid is stored in a reservoir at each wheel. Small motors are used to apply pressure to the hydraulic fluid when the brake pedal is pressed. The resulting force stops the vehicle. In a fully electronic braking system, there is no need for hydraulic fluid. Instead, an electric motor with a small gearbox applies the braking force.

Automakers have already introduced electrohydraulic brakes. One design under consideration has no mechanical linkage between the brake pedal and the master cylinder. Pedal position sensors and wires signal the vehicle computer when the brakes are applied.

There are several benefits to this setup. Brake pedal feel can be fine-tuned to the needs of a specific vehicle type or an individual's driving style. Also, the feedback that is felt when the anti-lock brakes (ABS) are activated can be removed. This feedback sometimes startles drivers. Because of this, a driver may not use the ABS properly.

AUTOMOTIVE EXCELLENCE
TEST PREP

Answering the following practice questions will help you prepare for the ASE certification tests.

1. Technician A says some vacuum boosters have a single diaphragm. Technician B says tandem vacuum boosters have two diaphragms. Who is correct?
- **a** Technician A.
- **b** Technician B.
- **c** Both Technician A and Technician B.
- **d** Neither Technician A nor Technician B.

2. A vacuum booster housing is a steel canister separated by:
- **a** The power piston.
- **b** A rubber diaphragm.
- **c** A spool valve.
- **d** A reaction disc.

3. What does the reaction disc do in a vacuum booster?
- **a** Helps the driver with reaction time.
- **b** Slows down reaction time to prevent brake lock-up.
- **c** Provides brake "feel" feedback to the driver.
- **d** Is not a part of the vacuum booster.

4. What types of boosters can be used in a car with a diesel engine?
- **a** A hydraulic brake booster.
- **b** A vacuum brake booster with a vacuum pump.
- **c** Both a and b.
- **d** Neither a nor b.

5. Which of these is not a part of a vacuum booster system?
- **a** Diaphragm.
- **b** Brake pedal pushrod.
- **c** Vacuum check valve.
- **d** Accumulator.

6. The pressurized fluid used in a hydro-boost booster system is:
- **a** Power steering fluid.
- **b** Automatic transmission fluid.
- **c** DOT 3 brake fluid.
- **d** Petroleum-based fluid.

7. Technician A says the accumulator on a hydro-boost system contains a spring. Technician B says the accumulator is gas filled. Who is correct?
- **a** Technician A.
- **b** Technician B.
- **c** Both Technician A and Technician B.
- **d** Neither Technician A nor Technician B.

8. Technician A says the Powermaster uses brake fluid for hydraulic power. Technician B says the master cylinder is partitioned to accept both brake fluid and hydraulic oil. Who is correct?
- **a** Technician A.
- **b** Technician B.
- **c** Both Technician A and Technician B.
- **d** Neither Technician A nor Technician B.

9. Technician A says power steering fluid should be changed routinely in a hydro-boost system. Technician B says a special fluid is used and never needs to be changed. Who is correct?
- **a** Technician A.
- **b** Technician B.
- **c** Both Technician A and Technician B.
- **d** Neither Technician A nor Technician B.

10. A car equipped with a hydro-boost booster has a very hard brake pedal with the engine running. What should be checked?
- **a** Vacuum supply to the booster.
- **b** Electrical supply to the pump.
- **c** The power steering pump belt.
- **d** The booster check valve.

CHAPTER 6

Diagnosing & Repairing Parking Brakes

You'll Be Able To:

- ⊗ Identify the two kinds of parking brake systems.
- ⊗ List the components of a parking brake system.
- ⊗ Inspect the parking brake warning light system.
- ⊗ Check parking brake operation and adjust as needed.
- ⊗ Inspect parking brake cables and components for defects and service as needed.

Terms to Know:

auxiliary parking brake

integral parking brake

parking brake equalizer

The Problem

Todd Parker has a car with an automatic transmission. He rarely uses the parking brake. Todd's friend, Antonio, asked to borrow his car. Antonio drove the car to San Francisco for a long weekend.

When Antonio returned Todd's car, Todd thought something was wrong. On his commute to work, he smelled something burning. Todd also found that he has to press the accelerator further to keep his car moving at the same speed. He says it feels as though the brakes are partially applied or dragging.

Your Challenge

As the service technician, you need to find answers to these questions:

❶ Is the brake warning light illuminated?

❷ Has Todd had the parking brake mechanism serviced recently?

❸ Did Antonio use the parking brake when he borrowed the car?

Section 1

Parking Brake Basics

The parking brake is designed to hold a parked vehicle stationary. It relies on static friction to keep the wheels from turning. In some cases, a parking brake uses the wheel brake components of the service brake system. In other cases, the parking brake is totally separate.

Parking Brake Components

Unlike service brakes, parking brakes are not operated by a hydraulic system. Parking brakes have a mechanical operating system.

Braking Mechanisms   Most parking brakes use the existing rear drum brake shoes, or rear disc pads, of the vehicle's wheel brake mechanism. Some vehicles, such as the Corvette, use a separate set of drum brakes. In these vehicles the parking brake shoes are applied independently of the rear disc brakes.

The two types of parking brake systems are integral parking brakes and auxiliary parking brakes. The **integral parking brake** is a parking brake that uses the same brake shoes or pads as the service brakes. The **auxiliary parking brake** is a parking brake that uses a separate set of brake linings.

Actuating Mechanisms   Two kinds of actuating mechanisms are used on parking brakes. Some systems use pedals. Other systems use hand levers. Most parking brake systems use cables to carry motion from the brake pedal or hand lever to the braking mechanism.

Brake Warning Light   All parking brakes have a warning light switch that closes when the parking brakes are applied. **Fig. 6-1.** When the switch closes, a red light on the instrument panel glows. It is usually the same light that glows if there is a service brake failure. The light warns the driver that the parking brakes are on.

If the brake warning light stays on when the parking brake hand lever or pedal is not applied, verify that the lever or pedal is returning fully to the released position. Also check the switch at the parking brake hand lever or pedal. Adjust and service as needed. If this light will not turn on and

Safety First   **Personal** Parking brakes are often mistakenly called emergency brakes. Applying the parking brakes will slow the vehicle. However, this will not bring the vehicle to a safe stop in a short distance.

it is not burned out, check and repair the electrical circuit. Refer to the vehicle service manual.

Parking Brake Construction and Operation

Three basic types of parking brake systems are commonly used on vehicles:
- Rear drum integral parking brake assemblies.
- Rear disc integral parking brake assemblies.
- Auxiliary parking brake assemblies.

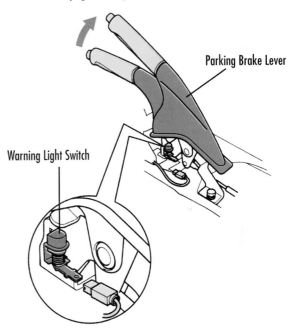

Parking Brake Lever

Warning Light Switch

Fig. 6-1 Parking brake warning light switch on a vehicle with a hand lever brake. *What is the purpose of the warning light switch?* (American Honda Motor Company)

Rear Drum Integral Parking Brakes

Rear drum brake assemblies are integral systems. This means they use the same brake shoes and drums as the service brakes.

This is the most common system. It is also very reliable. Drum brakes make good parking brakes because of their superb static friction. The service brake shoes double as parking brakes.

Rear Drum Integral Parking Brake Construction Rear drum brakes with integrated parking brake capability have additional brake hardware. This hardware consists of the parking brake cable, parking brake lever, and parking brake link. **Fig. 6-2.**

Rear Drum Integral Parking Brake Operation The parking brake is activated when the driver presses the parking brake pedal or pulls the parking brake hand lever. This pulls the parking brake cable, activating the mechanical parking brake components in the rear wheel brake assembly.

When the parking brake hand lever or pedal pulls the parking brake cable, it pivots the brake lever. This forces the rear brake shoe into contact with the drum. At the same time, the parking brake link pushes forward. This forces the front brake shoe

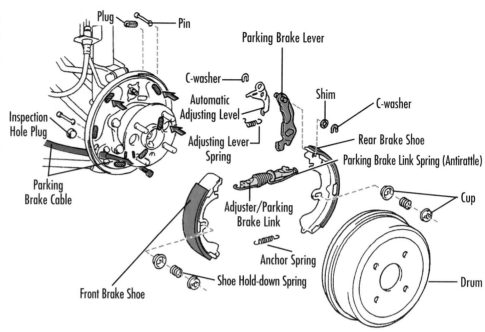

Fig. 6-2 Parking brake hardware is an integral part of the drum-style wheel brake mechanism. *Which major components of the wheel brake mechanism shown are specific to the parking brake system?*

into contact with the drum, applying the parking brakes. **Fig. 6-3.**

Because most rear drum brakes have the duo-servo design, any forward movement of the car increases the application effort of the brakes. In reverse, however, rotation of the drum tends to decrease the holding power of the parking brakes.

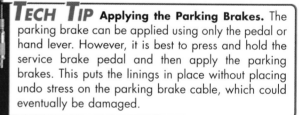

TECH TIP Applying the Parking Brakes. The parking brake can be applied using only the pedal or hand lever. However, it is best to press and hold the service brake pedal and then apply the parking brakes. This puts the linings in place without placing undo stress on the parking brake cable, which could eventually be damaged.

Rear Disc Integral Parking Brakes

Cars equipped with rear disc brakes and floating calipers often use the service brake pads for the parking brake function. Three basic types of mechanisms are used to mechanically actuate the parking brakes:

- Screw-nut-cone.
- Ball-ramp.
- Cam-rod.

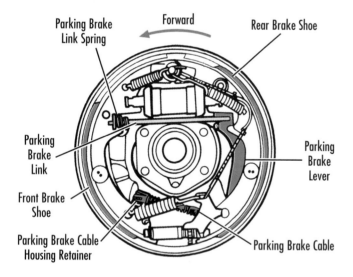

Fig. 6-3 A duo-servo drum brake. Applying the parking brake transmits force mechanically through the parking brake cable. This mechanical force actuates the parking brake lever. This pivots and forces the front and rear brake shoes against the drum. *Which component of the parking brake hardware transfers motion from the brake lever to the forward brake shoe?* (Ford Motor Company)

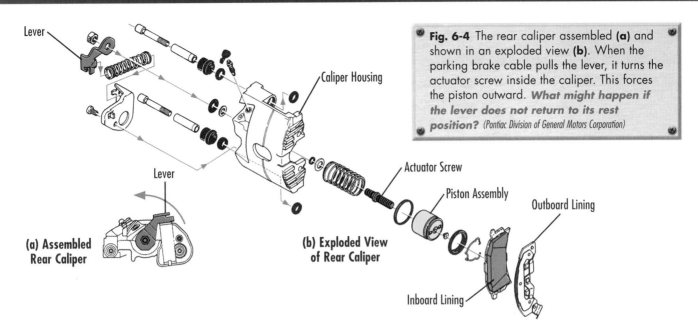

Fig. 6-4 The rear caliper assembled **(a)** and shown in an exploded view **(b)**. When the parking brake cable pulls the lever, it turns the actuator screw inside the caliper. This forces the piston outward. *What might happen if the lever does not return to its rest position?* (Pontiac Division of General Motors Corporation)

Lever

Caliper Housing

Lever

Actuator Screw

Piston Assembly

Outboard Lining

Inboard Lining

(a) Assembled Rear Caliper

(b) Exploded View of Rear Caliper

TECH TIP   Use the Parking Brake. Advise the vehicle owner to use the parking brake regularly. This is necessary to keep the brakes properly adjusted. It will also help to keep the parking brake mechanism operating correctly. When not used for long periods of time, parts may seize and cause the parking brake to not properly apply or release.

Screw-Nut-Cone Parking Brakes   Screw-nut-cone parking brakes are common on General Motors vehicles. This parking brake mechanism is also used to keep the rear disc brakes adjusted.

When the driver applies the parking brake, the cable pulls a lever on the outside of the caliper housing. The lever turns the actuator screw inside the caliper. A nut follows the threads of the screw and jams into the clutch surface of the caliper piston. **Fig. 6-4.**

The piston cannot turn because it is locked into place by tabs on the inboard brake pad. Because the piston cannot turn, it is forced to move outward. As it moves outward, it forces the piston against the brake pad, which contacts the rotor. The parking brake is released by removing tension from the parking brake cable.

After servicing screw-nut-cone rear calipers, the technician must cycle the parking brakes many times. This is the only way to move the nut along the screw threads sufficiently to set the initial brake adjustment. The screw-nut-cone parking brake piston assembly is nonserviceable.

Ball-Ramp Parking Brakes   The ball-ramp parking brake design is commonly used on Ford vehicles. Ball-ramp mechanisms are one-way clutches.

When the parking brake is applied, the cable pulls a lever on the outside of the caliper. The lever is attached to the operating shaft, which has an end plate. **Fig. 6-5.**

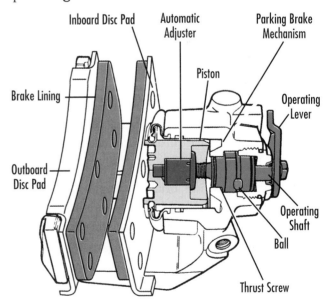

Inboard Disc Pad

Automatic Adjuster

Parking Brake Mechanism

Piston

Brake Lining

Operating Lever

Outboard Disc Pad

Operating Shaft

Ball

Thrust Screw

Fig. 6-5 A ball-ramp caliper assembly. When the operating lever moves, the balls ride up ramps. This forces the piston out and against the inboard disc pad. *What happens when the parking brake mechanism turns the nut inside the caliper piston?* (Ford Motor Company)

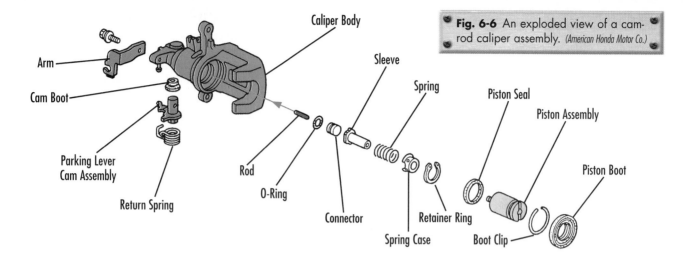

Fig. 6-6 An exploded view of a cam-rod caliper assembly. (American Honda Motor Co.)

As the lever turns, three steel balls between the operating shaft plate and thrust screw plate ride up ramps. They try to push the two plates apart.

Because the operating shaft face is against the caliper housing, it cannot move. As a result, the thrust screw must move. When it does, it turns the automatic adjuster nut inside the piston. This forces the caliper piston outward. As the piston moves outward, it pushes the brake pad against the rotor and applies the brakes.

Brakes are adjusted each time the service brakes are applied. Cycling the parking brake several times following a brake job, however, is necessary to make the initial adjustment.

Cam-Rod Parking Brakes *The cam-rod parking brake actuator* is also called the eccentric shaft and rod design. The operation is the same. Cam-rod parking brakes are more common on Asian imports. **Fig. 6-6.**

When the parking brake is applied, the cable pulls an arm (lever) located on the outside of the caliper. As the arm moves, it turns the parking lever cam assembly or eccentric shaft. The cam assembly or eccentric shaft pushes on a rod positioned between the cam and the actuator. As the rod pushes on the actuator, the actuator pushes on the piston. The piston then pushes the brake pads against the rotor.

Springs push the rod back when the parking brake is released. Like the screw-nut-cone design, the brakes are adjusted whenever the parking brakes are applied.

Auxiliary Parking Brakes

A growing number of vehicles with rear disc brakes use auxiliary drum brakes for parking only. This system used to be most common on systems with fixed calipers. However, it is now being used on systems with floating calipers as well.

Auxiliary Parking Brake Construction On a caliper system with an auxiliary parking brake, the rotor is shaped like a hat. The service brakes contact the brim of the hat. The parking brake is applied to the inside crown of the hat. **Fig. 6-7.**

The parking brake is simply a drum brake with a set of small shoes. The small shoes have soft linings with a high coefficient of friction. Even so, the soft linings will not wear out if properly used. This is because they provide only static friction.

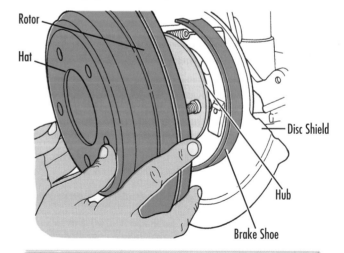

Fig. 6-7 The auxiliary drum parking brake consists of small brake shoes that act on the "hat" section of the rear brake rotor. *Why do the soft shoe linings not wear out if properly used?* (DaimlerChrysler)

Reading Exploded Views

Often, a technician has to "look" into a vehicle's system. A vehicle has many parts. They may be stacked one on top of another, as in an engine. Many small parts cannot be seen inside large components or assemblies. Even some large components cannot be seen clearly on the vehicle.

Service manuals and textbooks provide pictures that allow the technician to see into an assembled system. These pictures are called *exploded views.* Exploded view diagrams can clarify information that would be very difficult to explain using only words. These views show all pieces of a system. Even very small pieces such as O-rings, springs, nuts, screws, and bolts are shown. The picture indicates how the parts fit together to form components. Some of these parts may fit inside each other when the component is assembled.

The technician can refer to exploded view diagrams when disassembling or reassembling vehicle systems and components. Every part may be identified by name and/or part number. The technician can use the part name or number to order replacements.

Apply it!

Meets NATEF Communications Standards for adapting reading strategies and comprehending written information.

❶ Study **Fig. 6-6** until you understand how the parts between the caliper body and the piston assembly fit together.

❷ Write a sentence describing exactly how the piston is connected to the caliper body.

❸ Did you find it difficult to describe how the parts fit together? Explain.

Auxiliary Parking Brake Operation Most designs have two brake shoes. There is an adjuster screw at the bottom and a cam at the top where the shoes meet. When the parking brake is applied, a lever rotates the cam. This forces the shoes apart and against the drum.

General Motors uses a single shoe that is almost a circle. An actuator at the bottom expands this circular shoe, or band, whenever the parking brake is applied. Adjustments are made at the parking brake cable. **Fig. 6-8.**

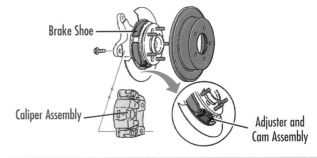

Fig. 6-8 This parking brake is a single, nearly circular band with sections of friction material bonded to it. There is an adjuster with an actuating cam at the bottom. *Where are single-shoe parking brake adjustments made?* (Pontiac Division of General Motors Corporation)

SECTION 1 KNOWLEDGE CHECK

❶ On what type of friction do parking brakes rely?

❷ What three types of parking brakes are used on vehicles?

❸ Which type of rear disc integral parking brake system is used to keep the service brakes adjusted?

❹ Which type of rear disc integral parking brake is most common on Asian import cars?

❺ Why do the shoes used in auxiliary parking brakes have soft linings?

Section 2

Parking Brake Controls

Parking brake controls consist of connected mechanical components. Components typically include the following:

- A hand lever or pedal to activate the parking brake.
- Cables to transfer force and motion from the parking brake hand lever or pedal to the parking brake mechanism at the wheels.
- A parking brake equalizer that distributes forces evenly. The **parking brake equalizer** is a device that balances the braking forces so that both rear brakes are applied evenly.

> **TECH TIP Cable Maintenance.** Parking brake cables transmit the force applied by a lever or foot pedal to the brakes at the wheels. Hanging under the vehicle, they are subject to road debris and water. Since the cable is made of woven steel wire, it can rust. A buildup of rust inside the cable housing may prevent the cable from moving. To prevent this, the parking brake cable should occasionally be lubricated with light oil where it enters the cable housing.

Parking Brake Hand Levers

Parking brake hand lever assemblies are simple ratchets with a pawl. The pawl engages the teeth of the stationary ratchet stop. When the hand lever is engaged, it pulls the parking brake cable. **Fig. 6-9.**

A release button on the end of the hand lever is pushed to release the parking brake. This disengages the pawl. The hand lever can be returned to the rest position. The cable should then become slack.

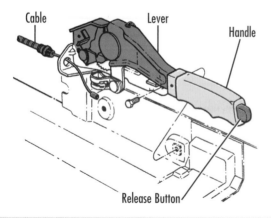

Fig. 6-9 Most parking brake hand levers have a simple ratchet and pawl design. *How are the parking brakes released?* (Pontiac Division of General Motors Corporation)

Parking Brake Pedals

Parking brake pedal mechanisms operate on the same ratchet principle as hand lever mechanisms. Pressing the brake pedal causes the ratchet to engage. **Fig. 6-10.**

In some cases the brake is released by pulling on a handle. In others the brake is released by pressing the pedal a second time. Some systems use a vacuum actuator to automatically release the brakes when the car is put in gear.

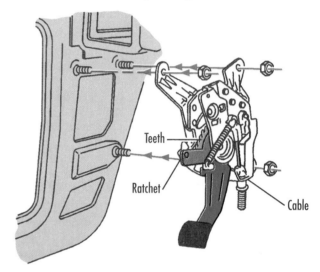

Fig. 6-10 Pedal-activated parking brakes also use a ratchet mechanism. *How are pedal-activated brakes released?* (Pontiac Division of General Motors Corporation)

Parking Brake Linkage

The parking brake linkage that connects the pedal or hand lever to the rear brakes is usually a cable. In the past, rods were used on some vehicles. They are still used on some trucks.

The cable is a braided wire rope covered with a metal or plastic sleeve. Most cars have at least two cables. The first leads from the parking brake pedal or hand lever to the equalizer. Another cable runs between the rear wheel brakes and through the equalizer.

Unless the parking brakes are used on a regular basis, the cables may rust. When this happens, the cable may not move in its housing. Cables should be lubricated occasionally to prevent corrosion.

If a cable seizes, each of the cables may be replaced individually. Do not attempt to repair a failed brake cable.

Levers in Braking Systems

Cables play an important part in parking brake systems. They "pull" on the integral or auxiliary pads or shoes to set the parking brakes. Can cables also "push"? At first glance, the question may seem strange. How can a flexible cable push?

In this activity you will make a working cable brake system. Using a bicycle caliper brake, you will show that the answer to the above questions depends on the path of the cable.

Apply it!

Testing a Brake Cable

Meets NATEF Science Standards for understanding energy, force, and levers.

Materials and Equipment
- Flat board, about 8" wide and 24" long [20 cm x 60 cm]
- Bicycle handlebar brake lever assembly
- Bicycle front brake cable
- Bicycle side-pull brake caliper assembly
- 6" [15 cm] piece of wooden dowel or broomstick, the same diameter as a bicycle handlebar
- Nails or screws to fasten the dowel to the board

1. Clamp the brake lever to the dowel. Fasten the dowel to the board with screws or nails. **Fig. A.**

2. Drill two holes in the board at positions X and Y. The center shaft of the caliper assembly must fit tightly into each hole (see illustrations).

3. Put the center shaft of the brake caliper assembly into hole X. **Fig. A.**

4. Cut the cable sheath and cable so that the cable will run in a straight line between the lever and the caliper. Thread the cable. Install the cable so that the caliper is slightly closed. **Fig. A.**

5. Push the brake lever down. Notice which part of the caliper moves. Hold the part that does not move with your fingers while you push the lever down. Do you feel that part trying to move?

6. Move the caliper to position Y. **Fig. B.** Press the lever down. What parts of the caliper move? Try holding one part of the caliper still. Observe what happens. Then hold the other part of the caliper. What happens?

Results and Analysis

This experiment demonstrates that a cable system can sometimes both pull on one part and push on another part. Look at **Figs. A** and **B.** What changed to allow the cable system to "push?" What special conditions are needed for a cable to act this way?

Notice the similarity between the bicycle brake mechanism and the disc brakes on a vehicle. Which do you think came first? Research the history of bicycle technology to find when caliper-rim brakes were first used.

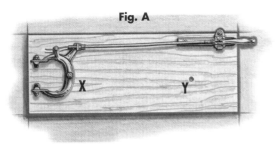

Fig. A

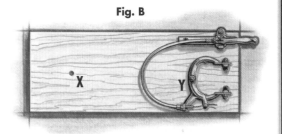

Fig. B

Parking Brake Equalizer

There are several different types of parking brake equalizers. The most common equalizer is a crescent-shaped cable guide attached to a threaded stud. The threaded stud provides tensioning adjustments for the parking brake cable assembly. **Fig. 6-11.**

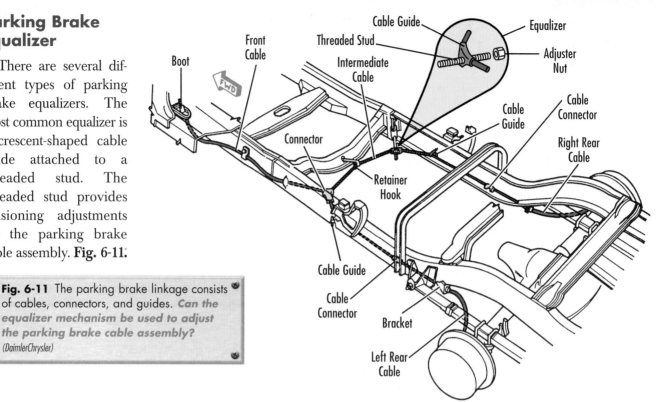

Fig. 6-11 The parking brake linkage consists of cables, connectors, and guides. *Can the equalizer mechanism be used to adjust the parking brake cable assembly?* (DaimlerChrysler)

SECTION 2 KNOWLEDGE CHECK

❶ What are two methods of activating parking brakes?

❷ How are lever-activated parking brakes released?

❸ What are two methods used to release parking brakes that are activated by a parking brake pedal?

❹ How are most parking brake cables constructed?

❺ What is the purpose of the parking brake equalizer?

Section 3

Parking Brake Diagnosis and Repair

There are two ways parking brakes can fail:
• The brakes can fail to hold.
• The brakes can fail to release.

Parking Brake Fails to Hold

If the brake fails to hold, the service brakes could be malfunctioning. The linings can be excessively worn or out of adjustment. Grease or oil on the linings can also cause failure.

If the service brakes are acceptable, the parking brake cable may need to be adjusted.

Parking Brake Fails to Release

If the parking brake fails to release, it is often due to a corroded cable. Many motorists, especially those who have cars with automatic transmissions, do not use their parking brake regularly. Unless the driver uses the parking brake regularly, corrosion can build up in the cable housing. Parking brake cables should be lubricated on a yearly basis. Apply lube spray at the cable housing and at the parking brake hand lever or pedal. Work the parking brake hand

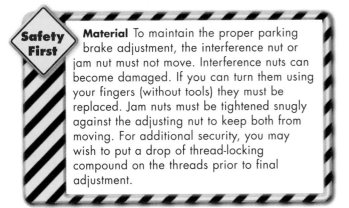

Material To maintain the proper parking brake adjustment, the interference nut or jam nut must not move. Interference nuts can become damaged. If you can turn them using your fingers (without tools) they must be replaced. Jam nuts must be tightened snugly against the adjusting nut to keep both from moving. For additional security, you may wish to put a drop of thread-locking compound on the threads prior to final adjustment.

lever or pedal several times to move the spray lube down and through the cable.

Applying the brake creates enough force to move the cable. But when the brake is released, the return springs may not be strong enough to overcome the friction in the cable. Frayed cables can also hang up and fail to release.

On cars with rear-caliper parking brakes, the cables could be overadjusted. The levers should travel fully to the stops (bosses) on the caliper housing when the brakes are released.

Often the brake warning light on the instrument panel will glow if the brakes do not fully release. The cause is usually a hand lever or pedal that does not fully retract or a bad brake warning light switch.

Adjusting Parking Brake Cables

Cables will stretch over time with use. Their effective length can be adjusted. Some cable adjusters are under the car, at the equalizer. Others are at the parking brake hand lever. On some systems there is an *interference nut,* a self-locking nut that will not loosen. On others there is a *jam nut.* This is a second nut that is tightened against the adjusting nut to keep it from moving.

Most lever-actuated cables are adjusted at the lever itself. Adjust by tightening a nut or turning an adjustment screw.

Replacing Parking Brake Cables

Replacing parking brake cables is straightforward. Most cables have a ball or barrel connector on the ends. This slips through a slot and seats in the component to which it connects.

TECH TIP **Do Not Twist the Cable.** If the cable is twisted, it can be damaged and may fail prematurely. To prevent twisting while adjusting, hold the cable's adjustment rod with a pair of locking pliers.

Replacing the Primary (Front) Cable To replace the primary cable:

1. Loosen the adjuster at the equalizer until there is sufficient slack to remove the cable from the equalizer. **Fig. 6-12.**

2. Remove the cable from the equalizer and then remove it from the pedal assembly or hand lever inside the car. **Fig. 6-13.**

3. Install the new cable by performing these steps in reverse order.

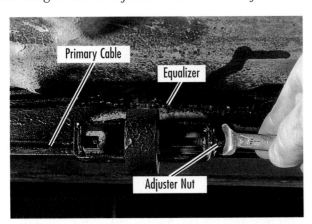

Fig. 6-12 At a point in front of the equalizer, hold the cable with a wrench. Then loosen the adjuster nut with another wrench. *How is the primary cable connected to the equalizer?* (Terry Wild Studio)

Fig. 6-13 The front cable (not visible) attaches to the brake pedal near the firewall and floorboard. *Where is the other end of the front cable attached?* (Terry Wild Studio)

Parking on a Hill

The only parking space you can find is on a sharp incline. You pull into that space and set the parking brake. You wonder if it will hold. What is the relationship between the angle of incline and the frictional force of the wheels that keeps the car from sliding?

The drawing below shows an incline with an angle, z. The rise, or height, of the incline is labeled y. The run, or horizontal distance, is labeled x.

The *slope* is the ratio of rise (y) to run (x) or y/x. The car will slip when the slope on which it is parked is greater than the value of the coefficient of static friction. A typical coefficient of static friction for rubber tires on dry concrete is 0.60.

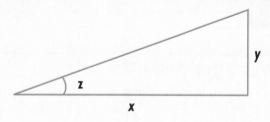

For example if the angle, z, is 5° with a run of 150 ft, the rise is 13.12 ft. The ratio of rise to run is:

$$\frac{y}{x} = \frac{13.12}{150} = 0.0875$$

A vehicle on a 5° slope will not slip if the coefficient of static friction is greater than 0.0875. The table shows the rise for a run of 150 feet, with angles of incline between 0° and 45°.

Apply it!

Meets NATEF Mathematics Standards for interpreting tables and using angles and geometric figures.

The table shows rise and run for angles of incline between 0° and 45°.

❶ Calculate the ratio of rise to run for the entries in the table.

❷ The coefficient of static friction for rubber tires on dry concrete is 0.60. Analyze the ratios you have calculated. Determine the approximate angle at which a vehicle will begin to slip.

❸ Assume that the street is wet. The coefficient of static friction for rubber tires on a wet street is only about 0.20. At what angle will the vehicle begin to slide?

Angle (z)	Rise (y)	Run (x)	y / x
0	0	150	0.0
5	13.12	150	0.0875
10	26.45	150	
15	40.19	150	
20	54.60	150	
25	69.95	150	
30	86.60	150	
35	105.03	150	
40	125.86	150	
45	150.00	150	

Replacing the Transfer (Rear) Cable(s) The *transfer cable(s)* runs between the rear wheels or from the equalizer to each rear wheel. This cable is generally replaced from under the car. To replace the transfer cable:

1. Disconnect the cable from the equalizer. **Fig. 6-14.**

2. Remove the cable(s) from the brackets. **Fig. 6-15.**

3. Once the cable(s) is free of the equalizer and brackets, it can be removed from the brake shoe or caliper actuator levers.

Most transfer cables are held to the backing plate with fingers. These fingers must be compressed before the cable can be removed from the backing plate. Install the replacement cable by performing these steps in reverse order. **Fig. 6-16.**

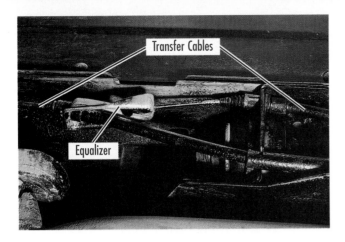

Fig. 6-14 The transfer cables attach to the equalizer. *What is the purpose of the equalizer?* (Terry Wild Studio)

Fig. 6-15 Brackets support the transfer cables as they are being routed to the rear brakes. (Terry Wild Studio)

Auxiliary Parking Brake Shoes

Adjusting   To adjust auxiliary shoes, turn a star wheel as you would when adjusting service drum brakes. There is usually an access hole in the rotor shield.

Replacing   Parking brake shoe replacement is similar to service brake shoe replacement. Check the vehicle service manual for specific instructions.

In general, the auxiliary parking brake shoes are much smaller than normal drum brake shoes. They are attached to the backing plate in a similar manner. Pins pass through the steel web of the shoe. The shoe is held in place by a spring.

TECH TIP   Compressing Cable Fingers. A useful trick for compressing the cable fingers is to use a small, worm-type hose clamp. Place the clamp around the cable fingers and then tighten by turning the screw.

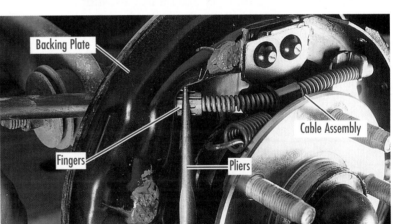

Fig. 6-16 Squeeze the fingers when removing the transfer cable from the backing plate. *How is the transfer cable accessed for replacement?* (Terry Wild Studio)

SECTION 3 KNOWLEDGE CHECK

1. What are two ways that parking brakes can fail?
2. What is frequently the problem when a parking brake fails to release?
3. Where are the adjusters for most hand lever activated parking brake systems?
4. The primary cable connects which two components of the parking brake system?
5. What is a good way to compress the fingers on the transfer cable when you need to remove it from the backing plate?

CHAPTER 6 REVIEW

Key Points

Meets the following NATEF Standards for Brakes: inspection and servicing of parking brakes and the parking brake warning light system.

- There are two basic types of parking brake systems: one uses the same brake shoes or pads as the service brakes; the other uses a separate set of brake linings.
- In some cases, parking brakes use components from the service brake system. In other cases, the parking brakes have separate components.
- The parking brake warning light system warns the driver that the parking brake is on.
- Screw-nut-cone parking brakes are common on General Motors vehicles. Ball-ramp parking brakes are used on Ford vehicles. Cam-rod parking brakes are more common on Asian imports.
- Parking brake hand lever assemblies are simple ratchets with a pawl.
- Parking brake pedal mechanisms operate on the same ratchet principle as hand lever mechanisms.
- Parking brake failure is usually due to corroded, worn, or out-of-adjustment cables.
- Servicing parking brakes includes cable adjustment and replacement.

Review Questions

1. What are the two kinds of parking brake systems?
2. What are the main components of a parking brake system?
3. What two kinds of actuating mechanisms are found on parking brakes?
4. Describe the inspection procedure for the parking brake warning light system.
5. What problems can occur in parking brake operation? What adjustments can be made?
6. What parking brake inspections should be done on a yearly basis?
7. Describe the process for inspecting and servicing parking brake cables and components for defects.
8. **Critical Thinking** Why are parking brakes needed? What might happen if they fail and the car is parked on a steep incline?
9. **Critical Thinking** Explain why motorists should use their parking brakes regularly.

TECHNOLOGY FORECAST FOR AUTOMOTIVE EXCELLENCE

Push-Button Parking Brakes

Activating a vehicle's parking brake will soon be as easy as pushing a button. With the switch to electrohydraulic and electronic brakes, today's mechanical parking brake systems will no longer be needed.

Braking action will be applied by an electric motor with a gearbox assembly. This device will be turned on and off by a dashboard- or console-mounted button.

With no more cumbersome levers to push or pull, people will be more likely to use their parking brakes—an important safety advantage. Vehicles will be less likely to roll if a transmission fails or if a driver fails to put a vehicle in gear properly—or forgets to do so at all.

An electronically controlled parking brake saves weight and is more efficient than today's mechanical designs. It also is easier for elderly and physically challenged drivers to use. Such people may not have the strength required to engage the parking brake correctly. With push-button operation, the parking brake can be safely applied by everyone.

A step beyond the push-button parking brake would be the automatic parking brake. Such a system would use software that would turn the parking brake on and off automatically.

AUTOMOTIVE EXCELLENCE
TEST PREP

Answering the following practice questions will help you prepare for the ASE certification tests.

1. Technician A says the parking brakes will fail if there is a leak in the master cylinder. Technician B says parking brakes are often mistakenly called emergency brakes but they will not bring a car to a stop in a short distance. Who is correct?
   - a Technician A.
   - b Technician B.
   - c Both Technician A and Technician B.
   - d Neither Technician A nor Technician B.

2. Parking brakes may be activated by:
   - a Engine vacuum.
   - b Placing the gear selector in PARK.
   - c A hand lever or pedal.
   - d The brake equalizer.

3. Rear-drum integral parking brakes:
   - a Use a cam-rod actuator.
   - b Use the same brake shoes and drums as the service brakes.
   - c Use a separate set of brake linings.
   - d Use a ball-ramp actuator.

4. Technician A says caliper-mounted parking brakes are used on floating calipers. Technician B says caliper-mounted parking brakes are a type of integral parking brake. Who is correct?
   - a Technician A.
   - b Technician B.
   - c Both Technician A and Technician B.
   - d Neither Technician A nor Technician B.

5. All of the following may be basic types of integral disc parking brakes *except*:
   - a Cam-rod.
   - b Strut.
   - c Screw-nut-cone.
   - d Ball-ramp.

6. Auxiliary parking brakes are used:
   - a To back up the regular parking brakes.
   - b Only on full-size sedans.
   - c Exclusively on imported cars.
   - d On vehicles with fixed or floating calipers.

7. Technician A says the auxiliary parking brake shoes have a low coefficient of friction. Technician B says auxiliary parking brake shoes wear rapidly. Who is correct?
   - a Technician A.
   - b Technician B.
   - c Both Technician A and Technician B.
   - d Neither Technician A nor Technician B.

8. Technician A says auxiliary parking brakes are self-adjusting. Technician B says integral parking brakes must be adjusted at every oil change interval. Who is correct?
   - a Technician A.
   - b Technician B.
   - c Both Technician A and Technician B.
   - d Neither Technician A nor Technician B.

9. Parking brake hand levers are released by:
   - a Depressing a button on the hand lever.
   - b Depressing the brake pedal.
   - c Pulling the release handle.
   - d Backing the car a short distance.

10. Technician A says the parking brakes may not hold if there is grease on the linings. Technician B says the brakes may not hold if the linings are excessively worn. Who is correct?
    - a Technician A.
    - b Technician B.
    - c Both Technician A and Technician B.
    - d Neither Technician A nor Technician B.

CHAPTER 7

Diagnosing & Repairing Anti-lock Brakes

You'll Be Able To:

- ✪ Describe how an anti-lock brake system (ABS) works.
- ✪ Describe the differences between an integral ABS and a nonintegral ABS.
- ✪ Determine whether the ABS warning light is functioning normally.
- ✪ Bleed ABS hydraulic circuits following the manufacturer's procedures.
- ✪ Describe the purpose and operating principles of traction control systems (TCS).

Terms to Know:

integral braking system

nonintegral braking system

oversteer

tone wheel

understeer

The Problem

Mr. Blackeagle comes into your shop on a rainy afternoon. He says he had to make a sudden stop last night when a deer darted out in front of his vehicle.

Mr. Blackeagle says when he slammed on the brakes, the brake pedal started jerking and he heard buzzing and clicking noises. They seemed to be coming from behind the instrument panel.

This is the first time Mr. Blackeagle has had this happen. He is afraid that his brakes might be failing. He wants you to check them out right away.

Your Challenge

As the service technician, you need to find answers to these questions:

1. Was the road slippery when Mr. Blackeagle hit the brakes?
2. Did his vehicle slide when he applied the brakes?
3. Was he able to maintain control and steer away from the deer?

Section 1

Anti-lock Brake System Operation

An anti-lock brake system (ABS) improves the driver's ability to control the vehicle while braking in panic stops or in emergencies. Stopping a vehicle quickly on a slick road can be dangerous. Tires skid when they lose traction on the road surface. This loss of traction happens when the driver applies the brakes so hard that the wheels "lock". Wheel lock means that the wheels stop rotating.

Wheel lock occurs when the force (speed) of the vehicle is greater than the friction between the tires and road surface. Wheels that are locked begin to skid. This breaks traction at the road surface.

A friction rating of a road surface depends on the construction of the road. Gravel has a very low friction rating. Dry asphalt has a high friction rating. Water on an asphalt road decreases its friction rating. When the friction rating of a road surface goes down, the stopping distance of a vehicle on that surface goes up. The stopping distance of a vehicle depends on the:
- Speed of the vehicle.
- Type and condition of tires.
- Driver's use of the vehicle's brakes.
- Road condition (dry, wet, snow, gravel).

Vehicle design also affects braking action. You need to consider whether the vehicle has the engine in the front or rear. You also need to consider whether it has rear-wheel drive, front-wheel drive, or four-wheel drive.

ABS was first developed in the United States for aircraft. Vehicle manufacturers adapted the system first for heavy-duty trucks and later for passenger vehicles.

Manufacturers soon realized the benefits of equipping vehicles such as vans and pickup trucks with ABS. These vehicles are harder to stop safely. Even the simpler and less expensive rear-wheel ABS greatly improved their stopping power.

Benefits of an Anti-lock Brake System

ABS is based on the concept of braking while maintaining control. **Fig. 7-1.** When a driver brakes and the wheels lock, the wheels skid. Because the wheels are not rotating, the driver cannot steer the vehicle. ABS allows the driver to brake the vehicle up to the point of wheel lockup. Thus, ABS allows the driver to decelerate, or slow down, and still steer the vehicle.

ABS benefits the average driver. In a panic situation, the average driver lacks the vehicle handling skills to apply the brakes to the point just prior to lockup.

During a skid, an experienced driver can pump the brakes and steer into the skid. This may help minimize a bad situation. ABS provides a better way to maintain vehicle steering and stopping control. ABS prevents the brakes from locking the wheels, regardless of most road conditions.

ABS helps drivers cope with the ability of rear wheels to maintain traction. A truck without ABS is designed to stop efficiently when fully loaded. It relies on the *proportioning valve* to balance front-to-rear braking effect. When the truck is empty and the vehicle's traction is low, the rear wheels may tend to slide.

(a) Without Anti-lock Brakes

(b) With Anti-lock Brakes

Fig. 7-1 Anti-lock brakes allow the driver to brake while maintaining control. *What affects the stopping distance of a vehicle?*

Personal Drivers new to ABS must learn the differences between ABS and standard brakes to realize their full benefit. With ABS, hard braking or panic stops will cause rapid pulsing, or modulation, of the brake pressure. Drivers will likely feel this through the brake pedal. Drivers describe this as if the brake pedal is "kicking" the driver's foot away from the pedal. Drivers need to know that this is normal with ABS. They should continue to apply the brakes and steer the vehicle as required to avoid an accident.

A manufacturer's decision to equip a vehicle with only rear-wheel ABS or all-wheel ABS depends on the vehicle's design. **Fig. 7-2.** It also depends on how much the customer is willing to pay for the option. More expensive vehicles are likely to have the more sophisticated ABS systems. These can include *yaw sensing* (sensing that the vehicle is sliding or spinning), *deceleration sensing*, and traction control (discussed later in this chapter).

The Role of Sensors

On most vehicles ABS provides normal braking action during regular braking. However, during hard braking ABS prevents wheel lockup. ABS functions whether the vehicle is being driven on snow, ice, gravel, or even wet leaves. ABS allows the brakes to slow the vehicle until the wheels are near lockup and the tires are starting to slide.

When wheel lockup is detected, ABS takes over. It stops further application of hydraulic brake pressure to the locking wheel. The brake fluid pressure is reduced, or even released, depending on the situation. This allows the wheel to resume rolling. Brake pressure is then reapplied, and the process starts over. The ABS process is similar to the rapid pumping action of the brakes. However, with ABS the process is done more quickly and with greater precision. Each wheel is controlled individually by the ABS control module.

Rapidly pumping the brake pedal will adjust the hydraulic brake pressure to all wheels at the same time. But with ABS a single wheel or a combination of all four may be controlled. The ABS rapidly pumps the brakes to keep the rate of wheel deceleration below the point at which the wheels lock.

The *wheel speed sensor* inputs wheel speed to the ABS control module. There are wheel speed sensors

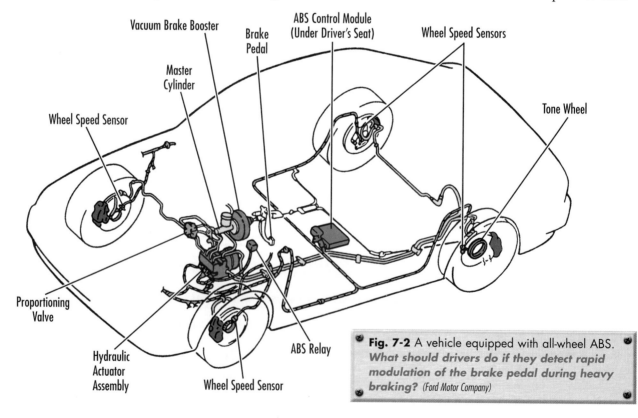

Vacuum Brake Booster
ABS Control Module (Under Driver's Seat)
Brake Pedal
Wheel Speed Sensors
Master Cylinder
Wheel Speed Sensor
Tone Wheel
Proportioning Valve
Hydraulic Actuator Assembly
Wheel Speed Sensor
ABS Relay

Fig. 7-2 A vehicle equipped with all-wheel ABS. *What should drivers do if they detect rapid modulation of the brake pedal during heavy braking?* (Ford Motor Company)

Interpreting Technical Illustrations

Illustrations and diagrams of parts, components, and systems are very useful. You have seen that an *exploded view* shows how the parts of a system fit together. **Figure 7-3(b)** is an example of a simple exploded view illustration. You also need to know how a system is related to other automotive components, including the chassis. One excellent way to view these relationships is to use an *outline* or *shadow* to represent the vehicle or a specific section of it. **Fig. 7-2.** The whole system can be pictured within the outline or shadow. You can then see the relationship of a system to other components and the chassis. You can also see how parts of the system relate to each other.

An illustration can show clearly and simply an object that might be hard to explain in words.

Apply it!

Meets NATEF Communications Standards for using study habits and methods.

Refer to **Fig. 7-2.**

❶ Follow the route of the hydraulic lines shown in yellow.

❷ What color is used to show the ABS?

❸ Where is the ABS control module?

❹ How does an outline or shadow view differ from an exploded view?

---

at each wheel. They continuously monitor wheel rotation speed. For each ABS-controlled wheel, a built-in toothed *tone wheel,* or *reluctor wheel,* rotates as part of the vehicle's wheel. **Fig. 7-3.**

The wheel speed sensor consists of a permanent magnet and a winding. This sensor is a permanent magnet sensor. The rotating toothed tone wheel is near the stationary wheel speed sensor. The sensor generates an alternating current (AC) signal as each tooth passes through the magnetic field of the wheel speed sensor. The faster the tone wheel turns, the stronger the sensor's output voltage and the higher its frequency. The AC signal is sent to the ABS control module through a shielded cable. The control module uses the signals to measure each wheel's speed.

ABS takes action when the brake light switch signals the control module that the brake pedal has been depressed. The control module action is based on wheel sensor input information and information stored in the on-board computer. The control module may determine that one or more wheels is nearing a locked condition. It then signals the *hydraulic actuator.* The actuator reduces, or modulates, brake pressure to the slowing wheel(s). This action prevents wheel lockup.

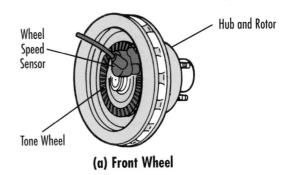

(a) Front Wheel

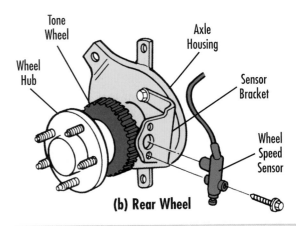

(b) Rear Wheel

Fig. 7-3 The wheel speed sensors generate signals. The ABS control module receives these signals. It uses them to determine whether the wheel is about to stop turning. The ABS control module then activates the ABS. *What device does the ABS control module use to determine wheel speed?* (Ford Motor Company)

Integral and Nonintegral Anti-lock Brake Systems

There are two types of ABS systems–integral and nonintegral (or nonintegrated). In an **integral braking system,** the brake booster, master cylinder, pump, accumulator, and pressure modulator may be combined as a single unit. Servicing an integral system can be more costly. Its components are more expensive. These units are not in wide use on current models.

A **nonintegral braking system** uses traditional brake system components, such as the master cylinder and brake booster. The ABS components include the ABS control module and hydraulic actuator. This is the most widely used unit today.

If the brake fluid reservoir is low on fluid in either an integral or nonintegral braking system, follow these steps:

1. Inspect the brake system for leaks.

2. Inspect the brake pads for wear.

3. Refer to the vehicle service manual to correct any problem in the brake system. Add brake fluid that meets the vehicle manufacturer's specifications. Follow the procedure described in the vehicle manual. In an ABS, do not add brake fluid without first checking the system for other problems. Some ABS systems require special procedures to check or adjust fluid levels.

SECTION 1 KNOWLEDGE CHECK

1. What are the benefits of an ABS over a conventional braking system?

2. What does the driver feel through the brake pedal during a panic stop with an anti-lock brake system?

3. What are the basic operating principles behind an ABS?

4. What component provides wheel speed input to the ABS control module?

5. Which system combines brake components in a single unit?

Section 2

Understanding ABS Channels

Anti-lock brake systems come in a variety of forms. Some operate on only the rear wheels. Some operate on all the wheels. Some activate a pair of wheels together and the other two wheels independently. Still others activate each wheel independently. These are called one-, two-, three-, and four-channel systems.

One-Channel ABS

The *one-channel ABS* has ABS only on the rear wheels. For economy and simplicity, many vans and pickup trucks have only rear-wheel ABS. In some cases these *rear-wheel anti-lock systems* (GM and DaimlerChrysler's *RWAL*) or *rear anti-lock brake systems* (Ford's *RABS*) control both rear wheels at the same time. **Fig. 7-4.**

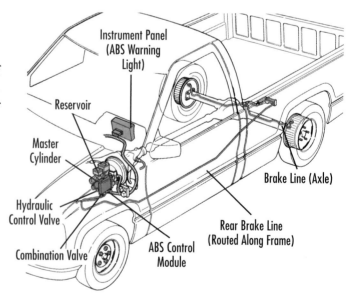

Fig. 7-4 The RWAL system is found on some GM and Dodge trucks. *Are the rear wheels controlled individually or together?*
(General Motors Corporation)

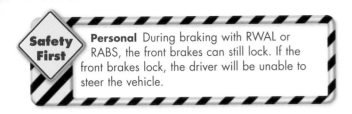

Wheel lockup is sensed either at the *differential carrier* or, as in GM, at the *transmission output shaft*. In these systems, hydraulic application of ABS affects control of both rear wheels together. A single hydraulic brake line supplies pressure to both rear-wheel cylinders. **Fig. 7-5.**

A vehicle equipped with one-channel ABS may have longer stopping distances when ABS is active. However, ABS reduces the possibility of vehicle spinout due to rear-wheel lockup.

A comparison of the front and rear brake pressures in this situation can be shown. **Fig. 7-6.** First the isolation valve closes. This prevents any increase in pressure from reaching the rear brakes.

If additional modulation is needed, the control module briefly opens the dump valve. This relieves

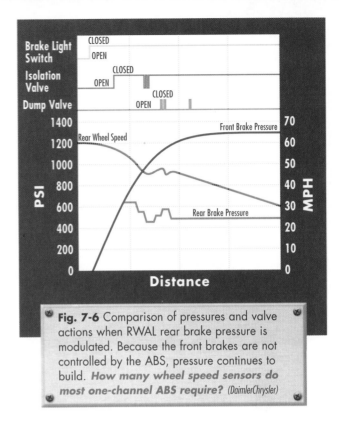

Fig. 7-6 Comparison of pressures and valve actions when RWAL rear brake pressure is modulated. Because the front brakes are not controlled by the ABS, pressure continues to build. *How many wheel speed sensors do most one-channel ABS require?* (DaimlerChrysler)

the pressure in the rear brakes by allowing fluid to flow from the brake circuit into the accumulator.

Three-Channel ABS

In three-channel ABS, both front wheels of the vehicle are independently controlled. If either front wheel nears lockup, the ABS can control only that wheel. This allows for improved braking and steering control. A single circuit controls both rear wheels at the same time.

The three-channel ABS combines the advantage of the rear-wheel-only system with the addition of independent lock-up control of the front wheels.

Fig. 7-5 The wheel speed sensor for the rear-wheel anti-lock brakes on many Dodge and Ford trucks is mounted in the differential carrier. *What is the advantage of one-channel ABS if the vehicle can experience longer stopping distances?* (DaimlerChrysler)

Interpreting ABS Graphs

The computer in an anti-lock brake system (ABS) prevents lockup during braking. It does this by monitoring the speed of each wheel compared to the vehicle speed.

To prevent lockup, the computer may choose one of three options. It may:
• Reduce braking pressure.
• Increase braking pressure.
• Hold the pressure constant.

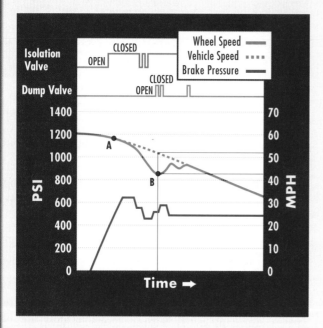

The computer does these three things by controlling the isolation and dump valves in the brake hydraulic system. Look at the graph of wheel speed, vehicle speed, and brake pressures. The dotted line represents overall vehicle speed. Under optimum conditions the wheel speed and the vehicle speed should remain the same.

At Point A the wheel speed has begun to decrease more quickly than the vehicle speed. The wheel starts to lockup. The ABS starts to control the hydraulic pressure to prevent lockup. At Point B the ABS has begun to overcome the lockup and allows the wheel to continue rotating. This begins to bring the wheel speed back where it belongs.

The ABS computer is programmed to calculate the *slip ratio*. This ratio can be written as:

$$\frac{\text{vehicle speed} - \text{wheel speed}}{\text{vehicle speed}} \times 100$$

This ratio (a fraction) is usually shown as a percentage by multiplying it by 100. The size of the slip ratio determines how the ABS controls the valves. The ideal slip ratio is zero. This means that the vehicle speed and the wheel speed are the same.

Apply it!

Meets NATEF Mathematics Standards for interpreting and using graphs.

❶ Follow the horizontal lines to the speed scale on the right of the graph. Write down your estimates of the speeds indicated there.

❷ Calculate the slip ratio at Point B. Use the slip ratio formula given above.

❸ Did the ABS keep the slip ratio at less than 30 percent as it was programmed to do? Explain.

Four-Channel ABS

A vehicle with four-wheel independent ABS control is equipped with a *four-channel ABS*. This system improves steering control at the front and slide prevention at the rear. It also paves the way for traction control systems. These systems rely on many of the ABS components.

In a three- or four-channel ABS system, solenoids in the system close or open one or more of the brake hydraulic circuits. Solenoids will either hold or release hydraulic pressure to one or more of the wheel brakes. This depends on which wheels are about to lock up.

SECTION 2 KNOWLEDGE CHECK

❶ How many basic ABS systems are there and what are they called?

❷ Which ABS is commonly found on vans and pickup trucks?

❸ What is the advantage of a two-channel ABS over a one-channel ABS?

❹ Which wheels are individually controlled in a three-channel ABS? What controls both rear wheels at the same time?

❺ What are two main advantages of four-channel ABS? How is a four-channel system similar to a three-channel system?

Section 3

Anti-lock Brake System Components

There are several components in an anti-lock brake system. Some components supply information to the ABS control module. Other components receive commands from the control module. Still others respond to hydraulic activity in the system. **Fig. 7-7.**

Anti-lock Brake System Control Module

The brain of the ABS is the control module. Like any electronic control module, it receives input data, processes it, and provides output data signals, or commands.

Input data comes from sensors. The ABS control module takes the data and reviews the data stored in its memory. The module then decides when and how action should be taken.

The control module changes output data to signals. These signals are sent to actuators. These actuators include the hydraulic control unit, the pump, and the brake and ABS lights in the instrument cluster. The control module decides which wheels should be ABS controlled under a variety of driving conditions. These driving conditions include:

• Hard braking.
• Hard cornering.
• Driving with four-wheel drive engaged.

The control module also initiates a self-diagnosis each time the vehicle is started. Depending on the vehicle, this self-test occurs between 5 and 25 mph [8–40 kph]. If the control module determines, after reviewing its stored data, that operating conditions are outside of specifications, it shuts down the ABS. The module then turns on the ABS warning light. It also stores a *diagnostic trouble code (DTC)*. This DTC can be retrieved later by a technician.

Hydraulic Actuator

In RWAL systems the ABS control module signals the *ABS relay* to power the *hydraulic actuator*. Either a two- or three-position flow control valve in the hydraulic actuator controls pressure to both rear wheels.

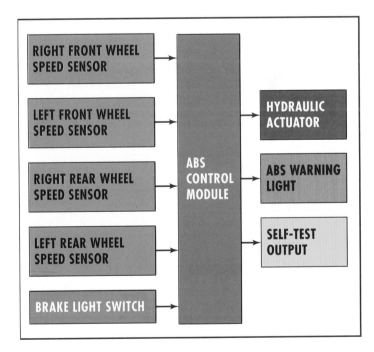

Fig. 7-7 Based on the input it receives, the ABS control module makes decisions and sends outputs. *How do ABS components function in three different ways?* (Ford Motor Company)

In a three-channel system, each wheel may have a wheel speed sensor. Although each front wheel brake is controlled individually by the actuator, a single three-position solenoid controls the pressure to both rear brakes. **Fig. 7-8.**

During normal braking, operating the brake pedal causes a normal increase of brake pressure at each wheel. **Fig. 7-9(a).** This changes when the system determines that a wheel lockup is about to occur. The ABS control module signals the solenoid valve in the hydraulic unit to close the spring-loaded check ball (right check ball). This prevents additional pressure from entering the brake line to the wheel. Instead of increasing with further application of the brakes, hydraulic pressure is held constant. This is the *hold-pressure position.* **Fig. 7-9(b).**

If the wheel still tends to slow down too fast or lock up, the ABS goes into the *reduce-pressure position.* **Fig. 7-9(c).** The ABS control module signals the

TECH TIP ABS **Component Names.** ABS hydraulic control and routing components have a number of names. These names include the hydraulic actuator, modulator, control module, control valve assembly, control unit, valve block, and valve body assembly. Names vary depending on the system and its manufacturer. Always refer to the vehicle service manual when working on an ABS-equipped vehicle. An ABS is too complex to repair without referring to a service manual.

solenoid valve to move the other check ball (left check ball) to open a passage that allows fluid to flow back from the wheel brake line to the reservoir in the hydraulic unit.

During ABS operation, this apply-hold-release cycling action causes the brake pedal to pulsate. This happens very rapidly and feels like a vibration. Under certain conditions with some systems, the solenoid valves can also allow the brake fluid pressure to increase.

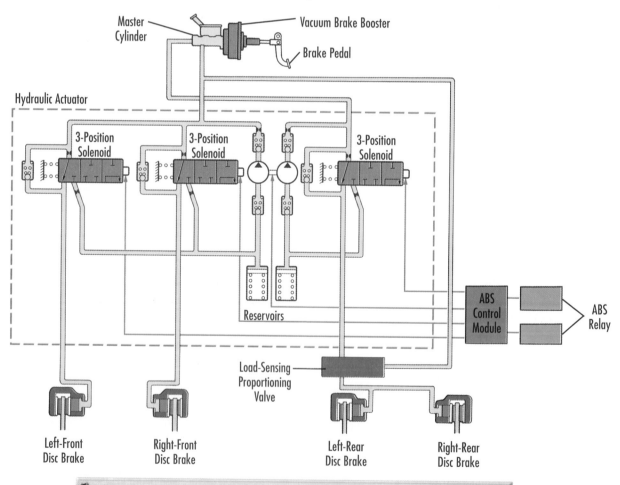

Fig. 7-8 In this three-channel system, the front brakes are controlled separately. The rear brakes are controlled together. *What controls brake pressure to the rear brakes?*

Pump, Accumulator, and Reservoir

The integral ABS must have high pressure available to be able to go into action. The pump builds up a reserve of brake fluid pressure. This reserve is stored in the *accumulator*. When accumulator brake fluid pressure drops to approximately 1,500 psi [10,343 kPa], a low-pressure switch turns on the pump. This rebuilds the pressure reserve.

A high-pressure switch shuts off the pump at around 2,500 psi [17,238 kPa]. Repeated braking can use up the high pressure in the accumulator. If this happens, the driver can still brake normally. However, the brake pedal will require greater than normal force.

In many non-integral systems, brake fluid vented from the wheel circuits is stored in a reservoir. This is separate from the one on the master cylinder. The reservoir is not pressurized.

The next time the brakes are applied, the motorized pump pressurizes the fluid. The pump sends the fluid back to the master cylinder.

Speed Sensors

In the GM RWAL system, a sensor at the output shaft of the transmission generates the speed signal. The signal is routed to the *digital ratio adapter controller (DRAC)*. There it is converted to a digital signal for the ABS control module.

DRAC output signals are also routed to the cruise control module, the *powertrain control module (PCM)*, and the instrument panel cluster (for the speedometer).

The ABS control module continuously monitors the speed of the wheels (or driveshaft or differential ring gear with RWAL). As long as all wheels are turning at about the same speed during a hard stop, the ABS control module takes no action.

The ABS is sensitive. It can be affected by a number of factors:

- Mismatched tires or badly worn tires in the same set.
- The mounting of larger- or smaller-than-stock tires. This will upset the preprogrammed functions of the ABS. Such tire conditions could cause dissimilar wheel speed signals. These signals can confuse the ABS control module, causing it to set a DTC.

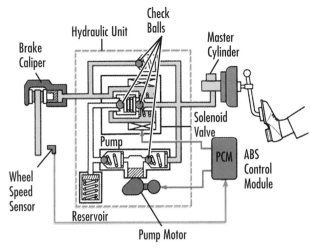

(a) Normal Braking Pressure

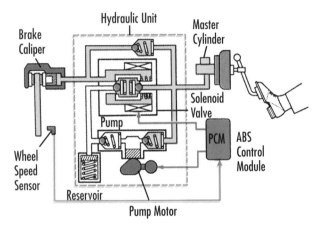

(b) Hold-Pressure Position

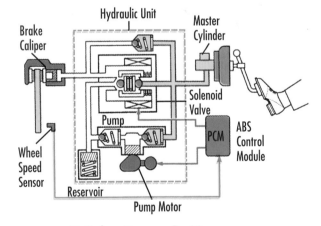

(c) Reduce-Pressure Position

Fig. 7-9 (a) During normal brake operation, fluid flow is normal and ABS is not activated. **(b)** If the wheel is decelerating and near lockup, pressure to the wheel is held. **(c)** If wheel lockup is imminent, the fluid pressure is reduced. *What happens when the ABS control module signals the solenoid valve to close the right check ball?* (DaimlerChrysler)

• If the vehicle uses a tone wheel mounted in the rear differential, changing the gear ratio can upset the ABS control module.

Check the vehicle service manual for the proper tires and gearing if the vehicle continues to set a DTC.

Deceleration Sensor

Some vehicles have an additional input to the ABS control module. It is called a *deceleration sensor,* or *G-sensor*. It tells the control module whether the vehicle is moving or stopped.

If a vehicle were on ice, pressing the brake pedal could cause a four-wheel lockup condition. Signals from the wheel speed sensors would mistakenly indicate that the vehicle had stopped. However, the G-sensor signals the ABS control module that the vehicle is still moving.

By comparing incoming data from the wheel sensors and the G-sensor, the ABS control module senses that all four wheels have stopped turning, but the vehicle is still moving. As a result, ABS is immediately enabled.

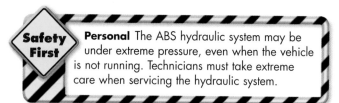

Safety First **Personal** The ABS hydraulic system may be under extreme pressure, even when the vehicle is not running. Technicians must take extreme care when servicing the hydraulic system.

ABS Warning Light

Many vehicles with anti-lock brake systems have a yellow or amber *anti-lock warning light* on their instrument panels. The warning light includes the symbols and letters (ABS). **Fig. 7-10.** The light glows during a self-test when the vehicle is started. It also glows if the ABS control module detects trouble in the anti-lock brake system. The light may also turn on to indicate that the anti-lock system is activated. In certain systems, DTCs stored by the ABS control module may be retrieved using the ABS warning light. The warning light can "blink out" the DTC in timed flashes. The flash sequence corresponds to a DTC listed in the vehicle service manual.

Brake Warning Light

The fluid reservoir that is part of the hydraulic unit includes a *low-fluid level sensor*. If the brake fluid is low, the sensor turns on a red instrument panel warning light or low brake fluid display. **Fig. 7-10.**

The red *brake warning light* will glow if there is low pressure in any section of the brake hydraulic system. This could be due to low brake fluid caused by normal brake pad wear. It could also be caused by a leak or rupture in the system. Both lights will glow briefly during the self-test of the ABS.

Lateral Acceleration Sensor

At least one manufacturer uses a *lateral acceleration sensor* in the ABS. This sensor senses hard

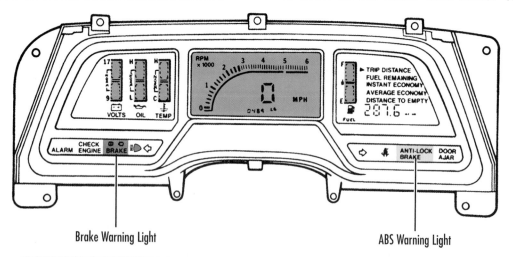

Brake Warning Light

ABS Warning Light

Fig. 7-10 The anti-lock warning light may indicate (1) a malfunction, (2) that the ABS is active, and/or (3) DTCs. *What is different about the warning light when it is indicating DTCs versus indicating that there is a malfunction or that the ABS is active?* (Ford Motor Company)

cornering during braking. Mercury switches or a Hall-effect lateral accelerometer detect high G-forces during moderate to hard turns. The sensor provides input to the ABS control module.

The result is that the ABS is enabled for better braking control–even under less-than-lockup conditions. This includes conditions such as braking on a curved exit ramp, where wheel lockup could cause the vehicle to slide or spin.

Safety First

Personal Without adequate training, drivers may be caught off guard by ABS pedal pulsations. Drivers might momentarily release the brake pedal. This will disengage ABS and increase the stopping distance. The pulsations can occur as rapidly as eighteen times per second. This is much faster than even a professional driver can pump the brakes. Practice forcing the ABS to operate under controlled conditions, such as in a vacant parking lot. With ABS activated, practice steering the vehicle as if to avoid hitting an object. A big advantage of ABS is that steering remains possible during hard braking. Drivers must remember the limits of their vehicles. Drivers should avoid a false sense of security with ABS. Yet they should remember to do more than panic and "stand on the brakes."

Four-Wheel-Drive Switch

For handling reasons, ABS on four-wheel-drive vehicles is disabled when the vehicle is in four-wheel drive. Only those vehicles that use a viscous-clutch differential can use ABS while in four-wheel drive

ABS is also disabled if the limited-slip front differential locks the front wheels together. The wheels are effectively locked through the drive train. In this situation the ABS cannot operate properly.

Diagnosing and Repairing Anti-lock Brakes

Only service personnel trained on the specific ABS should attempt to repair that system. Hydraulic fluid pressures in ABS systems are extremely high. Improper servicing procedures can result in personal injury and material damage. When servicing an ABS, consult the vehicle service manual. The designs of these complex systems vary from one vehicle manufacturer to another.

Diagnosing Anti-lock Brakes

Technicians must understand the function of each ABS component. Only then is it possible to determine if a component is malfunctioning. Diagnosis can be made using the ABS control module and the ABS *diagnostic connector*. **Fig. 7-11.** The diagnostic connector allows technicians to access the control module using a scan tool designed for the vehicle being serviced. The scan tool "reads" the DTCs stored in the control module. The DTCs provide the information needed to find problems in the ABS circuit.

ABS Diagnostic Connector There are many different styles of ABS diagnostic connectors in vehicles equipped with anti-lock brakes. The ABS diagnostic connector is the point at which the scan tool is connected. The scan tool provides the data readout on the status of the ABS system. Each connector is designed to interface with a product-specific scan tool to troubleshoot ABS. The ABS diagnostic connectors are under the dash, behind the glove box, under the hood, or under the seat.

Beginning with the 1996 model year, all ABS data are routed through the 16-pin data link connector (DLC). The DLC connector provides access to vehicle information, operating conditions, and diagnostic information.

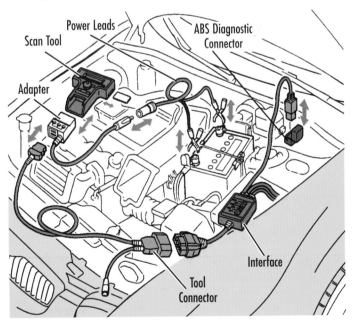

Power Leads
Scan Tool
ABS Diagnostic Connector
Adapter
Tool Connector
Interface

Fig. 7-11 Technicians use a scan tool to read the DTCs stored in the ABS control module. *What is the function of the diagnostic connector?* (Mazda Motors of America, Inc.)

Safety First

Personal Anti-lock brake systems have very high system pressures. Follow the procedure to "bleed down" the system pressure before attempting any repair requiring the removal of any system part. Improper repair could result in ABS malfunction. This could lead to injury to the vehicle occupants or to property damage. Always consult the service manual before doing any work on an ABS-equipped vehicle.

Bleeding Anti-lock Brakes

The procedure for bleeding anti-lock brakes may not totally purge air from the valves. Some vehicles require a special scan tool to perform a complete brake bleed. Refer to the vehicle service manual for specific procedures before bleeding. General guidelines for bleeding non-integral anti-lock brakes are as follows:

- With the engine and ignition OFF, pump the brake pedal several times. This will expel any vacuum remaining in the power brake booster.
- Remove the master cylinder reservoir cover. Fill the reservoir to the full level. Place the cover loosely over the opening.
- Pour a little brake fluid into a catch container– enough to submerge the end of the clear tubing during the bleeding process. This makes it easier to see the air bubbles being expelled from the brake system.
- The bleeding process sequence depends on whether the brake system is rear-split or diagonally split. With a front-rear split system, the sequence is right rear, left rear, right front, left front. With a diagonally split system, the sequence depends on the split. Always check the vehicle service manual for the correct order. When bleeding

brakes, check the brake fluid level in the master cylinder reservoir. If needed, refill the reservoir after completion of the bleeding of a caliper or wheel cylinder. Refill more often if needed.

- Loosen the bleeder valve on the wheel to be bled. Then tighten it only enough that it can be loosened and tightened easily.
- Install one end of the clear tubing over the bleeder valve. Leave room for a wrench to fit over the flats of the valve. Place the other end of the tubing into the catch container so the end is submerged near the bottom.
- Have an assistant pump the brake pedal twice and hold the pedal down. Open the bleeder valve and allow fluid and air to escape. Close the bleeder valve just before the fluid stops flowing. Have your assistant release the pedal, pump twice, and then hold the pedal down. Open the bleeder again and release more fluid and air. Continue this operation until there are no more air bubbles in the fluid.
- Securely tighten the bleeder valve. Remove the tube and container. Move on to the next caliper or wheel cylinder.
- If needed, add brake fluid to the master cylinder reservoir.
- Repeat the five steps above for the remaining calipers or wheel cylinders in the sequence specified.
- After bleeding all four calipers or wheel cylinders, top off the fluid level in the master cylinder reservoir. Secure the master cylinder reservoir cover.
- Check the operation of the brakes. Brake pedal travel should be normal. The pedal should be firm when depressed, not spongy.

SECTION 3 KNOWLEDGE CHECK

❶ Which of the ABS components provides output data signals, or commands?

❷ What device powers the hydraulic actuator?

❸ Which device in the ABS holds a reserve of pressurized hydraulic fluid?

❹ Where is the wheel speed sensor located on GM vehicles with RWAL?

❺ Which device in the ABS tells the ABS control module whether the vehicle is moving or stopped?

Section 4

Traction and Stability Control

When a wheel is given more torque than it can transfer to the road, the tire loses traction and spins. This occurs most often during heavy acceleration or on slippery surfaces. To prevent wheel spin, some vehicles with ABS also incorporate a *traction control system (TCS)*.

Traction Control

ABS components can also control wheel spin. As a result, many high-end cars come equipped with ABS and TCS. **Fig. 7-12.** When a wheel is about to spin, the TCS, using ABS components, applies the brake only at that wheel. This slows the wheel so wheel spin does not occur.

If a driver accelerates on a slippery surface, the TCS helps avoid wheel spin. This helps prevent momentary loss of vehicle control. It also reduces tire wear. With additional electronics it can also avoid loss of traction from "drag torque" when downshifting.

An ABS and a TCS share some components, such as wheel speed sensors and hydraulic actuators. A combined ABS and TCS control module is called an *anti-lock brake system-traction control module (ABS-TCS)*.

With TCS, when engine torque causes a drive wheel to break loose and spin, the ABS causes brake pressure to be applied to that wheel. This slows the wheel to prevent wheel spin.

If braking alone does not prevent wheel spin, the ABS control module calls for reduced engine speed and torque. It does this by signaling the powertrain control module (PCM). The PCM receives the request for less torque and retards ignition timing. If this does not stop wheel spin, the amount of fuel delivered to the engine is reduced.

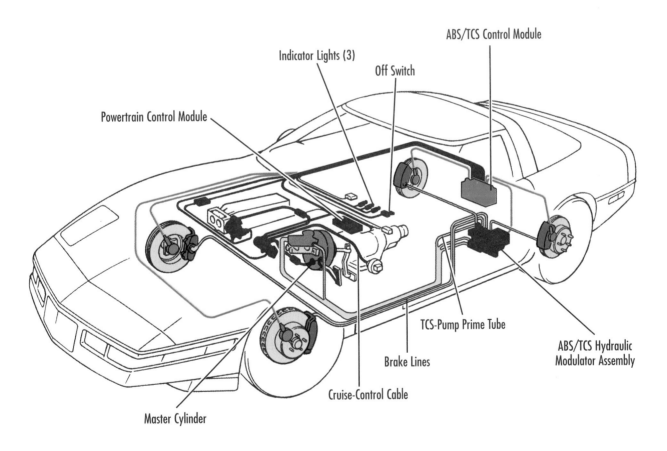

Fig. 7-12 A vehicle with an ABS and a traction control system (TCS). *What is the purpose of a TCS?* (General Motors Corporation)

Converting Energy in an ABS

A moving vehicle has *kinetic energy*. Kinetic energy is the energy of motion. The amount of kinetic energy depends on the mass and speed of the vehicle.

To stop a vehicle, this energy must be dissipated or dispersed. It must be changed into another form as quickly and efficiently as possible. This energy conversion is made by the braking system. The brakes convert kinetic energy into heat. They use the friction of the brake shoes or pads on the drums or discs.

A locked wheel and a skidding tire are certainly not the best ways to convert kinetic energy into heat. The heat created by the friction begins to melt the tire. Sliding on melting rubber, the tire has a very low coefficient of friction.

The ABS is designed to keep as much of the friction as possible inside the brakes, not on the road. The friction is as high as possible in the brakes. The kinetic energy of the vehicle goes into heating the brake drums or discs, not the tires and pavement.

Apply it!

Converting Energy to Heat

Meets NATEF Science Standards for understanding kinetic energy, friction, and heat.

Materials and Equipment
- Rubber wheel from a toy car, with its axle (a nail will do)
- Strip of coarse sandpaper
- Flat surface; such as a table with a tablecloth

To understand the conversion of kinetic energy into heat, try the following experiment. As you perform this experiment, try to keep the downward pressure on the wheel even and equal in each step.

1. Hold the wheel and its axle so that the wheel can easily turn. Roll the wheel across the flat surface. The very small force needed to do this comes from the low coefficient of rolling friction. This is why a vehicle, without braking, takes so long to coast to a stop.

2. Lock the wheel with your finger. Slide the wheel rapidly across the flat surface many times. The force needed depends on the coefficient of sliding friction. More force will be needed than in the previous step. Feel the wheel where it has been sliding. It's warm. This is the energy you supplied, converted to heat.

3. Roll the wheel across the flat surface while you hold the sandpaper very gently onto the top of the wheel. Let the sandpaper slide over the wheel. Do not let the wheel skid! This should require more force to turn the wheel on its axle. You have now moved the friction to where it is the greatest. You have spread the heat energy over the entire system.

4. Think carefully about what you have observed in this experiment. Do you see how an ABS is an energy manager for the conversion of kinetic energy into heat? Write a paragraph describing where the energy of a moving vehicle goes under the following conditions:

- Coasting to a stop.
- Locking up the brakes and skidding.
- Making an emergency stop with ABS.

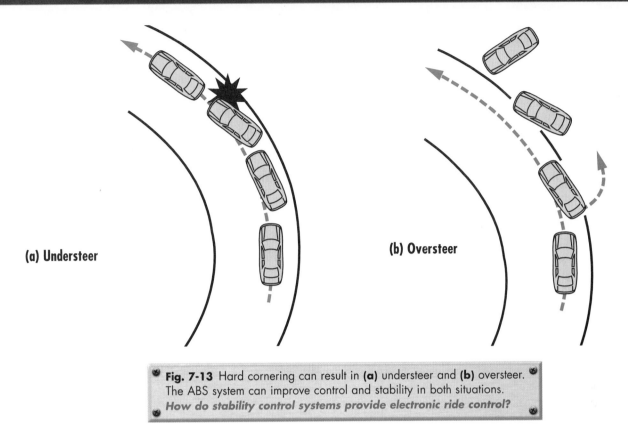

(a) Understeer   **(b) Oversteer**

Fig. 7-13 Hard cornering can result in **(a)** understeer and **(b)** oversteer. The ABS system can improve control and stability in both situations. *How do stability control systems provide electronic ride control?*

Stability Control

An ABS can also improve vehicle handling and stability. An advanced stability control system can prevent tire skid and vehicle instability when cornering.

Understeer is a condition that results when the front tires lose adhesion during cornering. The vehicle wants to push, instead of turning. **Fig. 7-13(a).** If understeer is detected, the ABS will apply braking force (drag) to the rear wheel opposite to the push. For example, if the vehicle is pushing to the right, while trying to turn left, ABS will drag the left rear wheel. The dragging action brings the vehicle back under control.

Oversteer is a condition that results when the rear tires lose adhesion during cornering. The rear end of the vehicle wants to swing out, or is loose. **Fig. 7-13(b).** If oversteer is detected, the ABS will apply braking force to the outside front wheel. This braking action will correct the oversteer, and prevent the rear end of the vehicle from swinging around.

The ABS may also use inputs from other motion and speed sensors to determine if the vehicle is under control. The stability controls are able to determine when a vehicle has reached its limit of directional change. The system can detect when a vehicle is about to skid out of control. The ABS may apply one or more of the wheel brakes to regain vehicle stability and directional control.

SECTION 4 KNOWLEDGE CHECK

❶ What causes a wheel to lose traction during acceleration?

❷ What is the TCS and what is its purpose?

❸ What components do the ABS and the TCS share?

❹ What can the ABS-TCS control module do if braking alone does not prevent wheel spin?

❺ What is a condition called "oversteer"? How is it prevented?

CHAPTER 7 REVIEW

Key Points

Meets the following NATEF Standards for Brakes: diagnosing the function of the ABS warning light; bleeding anti-lock brake systems.

- ABS offers an electronic method of pumping the brakes to avoid wheel lockup. It monitors wheel speed, and then lowers the brake fluid pressure to the brake on the wheel that is approaching lockup.
- In a nonintegral ABS, traditional brake system components such as the master cylinder and brake booster are used.
- The ABS brake warning light glows briefly during the self-test of the system.
- Bleeding an anti-lock brake system may not totally purge air from valves. Check the vehicle service manual. The use of a scanner may be required.
- A traction control system (TCS), which is a spin-off of ABS technology, prevents unwanted wheel spin.
- Some traction control systems can signal the PCM to reduce engine speed and torque.

Review Questions

1. Describe how an ABS works.
2. Describe the differences between an integral ABS and a nonintegral ABS.
3. How can you determine if the ABS warning light is functioning normally?
4. When bleeding an anti-lock brake system, why must you always follow manufacturer's procedures?
5. Identify the four different ABS systems.
6. Describe the purpose and operating principles of a traction control system (TCS).
7. Describe how an ABS is used to provide stability control.
8. (Critical Thinking) What is an ABS-TCS control module and how does it work?
9. (Critical Thinking) What could be done to four-wheel-drive vehicles so they can use ABS, even when in full-time four-wheel drive?

TECHNOLOGY FORECAST
FOR AUTOMOTIVE EXCELLENCE

ABS Moves into the Future

Anti-lock brake systems (ABS) have moved far beyond their original purpose. First used on airplanes to prevent wheel lockup, ABS was soon added to cars and trucks. It wasn't long before engineers developed new features, such as traction control and stability enhancement.

Future ABS units will continue to prevent wheel lockup. They will become standard on more vehicles. They may eventually be required by the government on all vehicles.

New ABS techniques will make braking more efficient. Electronic brake proportioning will help to even out the wear on tires and brakes by distributing braking force. Some engineers believe the system can also stop a vehicle more quickly.

Here's how electronic brake proportioning works. In a typical vehicle, weight shifts forward as the brakes are applied. The rear end becomes light, and most of the braking is done on the front wheels. With electronic brake proportioning, wheel speed sensors are located at each wheel. They measure wheel slip, no matter how small. This information is used by the computer to control braking pressure at the front and rear wheels.

Another new design uses sensors that monitor changes in the pavement. With this information, ABS can be better applied over bumps, turns, and potholes. This will make for safer braking in even the most extreme road conditions.

AUTOMOTIVE EXCELLENCE
TEST PREP

Answering the following practice questions will help you prepare for the ASE certification tests.

1. Technician A says an ABS helps stop a vehicle quickly without loss of control. Technician B says a vehicle without an ABS can sometimes stop faster on dry pavement with an expert driver. Who is correct?
   - ⓐ Technician A.
   - ⓑ Technician B.
   - ⓒ Both Technician A and Technician B.
   - ⓓ Neither Technician A nor Technician B.

2. During normal braking, the ABS:
   - ⓐ Has no effect on the brakes.
   - ⓑ Reduces brake hydraulic pressure equally at all wheels.
   - ⓒ Increases the brake hydraulic pressure equally at all wheels.
   - ⓓ Reduces the hydraulic pressure only at the rear wheels.

3. Technician A says, on an integral ABS, the brake booster is separate from the other components. Technician B says an integral ABS is more expensive to replace. Who is correct?
   - ⓐ Technician A.
   - ⓑ Technician B.
   - ⓒ Both Technician A and Technician B.
   - ⓓ Neither Technician A nor Technician B.

4. Technician A says the speed sensor for an RWAL system may be located in the rear differential. Technician B says the speed sensor may be located in the transmission. Who is correct?
   - ⓐ Technician A.
   - ⓑ Technician B.
   - ⓒ Both Technician A and Technician B.
   - ⓓ Neither Technician A nor Technician B.

5. Technician A says that in three-channel systems both front wheels of the vehicle are not controlled independently. Technician B says that in such systems both front wheels are controlled independently. Who is correct?
   - ⓐ Technician A.
   - ⓑ Technician B.
   - ⓒ Both Technician A and Technician B.
   - ⓓ Neither Technician A nor Technician B.

6. Technician A says a three-channel ABS combines the lower cost of an RWAL system with the added benefits of better steering control. Technician B says a three-channel ABS has a separate brake circuit to each rear wheel and a common circuit for the front wheels. Who is correct?
   - ⓐ Technician A.
   - ⓑ Technician B.
   - ⓒ Both Technician A and Technician B.
   - ⓓ Neither Technician A nor Technician B.

7. The ABS component that processes input signals, makes decisions, and outputs signals is the:
   - ⓐ PCM.
   - ⓑ ECM.
   - ⓒ BCM.
   - ⓓ ABS control module.

8. The hydraulic actuator can do all of the following, *except*:
   - ⓐ Hold brake pressure.
   - ⓑ Release brake pressure.
   - ⓒ Measure brake pressure.
   - ⓓ Increase brake pressure.

9. What type of signal does the wheel speed sensor generate?
   - ⓐ An AC signal.
   - ⓑ A DC signal.
   - ⓒ A digital signal.
   - ⓓ A pulsed DC signal.

10. Technician A says a lateral acceleration sensor tells the ABS control module that the vehicle is in a hard-cornering maneuver. Technician B says the lateral acceleration sensor tells the ABS control module the vehicle's forward velocity. Who is correct?
    - ⓐ Technician A.
    - ⓑ Technician B.
    - ⓒ Both Technician A and Technician B.
    - ⓓ Neither Technician A nor Technician B.

Electrical & Electronic Systems

AUTOMOTIVE CUSTOMER RELATIONS

Immediate opening for a "GREETER" in Service Department. Consists of greeting customers, filing, follow-up phone calls, and cashiering. Competitive wages and advancement for that special person. Apply in person.

MAINTENANCE TECHNICIAN

A local plastics company is searching for an individual to join our firm as a Maintenance Technician.

The successful candidate will possess a minimum of three years job-related experience, a high school diploma or GED equivalent; demonstrated mechanical, electrical, pneumatic, and hydraulic aptitude. Must have ability to diagnose, troubleshoot, and repair machinery. The ability to to read and interpret blueprints and schematic drawings is a must. Industrial electrical experience and PLC background preferred.

We offer a competitive wage and benefit package.

An Equal Opportunity Employer M/FH/V

Automotive Service Advisor

EXTREMELY BUSY SERVICE DEPARTMENT IS SEEKING CAREER-MINDED PERSON TO JOIN OUR MANAGEMENT TEAM. INDIVIDUAL MUST BE SENSITIVE TO CUSTOMER NEEDS AND CAPABLE OF OVERDELIVERING ON CUSTOMER EXPECTATIONS. EARNING POTENTIAL IS IN THE 40K-50K RANGE. IF YOU FEEL YOU'RE THE BEST AND WOULD LIKE TO JOIN A WINNING TEAM, CONTACT US TODAY.

ELECTRONICS TECHNICIAN

Must be able to troubleshoot and repair electronic controls. Training in digital AC and DC drive is required. A minimum 2 years of experience in industrial controls and electronics is necessary. Knowledge of technology a plus.

Career Focus Activity

After reading the above job descriptions, do the following:
- Imagine that you are applying for one of the positions. Write a letter of application explaining your qualifications.
- Choose one of the careers and research its availability in your city, in your state, and in the nation.
- What local business or industry could provide you with an opportunity to gain two years of experience in industrial controls?

CHAPTER 1

Electrical System Operation

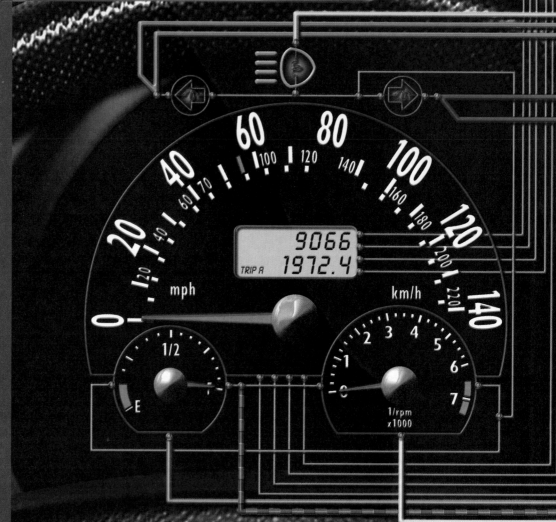

You'll Be Able To:

- Describe how current flows through a conductor.
- Define the relationships between voltage, current, and resistance.
- Apply Ohm's law.
- Read a wiring diagram.
- Use test equipment to locate a defective circuit.
- Safely handle electronic components.

Terms to Know:

conductor

current flow

fusible link

generator

multimeter

resistance

voltage

wiring diagram

The Problem

The automotive electrical system uses a generator and battery to supply electrical energy to the vehicle. Electrical and electronic circuits distribute and use the electrical energy to perform a variety of tasks. Electricity is used in powertrain control, cranking, lighting, safety, and convenience systems. It is important for the technician to know how electricity is generated, stored, distributed, and used in a vehicle's electrical system. If you have a basic knowledge of electrical and electronic fundamentals you will be better prepared to diagnose and repair problems in vehicles that are brought to you for service.

Your Challenge

As the service technician, you need to find answers to these questions:

1. How is electricity stored in a vehicle?
2. How is a wiring diagram used when diagnosing electrical problems?
3. What devices can be used to diagnose electrical problems?

Section 1

The Nature of Electricity

Understanding electricity requires a basic knowledge of physics. Everything we see, touch, feel, or smell is composed of matter. All matter is made of atoms, and all atoms have an atomic structure.

The atomic structure of an atom includes three basic particles: protons, neutrons, and electrons. Protons and neutrons are found in the nucleus, at the center of the atom. Protons have a positive (+) electrical charge. Neutrons have no electrical charge. Electrons orbit the nucleus of an atom much like a planet orbits the sun. Electrons have a negative (−) electrical charge. **Fig. 1-1.**

The atomic structure of the copper atom has 29 protons, 29 neutrons, and 29 electrons. The positive charge of the protons in the nucleus is exactly balanced by the negative charge of the electrons. Electrons circle the nucleus in different orbits, depending on their energy levels. These orbits are called shells.

The copper atom has 2 electrons in the shell closest to the nucleus. There are 8 electrons in the second shell and 18 electrons in the third shell. That leaves 1 electron in the outer shell. **Fig. 1-2.**

The electron or electrons located in the outer shell are the important ones in the study of electricity. In the copper atom, the one electron in the outer shell is a "free" electron. Free electrons can move from one atom to another. Other electrons are held tightly by the nucleus and do not move.

Conductors

An electrical **conductor** is a material that contains free electrons and easily conducts electrical current. The movement of free electrons through a conductor creates current flow. **Current flow** is the movement of electrical energy through a conductor.

Several materials are used as conductors in motor vehicles.
- Copper is an excellent conductor used in automotive wiring systems.
- Iron and steel, although not as good as copper, are used in a vehicle's body and frame as electrical grounds.
- Tin and lead, used in solder connections, are also conductors but not quite as good as iron and steel.
- Gold and silver are good conductors. They are used in the microchips in a vehicle's computerized control systems.

Insulators

Some materials do not have free electrons in their atomic structure. They hold their electrons very tightly. The lack of free electrons makes them poor electrical conductors. These materials are called *insulators*.

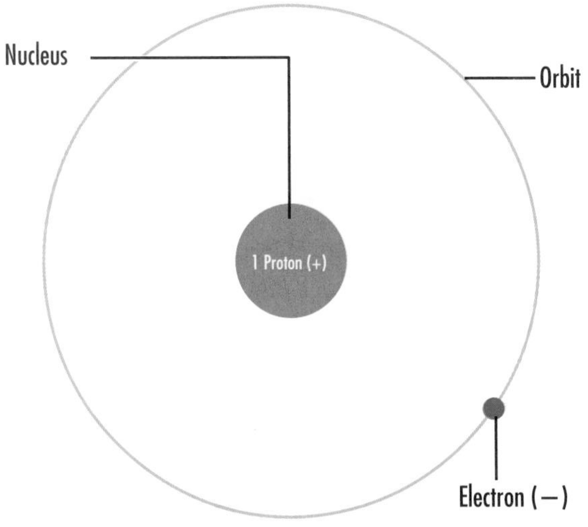

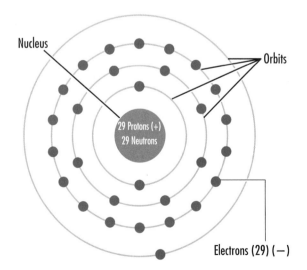

Nucleus — Orbit

1 Proton (+)

Electron (−)

Fig. 1-1 The hydrogen atom has the simplest structure. It has a single proton in the nucleus and no neutrons. It has one electron in orbit around the nucleus. *Which particle has a positive charge? Which particle has a negative charge?*

Nucleus — Orbits

29 Protons (+)
29 Neutrons

Electrons (29) (−)

Fig. 1-2 A copper atom has 29 protons and 29 neutrons in its nucleus, and 29 electrons in orbit around the nucleus. *With one electron in its outer shell, would copper be a good conductor?*

Using Ohm's Law

Ohm's law states that, in a circuit, the electromotive force *(E)* is equal to the product of the current *(I)* and the resistance *(R)*. Electromotive force is often referred to as voltage.

As a formula, Ohm's law is written as $E = I \times R$. Use this formula to calculate the electromotive force, or voltage, in a circuit when the current and resistance are known.

The unit for voltage *(E)* is the volt *(V)*. The unit for current *(I)* is the ampere *(A)*. The unit for resistance *(R)* is the ohm *(Ω)*.

If the current in a circuit is 15 A and the resistance is 0.4 Ω, what is the voltage?

Use $E = I \times R$, then $E = 15\ A \times 0.4\ Ω = 6\ V$.

If voltage and current are known, what formula could you use to determine resistance?

You know that $E = I \times R$. If both sides of the equation are divided by the current *(I)*, the relationships between the voltage, current, and resistance will be preserved.

You will write:

$$E = I \times R$$

Then you will divide both sides by I:

$$\frac{E}{I} = \frac{I \times R}{I}$$

$$\frac{E}{I} = R$$

$$R = \frac{E}{I}$$

Resistance is voltage divided by current.

Apply it!

Meets NATEF Mathematics Standards for using symbols and converting formulas to equivalent forms.

If voltage and resistance are known, what formula could you apply to determine the current? Rewrite $E = I \times R$ in the form "$I =$ _____."

Three materials are commonly used as insulators in motor vehicles.
- Plastic is used as an insulator around wires and plugs.
- Rubber is used in insulating mounts for electrical equipment.
- Glass is used to insulate and enclose electrical components like lightbulbs and fuses.

The Flow of Electricity

Traditional theory describes electricity as the flow of current from a positive source to a negative source. Electron flow theory describes electricity as the flow of electrons from a negative source to a positive source.

Both theories work well and are accepted by scientists. However, when using either theory, the complete set of rules for that theory must be used.

Voltage

Voltage is a measurement of the pressure that causes electrical energy (current) to flow. The flow of electrons through a wire is much like the flow of water through a pipe. A wire has free electrons that can flow. A pipe is filled with water that can flow. No movement takes place unless a force or pressure is applied. In the case of electrons, voltage supplies the pressure to cause electrical energy to flow. In the case of water, a force as basic as gravity can supply the pressure to cause water flow.

Voltage refers to the electromotive force (EMF) that moves electrons. Electrical pressure is measured in volts and uses the symbol *E*.

Current

Current flow is created when voltage moves electrons through a conductor. This flow is measured in amperes (symbol *A*). A flow of 1 ampere (or amp) means that many billions (6.25×10^{18}) of electrons are passing a defined point each second. This is a very large number. But the electrical charge of a single electron is extremely small.

There must be a complete circuit before current can flow. Even though a voltage may be present, current cannot flow unless there is a return path to the current source. There are two types of current, direct current (DC) and alternating current (AC).

Direct Current *Direct current* flows in a single direction. The direction of flow depends on the polarity of the applied voltage. Most automotive vehicles' electrical circuits use DC current.

Alternating Current *Alternating current* changes its direction of flow in a regular and predictable way. AC current moves back and forth in the circuit, reversing direction whenever the polarity of the applied voltage changes. The current in your house or apartment is AC.

Resistance

Resistance is the opposition to current flow. The amount of resistance is measured in units called *ohms*. The symbol R or the Greek symbol Ω is used to represent resistance.

No material is a perfect conductor of electricity. Even copper, a very good conductor, has some resistance.

The resistance of a conductor depends on the:
• Type of material from which it is made.
• Size of the material (for example, wire gauge).
• Length of the conductor.
• Temperature of the conductor. The lower the temperature, the less the resistance.

Battery cables are made of copper, have a large size, and are short. At normal operating temperatures, battery cables have very low resistance.

Resistance can also be designed into a circuit using devices called *resistors*. They have limited conductivity.

Ohm's Law

The three electrical values—voltage, current, and resistance—are mathematically related. This relationship is defined in Ohm's law. *Ohm's law* makes it possible to calculate any one of the three values if the other two values are known.

In Ohm's law, E = Voltage, I = Current, and R = Resistance.
• Voltage is equal to the current times the resistance ($E = I \times R$).
• Current is equal to the voltage divided by the resistance ($I = \frac{E}{R}$).
• Resistance is equal to the voltage divided by the current ($R = \frac{E}{I}$).

An Ohm's law pie chart can be used to find the unknown values. Covering the unknown value determines whether the other two values are multiplied or divided to find the unknown. **Fig. 1-3.**

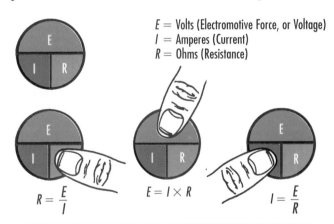

E = Volts (Electromotive Force, or Voltage)
I = Amperes (Current)
R = Ohms (Resistance)

$$R = \frac{E}{I} \qquad E = I \times R \qquad I = \frac{E}{R}$$

Fig. 1-3 Ohm's law states the relationships between voltage, current, and resistance. *What is the voltage if the current is 2 amperes and the resistance is 100 ohms?* (Ford Motor Company)

SECTION 1 KNOWLEDGE CHECK

❶ A 12-volt lightbulb has a resistance of 3 ohms. What is the current through the lightbulb?

❷ Why is glass a poor electrical conductor?

❸ A relay winding draws two amps of current when tested at 12 volts. What is the resistance of the winding?

❹ How many of the 29 electrons in the copper atom can be moved to create current flow?

❺ What voltage would it take to force 5 amps of current through a 5-ohm load?

Section 2

Electrical Automotive Components

The automotive electrical system produces and controls the flow of electricity in the vehicle. Electrical energy operates many systems in the vehicle, including the:

- Starting system.
- Computerized engine control system.
- Lighting system.
- Power accessories (such as electric windows and door locks).
- Climate control system (heating and air-conditioning).
- Audio system.

Wiring harnesses, fuse blocks, printed circuits, and other components are used throughout the vehicle to control the flow of electricity.

Automotive electrical circuits can contain several basic components. They are:

- A battery–stores electrical energy.
- A generator–generates electric current while the engine is running.
- Conductors–connect components in electrical circuits.
- Terminals (plug/unplug type)–allow wires and conductors to be connected or disconnected from a circuit.
- Fuses and circuit breakers–protect electronic circuits from excessive current flow.
- Connectors–keep terminals connected.
- Switches–used to open and close circuit paths.
- Loads–lights, motors, and other devices that use current flow to do useful work.
- Ground–path that returns electrical flow to the energy source.

Battery

A *battery* is an electrochemical device that stores electric current. The battery supplies current to the starter and ignition system when a vehicle is started. It can supply current to the electrical system when generator output is low. The battery also supplies current to maintain electronic memory circuits when the ignition is turned off.

Safety First

Personal Batteries contain an acid and explosive gases. Always wear safety glasses and avoid sparks and open flames when working with batteries. If safety precautions are not followed, personal injury and serious damage to the vehicle may occur.

Generator

The vehicle's **generator** is the device that converts mechanical energy into electrical energy. In motor vehicles, current is produced by mechanical generators, chemical generators, *photoelectric converters*, and *piezoelectric converters*.

Mechanical Generators A mechanical generator converts mechanical energy to electrical energy. The generator is driven by the engine. In operation, magnetic fields created in the *rotor* intercept the wires of the *stator* coil. This motion between the rotor magnetic field and the stator coil wires generates voltage. See Chapter 4, "Diagnosing & Repairing the Charging System," for more information.

Chemical Generators A chemical generator uses a chemical reaction to create current flow. This is how batteries work. See Chapter 2, "Diagnosing & Servicing the Battery," for more information on batteries.

Photoelectric Converters Photoelectric converters, or *photocells,* use light energy to create current flow. Photocells convert light into electricity. *Photo* means "light."

Piezoelectric Converters Piezoelectric converters use physical pressure to create current flow. This method is used in piezoelectric sensors.

Conductors

A material that conducts electrical current is called a conductor. Paths for current flow can be created in several ways.

Automotive Wire Most conductors are made of stranded copper wire covered with plastic insulation.

The size of the wire used is determined by the amount of current the wire will carry. The thickness of the insulation depends on the amount of voltage in the circuit.

General-purpose vehicle wiring conducts about 15 volts and has a thin layer of insulation. By comparison, spark plug wires must conduct several thousand volts. They have thick insulation to keep the high voltage from arcing to ground.

Groups of wires bundled together form a wiring harness. A harness provides a convenient way to route and protect wires in a confined area. Wiring harness illustrations show the wiring location on the vehicle. **Fig. 1-4.**

Printed Circuits Electronic circuits and dashboard wiring often use circuits printed on plastic boards. These printed circuits are called *circuit boards*. Printed circuits are more compact and easier to install. One printed circuit can replace many wiring harnesses and connectors. Printed circuits require less space and are more reliable than single wire connections. **Fig. 1-5.**

Terminals

Terminals are mechanical stampings made of tin-coated brass or steel. One end of the terminal is crimped or soldered onto the end of a wire. The other end is designed to mate with a matching terminal to form a connection to a second wire or a component. The terminal is designed to make a good electrical connection that can be easily separated and reconnected.

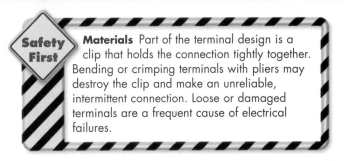

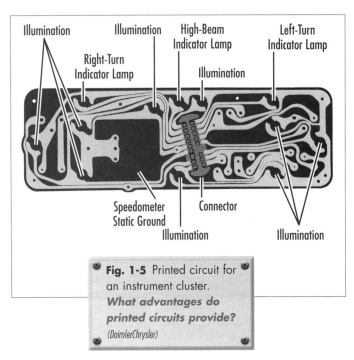

Fig. 1-5 Printed circuit for an instrument cluster. *What advantages do printed circuits provide?* (DaimlerChrysler)

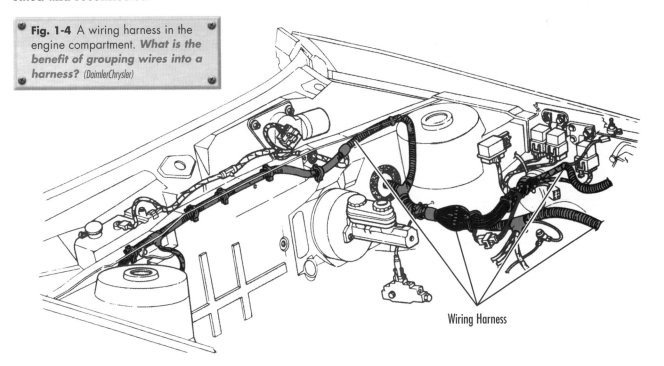

Fig. 1-4 A wiring harness in the engine compartment. *What is the benefit of grouping wires into a harness?* (DaimlerChrysler)

Wiring Harness

Fuses, Fusible Links, and Circuit Breakers

Fuses, fusible links, and circuit breakers are safety devices used to protect vehicle electrical systems. Electrical shorts and defective electrical components can quickly create high current flow. If the current is too high, circuits, components, or wiring will be damaged.

These safety devices work by opening the circuit to stop current flow.

Fuses Fuses contain a metallic element in a glass or plastic package. **Fig. 1-6.** The element is calibrated to melt when the current level in the circuit exceeds the rating of the fuse. Typical fuse ratings used in automobiles include 3, 5, 10, and 20 amps.

(a) Fuse Panel

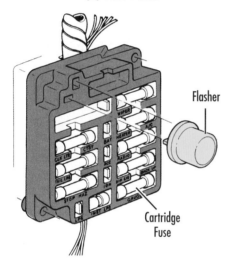

Flasher

Cartridge Fuse

(b) Fuses

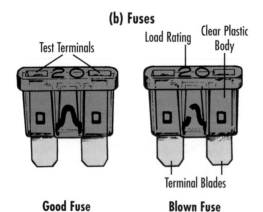

Test Terminals

Load Rating

Clear Plastic Body

Terminal Blades

Good Fuse   **Blown Fuse**

Fig. 1-6 (a) Fuse block or fuse panel with cartridge fuses in place. **(b)** A good and a blown blade fuse. *What might happen if you replace the blown fuse with one having a higher rating?* (Chevrolet Division of General Motors Corporation)

Materials Never replace a blown fuse with a fuse having a higher rating. A blown fuse indicates that a higher-than-normal current flow has occurred. A higher-rated fuse may not blow before damage is done to the circuit. Excess current will damage other devices in the circuit, melt insulation, or cause a fire.

When the fuse element melts (blows), the circuit opens. This quickly stops current flow and prevents damage. A fuse is a one-time use device that must be replaced after the fault is corrected.

Fusible Link A **fusible link** is a short length of insulated wire connected in the circuit. Typically it is four gauge sizes smaller than the wires it is protecting. Like a fuse, it is designed to melt when current flow exceeds the rating for the circuit. Unlike a fuse, it may not show visible signs of having blown. Check its continuity with a digital multimeter.

Some fusible links are easily replaced, similar to a cartridge or blade fuse. **Fig. 1-7.** Others are permanently installed as a part of the wiring harness.

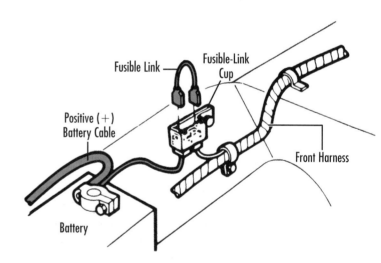

Fusible Link

Fusible-Link Cup

Positive (+) Battery Cable

Front Harness

Battery

Fig. 1-7 One type of fusible link, which uses spade lug connectors. *What do the colors used on fusible links indicate?* (Ford Motor Company)

To replace a fusible link, cut it out as shown at the top. **Fig. 1-8.** Strip the insulation. Splice the wires with a splice clip and solder in the new fusible link. Soldering the connection provides a strong permanent connection. Tape the splice with a double layer of electrical tape.

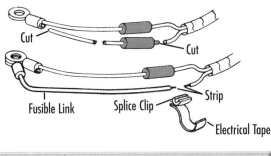

Fig. 1-8 This fusible link is part of the wiring harness. *Why is the fusible link connection soldered?* (General Motors Corporation)

Replacement fusible links may be color-coded to assure that the correct fusible link is used to protect the wire. Brown and black fusible links are used to protect 10- and 12-gauge wires, respectively. Green and orange are used to protect 14- and 16-gauge wires, respectively. The size and location of fusible links can be found in the vehicle's electrical and wiring diagrams.

Circuit Breakers A *circuit breaker* serves the same function as a fuse or fusible link. **Fig. 1-9.** The circuit breaker opens the circuit when it detects high current flow.

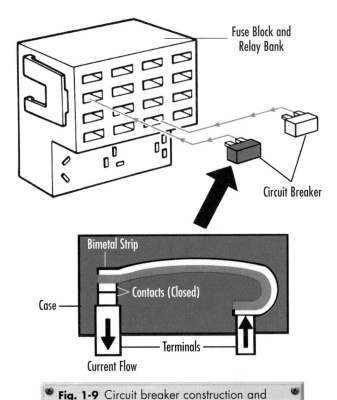

Fig. 1-9 Circuit breaker construction and placement. *Why are circuit breakers not considered one-time devices?* (DaimlerChrysler)

A circuit breaker is a reusable device. It can be reset. A manual circuit breaker must be reset by hand. An automatic circuit breaker resets itself when the bimetal strip cools down.

Connectors

Connectors are usually plastic-bodied connector blocks. They are used to connect and disconnect multiple wires. Connector blocks hold several male and/or female terminals. In this way entire wiring harnesses can be plugged together during assembly.

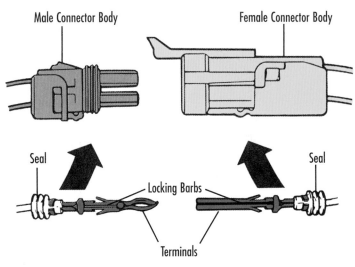

(a) Weather-Pack Connector

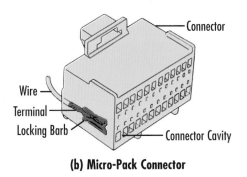

(b) Micro-Pack Connector

Fig. 1-10 Two types of electrical connectors. **(a)** A weather-pack connector. **(b)** A micro-pack connector. Metal terminals attach to the ends of the conductors to make the electrical connection. *Why are locking barbs used on connectors?* (Chevrolet Division of General Motors Corporation)

Connectors usually have rubber seals and outer coatings to protect the terminals. Connectors also have plastic locking barbs to secure the connector block halves together. This prevents mechanical strain on the terminals. **Fig. 1-10.**

Switches

Many types of switches are used in vehicles. Manually activated switches depend upon the driver's action to activate them.

Position switches are turned on or off by the movement of some part of the vehicle. The switch that controls the interior lights when a door is opened or closed is a position switch.

Load Devices

The purpose of an electrical system is to provide electrical energy where it is needed to do useful work. Motors convert electrical energy into rotary motion. Headlights use electrical energy to create light. Solenoids use electrical energy to create back and forth motion.

Electrical Circuits

Several types of electrical circuits are found in automotive wiring. The type of circuit used depends on how and when the electrical components are supplied with power.

Series Circuits

In a series circuit, individual components are connected end to end to form a single path for current flow. **Fig. 1-11(a).** Because there is only one path, the same amount of current flows through every part of the circuit. In addition, any resistance added to or removed from the circuit affects the entire circuit.

Series circuits have two major disadvantages. First, when connected in series, each circuit has to have its own switch and protective device. This is impractical because of the number of components and wires that would be needed. A second disadvantage is that in a series circuit if one component is open, the entire circuit is disabled. For example, if all exterior lighting was connected in series and one lightbulb burned out, none of the other lights would light. For both of these reasons, most automotive circuits are connected in parallel.

Parallel Circuits

In a parallel circuit, two or more loads are connected in separate branches. In most cases the parallel branches are connected in series with a common switch and protective device. **Fig. 1-11(b).** Equal voltage is applied to each branch of a parallel circuit. Current flow divides as it reaches the parallel path. The amount of current flowing through each branch depends on the resistance of that path only. Resistance in one branch of a parallel circuit does not affect current flowing through the other branches. Total current flow in a parallel circuit is equal to the sum of the current in each separate branch.

The backup light circuit is an example of a parallel circuit. Assuming that the bulbs are the same, an equal amount of current will flow through each bulb. The circuits for headlights, taillights, and turn signals are other examples of parallel circuits.

It is common to have both series and parallel connections in the same circuit. A backup light switch is wired in series with its fuse and in parallel with the two backup lights.

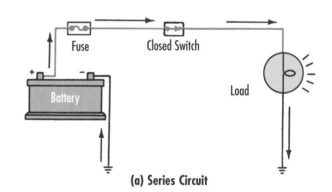

(a) Series Circuit

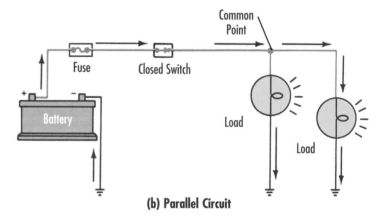

(b) Parallel Circuit

Fig. 1-11 (a) A switch in series with a load. **(b)** A switch in series with two loads connected in parallel. *Must the two loads in the parallel circuits have the same resistance?* (Ford Motor Company)

SECTION 2 KNOWLEDGE CHECK

❶ Name the electricity-producing devices that might be found in a motor vehicle.

❷ What advantages are provided by printed circuits?

❸ How do fuses and circuit breakers differ?

❹ How are fuses and fusible links similar?

❺ Name one advantage of a parallel circuit over a series circuit.

Section 3

Wiring Diagrams and Electrical Symbols

Electrical devices are shown on a printed page as symbols. Electrical circuits are shown as diagrams.

Wiring Diagrams

A **wiring diagram** is a drawing that shows the wires, connectors, and load devices in an electrical circuit. **Fig. 1-12.** A wiring diagram, sometimes

Fig. 1-12 A wiring diagram, or schematic, shows the wires, connections, and components in an electric circuit or system. The components are represented by symbols. *Are the backup lights connected in series or parallel?*

(Ford Motor Company)

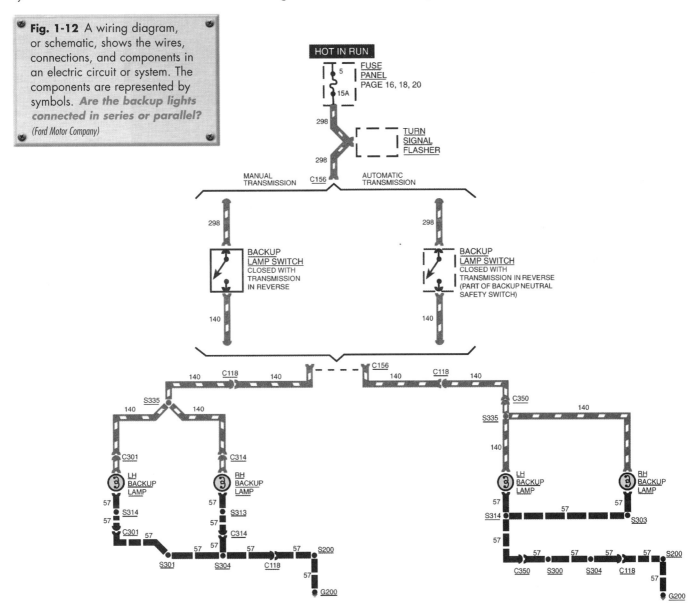

TYPICAL SYMBOLS USED IN ELECTRICAL WIRING DIAGRAMS

Symbol	Name	Symbol	Name
+	Positive	⤙	Connector
−	Negative	→	Male Connector
⏚	Ground	⤙	Female Connector
⊸◠⊸	Fuse	⬇⬇⬇	Multiple Connector
⊸⌒⊸	Circuit Breaker	─S	Denotes wire continues elsewhere
⊶⊣⊢⊷	Capacitor	→•→	Splice
Ω	Ohms	◇ J2	Splice Identification
⊶⋀⋀⊷	Resistor	◆◇	Optional wiring with / wiring without
⊶⋀⋀⊷	Variable Resistor	⊸⌁⌁⊷	Thermal Element (Bimetal Strip)
⋀⋀⋀⋀	Series Resistor	⅄	Y Windings
⊶◡◡◡⊷	Coil	⋁	Delta Windings
Stepup Coil	Stepup Coil	88:88	Digital Readout
Normally Open Contact	Normally Open Contact	⊷⊶	Single Filament Lamp
Normally Closed Contact	Normally Closed Contact	⊷⊶	Dual-Filament Lamp
⊸•⊶⊷	Closed Switch	⊷⊶	Light Emitting Diode (LED)
⊸⁄⊷	Open Switch	⊸⋀⊷	Thermistor
Closed Ganged Switch	Closed Ganged Switch	Gauge	Gauge
Open Ganged Switch	Open Ganged Switch	TIMER	Timer
Two-Pole Single-Throw Switch	Two-Pole Single-Throw Switch	Motor	Motor
Pressure Switch	Pressure Switch	Armature and Brushes	Armature and Brushes
Solenoid Switch	Solenoid Switch	Denotes wire goes through grommet	Denotes wire goes through grommet
Mercury Switch	Mercury Switch	Denotes wire goes through disconnect	Denotes wire goes through disconnect
⊸⊩⊷	Diode or Rectifier	STRG COLUMN	Denotes wire goes through steering column connector
⊸⊩⊷	Bidirectional Zener Diode		

Fig. 1-13 Some of the more common symbols used in electrical wiring diagrams. *Why are symbols used?* (DaimlerChrysler)

TECH TIP Finding Current Flow Direction. In many wiring diagrams, the current is assumed to flow from the power source at the top of the diagram to the ground at the bottom of the diagram.

called a schematic, shows where the various components are connected in the circuit. It shows wiring harness connections and color codes. Wiring diagrams do not show what the components actually look like or where they are located on the vehicle. A wiring diagram is very helpful to technicians when troubleshooting electrical problems.

Electrical Symbols

Electrical symbols are much like the icons used on personal computer displays. The symbols represent components. Symbols used by different manufacturers are similar but not necessarily identical. Charts are usually included in the vehicle service manual to explain each symbol. **Fig. 1-13.** Symbols for complicated or unique subassemblies may be shown as a block diagram. **Fig. 1-14.**

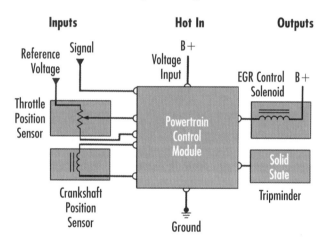

Fig. 1-14 A block diagram made up of boxes labeled with the circuit and device functions. Lines between the boxes represent the wiring that connects the devices. *When are block diagrams used?* (Ford Motor Company)

Electrical Test Equipment

There is a variety of electrical test and measurement equipment available to technicians. This equipment can be used to isolate the cause of electrical problems. General-purpose testers measure voltage, current, and resistance. Special-purpose testers may read engine trouble codes or display spark signal waveforms or other unique information.

Test Lights

Two types of test lights are in common use. These devices are used to check for voltage or for electrical continuity. They can be used to check for open circuits in motor and coil windings. They can check fuses, fusible links, and open circuits.

Circuit-Powered Test Lights Circuit-powered test lights are powered by the circuit being tested. They have two test leads; one lead connects to ground, and the other has a probe tip. A 12-volt bulb is connected between the ground lead and the probe tip. **Fig. 1-15.**

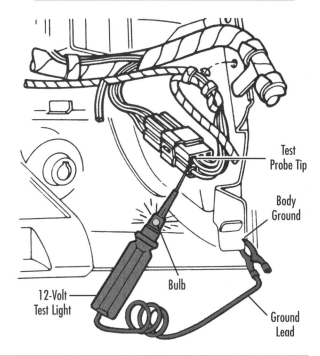

Fig. 1-15 A circuit-powered test light. Voltage from the circuit being tested powers a 12-volt test light. *If the light does not operate when testing a circuit, what conclusion can you draw?* (Ford Motor Company)

Test Probe Tip
Body Ground
12-Volt Test Light
Bulb
Ground Lead

Using Electrical Symbols

We use symbols every day. For example, letters are symbols for sounds. Symbols are often used as shortcuts for words, things, or ideas. We understand symbols when we learn what they mean. We know what a stop sign or a yield sign means without reading the words. Many symbols are accepted internationally. Then people who speak different languages can understand and use the same symbols.

Schematic wiring diagrams, or circuit diagrams, use symbols to indicate the components of electrical circuits. When you consult a vehicle service manual to read a schematic wiring diagram, be aware that not all manufacturers use the same symbols. You may need to find the chart of electrical symbols used in the manual if you see a symbol with which you are not familiar.

Apply it!

Meets NATEF Communications Standards for using text resources to gather data.

Study **Figs. 1-12** and **1-13** in this chapter.

❶ What companies publish these diagrams?

❷ On a sheet of paper, make a comparison chart of three electrical symbols. Draw the symbols for a single filament lamp or load, an open switch, and a ground, as shown in each of the figures.

❸ Compare the versions for each electrical component. Write down your observations based on your drawings of these symbols.

❹ Check several service manuals. Find out if the symbols used in their schematic diagrams differ from the ones you have drawn.

If the test probe contacts a 12-volt source, the bulb lights. Circuit-powered test lights can indicate whether voltage is available at a connection or test point.

Self-Powered Test Lights Self-powered test lights are also called *continuity testers.* A continuity tester checks for a complete circuit path, or continuity from one point to another. It has its own power source, usually a 1.5-volt battery.

One lead from the tester is connected to one point in a circuit. The test probe is connected to another point. If the circuit is complete, that is, if the wire has continuity, the bulb in the tester will light. **Fig. 1-16.**

A continuity tester cannot be used when more than one path exists between the two points being checked. If one path is open, the other path will still provide continuity. The open path will not be identified.

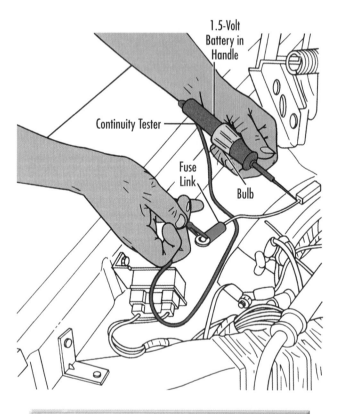

Logic Probes

Logic probes are sometimes called *electronic circuit testers.* They can be used to safely test electronic circuits. They use light-emitting diodes (LEDs), either singly or in combination, to display test results. Logic probes do not display values. The LEDs only indicate the presence of voltage, resistance, or polarity at the test point. **Fig. 1-17.**

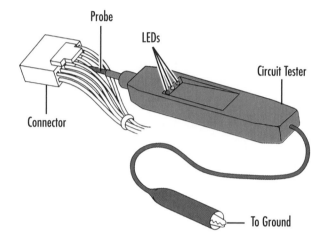

Volt-Ohm-Meters

Volt-ohm-meters (VOMs) are the most common general-purpose testers used by a technician. They are also called multimeters because they perform multiple functions. A **multimeter** measures voltage, current, and resistance. Many models have additional features, such as diode testing and frequency measurement.

A typical multimeter has two test leads with metal probes or test clips on the end. The leads must be connected to the proper terminals (jacks) on the front of the meter. It may be necessary to move the test leads to other jacks to measure other electrical units. The positive test lead is usually red. The negative lead is usually black. The leads are used to connect the meter to a circuit or to a component to make specific measurements. The leads should be connected with the correct *polarity*.

The rotary switch on the front of the meter is used to select the electrical value to be measured. In many cases the meter range (scale) is selected by the same switch. Accurate testing requires the use of the correct range. Refer to the instruction manual for the meter for specific instructions regarding its use.

Safety First **Materials** When using probes to make electrical measurements, do not damage wiring insulation or distort terminals. This could damage otherwise good components or connections. If it is necessary to penetrate insulation, be sure to reseal the penetration with electrical tape or RTV sealant. Failure to reseal could result in a short circuit.

Analog Multimeters Analog multimeters use a moving needle to indicate the value being measured. **Fig. 1-18.** Most of these meters have several ranges of measurement. Select the correct range before making a measurement. Do not use analog multimeters to test electronic components and circuits.

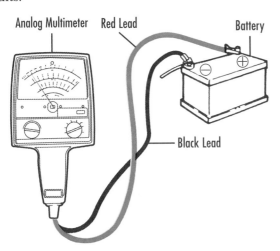

Fig. 1-18 An analog multimeter can be used as a voltmeter, ammeter, or ohmmeter. The relative position of the needle or pointer to the background scale indicates the reading, or measurement. *Why is a digital meter used for testing electronic components and circuits?* (Ford Motor Company)

Digital Multimeters Digital volt-ohm-meters (DVOMs) display readings numerically. The voltmeter on a DVOM has very high impedance (resistance), usually at least 10 million ohms (10 megohms). With high impedance, the current flow through the voltmeter will be very low and the effect of the meter on the circuit will be minimal. High impedance meters protect circuit components and ensure accurate readings while measurements are being made. Use only digital meters when testing electronic components and circuits. **Fig. 1-19.**

Fig. 1-19 A typical digital multimeter. *What unit is used to signify resistance?* (Fluke Corporation)

Scan Tools

Scan tools are devices used to read a vehicle's data stream and trouble codes. Many computerized systems in the automobile have self-diagnostic capabilities. The system computer can test circuits to determine those that are not working correctly. The computer (PCM) stores a *diagnostic trouble code (DTC)* for all circuits that fail the test. Scan tools, sometimes called scanners, can give a direct digital readout of these stored trouble codes. **Fig. 1-20.** Most scanners allow you to read data while the vehicle is operating.

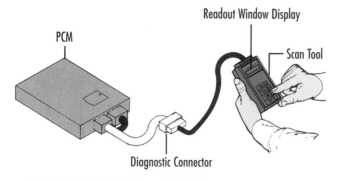

PCM

Readout Window Display

Scan Tool

Diagnostic Connector

Fig. 1-20 Scan tool retrieving trouble codes through a diagnostic connector. *What device stores trouble codes that are read out using scan tools?* (OTC Tool & Equipment Division of SPX Corporation)

Oscilloscopes

Oscilloscopes are devices that display voltage changes within a certain time period. Many voltages found in the automobile are not simply AC or DC. The voltages associated with fuel injectors and ignition coils have unique waveforms. If a component is not operating correctly, the waveform will change. Analyzing a waveform can aid in diagnosing electrical problems. **Fig. 1-21.**

Jumper Wire

A jumper wire is a short piece of wire used as a temporary connection between two points in a circuit. It may have alligator clips or other terminals on each end. Some jumper wires include a fuse or circuit breaker. Jumper wires should be used only to bypass non-resistive components such as switches and wires. Do not use a jumper wire to bypass circuit protection devices or load components such as lightbulbs or motors. **Fig. 1-22.**

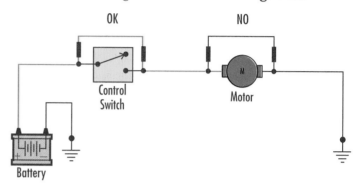

OK

NO

Control Switch

Motor

Battery

Fig 1-22 Correct and incorrect use of a jumper wire. *What will happen if a jumper wire is connected across a load device?* (Ford Motor Company)

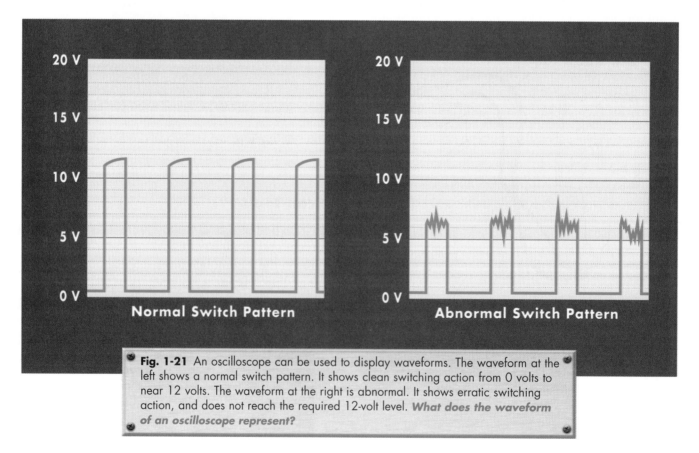

Normal Switch Pattern

Abnormal Switch Pattern

Fig. 1-21 An oscilloscope can be used to display waveforms. The waveform at the left shows a normal switch pattern. It shows clean switching action from 0 volts to near 12 volts. The waveform at the right is abnormal. It shows erratic switching action, and does not reach the required 12-volt level. *What does the waveform of an oscilloscope represent?*

Measuring Circuit Resistance

You can use Ohm's law to calculate the current and voltage in an automotive circuit. This will help you diagnose problems. However, unless you are careful, the values you calculate may not be correct.

Some circuits need to be analyzed before applying Ohm's law. Here's a simple example. The lamp filament of a 3496 replacement lamp for brakelights and taillights will measure about 0.6 ohms. Ohm's law predicts a current of about 20 amps:

$$\frac{12 \text{ volts}}{0.6 \text{ ohms}} = 20 \text{ amps}$$

But when the light is on, the filament glows brightly and becomes very hot. When this happens the resistance increases. The bright filament will actually be about 5 ohms. It will draw about 2.4 amps.

Power is the rate at which electrical energy is delivered to the circuit. The watt is the basic unit of power. The metal base of the lamp will show a 27-watt rating for the lamp. You can use the power formula to estimate the current drawn when the lamp is on.

$$P = E \times I \text{ or } I = \frac{P}{E}, \text{ where } P \text{ is the wattage}$$

$$\text{So, } \frac{27 \text{ watts}}{12 \text{ volts}} = 2.25 \text{ amps}$$

Use Ohm's law to estimate the actual resistance of the lamp in operation. You will find that the resistance is about 5 ohms.

Here is what happens. Before you apply Ohm's law to a *working circuit*, think about whether the current flow will cause anything to change. In our example, it did. Current flow through the filament increased its resistance. The "cold" resistance measured with the DVOM is the incorrect value to use in Ohm's law.

Assume that you measure the resistance of a motor with a DVOM. If you then use that measured resistance to estimate the current drawn in operation, you will be misled.

Some parts of a motor have a different resistance when the motor is spinning. That makes the actual resistance less than expected. During operation, the rotating armature of a motor generates a small countervoltage. Countervoltages have the opposite polarity to applied (battery) voltage. In an operating motor, battery voltage must overcome not only the resistance of the circuit, but the countervoltage as well. For this reason, using only a resistance measurement to calculate current flow in a motor circuit will lead to incorrect answers.

Apply it!

Applying Ohm's Law

Meets NATEF Science Standards for Ohm's law and the relationship of heat and resistance in electrical systems.

A 55-watt headlight measures about 0.5 ohms when cold.

❶ Use the power formula to estimate the current this light will draw when turned on.

❷ Use Ohm's law to estimate the light's "hot" resistance.

Results and Analysis

This problem gives another example of the importance of understanding the construction and operation of the components and circuits you work on.

Remember that Ohm's law applies *only* to loads that do not change their resistance under any condition. Wires, connectors, switches, and relay contacts should behave as predicted by Ohm's law. Always ask yourself, "Does the circuit I am testing contain a lamp or a motor that may alter the reading?"

Safety First

Materials Never use jumper wires to bypass load components such as motors, solenoids, or lightbulbs. Using jumper wires will cause excess current to flow through the circuit. Excess current will damage other components in the circuit and may be high enough to melt insulation or cause a fire. If in doubt, use only a fused jumper wire. This type of jumper wire has a fuse to protect against excessive current.

Electrical Circuit Problems

Troubleshooting an electrical circuit begins with knowing how the circuit should work. A wiring diagram is a guide to a current flow path. In addition to the wiring diagram, troubleshooting charts and basic system checks are available in the vehicle service manual. Use these to help locate the problem.

Circuit and wiring problems are most likely to occur at connection locations. Check ground and wiring harness connections, and make sure the terminals are tightly connected to the wire and are not damaged or corroded.

There are several basic steps to good electrical troubleshooting.

• Verify the customer complaint. (Identify the problem you need to solve.)
• Based upon the verified symptoms, make a preliminary diagnosis about what part of the system or what component is not working correctly.
• Use the vehicle service manual and your technical knowledge to determine how the system is supposed to function.
• Carefully inspect the affected system for damage or obvious causes of the problem.
• Make sure that all of the conditions are met for the system to operate.
• Perform tests to see if the problem is in the component or the circuit.

Open Circuit

An *open circuit* occurs when the electrical path is broken. Common causes of an open circuit are a burned-out fuse or lightbulb, a broken wire, or a defective switch. An open circuit stops current flow and prevents operation of devices in the circuit. **Fig. 1-23.** A jumper wire can be used to locate an open circuit. As an example, if a circuit does not work when the switch is turned on, connect a jumper

wire from the battery side of the switch to the load side. If the circuit now works, the switch is defective. Open circuits can also be found by visually inspecting wiring and connectors and by testing a circuit for voltage with a voltmeter or test light. An ohmmeter or self-powered test light can be used to test parts of a circuit and individual components for continuity.

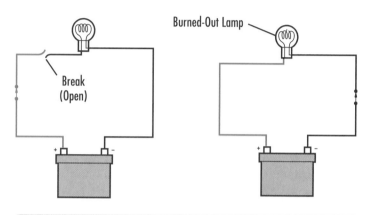

Burned-Out Lamp

Break (Open)

Fig 1-23 An open circuit is a break in the current flow path. Causes include a broken wire or a burned-out lamp. *Name two devices used to test for open circuits.* (Ford Motor Company)

Short Circuit

A *short circuit* occurs when two (or more) conductors touch each other where no connection is intended. In most cases, the insulation has been damaged. As a result, the bare wires come in contact with each other. A "short to power," or "wire-to-wire," short occurs when a conductor that is not powered comes in contact with one that is. This problem causes circuits to work when they are not turned on. For example, a short to power between horn wiring and turn signal wiring could cause the horn to sound when the turn signals are used. A short circuit in a coil of wire, such as a relay or solenoid, will bypass part of the normal resistance of the circuit, causing excessive current flow. This type of short circuit frequently causes a fuse to burn out or a circuit breaker to cycle on and off.

To locate a short circuit in wiring, note the unintended operation of a circuit. Use a wiring diagram to determine where the problem is most likely to occur. Inspect wires and wiring harnesses for physical damage. Short circuits can be found by measuring resistance or current draw. If the resistance is lower or the current draw is higher than specified, a short circuit is indicated. **Fig. 1-24.**

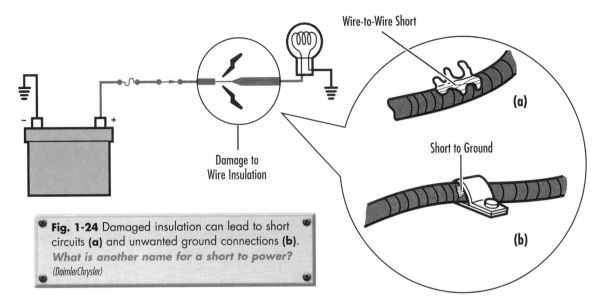

Fig. 1-24 Damaged insulation can lead to short circuits **(a)** and unwanted ground connections **(b)**. *What is another name for a short to power?* (DaimlerChrysler)

Grounded Circuit

Most automotive circuits are connected to "ground" as the return circuit to the battery and generator. When an unintended ground connection occurs, the circuit cannot operate normally. Unintended grounds are usually "copper to iron" connections. These are caused by damage to wire insulation. For example, a powered wire (copper) touches a metal surface of the engine, body, or chassis (usually iron). Current in the circuit returns to the battery or generator at that point instead of flowing through the intended load. The result is usually an open fuse or circuit breaker and no system operation. This problem is often referred to as a "short to ground." A short to voltage and a short to ground are not the same. While both problems may be referred to as a "short," the causes and testing procedures for each are different.

To test for unwanted ground connections, turn off the power. Disconnect any intentional grounds. Use a self-powered test light or ohmmeter to check for continuity between the conductor and a nearby clean metal surface. If there is continuity, the circuit is grounded.

Excessive Circuit Resistance

Excessive resistance in a circuit is one of the most common causes of electrical and electronic system failures. **Fig. 1-25.** The problem occurs when a circuit has more resistance than intended. Excessive resistance may be caused by:
• Loose or dirty connections.
• Undersized wires or cables.
• Broken strands of wire.
• Excessive heat buildup.
• Burned switch contacts.
• Excessively long wires.

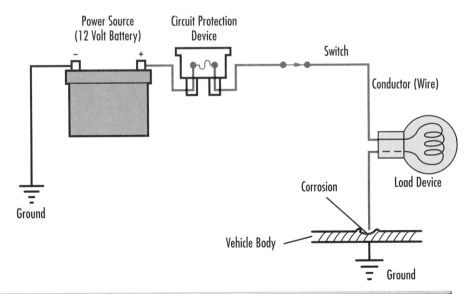

Fig. 1-25 Excessive resistance can cause a circuit to operate at less than normal efficiency. *How is excessive resistance found in a circuit?* (Ford Motor Company)

Excessive resistance causes a circuit to operate at less than normal efficiency. For example, a rusted ground connection in the headlight circuit may cause one headlight to be dim. If the connection is bad enough, the circuit will be open and the headlight will not come on at all.

Excessive resistance is found by measuring *voltage drop*. Connect a DVOM across the portion for the circuit to be tested. Operate the circuit so that there is current flow. If the voltage drop is higher than specified for that circuit, the resistance is higher than normal.

SECTION 3 KNOWLEDGE CHECK

❶ What does a wiring diagram show?

❷ A multimeter shows measurements of what three electrical quantities?

❸ Where are trouble codes stored?

❹ What information can an oscilloscope provide that a multimeter cannot?

❺ Name three types of circuit faults.

Section 4

Electronic Automotive Components

The term "electricity" is normally used to describe the flow of electrons through conducting materials such as copper, silver, and iron. The term "electronics" deals with the behavior and effects of electrons and with electronic devices.

Uses of Semiconductors

Pure silicon is neither a good conductor nor a good insulator. Semiconductors are manufactured by adding small amounts of certain elements to a thin silicon wafer, or "chip." The process of adding these elements is called "doping." One side of the chip is doped so that extra electrons will be present. The extra electrons make this layer of silicon negative. The other side of the chip is doped with a different element, causing fewer electrons to be present. This side will be positive.

Doping creates a positive (P) and a negative (N) layer on the same silicon chip. The two layers are known as a "PN junction." This is the basis for a diode, a common component of electronic circuits. A *diode* is a solid-state electronic device that allows the passage of an electric current in one direction only.

The basic PN junction can be combined in various ways to produce many electronic devices. Two of the more common uses of PN junctions are in *light-emitting diodes (LEDs)* and *transistors*. Silicon chips can be designed to control the flow of electrons in simple and complex circuits. Thousands of tiny electronic devices can be combined on a single silicon wafer to form an integrated circuit (IC). **Fig. 1-26.** These ICs are used in circuits such as *powertrain control modules (PCMs),* radios, electronic climate controls, and anti-lock brake control modules.

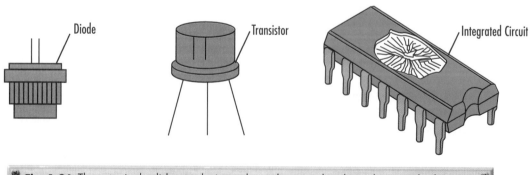

Diode   Transistor   Integrated Circuit

Fig. 1-26 These typical solid-state devices, shown here much enlarged, are a diode, a transistor, and an integrated circuit. *What is the most common material used to make semiconductor devices?*

Because they have no moving parts, electronic components are often referred to as *solid-state* devices. With no moving parts, electronic components cannot wear out. Other advantages of solid-state devices are small size, low cost, and very high operating speeds. Whenever possible, mechanical components on motor vehicles have been replaced with solid-state devices that can do the same job more efficiently.

Solid-State Components

A number of solid-state components can be made using combinations of the basic PN junction. Among the more common devices are diodes, Zener diodes, light-emitting diodes, transistors, and thermistors.

Diodes Diodes are one-way electronic check valves for current. **Fig. 1-27.** A diode allows conventional current flow (+ to −) only in the direction of the arrow in its symbol.

A diode "turns on" and conducts only when the correct voltage is applied to its terminals. If the applied voltage is too low or of the wrong polarity, the diode blocks current. Diodes can be tested with the DIODE TEST position on a DVOM. Connect the test leads to the diode and observe the meter. Reverse the leads and test again. The diode should conduct in one direction only. Diodes are commonly

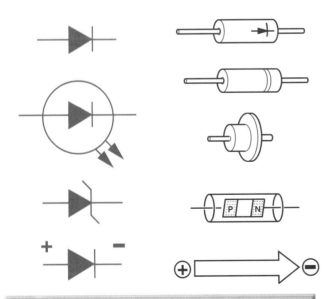

Fig. 1-27 Diodes can be represented in several ways. *In which direction will conventional current flow through a diode?* (General Motors Corporation)

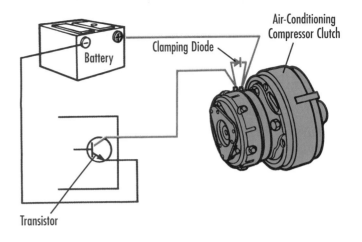

Fig. 1-28 A clamping diode protects against surge voltage. *How is the diode connected?* (General Motors Corporation)

used as rectifiers. Six diodes are used in an automotive generator to rectify (change) AC to DC.

Another common use for a diode is to prevent voltage spikes (momentary high-voltage surges). Voltage spikes can damage other electronic components. In this case a diode is connected in parallel with a winding. **Fig. 1-28.** When current flow through the winding stops, a high voltage is induced in the winding. This voltage can damage the module that controls the circuit. The diode provides a path to use up the unwanted voltage without having the voltage go through the module. Used in this way, the diode is called a *clamping diode* because it "clamps down" on induced voltage spikes. **Fig. 1-28.**

 TECH TIP Clamping Diodes. A defective module may have been caused by an open clamping diode. If a clamping diode is used in the circuit, check it before installing a new module.

Zener Diodes Zener diodes are a special type of diode. They have been doped so they are able to conduct in a reverse direction without being damaged. The zener diode symbol is similar to that of a regular diode except for the small lines added to the vertical line representing the cathode. Zener diodes are most often used as voltage sensors. Electronic voltage regulators use a zener diode to determine when voltage regulation occurs.

Light-Emitting Diodes Light-emitting diodes (LEDs) are similar to regular diodes except they are designed to give off light. They are produced by doping a silicon chip with different elements to produce red, green, orange, yellow, or blue light. There is a small lens on the diode housing to make the light visible. The symbol for an LED has arrows pointing away from the basic diode symbol. This indicates that light is being given off.

LEDs can be made very small and use very little current. They are used as indicating lamps on dash displays and in optical-type sensor circuits. Large LEDs have been used as stop lamps in some applications, such as in the Chrysler LHS.

Transistors Transistors are also constructed from layers of P and N material. They have three terminal connections instead of two. There are several types of transistors, all operating in a somewhat similar manner. **Fig. 1-29.** A transistor is often used as an electronic switch or relay. When the transistor is off, no current can flow. When the correct volt

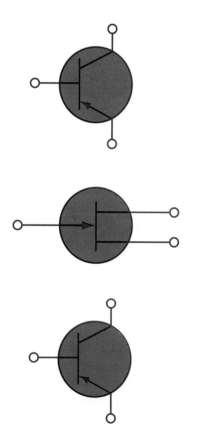

Fig. 1-29 Several different symbols are used to indicate a transistor. *Transistor operation is similar to what electrical device?*

age is applied to a control circuit, the transistor turns on and conducts current.

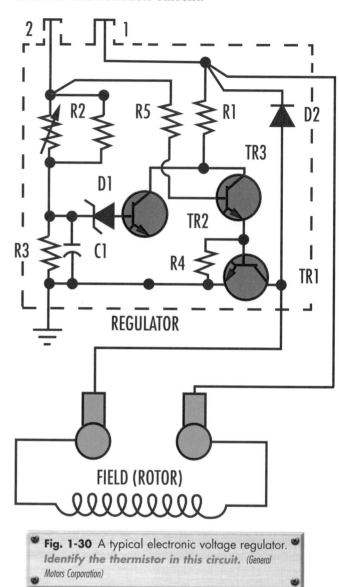

Fig. 1-30 A typical electronic voltage regulator. *Identify the thermistor in this circuit.* (General Motors Corporation)

With no moving parts, the transistor can be used as a high-speed switch, turning a circuit on and off thousands of times per second. A transistor can also be used as an amplifier. When a transistor is used in this way, a small amount of control current regulates a much larger amount of output current.

Thermistors *Thermistors* are small solid-state devices used for temperature sensing. The resistance of a thermistor may change from a few ohms to several million ohms as temperature changes. One symbol for a thermistor is the sawtooth resistor symbol with an arrow drawn through it. **Fig. 1-30.**

Also used is the resistor symbol with the letter *T* next to it. Thermistors are used as engine coolant temperature (ECT) sensors and intake air temperature (IAT) sensors. They are also used to provide temperature compensation in electronic voltage regulators.

Handling Solid-State Components

Technicians must use caution to avoid damaging sensitive electronic parts when servicing them. High temperatures, physical impact, high voltage, reverse polarity, and static electricity may damage solid-state components.

High temperature problems are minimized by mounting electronic components on an aluminum housing. Aluminum transfers heat more effectively than most materials. This protects the electronic components from heat related damage. In some cases special heat-transfer grease must be used between a component and the surface on which it is mounted.

The silicon chip that is the basis for all electronic components is a brittle, crystalline material. Internal connections may be broken if a component is exposed to sudden impact. Avoid striking or dropping components containing solid-state circuits.

In general, electronic circuits are designed to operate within the voltage range produced by the vehicle's electrical system. The technician must use care to avoid creating high-voltage spikes while working on the vehicle. High voltage in the electrical system may be caused by:
• Disconnecting a component or its wiring while the circuit is in operation.
• Disconnecting the battery while the engine is running.
• Improper use of a battery charger.

Damage to electronic components is frequently caused by reversed polarity connections. Make sure the cables are connected correctly when installing a battery, using jumper cables, or charging a battery.

A form of static electricity known as electrostatic discharge (ESD) can damage some electronic modules. The static energy builds up during normal work activities such as sliding across a car seat. ESD can damage a module without the technician's knowledge. Circuits containing modules that may be damaged by ESD are marked with a special symbol. **Fig. 1-31.**

To avoid damaging ESD sensitive modules, follow these steps:
• Read and follow the directions in the shop manual and on packaging.
• Leave a new module in its protective packaging until ready for installation.
• Do not touch any terminals on the module.
• Before handling a module, touch a nonsensitive metal surface on the vehicle. This will discharge any static voltage present.
• Anti-ESD wrist straps and work mats are available. Use them when recommended.

Fig. 1-31 A warning label for circuits containing components that may be damaged by electrostatic discharge. *What is another name for electrostatic discharge?* (General Motors Corporation)

SECTION 4 KNOWLEDGE CHECK

❶ What is the most common element used in making semiconductors?

❷ What automotive component uses diodes as rectifiers?

❸ What is the primary use of a Zener diode?

❹ What electronic device works like a relay or switch?

❺ Name two things a technician can do to avoid creating spike voltages when servicing a vehicle.

CHAPTER 1 REVIEW

Key Points

Meets the following NATEF Standards for Electrical/Electronic Systems: using wiring diagrams; checking and measuring electrical circuits.

- All matter is made of atoms. Atoms have three components—protons, neutrons, and electrons.
- Electrical current is carried through a conductor by "free" electrons.
- Ohm's law defines the relationships among voltage, current, and resistance.
- Ohm's law states that the voltage in a circuit is equal to the current multiplied by the resistance: $E = I \times R$.
- Fuses, fusible links, and circuit breakers protect electrical circuits by opening if excessive current occurs.
- Wiring diagrams show how circuit components are connected.
- Volt-ohm-meters, test lights, scan testers, and oscilloscopes are used to test electrical circuits.
- A test light can be used to locate a grounded circuit.
- Electronic modules can be damaged by electrostatic discharge (ESD).

Review Questions

1. How do "free" electrons carry current through a conductor?
2. According to Ohm's law, what is current (I) equal to?
3. What is the voltage (E) if the current (I) is 2 amperes and the resistance (R) is 6 ohms?
4. Is it possible for the same circuit to have both series and parallel branches? Explain.
5. What type of test instrument would you choose to measure voltage in sensitive electronic circuits?
6. What tool would you use to read out the trouble codes on a motor vehicle?
7. What steps would you follow to avoid damaging ESD sensitive modules?
8. **Critical Thinking** While checking a vehicle wiring problem, you notice some burned, brittle, and cracked insulation. What type of circuit problem would you suspect? What condition would you look for?
9. **Critical Thinking** How would you use a test light to locate an "open" in a simple fused circuit?

TECHNOLOGY FORECAST
FOR AUTOMOTIVE EXCELLENCE

42 Volts of Power!

A new automotive system is on its way. What's in it for car owners? It has what drivers today want—more power.

A 42-volt system could soon take the place of the 12-volt system used in vehicles for the past few decades. The new system is needed because the current 12-volt design can't handle the electronic gadgets car owners love.

Today's cars already have all kinds of electronics. A 42-volt system will provide plenty of power for these devices. It will also power such new devices as electronic valves, a flywheel starter/alternator, and electric brakes.

And that's not all. The new system will allow smaller wires and parts. This will mean less weight, more efficiency, lower cost, and better packaging.

Plus, mechanical relays should no longer be needed. This will make diagnosis of systems with solid-state switches much easier. Circuits can be protected without traditional fuses. Multiplexing can be more widely used. Multiplexing is sending multiple signals over one wire.

An interesting fact is that the 42-volt system will likely require two batteries. One would be a small 36-volt design for starting the engine. The other would be rated at 12 volts and provide electricity after the engine is started.

AUTOMOTIVE EXCELLENCE TEST PREP

Answering the following practice questions will help you prepare for the ASE certification tests.

1. Higher than normal current draw in a circuit could be caused by:
   - a A burned-out lightbulb.
   - b A short circuit.
   - c Poor connections.
   - d An undersized wire.

2. Technician A says making a voltage measurement requires connecting the multimeter in series with the power supply. Technician B says the multimeter should be across or in parallel to the power supply. Who is correct?
   - a Technician A.
   - b Technician B.
   - c Both Technician A and Technician B.
   - d Neither Technician A nor Technician B.

3. Excessive resistance in a circuit could be caused by:
   - a A short circuit.
   - b A burned-out fuse.
   - c A missing lightbulb.
   - d A defective switch.

4. The correct replacement for a battery cable is not available. Technician A says a heavier cable can be used if it fits properly. Technician B says a lighter gauge cable might overheat. Who is correct?
   - a Technician A.
   - b Technician B.
   - c Both Technician A and Technician B.
   - d Neither Technician A nor Technician B.

5. Which of the components listed produces a voltage?
   - a The wiring harness.
   - b A fusible link.
   - c The generator.
   - d A printed circuit.

6. The check of an electrical circuit shows the fusible link is open and the outer cover blistered. Technician A says the link could have broken from mechanical vibration. Technician B says the cause is likely to be a short circuit that melted the link. Who is correct?
   - a Technician A.
   - b Technician B.
   - c Both Technician A and Technician B.
   - d Neither Technician A nor Technician B.

7. On a vehicle with four taillights, one bulb does not light. This could be caused by:
   - a A burned-out fuse.
   - b A defective headlight switch.
   - c A defective light socket.
   - d Low battery voltage.

8. Technician A says a digital volt-ohm-meter can replace both a circuit-powered test light and a self-powered test light. Technician B says a scan tool is needed to get a direct readout of stored trouble codes. Who is correct?
   - a Technician A.
   - b Technician B.
   - c Both Technician A and Technician B.
   - d Neither Technician A nor Technician B.

9. Technician A says electrical symbols in a wiring diagram show how the components of the electrical system are connected. Technician B says the wiring diagram is used to show how the components operate. Who is correct?
   - a Technician A.
   - b Technician B.
   - c Both Technician A and Technician B.
   - d Neither Technician A nor Technician B.

10. A defective electronic module may have been caused by an open:
    - a Transistor.
    - b Thermistor.
    - c Clamping diode.
    - d Inductive winding.

Diagnosing & Servicing the Battery

You'll Be Able To:

- Explain the purpose, construction, and operation of the battery.
- Test the state of charge of a battery.
- Conduct a battery load test and an electronic battery test.
- Demonstrate how to slow charge and fast charge a battery.
- Explain how to service and replace a battery.
- Demonstrate the safe use of jumper cables.

Terms to Know:

battery jumper box

cold-cranking amps (CCA)

electrolyte

hydrometer

key-off loads

load test

reserve capacity

The Problem

Cecilia's vehicle is towed to your service center. She tells the service writer that the battery in her car was dead when she tried to start it this morning.

The service writer asks Cecilia if this problem has ever happened before and under what circumstances. Cecilia said the first time it happened was in the parking lot at the airport. The battery was dead when she returned from a week-long trip. The second time was in the driveway at her home after the car had not been driven for a few days.

Your Challenge

As the service technician, you need to find answers to these questions:

1. What could cause a battery to discharge after several days of standing?
2. What are the most likely causes of the problem?
3. What diagnostic checks can you perform to isolate the problem?

Section 1

Battery Construction and Operation

The automotive battery:

- Stores the electrical energy needed to operate the starter, ignition system, and fuel system during cranking.
- Supplies "key-off" power for lights and accessories.
- Supplies the power needed to operate various key-off loads. **Key-off loads** are devices that draw current even when all switches are turned off. Examples are computer and radio memory circuits. Key-off loads are sometimes referred to as *parasitic drains*.
- Provides some of the power for electrical loads during normal operation. This is necessary when the load exceeds the generator's output.
- Serves as a load regulator to dampen *transient voltages* that occur in the electrical system. These voltage spikes might otherwise damage sensitive electrical and electronic components.

Safety First **Personal** Always wear safety glasses when working around batteries. Batteries contain sulfuric acid. Do not allow the acid to touch any part of your body or clothing. Use clean water to flush any area that comes in contact with battery acid. See a doctor if needed.

Automotive batteries use lead and acid to store energy in chemical form. The chemicals used in the battery determine how much voltage it will produce. The amount of lead and acid in the battery determines the amount of energy (current) the battery can store. Drawing current out of the battery causes it to be discharged. An advantage of the lead-acid type of battery is that it may be recharged easily. Recharging restores the electrical energy lost during discharge.

Figure 2-1 shows the key parts of a typical automotive battery.

- The plastic case and cover serve as a container for the battery components.
- Vent caps cover cell openings. They are non-removable on maintenance-free batteries. Conventional batteries have removable vent caps. The vent allows the gases formed during battery operation to escape to the atmosphere.

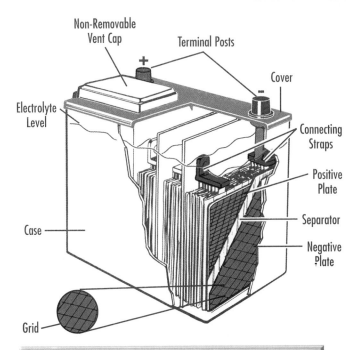
Fig. 2-1 Cutaway view of a maintenance-free battery. Major components are identified. *What two liquids make up the electrolyte?* (Ford Motor Company)

- The active materials are lead peroxide and sponge lead pastes that become the positive and negative *plates*.
- The grid is the lead alloy framework that holds the active plate materials.
- Separators are porous insulators used to separate the plates while allowing the electrolyte to flow between the plates.
- An **electrolyte** is a compound that conducts an electrical current in a water solution. In an automotive battery, the electrolyte is a mixture of sulfuric acid and water.
- In each cell, connecting straps connect plates of the same polarity. They also connect the cells together to form the battery. The end straps are connected to the battery terminals. Connecting straps are made of a lead alloy.
- Terminal posts are the connection points for battery cables. Some batteries have top posts; others have threaded side terminals. Some have both.

Safety First **Personal** Batteries can produce very high current. Remove all jewelry, including watches, necklaces, and similar items before working around batteries. Remove or put tape over rings. Keep tools and other metallic objects that could cause an arc away from battery terminals.

AUTOMOTIVE
MATHEMATICS
EXCELLENCE

Applying Ohm's Law to Series Circuits

A 12-volt automotive battery has six cells. Each cell has a voltage of about 2 volts. When the cells are connected in series, the combined voltage is higher than the voltage of any single cell by itself. The total voltage in a series circuit is the sum of the individual voltages. The formula for the total voltage, V_T, of the 12-volt battery is:

$$V_T = V_1 + V_2 + V_3 + V_4 + V_5 + V_6$$

In general, the total resistance in a series circuit is the sum of the individual resistances. You can write the formula for the total resistance, R_T, in a series circuit if there are three individual resistors, R_1, R_2, and R_3, as:

$$R_T = R_1 + R_2 + R_3$$

Ohm's law still applies in a series circuit. The total resistance, R_T, and the total voltage, V_T, are used in the formula for the circuit.

$$V_T = I \times R_T$$

You can find R_T and the current if you know the voltage and the value of each resistance in a series circuit. Consider a series circuit with a 12-volt battery and three resistors, whose values are 1.0 Ω, 2.5 Ω, and 2.5 Ω.

First, find the total resistance:

$$R_T = 1.0\ \Omega + 2.5\ \Omega + 2.5\ \Omega = 6\ \Omega$$

Use Ohm's law to find the current.

$V = I \times R_T$. Thus:

$$I = \frac{V}{R_T} = \frac{12\,V}{6\ \Omega} = 2\ \text{amps}$$

Apply it!

Meets NATEF Mathematics Standards for calculating algebraic expressions using addition and division.

❶ What is the total resistance and the current in a series circuit with a 6-volt battery and two resistances of 6 Ω and 4 Ω?

❷ What is the current in a circuit with two 6-volt batteries and two resistances of 0.15 Ω and 0.35 Ω, all connected in series?

Battery Construction

Automotive batteries have six cells connected in series. Each cell produces about 2.1 volts. When connected in series, they produce 12.6 volts (6 × 2.1). Connected in series means the positive (+) strap of the first cell is connected to the negative (−) strap of the second cell. The "+" of the second cell is connected to the "−" of the third cell, and so on. The cell wall seals around the lead straps that connect the cells. This keeps each cell separate. The first negative and the last positive connecting straps are connected to the external battery terminals.

The plastic case of the battery separates the cells and supports the plates, connecting straps, and terminals. **Fig. 2-2.** The case also holds the acid and water used as the electrolyte. Each cell is composed of a number of positive and negative plates. Typically there are nine to thirteen or

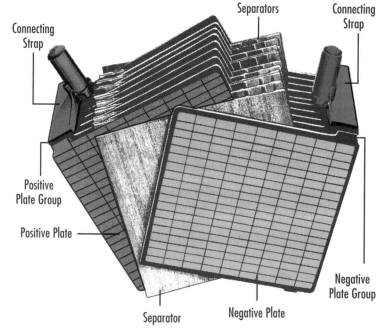

Connecting Strap

Separators

Connecting Strap

Positive Plate Group

Positive Plate

Separator

Negative Plate

Negative Plate Group

Fig. 2-2 Partially assembled battery components. *Why is it necessary to have separators between the positive and negative plates? (General Motors Corporation)*

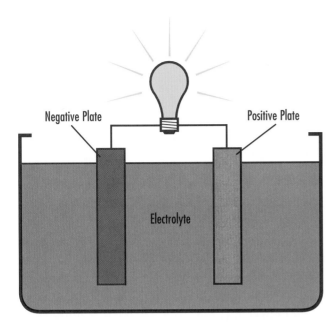

more of each. The amount of current a battery can provide depends on the amount of plate material in contact with the electrolyte. **Fig. 2-3.**

Battery plates are formed on a lead alloy grid. The grid structure supports the active plate materials of the battery. The positive plates are made of brown lead peroxide. Negative plates are composed of gray sponge lead. These materials are made into pastes. The pastes are pressed into and supported by the grid structure. Porous separators keep the plates from touching each other, while allowing the electrolyte to flow freely between the plates.

The electrolyte provides a conductive path for electron travel between the plates. In a fully charged battery, the electrolyte is approximately one-third sulfuric acid and two-thirds water. In a fully charged battery, the negative plates contain free electrons. A difference of potential exists between the positive and negative plates in each cell. When a load is connected to the battery, current flows to equalize the difference in potential between the plates.

When current flows from the battery, a chemical reaction occurs between the electrolyte and the

plates. As the battery discharges, the chemical reaction depletes the acid in the electrolyte. The electrolyte eventually becomes weaker because it contains a smaller amount of acid and a larger amount of water. A weak electrolyte cannot support a high level of current flow. The relationship between the amount of acid and water affects the density (specific gravity or weight) of the electrolyte. An electrolyte with a low density has a low level of acid. The density of the electrolyte provides an indication of the state-of-charge of the battery.

Battery Operation

To produce current, a chemical reaction must occur between the electrolyte and the active plate materials. The plates are slowly converted to *lead sulfate*. The formation of lead sulfate releases electrons. These electrons make up the current that flows from the battery. When the battery is fully discharged, most of the acid has left the electrolyte and entered the plates. The electrolyte is now mostly water. As a result, discharged batteries can freeze and be permanently damaged in cold weather.

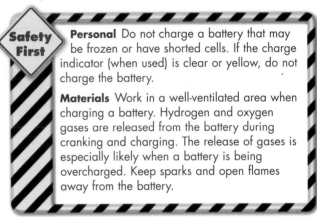

Automotive batteries are called *secondary cells* because, unlike some flashlight batteries (primary cells), they can be recharged. Charging reverses the chemical reaction of discharge by forcing electrons back into the battery. This converts the lead sulfate back into sulfuric acid and restores the plates to their original condition.

Part of the chemical reaction involved in charging a battery separates water into hydrogen and oxygen gases. In conventional batteries the hydrogen and oxygen gases escape through the vents and some water is lost. In maintenance-free batteries these gases are collected and recombined, so less water is lost. The use of lead-calcium alloys to make grids and straps reduces the amount of hydrogen produced. This further reduces water loss in maintenance-free batteries.

Battery Performance All batteries produce current through chemical reactions. These reactions are affected by temperature. As the battery gets colder, less current output is available. **Fig. 2-4.**

Many batteries fail in cold weather because their reduced output is not enough to crank the engine. As the battery gets hotter, harmful reactions such as self-discharge and grid corrosion become a problem. Battery life is usually shorter in hot climates.

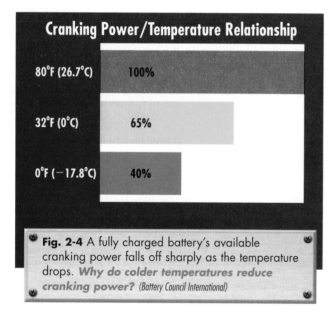

Fig. 2-4 A fully charged battery's available cranking power falls off sharply as the temperature drops. *Why do colder temperatures reduce cranking power?* (Battery Council International)

Poor maintenance and battery age also affect output. Batteries should be kept as fully charged as possible. A discharged battery means that lead sulfate remains on the plates *(sulfation)*. Eventually the sulfate hardens, making charging difficult or impossible. When a battery is not held securely in the tray, material can shake from the plates and connections may crack. Such conditions will reduce battery capacity and life.

Even well-maintained batteries eventually fail. Common causes of failure are hard sulfation, shedding of plate material, grid corrosion, and shorting between plates. In many cases a combination of these problems causes battery failure.

A battery has a service life. The service life of a battery is usually indicated in months. Ratings of 48, 60, and 75 months are typical. Batteries that are

at or near the end of their service life should be tested as a normal maintenance item.

Battery Ratings Capacity ratings measure a battery's ability to supply current under specified conditions. The two most widely used ratings are cold-cranking amps and reserve capacity.

Cold-Cranking Amps (CCA) The amount of current a fully charged battery can supply at 0°F [−17.8°C] for 30 seconds is known as **cold-cranking amps (CCA).** For example, let's look at a battery with a CCA rating of 400. The battery is able to supply 400 amps at 0°F (−17.8°C) for 30 seconds without the battery voltage falling below 7.2 volts. Most automotive batteries have CCA ratings between 400 and 1000 amps. This rating is especially important when cold starting is a concern.

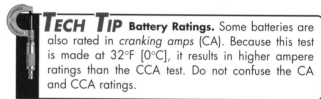

TECH TIP **Battery Ratings.** Some batteries are also rated in *cranking amps* (CA). Because this test is made at 32°F [0°C], it results in higher ampere ratings than the CCA test. Do not confuse the CA and CCA ratings.

Reserve Capacity The measure of how many minutes a battery can supply a load of 25 amps at 80°F [27°C] is known as **reserve capacity.** The minimum terminal voltage on this test is 10.5. Reserve capacity relates to how well the battery can handle key-off loads. A typical value for reserve capacities is 125 minutes. This means the battery can supply 25 amperes for 125 minutes at 80°F [27°C].

Table 2-A shows typical CCA and reserve capacities for batteries. It is not practical to convert from one of these rating systems to the other. Make sure any replacement battery has the minimum CCA and reserve capacity ratings specified by the vehicle's manufacturer.

Terminal Voltage The voltage measured across a battery under specified conditions is known as *terminal voltage.* The open circuit, or at rest, voltage of a fully charged battery is 12.6 volts. In use, however, the actual measured voltage will differ.

- During cranking, the voltage may drop to 10 volts or less, depending on load and temperature.
- With the engine running, charging-system voltage will usually be in the range of 13.5–15.0 volts, depending on temperature. Voltage outside this range usually means a charging system problem.

Table 2-A	BATTERY SPECIFICATIONS	
Test Load Amps	**Cold- Cranking Amps**	**Reserve Capacity Minutes**
200	405	75
210	430	90
250	500	90
310	625	90
310	630	105
330	770	115
360	730	115

SECTION 1 KNOWLEDGE CHECK

❶ What are key-off loads?

❷ What material is used to make the plate grids?

❸ What material is used as the active material on positive plates and on negative plates?

❹ What is the name of the acid used in a battery's electrolyte? What happens to this acid as the battery is discharged?

❺ What does the CCA rating of a battery indicate?

Section 2

Battery Inspection and Testing

Batteries are tested to determine their state of charge and condition. A thorough inspection of the battery helps determine what procedure to follow.

Battery Inspection

Signs of possible battery problems can often be found by carefully inspecting the battery, battery tray and hold-down clamps, and cables. Inspect the following:
- Battery tray and hold-down clamps—make sure the battery is held securely in place.
- Case and cover—look for loose posts and cracks or other physical damage. Clean surfaces to prevent small leakage currents from occurring across dirt and moisture.
- Terminal connections—make sure they are clean and tight.
- Cables—replace if damaged. Make sure replacement cables are the correct wire gauge (diameter).
- Electrolyte level—on batteries with removable vent caps only, remove the vent caps. Add water to bring electrolyte to indicated level. Distilled water is preferred.

Battery Testing

A number of tests are available for determining a battery's state of charge and condition. State-of-charge testing is done using the charge indicator built into the battery, when available. Regular hydrometers can be used on batteries with removable vent caps. An *open-circuit voltage (OCV) test* can determine the state of charge of any battery. A load test or an electronic test can determine if a battery is good or bad.

Using a Built-in Hydrometer The electrolyte in a fully charged battery is about one-third sulfuric acid and two-thirds water. As the battery discharges, the percentage of acid in the electrolyte decreases. This means the amount of acid remaining in the electrolyte is an accurate measure of the battery's state of charge. Sulfuric acid is heavier or denser than water.

A **hydrometer** is an instrument that measures the density of a liquid. This measurement of density is known as the *specific gravity* of a liquid. Specific gravity compares the density of any liquid to the density of an equal volume of water. A different type of hydrometer is used to test the anti-freeze protection of engine coolants.

Some maintenance-free batteries have a built-in hydrometer called a charge indicator. **Fig. 2-5.** This hydrometer has a small green ball in a cage at the bottom of a clear plastic tube. When the battery is adequately charged, the ball will float and appear as a green dot in the charge indicator. **Fig. 2-6.** The presence of the green dot does not necessarily mean that the battery is in good condition. It means only that the battery state of charge is acceptable. Only a load test or an electronic test can determine if a battery is good or bad.

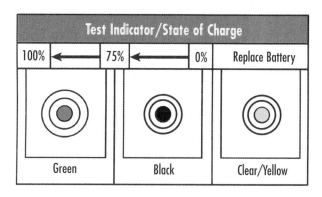

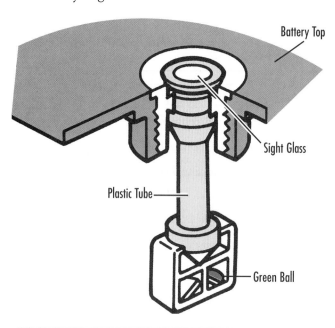

Fig. 2-5 Construction of the built-in hydrometer, or charge indicator. *If the indicator is green, does the battery still need charging?*

If the battery is discharged, the weaker electrolyte cannot float the green ball and the indicator will be dark. This means the battery must be charged and then tested. A clear or yellow indicator means the electrolyte level is too low, and the battery should be replaced. Do not attempt to test, charge, or jump start a battery that is too low on electrolyte. It could explode.

Some manufacturers use a slightly different color code for the built-in hydrometer. In these batteries, readings show:
• Green—a charge of 75 percent or more.
• Black—a charge between 50 percent and 75 percent.
• Red—a charge less than 50 percent.
• Yellow—a low electrolyte level.

Fig. 2-6 Appearance of the charge indicator under various battery conditions. *What will be the color of the charge indicator if the battery is low on electrolyte?* (DaimlerChrysler)

Refer to the vehicle service manual for specific information.

Using a Regular Hydrometer A regular hydrometer may be used on batteries with removable vent caps. **Fig. 2-7.**

A regular hydrometer and a built-in hydrometer operate on the same principle. A glass float or a number of colored balls float inside a glass tube. The float is calibrated to read in the range of about 1.1 to 1.3 specific gravity. A reading in this range means the liquid being tested is from 1.1 to 1.3

Fig. 2-7 Measuring the specific gravity of electrolyte in a battery cell using a regular hydrometer. (Jack Holtel)

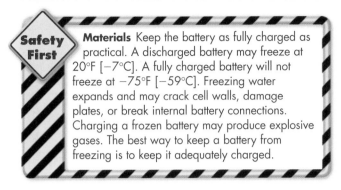

Materials Keep the battery as fully charged as practical. A discharged battery may freeze at 20°F [−7°C]. A fully charged battery will not freeze at −75°F [−59°C]. Freezing water expands and may crack cell walls, damage plates, or break internal battery connections. Charging a frozen battery may produce explosive gases. The best way to keep a battery from freezing is to keep it adequately charged.

times denser than water. The increased density is due to the acid present. Higher specific gravity readings mean there is more acid in the electrolyte. When colored balls are used instead of the glass float, the same principle applies. More floating balls indicate more acid and a higher state of charge.

To use a regular hydrometer:

1. Draw enough electrolyte from one cell to cause the float or colored balls to move up in the tube. Try to keep the tip of the hydrometer in the cell to prevent acid damage to nearby surfaces.

2. Note the number of floating balls or the numbers on the float.

3. If all of the balls float, the battery is fully charged. The battery is less than fully charged if only some of the balls float or if the glass float reads less than about 1.230 specific gravity. **Fig. 2-8.** (Note: It may be impossible to charge a used battery to 1.265 specific gravity.)

4. Rinse the hydrometer with water after each use.

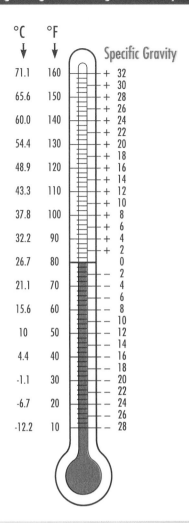

Fig. 2-9 Temperature correction of a hydrometer reading involves adding 4 points (0.004) of specific gravity for each 10°F [5.5°C] that the electrolyte temperature is above 80°F [26.7°C]. If the electrolyte is below 80°F, subtract 4 points for each 10°F. The correction chart may be part of the hydrometer. *The specific gravity reading is 1.250. The electrolyte temperature is 20°F [−7°C]. What is the corrected specific gravity rating?*

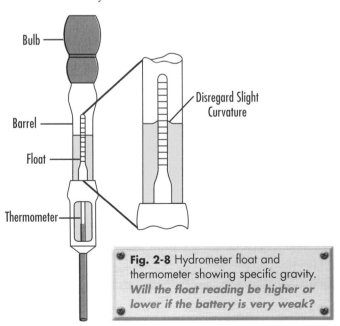

Fig. 2-8 Hydrometer float and thermometer showing specific gravity. *Will the float reading be higher or lower if the battery is very weak?*

Because the cells in a battery are separated from each other, it is necessary to perform the hydrometer test on each cell. The cell readings should be within 50 (0.050) specific gravity points of each other, regardless of the state of charge. If not, replace the battery.

The temperature of the electrolyte will affect the accuracy of hydrometer readings. Because of its size and weight, it may take several hours for a battery to warm to room temperature when brought inside. Many hydrometers have temperature correction values built in. The correction is necessary to obtain accurate readings. **Fig. 2-9.**

Open-Circuit Voltage (OCV) Test An open-circuit voltage (OCV) test may be used to determine the state of charge of a battery. It is the only way to check the state of charge on a battery that does not have a built-in hydrometer or removable vent caps. Before making the OCV test, turn on the headlights for 3–5 minutes. This will remove any surface charge from the plates. The surface charge may also be removed by cranking the engine for 15–30 seconds with the ignition disabled. Turn off all lights and accessories. Use a digital volt-ohm-meter (DVOM) to measure voltage at the battery terminals. **Fig. 2-10.**

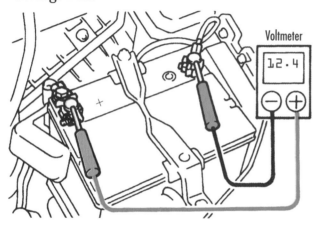

Voltmeter

Fig. 2-10 Open-circuit voltage is measured with all lights and accessories off. *Why must all loads be removed before conducting this test?*

There is a relationship among specific gravity, OCV, and battery state of charge. **Figure 2-11** can be used to determine battery state of charge using OCV.

Battery Load Test The battery load test is also called the *high-rate discharge test*. The **load test** is a test that measures terminal voltage while the battery is supplying a large current for 15 seconds. The load used in the test is approximately what starter current draw would be at normal temperature. Typical load testers include a voltmeter, ammeter, and an adjustable carbon pile resistor. Adapters are often required when testing or charging side terminal batteries. **Fig. 2-12.**

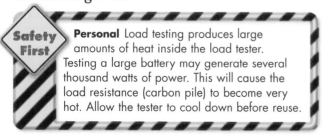

Safety First **Personal** Load testing produces large amounts of heat inside the load tester. Testing a large battery may generate several thousand watts of power. This will cause the load resistance (carbon pile) to become very hot. Allow the tester to cool down before reuse.

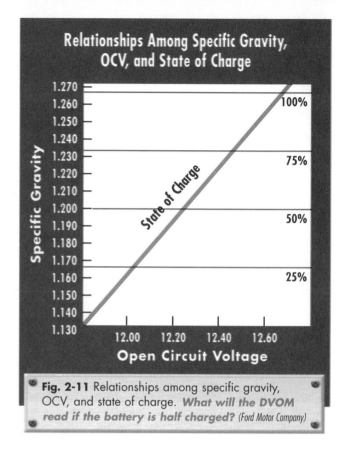

Relationships Among Specific Gravity, OCV, and State of Charge

Fig. 2-11 Relationships among specific gravity, OCV, and state of charge. *What will the DVOM read if the battery is half charged?* (Ford Motor Company)

The results of a battery load test are valid only when the battery is adequately charged. The battery is adequately charged only if one of the following occurs:

- The built-in charge indicator is green.
- The specific gravity reading (corrected for temperature, if needed) is at least 1.230.
- The OCV is at least 12.4 volts.

A battery that is not adequately charged will usually fail a load test. To perform a load test:

1. Make sure the load control is OFF.
2. Connect the tester as shown in the vehicle service manual or the tester instruction manual.
3. Remove the surface charge if the battery has been charged before the test is made. This can be done by turning on the headlights or cranking the engine.
4. Apply a load equal to one-half the CCA rating of the battery for 15 seconds. (Some batteries have the correct load value marked on a label or on the battery case.)
5. After 15 seconds, note the voltage and turn off the load.
6. Disconnect the tester.

Investigating Electrolytes

The electrical energy needed to start an automobile is stored in the battery in a chemical reaction. An automotive battery is an example of a wet-cell battery. Flashlight batteries are dry-cell batteries. A dry-cell battery contains a paste electrolyte.

In a wet-cell battery, the cells contain plates of two different materials. In the automotive battery, these plates are lead (Pb) and lead peroxide (PbO_2). A liquid electrolyte surrounds the plates. The electrolyte in an automotive battery is a mixture of about one-third sulfuric acid (H_2SO_4) and two-thirds water (H_2O).

A battery is discharged by turning on the ignition switch or using electrical accessories. The electrolyte reacts with the lead and lead peroxide. This chemical reaction moves electrons from the electrolyte to the plates. Electricity is produced when electrons are moved from one place to another in a closed circuit.

- The lead plates attract sulfate ions, along with their electrons, from the sulfuric acid. An *ion* is a charged atom in which the negative and positive charges are not equal.
- The hydrogen ions in the electrolyte "pull" oxygen and its electrons from the oxygen in the lead peroxide and replace them with sulfate ions to make lead sulphate ($PbSO_4$).
- The oxygen ions combine with hydrogen ions from the sulfuric acid to make more water.

Eventually both plates contain lead sulfate ($PbSO_4$). The chemical reaction stops. No electrons are moving, no more electricity is produced.

When a battery is recharged, the chemical reactions are reversed.

- The sulfate ions leave the plates and combine with hydrogen ions from the water to make more sulfuric acid.
- The oxygen ions that are left behind are now attracted to the lead peroxide plate. The plates are lead and lead peroxide again.

Apply it!

Making a Wet-Cell Battery

Meets NATEF Science Standards for laboratory safety and electrochemical reactions in batteries.

Materials and Equipment
- Safety glasses and rubber gloves
- Diluted sulfuric acid
- DVOM

- Wet-cell battery kit available from scientific supply stores. (Kit contains plastic tumbler, holder for metal strips, and terminals for measuring volts and current)

- 6″ piece of insulated copper wire with 1″ insulation stripped from each end
- 2 strips of zinc
- 2 strips of copper

Safety First **Personal** Sulfuric acid is corrosive. Do not get acid on skin or clothes. Wear safety glasses and rubber gloves during this experiment. Reread the safety features in the chapter.

❶ Fill the tumbler half-full with diluted sulfuric acid.

❷ Place one zinc strip in the acid. Is there a clue that a chemical reaction is taking place? Remove and rinse the strip in water.

❸ Repeat the above procedure with one of the copper strips. What did you observe?

❹ Attach the copper wire between the terminals. Attach the copper strips in the holders.

❺ Measure the voltage between the two terminals with the DVOM. What is the reading?

❻ Replace the copper strips with zinc strips. Repeat the procedure. What is the reading?

❼ Remove one of the zinc strips and rinse it in water. Replace it with a copper strip. Measure the voltage. What is the reading?

❽ Connect the DVOM to measure current. Place the metal strips as far apart as possible. There should be a small measurable current flow.

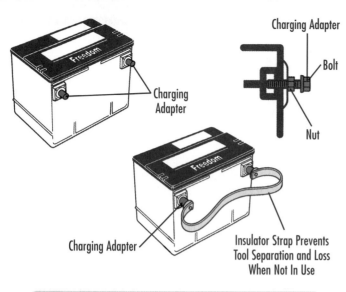

Fig. 2-12 Typical adapter tools. *When would it be necessary to use adapters?* (General Motors Corporation)

If the voltage is 9.6 volts or higher, the battery is good. Battery temperature will affect the voltage reading. If the battery is not at room temperature or higher, use the temperature compensation chart found in the tester's instructions or in the vehicle service manual. **Table 2-B.** If the voltage is less than noted for a given temperature, the battery is defective and should be replaced.

Table 2-B	TEMPERATURE COMPENSATION	
Minimum Battery Voltage	**Temperature**	
	°F	**°C**
9.6	70	21
9.5	60	16
9.4	50	10
9.3	40	4
9.1	30	−1
8.9	20	−7
8.7	10	−12
8.5	0	−18

Electronic Battery Testing An electronic battery tester may also be used to determine battery condition. **Fig. 2-13.** These testers are microprocessor controlled. The controller sends a low-voltage AC signal through the battery and measures the return pulse. The controller compares the returned signal to data in its memory to determine battery condition and state of charge. Electronic battery testers are reliable and easy to use. They can be used even

if the battery is not adequately charged. They can also identify batteries with internal damage that may be dangerous to charge.

To perform an electronic battery test:

1. Turn off all electrical loads.

2. Make sure the test leads make good contact with the battery terminals.

3. Enter the battery's CCA rating in the display window.

4. Press the START button.

The results of the test will be displayed in a few seconds. **Fig. 2-13.** Refer to the tester instruction manual for additional information.

Fig. 2-13 An electronic battery tester. A reading from an electronic battery tester is shown at the right. *What battery rating must be entered to conduct this test?* (Midtronics)

Key-Off Loads (Parasitic Drains) Testing An undercharged or dead battery may result from higher than normal key-off loads. These loads are also called *parasitic drains*. **Table 2-C.** They represent the current needed to maintain memory circuits in devices such as computers, radios, and climate controls. On most vehicles, normal key-off drain will be less than 30–50 *milliamps*. The actual drain depends on the number of electronic devices on the vehicle. Refer to the vehicle service manual for specifications. To measure parasitic drain:

1. Run the vehicle for a few minutes, making sure all electronic systems are turned on.

2. Turn off the ignition and all other switches. Make sure all lights are off and the doors are closed.

3. Without breaking the circuit, connect the highest ammeter scale of a DVOM between the negative

battery terminal and the negative cable. On side-terminal batteries, this requires a test adapter. **Fig. 2-14(a).**

4. On post-type batteries, hold the meter leads on the post and cable clamp. While maintaining this connection, carefully slide the clamp from the

post. **Fig. 2-14(b).** Change to the milliamp scale when the current flow is low enough to do so.

Very high parasitic drain is often caused by underhood, glovebox, or trunk lamps that do not shut off. If no obvious cause is found, follow instructions in the vehicle service manual to locate the cause of the high parasitic drain.

Table 2-C	CAUSES OF LOW OR DISCHARGED BATTERY	
Drains on the Battery		**Faulty Charging System**
• Batteries being less than adequately charged when received		• Generator not supplying the required amperage to recharge the battery
• Extended storage of vehicle (self-discharge and parasitic loads)		• Low charging voltage
• Drains caused by trunk or other lights that stay on		• Broken generator drive belt
• Parasitic drains		• Dirty or damaged wiring or connectors
Faulty Starter System		**Owner Cause**
• Excessive current draw from starter motor		• Extended cranking periods due to hard- or no-start condition
		• Accessory left on, such as headlights or interior lights

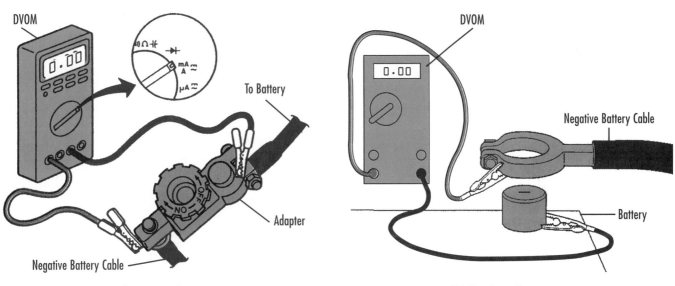

(a) Side-Terminal Battery

(b) Top Post Battery

Fig. 2-14 An adapter is needed when measuring parasitic drain on a side-terminal battery. The DVOM leads are connected to terminals on the adapter. *What DVOM scale is used to measure the key-off drain?* (General Motors Corporation)

SECTION 2 KNOWLEDGE CHECK

❶ Why is it important to visually inspect a battery?

❷ What does a hydrometer test indicate about the battery?

❸ What test is used to determine the state of charge of a battery with no built-in charge hydrometer and no vent caps?

❹ What test discharges the battery at a high rate while measuring the terminal voltage?

❺ How is key-off load measured?

Section 3

Battery Servicing

All batteries, even maintenance-free ones, benefit from periodic checks.

Routine battery maintenance includes:
- Inspecting for physical damage and corrosion.
- Cleaning the battery, cable connections, and tray.
- Testing.
- Charging when needed.
- Replacing if defective.

Battery Cleaning

Corrosion around the battery is caused by small amounts of acid that leave the cells in the escaping gases. Neutralize the acid with a mixture of baking soda and water. Flush clean with water. Do not allow the cleaning solution to enter the cell openings.

Cleaning a battery clamp or terminal requires a wire brush to remove corrosion and oxidation. Special brushes are available for this purpose. **Fig. 2-15.** Cable-clamp bolts may corrode and become weakened. Corroded bolts should be replaced.

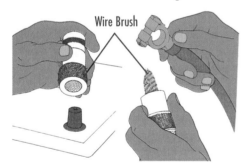

Wire Brush

Fig. 2-15 Using a battery terminal brush to clean the battery terminal posts and clamps. *Why is it important to clean the connections?*

To prevent corrosion, install chemically treated felt washers before reconnecting the battery cables. Protective spray coatings may be used after reconnecting battery cables. Lubricating grease is not recommended because it may soften and run off.

Clean the battery tray as necessary. Replace it if it is damaged. After removing any rust and corrosion, repaint the tray with rust-inhibiting paint. Be careful to avoid overspray that may damage or mar other surfaces. Replace hold-down brackets or bolts if they are damaged or missing.

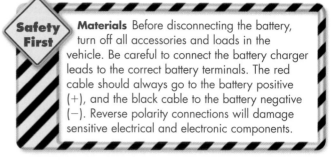

Safety First

Materials Before disconnecting the battery, turn off all accessories and loads in the vehicle. Be careful to connect the battery charger leads to the correct battery terminals. The red cable should always go to the battery positive (+), and the black cable to the battery negative (−). Reverse polarity connections will damage sensitive electrical and electronic components.

Battery Charging

There are several methods for charging a battery. Slow charging uses a current of 3–10 amps and may take 24 hours or more. Medium charging is done with a charging current of 15–25 amps, usually for 3–5 hours. Fast charging uses a current of 30–50 amps for up to one hour. **Fig. 2-16.**

Normally, batteries are charged until they are three-fourths charged or more. Avoid overcharging by checking the state of charge frequently. Reduce

Fig. 2-16 Battery charger connected to a battery in a vehicle. Disconnect the negative (ground) cable from the battery before connecting the charger cables. *Why is it important to disconnect the negative battery cable before connecting the battery charger?* (Jack Holtel)

the charge rate if the battery releases excessive gases or feels unusually warm to the touch. When charging a battery in the vehicle, disconnect the negative battery cable to avoid damaging sensitive electronic components. Never attempt to charge a battery that is frozen or that has a shorted cell.

A slow charge rate is best for the battery. The faster the charge rate, the greater the internal heating. Battery charging should always be done in a well-ventilated area.

Battery Replacement

The three common mistakes made during battery replacement are:
• Installation of a battery with too little capacity.
• Failure to properly secure the battery to the battery tray.
• Improper tightening and protection of the battery terminals.

Safety First — **Personal** Before disconnecting a battery, turn off all electrical loads. This will prevent a spark from occurring when the cable is removed. Always remove the ground cable first. This prevents arcing if a wrench touches another metal surface while touching a battery terminal.

Battery Removal To remove a battery:

1. Connect a memory holder to the cigarette lighter receptacle. **Fig. 2-17.** Memory holders use a small 9- or 12-volt battery to supply power to memory circuits while the vehicle battery is disconnected. This prevents the loss of stored information such as engine operating adjustments and radio and climate-control presets.

Fig. 2-17 A memory holder prevents the loss of electronic memory while the battery is disconnected. *What will happen if you open a door to the vehicle while only the memory holder is connected?* (Sears, Roebuck & Co.)

Do not open doors or use any electrical devices in the vehicle while using the memory holder. The small batteries are not capable of supplying current for normal electrical loads.

2. Remove the negative or ground cable from its battery terminal. Avoid damaging the posts when removing the cable clamps. Use cable pullers and battery pliers to prevent battery damage. **Fig. 2-18.**

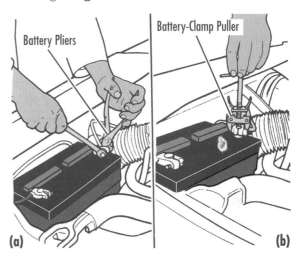

Battery Pliers — Battery-Clamp Puller

(a)   (b)

Fig. 2-18 Loosening nut **(a)** and using a battery-clamp puller **(b)** to remove the cable clamp from the battery post. This technician is using battery pliers to hold the nut. *(Ford Motor Company)*

3. Remove the positive battery cable from the battery. Use the same method as for the negative cable.

4. Remove the battery hold-down clamps from the battery and the battery tray.

5. Carefully remove the battery from the battery tray. Use a clamp-type battery carrier if needed. **Fig. 2-19.**

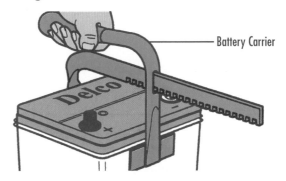

Battery Carrier

Fig. 2-19 A lifting tool makes it easier to remove batteries from tight spaces. *What holds the battery in the battery tray?* (General Motors Corporation)

TECH TIP **Battery Carrier.** Batteries can be heavy and difficult to remove. A clamp-type battery carrier may be helpful in removing the battery when clearances are small.

Battery Installation Before installing a new battery, make sure you have the correct replacement. The size of new battery should be the same as the original. It should also have the same terminal arrangement. The CCA or reserve capacity rating should be the same or higher. The battery tray should be clean. The drain holes should be open. All mounting hardware should be in good condition. Clean, repair, or replace these items as needed. To install a battery:

1. Place the battery in the tray with the terminals in the proper position.

2. Install the battery hold-down clamps. Make sure they are secure.

3. To prevent corrosion, install chemically treated felt washers on both terminals.

4. Connect the positive battery cable. Tighten the clamp or bolt securely, but do not overtighten.

5. Connect the negative or ground cable. Tighten the clamp or bolt securely, but do not overtighten.

6. Disconnect the memory holder from the cigarette lighter.

7. Remove all tools and equipment from the vehicle.

8. Make sure the vehicle starts and runs normally.

9. Check the charging system dash indicator for normal operation.

Writing a Memo

Your ability to communicate with others on the job is very important. You will need to tell others what you have learned and what you have done to solve your customer's problems. You will often need to give this information to the service writer. One of the best ways to do this is to write a memo.

A memo is an informal written note containing information, directions, or suggestions. A memo usually begins with four lines of information:

DATE: (date when you write the memo)

TO: (person receiving the memo)

FROM: (person sending the memo)

RE: (subject of the memo)

The abbreviation *RE* is short for "regarding."

Information you should put in a memo includes:

• The customer's complaint.
• The vehicle's symptoms.
• What you have done, including diagnostic tests, repairs, or adjustments.
• Your recommendations.

In a memo, include all the information you think is necessary. However, the memo should be as short and clear as possible. It will become an important part of a vehicle's repair file.

Apply it!

Meets NATEF Communications Standards for organizing written information and adapting style and structure for a written report.

❶ Reread *The Problem* at the beginning of the chapter. Assume you are the technician who checked out the vehicle.

❷ Write a sample memo to the service writer. In your memo include:
• The DATE, TO, FROM, and RE lines. Make up a name for the service writer.
• All the details you think he or she will need to identify the vehicle. Include its mileage, make, model, year, and VIN (vehicle identification number).
• The warranty information for the battery.
• What you have done to check out the vehicle.
• Your conclusions and recommendations.

Battery Damage

The following will reduce battery capacity and life:

- Overcharging causes loss of electrolyte and damage to the plates and grids. Charge batteries as slowly as practical. Turn off the charger when the battery nears full charge. If overcharging occurs during normal operation, check the charging system.

- Hard sulfation occurs when the battery remains in a discharged condition for a month or more. The lead sulfate on the plates hardens, making it difficult or impossible to charge the battery. Sulfation is one of the leading causes of battery failure.

- Deep cycling occurs when a deep discharge and charge cycle is repeated often. This results in excessive shedding of active material from the plates. The material will settle to the bottom of the battery case, where it can short between plates.

- Freezing damage occurs when weak batteries freeze and damage the plates and possibly the case.

- Low electrolyte level allows uncovered area of plates to dry out.

- Excessive vibration is caused by the improper mounting of the battery to the case. The vibration causes internal damage to the plates and connectors.

Safety First

Personal If you encounter a vehicle with a battery that is hot and spewing electrolyte, be very careful! Shut off the engine to prevent additional charging. Make sure the area around the car is well ventilated. Do not allow sparks or open flames near the battery. Stay away from the battery until it cools. This may take an hour or more. After the battery has cooled, check it for internal shorts. A battery with an internal short will fail a load test or an electronic battery test. If the battery tests okay, check the charging system. It may have caused the overcharging problem.

Jump Starting

Jump starting is the process of starting the engine in one vehicle by connecting it to the battery in another vehicle. **Fig. 2-20.** Jump starting can be a dangerous process if not done correctly. Whenever possible, check the condition of the bad battery before making any connections. Batteries that are completely discharged or low on electrolyte can explode. Avoid causing arcs when making the connections. If not done properly, the resulting sparks could ignite fuel vapors or battery gases. They can also damage the terminals or components where the arc occurs. Replacement of damaged components can be very costly.

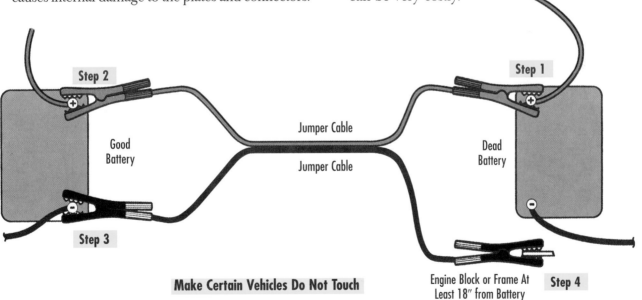

Make Certain Vehicles Do Not Touch

Fig. 2-20 The four basic steps in connecting jump starting cables. **Step 1:** Connect the (+) cable to the (+) terminal of the dead battery. **Step 2:** Connect the opposite end of the (+) cable to the (+) terminal of the good battery. **Step 3:** Connect the (−) cable to the (−) terminal of the good battery. **Step 4:** Connect the opposite end of the (−) cable to a good ground connection away from the battery in the vehicle with the dead battery. *(General Motors Corporation)*

Safety First

Personal Batteries contain explosive hydrogen gas and corrosive acid. Always protect your eyes and exposed skin.

Materials Connecting batteries for jump starting can create dangerous voltage and current surges. These surges can permanently damage expensive electronic components in the vehicle. Always observe proper polarity and connection practices when jump starting.

Some vehicles may have convenient connection points under the hood for both positive and negative terminals. **Fig. 2-21.** Using these terminals keeps dangerous sparks away from the battery.

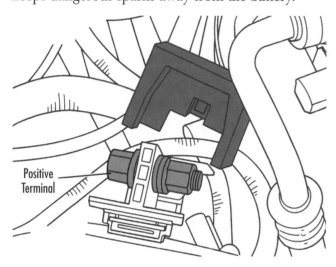

Positive Terminal

(a) Remote Positive Battery Terminal

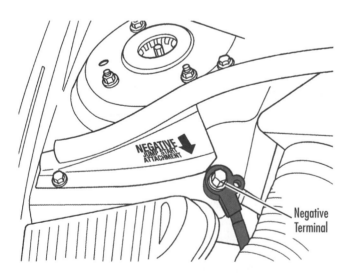

NEGATIVE JUMP START ATTACHMENT

Negative Terminal

(b) Remote Negative Battery Terminal

Fig. 2-21 Connection points found on some vehicles. (DaimlerChrysler)

If you must jump start a vehicle, use extreme caution. Follow the instructions in the vehicle owner's manual. Be sure to follow all safety precautions.
• Wear safety goggles.
• Avoid sparks and open flames.
• Do not allow the vehicles to touch. If the vehicles touch, a ground connection may be made.

The jump starting procedure requires a set of jumper cables of adequate wire gauge (diameter). Always connect cables in the proper sequence, observing correct polarity.

1. Make sure both vehicles have a 12-volt negative ground electrical system.

2. Make sure the vehicles aren't touching each other.

3. Set the parking brakes on both vehicles. Place transmissions in PARK or NEUTRAL.

4. Turn off all lights and accessories. This will avoid accidental damage to these components.

5. Connect one end of the red positive (+) cable to the positive terminal of the dead battery.

6. Connect the other end of the positive (+) cable to the positive (+) terminal of the good battery.

7. Connect one end of the black negative (−) cable to the negative (−) terminal of the good battery.

8. Connect the other end of the negative (−) cable to the engine block or frame of the vehicle with the dead battery. To avoid the danger of sparks igniting battery gases, make sure this connection is at least 18″ [46 cm] from the dead battery.

9. Start the vehicle with the good battery. Turn on the blower motor of the vehicle with the good battery. This helps to prevent potential voltage surges, which can damage sensitive electronic systems.

10. Attempt to start the engine of the vehicle with the dead battery. To avoid starter damage, do not crank the engine longer than 30 seconds at a time.

11. After the engine starts, remove the jumper cables in reverse order. Do not allow the positive and negative cables to touch. Do not allow the positive cable to touch the surface of any vehicle.

Using a Battery Jumper Box

Many shops use a battery jumper box instead of jumper cables. A **battery jumper box** is a portable power pack used for jump starting vehicles. **Fig. 2-22.** The jumper-box cables are connected in the same way as jumper cables. Connect the red positive lead to the battery positive terminal. Connect the black negative lead to a clean ground connection at least 18″ [46 cm] from the battery.

Safety First — **Materials** Some manufacturers may recommend additional precautions when using jumper cables. This information can be found in the owner's manual.

Using a Portable Battery Pack

A *portable battery pack* eliminates the need for jumper cables. These devices charge the battery through the vehicle's auxiliary power connector (cigarette lighter). The powerful yet compact packs can be carried in the trunk or inside the vehicle. Although designs may vary, general guidelines for use are:

1. Turn off all interior and exterior lights and accessories.
2. Plug the end of the connecting cord into the auxiliary power connector.

Safety First — **Personal** Batteries contain a sulfuric acid solution. Always keep a battery level when removing it. Corrosive acid can spill from the filler or vent caps if the battery is allowed to tip over. If the acid comes in contact with the skin, serious burns can result.

Materials Acid spills can damage paint and fabrics. Always flush or neutralize any acid spill to minimize any corrosive effects.

Fig. 2-22 A battery jumper box may be more convenient to use than jumper cables. *How are the cables connected?* (Jack Holtel)

3. Turn the battery pack on as indicated in the instructions for the unit.
4. If equipped, an indicator light should show that the unit is charging the battery. Charging times vary, depending on the size and type of battery pack used. Average times may range from 15 to 30 minutes.
5. When charging is complete, unplug the battery pack from the auxiliary power connector.
6. Always be sure to recharge the unit, following the instructions in the manual.

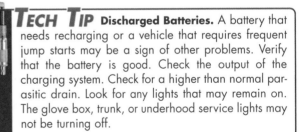

TECH TIP Discharged Batteries. A battery that needs recharging or a vehicle that requires frequent jump starts may be a sign of other problems. Verify that the battery is good. Check the output of the charging system. Check for a higher than normal parasitic drain. Look for any lights that may remain on. The glove box, trunk, or underhood service lights may not be turning off.

SECTION 3 KNOWLEDGE CHECK

1. Why is slow charging a battery recommended over fast charging?
2. What can be done to prevent corrosion on batteries?
3. When removing a battery, why is it important to use a memory holder?
4. When disconnecting a battery, which cable should be removed first?
5. Describe the jump starting procedure.

CHAPTER 2 REVIEW

Key Points

Meets the following NATEF Standards for Electrical/Electronic Systems: inspecting batteries; performing battery state-of-charge and load tests.

- The primary components of an automotive battery are the case, cover, vent caps, plates, separators, electrolyte, and terminals.
- A battery operates by storing electrical energy in chemical form. The formation of lead sulfate during discharge releases electrons that become battery current.
- A battery hydrometer, built-in hydrometer, or open circuit voltage test can be used to determine battery state of charge.
- A load test or a battery load or electronic battery test can be used to determine battery condition.
- The faster a battery is charged, the greater the internal heating. A slow charge rate is better.
- When replacing a battery, using a memory holder prevents the loss of stored information in memory circuits.
- Jump starting is the process of starting the engine in one vehicle by using the battery in another vehicle.

Review Questions

❶ Explain the tasks a battery performs in an automotive electrical system.

❷ What are the primary internal components of a battery?

❸ Describe how a battery produces voltage.

❹ What tests can be performed to test a battery's state of charge?

❺ Explain how to make a battery load (high-rate discharge) test and an electronic battery test.

❻ Explain the difference between slow and fast charging a battery.

❼ Describe the procedure for servicing and replacing a battery.

❽ Describe the process for the safe use of jumper cables.

❾ **Critical Thinking** Is it necessary to perform a load test on a battery that has passed an electronic battery test? Explain your answer.

TECHNOLOGY FORECAST FOR AUTOMOTIVE EXCELLENCE

Two Batteries Are Better Than One

Cars and trucks could soon have two batteries. One would be a traditional, lead-acid battery. The other would have a smaller, thin-foil design.

The two-battery design is better because a battery is asked to do two different jobs. First, it must provide a large burst of power to start the engine. Later, it may have to supply electricity for the lights, radio, and other accessories.

The new thin-foil battery will be used to start the engine. It will have lead-acid plates like today's batteries but will be smaller and lighter. Thin spirals of lead foil will replace the heavy-plate electrodes.

These spirals will have more surface area and be closer together. This arrangement will produce more power more quickly.

The second battery will be a normal lead-acid battery. It will provide the electricity needed after the engine is started.

This two-battery system could reduce weight and make better use of space. The setup also is safer. Should a driver leave the lights on and run down the larger battery, the smaller one can start the vehicle. The vehicle can then be taken to a service facility where the larger battery can be recharged.

AUTOMOTIVE EXCELLENCE
TEST PREP

Answering the following practice questions will help you prepare for the ASE certification tests.

1. Each cell in an automotive battery produces:
   a 1.5 volts.
   b 2.1 volts.
   c 2.6 volts.
   d 12.6 volts.

2. Technician A says batteries used in very hot climates may have a shorter life than batteries used in cold climates. Technician B says batteries fail more quickly in cold climates due to high cranking currents. Who is correct?
   a Technician A.
   b Technician B.
   c Both Technician A and Technician B.
   d Neither Technician A nor Technician B.

3. Technician A says a hydrometer reading measures how much cranking current is available to the starter. Technician B says the hydrometer reading indicates the state of charge in the battery. Who is correct?
   a Technician A.
   b Technician B.
   c Both Technician A and Technician B.
   d Neither Technician A nor Technician B.

4. Technician A says a hydrometer test can be made at any temperature. Technician B says this is true as long as corrections are made to the readings when needed. Who is correct?
   a Technician A.
   b Technician B.
   c Both Technician A and Technician B.
   d Neither Technician A nor Technician B.

5. Technician A says charging frozen batteries is dangerous. Technician B says overcharging a battery is dangerous. Who is correct?
   a Technician A.
   b Technician B.
   c Both Technician A and Technician B.
   d Neither Technician A nor Technician B.

6. Technician A says it is important to properly secure the battery to the battery tray. Technician B says the battery plates could be damaged if the battery is loose in the tray. Who is correct?
   a Technician A.
   b Technician B.
   c Both Technician A and Technician B.
   d Neither Technician A nor Technician B.

7. A digital volt-ohm-meter is used to perform:
   a Battery load testing.
   b Specific gravity testing.
   c Open-circuit voltage testing.
   d Reserve capacity testing.

8. A customer complains that the engine in her car cranks slowly. Technician A says the problem could be a bad battery. Technician B says it could be poor connections between the battery cables and terminals. Who is correct?
   a Technician A.
   b Technician B.
   c Both Technician A and Technician B.
   d Neither Technician A nor Technician B.

9. A sport utility vehicle is brought into the shop and found to need a new battery. The exact battery needed is not available. Technician A says any battery with the same or higher ratings can be used. Technician B says the replacement battery must fit the battery box, connect correctly to the cables, and have the same or higher rating. Who is correct?
   a Technician A.
   b Technician B.
   c Both Technician A and Technician B.
   d Neither Technician A nor Technician B.

10. A survey by battery manufacturers shows that most errors are made when:
    a Choosing the battery.
    b Tightening the terminals.
    c Improperly mounting the battery.
    d All of the above.

CHAPTER 3

Diagnosing & Repairing the Starting System

You'll Be Able To:

- ⊗ Explain the operation of a starting system.
- ⊗ Identify the major internal components of a starter.
- ⊗ Explain the function of an overrunning clutch.
- ⊗ Explain the purpose of the ring gear and pinion.
- ⊗ Test solenoid windings with an ohmmeter.

Terms to Know:

armature
commutator
electromagnetic field
overrunning clutch
starter
starter relay
starter solenoid

The Problem ⟩⟩⟩⟩⟩

Luci Hinks has had her vehicle towed to your service center. She tells you that she has had starting problems for the past week.

She says, "Sometimes I have to turn the ignition switch to START several times before the engine cranks. But sometimes it starts on the first try." This morning her vehicle would not start at all. To verify her complaint, you turn the ignition switch to the START position, and the engine cranks.

"See, that's what it's been doing," she says. "Sometimes it starts, sometimes it doesn't."

Your Challenge

As the service technician, you need to find answers to these questions:

❶ What information can Luci supply to help isolate the problem?

❷ What are the most likely, and least likely, causes?

❸ What diagnostic checks would you perform to isolate the problem?

Section 1

The Need for a Starting System

The vehicle starting system provides an efficient and reliable method to crank an engine. An internal combustion engine needs air, fuel, and an ignition source to operate. The *engine crankshaft* must rotate to draw in and compress an air/fuel mixture in the engine's cylinders. The starting system converts electrical energy from the battery into mechanical energy from the starter. This energy is then used to crank the engine.

The major parts of the typical starting system are:
- **Battery**–an electrical storage device and the source of current for the starting system.
- **Ignition switch**–the main control device for the starting system.
- **Starter**–a high-torque electric motor that cranks the engine.
- **Starter relay**–an electrical device that opens or closes a circuit in response to a voltage signal.
- **Starter solenoid**–an electromechanical device that, when connected to an electrical source such as a battery, produces a mechanical movement. Some systems use both a relay and a solenoid.
- **Park/neutral position switch**–a switch that prevents the starter relay or solenoid from closing when the vehicle is in gear.

These parts are used in two separate electrical circuits, a high-current motor circuit (red) and a low-current control circuit (blue). **Fig. 3-1.** The high current needed to operate the starter (150 amps or more) requires a heavy-gauge cable. It is not practical or safe to route this high-current circuit through the ignition switch and dashboard. Instead, the ignition switch controls a circuit that supplies low current to the starter relay or solenoid. The contacts of the relay or solenoid control the high-current path to the starter. When the contacts close, the starter connects directly to the battery.

A park/neutral position switch in the low-current circuit controls starter operation. This switch prevents the relay or solenoid from operating with the vehicle in gear. The gear selector of an automatic transmission must be in PARK or NEUTRAL. A clutch-pedal operated switch is used with a manual transmission.

How Does a Starting System Work?

The battery is the source of electric current to operate the starter. The low- and high-current circuits control where and when the current flows. The starter converts this current into the mechanical *torque* (turning power) needed to crank the engine.

With the ignition switch in the START position, low current flows from the battery. It flows through the ignition switch and the PARK/NEUTRAL position switch to the starter relay or solenoid. This closes the contacts in the relay or solenoid and completes a high-current path from the battery to the starter. With current applied, the starter cranks the engine.

Crankshaft rotation brings air and fuel into the engine's cylinders. A high-voltage spark, created at the spark plugs, ignites the air/fuel mixture in the cylinders. Power from the burning fuel in the cylinders provides the energy to maintain engine operation.

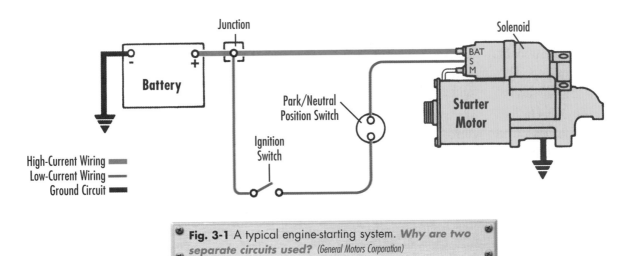

Fig. 3-1 A typical engine-starting system. *Why are two separate circuits used?* (General Motors Corporation)

Starter Construction

A starter is a high-torque electric motor that converts electrical energy from the battery into mechanical energy to crank the engine. Advances in technology have changed the construction of some starters. Their new design makes them smaller, lighter, and more powerful. They require less current by using permanent magnets instead of field coils. They use internal *gear reduction* to increase cranking torque.

 ***TECH TIP* Electrical Connections.** Poor electrical connections to the starter or the engine block are often a source of trouble. Poor connections create resistance to current flow through the circuits of the starting system. Low-current flow results in poor cranking performance. Battery terminals may develop a layer of corrosion. Remember to check all cable connections. Terminals must be clean and tight.

Field Coil Starter

A field coil starter contains two basic assemblies: an armature and field coils. **Fig. 3-2.** The field coils are mounted in the starter housing.

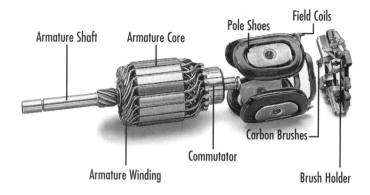

Fig. 3-2 Components of the field coil starter: armature shaft, armature winding, armature core, commutator, pole shoes, field coils, carbon brushes, and brush holder. *(Robert Bosch GmbH)*

An electromagnet consists of a soft iron pole shoe (core) wrapped with a coil of copper wire. The field coil is an *electromagnet* that produces a stationary magnetic field. The pole shoe guides and intensifies the magnetic field so that it reacts strongly with the armature. Field coil starters contain either four or six field coils.

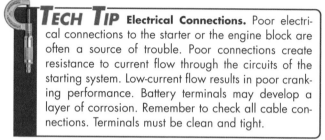

Ask the Right Questions

Sherlock Holmes, the famous fictional detective, had great powers of deduction. He was able to solve a case by reasoning from known facts. Unlike Holmes, you have a willing witness—the customer. The customer has firsthand knowledge about the vehicle's problem. To translate your customer's information into clues, you need to ask the right questions. The answers to these questions will help you diagnose the vehicle's problem.

Successful technicians are skilled at asking questions that lead to the source of the problem. Specific questions related to Who, What, When, Where, and How should be the basis for your investigation. When you read *The Problem* at the beginning of this chapter, did you notice that you needed to find out when the vehicle does not start? You will need to ask questions, such as,

"Does this happen all the time?" Finally, you will have enough information to begin your diagnosis.

Apply it!

Meets NATEF Communications Standards for speaking, verbal cues, and oral information.

Read again *The Problem* at the beginning of this chapter.

❶ Make a list of questions you might ask the customer to help you diagnose the problem with her vehicle.

❷ Share your list with your team members. Ask for their comments on your questions.

❸ Make a list of key questions to ask customers. This will be a valuable resource for you on the job.

The **armature** is the part of the starter that rotates. It contains many individual windings (coils). A nonconducting material electrically separates the coils. Current flowing through the coils creates magnetic fields. A laminated iron core increases the strength of the magnetic fields in the armature.

The armature mounts on a shaft located inside the starter housing. Bearings support the armature at each end. The smaller the space between the armature and the field magnets, the stronger the magnetic reaction. However, the armature must be able to rotate freely without touching the field magnets.

The **commutator** is a series of copper segments placed side by side to form a ring around the armature shaft. Nonconducting material separates each segment. Two segments, appropriately placed, connect to each end of a coil in the armature. As the armature rotates, carbon brushes ride against the commutator segments and supply current to the armature coils.

Permanent Magnet Starter

Many starters use permanent magnets. A *permanent magnet* is a ceramic magnetic material. It does not require current flow through a field coil to create its magnetic field. Permanent magnets replace the electromagnetics.

Starter Operation

Electric motors work on the principle of electromagnetism. *Electromagnetism* occurs when current flowing through a conductor creates an electromagnetic field. An **electromagnetic field** is the space around an electromagnet that is filled with invisible lines of force. The strength of the field depends on the amount of current flow and the number of wires in the coil. Placing an iron core inside the coil also increases field strength.

A basic law of magnetism states that like magnetic poles repel, unlike magnetic poles attract. In a field coil starter, current passing through the field coils creates an electromagnetic field. **Fig. 3-3.** The same current passes through one brush, into the commutator segment, through the armature coil, out through the other segment, and through another brush to ground.

Current passing through the armature coils creates another electromagnetic field. As the armature rotates, the brushes and the commutator reverse the direction of the current flow in the armature coils. Reversing the direction of current in each coil reverses the direction of the magnetic field. The magnetic fields in the armature are attracted and repelled by the magnetic field created by the field coils. The action of the magnetic fields makes the armature rotate.

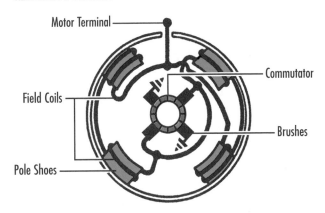

Fig. 3-3 The starter has four field coils, four brushes, and multiple armature windings. *What component is not needed on a permanent magnet starter?* (General Motors Corporation)

In a permanent magnet starter, current flows only through the armature windings. The permanent magnets provide the stationary magnetic field. The magnetic fields in the armature are attracted and repelled by the magnetic fields created by the permanent magnets.

High current flows to the starter only when the ignition switch is in the START position. As long as current flow exists, the armature continues to rotate. When the ignition switch returns to the RUN position, the contacts of the starter relay or solenoid open. No current flows to the starter. Without current flow, the starter armature stops rotating. The starter stops cranking the engine.

Solenoid Starter

The solenoid in a solenoid-operated starter has two functions. **Fig. 3-4(a).** It closes the high-current contacts connecting the battery to the starter. It also moves the starter pinion gear assembly into mesh with the ring gear. The ring gear is located on the flywheel or drive plate of the engine. The low-current control circuit energizes the coils of the solenoid. The magnetic field that is created pulls the plunger into the solenoid. **Fig. 3-5(a).** The end of the shift lever, attached to the plunger, moves in the same direction. As the plunger moves, the shift lever moves the pinion gear assembly on the armature shaft toward the ring gear. **Fig. 3-5(b).** Electrical contacts at the base of the plunger close when the pinion and ring gears fully mesh. **Fig. 3-5(c).** The closed contacts provide a current path between the battery and the starter. The starter begins to crank the engine.

Moveable Pole Shoe Starter

The moveable pole shoe starter is an older design. It does not use a solenoid. **Fig. 3-4(b).** It requires a separately mounted relay to supply

current to the starter. **Fig. 3-6.** The low-current control circuit connects the relay coil to the battery. Current applied to the coil energizes the relay. The energized coil pulls the relay plunger into the center of the coil. Contacts on the plunger connect the battery and the starter. The closed contacts provide a current path between the battery and the field coils in the starter.

Two windings in this starter act as solenoid windings. The first is grounded to the case. The second is grounded through a set of copper contacts that are normally closed. When current is first applied to the starter, most of it will flow through the two windings. This creates a strong magnetic field that pulls the moveable pole shoe down. An attached shift lever moves the pinion into engagement. When the pole shoe is pulled down, a small arm opens the copper grounding contacts. This opens the "pull-in" circuit. The starter now operates in a normal manner.

Moving the ignition switch to the RUN position removes current from the relay or solenoid. The return spring on the pole shoe moves the shift lever and pinion gear assembly out of engagement.

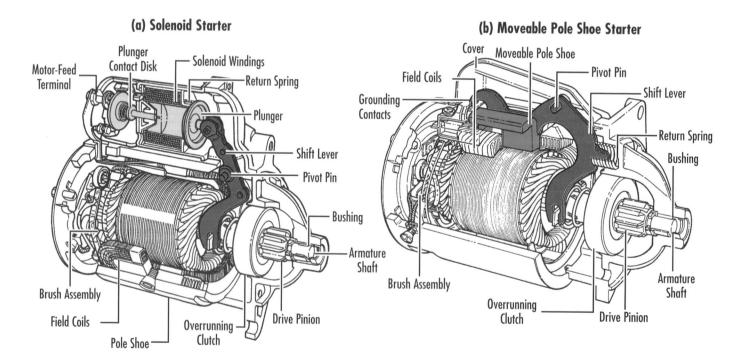

Fig. 3-4 Cutaway views of two starters. A solenoid starter **(a)** and a moveable pole shoe starter **(b)**. *What are the functions of the solenoid and the pole shoe?* (Ford Motor Company)

(a) Pinion Gear Disengaged

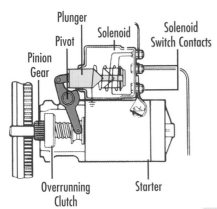

(b) Pinion Gear Partially Engaged

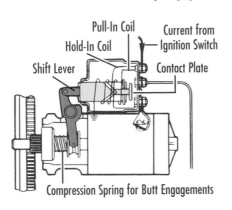

(c) Pinion Gear Fully Engaged and Starter Cranking

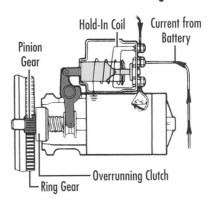

Fig. 3-5 Action of the starter solenoid. *At which point is cranking current supplied to the starter?* (Delco-Remy Division of General Motors Corporation)

Relay and Solenoid Starters

Some manufacturers use both a relay and a solenoid. **Fig. 3-7.** Low current applied to the relay coil pulls the relay plunger into the center of the coil. Contacts on the relay plunger connect low current to the solenoid coil.

Low current applied to the solenoid coil energizes the solenoid. The energized coil pulls the solenoid plunger into the center of the solenoid coil. Electrical contacts connected to the base of the solenoid plunger close when the pinion and ring gear mesh. The closed contacts provide a high current path between the battery and the starter.

Typical service problems with relays and solenoids may include burned or pitted contacts and binding of the plunger assembly. Defective relays and solenoids are replaced. A chattering or buzzing from a relay, solenoid, or starter is an indication of low battery current or high resistance in the circuit.

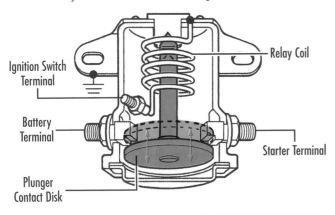

Fig. 3-6 The internal construction of one type of starter relay. *Which parts are similar to the parts in a starter solenoid?* (Ford Motor Company)

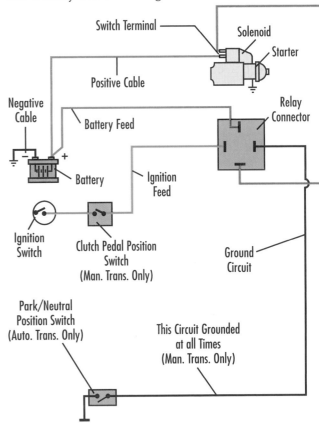

Fig. 3-7 A combination relay and solenoid circuit. *Which component is energized first, the relay or the solenoid?* (DaimlerChrysler)

It's All About Magnetism!

You turn the key, and the engine starts. It sounds simple, but is it? The starting system–battery, solenoid, and starter–makes it happen, using electricity and magnetic force.

Electric current flows through the coils of the solenoid. This creates a magnetic field that produces movement that closes contacts to complete a circuit. The current from the battery can now crank the engine.

Let's see how an electromagnetic solenoid works.

When electric current flows through a wire (a conductor), it creates a magnetic field around the wire. To find out the field's direction, use the *right-hand rule*. Hold your right hand directly in front of you. Point your right thumb in the direction in which electric current is flowing and curve your fingers. The curve of your fingers shows the direction of the magnetic field! Try it.

Figure A shows current flowing through a single loop, or coil, of wire. Use the right-hand rule to find the direction of the magnetic field at different points along the loop. Note that the lines of magnetic force all flow in the same direction in the center of the loop. That's where the field is concentrated.

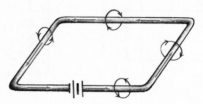

Fig. A

If you wind the wire into a coil of many loops to make a solenoid, the magnetic fields are added together. The more loops in a solenoid's coil, the stronger its magnetic field. An iron core in the coil's center will increase magnetic strength.

When the solenoid is energized, magnetic forces will attract or repel this core, or plunger, causing movement. This movement will close contacts in the starter circuit. It will also operate the shift lever to move the pinion gear into position.

Apply it!

Constructing an Electromagnet

Meets NATEF Science Standards for electrical circuitry, grounding, voltage, current, resistance, and magnetism.

Materials and Equipment
- Digital multimeter
- No. 36 copper wire (plain enamel magnet wire), 25–30' [8–9 m]
- DC power supply: regulated and fused, low voltage, variable (0–12 volts, 0.5 amps)
- Magnetic compass
- Coil core: a large iron nail or steel bolt
- Insulating tape

❶ Wrap a layer of insulating tape around the coil core. Wind the insulated magnet wire around the coil core in closely spaced turns. Avoid overlapping the windings.

❷ Tape the first and last turn to the coil core. Strip the insulation from each end of the wire.

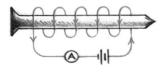

Fig. B

❸ Make sure the power supply is turned off. Connect the digital multimeter, Ⓐ, in series with the coil and the power supply. **Fig. B.** Set the multimeter to measure current.

❹ Hold a compass near the end of the coil. What happens?

❺ Turn on the power supply. Repeat Step 4 with the voltage set at 1 volt. Observe the magnetic force of the electromagnet on the compass needle.

❻ Reverse and increase the current. Observe. What happens to the compass needle?

Fig. 3-8 The pinion gear meshes with and drives the ring gear. *Where is the ring gear located?* (Robert Bosch GmbH)

Pinion Gear

Ring Gear

Starter Drive Assembly

The starter drive assembly engages and disengages the pinion gear and the ring gear.

Pinion Gear The *pinion gear* is the smallest gear in a gear set. The pinion gear on the armature shaft meshes with the larger ring gear on the flywheel or drive plate of the engine. **Fig. 3-8.** The ring gear has about fifteen times as many teeth as the pinion gear. The pinion gear must rotate fifteen times to rotate the ring gear one time. If the starter operates at 3,000 rpm, it will crank the engine at 200 rpm.

Overrunning Clutch The **overrunning clutch** prevents the engine from driving the starter. **Fig. 3-9.** As the engine starts, its speed may rapidly increase to 2,000 rpm or higher. If the pinion gear remained engaged with the ring gear, the starter's armature would rotate at 30,000 rpm! The armature would break apart at such speed. The overrunning clutch transmits torque only in one direction. It rotates

freely in the opposite direction, preventing armature damage.

During starter operation, rotation of the armature forces the rollers into the small end of the notches in the clutch shell. This action locks the clutch to transmit cranking torque to the engine. As the engine starts, ring gear rotation turns the pinion and clutch shell faster than the armature. The roller springs force the rollers into the large end of the shell notches. The overrunning clutch unlocks, removing cranking torque, allowing the pinion gear to spin freely until it disengages.

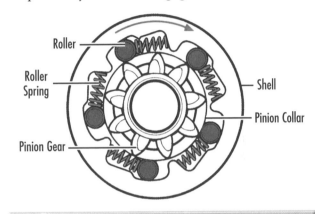

Roller

Roller Spring

Pinion Gear

Shell

Pinion Collar

Fig. 3-9 A cutaway view of an overrunning clutch with pinion gear. Note the notches in the shell. *What makes the overrunning clutch unlock?* (Delco-Remy Division of General Motors Corporation)

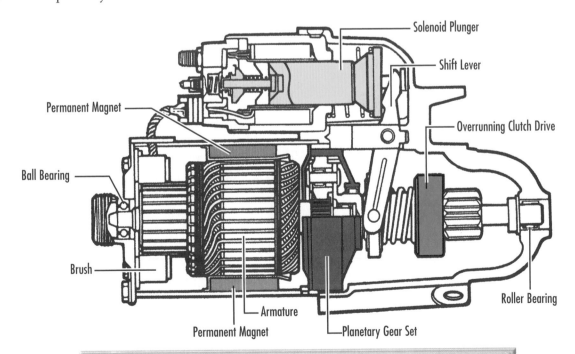

Solenoid Plunger

Shift Lever

Permanent Magnet

Overrunning Clutch Drive

Ball Bearing

Brush

Armature

Permanent Magnet

Planetary Gear Set

Roller Bearing

Fig. 3-10 A cutaway view of a permanent magnet starter with internal gear reduction. *Why does this starter have a gear ratio of 45:1?* (General Motors Corporation)

Gear Reduction and Cranking Torque The forces of friction and compression require high torque to crank the engine. Torque is affected by starter design and by the gear ratio between the pinion and the ring gear. *Gear ratio* is the number of rotations a pinion gear must turn to rotate a driven gear one time. If the starter connects directly to the crankshaft, the gear ratio is 1:1. A direct connection does not supply enough torque to crank the engine.

Gear reduction increases starter torque by increasing the gear ratio. In our example starter, the pinion gear and ring gear form an external reduction gear set. The pinion gear must rotate fifteen times to rotate the ring gear one time, for a ratio of 15:1.

Some starters use a combination of internal and external gear reduction to increase torque. **Fig. 3-10.** Internal gear reduction is supplied by *planetary gears.* **Fig. 3-11.** External gear reduction is supplied by the starter pinion gear and the engine ring gear. The combination of internal and external gear reduction provides a much higher gear ratio, for example 45:1. The pinion gear must rotate forty-five times to rotate the ring gear one revolution.

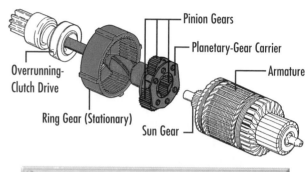

Fig. 3-11 An exploded view of a starter using planetary gear reduction. *Why is gear reduction used?* (Delco-Remy Division of General Motors Corporation)

SECTION 1 KNOWLEDGE CHECK

❶ Explain the purpose of a starting system.

❷ Describe the differences in construction between a field coil starter and a permanent magnet starter.

❸ Describe the function of a moveable pole shoe.

❹ Explain the function of an overrunning clutch.

❺ Describe how gear reduction improves cranking torque.

Section 2

Diagnosing the Starting System

The three basic starting system complaints are:
• The engine does not crank.
• The engine cranks slowly but does not start.
• The starter operates but does not crank the engine.

Many starting system problems are diagnosed by looking and listening when cranking the engine. **Table 3-A.**
• Does the relay or solenoid chatter?
• Do the headlights or interior lights dim heavily or go out?
• Does the engine crank at a normal speed?

Cranking Voltage/Current Draw Test

The *cranking voltage/current draw test* measures battery voltage and current during engine cranking. **Fig. 3-12.** Make sure the battery is fully charged

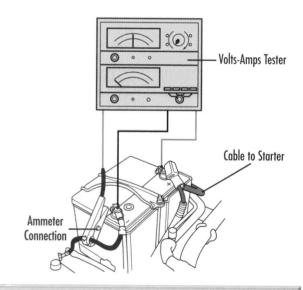

Fig. 3-12 The cranking voltage/current draw test measures battery voltage and current during engine cranking. *Why must the ignition or fuel system be disabled?* (American Honda Motor Co.)

Table 3-A	DIAGNOSING THE STARTING SYSTEM	
Complaint	**Possible Cause**	**Check or Correction**
1. No cranking, lights stay bright	a. Open circuit in ignition switch b. Open circuit in starter c. Open in control circuit d. Open fusible link	a. Check switch contacts and connections b. Check commutator, brushes, and connections c. Check solenoid or relay, switch, and connections d. Correct condition causing link to blow; replace link
2. No cranking, lights dim heavily	a. Trouble in engine b. Battery low c. Very low temperature d. Frozen armature bearing, short in starter	a. Check engine to find trouble b. Check, recharge, or replace battery c. Battery must be fully charged, with engine, wiring circuits, and starter in good condition d. Repair or replace starter
3. No cranking, lights dim slightly	a. Faulty or slipping drive b. Excessive resistance or open circuit in starter	a. Replace parts b. Clean commutator; replace brushes; repair poor connections
4. No cranking, lights go out	a. Poor connections, probably at battery	a. Clean cable clamp and terminal; tighten clamp
No cranking, no lights	a. Battery dead b. Open circuit	a. Recharge or replace battery b. Clean and tighten connections; replace wiring
5. Engine cranks slowly but does not start	a. Battery run down b. Very low temperature c. Starter defective d. Undersized battery cables or battery e. Mechanical trouble in engine f. Driver has run down battery trying to start vehicle	a. Check, recharge, or replace battery b. Battery must be fully charged, with engine, wiring circuits, and starter in good condition c. Repair or replace starter d. Install cables or battery of adequate size e. Check engine f. See item 7
6. Engine cranks at normal speed, but does not start	a. Ignition system defective b. Fuel system defective c. Air leaks in intake system d. Engine defective	a. Make spark test; check timing and ignition system b. Check fuel pump, fuel line, or fuel delivery system c. Tighten mountings and fittings; replace gaskets as needed d. Check compression, valve timing, etc.
7. Relay or solenoid chatters	a. Relay or solenoid defective b. Low battery c. Burned relay or solenoid contacts	a. Replace relay or solenoid b. Charge battery c. Replace relay or solenoid
8. Pinion disengages slowly after starting	a. Sticking solenoid plunger b. Overrunning clutch sticks on armature shaft c. Overrunning clutch defective d. Shift lever return spring weak	a. Repair or replace solenoid b. Clean armature shaft and clutch sleeve c. Replace clutch d. Repair or replace starter
9. Unusual noises	a. High-pitched whine during cranking (before engine starts) b. High-pitched whine after engine starts as ignition key is released c. Loud buzzing or siren sound after engine starts d. Rumble, growl, or knock after engine starts, and starter is coasting to a stop	a. Too much clearance between pinion gear and ring gear b. Too little clearance between pinion gear and ring gear c. Defective overrunning clutch d. Bent or unbalanced armature

before beginning the test. Low cranking voltage may indicate a bad battery, bad starter, or poor cable connections. To perform the test:

1. Disable the ignition or fuel system to prevent starting. Refer to the manufacturer's service manual.

2. Connect a voltmeter across the battery terminals.

3. Connect the ammeter current probe around the cable to the starter.

4. Turn the ignition switch to the START position.

5. Crank the engine 5–10 seconds and note the voltage and current readings.

6. Release the ignition switch.

Cranking voltage is normal if:

• The voltage reading is 9 volts or higher at 70°F [21°C].

• The engine cranks normally.

Cranking voltage is not normal if:

• The voltage is below 9 volts.

Cranking current is normal if:

• The current draw is between 140–200 amps.

The readings vary with starter size, engine size, and engine compression ratio. Consult the manufacturer's service manual for specified values. Readings higher or lower than specified may indicate a faulty starter.

If abnormal results are obtained:
• Check for resistance in the high-current circuit, the electrical cables, battery terminals, and ground connections.

Voltage Drop Test

The *voltage drop test* is performed while cranking the engine. This test checks for high resistance across a cable, component, or connection. Resistance reduces voltage. A voltage drop across

any component reduces the available cranking voltage. **Fig. 3-13.** Normal voltage drop readings should be less than 0.5 volts. Excessive voltage drop indicates high resistance at the point of measurement. Resistance may be due to loose or damaged connections, undersized cables, or burned contacts. Consult the manufacturer's service manual for specified values.

To perform the voltage drop test:

1. Disable the ignition or fuel system. Refer to the manufacturer's service manual.

2. Connect the voltmeter leads across each cable, switch, component, and connection in the circuit.

Calculating Resistance

Mr. Jackson's truck does not start. You turn on the headlights, and they appear normal. When you try to crank the engine, it turns over slowly. You check the battery connections, and they feel solid, even though there is a layer of corrosion. Because the headlights burned brightly, you feel the battery is okay.

You suspect the problem is excessive resistance somewhere in the circuit. This could be caused by a loose or corroded connection. You decide to perform a voltage drop test for excessive resistance. You will measure the voltage drop across each connection while trying to start the engine. Then you decide to calculate the resistance and the power loss at each connection.

You perform the voltage drop test by measuring the voltage difference between the terminal post and the cable clamp while cranking the engine. You also connect an ammeter clamp around the cable to the starter. With a cranking current of 150 amps, you measure 1 volt between the terminal post and the cable clamp.

You calculate the resistance across that connection using Ohm's law.

$$R = \frac{E}{I} = \frac{1 \text{ volt}}{150 \text{ amps}} = 0.006 \text{ ohms}$$

For a given current through the connection, the greater the voltage drop, the greater the resistance in the connection.

You can calculate the power being lost in the connection using the formula:

$$\text{Power} = I \times E = 1 \times 150 = 150 \text{ watts}$$

This 150 watts of power will appear as heat in the resistance of the connection. This heat may further deteriorate the already poor connection. The heat is power that did not reach the starter. It may be part of the reason for the engine's slow cranking.

Apply it!

Meets NATEF Mathematics Standards for formulas, decimals, symbols, estimating, and determining exact values.

When you test the same connection in a van, you measure a voltage drop of 0.15 volts for a 150–amp starting current.

❶ Use the measured voltage for the van to calculate the resistance across the connection. Compare your result for the van with the calculated resistance for Mr. Jackson's truck.

❷ Check the instructions for the voltage drop test in this chapter. Which vehicle's voltage drop reading indicates excessive resistance?

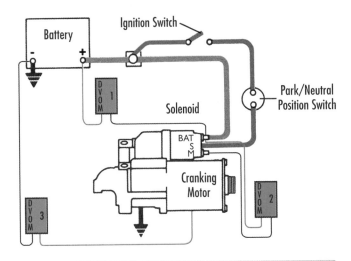

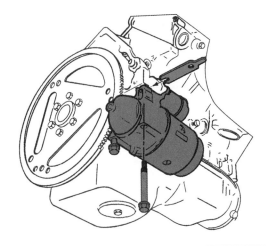

3. Hold the ignition switch to the START position.

4. Note the voltage reading across each component.

5. Compare with manufacturer's specifications.

Starter No-Load Bench Test

A starter no-load bench test can be made on some starters after the starter has been removed from the vehicle. To remove the starter:

1. Disconnect the negative battery cable.

2. Raise the vehicle if necessary. Use wheel blocks and jackstands to prevent the vehicle from falling.

3. Disconnect any braces, shields, or parts that interfere with removal.

4. Disconnect the wiring from the starter.

5. Support the starter.

6. Remove the mounting bolts.

7. Notice the location and number of any shims between the starter and the engine. The shims determine the clearance between the pinion and the ring gear. **Fig. 3-14.** Failure to replace shims may result in improper meshing of the gears.

8. Remove the starter from the vehicle.

To replace the starter, hold it in the proper position. Reinstall any shims that were in place when the starter was removed. Install and tighten starter mounting bolts. Connect the solenoid wires, and then reconnect the negative battery cable.

The no-load test checks for normal armature speed and current draw when a specified voltage is applied.

The test requires the following:
- An ammeter, a voltmeter, or a combination volt-amps tester (VAT).
- A remote starter switch.
- A battery.
- A carbon pile to adjust voltage to the starter (may be part of the VAT).

To perform the starter no-load bench test:

1. Connect the components. **Fig. 3-15.**

2. Close the remote starter switch.

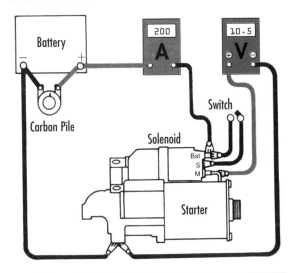

3. Adjust the carbon pile to obtain the manufacturer's specified no-load voltage.

4. Note the current flow and listen for normal armature rotation.

5. Compare the readings with those found in the manufacturer's specifications. Possible test results:
   - Rated current draw and speed indicate normal operation.
   - Low speed and high-current draw indicate internal friction or shorted armature.
   - No speed with high-current draw indicates a grounded component or frozen armature.
   - No speed with no current draw indicates an open circuit in the starter or solenoid.

Testing Relays and Solenoids

Failure of an engine to crank when the key is turned to START may be due to a defective relay or solenoid. If no movement of the relay or solenoid plunger can be detected, follow these steps:
- Make sure the battery is good and properly charged.
- Inspect the battery cables and wiring to the starter for damage.
- Make sure the park/neutral position switch is closed. (The shift lever must be in PARK or NEUTRAL or the clutch pedal fully depressed.)
- Have someone attempt to crank the engine. Use a DVOM or test lamp to check for voltage at the switch (S) terminal of the relay or solenoid.

- If no voltage is present, check the wiring circuit to the switch terminal. This will include the ignition switch and the park/neutral position switch or clutch switch. Repair this circuit as needed.
- If voltage is present, but the plunger does not move, the relay or solenoid is defective. Check the relay winding with an ohmmeter. Disconnect the wire at the switch terminal. Connect one ohmmeter lead to the switch terminal. Connect the other to a bare metal surface on the relay. If the ohmmeter indicates an open circuit, replace the relay.
- A starter solenoid has two windings. Each must be checked separately. This is most easily done with the starter on the bench. Disconnect the heavy connector between the solenoid and the starter. Connect one ohmmeter lead to the switch (S) terminal. Connect the other to the terminal that was connected to the starter. **Fig. 3-16(a).** Next connect the ohmmeter leads from the "S" terminal to the solenoid case. **Fig. 3-16(b).** If either ohmmeter reading indicates an open circuit, replace the solenoid.

If the relay or solenoid plunger does move, but cranking is slow, the problem may be high resistance in the main cranking circuit. Measure voltage drop to determine if high resistance is present. If the voltage drop across the relay or solenoid contacts is higher than specified (usually about 0.3 volt), replacement is required.

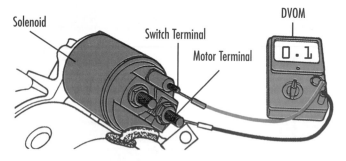

(a) Continuity Test Between Switch Terminal and Motor Terminal

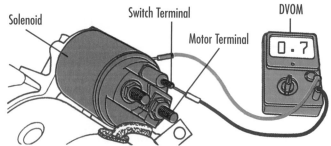

(b) Continuity Test Between Switch Terminal and Solenoid Case

Fig. 3-16 Using a DVOM to test resistance or continuity of solenoid windings. *What would you do if either winding has an open circuit?* (DaimlerChrysler)

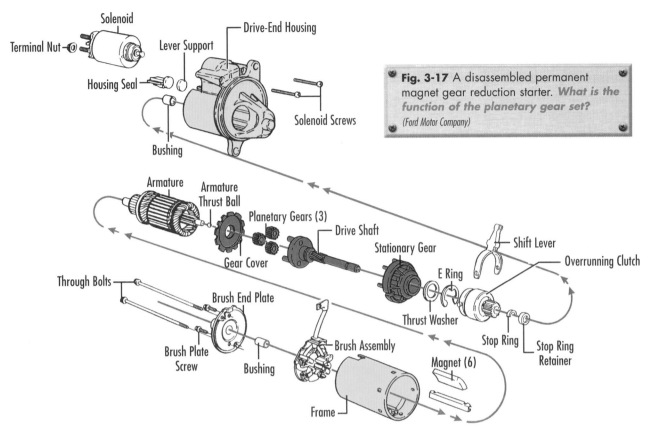

Fig. 3-17 A disassembled permanent magnet gear reduction starter. *What is the function of the planetary gear set?* (Ford Motor Company)

Starter Service

Most starters are relatively maintenance-free. However, they may fail with use. Most service facilities install a new or remanufactured starter when service is needed. Rebuilding starters is not a common service practice. However, if replacement is not an option, it may be possible to rebuild a starter. **Fig. 3-17.**

Major steps in rebuilding a starter may include the following:
- Replacing armature bearings and bushings.

- Testing the armature and field coils.
- Resurfacing the commutator.
- Replacing defective field coils.
- Replacing worn brushes.
- Replacing the overrunning clutch.
- Replacing electrical contacts in the solenoid.
- Lubricating the splines, shafts, and bearings.
- Checking the pinion clearance.
- Bench testing performance.

SECTION 2 KNOWLEDGE CHECK

❶ Describe the differences between a cranking voltage test and voltage drop test.

❷ Explain what could cause abnormal cranking current draw test results.

❸ Describe how to perform a voltage drop test. Why is it important that connections be clean and tight?

❹ Explain how to test solenoid windings with an ohmmeter.

❺ When replacing a starter, why is shim placement important?

CHAPTER 3 REVIEW

Key Points

Meets the following NATEF Standards for Electrical/Electronic Systems: starter testing; starter removal and replacement.

- The battery supplies electrical current to the starter through the relay or solenoid contacts.
- Though similar in operation, field coil and permanent magnet starters differ in construction.
- The starter armature rotates because the magnetic field of the field coils or permanent magnets repels the magnetic field of the armature.
- The pinion gear and ring gear provide a gear ratio of 15:1.
- An overrunning clutch transmits starter torque in only one direction.
- Relay and solenoid windings can be checked with an ohmmeter.
- The cranking voltage/current draw test measures battery voltage and current during engine cranking.
- The voltage drop test checks for high resistance across a cable, component, or connection.

Review Questions

1. Explain the operation of a starting system.
2. Identify the major internal components of a starter.
3. What component prevents starter operation when the vehicle is in gear?
4. What is the purpose of the ring gear and pinion in starter operation?
5. What two functions are performed by a starter solenoid?
6. Describe the operation of an overrunning clutch.
7. What is gear ratio?
8. Explain the procedure for testing solenoid windings with an ohmmeter.
9. **Critical Thinking** You have the option to rebuild the starter or install a replacement. Which option would you choose? Why?

TECHNOLOGY FORECAST
FOR AUTOMOTIVE EXCELLENCE

The Changing Role of Starters

Modern starters have come a long way from the old-fashioned hand crank. And they keep on changing. They are lighter, stronger, and more efficient than ever.

While basic starter motor designs may not change soon, their control systems may. They may depend more on electronics and less on electro-mechanical parts, such as solenoids. Designers may change some metal parts to make them weigh less. They may replace other metal parts with lighter plastic ones.

The creation of new powertrains will result in major changes in starter design. New powertrains may include hybrid (mixed) designs that use two or more energy sources. For example, these

designs may blend power from an internal combustion engine with battery power.

Designers might combine starters with other powertrain parts. For example, they might combine such a starter with a generator and mount it on the engine flywheel.

By design, engines with hybrid powertrains would run only part of the time. The engine would stop and restart in response to changes in speed. For such a system to work, the starter would have to work with the engine in a fully integrated way. Engineers might also design still other types of starting systems.

Who knows? The only thing certain is change.

AUTOMOTIVE EXCELLENCE TEST PREP

Answering the following practice questions will help you prepare for the ASE certification tests.

1. Technician A says the high current for the starter comes from the battery. Technician B says high current comes from the ignition switch. Who is correct?

   a Technician A.
   b Technician B.
   c Both Technician A and Technician B.
   d Neither Technician A nor Technician B.

2. A truck engine with a manual transmission does not crank when the key is turned to START. The battery is good. The clutch is depressed. Which of the following is the most likely cause?

   a Low resistance in the high-current circuit.
   b The clutch pedal position switch is bad.
   c The truck is in high gear.
   d All of the above.

3. A starter current draw test shows a reading of 100 amps. The manufacturer's specifications indicate the reading should be 190 amps. Which of the following can cause the problem?

   a Corroded battery terminals.
   b A bad ground connection to the engine block.
   c Worn brushes in the starter.
   d All of the above.

4. Sometimes an engine starts right up. Sometimes it cranks slowly. Technician A says it may be high resistance in the high-current circuit. Technician B says it may be low resistance in the low-current circuit. Who is correct?

   a Technician A.
   b Technician B.
   c Both Technician A and Technician B.
   d Neither Technician A nor Technician B.

5. Technician A says a gear-reduction starter gives better cranking performance. Technician B says it provides greater torque by using a higher gear ratio. Who is correct?

   a Technician A.
   b Technician B.
   c Both Technician A and Technician B.
   d Neither Technician A nor Technician B.

6. Higher than normal cranking current draw could result from:

   a A bad battery.
   b A starter or engine mechanical problem.
   c Poor cable connections.
   d All of the above.

7. A car has a grinding noise during cranking. Technician A says the starter drive mechanism may be damaged. Technician B says the starter shims may be missing. Who is correct?

   a Technician A.
   b Technician B.
   c Both Technician A and Technician B.
   d Neither Technician A nor Technician B.

8. A cranking current draw test shows a reading twice as high as specifications, and the engine does not crank. Technician A says the cause could be a defective starter. Technician B says it could be a seized engine. Who is correct?

   a Technician A.
   b Technician B.
   c Both Technician A and Technician B.
   d Neither Technician A nor Technician B.

9. Technician A says the overrunning clutch prevents the armature from being driven by the ring gear once the engine starts. Technician B says the clutch is used to disengage the pinion gear from the ring gear once the engine starts. Who is correct?

   a Technician A.
   b Technician B.
   c Both Technician A and Technician B.
   d Neither Technician A nor Technician B.

10. Technician A says worn brushes in the starter can cause high current flow. Technician B says worn bearings or bushings in the motor can cause high current flow. Who is correct?

   a Technician A.
   b Technician B.
   c Both Technician A and Technician B.
   d Neither Technician A nor Technician B.

CHAPTER 4

Diagnosing & Repairing the Charging System

You'll Be Able To:

- ⊗ Explain how a generator works.
- ⊗ Diagnose charging system problems.
- ⊗ Inspect and test a generator and adjust the drive belt.
- ⊗ Inspect and test a voltage regulator.
- ⊗ Remove and replace a generator.

Terms to Know:

brush
diode
rotor
slip rings
stator windings
voltage regulator

The Problem 〉〉〉〉〉 Your Challenge

Your customer has a van with an electrically operated wheelchair lift. Most of her driving is within three miles of her home.

Today, she had the van towed in for service again. The battery was dead this morning when she tried to start it. This has happened several times before. In the three years she has owned the van, the battery has been replaced twice and the generator once. Your customer is very unhappy. Whatever the problem is, she wants it corrected this time.

As the service technician, you need to find answers to these questions:

❶ What additional load is on this battery?

❷ What are the most likely and least likely causes of this problem?

❸ What diagnostic checks should you perform to isolate the problem?

Section 1

Charging System Construction and Operation

The charging system supplies all current needed for vehicle operation, lights, and accessories while the engine is running. It should also supply enough current to keep the battery charged. **Fig. 4-1.**

When the engine is not running, the battery alone supplies the current needed for cranking, ignition, fuel injection, and all other electrical loads. The battery also supplies current when the engine is running if electrical loads are higher than the generator output. If the generator fails, the battery is the only source of current.

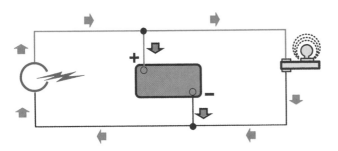

Fig. 4-1 Generator supplying load current and charging battery. *Where does the electrical power come from if the generator is not working?* (General Motors Corporation)

The charging system includes a generator, voltage regulator, system wiring, and the battery. A gauge or warning light on the instrument panel allows the driver to monitor the system. **Fig. 4-2.**

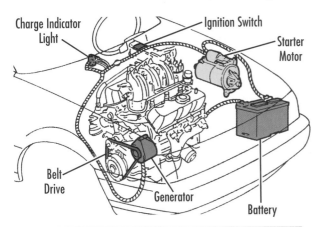

Charge Indicator Light — Ignition Switch — Starter Motor — Belt Drive — Generator — Battery

Fig. 4-2 A typical charging system. *Which parts shown are not part of the charging system?* (Ford Motor Company)

Today's charging systems use an alternating current (AC) generator. At one time they were called alternators. Before the early 1960s most vehicles had direct current (DC) generators. DC generators are still found on older vehicles.

Charging System Construction

The engine supplies the mechanical energy to drive the generator. Serpentine belts (or the older V-belts) transfer power from the engine to the generator. **Fig. 4-3.** These belts are made of rubber reinforced with fiberglass or nylon cords for strength. Idler, or tension, pulleys maintain a constant tension on the belt to prevent slippage under load.

The drive ratio between the pulleys is about 3:1 (three to one). This means the generator rotor turns three revolutions for one crankshaft revolution. The higher rotor speed allows the generator to supply electrical power even at low engine speeds. If the drive ratio was too high, the generator would self-destruct at high engine speeds.

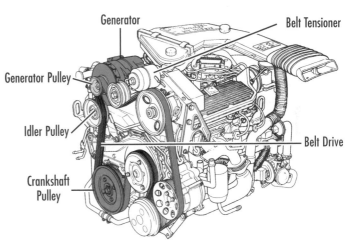

Generator — Belt Tensioner — Generator Pulley — Idler Pulley — Belt Drive — Crankshaft Pulley

Fig. 4-3 Generator mounting and drive-belt arrangement on an engine. Notice the difference in diameter between the crankshaft pulley and the generator pulley. *Why is it important that the generator rotor turn faster than the engine?* (General Motors Corporation)

Generator Construction A generator works on the principle of *electromagnetic induction*. When a moving magnetic field cuts across stationary conductors, an electrical pressure (voltage) is induced in the conductors. This voltage will produce current flow if there is a complete circuit.

Demonstrating Generator Action

You are driving–and suddenly the generator fails. The charging system warning light comes on! What happens if you ignore these warnings? Eventually, the battery goes dead and the engine stops running! A dead battery with no generator to charge it cannot supply the voltage to run the engine. Your generator is necessary for continued vehicle operation.

A generator is basically the opposite of a starter motor. In a starter motor, a magnetic field and a coil (armature) convert current into a rotational torque that cranks the engine. In an automotive generator, a magnetic field and a coil (rotor) convert the torque of the engine into a voltage (and current) in another coil (stator). The voltage in the stator coil produces current flow in the charging system.

Here is a simple experiment you can do to demonstrate generator action.

Apply it!

Inducing a Current in a Coil

Meets NATEF Science Standards for generators.

Materials and Equipment
- 50 feet [15m] of 24-gauge insulated copper wire
- Multimeter with DC milliamp/microamp scale
- Ceramic block magnet (about 0.4″ x 0.9″ x 1.9″ [about 9 mm x 22 mm x 48 mm])
- Electrical tape

❶ Set the multimeter on the most sensitive DC-current scale.

❷ Loosely wind the wire around your hand in a coil of about 3″ [7.6 cm] in diameter. Lay the coil of wire flat on a table.

❸ Wrap a length of tape around the right side of the coil to keep it wound. Wrap another length of tape around the left side of the coil.

❹ Scrape about 1″ [2.5 cm] of the insulation from each end of the wire. Attach the multimeter leads to the ends of the wire.

❺ Hold the magnet very near one side of the coil without moving it. Make sure the magnet is either face down or face up. Observe that no current is induced as long as the magnet is stationary.

❻ Quickly move the magnet to the right. Note that a small amount of current is generated.

❼ Stop the magnet. Notice that again no current is generated.

❽ Move the magnet back to the left. Notice that current is generated, but the needle moves in the opposite direction.

The amount of voltage induced in the conductors depends on three factors:
- The strength of the magnetic field.
- The number of stationary conductors.
- The speed at which the magnetic field cuts across the conductors.

Rotor Assembly The **rotor** is the part of the generator that is rotated by the drive belt. It creates the rotating magnetic field of the generator. At the center of a rotor is an iron shaft and core. The shaft is supported on both ends by bearings in the generator housing. Many loops (turns) of small wire are wound around the iron core. This winding is called the rotor field coil because current flow through the coil creates a magnetic field.

The ends of the rotor field coil windings are connected to the slip rings. **Slip rings** allow current to flow through the field coil while it rotates. Small carbon brushes ride on the slip rings. A **brush** is a block of conducting material, such as carbon. It

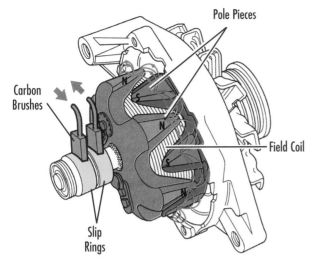

Fig. 4-4 A typical generator rotor assembly. *What are the ends of the field-coil windings connected to?* (General Motors Corporation)

rests against the rotating ring to form a continuous electric circuit. These brushes connect the field circuit wiring to the field coil.

Soft iron *pole pieces* are assembled over the ends of the field coil. The iron core and pole pieces concentrate and direct the magnetic field produced by the field coil. A typical rotor may have six north and six south poles. When assembled, every other pole has the opposite magnetic polarity. **Fig. 4-4.**

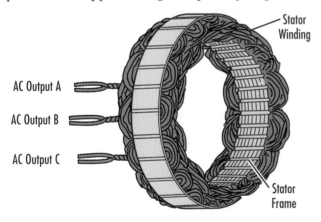

Fig. 4-5 A typical stator assembly. *How many windings (phases) are used in a typical stator?* (General Motors Corporation)

Stator Assembly   The **stator windings** are the stationary conductors in a generator. The rotor spins inside the stator assembly. Stators usually have three separate windings, or phases. The three windings are

equally spaced inside an iron core called the *stator frame.* **Fig. 4-5.**

Stator windings may be connected in two different ways. The wiring diagram for a wye-connected stator resembles the letter *Y*. The three windings have one common connection. The wiring diagram for a delta-connected stator is a triangle. In this case there is no common connection. However, pairs of windings are connected at three points. The two types of stators operate in the same manner. Three windings provide a stable voltage output from the generator. **Fig. 4-6.**

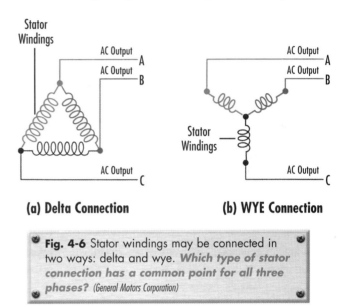

(a) Delta Connection   **(b) WYE Connection**

Fig. 4-6 Stator windings may be connected in two ways: delta and wye. *Which type of stator connection has a common point for all three phases?* (General Motors Corporation)

Diodes   Automotive vehicles operate on DC. The output of an AC generator must be rectified (changed) to DC. This is the job of the diode rectifier. A **diode** is an electrical device that allows current to flow in only one direction. **Fig. 4-7.** By using two diodes per winding, the AC output of the stator winding is changed to DC. Diodes are either positive or negative, depending on the polarity of the current they conduct.

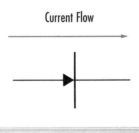

Current Flow

Fig. 4-7 A diode permits current flow in only one direction. *How many diodes are used for each winding?* (General Motors Corporation)

The three positive diodes are mounted in a small aluminum casting called the positive rectifier (diode) bridge. This bridge is connected to the output (BAT) terminal of the generator. A capacitor is also connected from the bridge to ground. It reduces radio noise and helps to smooth out the voltage pulses from the stator. Three negative diodes are mounted in the negative rectifier bridge. This bridge is grounded at the generator housing. The positive and negative rectifier bridges are assembled together as a unit but they are insulated from each other. **Fig. 4-8.**

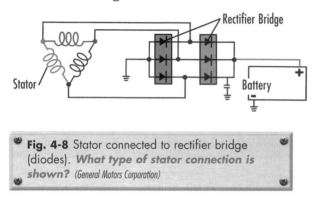

Fig. 4-8 Stator connected to rectifier bridge (diodes). *What type of stator connection is shown?* (General Motors Corporation)

Generator Housing The generator housing consists of two aluminum end frames assembled together. The rotor, stator, diodes, bearings, and capacitor are located inside the generator housing. **Fig. 4-9.**

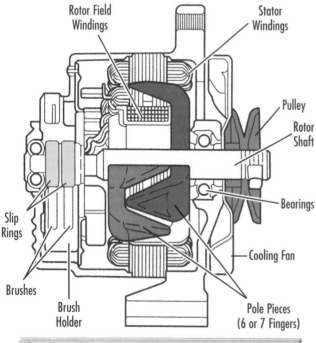

Fig. 4-9 Cutaway view of a generator. Some components are not visible in this view. *How is the rotor shaft supported?* (General Motors Corporation)

Some generators have a voltage regulator located inside or on the back of the housing. The drive pulley and cooling fan are mounted on the rotor shaft at the drive end of the generator.

Generator terminals are located at the slip-ring end of the housing. The output (battery or BAT) terminal is connected by a fairly large wire that leads to the positive battery terminal. Other terminals are used for the field circuit and to monitor or control charging system operation. **Fig. 4-10.**

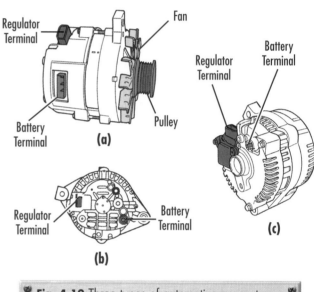

Fig. 4-10 Three types of automotive generators. *How is the output terminal usually marked?* (Ford Motor Company)

Charging System Operation

The generator and voltage regulator work together to produce the current needed by the electrical system. The generator produces the output current. As with a bar magnet, the magnetic fields produced in the rotor of the generator have polarity (north and south). When the magnetic field is not cutting across a stator winding, no voltage is created. See the 0° point in **Fig. 4-11.** When the rotor moves so that a south pole nears one side of a stator winding, a north pole will near the other side of the same winding.

The magnetic fields will induce a voltage that rises from 0 to a maximum value. This happens at 90° of rotation. As the two magnetic poles move past this winding, the induced voltage falls back to 0. At 180° of rotation, no voltage is being induced.

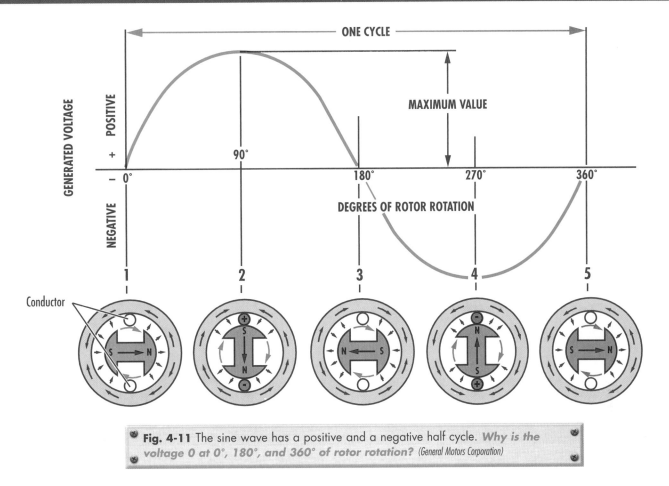

Fig. 4-11 The sine wave has a positive and a negative half cycle. *Why is the voltage 0 at 0°, 180°, and 360° of rotor rotation?* (General Motors Corporation)

This is because the magnetic fields are not cutting across the stator winding in this position.

As rotation continues, the opposite poles will near the same stator winding. Now the pattern will be repeated. However, the induced voltage will be of the opposite polarity. The rotor has turned 270°. With one full revolution, the rotor is back at its starting point. The voltage is again at 0, as it was at 180°. Each movement of a pair of rotor poles past a stator winding will produce a *sine wave* (AC) voltage pattern.

Because the stator has three separate windings, it produces three separate sine waves or phases. These sine waves are spaced 120° apart. With this arrangement the voltage never drops to 0. As voltage from one phase begins to drop, voltage from the next phase is increasing to the peak value. **Fig. 4-12.** This provides a more constant voltage output from the stator.

Rectifying AC Voltage   The AC voltage induced in the stator is changed to DC by the diode rectifier. This process is called *rectification*. The diodes allow current flow in one direction only. They route negative voltage pulses through the stator circuit. This

ensures that all current at the output terminal flows in the same direction. Current induced in the stator windings is always AC. Current flowing from the output terminal is always DC. The diodes permit both halves of the sine wave to be used. This is called *full-wave rectification*.

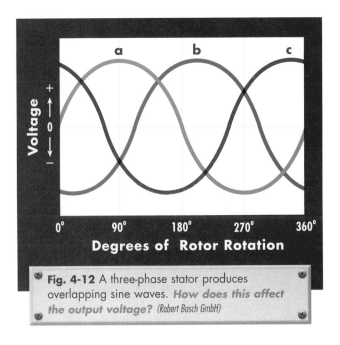

Fig. 4-12 A three-phase stator produces overlapping sine waves. *How does this affect the output voltage?* (Robert Bosch GmbH)

By using 12 magnetic poles in the rotor, three-phase windings in the stator, and full-wave rectification, the output of the generator is almost a smooth DC voltage. The battery and a capacitor in the generator smooth out most of the remaining voltage ripple. **Fig. 4-13.**

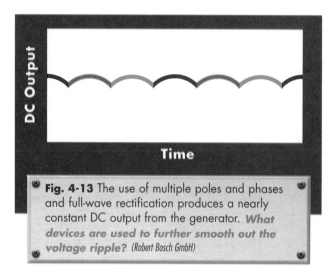

Fig. 4-13 The use of multiple poles and phases and full-wave rectification produces a nearly constant DC output from the generator. *What devices are used to further smooth out the voltage ripple?* (Robert Bosch GmbH)

Regulating the Generator Generators can produce enough voltage to seriously damage the electrical system. Voltage output must be limited to prevent overcharging the battery and damaging electrical and electronic components. An electronic **voltage regulator** is a device used to control the generator output voltage. Voltage regulators can be mounted inside the generator, on the back of the generator, or under the hood. **Fig. 4-14.** On some vehicles the voltage regulator is a part of the powertrain control module (PCM).

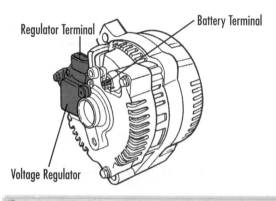

Fig. 4-14 A generator with the regulator mounted on the rear of the housing. *Name two other places the voltage regulator may be located.* (Ford Motor Company)

With the engine running, the voltage regulator senses charging system voltage. The normal range is from about 13.5 to 15.0 volts, depending on the underhood temperature. As long as the output is below this range, no regulation is required. When operating conditions cause generator output to reach the maximum desired voltage, the regulator begins to control rotor field current. Regulating field current controls the strength of the rotor's magnetic field. Reducing field current reduces the strength of the magnetic field. With a weaker magnetic field, voltage output is reduced.

Electronic voltage regulators control field current with a transistor. **Fig. 4-15.** Turning the transistor on and off very rapidly controls the amount of current flowing through the rotor.

Some regulators control field current by varying the frequency of the current pulses to the rotor. Other regulators use a fixed frequency but vary the "on" time (duty cycle) of the current. In either case,

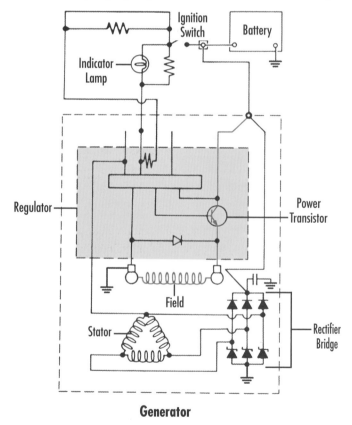

Generator

Fig. 4-15 Diagram of a generator with a built-in (integral) voltage regulator. The transistor in the regulator is in series with the field winding. *What is the purpose of the transistor?* (General Motors Corporation)

the voltage regulator controls the generator's output by varying the amount of current flowing in the rotor coil.

With high engine speed and low electrical loads, the generator voltage tends to be too high. During these conditions rotor current flow can be on for as little as 10 percent of the time and off for 90 percent. Average field current is low.

During low-speed, high-load conditions, the reverse is true. Voltage output will be less than the maximum allowed. Current will flow through the rotor for 90 percent of the time and be off for 10 percent. Average field current is high. **Fig. 4-16.**

Temperature Compensation The voltage required to charge a battery varies with battery temperature. It may take almost 16 volts to charge a very cold battery. When the battery is hot, the required voltage is much less. A temperature-sensitive resistor, called a thermistor, is used in most charging systems. It adjusts regulator voltage according to the underhood or battery temperature.

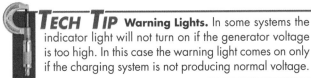
TECH TIP **Warning Lights.** In some systems the indicator light will not turn on if the generator voltage is too high. In this case the warning light comes on only if the charging system is not producing normal voltage.

The thermistor may be located in the voltage regulator or near the battery. The normal range of charging voltage is from about 13.5 volts (hot) to over 15.0 volts (cold).

Instrument Panel Charge Indicators

Displays on the instrument panel inform the driver when the charging system is not operating normally. These indicators can also be helpful to the technician when diagnosing problems in the system.

Indicator Light Some vehicles are equipped with a charging system indicator light. The light is operated by the difference between battery voltage and charging voltage. The charging indicator light should come on when the ignition switch is turned on. With the engine running, the lamp will go out if the generator is producing normal voltage output. If generator output is low or zero, the light will be on. On some vehicles the light will also be on if the voltage is higher than normal. The light cannot warn the driver if the voltage is within the normal range but generator current output is lower than it should be. Indicator lights have the advantage of being simple and easy to understand. **Fig. 4-17.**

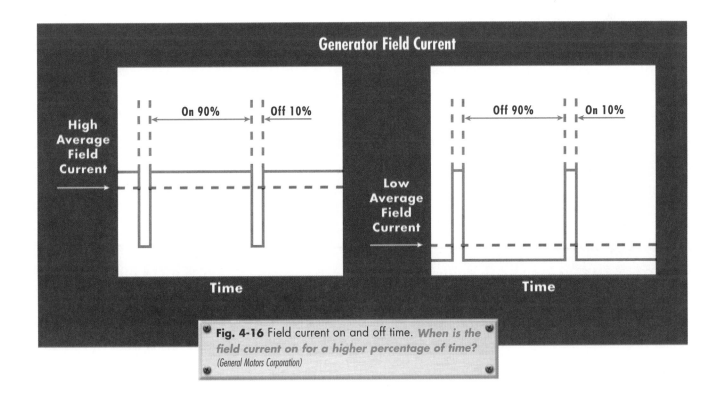
Fig. 4-16 Field current on and off time. *When is the field current on for a higher percentage of time?*
(General Motors Corporation)

Analyzing Sine Waves

Refer to the graph in **Fig. 4-11.** Ask these questions to better understand the graph.

1. What is the initial voltage?

2. What is the voltage at 180° of rotor rotation? At 360°?

3. What is the polarity of the voltage values that are found between 0° and 180°?

4. What is the polarity of the voltage values that are found between 180° and 360°?

5. Would the voltage between 180° and 360° power a battery?

The answers:

1. The initial voltage is 0 V.

2. At 180° and 360°, the graph indicates the voltage is zero.

3. Between 0° and 180°, the voltage is positive.

4. The graph indicates the voltage is negative between 180° and 360°.

5. The voltage is negative. The voltage of the battery must be positive to be used.

Figure A shows the result of adding another waveform starting when the rotor rotation is at 120°.

Now, with the second waveform, there is a positive voltage at 180°. Which areas now represent positive voltage? Which represents a negative voltage?

The areas A and B represent positive voltage. Areas C, D, and E represent negative voltage.

Apply it!

Meets NATEF Mathematics Standards for interpreting graphs, understanding geometric figures, and drawing conclusions.

❶ If a third waveform starts 120° after the second wave, the graph in **Fig. 4-12** would result. Is there a positive voltage corresponding to each degree of rotor rotation?

❷ In addition, the voltage varies less. The voltage is kept closer to what value?

The graph shown in **Fig. B** illustrates the net effect of generating the three alternating waveforms. Note that the waves are examples of sine waves generated by a sine function. A complete analysis of a sine function requires the study of trigonometry, which is not covered in this text.

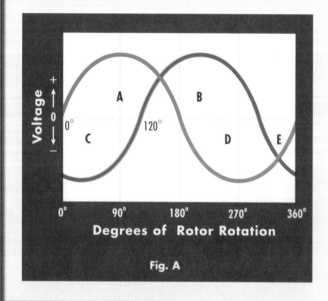

Fig. A

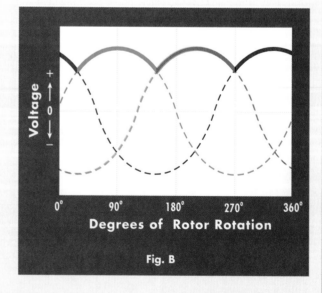

Fig. B

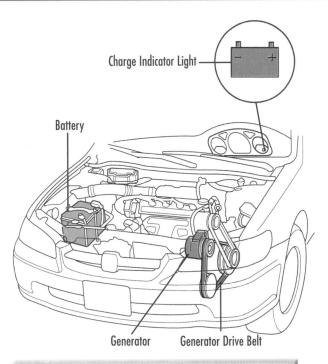

Charge Indicator Light

Battery

Generator   Generator Drive Belt

Fig. 4-17 A typical charge indicator light showing the international charging system symbol. *What is the advantage of an indicator light over a meter?* (American Honda Motor Company)

Voltmeter An instrument panel voltmeter is connected so that it displays voltage in the electrical system. The voltmeter provides a more accurate reading of system activity than does an indicator light.

- With the key off, a dash voltmeter needle will be at the low end of the scale. The meter is not connected until the ignition switch is turned on.
- With the key on and engine off, the voltmeter needle indicates battery voltage. Normally this is between 12 and 13 volts.
- While the engine is cranking, the voltmeter needle may drop to 11 volts or less.
- With the engine running, the voltmeter needle indicates charging system voltage. The normal range is between 13.5 and 15.0 volts, depending on battery condition and underhood temperature. Some dash voltmeters do not display actual voltage. Instead, they show a range of low, normal, and high values.

TECH TIP **Dash Ammeters.** The dash ammeter does not indicate total generator output. Current used by the engine operating systems, lights, and accessories is not included in the ammeter display.

Ammeter Dash ammeters are connected so they display only current going into or out of the battery (except for cranking current). The ammeter indicates whether current is flowing into the battery (C, or "charging") or out of the battery (D, or "discharging").

Immediately after an engine is started, the ammeter normally indicates that the battery is being charged. After the engine has been running for some time, the ammeter shows a lower charge rate. The charge rate drops as the battery is charged by the generator.

Some vehicles equipped with charge indicator lights also have a second warning system. The PCM monitors charging system voltage. It will turn on the *malfunction indicator lamp (MIL)* if certain charging system problems occur. (This lamp may be labeled CHECK ENGINE.) For example, the lamp is turned on and a *diagnostic trouble code (DTC)* is stored if system voltage is above 16.9 volts for more than 50 seconds. Lamp operation and diagnostic trouble code storage also occur for other charging system problems.

SECTION 1 KNOWLEDGE CHECK

❶ What is the primary source of electrical power in a motor vehicle when the engine is running?

❷ What three factors determine the amount of voltage induced in a winding?

❸ What component creates the magnetic field needed to induce voltage in an AC generator?

❹ What is the name of the stationary conductors in a generator?

❺ Why is it necessary to rectify the generator's output?

Section 2

Charging System Testing

You should perform charging system testing and diagnosis in a logical order. Without a test sequence in mind, you may overlook important clues to system failures.

Preliminary Checks

Before testing the charging system itself, make several checks to determine the overall condition of the charging circuit. Failure to make these checks prior to testing may cause misleading results during testing.

1. Inspect the drive belt and check its tension.
2. Start the engine. Observe the charging system indicator on the instrument panel.
3. If the MIL is on, check for charging system related DTCs.
4. Inspect the battery and charging system cables, wires, and connections. Repair as needed.
5. Test the battery for condition and state of charge. Charge or replace if needed.

Check the generator drive-belt condition and tension before testing the output of the charging system. A worn or loose belt will slip, causing generator output to be lower than normal. A damaged belt may break, disabling the water pump, generator, and power steering pump. The air conditioning compressor and cooling fan may also be affected.

Replace any belts that are frayed, torn, or have parts of the V-ribs missing. **Fig. 4-18.** Also replace belts that slip or squeal even after proper adjustments have been made. Replace V-belts if they are badly glazed (shiny) or worn. A worn belt rides on the bottom instead of the sides of a pulley. Ignore minor cracks on a serpentine drive belt. Cracks are normal and do not interfere with the proper operation of the belt. Belt dressings intended to improve the friction or grip of V-belts should not be used on serpentine belts.

Drive-Belt Tension

It takes several horsepower to drive a generator at maximum output. Even a good or new belt cannot transmit enough power if it slips due to insufficient tension. If belt tension is too high, there will be higher than normal loads on the bearings of the parts being driven. This could cause early bearing failure. The best way to check belt tension is with a belt-tension gauge. This gauge measures the amount of belt deflection under a specified load.

In **Fig. 4-19,** belt tension is adjusted by first loosening the mounting bolt and lock nut. The adjusting bolt is then adjusted to obtain the desired tension. Other vehicles are similar, but some do not use adjusting bolts. In vehicles without adjusting bolts, the tension is adjusted by carefully prying against the stator frame in the middle of the generator. Do not pry against the end frames. Refer to a service manual for specific adjustment procedures.

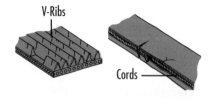

Fig. 4-18 A damaged serpentine belt. The cord structure in the belt gives it strength. The V-ribs give it surface contact area for gripping the pulleys. *What will happen if pieces of the V-ribs are missing?*

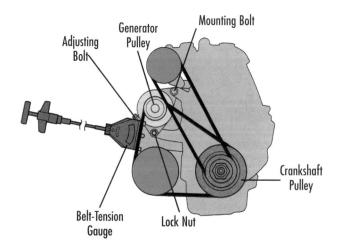

Fig. 4-19 The belt-tension gauge measures belt deflection under specified loads. *Why is it important that the generator's drive belt be properly tensioned?* (American Honda Motor Company)

Some vehicles use an automatic belt tensioner to keep the belt properly adjusted. **Fig. 4-20.** In normal use, periodic belt-tension adjustments are not necessary. To remove the drive belt, you must release the tension applied by the tensioner. Tension is released by rotating against spring pressure. Check for free movement of the tensioner after the belt is replaced.

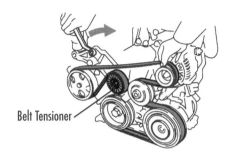

Belt Tensioner

Fig. 4-20 Automatic belt tensioners are used on some vehicles. *When is it necessary to release the tension from the belt?*

Charging System Testing

Four tests may be made to determine if the charging system is operating normally. They are:
- Charging voltage test.
- Charging current test.
- Voltage regulator test.
- Voltage drop test.

Charging Voltage Test The charging voltage test is a simple, quick test. It determines whether generator voltage is within the normal range. It does not indicate whether generator current output is normal. To measure charging system voltage:

1. Connect a voltmeter to the battery terminals. **Fig. 4-21.**

2. Start the engine and run it at about 2,500 rpm. Observe the voltage at the battery. It should be in the range of 13.5 to 15.0 volts, depending on temperature. Voltage may be higher if the charging system is very cold.

3. If the voltage is between 13.5 and 15.0, generator voltage output is normal. (Check current output if needed.)

4. If the voltage is less than 13.5, generator voltage output may be low. Perform a current output test to verify that the generator is defective.

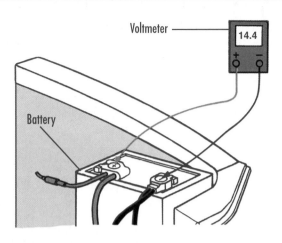

Voltmeter — 14.4

Battery

Fig. 4-21 The charging voltage test measures charging system voltage. *At what speed should the engine be running during this test?* (DaimlerChrysler)

5. If the voltage is over 15.0 (at normal temperature), the voltage regulator is probably defective. In most cases, it is not practical to bench test a voltage regulator.

Charging Current Test The charging current test measures the maximum current output of the generator at a specified voltage. It is the only test that can determine if the current output is within the manufacturer's specifications. Measure charging system current output using a volts-ampere tester (VAT). **Fig. 4-22.** This tester has a built-in carbon-pile load and inductive (clamp-on) ammeter lead.

To perform the charging current test:

1. Make sure the tester load control is turned off. Make sure that all switches are in the proper positions.

2. Connect the tester leads as follows: Connect the heavy load test leads and voltmeter leads to the proper battery terminal. (Voltmeter leads do not have to be connected if the voltmeter selector switch is placed on INTERNAL.)

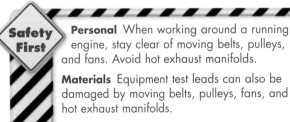

Safety First

Personal When working around a running engine, stay clear of moving belts, pulleys, and fans. Avoid hot exhaust manifolds.

Materials Equipment test leads can also be damaged by moving belts, pulleys, fans, and hot exhaust manifolds.

Fig. 4-22 The VAT-45 combines an ammeter, voltmeter, and load device into a single tester. *What is the inductive pick-up lead used for?* (Snap-on Tools Company)

3. Clamp the inductive ammeter lead over the heavy wire connected to the BAT terminal on the generator. Make sure the arrow on the inductive pickup is pointed in the correct direction.

4. Start the engine and run it at about 2,500 rpm.

5. Turn the load control knob clockwise to get the maximum ammeter reading without the voltage dropping below 13.0 volts.

6. Turn off the load control as quickly as possible. The carbon-pile load will overheat if left on too long.

7. The generator is good if the ammeter reading is within 15 amps of rated output. This value may be marked on the generator or found in the service manual.

8. If the output is low or zero, use a voltmeter to see if the correct voltage is present at all generator terminals. Refer to the service manual or wiring diagram for specified values. If the voltages are normal, remove the generator for bench testing or replacement.

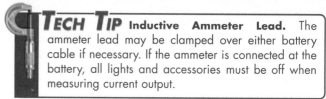

TECH TIP **Inductive Ammeter Lead.** The ammeter lead may be clamped over either battery cable if necessary. If the ammeter is connected at the battery, all lights and accessories must be off when measuring current output.

Voltage Regulator Test   The VAT can also be used to check the voltage regulator. If the generator current output is normal, check the charging voltage. With the engine still running under test conditions, note the ammeter reading. If the charge rate is less than 20 amps, read the voltmeter. This will be the regulated voltage. If the charge rate is above 20 amps, the generator may not be producing maximum voltage. In this case, charge the battery (with the generator or a charger) until the ammeter reads less than 20 amps during the test. Then read the voltmeter again.

Some generators can be tested with the voltage regulator bypassed. This will determine if the generator or regulator is at fault. Bypassing the voltage regulator is known as "full fielding" the generator. The procedure requires either a voltage or ground connection to the proper field terminal, depending on the type of generator. **Fig. 4-23.** If low generator output increases to normal when the regulator is bypassed, the regulator is defective.

Not all manufacturers recommend (or permit) full-field testing. With the regulator bypassed, there is no control over system voltage. If the voltage becomes too high, sensitive electronic components

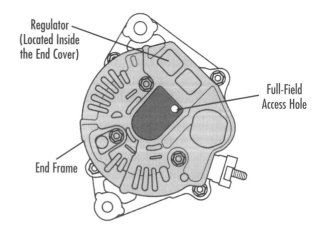

Regulator (Located Inside the End Cover)

Full-Field Access Hole

End Frame

Fig. 4-23 A generator with an access hole for full fielding. A terminal inside the opening is connected to ground to bypass the voltage regulator. *Why is the full-fielding procedure used on some generators?* (American Honda Motor Company)

will be damaged. Refer to the charging system section of the service manual to determine if and how full-field testing can be done on a generator.

Voltage Drop Test Excessive resistance in the charging system is found by measuring voltage drop. Perform this test after performing the charging current and voltage regulator tests with a VAT. With the VAT still connected and the engine running, follow these steps:

1. Turn the voltmeter selector switch to the EXTERNAL 3 VOLT position.

2. Connect the positive voltmeter lead to the BAT terminal on the back of the generator. Connect

Materials System damage may occur if voltage is allowed to exceed 16.5 on the full-field test.

the negative voltmeter lead to the positive battery terminal.

3. Adjust engine speed so the ammeter reads between 10 and 20 amps. Read the voltmeter. This will be insulated-circuit voltage drop.

4. Move the positive voltmeter lead to the negative terminal of the battery. Move the negative voltmeter lead to the generator housing. Read the voltmeter. This will be ground-circuit voltage drop.

Check the manufacturer's specifications in the service manual. If voltage drop is higher than specified, there is excessive resistance in that circuit. Find and correct the problem.

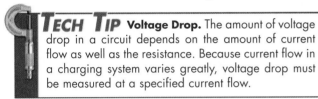

TECH TIP **Voltage Drop.** The amount of voltage drop in a circuit depends on the amount of current flow as well as the resistance. Because current flow in a charging system varies greatly, voltage drop must be measured at a specified current flow.

SECTION 2 KNOWLEDGE CHECK

❶ What happens to charging system output if the drive belt slips?

❷ What does a belt-tension gauge measure?

❸ What is the normal range for charging system voltage?

❹ What test is performed to determine if a generator can produce its rated output?

❺ What is a voltage drop test used for?

Section 3

Charging System Diagnosis

Charging system problems can be caused by the generator, voltage regulator, or battery. Defective wiring, poor connections, and slipping drive belts can also cause problems.

Undercharged Battery

An undercharged battery usually results in slow cranking. Possible causes for this problem include:
- Low generator current output.
- Low charging voltage.
- High resistance in the charging circuit.
- Lights or accessories used with engine off.
- Frequent short-trip driving.
- High key-off (parasitic) drain.
- Hard sulfation on battery plates.

- Use of high-current accessories added after the vehicle was manufactured.

If the battery and charging system are tested and are operating normally, check for the other possible causes of the undercharged battery.

Overcharged Battery

Overcharging the battery can produce dangerous hydrogen gas that may explode. Sulfuric acid vapors usually spew from battery vents during overcharging. If the overcharging is caused by high generator voltage, damage to the electrical system may occur. Possible causes of overcharging include:
- A defective voltage regulator.
- A grounded rotor coil (in some generators).
- A defective battery (shorted cell).
- High resistance in voltage regulator sensing circuit.

Charging System Noise

Generator noise may result from either mechanical or electrical causes. Mechanical noise may be caused by defective bearings, bent pulleys, defective belts, or bent cooling fan blades. Defective brushes and slip rings may also produce noise. Radio noise (static) may be caused by a defective generator.

 TECH TIP **Noisy Generator.** To determine if the cause of a noisy generator is mechanical or electrical, disconnect the field circuit and start the engine. If the noise is gone, the problem is electrical.

Common System Problems

Generator components most likely to fail are the bearings, brushes, and rectifiers. Poor connections in the generator or in the circuit wiring are also common causes of charging system failure.

Table 4-A, Diagnosing the Charging System, identifies common charging system problems, possible causes, and suggested checks and corrections.

Safety First

Personal The output (BAT) terminal on the generator is connected directly to the battery. Accidentally shorting this terminal to ground with tools or wires can cause very high current flow. This may result in serious burns as well as damage to the vehicle wiring. Whenever possible, disconnect the negative battery cable before working on generator wiring.

Materials Do not disconnect charging system wiring or the battery while the generator is operating. Doing so may produce voltage spikes that could damage electronic components.

Table 4-A	DIAGNOSING THE CHARGING SYSTEM	
Condition	**Possible Cause**	**Check or Correction**
1. Battery does not stay charged—engine starts okay	a. Battery defective b. Loose or worn generator belt c. Damaged or worn wiring or cables d. Generator defective e. Regulator defective f. Other electrical system malfunction	a. Test battery; replace if necessary b. Adjust or replace belt c. Repair as required d. Test and/or replace components as required e. Test; replace if necessary f. Check other systems for current draw; service as required
2. Generator noisy	a. Loose or worn generator belt b. Bent pulley flanges c. Generator defective d. Loose generator mounting	a. Adjust tension or replace belt b. Replace pulley c. Service or replace generator d. Tighten
3. Lights or fuses burn out frequently	a. Damaged or worn wiring b. Generator or regulator defective c. Battery defective	a. Service as required b. Test, service, and replace if necessary c. Test; replace if necessary
4. Charge indicator light flickers after engine starts or comes on while driving	a. Loose or worn generator belt b. Generator defective c. Field-circuit ground defective d. Regulator defective e. Light circuit wiring or connector defective	a. Adjust tension or replace b. Service or replace c. Service or replace wiring or connection d. Test; replace if necessary e. Repair as required
5. Charge indicator light flickers while driving	a. Loose or worn generator belt b. Loose or improper wiring connections c. Generator defective d. Regulator defective	a. Adjust tension or replace belt b. Service as required c. Service or replace d. Test; replace if necessary
6. Charge indicator meter shows discharge	a. Loose or worn generator belt b. Damaged or worn wiring (grounded or open between generator and battery) c. Field-circuit ground defective d. Generator defective e. Regulator defective f. Meter wiring or connections defective g. Damaged or defective meter h. Other electrical system malfunction	a. Adjust tension or replace belt b. Repair or replace wiring c. Repair or replace wiring d. Service or replace e. Test; replace if necessary f. Service as required g. Replace h. Service as required

SECTION 3 KNOWLEDGE CHECK

1 How is an undercharged battery usually noticed?

2 Name three possible causes of an undercharged battery.

3 What safety hazard occurs when a battery is being overcharged?

4 An overcharged battery may result from the failure of which device in the charging system?

5 What are some causes of mechanical noise in generators?

Section 4

Generator Removal, Service, and Installation

The generator and its wiring may be accessed from the top, bottom, or side of the engine compartment. You may need to remove parts or accessories to access the generator. Consult the service manual for specific instructions.

Removing the Generator

To remove the generator:

1. Connect a memory holder to the cigarette lighter receptacle.

2. Disconnect the negative battery cable.

3. Disconnect and remove the wiring harness from the generator.

4. Release tension from the drive belt and remove it from the generator pulley.

5. Remove the lock bolt, mounting bolt, and adjusting bolt that mounts the generator. **Fig. 4-24.**

6. Remove the generator from the vehicle.

Generator Service

In many cases a defective generator will be replaced with a new or rebuilt unit. Some manufacturers do not recommend generator repairs in

TECH TIP **Identifying Wires.** In some cases it may be necessary to identify individual wires before you remove them. This makes it easier to get the wires back on the correct terminal when reinstalling the generator.

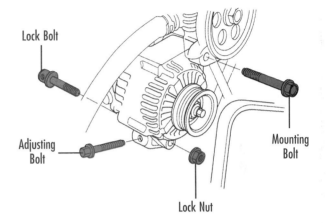

Lock Bolt

Adjusting Bolt

Mounting Bolt

Lock Nut

Fig. 4-24 Generator removal. *When is it necessary to remove the wiring harness after removing the generator from its mounting?* (American Honda Motor Company)

the service shop. When replacement is not practical, it may be possible to bench test and repair the defective unit.

Generator Disassembly To disassemble the generator:

1. Make reference marks on the two end frames so they can be correctly reassembled.

2. Remove the through bolts that hold the drive end and slip-ring end frames together.

3. Carefully separate the end frames.

Generator Bench Testing To bench test a generator:

1. Clean the parts with a dry or slightly damp cloth. Do not place electrical parts in a cleaning tank.

2. Inspect the brushes, slip rings, and bearings.

3. Inspect and test the rotor and stator. Refer to the service manual for test procedures and specifications. **Fig. 4-25.**

4. Check if the diodes can be easily disconnected from the stator. If so, they can be tested using the DIODE TEST position on a DVOM. Special testers are required if the diodes and stator cannot be separated.

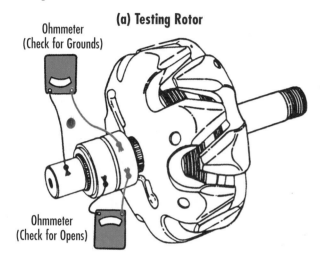

(a) Testing Rotor

Ohmmeter
(Check for Grounds)

Ohmmeter
(Check for Opens)

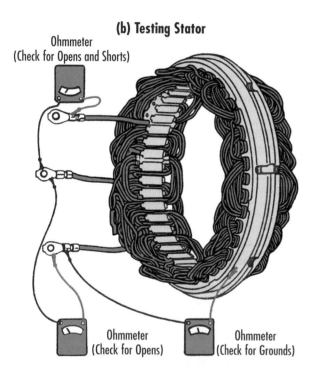

(b) Testing Stator

Ohmmeter
(Check for Opens and Shorts)

Ohmmeter
(Check for Opens)

Ohmmeter
(Check for Grounds)

Fig. 4-25 Bench testing generator components. *Where will you find specifications for rotor-coil current draw?* (General Motors Corporation)

Repairing the Generator To repair the generator:

1. Replace defective or worn parts.

2. Repair wires or connections if needed.

3. Replace the voltage regulator if it does not pass the tests for output or voltage regulation.

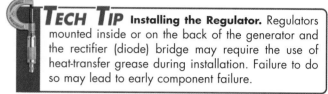

TECH TIP Installing the Regulator. Regulators mounted inside or on the back of the generator and the rectifier (diode) bridge may require the use of heat-transfer grease during installation. Failure to do so may lead to early component failure.

Generator Reassembly To reassemble the generator:

1. Make sure all parts are properly installed.

2. Make sure the brushes are properly retained in the brush holder.

3. Carefully reassemble the two end frames using the reference marks made earlier.

4. Install and tighten the through bolts.

5. Remove the brush retainer, if used.

6. Check the generator with an ohmmeter before reinstalling it. The output terminal should not be grounded. The field circuit should be tested for continuity, if possible.

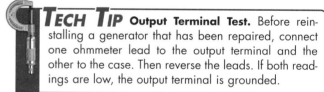

TECH TIP Output Terminal Test. Before reinstalling a generator that has been repaired, connect one ohmmeter lead to the output terminal and the other to the case. Then reverse the leads. If both readings are low, the output terminal is grounded.

Installing the Generator

Replacement generators may have a variety of output ratings, pulley types, voltage regulator calibrations, and mounting positions. Check part numbers to be sure you are installing the correct generator. Install the proper drive pulley on the generator if necessary. Follow the instructions in the service manual for pulley removal and installation.

To install a generator:

1. Connect the generator wiring harness first if necessary.

2. Position the generator in its mounting bracket.

Using Electrical Schematics

Electrical schematics, or wiring diagrams, are the "road maps" for current flow. These road maps show how electricity travels through a circuit. These diagrams can be as simple as **Fig. 4-1.** They can also be quite complex as in **Fig. 4-15,`** which shows many components.

Being able to read an electrical schematic is vital to your job as an automotive technician. The electrical and electronic systems on a vehicle are becoming more complex every year. These changes result in new schematics.

These schematics are printed in service manuals and textbooks so that you can "read" the electrical paths. Technical terms are used to identify parts in schematics. Electrical symbols tell what is happening to the electrical current. You may encounter a symbol you do not recognize. Refer to the key in the book or manual for an explanation. Keys are usually provided to explain what the symbols represent.

Understanding these terms and symbols will help you successfully diagnose and solve vehicle problems.

Apply it!

Meets NATEF Communications Standards for comprehending and using written information to determine technical sequences.

Refer to **Fig. 4-15.** This figure shows the electrical schematic for a generator with a built-in voltage regulator.

❶ On a sheet of paper, list the technical terms for the parts shown in **Fig. 4-15.** Explain the function of each part in a sentence.

❷ Identify and draw the electrical symbols used in the figure. Explain what each one represents.

❸ Follow the path of the electrical current from the generator to the battery.

❹ Write a paragraph explaining this wiring diagram.

---

3. Install the pivot and clamp bolts. Do not tighten.

4. Install the drive belt and place it over the pulley.

5. Allow the automatic tensioner to apply tension to the drive belt. If a tensioner is not used, adjust the belt using a tension gauge.

6. Tighten the pivot and clamp bolts using specified torque values from the vehicle service manual.

7. Connect any wires not previously connected.

8. Reconnect the negative battery cable.

9. Disconnect the memory holder.

10. Start the vehicle and verify proper charging system operation.

SECTION 4 KNOWLEDGE CHECK

❶ What should you connect to the vehicle before removing the negative battery cable?

❷ What should you do to the generator before disassembly?

❸ What device can be used to test diodes that are disconnected from the stator?

❹ What two tests should be made on a repaired generator before it is reinstalled?

❺ After servicing the charging system, what is the last thing you should do before returning the vehicle to the owner?

CHAPTER 4 REVIEW

Key Points

Meets the following NATEF Standards for Electrical/Electronic Systems: diagnosing charging system problems; inspecting and adjusting drive belts.

- Generators use electromagnetic induction to generate voltage.
- The major components inside a generator are the rotor, stator, and rectifier (diode) bridge.
- Diodes are used to rectify (change) the AC voltage in the stator to DC voltage at the output terminal.
- The charging system indicator light is operated by the difference between battery voltage and charging voltage.
- On-the-vehicle charging system tests include the charging voltage test, charging current test, voltage regulator test, and voltage drop test.
- The only complete test of a generator is the current output test.
- Some generators can be tested for current output with the regulator bypassed.
- Checking and adjusting drive-belt tension is an important part of charging system service.
- Always check part numbers when replacing a generator or regulator.

Review Questions

1. Describe the theory and operation of a generator.
2. Explain how the AC current in the stator is rectified to DC current at the output terminal.
3. Explain how to diagnose charging system problems.
4. Describe the procedures for inspecting and testing a generator and adjusting the drive belt.
5. What is the procedure for inspecting and testing a voltage regulator?
6. When performing the voltage regulator test, what does a charge rate above 20 amps indicate? What procedure should be followed if the charge rate is above 20 amps?
7. Describe the process for removing and replacing a generator.
8. **Critical Thinking** Why would a dash ammeter show a high rate of charge immediately after starting the engine?
9. **Critical Thinking** Explain why it is important to turn off all lights and accessories when the inductive ammeter lead from a VAT is connected over a battery cable.

TECHNOLOGY FORECAST
FOR AUTOMOTIVE EXCELLENCE

A New Generation Generator

Today's cars and trucks are filled with electronic options. We enjoy heated seats, CD players, and quick-defrost windows. These features make drive time more pleasant, but they take a toll on the generator. Not far into the future are electrohydraulic brakes and steering systems, heated catalysts, and more advanced engines. These new systems will need a higher charging capacity than vehicles have today.

Engineers are working on new generators to replace present generators. Generators have been in cars and trucks for nearly 40 years. The new generators will be up to 20 percent more efficient.

They will be far quieter than today's models. They'll also produce at least 90 percent of their available amperage at idle.

Several new technologies are in the works. These include switch-reluctance designs and permanent magnet designs that are more efficient than today's generators. Direct-drive generators are another possibility. They would replace the drive belt and eliminate problems such as vibration. A combined flywheel generator/starter is also being studied, as well as liquid-cooling and load-following designs.

AUTOMOTIVE EXCELLENCE
TEST PREP

Answering the following practice questions will help you prepare for the ASE certification tests.

1. Which type of stator has each phase winding connected to a common point?
   - ⓐ A wye-connected stator.
   - ⓑ A delta-connected stator.
   - ⓒ Both a and b.
   - ⓓ Neither a nor b.

2. Technician A says the diodes in the rectifier bridge rectify DC voltage produced in the stator into AC voltage. Technician B says the diodes rectify AC voltage to DC voltage. Who is correct?
   - ⓐ Technician A.
   - ⓑ Technician B.
   - ⓒ Both Technician A and Technician B.
   - ⓓ Neither Technician A nor Technician B.

3. The output of a generator can be described as:
   - ⓐ Alternating current.
   - ⓑ Direct current.
   - ⓒ Pulsed alternating current.
   - ⓓ Half-wave rectified.

4. Technician A says the stator has three different coils of wire. Technician B says current from the stator controls the field current. Who is correct?
   - ⓐ Technician A.
   - ⓑ Technician B.
   - ⓒ Both Technician A and Technician B.
   - ⓓ Neither Technician A nor Technician B.

5. The use of two diodes for each phase winding of the stator:
   - ⓐ Causes generator output to be half-wave rectified.
   - ⓑ Eliminates ripple voltage.
   - ⓒ Uses both positive and negative voltage pulses.
   - ⓓ Eliminates electrical noise from audio equipment.

6. Technician A says the output of the generator increases as engine speed increases. Technician B says the voltage regulator controls the voltage of the generator regardless of engine speed. Who is correct?
   - ⓐ Technician A.
   - ⓑ Technician B.
   - ⓒ Both Technician A and Technician B.
   - ⓓ Neither Technician A nor Technician B.

7. A charge indicator light that is on may be an indication of which of the following problems?
   - ⓐ Generator voltage is low.
   - ⓑ Generator current is high.
   - ⓒ Both a and b.
   - ⓓ Neither a nor b.

8. Technician A says an instrument panel ammeter indicates the amount of current going to or from the battery. Technician B says starter motor current does not go through the ammeter. Who is correct?
   - ⓐ Technician A.
   - ⓑ Technician B.
   - ⓒ Both Technician A and Technician B.
   - ⓓ Neither Technician A nor Technician B.

9. Technician A says a full-field test bypasses the rectifier bridge to see if the generator can produce maximum output. Technician B says the full-field test is recommended for all generators. Who is correct?
   - ⓐ Technician A.
   - ⓑ Technician B.
   - ⓒ Both Technician A and Technician B.
   - ⓓ Neither Technician A nor Technician B.

10. When removing a generator from a vehicle, the first step is to:
    - ⓐ Disconnect the positive battery cable.
    - ⓑ Disconnect the negative battery cable.
    - ⓒ Loosen the mounting bolts.
    - ⓓ Connect a memory holder.

CHAPTER 5

Diagnosing & Repairing Lighting Systems

You'll Be Able To:

- Explain how the exterior lighting circuit works.
- Explain how the brake, turn signal, and hazard warning light circuits are wired.
- Explain the operation of a brake warning light system.
- Diagnose lighting circuit problems.
- Explain how headlights are aimed.
- Explain how an electromagnetic gauge works.

Terms to Know:

electromagnetic display

halogen lamp

high-intensity discharge (HID) lamp

incandescent bulb

liquid crystal display (LCD)

vacuum tube fluorescent (VTF) display

The Problem

Bob McCue has brought in his pickup for an oil change and lubrication. He tells the service writer that when he tows his boat trailer with the truck, the turn signals flash much faster than normal. He says this doesn't happen when he tows the same trailer with his Lincoln.

Why does this happen only with the truck? Is this normal? Could the wiring of the trailer be shorting out something in the truck? Could this damage the turn signals?

Your Challenge

As the service technician, you need to find answers to these questions:

❶ How does the turn signal system work?

❷ What determines the rate at which the lights flash?

❸ What problems in the electrical system could cause the unusual flash rate?

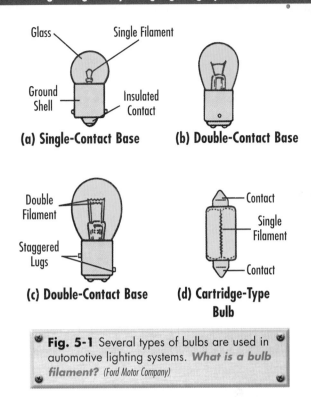

(a) Single-Contact Base   **(b) Double-Contact Base**

(c) Double-Contact Base   **(d) Cartridge-Type Bulb**

Fig. 5-1 Several types of bulbs are used in automotive lighting systems. *What is a bulb filament?* (Ford Motor Company)

Section 1

Lighting Systems

The lighting system conducts power to the exterior and interior lights on a vehicle.

Exterior lights generally include the head, parking, tail, side marker, and license plate lights. Separate circuits control the remaining exterior lights, including brake, turn signal, backup, and hazard warning lights. *Daytime running lights (DRLs)* are a separately controlled safety feature using the headlights. Fog light circuits have a separate on/off switch but may be powered through the headlight switch.

Interior lights include instrument panel lights, courtesy lights, and the dome/map light. Glove box, trunk, underhood, and vanity mirror lights are individually controlled.

Sources of Light

All lights in a vehicle are electrically operated. Several different technologies produce light.

Incandescent Bulbs A bulb that uses a tungsten filament placed in a vacuum inside a glass bulb is known as an **incandescent bulb.** The *filament* is the small wire-like conductor inside the bulb. Current flowing through the filament causes it to glow white-hot and give off light and heat. The vacuum in the glass bulb prevents the filament from burning up due to the oxygen in air. These bulbs are typically used for side marker, license plate, and interior lights. **Fig. 5-1.**

Some bulbs have two filaments sharing a common ground connection. Dual-filament bulbs are used in brakelights and taillights and in some high-beam/low-beam headlights. Each filament has its own terminal. This allows the unit to operate as two separate bulbs. **Fig. 5-2.**

Gas-Filled Bulbs Some lamps use a gas instead of a vacuum inside the bulb. The gas may protect the filament or be used in place of it. A **halogen lamp** is a bulb filled with halogen gas. Halogen is an inert (chemically inactive) gas. This protects the filament from burnout and allows it to operate at a higher temperature. The higher temperature of

the filament changes the color and the intensity of the light. Halogen bulbs produce a whiter light. It is about 25 percent brighter than the light from a conventional lightbulb.

Mercury vapor or xenon gas is used in high-intensity discharge (HID) lamps. A **high-intensity discharge (HID) lamp** is a lamp in which light is produced when high voltage creates an arc between two electrodes. The light source (lamp) is also referred to as a *burner.* A device called a *ballast module* produces the high voltage required for

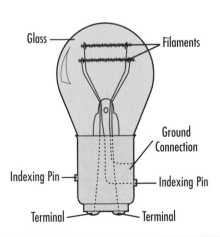

Fig. 5-2 Some automotive lightbulbs have dual filaments. *Give an example of a dual-filament bulb used in automotive lighting.*

Calculating Wattage

The *watt* is the basic unit of power. *Power* is the rate at which electrical energy is delivered to a circuit. "How many watts?" is a question often asked about speakers, sound systems, hair dryers, and lightbulbs. A sound system may be rated at 20 watts and a hair dryer at 1,250 watts. Household bulbs may be rated at 40, 100, or 150 watts. Some automotive bulbs are rated at less than 5 watts.

Ohm's law states that voltage is the product of current and resistance. If you know two of the values, you can calculate the third. The power formula is related to Ohm's law. Use the power formula to calculate the current flow when a lamp is on. The power formula is stated as $P = E \times I$, where:

P = power in watts (W)

E = operating voltage

I = current in amperes

What is the power for a lightbulb with a voltage of 12.6 V and a current of 1.04 A?

$$P = E \times I$$
$$P = 12.6 \text{ V} \times 1.04 \text{ A} = 13.1 \text{ W}$$

Round wattage to tenths, voltage to tenths, and current to hundredths.

Rewrite the power formula in terms of the current. Solve for I.

Start with $P = E \times I$. Divide both sides by E:

$$\frac{P}{E} = \frac{E \times I}{E}$$
$$I = \frac{P}{E}$$

Calculate the current in a bulb where the voltage is 14.5 V and the power is 4.9 W:

$$I = \frac{P}{E} = \frac{4.9 \text{ W}}{14.5 \text{ V}} = 0.34 \text{ A}$$

A 1154 bulb has a double filament. The operating voltage is 6.4 V and the currents are 2.63 A and 0.75 A.

What are the power values for each filament?

For the lower-current filament:

$$P = E \times I = 6.4 \text{ V} \times 0.75 \text{ A} = 4.8 \text{ W}$$

For the higher-current filament:

$$P = E \times I = 6.4 \text{ V} \times 2.63 \text{ A} = 16.8 \text{ W}$$

Apply it!

Meets NATEF Mathematics Standards for converting formulas to equivalent forms.

1. In a 1157 bulb, the wattage is 26.9 W and 7.6 W. The voltage is 12.6 V. What are the corresponding current values?

2. In a 1176 bulb, the voltage is 14.5 V and the currents are 1.34 A and 0.59 A. What are the corresponding wattages?

HID operation. The ballast module may be a separate assembly or part of the lamp housing. HID lamps produce a slightly bluish light. It is several times brighter than the light from halogen bulbs. HID lamps have no filaments to break. They last much longer than conventional or halogen bulbs. **Fig. 5-3.**

Neon gas can also be used for vehicle lighting. Glass tubes filled with neon are energized by high voltage produced by a ballast device. The motion and energy of the neon gas produce light without the need for an internal arc or filament.

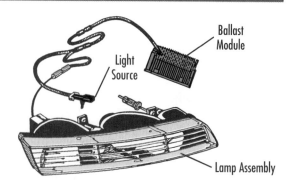

Fig. 5-3 High-intensity discharge (HID) lamps require a ballast module. *What does the ballast module do?* (Ford Motor Company)

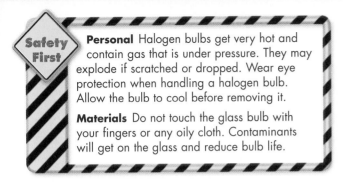

Neon-filled tubes are used in some brake lights because they light up quickly when energized. This gives drivers more time to react to the brake light signals.

Light-Emitting Diodes Light-emitting diodes (LEDs) are commonly used in automotive interior and exterior lighting. The most common LED color is red, which is used for brake lights and taillights. **Fig. 5-4.**

LEDs are also used as instrument panel warning lamps. Having no filament, they are less likely to require replacement than conventional bulbs.

Basic Lighting Circuits

For a lighting system to perform as intended, the driver must be able to control system operation. This is possible through the use of a circuit. The circuit may include switches, relays, and modules as well as the necessary lamps. **Fig. 5-5.**

Headlight and Related Circuits The headlight circuit controls the most often used exterior lighting on a vehicle. This circuit also controls instrument panel lighting. The major components of a basic headlight circuit are the:

- Headlight switch.
- Circuit protection device.
- Dimmer switch.
- Headlights.
- Front parking lights.
- Rear parking lights/ taillights.
- Side marker lights.
- License plate light(s).

A typical headlight switch has three positions:

- OFF—all lights are off.
- PARK—headlights are off; all other lights are on.
- HEAD—all lights are on.

The headlight switch can be a push/pull, rocker, or rotary type of switch. When the exterior lights are on, the headlight switch also provides power to the instrument panel lights. A *rheostat* is either part of the headlight switch or is mounted nearby.

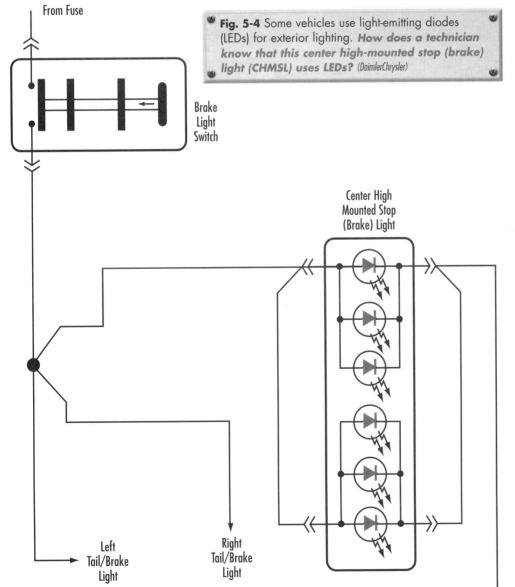

Fig. 5-4 Some vehicles use light-emitting diodes (LEDs) for exterior lighting. *How does a technician know that this center high-mounted stop (brake) light (CHMSL) uses LEDs?* (DaimlerChrysler)

From Fuse

Brake Light Switch

Center High Mounted Stop (Brake) Light

Left Tail/Brake Light

Right Tail/Brake Light

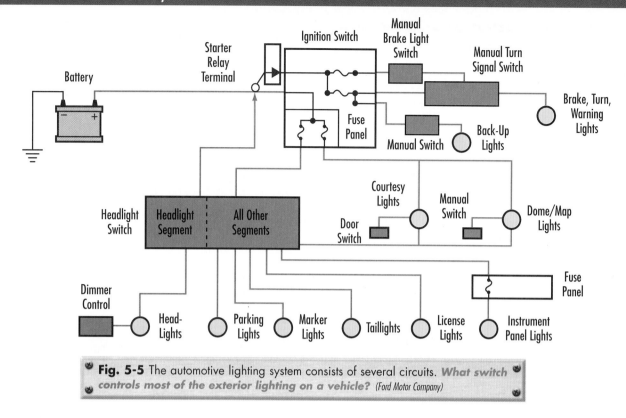

Fig. 5-5 The automotive lighting system consists of several circuits. *What switch controls most of the exterior lighting on a vehicle?* (Ford Motor Company)

Rotating the rheostat knob varies the current flow through the panel light circuit. This rheostat controls the brightness of the instrument panel lights. Full rotation of the rheostat in one direction closes a switch that turns on the vehicle's interior lights. **Fig. 5-6.**

Many headlight switches have built-in circuit breakers to protect the headlight circuits against electrical overloads. A circuit breaker is used instead of a fuse or fusible link so that the headlights do not suddenly go out if an overload occurs. Instead, the headlights will pulse on and off, providing some road illumination. A circuit breaker protects only the headlight circuit. Fuses protect the other light circuits controlled by the headlight switch.

Some headlight circuits do not use circuit breakers. Instead, there is a separate fuse for each headlight. If one fuse burns out, the other light(s) are not affected.

A headlight *dimmer switch* is usually part of the *multifunction switch* on the steering column. The multifunction switch is used to select either high- or low-beam headlights. It may also control the windshield wipers and turn signals. When the high-beams are selected, a high-beam indicator lamp on the instrument panel is turned on. **Fig. 5-7.** Some dimmer switches have a "flash to pass" feature. This circuit turns on the high-beam lights only while the

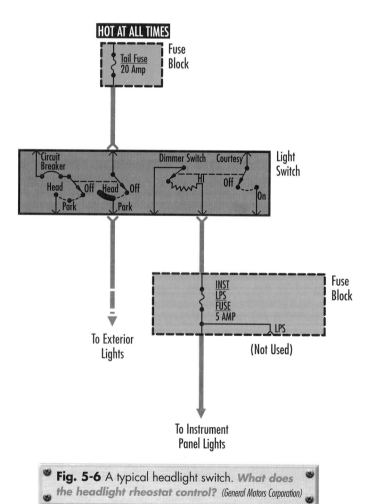

Fig. 5-6 A typical headlight switch. *What does the headlight rheostat control?* (General Motors Corporation)

switch is held in the flash to pass position. When the switch is released, the lights return to low beam.

A variety of headlight systems have been used in recent years. Sealed-beam headlights are either incandescent or halogen. They are either round or rectangular. They are used in two-headlight and four-headlight combinations. Sealed-beam bulbs contain the filament(s), reflectors, and lenses in one unit. They are identified as type-1 or type-2 bulbs, depending on the number of filaments used. Type-1 bulbs have one filament (either high- or low-beam). Type-2 bulbs have two filaments, one high beam and one low beam. The letter H or the word HALOGEN appears on the lenses of halogen sealed-beam units.

Many vehicles use composite (aerodynamic) headlights. These lights have a plastic housing that includes a reflector and lens. A replaceable halogen or HID bulb fits into the back of the housing. **Fig. 5-8.**

The position of the headlights is adjustable. This allows proper placement of the light beam on the road in front of the vehicle.

Parking/Tail/License/Side Marker Lights When the headlight switch is in the PARK or HEAD position, power is supplied to lights at the front, side, and rear of the vehicle. Some parking lights (front) and taillights (rear) use dual-filament bulbs. One filament is powered by the headlight switch for parking light and taillight operation. The second filament has less resistance. This permits more current flow, producing brighter light. This filament is used for brake light and turn signal light operation. The only

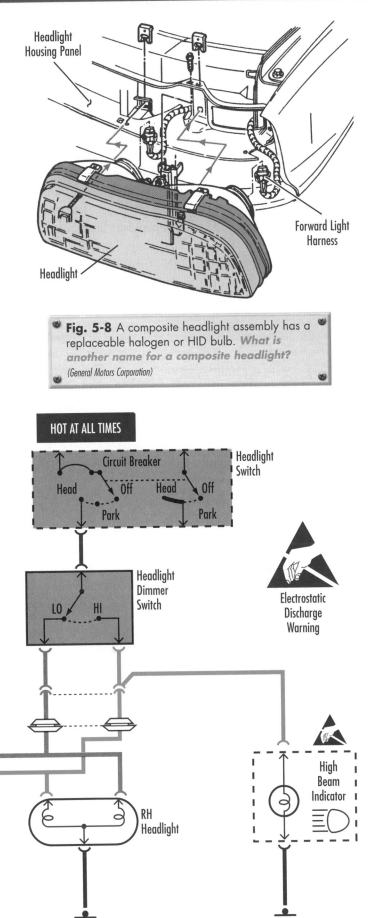

Fig. 5-8 A composite headlight assembly has a replaceable halogen or HID bulb. *What is another name for a composite headlight?* (General Motors Corporation)

Fig. 5-7 The dimmer switch controls power to two separate sets of headlights. *What additional light comes on when the high-beam headlights are used?* (General Motors Corporation)

connection between the headlights and the other exterior lights is the headlight switch. One or more fuses usually protect exterior light circuits.

Instrument Panel Lighting Turning on the headlight switch supplies power to the instrument panel lighting circuit. The panel lights illuminate the speedometer and other gauges, the radio, and the climate controls. In many applications this circuit is routed through the taillight fuse as well as the panel lights fuse. If the taillight fuse is burned out, the panel lights are also out. This should prompt the driver to get the necessary repairs. The panel light circuit is controlled by a rheostat in the headlight switch or mounted separately on the dash. The rheostat controls the brightness of the panel lights. **Fig. 5-9.**

Interior/Courtesy Lights Vehicle interior and courtesy lights are controlled by two different switches. When the instrument panel light rheostat is turned past its full brightness position, a switch closes to turn on the interior lights. The interior lights stay on until the rheostat switch is rotated out of the closed position. Door jamb switches also control interior lights. When any door on a vehicle is opened, a switch on the door jamb causes the interior lights to come on. The lights go out when all doors are closed. **Fig. 5-10.**

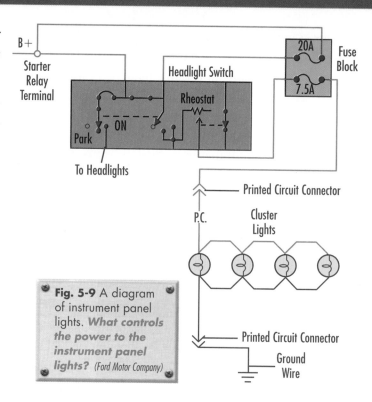

Fig. 5-9 A diagram of instrument panel lights. *What controls the power to the instrument panel lights?* (Ford Motor Company)

Some vehicles have illuminated entry systems. This system uses an electronic module to turn on the interior lighting before the doors are opened. The system can be activated by lifting a door handle or by using a *remote keyless entry* transmitter. In most cases the module will turn off the interior lights gradually after the doors are closed.

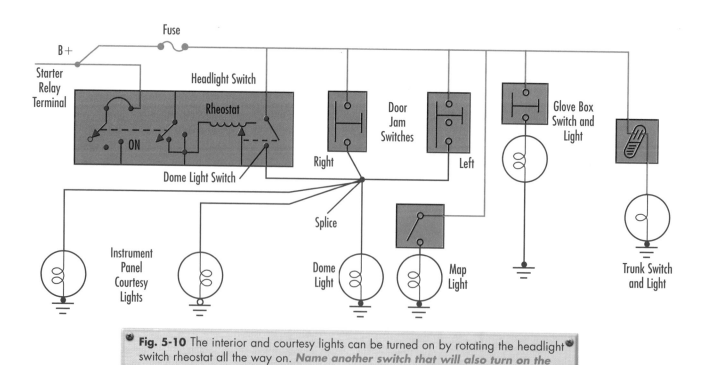

Fig. 5-10 The interior and courtesy lights can be turned on by rotating the headlight switch rheostat all the way on. *Name another switch that will also turn on the courtesy lights.* (Ford Motor Company)

Electronic Lighting Controls

Many lighting circuits use sensors, relays, and electronic modules to control some lighting functions.

Automatic Lighting Control Several versions of automatic lighting control are in use. On some vehicles the exterior lights are controlled automatically, depending on the amount of light sensed by an ambient light sensor. This sensor is usually mounted on the top of the instrument panel. When the light level is high (daylight), the exterior lights remain off. As darkness approaches, the light sensor signals a module to turn on the exterior lights. The exterior lights can also be turned on at any time with the headlight switch.

Automatic Headlight Dimming Some vehicles have circuits that use light sensors to signal the presence of oncoming traffic. If the high-beam headlights are on, a module switches the headlights to low beam when sufficient light strikes the sensor. The sensitivity of the circuit is usually adjusted with a rheostat on the instrument panel.

Lamp and Body Control Modules Some vehicles use lamp modules or *body control modules (BCMs)* to control vehicle lighting. The module receives inputs from various switches and supplies power to the appropriate lights through a relay or solid state switch. Electronic modules are commonly used to control automatic exterior lighting, headlight off delay, and illuminated entry systems. **Fig. 5-11.**

Daytime Running Lights The daytime running lights (DRL) module turns on the headlights at a slightly reduced power during daytime. This safety feature makes the vehicle more visible. Operating the lights at reduced power in daylight increases the life of the bulb. Some DRL circuits use an ambient light sensor mounted on the instrument panel. When the ambient light sensor "sees" a lower light level, the module supplies normal power to the lights for nighttime driving.

In general, the headlights operate in the DRL mode when the ignition switch is on and the transmission is in gear. Some DRL circuits do not use

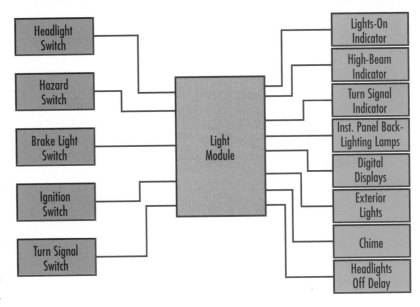

Fig. 5-11 Some vehicles use electronic modules to control lighting circuits. *Name two lighting functions that could be controlled by an electronic module.* (DaimlerChrysler)

light sensors. In such cases the headlights must be turned on manually for normal nighttime brightness. Other exterior lights, such as side marker and taillights, are not affected by DRL. They are controlled by the headlight switch.

Brake, Turn Signal, and Hazard Warning Lights

Brake, turn signal, and hazard warning lights are used to signal actions by the driver. They are not affected by headlight circuit operation. These circuits often share bulb filaments and wiring.

Brakelight Circuits A switch operated by the brake pedal controls the brakelights. The brakelight circuit includes lights on both sides at the rear of the vehicle and the *center high-mounted stop (brake) light (CHMSL).* The CHMSL is located at the rear of the vehicle at approximately driver's eye level for better visibility.

When the brakes are applied, the brakelight switch closes, supplying current to the brakelight bulbs. If the brakelights and turn signals share the same bulb filament, the brakelight circuit passes through the turn signal switch. This makes it possible for the rear turn signal filament to flash even when the brakelight circuit is activated.

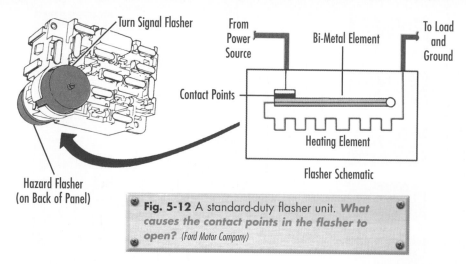

Turn Signal Flasher

From Power Source

Bi-Metal Element

To Load and Ground

Contact Points

Heating Element

Flasher Schematic

Hazard Flasher (on Back of Panel)

Fig. 5-12 A standard-duty flasher unit. *What causes the contact points in the flasher to open?* (Ford Motor Company)

Turn Signal Circuits

The turn signal circuit flashes lights on either side of the vehicle, signaling other drivers when the vehicle is turning. The circuit consists of the turn signal switch, flasher unit, turn signal bulbs (or filaments), and indicator lights. The flasher unit controls turn signal operation. A standard-duty flasher unit contains a set of normally closed electrical contacts, a bimetal element, and a heating element. **Fig. 5-12.**

The flasher unit acts like a self-resetting circuit breaker. When the turn signal switch is moved to either the left- or right-turn position, the circuit is activated. Current flows through the contacts and heating coil of the flasher to the bulbs. Heat causes the bimetal arm to bend, opening the contacts. With the contacts open, current flow stops, causing the turn signal bulbs to go out. The bimetal element cools and returns to its original position, closing the contacts. As this process is repeated, the turn signal bulbs and the turn indicator bulb(s) on the instrument panel flash on and off.

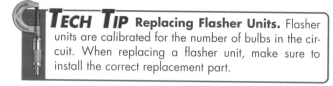

TECH TIP **Replacing Flasher Units.** Flasher units are calibrated for the number of bulbs in the circuit. When replacing a flasher unit, make sure to install the correct replacement part.

With the standard-duty flasher, the flash rate depends on the amount of current flow in the circuit. A burned-out bulb or excessive resistance in the circuit decreases the flash rate. Connecting trailer lights to the vehicle increases the flash rate. The flash rate can also be affected by the use of incorrect bulbs. With a standard-duty flasher, an abnormal flash rate indicates a problem in the

circuit. Continued use of a turn signal circuit with an increased flash rate could damage the flasher unit.

Flasher Units Heavy-duty flashers are available for vehicles wired for trailer pulling. This flasher has normally open contacts. Current flows through the heating element only when the circuit is first energized. Heat causes the bimetal element to close the contacts. The flash rate does not depend on the number of bulbs in the circuit. A disadvantage of a heavy-duty flasher is that the flash rate will not change if one or more bulbs are burned out.

Some vehicles use electronic flashers that have timing circuits to control power transistors. The transistor turns on and off at a uniform rate, regardless of the current flow in the circuit. A single electronic flasher unit may be used for both the turn signal and hazard warning circuits. In some applications the flasher function is included in the body control module or lighting module instead of being a separate assembly.

Hazard Warning Circuits The hazard warning circuit flashes all of the turn signals and the dash indicator lamps at the same time. This circuit has its own on/off switch, located on the steering column or on the instrument panel. A heavy-duty flasher is used due to the number of bulbs in the circuit.

Backup Light Circuit

The backup light circuit is controlled by the backup light switch. On vehicles with automatic transmissions, the backup light switch is part of the *park/neutral position (P/NP) switch.* Vehicles with manual transmissions use a switch operated by the shift linkage. When the transmission is in reverse, the switch is closed and the backup lights are turned on. If the backup lights are on when the transmission is not in reverse, the backup light switch is defective or out of adjustment. **Fig. 5-13.**

Diagnosing Lighting Problems

Bulbs either light or are burned out. When a light is dimmer than normal, it is because of a problem in the circuit. Brighter than normal bulb operation could be caused by high

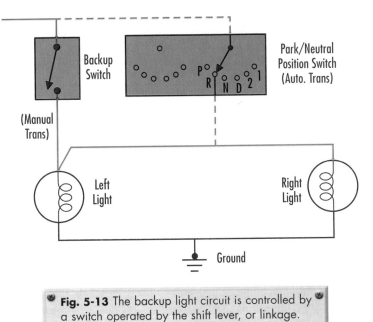

Fig. 5-13 The backup light circuit is controlled by a switch operated by the shift lever, or linkage. *What could cause the backup lights to be on in PARK or NEUTRAL?* (Ford Motor Company)

charging system voltage (all lights) or use of the wrong bulb. Intermittent light operation is almost always due to a loose or dirty connection in the circuit.

When diagnosing lighting problems, remember that lightbulbs are connected in parallel. If a light is out, check to see if other lights in the same circuit are on. If they are on, the power supply, circuit protection device, and switch for that circuit are working.

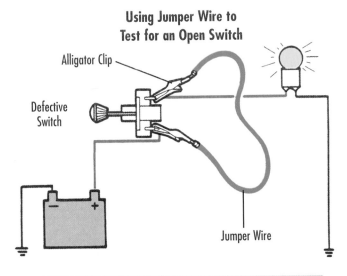

Fig. 5-14 A jumper wire can be used to bypass a switch when testing a circuit. *What two types of electrical devices should not be bypassed with a jumper wire?* (Ford Motor Company)

If an entire lighting circuit is not working, check the switch and circuit protection devices first. A jumper wire can be used to bypass a switch. If the circuit works with the jumper connected, the switch is defective. **Fig. 5-14.** Do not use a jumper wire to bypass circuit protection devices or electrical loads.

You can use a test light to test fuses without removing them from the fuse block. Before replacing a fuse, fusible link, or circuit breaker, check the circuit for shorts or grounds that may have caused an electrical overload. Repair the circuit before installing new parts.

If the switch and protection devices are good, use a test light or *digital volt-ohm-meter (DVOM)* to check for voltage at various points in the circuit until the problem has been found. **Fig. 5-15.**

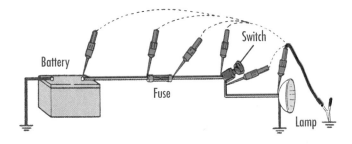

Fig. 5-15 You can use a test light to check an electrical circuit for voltage. *What test device can you use in place of a test light?* (Ford Motor Company)

If only one light is out, remove and inspect the bulb and its socket. A burned-out bulb may be discolored or have a loose filament visible inside. Check the socket for corrosion and loose connections.

Many lighting problems are caused by a poor ground connection. This is a frequent cause of one or more dim lights. Some lightbulb sockets are connected directly to ground. Others require a ground wire connection. When working on lighting problems, make sure all ground connections are clean and tight.

Bulb Replacement Light assemblies and bulb replacement methods vary considerably among vehicles. If the procedure for bulb removal is not

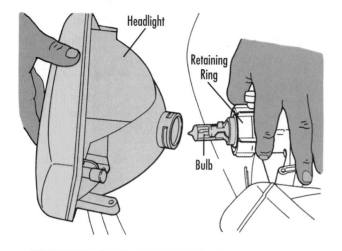

Fig. 5-16 A retaining ring must be rotated to remove the bulb from this headlight assembly. *What should you do if you are not sure how to safely remove a headlight bulb?* (DaimlerChrysler)

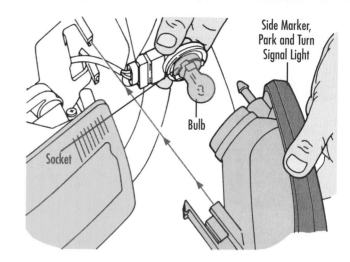

Fig. 5-17 Some small bulbs can be removed from the rear of the light housing. *How is the bulb socket removed from the housing?* (DaimlerChrysler)

clear, refer to the vehicle service manual for instructions. The following suggestions may be helpful.

• Headlight bulb replacement is usually made through the rear of the light assembly. This may require the removal of trim pieces and/or the light housing to gain access to the bulb.

• Halogen headlight bulbs are removed after turning or releasing a retaining ring. **Fig. 5-16.**

• Some fogging of the composite type of light assembly due to moisture is normal. Unless unusual conditions exist, the moisture escapes to the atmosphere through vents provided for that purpose.

• When replacing sealed-beam units, do not tamper with the headlight adjusting screws.

• In some cases, you can replace small bulbs by first removing the light socket from the back of the light housing. Carefully turn the socket to release it before pulling it out. **Fig. 5-17.**

• If you cannot remove the bulb socket from the housing, you will need to remove the lens from the light assembly to access the bulb.

• Some bulbs are removed from their sockets by pushing in and rotating the bulb. Other bulbs can be pulled straight out of the socket.

• Make sure to install the correct bulb. Bulb numbers are usually given in the owner's manual and vehicle service manual.

• Make sure the gaskets and seals around the light housing are in good condition and properly aligned.

Headlight Aiming

Headlight position is mechanically adjustable. The lights must be aimed to provide maximum visibility without blinding oncoming drivers. Headlights are aimed using the adjustment screws provided for that purpose.

Two methods of checking alignment are widely used. The screen method of aiming provides a screen (target) mounted on a wall at a specified height and distance from the front of the vehicle. With the vehicle properly positioned, the adjusting screws are turned until the headlight beam strikes the screen in the proper location. **Fig. 5-18.**

TECH TIP **Headlight Aiming.** Verify proper headlight operation. Check headlight aim on low beam. If necessary, verify the adjustment on high beam.

The required dimensions and other specifications for this procedure are found in the vehicle service manual.

The second alignment method uses headlight aimers temporarily attached to the headlight assembly.

Safety First **Materials** Do not operate the headlights with the aimers installed. Excessive heat buildup may damage the headlight lens.

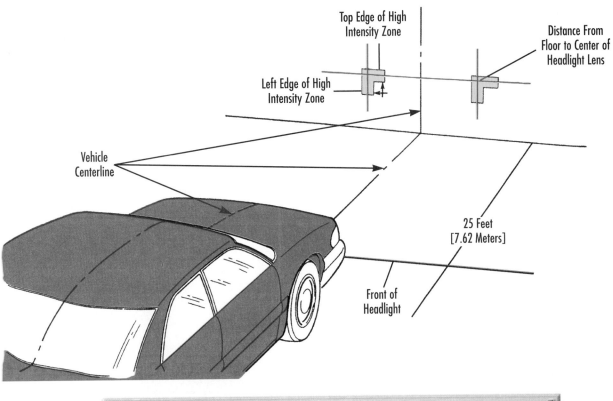

Fig. 5-18 Headlights can be aimed using the screen method. *What are the two important dimensions when placing a headlight aiming screen in front of a vehicle?* (DaimlerChrysler)

The aimers must be installed so that the adjustment rods contact the aiming pads on the headlight assembly.

Adapters are available for use with various types of lights. The headlight is adjusted to move a bubble level on the aimer to a specified position. Some vehicles have a bubble level built into the headlight assembly.

Before checking headlight aim, make these preliminary checks:
- Vehicle must be on a level floor.
- Fuel tank level should be as specified in the vehicle service manual.
- Tires should be properly inflated.
- Luggage area should not be heavily loaded.
- If vehicle is regularly used to pull a trailer, this load should be simulated.

SECTION 1 KNOWLEDGE CHECK

❶ Explain the difference between incandescent, halogen, and high-intensity discharge (HID) lamps.

❷ Describe the two methods of protecting a headlight circuit from electrical overloads.

❸ What switch is used to select either high- or low-beam headlights?

❹ How is the brightness of instrument panel lights controlled? Where is the brightness control located?

❺ Explain why you might replace a standard duty flasher with a heavy-duty unit on some vehicles. What is a disadvantage of a heavy-duty flasher?

Section 2

Instrument Panel Displays

Instrument panel displays provide important information concerning vehicle operating conditions. Gauges, lamps, and digital displays inform the driver about variables such as speed, oil pressure, and fuel level. Most of these displays are located in a section of the instrument panel known as the instrument cluster.

Types of Displays

Driver information displays are divided into three basic types: electromagnetic (analog) displays, digital displays, and warning lamps. Most vehicles use both electromagnetic (analog) displays and warning lamps. **Fig. 5-19.** Some vehicles use various combinations of the three types of displays.

Electromagnetic Displays A gauge that uses electromagnetism to move an indicating needle (pointer) is known as an **electromagnetic display**. Magnetic coils in the gauge unit are connected to a regulated voltage source. Ground for magnetic coils is through both the cluster ground connection and a variable resistance sending unit. The needle indicates the value of the condition being monitored. Electromagnetic gauges are also known as *analog displays* because the units indicated can vary between minimum and maximum values. Analog displays include temperature, fuel, and oil pressure gauges. **Fig. 5-20.**

When the gauge is in operation, voltage is applied to the magnetic coils. The amount of current flow through the coils depends on the resistance of the sending unit. The gauge circuit is calibrated to make the indicating pointer read high or low, depending on the condition being measured. Electromagnetic gauge circuits differ in design and operation. Some gauges read high when the sending unit resistance is high. Other gauges read low under the same conditions.

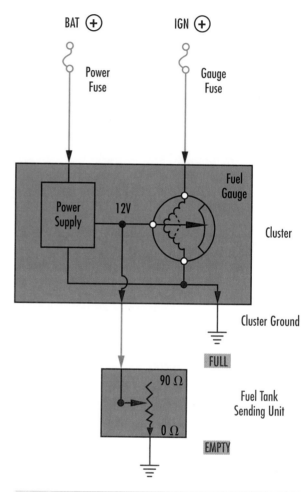

Fig. 5-20 A typical electromagnetic gauge circuit. *Where are the electromagnetic coils grounded?* (General Motors Corporation)

Fig. 5-19 An electromagnetic, or analog, information display. *Why are electromagnetic displays also called analog?* (DaimlerChrysler)

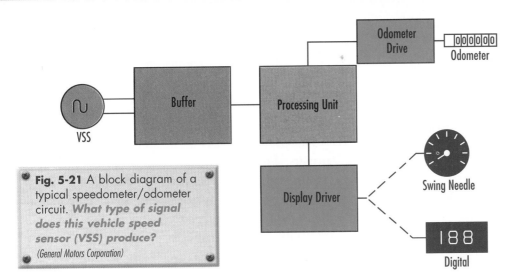

Fig. 5-21 A block diagram of a typical speedometer/odometer circuit. *What type of signal does this vehicle speed sensor (VSS) produce?* (General Motors Corporation)

Electronic circuits in the cluster convert the VSS signal into a speed display and operate a *stepper motor* odometer to record the distance traveled. **Fig. 5-21** and **5-22.**

Solid State Digital Displays Several types of digital (graphic) images are used in instrument panel driver information systems. The information may be displayed as letters, numbers, or bar graphs. Electronically controlled solid state light segments create the desired images. Each segment in a multisegment display is a separate light source. Digital displays are controlled by a microprocessor known as a *central processing unit (CPU)* in the instrument cluster. Based on inputs from sending units, the CPU turns on the appropriate light segments to form letters, numbers, or bar graphs.

Light-Emitting Diodes *Light-emitting diodes (LEDs)* are used in 7- or 11-segment displays to form letters and numbers. **Fig. 5-23.** LEDs can also be arranged to form bar graphs.

Vacuum Tube Fluorescent Displays A display that produces light in a manner similar to a television tube is known as a **vacuum tube fluorescent (VTF) display.** Electrons striking a coated surface (anode) cause it to give off a greenish light, which forms the display. Electronic switches at the back of the display control the segments to be illuminated to produce the images. Voltage pulses

In some applications sensor output will be connected to the *powertrain control module (PCM)* instead of the instrument cluster. In this way the PCM can route sensor information to other modules that require the same data. Refer to the vehicle service manual for information on sending unit circuits and gauge operation. You will need this information when diagnosing gauge problems.

Another analog display is the pointer-type speedometer. At one time speedometers were mechanically driven by a cable. Now most speedometers are electronically operated. As with other analog displays, the speedometer pointer is moved by electromagnetic fields. Instead of a variable resistance sending unit, the input signal for a speedometer comes from a *vehicle speed sensor (VSS)*. A typical VSS is a small AC signal generator driven by a shaft in the transmission. When the vehicle is moving, the VSS generates a voltage signal proportional to vehicle speed.

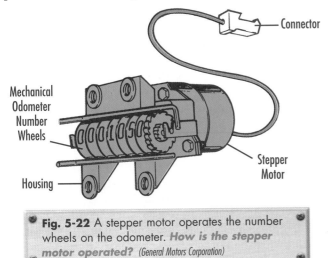

Fig. 5-22 A stepper motor operates the number wheels on the odometer. *How is the stepper motor operated?* (General Motors Corporation)

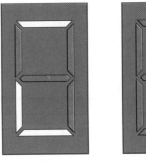

Fig. 5-23 A seven-segment LED display creates numerical images. *What controls which segments are turned on?* (General Motors Corporation)

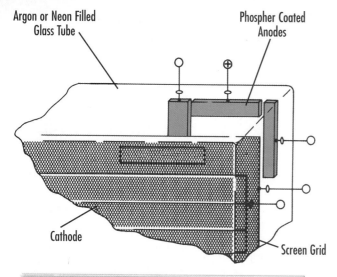

Argon or Neon Filled Glass Tube

Phospher Coated Anodes

Cathode

Screen Grid

Fig. 5-24 Coated surfaces in a vacuum tube fluorescent display give off a greenish light when struck by electrons. *What controls the voltage to the anodes?* (General Motors Corporation)

from the CPU control the light sources (anodes). **Fig. 5-24.**

Liquid Crystal Displays A panel that is placed in front of an incandescent or halogen lightbulb is known as a **liquid crystal display (LCD).** The LCD does not actually produce light. The LCD panel is a "sandwich" of a fluid between two layers of glass that have a conductive coating. Voltage signals from the CPU to the glass layers cause the fluid between the glass to either pass or block light from the bulb. The resulting light and dark segments form the images. **Fig. 5-25.**

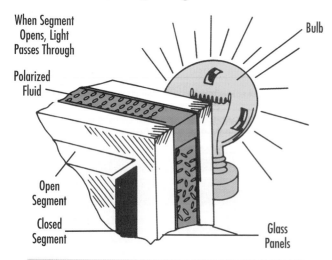

When Segment Opens, Light Passes Through

Bulb

Polarized Fluid

Open Segment

Closed Segment

Glass Panels

Fig. 5-25 Liquid crystal displays use a fluid to control the passage of light. *Why is a bulb used behind the LCD panel?* (General Motors Corporation)

LCD displays produce only black and white images. Color filters are often used in front of the display to produce a variety of colored images. In addition to instrument panel displays, LCDs are widely used in DVOMs, calculators, and gasoline pump displays.

Digital display circuits are similar to analog gauge circuits. **Fig. 5-26.** Each variable to be displayed requires an input signal from a sending unit. In general, the same types of sensors are used for both analog and digital displays. While the sending units may look the same, they are not necessarily interchangeable. Always check part numbers when replacing components.

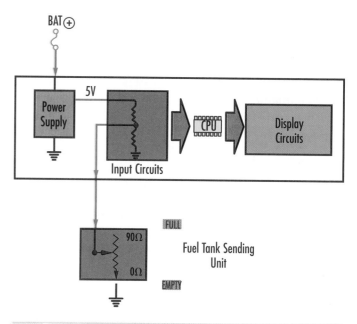

BAT ⊕

Power Supply

5V

Input Circuits

CPU

Display Circuits

90Ω

0Ω

FULL

EMPTY

Fuel Tank Sending Unit

Fig. 5-26 Digital display circuits are similar to analog gauge circuits. *Will this display read high or low if the sending unit lead wire is grounded?* (General Motors Corporation)

Digital odometers require a means of storing mileage data when the ignition switch is off. This information is stored in a nonvolatile random access memory (NVRAM) chip. As the vehicle is driven, accumulated mileage is stored in the NVRAM. When the ignition is shut off, the chip retains the information in its memory for display when the ignition is turned on again. If a new or rebuilt cluster is installed in a vehicle, the original NVRAM must be installed in the replacement unit.

Instrument Panel Warning Lights Instrument panel warning lights are less expensive than analog gauges or electronic displays. A driver may be

Table 5-A	BULB INDICATORS	
Indicator	**Color**	**Bulb #/LED**
Battery	Red	LED
Brake	Red	LED
Air bag	Red	LED
Low Oil	Red	LED
Seatbelt	Red	LED
Temperature	Red	LED
Door Ajar	Red	LED
Decklid Ajar	Red	LED
High Beam	Blue	74
Turn Signals	Green	194
Cruise	Green	74
Trac On	Green	74
Trac Off	Amber	LED
Check Engine	Amber	LED
ABS	Amber	LED
Low Fuel	Amber	LED
Low Washer Fluid	Amber	LED
Illumination	Bluegreen	194

more likely to notice a warning light than to notice an unusual operating condition on a gauge or other display. Red, amber, green, and blue indicators are often used as instrument panel warning signals. Warning lights are usually either incandescent bulbs or LEDs. **Table 5-A.**

As with other displays, warning lights require an input signal to operate correctly. The input signal is usually a simple switch that turns on the warning light. The switch can be on either the battery or ground side of the warning light.

In externally grounded warning circuits, voltage is supplied to the lights when the ignition switch is on. When the grounding switch is closed, the light will be on. When the switch is opened, the light will go out. Some of these circuits also have a bulb test feature. When the ignition switch is turned and held just past the RUN position, the switch grounds the light circuit. This is the bulb test section of the ignition switch. The bulb test is used in circuits where the warning light does not come on when the ignition switch is first turned on.

The ground switch for some warning lights is in an electronic module. The module monitors the operation of the system and grounds the circuit to turn on the warning light when appropriate. In some cases the module will cycle the warning light on and off to signal specific conditions or trouble codes.

Warning light circuits can also be internally grounded. In this case the ignition switch does not have to be in the circuit. Instead, a switch that controls the individual circuit is located between the battery and the light circuit. When the control switch is closed, the warning light comes on. The warning light stays on as long as the switch is on. Examples of this type of circuit include the high-beam indicator, hazard warning indicators, and seat belt warning light.

Safety First **Materials** Some older vehicles have thermoelectric gauges that use instrument voltage regulators (IVRs). Grounding the sending unit lead wire on these vehicles may damage the gauge. Refer to the vehicle service manual if in doubt about the type of gauges used.

Diagnosing and Servicing Instrument Panel Displays

When diagnosing instrument panel display problems, note how many displays are affected. For example, if only one gauge, display, or lamp is out, it is unlikely that the problem is a fuse or common ground connection. Also determine if the problem is failure to operate, intermittent operation, or an inaccurate display.

Gauges and Graphic Displays If only one gauge or digital display fails to respond normally, refer to a wiring diagram to determine how the circuit is powered and grounded. An open circuit in the sending unit or its wiring can cause this problem. With the ignition switch on, disconnect the lead wire at the sending unit. Observe the display while briefly touching the lead wire to a good ground connection. If the display changes when you do this, the sending unit is probably defective. If the display does not change, check the wiring between the sending unit and the instrument panel. Inspect the printed circuit board and connectors for damage. If the wiring is good, the gauge or display is defective.

When one gauge or display reads inaccurately, check for resistance in the sending unit circuit. You can use gauge testers to temporarily take the place of the sending unit. You can use them on both

Grounded!

Many vehicle electrical systems use the frame or chassis as one conductor in the circuit. For example, both the parking brake and brake fluid level warning switches connect one side of the warning light to the chassis ground.

This arrangement is economical and reliable. Using the chassis as one conductor saves wire, weight, complexity, and cost. Because the switches are frequently far from the warning lights, putting the switches in the positive, or "hot," wire of the circuit would require twice as much wire and many extra connectors.

This type of circuit is easy to build for light testing circuits. The short circuit of a switch wire to ground causes a warning. However, it does not blow a fuse or cause damage. Also, the chassis is a very reliable conductor. It cannot easily break and create an open circuit.

Apply it!

Constructing a Grounded Circuit

Meets NATEF Science Standards for understanding electrical grounds and fuses.

Materials and Equipment
- 12-volt battery or power supply
- In-line automotive fuse assembly, 1 amp or less
- Small 12-volt lamp and socket assembly
- 3 normally open push-button switches
- About 10′ [3 m] of hookup wire
- Piece of wood, about 6″ × 24″ x 1″ thick [about 15 cm × 60 cm × 3 cm]
- Strip of sheet metal, about 2″ [5 cm] wide and as long as the board
- Sheet metal screws, ½″–¾″ [1 cm–2 cm]

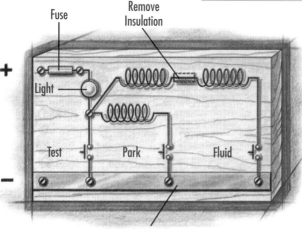

Metal Strip (Chassis)

❶ Construct the circuit shown here. This circuit demonstrates the three reasons for using the chassis as the "ground return" part of the system. The coiled hookup wires represent the distance from the dashboard to the parking brake and fluid level sensor switches.

❷ Test the circuit. Push the test switch to close it. Then release it. The warning light will come on, and then go off. This represents the light test when the vehicle is started.

❸ Simulate the situation where the parking brake is set and the fluid level is low. Press and hold the parking brake and fluid-low switches one at a time. This situation signals trouble by lighting the warning light.

❹ Remove the fuse. Then press and release the test switch. The warning light does not flash. This indicates a fault.

❺ Replace the fuse. Now simulate a wire being shorted to ground. To do this, pull down and press the bare part of the fluid-low wire to the metal chassis strip. The warning light comes on to indicate a fault, but the fuse does not blow.

❻ In this system, the fuse, light, and test switch are located near each other, as they are in a vehicle. The switches are located far away. Can you think of a vehicle electrical system in which the fuse and switch(es) are located together and the lights are far away?

analog and digital displays. The tester has one or more rheostats that allow you to set specific resistance values as listed in the vehicle service manual. Connect the tester to the sending unit lead wire. Adjust the resistance as required. If the display is not as specified, the sending unit is probably defective. Replace defective analog gauges individually. Replace digital display modules as a unit. Stepper motor odometers and digital odometer memory chips must be removed from the defective unit and reinstalled in the replacement part.

Warning Lights When warning lights fail to come on, check the fuse by observing if other lights on the same fuse are working. If the fuse is good, check the bulb if it is easily accessible. If the bulb is not easily accessible, it may be easier to check the circuit by applying voltage or ground. Refer to the vehicle service manual to determine if the circuit is internally or externally grounded.

Depending on the circuit, use a jumper wire to apply voltage, or disconnect and ground the lead wire to the switch. If the light does not light, the bulb is probably defective. If the light does light, repair the wiring or replace the control switch as needed. If a warning light stays on all the time, check for a grounded wire to the switch or a defective switch.

SECTION 2 KNOWLEDGE CHECK

❶ Explain how an electromagnetic gauge or display works.

❷ Describe the difference between a vacuum tube fluorescent display and a liquid crystal display.

❸ Describe the two ways in which warning light circuits are grounded.

❹ How do you determine if a gauge sending unit is defective?

❺ How do you diagnose a warning light that does not come on?

Section 3

Brake Warning Lights, Switches, and Sensors

The brake warning light is located in the instrument panel. When lit, it alerts the driver to a problem with one or more of the brake systems. It is usually a red light and it may display an icon or the word "brake." It may display both. **Fig.** 5-27.

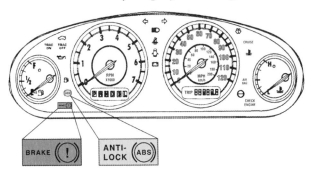

Fig. 5-27 The brake warning light will glow when there is a problem with brake systems or components. *How can you recognize the brake warning light? (DaimlerChrysler)*

The brake warning light will come on briefly when the driver turns the ignition switch to the ON position. This is the bulb check. If the light is not glowing, the bulb may be burned out, the socket could be loose, or a fuse could be blown. Inspect and replace any damaged parts. Refer to the vehicle service manual.

The brake warning light may also glow under the following conditions:
• If the parking brake is set.
• If the fluid level in the master cylinder is low.
• If the pressure-differential switch senses a difference in pressure between two hydraulic circuits.

The anti-lock brake system (ABS) warning light is different from other brake warning lights. First, it is usually amber, not red. Second, it defaults to ON and must be turned off by the ABS control module.

Safety First **Personal** If the brake warning light does not come on with the ignition key in the run position, repair the light system immediately. In this failure mode, the driver will not be alerted to a potentially life-threatening vehicle condition.

Recognizing Consequences

"Potentially life-threatening!" These are serious words. When a medical doctor speaks these words, they mean that a life might be ended by an illness. In an automotive context, these words are as serious as when they are spoken by a doctor. They, too, indicate a condition that could lead to the death of a driver, passenger, or pedestrian.

As a technician servicing a vehicle, you are responsible for the safety of the occupants of a vehicle. You need to read service manuals and safety bulletins very carefully. Keep your skills and knowledge up to date. Learn about new safety issues and new solutions by reading new information as it is released. You need to pay close attention to details. As a technician, you must accept the serious responsibilities of your job.

Apply it!

Meets NATEF Communications Standards for making inferences and predicting outcomes.

As you read Section 3, *Brake Warning Lights, Switches, and Sensors:*

❶ Fold a sheet of notebook paper in half lengthwise. Head one half "Warnings." Head the other half "Consequences If Not Serviced."

❷ As you read, list the "Warnings" given for the brake warning lights.

❸ Study each warning. In the "Consequences" column, write the possible results if the warnings are ignored. Be sure to explain why some situations are or could become life-threatening.

When the ignition is turned to the ON position, the amber ABS light will glow to show that it is working. Once the ABS performs its automatic self-check, the ABS control module opens the circuit to the warning lamp and turns it off.

The ABS will not function when the ABS warning light is illuminated. The ABS warning light glows whenever there is an electrical or hydraulic malfunction of the ABS.

Activating Switches and Sensors

Brake system switches and sensors are activated either by driver input or system input.

Brake Light Switches All vehicles have at least one brake light switch. Some have more than one. Depressing the brake pedal activates the brake light switch. On some vehicles the brake light switches' only function may be to turn on the brake lights. **Fig. 5-28.**

The brake light switch may also have multiple functions. It may be used to provide information to the ABS, the torque converter clutch, or the cruise control system. Many vehicles have additional independent switches for these purposes.

In older domestic vehicles, the brake light switch is usually wired into the same circuit as the turn signals and hazard flashers. This turns on the brake lights when the brakes are applied. The same filament is used for both brake lights and turn signals. Import vehicles and later-model domestic vehicles with amber rear turn-signal lights have separate circuits for rear turn signals and brake lights.

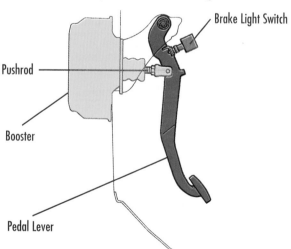

Fig. 5-28 All vehicles have simple brake light switches. *Do some vehicles have more than one brake light switch?*

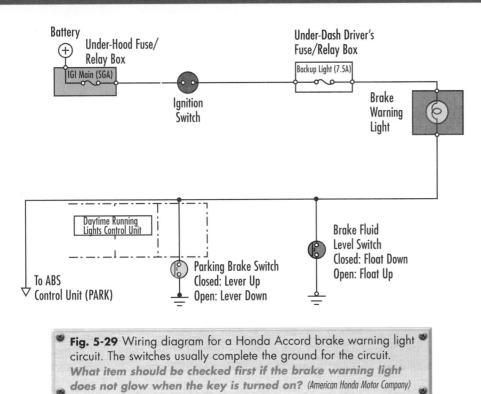

Fig. 5-29 Wiring diagram for a Honda Accord brake warning light circuit. The switches usually complete the ground for the circuit. *What item should be checked first if the brake warning light does not glow when the key is turned on?* (American Honda Motor Company)

driver that the brake fluid level in the master cylinder may be dangerously low.

The sensor may be located in the side of the reservoir. Or, it may be a part of the reservoir's cap.

On some vehicles, the warning lights receive battery power only after the voltage has passed through two fuses. Always check the fuses first if the warning lamp does not glow when the key is turned on. The wiring diagram also shows that the switches complete the ground side of the circuit to turn on the light. **Fig. 5-29.**

Parking Brake Switch The parking brake switch closes to complete the brake warning light circuit to ground. It turns on the brake warning light whenever the parking brake is applied. The brake warning light may also remain on if the parking brake is not fully released. This alerts the driver to fully release the parking brake before proceeding. Not fully releasing the parking brake may allow it to drag. This can lead to serious overheating of the wheel brake mechanism and shorten brake lining life.

Brake Fluid Level Sensors Most current vehicle models are equipped with low brake fluid level sensors. These sensors are simply float-operated switches. When the float drops because of low fluid level, the switch closes. When the switch closes, it completes the ground circuit and illuminates the brake warning light. This warning light alerts the

Brake Wear Indicator Light

Some vehicles, particularly the more expensive models, have electric brake pad wear indicators. A sensor wire is molded into the pad lining. When the lining wears to a predetermined depth, a wire contacts the brake rotor.

The brake rotor, acting like a switch, completes the circuit to ground. This illuminates the brake-wear indicator light on the instrument panel.

 **TECH TIP** **Checking Brake Warning Lights.** On some vehicles equipped with daytime running lights (DRLs), the brake warning lamp circuit passes through the DRL control module. If the brake warning light will not go out, the problem could be in this control module.

SECTION 3 KNOWLEDGE CHECK

❶ What is the purpose of the brake warning light?

❷ What functions, in addition to turning on the brake light, may the brake light switch provide?

❸ How does the ABS function when the ABS warning light is turned on?

❹ What is the purpose of the brake fluid level sensor light?

❺ What is the purpose of the brake wear indicator light?

CHAPTER 5 REVIEW

Key Points

Meets the following NATEF Standards for Electrical/ Electronic systems: inspecting, replacing, and aiming headlights; diagnosing incorrect turn signal operation.

- The headlight switch controls all regular exterior lighting.
- On vehicles with combination turn signal and brakelight bulbs, the brake light circuit goes through the turn signal switch.
- The flash rate for standard-duty flasher units depends on the amount of current flow in the circuit.
- Correct headlight aiming assures good visibility.
- The value displayed on an electromagnetic gauge depends on the resistance of the sending unit.
- Digital displays are controlled by a central processing unit (CPU).
- Warning light circuits may be switched at either the voltage or ground side of the circuit.
- The functions of the brake light switch can vary, in addition to turning on the brake lights.
- If the brake warning light is turned on, this may be an indication that the parking brake has not been released.

Review Questions

1. Identify the lighting circuits that come on when the headlight switch is in the HEAD position.
2. Explain the purpose and operation of daytime running lights (DRL).
3. Explain how brake, turn signal, and hazard warning light circuits are wired.
4. When diagnosing vehicle lighting problems, what two connections are required by all circuits?
5. How are headlights aimed?
6. How does an electromagnetic gauge work?
7. Describe the operation of a brake warning light and switch.
8. **Critical Thinking** A customer complains that only the left turn signals work. What would you do to diagnose this problem?
9. **Critical Thinking** In a vehicle with electromagnetic gauges, all gauges work but they are not accurate. What could be the cause of this problem?

TECHNOLOGY FORECAST
FOR AUTOMOTIVE EXCELLENCE

Headlights of the Future

Engineers are working on a new type of headlamp system. It is called a high intensity discharge (HID) headlamp. HID headlamps are different from most headlights used today, although they can be found on some luxury cars. You can tell an HID headlamp by its blue-white color.

HID headlights are brighter, smaller, and more efficient than standard headlamps. They make light from a bright arc that is produced when high voltage moves between two electrodes. HID headlights don't use a filament that could burn out, and they don't use halogen gas.

The benefits of HID systems are many. They produce more light than today's standard headlight, so drivers can see better and farther at night. An HID headlight's beam won't hurt the eyes of other drivers.

There are some problems with HID systems. They are expensive to make and repair. They don't have a high-beam setting. Engineers are working on ways to solve these problems.

Other lighting systems are also being studied. Neon gas, normally used in signs, can be used in taillights. Neon lights up quicker and more evenly than normal lightbulbs. There is no filament to replace. A neon light is expensive, but it should last a vehicle's lifetime.

AUTOMOTIVE EXCELLENCE
TEST PREP

Answering the following practice questions will help you prepare for the ASE certification tests.

1. Technician A says all automotive lightbulbs have filaments. Technician B says high-intensity discharge bulbs do not have filaments. Who is correct?
   - ⓐ Technician A.
   - ⓑ Technician B.
   - ⓒ Both Technician A and Technician B.
   - ⓓ Neither Technician A nor Technician B.

2. Which type of electronic display uses a bulb as its light source?
   - ⓐ Vacuum tube fluorescent.
   - ⓑ Light-emitting diode.
   - ⓒ Liquid crystal display.
   - ⓓ Non-volatile random access memory.

3. Which lights operate when the daytime running lights are on?
   - ⓐ Headlights only.
   - ⓑ Head and taillights only.
   - ⓒ Head, tail, and side marker lights.
   - ⓓ All exterior lights.

4. Technician A says a heavy-duty flasher unit can be installed when connecting trailer lights to a vehicle. Technician B says heavy-duty flashers are used for hazard warning light circuits. Who is correct?
   - ⓐ Technician A.
   - ⓑ Technician B.
   - ⓒ Both Technician A and Technician B.
   - ⓓ Neither Technician A nor Technician B.

5. What could cause one bulb in a circuit to be brighter than normal?
   - ⓐ High resistance in circuit from battery.
   - ⓑ High charging system voltage.
   - ⓒ Use of incorrect bulb.
   - ⓓ Both b and c.

6. Two technicians are discussing headlight aiming. Technician A says the vehicle must be on a level surface. Technician B says normal vehicle loads must be simulated. Who is correct?
   - ⓐ Technician A.
   - ⓑ Technician B.
   - ⓒ Both Technician A and Technician B.
   - ⓓ Neither Technician A nor Technician B.

7. The red brake warning light does not come on when the key is turned to the ON position. The cause may be:
   - ⓐ A blown fuse.
   - ⓑ A burned-out bulb.
   - ⓒ A loose bulb socket.
   - ⓓ All of the above.

8. Where are the coils for electromagnetic gauges grounded?
   - ⓐ At the sending unit.
   - ⓑ At the cluster ground.
   - ⓒ Both a and b.
   - ⓓ Neither a nor b.

9. Which of the following is not used as a gauge sending unit?
   - ⓐ A thermistor.
   - ⓑ A potentiometer.
   - ⓒ A rheostat.
   - ⓓ A power supply.

10. Technician A says some headlight circuits are protected by circuit breakers. Technician B says fuses protect some headlight circuits. Who is correct?
    - ⓐ Technician A.
    - ⓑ Technician B.
    - ⓒ Both Technician A and Technician B.
    - ⓓ Neither Technician A nor Technician B.

CHAPTER 6

Diagnosing & Repairing Accessory and Safety Systems

You'll Be Able To:

- ⊗ Explain how small direct current (DC) motor circuits work.
- ⊗ Describe the operation of windshield wiper and washer systems.
- ⊗ Explain how cruise control systems operate.
- ⊗ Identify the components of the supplemental restraint system.

Terms to Know:

air bag control module

crash (impact) sensor

cruise control vacuum servo

horn relay

inflator module

park switch

positive temperature coefficient (PTC) resistor

The Problem ⟩ ⟩⟩ ⟩⟩ ⟩⟩ ⟩⟩

Anna Lopez leaves her two-year-old passenger car outside your shop with a note on the instrument panel. The note says, "Please check the windshield wipers." She gives her name and a telephone number.

You notice that the wipers have stopped near the middle of the windshield. When you turn on the ignition switch, the wipers begin to operate. When you attempt to turn the wipers off, you notice that the switch is already in the OFF position.

Your Challenge

As the service technician, you need to find answers to these questions:

❶ How can the wiper motor run with the switch OFF?

❷ Why don't the wipers return to the park position?

❸ What is responsible for stopping the wipers in the correct position?

Section 1

Accessory Motor Circuits

Small direct current (DC) motors are widely used to power automotive accessory systems. Most of these motors are permanent magnet (PM) type motors. The motor armature rotates inside a permanent magnetic field. Depending on their use, many of these motors are wired to be reversing. Windshield wipers and washers, power windows and seats, and blower motors are examples of permanent magnet motors. A portion of the operating circuit for some small motors may be in the *body control module (BCM)*. The BCM uses inputs from a variety of sensors to control the operation of the motor.

Permanent Magnet Motor Principles

Permanent magnet motors contain four major components: an *armature, brushes, commutator,* and the *magnet(s)*. The armature is positioned between the poles of the magnet. The operation of the permanent magnet motor is similar to that of the starter motor. The brushes and commutator conduct current to the armature windings, creating an electromagnetic field. This magnetic field reacts with the permanent magnetic field, causing the armature to rotate. **Fig. 6-1.** The armature shaft is connected either directly or through gears to the device to be operated.

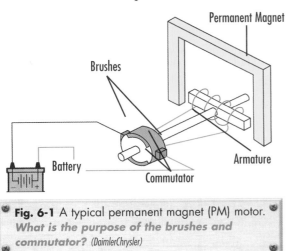

Fig. 6-1 A typical permanent magnet (PM) motor. *What is the purpose of the brushes and commutator?* (DaimlerChrysler)

Small motor circuits often use two protection devices. A fuse or circuit breaker in the fuse panel protects the wiring. In addition, a self-resetting circuit breaker is located inside the motor. **Fig. 6-2.** The circuit breaker in the motor can be either mechanically or electronically operated. Electronic circuit breakers

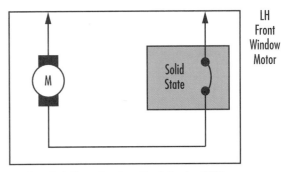

Note: Each Motor Contains a Circuit Breaker (PTC). It Resets Only After Voltage Is Removed from the Motor.

Fig. 6-2 Many small motors have an internal circuit breaker. *How is this circuit breaker reset?* (General Motors Corporation)

are positive temperature coefficient resistors. A **positive temperature coefficient (PTC) resistor** is a solid state device that opens a circuit when an overcurrent condition occurs. Both types of circuit breaker reset when power to the motor is shut off.

Nonreversing Motors

The polarity of an electromagnetic field depends on the direction of the current flow that creates the field. In a PM motor, the direction of armature rotation depends on the direction of current flow through the armature windings.

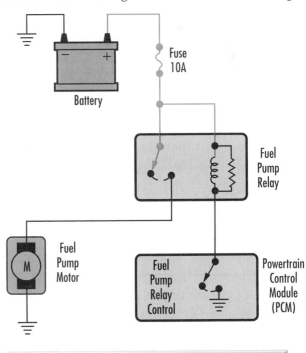

Fig. 6-3 The armature in nonreversing motors is connected directly to ground. *Why is this motor nonreversing?* (DaimlerChrysler)

When one brush of a motor is connected directly to ground, the motor is nonreversing. When power is applied, current always flows through the armature in the same direction. This means that the magnetic field always has the same polarity. The armature always rotates in the same direction. Nonreversing motors are used to operate such devices as fuel pumps, heater blowers, and windshield wipers. **Fig. 6-3.**

Reversing Motors

If the direction of current flow through an armature is reversed, its magnetic polarity is also reversed. This causes the armature to rotate in the opposite direction. The armature in a reversing motor is not connected directly to ground. Instead, the circuit is wired through a control switch. One control switch contact provides power while another contact supplies a ground connection. The switch contacts may be "ganged" (connected together) so that they move at the same time. By changing the position of the switch the direction of the current flow is changed. The armature can be made to rotate in either direction. Reversing motors are used to operate devices such as power windows, seats, and antennas. **Fig. 6-4.**

Some vehicles with power windows have an "express down" feature for the driver's window only. When the driver's window switch is momentarily moved to the express position, the window will move all the way down without the switch being held. An electronic module controls the express down circuit. **Fig. 6-5.** For safety reasons this express feature does not work when the window is being raised.

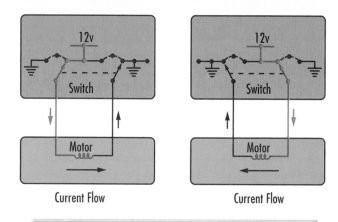

Current Flow   Current Flow

Fig. 6-4 Reversing motors have both power and ground connected through switch contacts. *Why does the armature reverse direction when the current flow is reversed?* (DaimlerChrysler)

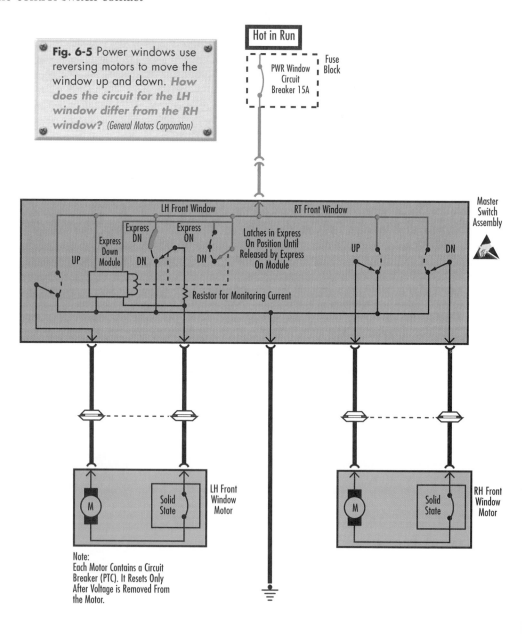

Fig. 6-5 Power windows use reversing motors to move the window up and down. *How does the circuit for the LH window differ from the RH window?* (General Motors Corporation)

Note:
Each Motor Contains a Circuit Breaker (PTC). It Resets Only After Voltage is Removed From the Motor.

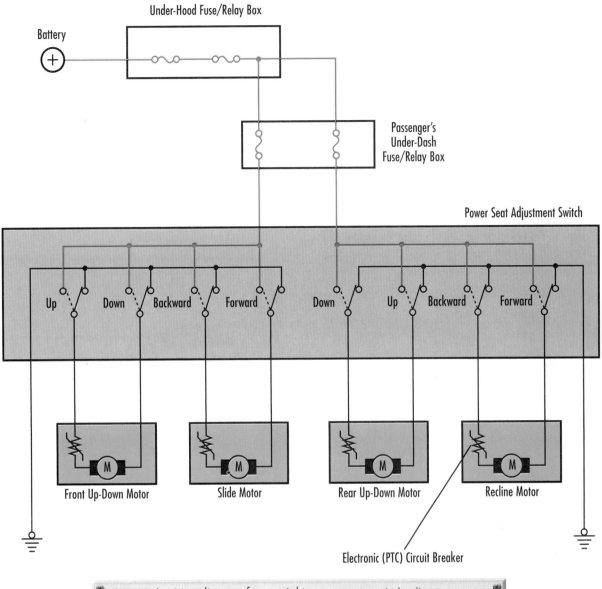

Battery

Under-Hood Fuse/Relay Box

Passenger's Under-Dash Fuse/Relay Box

Power Seat Adjustment Switch

Up   Down   Backward   Forward   Down   Up   Backward   Forward

Front Up-Down Motor   Slide Motor   Rear Up-Down Motor   Recline Motor

Electronic (PTC) Circuit Breaker

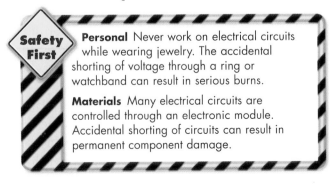

Fig. 6-6 A wiring diagram for an eight-way power seat circuit. *How many fuses and circuit breakers are used in this circuit?* (American Honda Motor Company)

Some accessory circuits use more than one reversing-type motor. Power seat circuits, for example use several motors, depending on the number of possible seat adjustments. **Fig. 6-6.**

Windshield Wiper Circuits Most windshield wiper circuits use nonreversing PM motors, designed for two-speed operation. They have three brushes instead of two. An insulated and a ground brush are located 180° apart on the commutator. A second insulated brush is located near the first. When the wiper switch completes the circuit to the first insulated brush, the wiper motor operates at normal speed. Moving the wiper switch to the HIGH position moves the circuit to the second insulated

brush. This causes the motor to operate at high speed. The wiper motor assembly includes a gearbox and attached linkages that convert the rotary motion of the motor armature to the back-and-forth motion of the wiper arms.

Safety First

Personal Never work on electrical circuits while wearing jewelry. The accidental shorting of voltage through a ring or watchband can result in serious burns.

Materials Many electrical circuits are controlled through an electronic module. Accidental shorting of circuits can result in permanent component damage.

Using Metric Prefixes

You will frequently use metric measurements in your work. It is important to understand the metric system. With a little practice, you will be surprised at how easy and convenient it is to use this system.

Metric units for specific measurements such as length, pressure, resistance, voltage, and current are related to each other by factors of ten. Prefixes indicate how many multiplications or divisions by ten are involved. In the table, the most commonly used metric prefixes are unshaded.

As you know, multiplying by ten can be done by moving the decimal point to the right:

$$2 \text{ kg} = 2{,}000 \text{ g} = 2{,}000{,}000 \text{ mg}$$

Dividing by ten can be done by moving the decimal point to the left:

$$360 \text{ mm} = 36 \text{ cm} = 0.36 \text{ m}$$

You might choose to use any one of the three ways of expressing these measurements. Your choice would depend on the situation.

It is easier to use factors of ten than it is to use many of the factors found in the English (customary) system. Factors such as 12, 16, 36, 128, or 5280 are harder to remember and use in multiplication or division. Using metric prefixes, you can do the mathematics more easily in your head.

Prefix	Symbol	Factor
giga-	G	× 1,000,000,000
mega-	M	× 1,000,000
kilo-	k	× 1,000
hecto-	h	× 100
deca-	da	× 10
deci-	d	÷ 10
centi-	c	÷ 100
milli-	m	÷ 1000
micro-	μ	÷ 1,000,000
nana-	n	÷ 1.000,000,000

Apply it!

Meets NATEF Mathematics Standards for understanding place values when using the metric system.

Convert the following to complete each statement.

❶ 0.0005 A = _?_ mA = _?_ μA

❷ 12 kΩ = _?_ Ω = _?_ MΩ

❸ 36 km = _?_ m = _?_ mm

❹ 0.0028 V = _?_ μV = _?_ mV

❺ 120,000 Pa = _?_ kPa = _?_ Mpa

Which of the three ways of writing each number do you think would be easiest to use? Why? Remember they all represent the same value.

The motor assembly also contains a park switch operated by a cam, relay, or both. The **park switch** is a switch that supplies power to the motor after the wiper switch is turned off. This allows the blades to travel to their normal park position at the base of the windshield before the motor stops. If the park switch does not close, the wiper blades stop where they are when the wiper switch is turned off. If the park switch contacts stick closed, the wipers do not stop when the wiper switch is turned off. **Fig. 6-7.**

Many windshield wiper circuits contain an electronic module to provide pulse, or intermittent, operation. When the wiper switch is in the pulse position, a timing delay circuit controls the voltage to the module. Varying the position of the switch controls the amount of delay that occurs between wiper strokes.

Windshield Washer Circuits Windshield washer circuits work in conjunction with the windshield wipers. A nonreversing PM motor drives the washer pump. The motor is controlled by a separate set of contacts in the wiper switch. In most cases when the washer circuit is activated, the wiper motor operates at low speed. The wiper completes several strokes after the washer shuts off, then returns to its previous status (operating or parked).

TECH TIP Avoiding Module Damage. If the motor circuit includes an electronic module, use only a digital voltmeter when measuring voltage. Use caution when handling the module to avoid damage due to electrostatic discharge (ESD).

Diagnosing Motor Circuits

Because all direct current motors use the same operating principles, the diagnostic techniques for

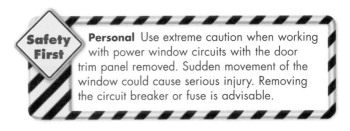

Safety First Personal Use extreme caution when working with power window circuits with the door trim panel removed. Sudden movement of the window could cause serious injury. Removing the circuit breaker or fuse is advisable.

all motor circuits are very similar. Before attempting to diagnose problems in the electrical circuit, make sure the motor is free to move. Check for mechanical interference that could prevent the motor from operating normally. If a wiper motor operates at only one speed, refer to a wiring diagram to determine the current path for the other brush circuit.

Checking Power Source Some motor circuits are powered only when the ignition switch is on. Other circuits are not controlled by the ignition switch (hot at all times). Make sure the ignition switch is in the correct position for the circuit you are working on. If the circuit has more than one switch controlling the motor, try all of the switches. If the motor can be operated from any switch, the power source and circuit protection devices are good. If the motor does not operate at all, check the fuse and circuit breaker. Circuit breakers inside a motor reset when the power source (control switch) is turned off. Check the circuit for an electrical overload before replacing an open fuse or circuit breaker. An overload may be caused by a motor that cannot rotate due to internal or external interference.

Fig. 6-7 Nonreversing permanent magnet windshield wiper and washer motors. *What two devices operate the park switch on the wiper motor?* (General Motors Corporation)

Some motor circuits use relays to connect both the power and ground side of the motor. In this case the control switches conduct only the low current used by the relay windings. With less current flow, there is less arcing at the switch contacts. The higher current flow required by the motor is conducted through the relay contacts. **Fig. 6-8.**

Use a voltmeter or test light to check for voltage from the power source to the motor terminal when the switch is on. For reversing motors, you need to refer to a wiring diagram to determine which side of the circuit is powered for a given direction of movement. If there is no voltage at the motor, check the switch, connectors, and wiring for an *open circuit.* If the circuit uses relays, check for voltage to and from the relays while the circuit is activated. Check the relay windings for continuity with a voltmeter or ohmmeter. If voltage is present at the battery side of the motor, check the ground circuit.

Checking the Ground Circuit Nonreversing motors are usually grounded through their cases. Make sure the motor mounting bolts are clean and tight. A loose or corroded mounting bolt cannot provide a good ground connection. Reversing motors are grounded through a separate set of contacts in the control switch. With the switch in the operating position, measure voltage drop from the motor to a ground connection. A voltage drop of less than 0.5 volts is usually

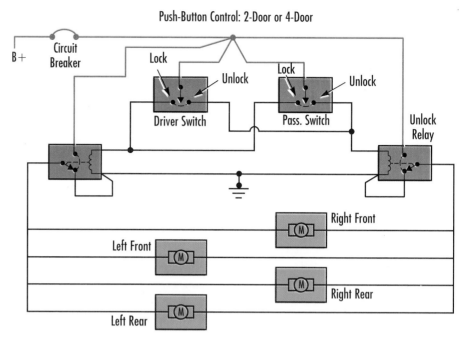

Push-Button Control: 2-Door or 4-Door

Fig. 6-8 Some motor circuits use relays to control current to and from the motor. *What is the advantage of using relays in the motor circuit?* (Ford Motor Company)

considered acceptable. If the voltage drop is higher, check the circuit for poor connections. If the ground connection is questionable, use a jumper wire to connect the motor case to a known good ground. If the motor operates with the jumper connected, the ground circuit is defective.

If the power source and ground circuit are good, the motor itself is usually defective. Most small motors are serviced by replacement only. Some replacement parts may be stocked for wiper motors. Check with a parts supplier for the availability of these parts.

SECTION 1 KNOWLEDGE CHECK

❶ What is a positive temperature coefficient resistor? How is it used in a permanent magnet motor?

❷ How is the direction of armature rotation reversed in a permanent magnet motor?

❸ How is the "express down" feature of the driver's side window controlled?

❹ What is the purpose of the second insulated brush in a windshield wiper motor?

❺ What is the purpose of the park switch in a windshield wiper circuit?

Section 2

Cruise Control Systems

Cruise (speed) control systems are designed to reduce driver fatigue by maintaining a constant cruising speed. When engaged, the system controls throttle position to maintain the driver-selected vehicle speed. Two types of cruise control systems are currently in use: the electronic/vacuum operated system and the electronic system.

Electronic/Vacuum Systems

Electronic/vacuum cruise control systems use an electronic circuit to regulate vacuum to the cruise control vacuum servo. A **cruise control vacuum servo** is a device that uses a vacuum operated diaphragm to hold the throttle linkage open. In addition to the diaphragm, the servo has an inlet and outlet valve and a servo position sensor. **Fig. 6-9.** When the system is disengaged, a return spring closes the throttle.

Components The major components of an electronic/vacuum cruise control system are the:
- Electronic control unit.
- Cruise control switch.
- Vacuum servo.
- Brake (and clutch, if used) switch.
- Vacuum release valve.
- Throttle cable.
- Vehicle speed sensor (VSS) input.
- Manifold vacuum source.

On some vehicles the control unit is a self-contained module located under the instrument panel. Other vehicles have the control unit function built into the powertrain control module (PCM) or body control module (BCM).

> **TECH TIP** **Reading Cruise Control System Information.** When the cruise control circuit is in the PCM or BCM, diagnostic trouble codes (DTCs) and other cruise system information can be read in the data stream.

The cruise control switches are often part of a multifunction switch on the steering column. The switches may also be located on the steering wheel or on the instrument panel. The vacuum servo is located under the hood and is connected by a cable to the throttle linkage. The brake and clutch switches disengage cruise operation when the pedal is depressed. Applying the brakes disengages the system in two ways. A vacuum release valve and an electric release switch are mounted on the brake pedal bracket.

When the brake pedal is depressed, the vacuum release valve opens to vent the vacuum from the servo. At the same time, the electric release switch opens, electrically disconnecting the system. Two brake release devices are used as a safety feature. If either device fails, the

Internal View

Diaphragm

Vacuum Solenoid and Valve (Normally Closed)

Servo Position Sensor

Coil

Vent Solenoid and Valve (Normally Open)

Vacuum Release Port (To Vacuum Brake Release Valve)

Vacuum Supply Port (To Vacuum Supply)

Fig. 6-9 A typical cruise control vacuum servo. *What opens the throttle when the cruise control is engaged?* (General Motors Corporation)

other will still disengage the system. Vehicles with manual transmissions may use an electric release switch operated by the clutch pedal in addition to the two brake release devices. **Fig. 6-10.**

When either the clutch pedal or brake pedal is depressed, the system is electrically disengaged.

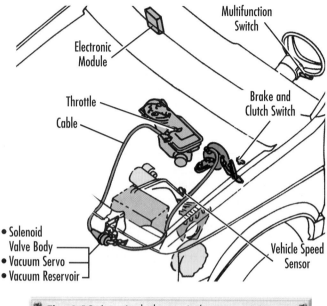

Fig. 6-10 A typical electronic/vacuum cruise control system. *How is the vacuum servo connected to the throttle lever?* (DaimlerChrysler)

System Operation The electronic module uses inputs from the cruise control switch, VSS, servo-position sensor, and brake/clutch release switches. **Fig. 6-11.** In most cases four modes of operation are possible: cruise, coast, resume/ accelerate, and tap up/tap down.

For safety reasons, the vehicle must be traveling above 25 mph [40 kph] to engage the system. When the vehicle slows below this point, the cruise control system automatically disengages.

With the cruise switch ON, momentarily pressing the SET button activates the system. The control unit opens the vacuum inlet valve and closes the vent valve on the servo. As vacuum builds in the servo, the diaphragm moves the throttle to a position that maintains the selected cruising speed. When the speed is reached, the control unit closes the inlet valve. This seals the vacuum in the servo vacuum chamber and maintains the set speed.

If the vehicle speed begins to decrease (going uphill, for example), the controller opens the inlet valve to admit more vacuum. This causes the servo to open the throttle farther to maintain the set speed. When vehicle speed increases (going downhill), the vent valve is opened to release some vacuum. This causes the throttle to close slightly to prevent over-speeding. During operation the controller opens and closes the servo valves as needed. This provides the throttle opening required to maintain the desired vehicle speed.

If the SET button is held in, the vehicle decelerates until the button is released. The speed at which the button is released is the new set speed. On many systems, if the SET button is pushed in and released immediately, the set speed decreases by 1 mph [1.6 kph]. This is the "tap down" feature.

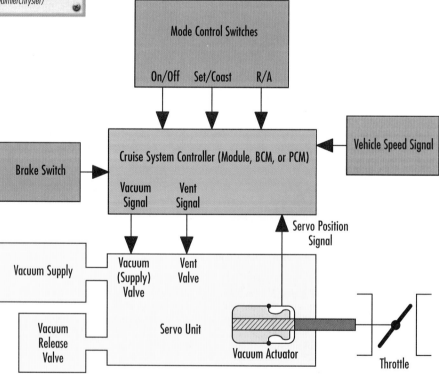

Fig. 6-11 A block diagram of a typical electronic/vacuum cruise control system. *What three input signals does the controller use to operate the system?* (General Motors Corporation)

The RESUME/ACCEL (R/A) switch serves three purposes. If the brake or clutch switch has disengaged the cruise system, the R/A switch causes the system to resume the previous speed. The R/A function also restores the previous speed if the system was disengaged because vehicle speed dropped below 25 mph [40 kph]. The resume feature does not work if the cruise or ignition switches have been turned off. If the R/A switch is held in the ON position while in the cruise mode, the vehicle accelerates until the switch is released. Moving the R/A switch to ON and releasing it immediately increases the set speed by 1 mph [1.6 kph]. This is the "tap up" mode.

Electronic Systems

Electronic (stepper motor) cruise control systems do not use a vacuum servo. Instead, an electronic controller and a *stepper motor* are combined into a single unit called a cruise control module. **Fig. 6-12.** The stepper motor replaces the vacuum diaphragm. It opens the throttle to maintain the desired cruising speed. In some applications the control module is still referred to as a servo.

Electronic System Components The cruise control module is located under the hood of the vehicle. In addition to the electronic controller and stepper motor, the module contains a solenoid-operated clutch and a drum gear and strap. **Fig. 6-13.** The drum gear strap is connected to the throttle cable.

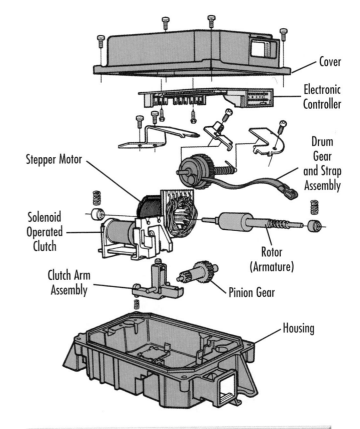

Fig. 6-13 The cruise control module contains the electronic controller, stepper motor, solenoid-operated clutch, and drum gear and strap. *What is the drum gear strap connected to?* (General Motors Corporation)

The cruise control switch and VSS are similar to those used for the electronic/vacuum cruise control system. Two brake electrical release switches are used. One switch serves as a backup for the other. A clutch electrical release switch is used on vehicles with manual transmissions.

Operation As with other cruise control systems, vehicle speed must be above 25 mph [40 kph] for the circuit to engage. With the cruise switch ON, pressing the SET button activates the system. The cruise control module energizes the solenoid-operated clutch. This connects the drum gear and strap to the stepper motor rotor (armature). The

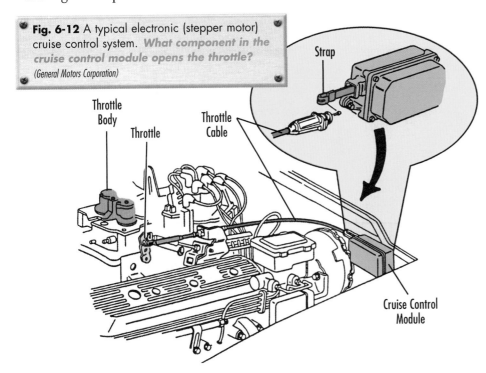

Fig. 6-12 A typical electronic (stepper motor) cruise control system. *What component in the cruise control module opens the throttle?* (General Motors Corporation)

Using Diagnostic Trees

Service manuals and books sometimes have problem solving trees. They are diagrams that rather resemble upside-down trees. The problem, or "root," is at the top. The "branches" represent testing methods. The "leaves" are outcomes of the tests. As you work your way down the tree, a YES or NO response gives you a service procedure or another test to perform.

These diagnostic trees are sometimes presented as tables. Shown below is a "tree" based on information from a 1998 General Motors service manual. Problems, testing, and outcomes are shown in this tree. Using diagnostic trees will give you experience in following directions. Trees will also help you diagnose automotive problems.

Apply it!

Meets NATEF Communications Standards for comprehending and using problem solving trees.

❶ What cruise control problem is shown in the diagnostic tree?

❷ What action do you take if the test light comes on?

❸ After what step might you replace the turn signal lever?

❹ In Steps 2 and 5 what answer to the questions about poor connections prompts you to replace a component?

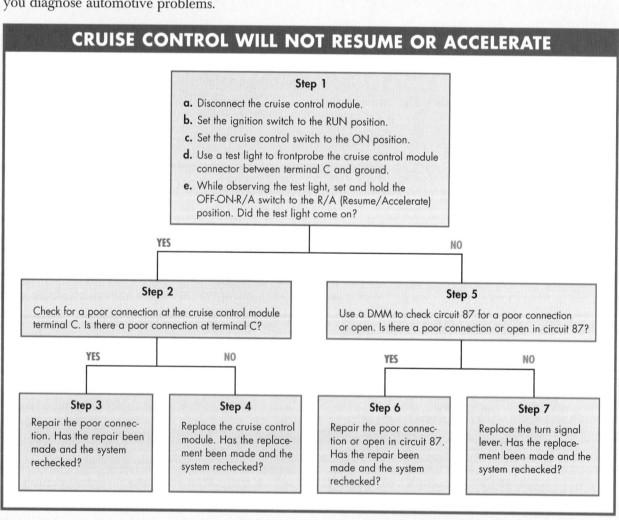

CRUISE CONTROL WILL NOT RESUME OR ACCELERATE

Step 1

a. Disconnect the cruise control module.

b. Set the ignition switch to the RUN position.

c. Set the cruise control switch to the ON position.

d. Use a test light to frontprobe the cruise control module connector between terminal C and ground.

e. While observing the test light, set and hold the OFF-ON-R/A switch to the R/A (Resume/Accelerate) position. Did the test light come on?

YES — NO

Step 2

Check for a poor connection at the cruise control module terminal C. Is there a poor connection at terminal C?

Step 5

Use a DMM to check circuit 87 for a poor connection or open. Is there a poor connection or open in circuit 87?

YES — NO

YES — NO

Step 3

Repair the poor connection. Has the repair been made and the system rechecked?

Step 4

Replace the cruise control module. Has the replacement been made and the system rechecked?

Step 6

Repair the poor connection or open in circuit 87. Has the repair been made and the system rechecked?

Step 7

Replace the turn signal lever. Has the replacement been made and the system rechecked?

control circuit moves the armature in small steps to open the throttle to the position necessary to maintain the set speed. Stepper motor movement causes the drum and strap to open or close the throttle.

As soon as the brake, clutch, or cruise switch opens the electrical circuit, the clutch is disengaged. This disconnects the drum gear from the motor armature and allows a return spring to close the throttle. When the system is disengaged, the armature returns to the closed throttle position.

The coast, resume/accelerate, and tap up/tap down modes of operation for this system are identical to that of the electronic/vacuum cruise control system.

Diagnosing and Repairing Cruise Control Systems

When diagnosing any cruise control problem, determine if the control circuit is in the PCM or BCM. If it is, use a *scan tool* to read any diagnostic trouble codes (DTCs) and other information related to the cruise control operation. Electronic testers are available for some cruise control systems. They plug into the wiring harness between the controller and servo. Refer to the manufacturer's instructions when using such a tester.

The most common problems with cruise control systems are failure to engage and failure to maintain a set speed.

Failure to Engage If the cruise control system does not engage, take the following steps:

1. Check the cruise control fuse.

2. Inspect the system wiring and connections.

3. Check for a vehicle speed input to the controller.

4. Referring to a wiring diagram, check for the correct voltage at the controller and servo terminals as the cruise control switch is operated.

5. Check the brake (and clutch) switches for proper position and operation.

> **Safety First** **Materials** Do not energize the servo solenoid valves with jumper wires while the engine is running. The engine over-speed condition that will result could cause severe engine damage.

The following checks apply to vacuum-operated systems only:

1. Start the engine. Check for vacuum at the vacuum line to the servo. Vacuum should be at least 10 in. [25 cm] of mercury at idle. **Fig. 6-14.**

2. Check for leaks in the vacuum release hose and valve.

3. Check the resistance of the servo solenoid windings. Refer to the vehicle service manual for specifications.

4. Check for a vacuum leak at the servo. (See the procedure below.) If the results of these checks are satisfactory, the controller is usually defective.

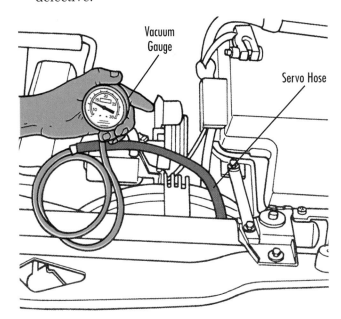

Vacuum Gauge

Servo Hose

Fig. 6-14 Using a vacuum gauge to check for a vacuum at the servo. *How much vacuum should be present?* (DaimlerChrysler)

Failure to Maintain a Set Speed If the cruise system fails to maintain the set speed, check the throttle linkage adjustment. Refer to the vehicle service manual for the correct procedure. If the linkage adjustment is good, check for a vacuum leak at the servo. With the engine off, use jumper wires to energize the servo solenoid valves. Apply vacuum with a hand-held vacuum pump. If the servo fails to hold vacuum, it must be replaced. If the linkage adjustment and servo are good, the controller is probably defective.

SECTION 2 KNOWLEDGE CHECK

❶ What two types of cruise control systems are currently in use?

❷ Name the three places where the control function for the electronic/vacuum cruise control system may be found.

❸ Explain the "tap up" feature available on many cruise control systems.

❹ On an electronic cruise control system, how is the drum gear disconnected from the motor armature when the system is to be disengaged?

❺ What are the two most common problems associated with cruise control systems?

Section 3

Supplemental Restraint Systems

Supplemental restraint systems (SRSs) are also called *air bag systems*. The original systems were designed to protect only the driver. In addition, the air bags inflated only if there was sufficient impact at the front of the vehicle. Newer systems also protect some passengers. They can be triggered by side as well as frontal impacts.

Air bag systems are intended to be used with seat belt restraints. An unbelted driver or passenger is not likely to receive maximum protection from the air bags. Even in relatively minor impacts, unbelted occupants have been seriously injured or killed by inflating air bags.

Air bag inflation (deployment) occurs when onboard sensors detect a sudden impact of sufficient severity. Current flow through an igniter circuit starts a chemical reaction in a solid propellant. The chemical reaction generates nitrogen gas that inflates the air bag. A few vehicles use compressed gas to inflate the air bag(s). The gas is stored in a cylinder in the air bag module.

Components of Air Bag Systems

Most air bag systems use similar components, but the terminology is not standardized. Each manufacturer has unique terms for some system components. Always refer to the service manual or training manual if you are unfamiliar with the components in the system you are working on. **Fig. 6-15.**

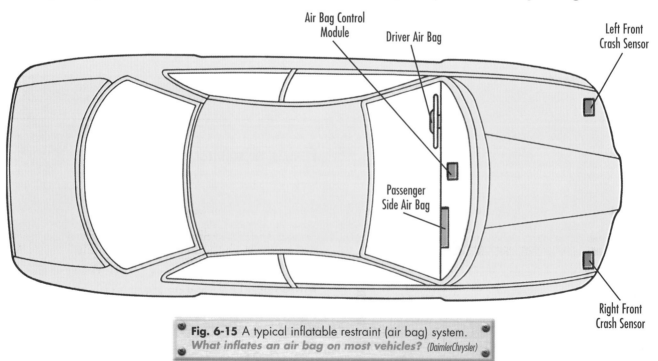

Air Bag Control Module

Driver Air Bag

Left Front Crash Sensor

Passenger Side Air Bag

Right Front Crash Sensor

Fig. 6-15 A typical inflatable restraint (air bag) system.
What inflates an air bag on most vehicles? (DaimlerChrysler)

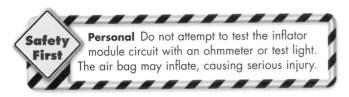

Safety First **Personal** Do not attempt to test the inflator module circuit with an ohmmeter or test light. The air bag may inflate, causing serious injury.

Inflator Module The **inflator module** is the module that contains the air bag, ignitor, solid propellant, and cover. The driver side inflator module is located in the center of the steering wheel. **Fig. 6-16.** The passenger side inflator module is mounted in the right side of the instrument panel. Side inflator modules are located in the door or seat assemblies.

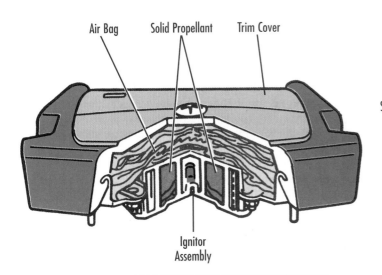

Air Bag   Solid Propellant   Trim Cover

Ignitor Assembly

Fig. 6-16 A driver-side air bag inflator module. *Where is the passenger side inflator module located?* (DaimlerChrysler)

Air Bag Control Module The **air bag control module** is the module that monitors system operation, controls the air bag warning light, and stores trouble codes. It is located in the passenger compartment of the vehicle. In some cases a backup power supply and a crash sensor may be located in this module.

Crash (Impact) Sensors The **crash (impact) sensors** are the sensors that close an electrical circuit when sufficient impact occurs. They are usually mounted under the hood and in the passenger compartment. The sensor in the passenger compartment may be located inside the air bag control module. The number and location of crash sensors varies among vehicles.

Most air bag systems use two crash sensors. One sensor is on the battery side of the circuit, while the other is on the ground side. The two sensors are calibrated to close at different impact levels. The sensors cannot be interchanged. Both must close at the same time for air bag inflation to occur. The sensors, along with the ignitor, power supply, and wiring make up the "deployment loop." A pair of identical sensors may be connected in parallel. If either one of the sensors closes, that part of the circuit is complete. **Fig. 6-17.**

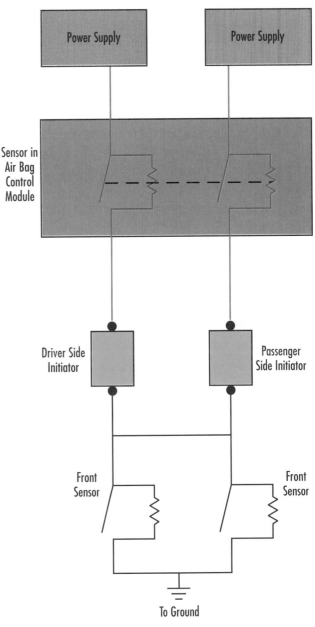

Power Supply   Power Supply

Sensor in Air Bag Control Module

Driver Side Initiator   Passenger Side Initiator

Front Sensor   Front Sensor

To Ground

Fig. 6-17 A typical air bag circuit deployment loop. *What components make up the deployment loop?* (DaimlerChrysler)

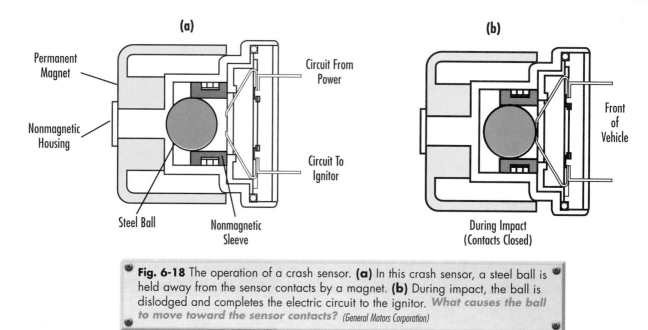

(a)

Permanent Magnet

Nonmagnetic Housing

Circuit From Power

Circuit To Ignitor

Steel Ball

Nonmagnetic Sleeve

(b)

Front of Vehicle

During Impact (Contacts Closed)

Fig. 6-18 The operation of a crash sensor. **(a)** In this crash sensor, a steel ball is held away from the sensor contacts by a magnet. **(b)** During impact, the ball is dislodged and completes the electric circuit to the ignitor. *What causes the ball to move toward the sensor contacts?* (General Motors Corporation)

A common type of crash sensor uses a gold-plated steel ball held in place by a magnet. If a crash generates enough force to move the ball away from the magnet, the circuit between the sensor contacts closes. If both sensors in the circuit are closed, the air bag(s) inflate. **Fig. 6-18.**

Reserve (Backup) Power Supply The reserve power supply stores a voltage high enough to deploy the air bags. This is in case the battery connection is lost in a collision before the crash sensors close. The backup power supply can be a stand-alone unit or located in the control module.

Safety First **Personal** The backup power supply stays energized for several minutes after the ignition is turned off. When disabling an air bag system, wait at least ten minutes (or the time specified) before beginning work on the air bag system.

Clockspring The clockspring consists of two or more wires coiled inside a plastic housing mounted at the top of the steering column. The clockspring maintains a "hard wired" connection between the steering column wiring and the rotating steering

Fig. 6-19 The clockspring assembly is mounted at the top of the steering column. *What is the purpose of the clockspring assembly?* (General Motors Corporation)

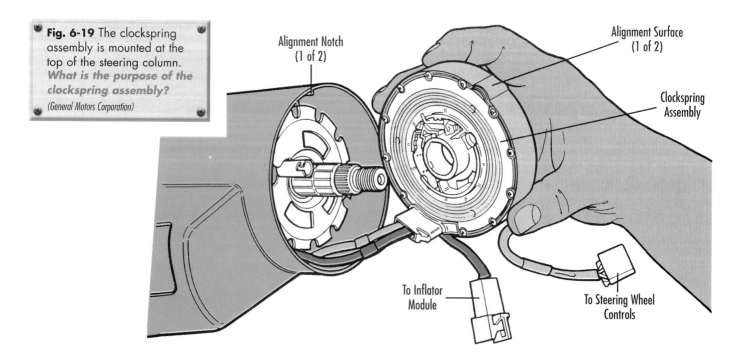

Alignment Notch (1 of 2)

Alignment Surface (1 of 2)

Clockspring Assembly

To Inflator Module

To Steering Wheel Controls

Using a Switching Transistor

Switching transistors are used in everything from supercomputers to sports cars. The electronic switch inside a computer-controlled lighting system is a transistor. Transistors have three connections:

- Collector.
- Base.
- Emitter.

The base electrode acts like a safety valve. A threshold base voltage triggers a large current flow from the collector to the emitter. The transistor switch is closed. All current flow from the collector to the emitter stops if there is no base current. The transistor switch is open. This switching action sends current through a small lamp. For larger lights, it sends current through a relay.

Apply it!

Demonstrating a Transistor's Switching Action

Meets NATEF Science Standards for electrical parameters and semiconductors.

Materials and Equipment
- 2N3903 general-purpose NPN transistor
- Multimeter
- 330-ohm resistor (R_C)
- Light-emitting diode (LED)
- 1.0 kilo-ohm potentiometer (R_{POT})
- 2.7 kilo-ohm resistor (R_B)
- Breadboard

You can demonstrate a transistor's switching action on a breadboard.

❶ Construct the circuit shown here.

❷ Adjust the potentiometer to set the base voltage to approximately 0.7 volts or greater. The light-emitting diode will start glowing. The switch is now turned on. The LED emits visible light when it is conducting.

❸ Decrease the base voltage to less than 0.6 volts and the LED stops glowing completely. This happens because the transistor has switched off. Current can no longer flow from the voltage supply to ground through the transistor.

❹ Explain how the transistor in this experiment is similar to a mechanical switch.

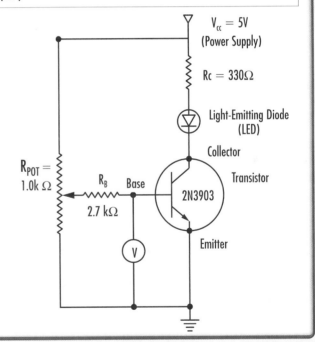

wheel. This ensures good electrical contact between the driver side inflator assembly and the rest of the deployment circuit. Additional wires in the clockspring assembly may be used for the horn, steering wheel controls, or other similar circuits. **Fig. 6-19.**

Indicator Light   Air bag systems use a warning light on the instrument panel to inform the driver about the status of the system. The light should be off when the ignition switch is off. During the first 10 seconds after the ignition switch is turned on, the light should either stay on or flash on and off. The light goes out if the system passes a self-test initiated by the air bag control module. If the warning light does not come on, does not stay on or flash as it should, or fails to go out, there is a fault in the system.

Operation of Air Bag Systems

Most air bag systems operate in a similar manner. The following is a general description of a typical air bag system.

Startup Routine When the ignition is turned on, the control module checks the system and stores a voltage in the reserve power supply. The indicator light comes on or begins flashing. If the system passes the diagnostic check, the light goes out after about 10 seconds. If there is a problem in the system, the light either stays on or begins flashing trouble codes.

The control unit continuously checks the air bag system for possible faults. If none are found, the indicator light remains off. If a fault is detected, the warning light is turned on and a fault code is stored in the module's memory. Fault codes can be read by counting the number of light flashes or by using a scan tool.

Air Bag Deployment If the vehicle experiences an impact severe enough to require air bag deployment,

two or more of the crash sensors close. This completes the deployment loop. Voltage from the battery or reserve power supply causes current to flow through the igniter. The ignitor starts a chemical reaction in the solid propellant that produces nitrogen gas. The gas inflates the air bag(s). **Fig. 6-20.**

Deployment Chemistry In addition to nitrogen gas, other materials are released during the chemical reaction. Small amounts of sodium azide, sodium hydroxide, and sodium bicarbonate may be found around a deployed air bag. Sodium azide is the primary propellant used in most air bag modules. Only very small amounts are present in a vehicle with a deployed air bag. Sodium hydroxide can irritate the eyes and skin.

Sodium bicarbonate is harmless. The "smoke" and dust released when an air bag deploys is talcum powder or cornstarch. They are used to keep the air bag surfaces from sticking to each other before deployment.

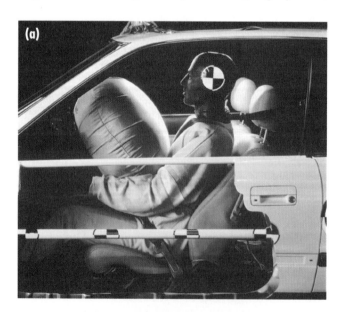
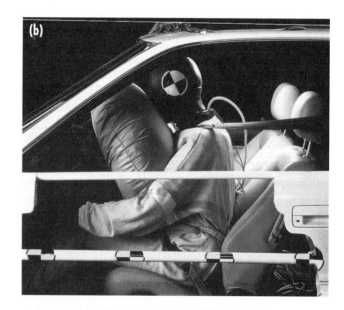

Fig. 6-20 Air bag deployment sequence. **(a)** Steering wheel at impact point, air bag begins deployment. **(b)** The air bag is fully deployed, protecting the driver from frontal impact. *How many crash sensors must sense the impact before the air bag deploys?* (National Highway Traffic Safety Administration)

Diagnosis and Repair Special procedures are required when diagnosing and repairing inflatable restraint systems. Failure to follow specific testing and servicing instructions can cause accidental deployment of the air bag(s). In addition, improper servicing and handling of component parts may prevent the system from deploying successfully. Procedures that apply to one system may not apply to another. Always read and follow the vehicle service manual instructions. Pay particular attention to the many safety warnings and precautions.

Reading and Erasing Trouble Codes Some inflatable restraint systems use the air bag warning light to display trouble codes. The system is put into diagnostic mode with a jumper wire connection or by cycling a switch. With the ignition on, the technician counts the number of light flashes to determine the trouble code(s) present. In some cases the codes are repeated several times with a slightly longer pause between displays. On other vehicles trouble codes are read with a scan tool connected to the *data link connector (DLC)* or a separate test connector for the air bag system. On some systems it is possible to read trouble codes using either the flash code or scan tool method.

Erasing trouble codes usually requires the use of a scan tool. With the scan tool connected and the ignition on, follow the instructions on the scan tool display. In some cases the codes will be erased only when the fault is corrected. On some systems, codes can be erased by cycling the driver's door light switch on and off at least five times within 7 seconds of turning the ignition on. Always refer to the appropriate service manual for specific information on reading and erasing trouble codes.

> **TECH TIP Warning Lights.** When diagnosing a problem, always observe the operation of system warning lights. If a light does not come on when it should, begin your diagnosis with the light circuit.

Disabling the Air Bag System A variety of complex air bag systems are in use. It is not possible to have a single disabling procedure that is suitable for all applications. The following procedures are often used in various combinations as steps in preventing an air bag from deploying accidentally. Use the procedures specifically recommended for the vehicle you are working on:

> **Safety First** **Personal** Disable the air bag(s) before working on the inflatable restraint system or on any components that require you to work near the air bags. This will prevent accidental deployment that could cause serious injuries.

- Turn off the ignition switch.
- Disconnect the battery; isolate the cable from the post.
- Disconnect the clockspring connector at the base of the steering column.
- Connect a shorting clip across the disconnected clockspring connector (if not already in place).
- Wait the recommended time for the reserve power supply to shut down.
- Remove the air bag fuse.
- Disconnect the reserve power supply.

Servicing Inflatable Restraint Systems Personal injury and extensive damage to the vehicle may occur if specific repair and replacement procedures are not followed exactly. Read the service manual carefully and perform the necessary steps in the order listed. Read all of the instructions before beginning the procedure. Follow the procedure(s) listed for disabling the system before beginning repairs.

Components of an inflatable restraint system cannot be repaired or adjusted. Defective or damaged parts must be replaced. In addition to the inflator module, some manufacturers require replacement of other deployment loop components if the bag has been deployed. The location and orientation (direction) of sensors are critical. Refer to the vehicle service manual for all service-related information and procedures.

Inflated air bags must be disposed of and documented according to specified regulations.

Seat Belts

Seat belts and air bags are designed to be used as a total passenger restraint system. Seat belts should be checked for wear and proper operation.

Inspecting Seat Belts

Normal use of seat belts may cause wear and deterioration. The seat belt attachment and retraction mechanism may also show signs of wear from extended use and collision damage.

When inspecting seat belts:

1. Extend the seat belt so its total length can be seen.

2. Check the belts for cuts, noticeable wear, and deterioration. If involved in a collision, the belt fabric may show signs of damage as the belt passed through the guide plates.

3. Attach the two portions of the buckle. Pull sharply and firmly on the belt to check buckle operation. Make sure the belt releases easily when the buckle release is pressed.

4. Check the belt anchor points for cracking or tearing. Look for sharp or exposed edges that may cut into the seat belt. Be sure the anchor brackets are firmly attached.

5. Check the retractor assemblies for proper operation. When the seat belt is released, it should retract fully, without binding or jamming.

Servicing Seat Belts

Seat belts are not normally repaired. Any seat belt that is damaged should be replaced with a new assembly. Some seat belts are automatically extended and retracted. Refer to the vehicle service manual for the proper removal and installation procedures.

When removing seat belts:

1. Replace any component that is damaged.

2. Wear protective gloves and clothing. The confined areas where the belts are attached may contain sharp edges.

3. Remove the anchor brackets and retractor mechanisms. Attachment bolts may have corroded. Be careful not to damage the bolt. If the bolts are frozen, apply a solvent to dissolve any corrosion.

4. Inspect all anchor points for corrosion and structural damage.

When installing seat belts:

1. Position the belt assembly to ensure the belt is not twisted or kinked.

2. Attach all mounting hardware with the correct hardened bolts.

3. Tighten all mounting bolts to the correct torque specifications.

4. Check the retractor assemblies for proper operation. When the seat belt is released, it should retract fully, without binding or jamming.

SECTION 3 KNOWLEDGE CHECK

❶ Why are air bag systems called "supplemental" restraint systems?

❷ What components make up the deployment loop?

❸ What is the purpose of the clockspring?

❹ What should the air bag warning light do when the ignition switch is first turned on?

❺ What are the two ways in which air bag trouble codes can be read?

Section 4

Horn Circuits

A typical horn circuit consists of either one or two horns, a horn relay, and a horn switch. **Fig. 6-21.**

Horns are usually located near the front of the vehicle under the hood. The horn relay may be located under the instrument panel or hood. The **horn relay** is a relay that controls current flow from the battery to the horns. Most horn switches are located on the steering wheel or on the steering column.

If the horn switch is located on the steering wheel, some provision must be made for continuous electrical contact while the wheel is rotated. Some vehicles use brushes and slip rings to maintain electrical contact in the circuit during wheel rotation. Vehicles with a supplemental restraint (air bag) in the steering wheel may use part of the air

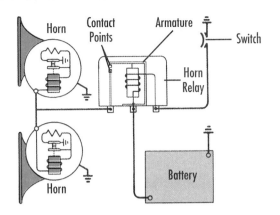

Fig. 6-21 Automotive horn circuits use a relay to supply current to the horns. *Where is the relay winding grounded?* (General Motors Corporation)

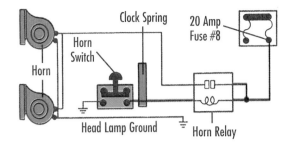

Fig. 6-22 On vehicles with air bag systems, the clockspring may be part of the horn switch wiring. *If a clockspring is not used, how is the circuit between the steering wheel and column connected?* (DaimlerChrysler)

bag wiring harness (clockspring) in the steering column for horn switch wiring. **Fig. 6-22.**

Closing the horn switch completes a ground path activating the horn relay. When the relay contacts close, battery voltage is connected to the horns. Current flow through a winding in each horn creates a magnetic field that causes an armature to move a diaphragm.

A pair of contact points inside the horn open when the armature movement reaches its limit, opening the circuit. A spring returns the armature and the diaphragm to its original position, closing the contacts. Repeating this cycle many times per second causes the diaphragm to vibrate. Movement of the diaphragm is similar to the cone of a speaker. **Fig. 6-23.**

Diagnosing Horn Problems

If the horns do not operate, check the wiring and connections. Check the fuse. If the fuse is good, use a voltmeter or test light to check for voltage at the relay. If voltage is present, have someone operate the horn switch while you check for voltage from the relay. If voltage is present at the relay output, there is an open circuit between this point and the horns.

Check the wiring, connections, and ground circuit. If there is no voltage at the relay output, use a jumper wire to ground the horn switch terminal at the relay. Make sure you ground the correct terminal. If the horn sounds, there is an open in the horn switch circuit. If the horn does not sound, the relay is defective.

Poor sound quality from horns can be caused by high resistance in the circuit, including the ground connection. Some horns can be adjusted. Connect an ammeter in series with one horn at a time. Operate the horn and turn the adjusting screw to obtain the specified current draw.

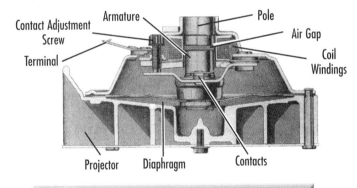

Fig. 6-23 Automotive horns use a vibrating diaphragm to produce sound. *What makes the horn diaphragm vibrate?* (General Motors Corporation)

SECTION 4 KNOWLEDGE CHECK

1. Where are horns usually located?
2. What electrical component is used to control current to the horns?
3. How is the horn relay activated?
4. How can you check the horn relay using a jumper wire?
5. How can some horns be adjusted?

CHAPTER 6 REVIEW

Key Points

Meets the following NATEF Standards for Electrical/Electronic Systems: diagnosing incorrect wiper operation; diagnosing incorrect operation of motor-driven accessory circuits.

- Many small motors have an internal electronic circuit breaker.
- Direct current motors can be reversing or non-reversing.
- Windshield wiper motors have an additional insulated brush to provide two-speed operation.
- Cruise control systems use either a vacuum diaphragm or a stepper motor to control throttle opening.
- Cruise control cannot be engaged below a minimum road speed designed into the system.
- Most inflatable restraint systems use a solid propellant to deploy the air bag(s).
- Inflatable restraint systems have a self-diagnostic capability.
- Air bags are designed to be used as a supplemental system with seat belts.
- The inflatable restraint system should be temporarily disabled when working on or near system components.

Review Questions

1. Name the four major components in a permanent magnet motor.
2. How do small direct current (DC) motor circuits work?
3. How do windshield wiper and washer systems operate?
4. Name the major components of an electronic/vacuum cruise control system.
5. How does a cruise control system operate?
6. How is cruise control diagnosis made easier if the electronic circuit for the system is in the PCM or BCM?
7. Explain how crash sensors are connected in the inflator module circuit.
8. What are the components of the supplemental restraint system?
9. **Critical Thinking** Describe the operation of a typical windshield wiper motor if the wire to the first insulated brush (low speed) is open.
10. **Critical Thinking** Why is it unsafe to drive a vehicle if the air bag warning light is on?

TECHNOLOGY FORECAST FOR AUTOMOTIVE EXCELLENCE

Night Vision Cameras

Soon, drivers will not have to be afraid of the dark. Night-vision systems will let them see farther and better. Making nighttime driving safe is important. About half of all fatal car accidents happen at night.

Night-vision systems use a camera that can "see" heat given off by people, animals, and objects. This heat is a kind of light called infrared light. Infrared light cannot be seen by the eye. A small image displayed on the windshield shows the driver what the night-vision camera detects.

With a night-vision system, the driver can see up to one-quarter of a mile away. That distance is three to five times farther than with low-beam headlights. It is up to three times farther than with high-beam headlights.

At normal cruising speeds, the driver will have more time to stop the vehicle or steer away from objects. Accidents and injuries can be avoided.

Night-vision cameras weren't built first for cars. The U.S. military made them so soldiers can see in the dark. They can see the terrain and identify objects and personnel. One day soon, night-vision systems may enable drivers to see better at night on America's highways.

AUTOMOTIVE EXCELLENCE TEST PREP

Answering the following practice questions will help you prepare for the ASE certification tests.

1. Which component is not used in a permanent magnet motor?
   - **ⓐ** An armature.
   - **ⓑ** Field coils.
   - **ⓒ** Brushes.
   - **ⓓ** A commutator.

2. Technician A says many small permanent magnet motors have an internal circuit breaker. Technician B says these circuit breakers have to be manually reset. Who is correct?
   - **ⓐ** Technician A.
   - **ⓑ** Technician B.
   - **ⓒ** Both Technician A and Technician B.
   - **ⓓ** Neither Technician A nor Technician B.

3. Technician A says wiper motors with three brushes are reversing-type motors. Technician B says the extra brush is used to operate the motor at high speed. Who is correct?
   - **ⓐ** Technician A.
   - **ⓑ** Technician B.
   - **ⓒ** Both Technician A and Technician B.
   - **ⓓ** Neither Technician A nor Technician B.

4. Pulse (intermittent) operation of a windshield wiper circuit is controlled by:
   - **ⓐ** A third brush.
   - **ⓑ** A two-way switch.
   - **ⓒ** An electronic module.
   - **ⓓ** An electronic circuit breaker.

5. Technician A says electronic cruise control systems do not use a vacuum servo. Technician B says electronic cruise systems use a stepper motor. Who is correct?
   - **ⓐ** Technician A.
   - **ⓑ** Technician B.
   - **ⓒ** Both Technician A and Technician B.
   - **ⓓ** Neither Technician A nor Technician B.

6. In an electronic/vacuum cruise control system, the throttle is opened by a:
   - **ⓐ** Permanent magnet motor.
   - **ⓑ** Stepper motor.
   - **ⓒ** Control module.
   - **ⓓ** Diaphragm.

7. Two technicians are discussing checking a vacuum servo for leaks. Technician A says if the servo does not hold vacuum, it may be necessary to replace the diaphragm. Technician B says the entire servo will have to be replaced. Who is correct?
   - **ⓐ** Technician A.
   - **ⓑ** Technician B.
   - **ⓒ** Both Technician A and Technician B.
   - **ⓓ** Neither Technician A nor Technician B.

8. Which module stores trouble codes for an inflatable restraint system?
   - **ⓐ** Inflator module.
   - **ⓑ** Air bag control module.
   - **ⓒ** Body control module.
   - **ⓓ** Powertrain control module.

9. Technician A says the clockspring is part of the inflator module. Technician B says it is located in the air bag control module. Who is correct?
   - **ⓐ** Technician A.
   - **ⓑ** Technician B.
   - **ⓒ** Both Technician A and Technician B.
   - **ⓓ** Neither Technician A nor Technician B.

10. Which procedure is not a step in disabling an air bag system?
    - **ⓐ** Disconnect the battery.
    - **ⓑ** Disconnect the clockspring connector at the base of the steering column.
    - **ⓒ** Disconnect the air bag control module.
    - **ⓓ** Wait for the reserve power supply to shut down.

Engine Performance

Automotive Managers & Assistant Managers

A busy automotive service center is now hiring. High-powered, energized individuals will be hired and trained to join and keep pace in this fast-growing industry.

Only Go-Getters Should Apply!

Starting wage is open for the right individual. Full-time benefits include paid sick leave, paid vacation, holiday pay, and profit sharing. If you have a high school diploma or GED and you're trainable, not afraid of hard work, and interested in a career opportunity, call for an immediate interview.

Exhaust Systems Analyst

Critical Skills and Knowledge: BS or MS in mechanical engineering; experience with computational fluid dynamics analysis (ICEM-CFD); some experience with systems analysis code preferred.

Key Tasks: perform analysis to optimize the engine air induction system and exhaust system in order to meet performance, fuel economy, and emissions targets.

TECHNICAL SPECIALIST

- Perform complex analysis, troubleshooting, test, repair, calibration, rework, and/or inspection duties in the technical specialty area of service.

- Must have demonstrated ability to work in a combination of multiple disciplines, including gases and chemicals, hydraulics, pneumatics, analog and digital circuitry, computerized systems, systems integration, robotics, vacuum technology, customer service, statistics, algebra, and geometry.

- Must also have good analytical, communication, and organizational skills.

- Requires Associate's degree in electronics/chemistry, electronic technology school diploma, or formal metrology training, with at least 5–8 years of related experience.

ENGINE DEVELOPMENT ENGINEER

If you have a strong work ethic and mechanical aptitude or some mechanical experience, here's an opportunity for you. We are currently seeking mechanical engineers to work on a cutting-edge engine development team. Top applicants will have solid experience with internal combustion engine design and development.

Career-Focus Activity

After reading the above job descriptions, do the following:

- Imagine that you are applying for one of the positions. Write a letter of inquiry.
- Choose one of the careers and research its projected growth in your state.
- What science and mathematics courses would best prepare you to apply for the Technical Specialist position above?

CHAPTER 1

Piston Engine Operation

The Problem

Engines are made in many sizes and power ratings. Four-cylinder engines power smaller vehicles, where fuel economy is more important. Larger and more powerful six- and eight-cylinder engines power most medium- and full-size vehicles. Ten- and twelve-cylinder engines put the performance in high performance sport models.

But how does the engine produce power? How is fuel burned in the engine? How is engine power used to move the vehicle? The answers to these questions are in this chapter.

Your Challenge

As the service technician, you need to find answers to these questions:

❶ How does an engine produce power?

❷ What is a piston?

❸ What is a crankshaft?

❹ What is a camshaft?

❺ What is a connecting rod?

Section 1

Internal Combustion Engines

Automotive engines are internal combustion engines. **Fig. 1-1.** An **engine** is a machine that turns heat energy into mechanical energy. An internal combustion engine burns fuel internally. The heat produced from burning a fuel creates the power that moves the vehicle.

Most automotive engines are called reciprocating engines because their pistons move up and down inside the cylinders. **Fig. 1-2.** A **piston** is a cylindrical plug that fits inside the cylinder. It receives and transmits motion as a result of pressure changes applied to it.

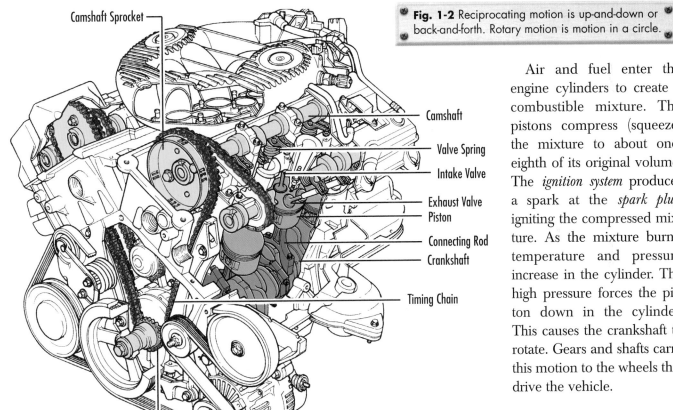

Camshaft Sprocket

Camshaft
Valve Spring
Intake Valve
Exhaust Valve
Piston
Connecting Rod
Crankshaft
Timing Chain

Crankshaft Sprocket

Fig. 1-1 A V-6 engine with dual overhead camshafts. *A spark ignition engine runs on what fuels?* (DaimlerChrysler)

There are two types of internal combustion piston engines: *spark-ignition* (gasoline) and *compression-ignition* (diesel).

Internal combustion piston engines differ in:
- The type of fuel they use.
- The way ignition of the *air/fuel mixture* occurs.

Spark-Ignition Engine

Most spark-ignition engines run on liquid fuels, such as gasoline, alcohol, or a gasoline/alcohol blend. Some spark-ignition engines run on gaseous fuels, such as propane or natural gas.

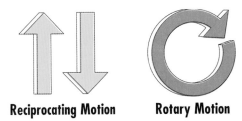

Reciprocating Motion   **Rotary Motion**

Fig. 1-2 Reciprocating motion is up-and-down or back-and-forth. Rotary motion is motion in a circle.

Air and fuel enter the engine cylinders to create a combustible mixture. The pistons compress (squeeze) the mixture to about one-eighth of its original volume. The *ignition system* produces a spark at the *spark plug,* igniting the compressed mixture. As the mixture burns, temperature and pressure increase in the cylinder. The high pressure forces the piston down in the cylinder. This causes the crankshaft to rotate. Gears and shafts carry this motion to the wheels that drive the vehicle.

Compression-Ignition Engine

A diesel (compression ignition) engine runs on a light fuel oil similar to kerosene. In this type of engine, the piston compresses only air. Compressing air to about one-twentieth of its original volume raises its temperature to 1,000°F [538°C] or higher. The fuel is injected (sprayed) into the cylinder, where it is ignited by the heated air. As the mixture burns, the pressure forces the piston down in the cylinder.

Engine Construction

Spark- and compression-ignition engines are similar in construction. Both have engine blocks and cylinder heads. Both have pistons that move up and down in the cylinders. The *cylinders*, or cylinder bores, are machined openings through the engine block. A *cylinder head* covers the top of the cylinders. The bottom of each cylinder is open. The pistons are connected through this opening to the *crankshaft.*

The two travel limits for a piston are defined as *top dead center (TDC)* and *bottom dead center (BDC)*. A piston stroke takes place when the piston moves from TDC to BDC or from BDC to TDC. **Fig. 1-3.**

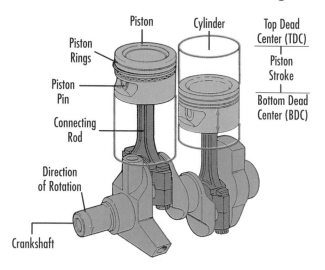

Fig. 1-3 The reciprocating action of pistons in the cylinders. *When does a complete piston stroke take place?* (General Motors Corporation)

The Engine Block

The *engine block,* also called the *cylinder block,* is a precision metal casting. **Fig. 1-4.** The block contains the:

• Cylinders, or cylinder bores.
• Pistons and connecting rod assemblies.
• Camshaft, for engines that do not have an overhead camshaft design.
• Crankshaft.

Figure 1-5 shows the events that take place in the cylinder of a spark-ignition engine. The piston has completed its intake stroke. It is at its lower limit of travel, bottom dead center. **Fig. 1-5(a).** The space above the piston contains the air/fuel mixture.

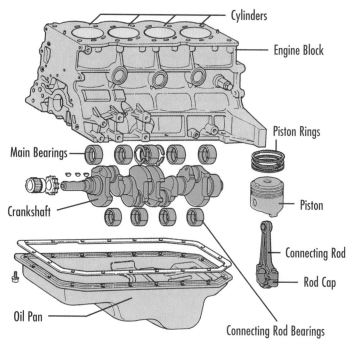

Fig. 1-4 The engine block and lower end parts. *Where does the cylinder head attach?*

Next, the piston moves up the cylinder toward top dead center. **Fig. 1-5(b).** This motion compresses the mixture. As the piston nears top dead center, an electric spark ignites the mixture. The mixture burns rapidly. This creates heat and high pressure that push the piston down in the cylinder. **Fig. 1-5(c).**

This downward movement creates power. At the bottom of the power stroke, the piston begins the exhaust stroke and moves up in the cylinder. The exhaust valves open, and the burned gases are pushed from the cylinder. **Fig. 1-5(d).**

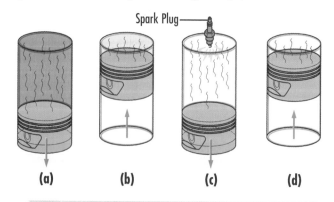

Fig. 1-5 Four views showing the piston movement in an engine cylinder. *How many piston strokes are needed for one power cycle?*

Pistons and Piston Rings Pistons are usually made of an aluminum alloy, which is aluminum mixed with other metals. **Fig. 1-6.** They are slightly smaller than the cylinders so that they can move up and down freely.

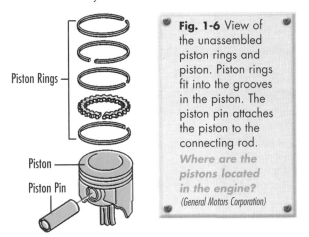

Fig. 1-6 View of the unassembled piston rings and piston. Piston rings fit into the grooves in the piston. The piston pin attaches the piston to the connecting rod. *Where are the pistons located in the engine?* (General Motors Corporation)

The small gap between the piston and cylinder wall is known as *piston clearance.* **Fig. 1-7.** Piston clearance provides the sliding fit. If not properly sealed, this gap allows some of the compressed air/fuel mixture and combustion gases to leak past the piston. This leakage is called *blowby.* Blowby reduces power, wastes fuel, and pollutes the air. The piston rings seal the gap between the piston and the cylinder wall. Each ring fits into ring grooves cut into the piston. There are two types of piston rings:

Compression rings form a sliding seal between the piston and the cylinder wall. They reduce or control blowby of combustion gases.

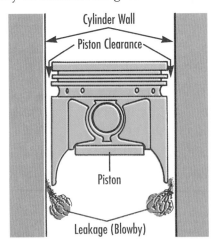

Fig. 1-7 Blowby occurs between the piston and cylinder wall. *What reduces the amount of blowby?*

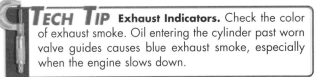

TECH TIP **Exhaust Indicators.** Check the color of exhaust smoke. Oil entering the cylinder past worn valve guides causes blue exhaust smoke, especially when the engine slows down.

Oil rings, or oil-control rings, scrape excess oil from the cylinder wall and return it to the crankcase.

Crankshaft The reciprocating motion of the pistons must be changed to rotary motion. Rotary motion is what turns the vehicle's drive wheels. The *connecting rods* and the crankshaft make this conversion possible. A piston pin connects each piston to the small end of the connecting rod. The connecting rod connects the piston to the crankshaft. **Fig. 1-8.**

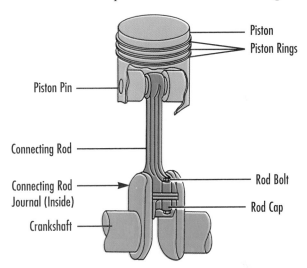

Fig. 1-8 Piston and connecting rod assembly attached to a connecting rod journal on the crankshaft. The piston is shown partly cut away to show how the piston is attached to the connecting rod. *What attaches the piston to the connecting rod?*

The *rod cap* and *rod bolts* attach the connecting rod to the *connecting rod journal.* The journal holds a split bearing (two halves), or *connecting rod bearing,* in place in the cap and rod. **Fig. 1-9.** A slight clearance allows the connecting rod journal to turn inside the bearing. Oil fills this clearance to lubricate the bearing and prevent metal-to-metal contact. As the crankshaft turns, the connecting rod journal moves in a circle. **Figs. 1-8** and **1-9.**

As the piston moves up and down in the cylinder its connecting rod journal moves in a circle around the centerline of the crankshaft. On the downstroke the connecting rod moves to one side, as its lower end follows the movement of the crankshaft rod journal. As the piston reaches BDC, the connecting rod journal continues to move in a circle. As the journal begins to move up the connecting rod pushes the piston up on the next stroke. In this way, the crankshaft changes the reciprocating motion of the piston to rotary motion at the drivetrain.

The Cylinder Head

The cylinder head is bolted to the top of the engine block. The cylinder head contains the:
- Intake valves, exhaust valves, and connecting parts.
- Camshaft for engines with overhead camshaft design.
- *Combustion chamber* (the upper portion of the cylinder located in the head).

Hot Gases Are Really Cool

Do you want to know why hot gases are so cool? A diesel engine would have no power without hot air to ignite the fuel within its cylinders. When a diesel engine piston moves up on the compression stroke, it compresses the air above it. This causes the air pressure and temperature to increase within the cylinder. The temperature increases so much that a light spray of fuel ignites as soon as it mixes with the highly compressed air in the top of the cylinder!

Can you explain the relationship between temperature and pressure for a gas? The answer to this question is very simple. Increasing the pressure for a gas (air) trapped in a container increases the temperature. Similarly, increasing the temperature increases the pressure. Gas pressure also decreases with decreasing temperature.

Safety First

Personal Use eye protection. Wear gloves and clothing with long sleeves.

Caution Do not heat an empty bottle in the microwave. Always put some water inside the bottle. Never heat the bottle with the top sealed.

Apply it!

Exploring Temperature and Pressure

Meets NATEF Science Standards for understanding the relationship between pressure and temperature and the effect of how adding heat causes vaporization.

Materials and Equipment
- 1-liter plastic bottle with a screw-on top, or plastic dish detergent bottle, or small plastic milk bottle
- Microwave oven and sink or water hose
- 4 tablespoons of water

Here's a simple experiment that you may want to try at home. It's just the reverse of what happens to the air within an engine's cylinder on the compression stroke.

❶ Remove the lid from the plastic bottle and rinse it thoroughly.

❷ Add about 4 tablespoons of water to the bottle. Heat it in a microwave oven for 1 minute.

❸ Remove the warm bottle from the oven. Screw on the lid tightly.

❹ Run cold water over the bottle for a minute or so. Observe what happens.

Results and Analysis As you probably observed, when the warm plastic bottle is cooled, it collapses. Can you explain why?

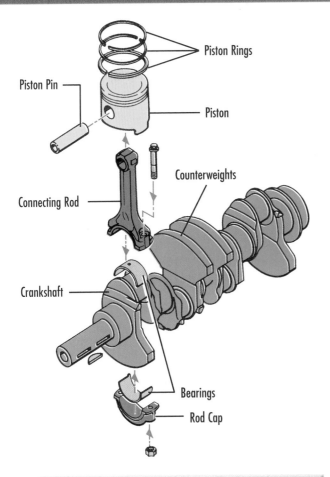

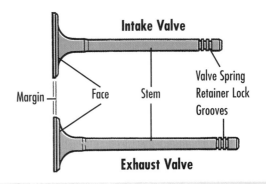

Fig. 1-10 An intake valve and an exhaust valve. The intake valve head diameter is larger than the exhaust valve head diameter. *In which part of the engine do these valves fit?* (DaimlerChrysler)

Fig. 1-9 Crankshaft with one piston and connecting rod assembly. This shows how the piston attaches through the connecting rod to the rod journal on the crankshaft. *What is the function of the crankshaft?*

Each cylinder has one or more *intake valves* and *exhaust valves.* **Fig. 1-10.** The intake valve controls the flow of the air/fuel mixture into the cylinder. The exhaust valve controls the flow of exhaust gas from the cylinder. The valves fit in the intake and exhaust ports of the cylinder head.

Most cylinders have two ports, or holes, in the combustion chamber area of the cylinder head. One port is the intake port; the other is the exhaust port. The air/fuel mixture enters the cylinder through the intake port. Burned gases leave the cylinder through the exhaust port. Some engines have multiple intake and exhaust ports and valves.

When a valve closes, it seals tightly against the valve seat. A **valve seat** is the surface against which the valve face comes in contact to provide a seal against leakage. In the closed position, the valve face and seat should form an air-tight seal. When a valve moves off its seat, the port is open. The air/fuel mixture or exhaust gas can then pass through the port.

The timing of valve opening and closing will vary with engine design. The intake valve opens before the intake stroke begins and closes after it ends. The exhaust valve opens before the exhaust stroke begins and closes after it ends. This is called *valve overlap.* Valve overlap improves engine "breathing," or the flow of air/fuel mixture and exhaust gases into and out of the cylinders.

SECTION 1 KNOWLEDGE CHECK

❶ What are the two major types of piston engines? What are the main differences between them?

❷ Name the two types of piston rings.

❸ Which parts of the engine convert reciprocal motion to rotary motion? How?

❹ What does the connecting rod connect in the engine?

❺ Where do the intake and exhaust valves fit in an engine?

Section 2

Engine Operation

An engine converts energy, in the form of a fuel, to motion. The fuel is burned and converted into heat. The heat develops pressure, which applies force to the engine's pistons. The pistons transfer this force, as reciprocating motion, to the engine's crankshaft. The crankshaft converts the reciprocating motion to rotary motion. The rotary motion is transferred through the *drive train* to provide motion to the vehicle's drive wheels.

Piston Action

The actions of the piston are divided into four strokes. A piston stroke is the movement of the piston from TDC to BDC or from BDC to TDC. The complete power cycle requires four piston strokes. **Fig. 1-11.**

- Intake.
- Compression.
- Power.
- Exhaust.

TECH TIP Compression Ratio. Carbon deposits inside the cylinder can increase the compression ratio. A cylinder compression test can show abnormal readings. When the compression ratio is too high, the air/fuel mixture may self-ignite.

This makes the engine a *four-stroke*, or *four-cycle*, *engine*. One complete four-stroke cycle requires two complete revolutions (720°) of the crankshaft.

Intake Stroke During the intake stroke, the piston moves down in the cylinder. The intake valve is open. The downward movement of the piston creates a partial vacuum in the cylinder. Atmospheric pressure forces the air/fuel mixture into the cylinder to fill the vacuum. As the piston moves from TDC to BDC, the crankshaft rotates 180°, or one-half turn.

Compression Stroke After the piston moves past BDC on the intake stroke, the *compression stroke* begins. The intake and exhaust valves are closed. As it moves up, the piston compresses the air/fuel mixture in the space between the top of the piston and the cylinder head. This space is the combustion chamber. The piston compresses the air/fuel mixture to about one-eighth of its original volume.

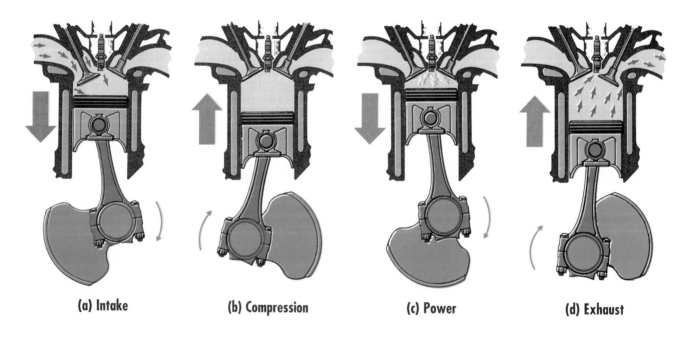

(a) Intake   (b) Compression   (c) Power   (d) Exhaust

Fig. 1-11 The four piston strokes. **(a)** Intake stroke, **(b)** Compression stroke, **(c)** Power stroke, **(d)** Exhaust stroke.

Calculating Engine Displacement

Someone may refer to a 350-cubic-inch engine (350 in³) or a 5.7-liter engine. They are using the engine's displacement to describe it. *Displacement* is the volume swept out when the piston moves from one end of the cylinder to the other.

You can calculate the displacement of an engine if you know the piston stroke length, the cylinder bore diameter, and the number of cylinders.

A Ford Mustang has an eight-cylinder 4.6-liter engine with a cylinder bore diameter of 3.55″ and a piston stroke length of 3.54″. You want to know the engine displacement in cubic inches. You can find this in two ways.

> ## Mathematics Toolbox
>
> $\pi = 3.1416$
>
> Radius of a circle: $r = \dfrac{\text{diameter}}{2} = \dfrac{d}{2}$
>
> Area of a circle: $A = \pi r^2 = \pi \left(\dfrac{d}{2}\right)^2 = \pi \dfrac{d^2}{4}$
>
> Volume of a cylinder: $V = \pi r^2 l$
>
> 1 in = 2.54 cm
> 1 in³ = (2.54 cm)³ = 16.39 cm³
> 1 liter = 1000 cm³

Method One

You can calculate the volume of each cylinder in cubic inches (in³) and multiply by the number of cylinders.

You first calculate the volume of one cylinder. Remember that:

$$r = \frac{d}{2} = \frac{3.55 \text{ in}}{2} \text{ and } r^2 = \frac{(3.55 \text{ in})^2}{4}$$

$$V = \pi r^2 l = 3.1416 \times \left[\frac{(3.55)^2}{4}\right] \times 3.54$$

$$V = 35.0389323 \text{ in}^3 \cong 35.04 \text{ in}^3$$

Then find the total volume of all eight cylinders:

Total $V = 35.04 \text{ in}^3 \times 8 = 280.32 \text{ in}^3$

The manufacturer's specs actually list the engine displacement as 281 in³. Your answer may differ from the manufacturer's value due to rounding off.

Method Two

You can convert 4.6 liters to in³.

The engine displacement is 4.6 liters, which is 4600 cm³.

1 in³ = 16.39 cm³, so 1 cm³ = $\left(\dfrac{1}{16.39}\right)$ in³.

The formula for total volume is:

$$V = 4600 \times \left(\frac{1}{16.39}\right) \text{ in}^3 = 280.66 \text{ in}^3$$

This figure is rounded off to 281 in³.

> ### Apply it!
>
> Meets NATEF Mathematics Standards for measuring volume and using both English and metric systems.
>
> A Chevrolet V-8 engine has a cylinder bore diameter of 4″ and a piston stroke length of 3.25″. Calculate the engine displacement in both in³ and liters.

Compression ratio is the volume in the cylinder with the piston at BDC divided by the volume in the cylinder with the piston at TDC. **Fig. 1-12.** It is the measure of how much the air/fuel mixture is compressed during the compression stroke. For example, if the mixture is compressed to one-eighth of its original volume, the compression ratio is 8 to 1 (written 8:1). As the piston moves from BDC to TDC, the crankshaft rotates 180°.

Compression ratios above 9.5:1 may create enough heat to self-ignite the air/fuel mixture without the aid of a spark. In a spark-ignition engine, such *detonation* can damage the pistons, piston rings, spark plugs, and valves.

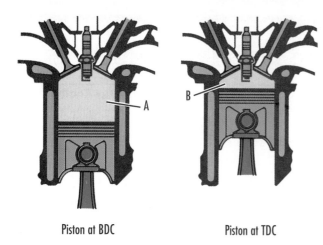

Piston at BDC   Piston at TDC

Fig. 1-12 Compression ratio is the volume in the cylinder with the piston at BDC divided by the volume in the cylinder with the piston at TDC, or (A) divided by (B). Note that (A) is the original volume. (B) is the compressed volume. *If the volume is compressed to one-eighth of its original volume, what is the compression ratio?*

Power Stroke As the piston nears TDC at the end of the compression stroke, both the intake and exhaust valves are closed. An electric spark jumps the gap at the spark plug. As the mixture burns, high temperatures and pressures are created in the combustion chamber. The force of the expanding combustion gases pushes down on the top of the piston. The connecting rod transmits this force to the crankshaft, which rotates another 180° from TDC to BDC.

Exhaust Stroke As the piston approaches BDC on the power stroke, the exhaust valve opens. After the connecting rod journal passes through BDC, the piston moves up, forcing the burned gases out through the open exhaust port. As the piston moves from BDC to TDC, the crankshaft rotates 180°.

Valve Action

The **valve train** is a series of parts that open and close the valves by transferring cam-lobe movement to the valves. One four-stroke cycle requires two revolutions of the crankshaft. This requires only one revolution of the camshaft. The crankshaft drives the camshaft through gears, sprockets, and a chain or through sprockets and a toothed timing belt. The sprocket driving the camshaft has twice as many teeth as the crankshaft sprocket.

Valve action starts at the camshaft. The **camshaft** is a shaft having a series of cams for operating the valve mechanisms. **Fig. 1-13.** Each *cam* is a round collar with a high spot, or lobe. Most camshafts have a cam lobe for each valve in the engine.

In an *overhead-camshaft engine*, the camshaft mounts in the cylinder head. **Fig. 1-14.** One end of a *rocker arm* contacts the tip of each valve stem. The other end of the rocker arm contacts the valve lifter. Valve lifters may be either solid metal or hydraulically operated. In either case, the lifter is used to maintain the desired clearance between the rocker arm and the valve stem. When the rotating camshaft moves a cam lobe in contact with a lifter, the rocker arm pivots to push the valve open. As the camshaft continues to rotate, the lobe moves away from the lifter. Spring tension now closes the valve.

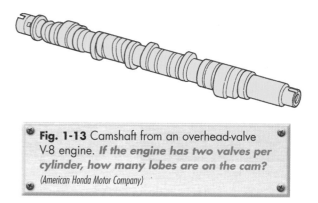

Fig. 1-13 Camshaft from an overhead-valve V-8 engine. *If the engine has two valves per cylinder, how many lobes are on the cam?* (American Honda Motor Company)

Power Flow

A single-cylinder four-cycle engine has one power stroke for every two rotations of its crankshaft. During the other three strokes, exhaust, intake, and compression, the piston does not deliver power. The single cylinder engine produces power only one-fourth of its running time. The crankshaft increases its rotational speed on the power stroke. It loses speed on the non-power strokes.

In general, an engine with multiple cylinders runs more smoothly. Automotive engines have four or more cylinders to provide a more even and smooth power flow. A complete engine cycle requires 2 complete rotations of the crankshaft, 720°. In a four-cylinder engine, a power stroke occurs every 180° of crankshaft rotation. A six-cylinder engine provides a power stroke

Explaining Things to Your Customer

The section you are reading is titled *Engine Operation*. This section uses technical language to provide information on the internal combustion engine.

During your years of servicing vehicles, you may be asked to share your knowledge with customers. A customer may ask, "How does the engine work?" You may wonder why you need to explain this to your customer. The reason is simple. Answering this question is one way to maintain good customer relations.

Because you are an automotive technician, you may think first in terms of technical processes. However, to make clear what you know, you will have to use terms that your customer can understand. Focus on clearly explaining the four piston strokes. This will help you to begin your explanation of how an internal combustion engine works.

Apply it!

Meets NATEF Communications Standards for adapting and using appropriate communication strategies and styles.

❶ Fold a sheet of notebook paper in half lengthwise. Head the left half of the sheet "Technical." Head the right half "Customer."

❷ In your own words, list on the "Technical" side the steps that occur in each piston stroke.

❸ On the "Customer" side, list the important steps that you believe the customer needs to know.

❹ Break into teams. Referring to the "Customer" side, explain to your team how the engine works. Ask them whether you have given them enough information. Ask also whether they think you have used words the customer can understand.

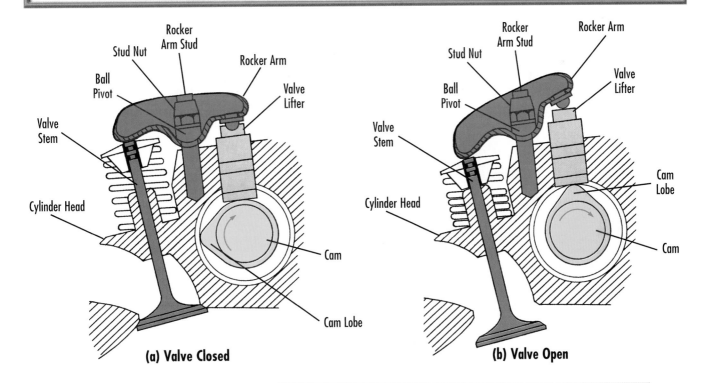

(a) Valve Closed

(b) Valve Open

Fig. 1-14 Valve operation on an overhead-camshaft engine. **(a)** The cam lobe is positioned away from the valve lifter and the valve is closed. **(b)** The cam lobe pushes the valve lifter and rocker arm up, pushing the valve down and opening the port. *How many times does the crankshaft turn for one turn of the camshaft?*

every 120 degrees. An eight-cylinder engine provides a power stroke every 90 degrees. With six or more cylinders, the power strokes follow each other very closely. Before the completion of a power stroke in one cylinder, a power stroke starts in another cylinder. This overlap results in a smoother-running engine.

Even when the power bursts overlap, the flow of power from the pistons to the crankshaft is not completely smooth. The crankshaft tends to speed up on each power stroke and slow down between power strokes. Without a way to store the energy from each power stroke, the engine still runs unevenly.

A *flywheel* is a mechanical device that is used to store energy. When energy sets a flywheel in motion, the weight or mass of the flywheel maintains that motion, storing the energy. A flywheel is mounted on one end of the crankshaft. The device on the other end of the crankshaft is known as the *damper*. **Fig. 1-15.**

The flywheel and damper store the energy produced by each power stroke. Between power strokes, the flywheel and damper transfer the stored energy back to the crankshaft. This action balances the energy between the power strokes. The flywheel and damper work together to produce a smoother-running engine.

Vehicles with automatic transmissions use a lightweight *drive plate* (or *flex plate*) and a fluid-filled *torque converter* in place of a flywheel. The weights offset any sudden change in the rotational speed of the crankshaft. They gradually absorb the power burst during the power stroke. They also resist slowdown of the crankshaft between the power strokes.

Both the flywheel and the drive plate have a ring gear mounted to them. When the vehicle is started, a *pinion gear* on the starter meshes with the *ring gear*. This turns the crankshaft to crank the engine.

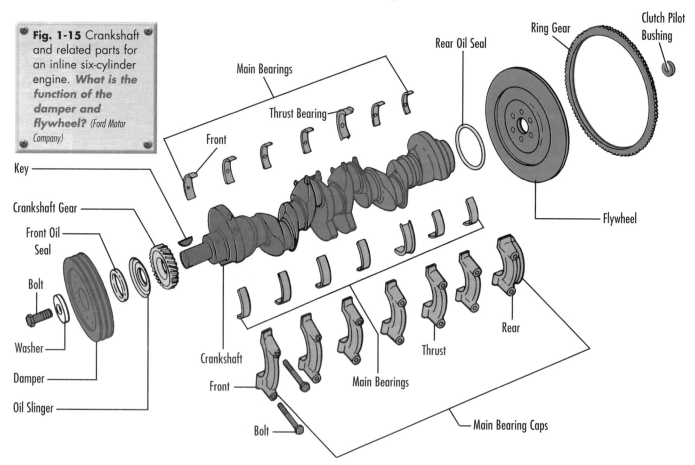

Fig. 1-15 Crankshaft and related parts for an inline six-cylinder engine. *What is the function of the damper and flywheel?* (Ford Motor Company)

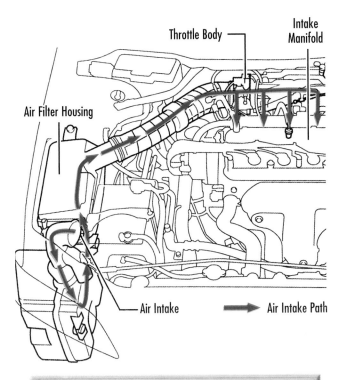

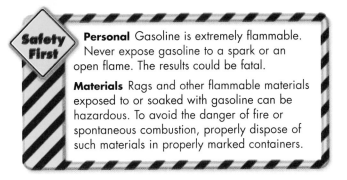

Fig. 1-16 Airflow through a remote-mounted air cleaner. The blue arrows show the air intake path. *Which part of an air induction system controls engine speed?* (American Honda Motor Company)

Fuel System

The fuel system delivers gasoline (or similar fuel) to the engine. **Fig. 1-17.** Air mixes with fuel to form a combustible air/fuel mixture, which burns quickly. This mixture fills each cylinder, where it is compressed and burned.

The fuel tank holds a supply of fuel. The fuel pump moves the fuel from the tank to the engine. On most fuel-injected vehicles, the fuel pump is located in the fuel tank.

Basic Engine Systems

A spark-ignition, fuel-injected engine requires six basic systems:
- Air induction system.
- Fuel system.
- Ignition system.
- Lubricating system.
- Cooling system.
- Exhaust system.

Air Induction System

The function of the *air induction system* is to direct clean, filtered air to the *intake manifold*. **Fig. 1-16.** Air entering the engine contains particles of dirt, which can damage the engine. An *air filter* in the air filter housing cleans the air. A *throttle body* contains a throttle blade that controls the engine speed by varying the amount of air entering the engine. The intake manifold carries air into the engine's cylinders. The intake manifold is located between the throttle body and the cylinder head.

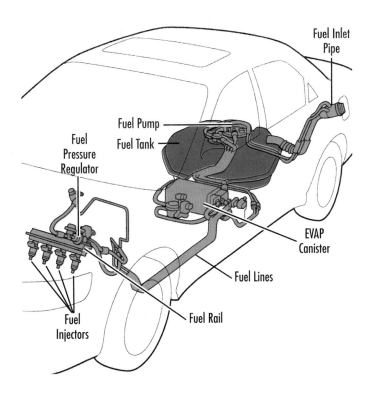

Fig. 1-17 The fuel system of a typical fuel-injected car. *Where is the fuel pump located on most fuel-injected vehicles?* (American Honda Motor Company)

Ignition System

The *ignition system* (except in diesel engines) uses one or more ignition coils to increase the low voltage of the battery to 20,000 volts or more. This voltage creates the spark that jumps the gap at the spark plug. The spark ignites the compressed air/fuel mixture, causing combustion to begin.

Lubricating System

An engine has many moving metal parts. If metal parts rub against each other, they wear quickly. To prevent this, engines have a *lubricating system* that coats moving parts with oil. **Fig. 1-18.** The oil film reduces the friction between the moving parts.

The lubricating system has an oil pan (sump) at the bottom of the engine that holds several quarts or liters of oil. An engine-driven *oil pump* circulates oil from the oil pan, through the engine, and back to the oil pan.

Cooling System

The *cooling system* keeps the engine at a safe and efficient operating temperature. Where there is combustion, there is heat. Burning the air/fuel mixture

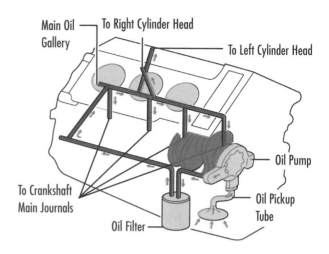

Fig. 1-18 Engine lubricating system, showing how the oil flows to the engine parts. *What circulates the oil through the engine?* (DaimlerChrysler)

raises the temperature inside the engine. Much of the heat leaves the engine through the exhaust gas.

The engine cooling system removes most of the remaining heat. **Fig. 1-19.** The engine has water jackets, which are open spaces surrounding the cylinders. Engine coolant, a mixture of water and *antifreeze*,

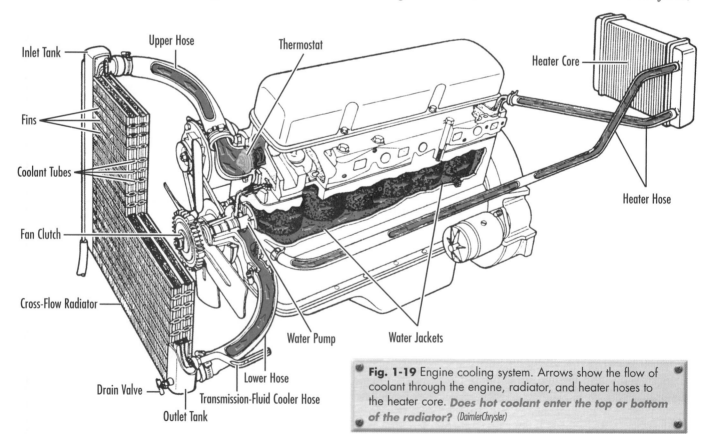

Fig. 1-19 Engine cooling system. Arrows show the flow of coolant through the engine, radiator, and heater hoses to the heater core. *Does hot coolant enter the top or bottom of the radiator?* (DaimlerChrysler)

moves through the water jackets. The coolant absorbs the heat and carries it to the *radiator*. Air passing through the radiator carries away the heat.

Exhaust System

The *exhaust system* carries exhaust gases away from the engine. **Fig. 1-20.** The exhaust gases pass from the cylinder, through the exhaust port, and into the *exhaust manifold*. The exhaust manifold connects to a *header pipe* that carries the exhaust gases to a *catalytic converter*.

The catalytic converter is located between the engine and the muffler. The converter changes harmful exhaust pollutants into harmless gases. The result is the exhaust gas leaves the catalytic converter with fewer pollutants.

Some vehicles use more than one catalytic converter. The number and location of the converters depends on engine used, the type of vehicle, and whether it has a single or dual exhaust system.

The converter may be equipped with heat shields to prevent unwanted heat from reaching the vehicle's floor pans. The floor pans may also be shielded and insulated from excessive heat.

The exhaust gases then enter the *muffler*. The muffler reduces the noise created by combustion of the air/fuel mixture. Some exhaust systems use one or more *resonators* to reduce exhaust noise. The *tail pipe* safely vents the exhaust gases away from the vehicle.

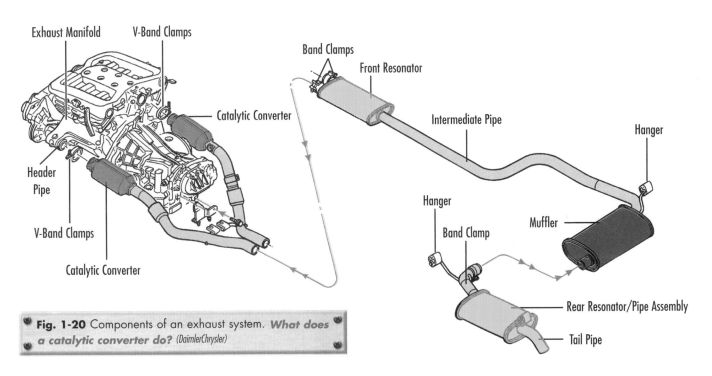

Fig. 1-20 Components of an exhaust system. *What does a catalytic converter do?* (DaimlerChrysler)

SECTION 2 KNOWLEDGE CHECK

❶ What are the four piston strokes? Explain the function of each piston stroke.

❷ What part of the engine causes the valves to open and close?

❸ Why do automotive engines have multiple cylinders?

❹ Which two engine parts smooth the power bursts from the pistons? Explain how this is done.

❺ Name the six basic engine systems in a spark-ignition, fuel-injected engine.

CHAPTER 1 REVIEW

Key Points

Satisfies the following NATEF Standards for Engine Performance: supporting knowledge for diagnosing engine mechanical problems.

- To be effective, the piston must slide smoothly in the engine cylinder.
- The four piston strokes of a spark-ignition engine take their names from their functions.
- The pistons, connecting rods, and crankshaft work together to convert heat energy to mechanical energy.
- Though similar in design, the intake and exhaust valves have different functions.
- Each piston stroke has a definite function in the power cycle.
- The engine has six basic supporting systems.
- The intake manifold carries air into the engine's cylinders.
- The engine cooling system controls engine operating temperature.

Review Questions

❶ Describe the function of the pistons.

❷ What would happen if the compression rings were missing or broken?

❸ Describe the purpose of the connecting rods and crankshaft.

❹ Describe how the intake and exhaust valves control the flow of the intake mixture and exhaust gases.

❺ Why is the compression ratio of a spark-ignition engine limited to about 9.5:1?

❻ Name the four piston strokes of a four-cycle engine.

❼ Explain the function of each piston stroke.

❽ (Critical Thinking) Why do you think diesel engines are not widely used in passenger vehicle applications?

❾ (Critical Thinking) What might happen if the manufacturer's recommended oil change intervals are always exceeded?

TECHNOLOGY FORECAST FOR AUTOMOTIVE EXCELLENCE

Powered by Hydrogen

The internal combustion engine has been the heart of the car for more than 100 years. But an electric motor could soon replace it.

Automakers want to make the change because electric motors produce fewer emissions than gasoline-powered piston engines. Electric motors are also more quiet. They might even require less maintenance.

Cars with electric motors are now being built. These motors need many improvements before they can replace the combustion engine. The driving range is too short. The batteries that power the electric motors are heavy and slow to recharge. There is another major drawback for consumers–they are expensive.

To provide the electric current these motors need, engineers are designing fuel cells. First used by NASA in the U.S. space program, fuel cells could easily power cars and trucks. A fuel cell creates electricity from hydrogen stored in the vehicle as either a gas or a liquid. Hydrogen also can be made from gasoline or methanol.

Once the hydrogen enters the fuel cell, it is split into protons and electrons. The electrons are used to run the electric motors. The protons pass through a thin membrane to combine with electrons and oxygen in the air. The result is water–a safe vehicle emission.

AUTOMOTIVE EXCELLENCE
TEST PREP

Answering the following practice questions will help you prepare for the ASE certification tests.

1. Two technicians are rebuilding an engine. Technician A says the clearance between the piston and the cylinder wall causes blowby. Technician B says piston movement requires a slight amount of clearance between the piston and the cylinder wall. Who is correct?
   ⓐ Technician A.
   ⓑ Technician B.
   ⓒ Both Technician A and Technician B.
   ⓓ Neither Technician A nor Technician B.

2. Technician A says some piston rings control blowby. Technician B says some piston rings control oil consumption. Who is correct?
   ⓐ Technician A.
   ⓑ Technician B.
   ⓒ Both Technician A and Technician B.
   ⓓ Neither Technician A nor Technician B.

3. The connecting rod performs which function in the engine?
   ⓐ Seals the piston to the cylinder wall.
   ⓑ Connects the intake valve to the camshaft.
   ⓒ Connects the camshaft to the crankshaft.
   ⓓ Connects the piston to the crankshaft.

4. The intake valve on most engines:
   ⓐ Has a larger head diameter than the exhaust valve.
   ⓑ Has a smaller head diameter than the exhaust valve.
   ⓒ Is closed on the intake stroke.
   ⓓ Is open on the exhaust stroke.

5. The exhaust valve on most engines:
   ⓐ Has a larger head diameter than the intake valve.
   ⓑ Has a smaller head diameter than the intake valve.
   ⓒ Is open on the intake stroke.
   ⓓ Is closed on the exhaust stroke.

6. Two technicians are diagnosing an engine that won't start. Technician A says the spark must occur when the piston nears TDC on the compression stroke. Technician B says both valves are closed on the compression stroke. Who is correct?
   ⓐ Technician A.
   ⓑ Technician B.
   ⓒ Both Technician A and Technician B.
   ⓓ Neither Technician A nor Technician B.

7. A piston stroke is piston movement from:
   ⓐ BDC to cylinder midpoint.
   ⓑ TDC to cylinder midpoint.
   ⓒ BDC to TDC or TDC to BDC.
   ⓓ Cylinder midpoint to cylinder top.

8. The engine produces power by the:
   ⓐ Rotation of the crankshaft.
   ⓑ Valve action.
   ⓒ Combustion pressure pushing on pistons.
   ⓓ Up-and-down movement of pistons.

9. Two technicians are discussing the exhaust system. Technician A says that the catalytic converter is located between the engine and the muffler. Technician B says that the converter is located between the muffler and the tailpipe.
   ⓐ Technician A.
   ⓑ Technician B.
   ⓒ Both Technician A and Technician B.
   ⓓ Neither Technician A nor Technician B.

10. Two technicians are trying to find the fuel pump. Technician A says the fuel pump may be located in the fuel tank. Technician B says some automotive engines don't need fuel pumps. Who is correct?
   ⓐ Technician A.
   ⓑ Technician B.
   ⓒ Both Technician A and Technician B.
   ⓓ Neither Technician A nor Technician B.

CHAPTER 2

Diagnosing Engine Problems

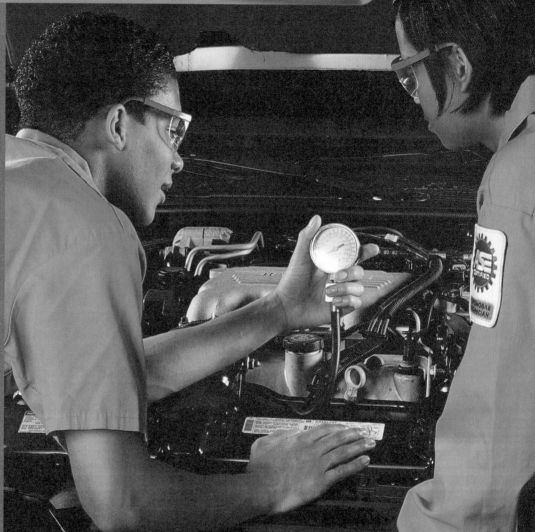

You'll Be Able To:

- ⊗ Form a diagnostic strategy to locate engine troubles.
- ⊗ Interpret the various colors of exhaust smoke.
- ⊗ Perform cylinder compression, cylinder leakage, and engine vacuum tests.
- ⊗ Diagnose engine noises.
- ⊗ Diagnose cooling system problems using various cooling system tests.

Terms to Know:

cooling system pressure test

cranking vacuum test

cylinder compression

cylinder leakage test

engine vacuum

service history

snap-throttle vacuum test

vacuum gauge

The Problem

The engine in John Tall Bear's vehicle is running rough. The roughness is most noticeable at idle. As the speed increases, the engine smoothes out.

You consult with Theresa, a fellow technician. You both agree that many things can be causing the problem, including low cylinder compression. However, you want to be sure before proceeding with any repairs. Using proper diagnostic practices and procedures will help isolate the problem.

Your Challenge

As the service technician, you need to find the answers to these questions:

1. Did the problem appear suddenly, or has it been slowly getting worse?
2. Has John's vehicle had any major engine service?
3. What test equipment might you use to isolate the problem?

Fig. 2-1 A vehicle's service history is a written record of its maintenance. *How can a service history provide valuable diagnostic information?* (Gary D. Landsman/The Stock Market)

Section 1

General Engine Diagnosis

Diagnosis is the detective work that answers the question, "What is wrong?"

An engine is a machine. As might any machine, it may develop mechanical problems. You may find the cause in the mechanical parts of the engine itself. The cause may also be in one or more of the engine support systems (air induction, fuel, ignition, lubricating, cooling, and exhaust).

Correctly identifying the cause of a problem is often the hardest part of the job. Make a visual inspection of the engine and its support systems. Check for fuel, oil, and coolant leaks. Check for loose or missing parts. If you do not find an obvious cause, form a diagnostic strategy.

Diagnostic Strategy

A diagnostic strategy is a planned, step-by-step procedure you can use to locate a problem. A diagnostic strategy has five basic rules:

- Know the system. Know the parts and how they work together. Know what happens if parts fail to work properly.
- Be aware of the common problems that occur most often.
- Determine the history of the complaint. Did the problem start suddenly or gradually? Does the problem occur with the engine warm or cold? Does it occur at all speeds?
- Look up the service history of the vehicle. The **service history** is a written service record for the vehicle. It is kept by the service facility. **Fig. 2-1.** How many miles are on the odometer? How old is the vehicle? What kind of treatment has it had? Has the vehicle been previously serviced for the same problem?
- Get as much information as possible from the driver. This may help direct your diagnostic strategy.

Diagnostic Tests

Worn engine parts cause:
- Engine noise.
- Loss of power
- Poor engine performance.
- Increased fuel and oil consumption.
- High exhaust emissions.

Worn engine parts may cause a loss of compression. Worn or broken pistons and rings, for example, cause increased blow-by and oil consumption. Damaged valves allow leakage between the valves and valve seats. A leaking *cylinder head gasket* lets compression and combustion gases leak from the cylinder.

Some engine problems can be diagnosed by checking the color and odor of the exhaust. An internal coolant leak into the combustion chamber produces heavy white smoke. Oil entering the combustion chamber past worn valve guides or piston rings produces blue smoke. Rich fuel mixtures may cause black exhaust smoke and a "rotten egg" odor.

Observing and listening will not always lead to a diagnosis. You may need to perform diagnostic tests.

Safety First

Personal Keep clothes and hands clear of rotating engine drive belts and pulleys. When using compressed air, always wear safety goggles. Serious injury can result when safety is not observed.

Materials Be sure the areas around exposed spark plug holes and the throttle body air intake are free of dirt, small materials, or loose parts. Small pieces can fall into the engine and cause severe engine damage when the engine cranks.

Cylinder Compression Test

Cylinder compression is the pressure developed in the cylinder as the engine cranks or runs. The average compression pressure during cranking is about 150 psi [1034 kPa]. Because compression raises the density and temperature of the air/fuel mixture, good compression is vital to proper combustion. When damaged or worn engine parts reduce compression, the engine loses power.

Cylinder compression tests measure compression in each cylinder as the engine cranks. **Fig. 2-2.** You may have to make two compression tests, a "dry test" and a "wet test."

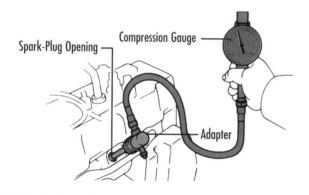

Spark-Plug Opening
Compression Gauge
Adapter

Fig. 2-2 In a compression test, the compression gauge is connected to the spark plug opening. *Why is compression so important to engine performance?*

A dry test is made by cranking the engine with a test gauge installed. Depending on results, it may be necessary to make the wet test. This is done after squirting a small amount of oil into the cylinder. The oil will help the piston rings seal against the cylinder walls.

TECH TIP Compression Readings. Compression pressure drops about 2 percent for each 1,000' [304.8 m] of altitude. When using the manufacturer's specifications, you must adjust the reading for altitude. Always know the approximate altitude of your service center. The local office of the National Weather Service has this information.

Dry Compression Test The *dry compression test* is always performed first. It is performed without adding oil to the cylinder.

To perform this test:

1. Make sure the engine cranks normally and is at normal operating temperature.

2. Disable the ignition or fuel system. Refer to the manufacturer's recommendation for the proper procedures.

3. Loosen the spark plugs about one turn. Use compressed air to blow dirt away from the plug wells.

4. Carefully remove the spark plugs. Be careful not to damage the cylinder head threads.

5. Thread the compression tester into the spark plug hole of the cylinder being tested.

6. Block the throttle plate open for better breathing.

7. Crank the engine at least four compression strokes. Crank the same number of strokes for each cylinder.

8. Record the test results for each cylinder.

Check the vehicle service manual for the specifications for normal compression pressure. Low readings in any cylinder may indicate leaking valves, worn or broken rings, or a defective cylinder head gasket. Low but equal readings in two adjacent cylinders may indicate a blown head gasket.

If the dry compression is within specs, remove the throttle block, and reinstall the spark plugs and the cables. Restore any circuit that was disabled to prevent starting.

If the dry compression test results in low readings, pressure is leaking from the cylinder. A wet compression test may provide additional information.

Wet Compression Test If the compression pressure is low on one or more cylinders, a wet compression test should be made.

Safety First **Materials** Never use the wet compression test on a diesel engine. Oil may fill the combustion chamber and cause a *hydrostatic lock*. This can crack the piston and bend the connecting rod. The oil might also ignite and damage the compression tester.

To perform a wet compression test:

Make sure the engine is still prepared for compression testing as noted in "Dry Compression Test."

1. Squirt about one tablespoon [15 cc] of engine oil through the spark plug hole.

2. Crank the engine for five seconds to distribute the oil to the cylinder walls and rings.

3. Thread the compression tester into the spark plug opening.

4. Recheck the compression as in the dry test and record the results.

5. Remove the throttle block when finished. Replace the plugs and cables. Restore any circuit that was disabled to prevent starting.

If the wet compression test gives a near normal reading, compression is probably leaking past worn piston rings. Other possible causes include worn pistons or scratched cylinder walls.

TECH TIP **Loose Carbon.** Carbon particles loosened by removing the spark plugs can lodge under a valve and cause a low compression reading. You may need to repeat the test to ensure an accurate reading.

Reinstall the spark plugs and start the engine. Allow the engine to reach normal operating temperature. Repeat the test.

If the compression remains low on the wet test, there may be leakage past the valves or head gasket. Possible causes of valve leakage include:
- Broken valve springs.
- Carbon deposits on the valve face or seat.
- Incorrect valve adjustment.
- Loose valve seat inserts.
- Burned, sticking, or bent valves.
- Incorrect valve timing.

Improper valve adjustment holds the valves slightly off their seats. Incorrect timing of the camshaft causes the valves to open too early or too late in the combustion cycle. "Engine Repair" in Volume 2 of *Automotive Excellence* discusses camshaft timing.

Common causes of low compression include wear or damage to the rings, pistons, cylinder walls, valves, or head gasket. Remove the cylinder head(s) to repair the problem, except when the cause is an improper valve adjustment or incorrect camshaft timing.

AUTOMOTIVE MATHEMATICS EXCELLENCE

Measuring Compression

A customer complains that her vehicle's engine seems to have lost power. You suspect that one or more of the four cylinders may have worn or broken rings or leaking valves. You decide to perform a compression test. You will compare the compression in each cylinder with the manufacturer's specifications in the service manual.

According to the manufacturer, the high-to-low pressure readings in the cylinders should not vary by more than 25 percent of the highest value. This means that the lowest reading must be at least 75 percent of the highest reading.

You perform a dry compression test on the four-cylinder engine. The readings are:

145 psi, 120 psi, 145 psi, and 156 psi

Are these readings within specifications? Note: To convert these readings to metric measurements, remember that 1 psi = 6.895 kPa.

The highest reading is 156 psi; 25 percent of 156 psi is 39 psi.

$$156 \text{ psi} - 39 \text{ psi} = 117 \text{ psi}$$

The lowest reading that is within specifications is 117 psi.

Since the lowest reading is 120 psi, and 120 psi is greater than 117 psi, all the readings are within specifications.

You can also find the lowest reading within acceptable range by calculating 75 percent of 156 psi, which would be 117 psi.

Apply it!

Meets NATEF Mathematics Standards for percentages and tolerances.

If the compression readings in the cylinders had been 145 psi, 142 psi, 123 psi, and 105 psi, would the readings have been within specifications? Explain your answer.

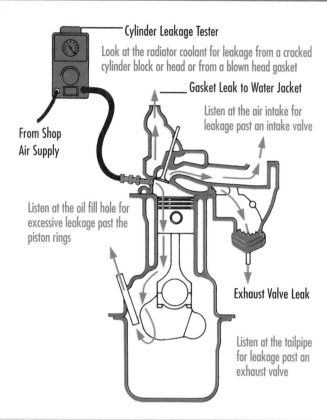

Cylinder Leakage Tester

Look at the radiator coolant for leakage from a cracked cylinder block or head or from a blown head gasket

Gasket Leak to Water Jacket

Listen at the air intake for leakage past an intake valve

From Shop Air Supply

Listen at the oil fill hole for excessive leakage past the piston rings

Exhaust Valve Leak

Listen at the tailpipe for leakage past an exhaust valve

Fig. 2-3 With the piston at TDC, the cylinder leakage test applies pressure to the cylinder through the spark plug hole. *What does a cylinder leakage test indicate that a compression test does not?* (Sun Electric Corporation)

Cylinder Leakage Test

A **cylinder leakage test** is another test that checks a cylinder's ability to hold pressure. Air pressure is applied to the cylinder with the piston at top dead center (TDC) on the compression stroke. In this position, the intake and exhaust valves are closed. **Fig. 2-3.** If the engine is in good condition, the percent of cylinder leakage will be less than the maximum specified.

To perform a cylinder leakage test:

1. Remove all spark plugs.

2. Remove the air cleaner.

3. Remove the oil-fill cap and dipstick.

4. Remove the radiator cap and fill the radiator to the proper level.

5. Block the throttle wide open.

6. Thread the leakage test adapter with the whistle into the spark plug hole of the number-one cylinder.

7. Crank the engine until the whistle sounds. This means that the number-one piston is moving up on the compression stroke.

8. Continue cranking the engine until the piston is at TDC. Consult the manufacturer's service manual for the proper procedure.

9. Disconnect the whistle from the adapter hose and connect the leak tester.

10. Calibrate the leakage tester as instructed by the manufacturer.

11. Apply the specified air pressure to the cylinder. Record the gauge reading. Observe the percentage of leakage from the cylinder. **Fig.** 2-4. Specifications vary, but a leakage of more than 20 percent is normally over acceptable limits.

12. Listen at the air intake, tailpipe, and oil-fill openings. Air leaking from the air intake indicates a leaking intake valve. Air leaking from the tailpipe indicates a leaking exhaust valve. Excessive air leaking from the oil-fill opening may indicate worn piston rings. If air escapes from the next spark plug hole, the engine has a blown head gasket between the cylinders. If air bubbles up through the radiator, the engine has a blown head gasket, a cracked cylinder block, or a cracked cylinder head.

13. Check each cylinder in the same way.

14. Remove the throttle block. Reinstall spark plugs and cables. Restore any circuit that was disabled to prevent starting.

Fig. 2-4 Performing a cylinder leakage test. *How can you determine if the reading is within acceptable limits?* (Jack Holtel)

Table 2-A   VACUUM GAUGE READINGS

Gauge Display	Reading	Diagnosis	Gauge Display	Reading	Diagnosis
	Average and steady at 15"–22".	Everything is normal.		Needle drops to low reading, returns to normal, drops back, and repeats this pattern at a regular interval.	Burned or leaking valve.
	Extremely low reading; needle holds steady.	Air leak at the intake manifold or throttle body; incorrect timing.		Needle drops to zero as engine rpm is increased.	Restricted exhaust system.
	Needle fluctuates between high and low reading.	Blown head gasket between two adjacent cylinders. Check with compression test.		Needle holds steady at 12"–16", drops to 0 and back to about 21" as throttle is opened and released.	Late ignition or valve timing. Leaking piston rings. Check with compression test.
	Needle fluctuates very slowly, ranging 4" or 5".	Incorrect air/fuel mixture.			
	Needle fluctuates rapidly at idle, steadies as rpm is increased.	Worn valve guides.			

Engine Vacuum

Engine vacuum can be a good indicator of how well an engine is performing. **Engine vacuum** is the low-pressure condition created as the crankshaft turns, pulling the rod and piston down in the cylinder. The **vacuum gauge** is a device that measures intake manifold vacuum in inches of mercury (Hg). **Fig. 2-5.** There will be some intake manifold vacuum anytime the engine is cranking or running. Note this is not always true if the engine is supercharged or turbocharged.

Engine Vacuum Test Engine speed and load, the position of the throttle plate, and possible engine defects affect engine vacuum. The *engine vacuum test* (also referred to as the *manifold absolute pressure test*) checks vacuum readings through a range of engine operating conditions.

To perform an engine vacuum test:

1. Make sure all vacuum hoses are properly connected and not leaking.

2. Connect the vacuum gauge to a source of intake manifold vacuum. Check the manufacturer's service manuals for proper connection points.

3. Start the engine and run it until it reaches normal operating temperature.

4. Record the readings at idle and at several other engine speeds.

The readings on the vacuum gauge provide valuable information. **Table 2-A.** Keep in mind that engine vacuum decreases about 1" [25 mm] of mercury (Hg) per 1,000' [304.8 m] increase in altitude.

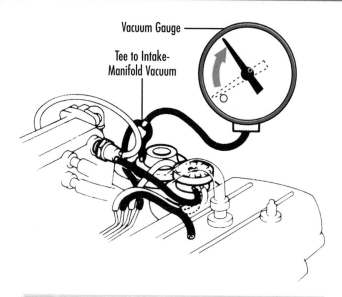

Vacuum Gauge

Tee to Intake-
Manifold Vacuum

Fig. 2-5 A vacuum gauge is a common diagnostic tool. *What unit of measure does a vacuum gauge display?*

You may need to adjust your readings. The following items describe common gauge readings.

- A high, steady reading between 15″ and 22″ [381 mm and 559 mm] of mercury means normal engine condition.
- A low, steady reading may mean late ignition timing, late valve timing, cylinder leakage, or an air leak at the intake manifold or throttle body.
- A reading that drops back to zero as engine speed steadily increases may mean a restricted catalytic converter, muffler, or exhaust pipe.
- A regular drop in the reading may mean that a single valve is leaking or stuck open.
- An irregular drop in the reading may mean a valve is sticking open part of the time.
- A slow drop and increase in the reading may mean an incorrect air/fuel mixture.

Snap-Throttle Vacuum Test A **snap-throttle vacuum test** is a test that shows the condition of the pistons and piston rings.

To perform a snap throttle vacuum test:

1. Perform steps 1–3 as noted in "Engine Vacuum Test."
2. Increase the engine speed and then quickly release the throttle.
3. Record the reading as the engine speed slows to idle.

A reading of 23″ to 25″ [584 mm to 635 mm] of mercury indicates normal operation.

Cranking Vacuum Test The **cranking vacuum test** is a test that measures engine vacuum while the engine is cranking. To perform a cranking vacuum test:

1. Connect the vacuum gauge to a source of intake manifold vacuum. Check the vehicle service manual for proper connection points.
2. With the engine at normal operating temperature, close the throttle and plug the positive crankcase ventilation (PCV) hose.
3. Disable the ignition or fuel system and crank the engine.

An even vacuum reading at normal cranking speed means the valve timing is correct and the engine is mechanically sound. An uneven reading means there is an air leak into one or more cylinders. Burned valves or poor valve seating may cause leakage.

A low or zero reading may indicate that the camshaft timing is incorrect or that the timing belt, gears, or chain has failed.

SECTION 1 KNOWLEDGE CHECK

❶ Why is a visual inspection the first step of the diagnostic process?

❷ Where do you find the service history of a vehicle?

❸ What does an irregular drop in the reading on the vacuum mean?

❹ What is indicated if the vacuum gauge reading drops back to zero as the engine speed steadily increases?

❺ If a dry compression test gives a low reading and a wet test gives a normal reading, what components could be at fault?

Section 2

Diagnosing Engine Noises

Some engine noises are normal. Others may point to serious problems that require prompt attention. Excessive engine noise should not be overlooked as an indication of a serious engine problem.

Using an Automotive Stethoscope

Engine noises can travel throughout the engine. They can be hard to isolate and detect. An *automotive stethoscope* helps isolate and amplify noises in a running engine. **Fig. 2-6.**

Valve Train Noises

Valve train noise is a regular clicking noise that occurs at one-half engine speed. When an engine is cold, some valve train noise is normal. Engines with mechanical *valve lash adjusters* usually produce more noise than do engines with *hydraulic lash adjusters.*

Too much clearance between the *rocker arm* (or *cam follower*) and the *valve stem* causes excess valve train noise. Other causes of noise include worn valve guides, sticking valves, and sticking or defective hydraulic adjusters. If excess valve train noise occurs when the engine is at normal operating temperature, the engine may need service.

More detailed information on valve train diagnosis and repair is found in Volume 2 of *Automotive Excellence.*

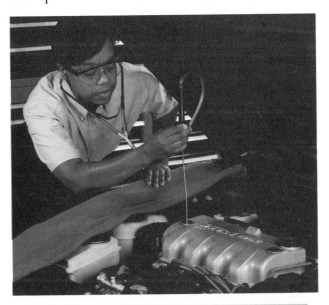

Fig. 2-6 To locate noises, technicians use automotive stethoscopes. Most stethoscopes are mechanical units. Some use electronics to amplify and clarify noises. *What makes the exact location of a noise so hard to detect?* (Jack Holtel)

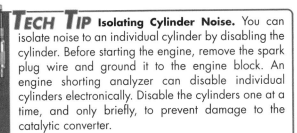

TECH TIP **Isolating Cylinder Noise.** You can isolate noise to an individual cylinder by disabling the cylinder. Before starting the engine, remove the spark plug wire and ground it to the engine block. An engine shorting analyzer can disable individual cylinders electronically. Disable the cylinders one at a time, and only briefly, to prevent damage to the catalytic converter.

Engine Block Noises

Engine block noises are produced by components within the engine itself. The heavy casting of the block may make some noises difficult to locate or diagnose.

Connecting Rod Noise Connecting rod bearing noise is usually a light knocking or pounding sound. It is most noticeable when engine speed is constant (not speeding up or slowing down). Extreme wear or poor lubrication causes rod bearing noise. This type of wear is also indicated by low oil pressure and high oil consumption.

Piston Pin Noise A loose piston pin creates a noise similar to valve train noise. However, piston pin noise has a unique, metallic double knocking sound.

Piston Slap *Piston slap* produces a muffled, hollow, knocking noise. When the engine is cold, piston slap is usually louder. Although a slight amount of piston slap is normal in many engines, loud, continuous piston slap means that the engine needs service. Overheating, poor lubrication, piston wear, and cylinder wear cause piston slap. Increased oil consumption may also occur.

Crankshaft Knock Worn *main bearings* cause a rumbling or knocking noise that is loudest when the engine is cold or under heavy load. Worn main bearings may cause low oil pressure and increased oil consumption.

Other Noises Some noises can sound like valve train or engine noises. A leaking exhaust manifold can sound like valve train noise. A loose flywheel can sound like worn main or connecting rod bearings. A loose timing belt or a dented flywheel cover or oil pan that interferes with moving parts can sound like crankshaft noises.

Before disassembling an engine, always verify that the source of the noise is not external. An effective diagnostic strategy can help ensure an accurate diagnosis.

Using the Scientific Method

Mechanical problems are often hard to diagnose. You can not afford to waste time when solving a problem. Time is money for you and for the customer. You need to use a systematic approach to finding the solution.

As a technician, you should be familiar with the scientific method. The *scientific method* is an approach to a problem in which researchers make an observation, collect information, form a hypothesis, perform experiments, and analyze the results. Think of the scientific method as a logical approach to problem solving. Using it can help you deal with a complicated problem in an efficient way. You can use these four simple steps to apply the scientific method.

1. **Observe and Gather Information** Observation means finding out as much information as you can about the problem. Part of this step is talking to the customer to find out what has been observed. Looking and listening for clues to the trouble are also ways to add to your observation.

2. **Hypothesize** You now use the information you have gathered and your knowledge and experience. You make a hypothesis, or educated guess, about what the problem is.

3. **Test** Here, you try different methods to see if your hypothesis is correct. Now is when you use the gauges and electronic equipment to help you prove or disprove your hypothesis.

4. **Analyze** The results of your testing give you another set of information. You use this new information to determine if you have found the problem. If you have not, you begin the process again with step one. However, having used the method once, you have additional information to add to step one to help you solve your customer's problem. You will have eliminated one possibility. Now you will be able to test other possibilities using what you have already learned.

Remember, use **OHTA: O**bserve, **H**ypothesize, **T**est, and **A**nalyze.

Apply it!

Meets NATEF Communications and Science Standards for problem solving using the scientific method.

1. Reread "The Problem" on the first page of this chapter.
2. Apply **OHTA**.
   - Write down your observations.
   - Write down your hypothesis.
   - Write down the tests you will use to prove your hypothesis.
   - Using the test results, analyze your information to see if your hypothesis is correct.
   - If your hypothesis was not correct, use what you learned to make a new hypothesis to test.

SECTION 2 KNOWLEDGE CHECK

1. What piece of test equipment helps to isolate engine noises?
2. Why do valve train noises occur at one-half engine speed?
3. What are the causes of excessive valve train noise?
4. What are two indicators of rod bearing wear?
5. How can you isolate engine noise to an individual cylinder?

Section 3

Diagnosing the Cooling System

Pressurizing a cooling system raises the boiling point of the coolant. The higher boiling point improves the coolant's ability to transfer heat and cool the engine. A high pressure also keeps the coolant from forming vapor bubbles at "hot spots" in the engine. Cooling system pressure usually runs between 14 and 18 psi [97 and 124 kPa].

Frequent need to add coolant indicates a cooling system leak. A compression test or a cylinder leakage test can diagnose internal engine cooling system leaks.

Cooling system tests limit pressure to no more than 18 psi [124 kPa]. Higher pressure can damage the radiator, heater core, and coolant hoses.

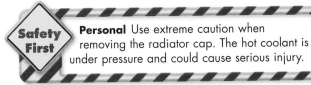

Safety First | **Personal** Use extreme caution when removing the radiator cap. The hot coolant is under pressure and could cause serious injury.

Cooling System Pressure Test

A **cooling system pressure test** is a test that diagnoses external cooling system leaks. **Fig. 2-7.**

1. Make sure that the coolant is at the correct level.

2. Attach the pressure tester to the radiator neck, using the proper adapter.

3. Adjust the pressure to the specified value.

4. Wait a few minutes before looking for leaks around the water pump shaft, coolant hoses, and *radiator core.*

5. Check inside the passenger compartment for leaks at the heater core.

If you do not see any external leaks, a cylinder leakage test may indicate a coolant leak in the cylinder head, head gasket, or engine block.

By using an adapter that connects the cap to the pressure tester, you can test the radiator pressure cap. The cap must hold the specified system pressure without leaking. If the cap leaks, the coolant can escape through the *radiator overflow tube.*

The radiator pressure cap has a pressure-relief valve to prevent excessive pressure. When the pressure goes too high, it opens the valve. Excess

Fig. 2-7 The cooling system pressure tester checks the cooling system for leaks. *What is the maximum pressure used for this test?* (Terry Wild Studio)

pressure and coolant then escape into the coolant recovery tank, also called the expansion tank.

To check a coolant hose, squeeze it. It should not collapse easily. Replace any hose that is soft, very hard, damaged, or swollen.

Cooling System Temperature Test

Engine temperature is controlled by engine coolant. The *cooling system thermostat* regulates the flow of coolant through the engine and cooling system. A properly operating thermostat keeps the engine within the correct operating temperature range. There are several different methods for testing thermostats. Refer to the service manual for details. In many cases it may be more practical to replace the thermostat if its operation is in question.

Modern engines operate at a coolant temperature of about 190°F [88°C]. If the thermostat maintains too low a temperature, efficient combustion cannot occur. There may be a build-up of moisture in the engine. If the thermostat maintains too high a temperature, the engine may overheat. Overheating can warp cylinder heads, cause blown head gaskets, score pistons, and cause spark knock (detonation).

Analyzing Fluids

We use the five senses of smell, taste, touch, sight, and hearing to pick up messages about what is going on around us. For example, our sense of hearing wakes us to our clock radio. Our sense of sight allows us to find our way around the house or to school. Touch may help us estimate temperature when we step out of our house. We use hearing to keep in touch with our friends when we talk with them. Smell may give us a clue as to what is for lunch. Our sense of taste will let us know if lunch was worth the wait.

Automotive technicians also use their senses. In this chapter you learn about diagnostic strategy. When automotive technicians put together a *diagnostic strategy* they rely on their senses. You learn how observing the color of exhaust can help you diagnose some engine problems. This chapter also explains how listening to engine noise can help a technician detect worn engine parts such as bearings and rocker arms. You are told that some obvious engine problems can be detected by fluid leaks. Let's see how you might do this.

Safety First

Personal In the past, some technicians detected coolant leaks by taste. Do not taste fluids! The main ingredient in coolant is ethylene glycol, which is poisonous.

Materials Rags saturated with any flammable fluid must be disposed of properly.

Fluid leaks are often one of the first clues that a technician has in identifying a possible engine problem. Such leaks can occur before the problem becomes serious enough to cause major damage to the engine. With just a little practice you can "fine tune" your senses to help you identify fluid leaks from an engine. This can be your first step in a diagnostic strategy.

You may use a white shop towel to wipe up fluids dripping from the engine or from another location under the vehicle.

Apply it!

Identifying Auto Fluid Leaks

Meets NATEF Science Standards for analyzing fluids, analyzing waste management, and applying safety procedures.

To help train your senses to identify auto fluid leaks, try this.

1. Look around the garage at home, the auto shop at school, or an auto parts store. Pick up samples of the following automotive fluids:
   - Brake fluid.
   - Coolant (antifreeze).
   - Engine oil (new and used).
   - Transmission fluid.
   - Power steering fluid.

2. Get a white paper towel or shop rag. Place a small spot (about the size of a quarter) of each of the fluids you have collected on the towel or rag. Use your senses to learn as much as you can about the properties of each spot. Important clues to look for are:
   - What color is it?
   - Does it have a particular odor?
   - How does it feel? Is it slippery or sticky, for example?

3. Once you have practiced using your senses on these automotive fluids, take one of your white rags out to a parking lot. Look for fluids on the ground. Dip your rag or towel into one of them. See if you can identify it.

4. Determine how to discard the rags and clean up these fluids in an acceptable manner.

A *pyrometer* is an instrument that checks temperature. **Fig. 2-8.** Pyrometers are available in two types, contact and non-contact. Contact types use a probe, which must touch the engine block or thermostat housing to check temperature. Non-contact pyrometers, or *infrared pyrometers*, check engine temperature by measuring heat radiated from the thermostat housing or engine block.

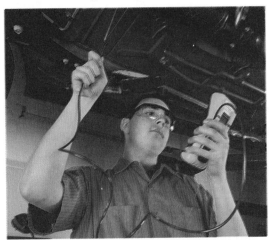

Fig. 2-8 The pyrometer checks engine temperature. *What is another name for a non-contact pyrometer?* (Jack Holtel)

Coolant Condition Check Coolant freezing point can be checked with a *float hydrometer* or a *ball hydrometer.*

Place the end of the float hydrometer's rubber tube into the coolant. Draw coolant up into the hydrometer. **Fig. 2-9 (a).** The position of the float in the coolant is determined by the concentration of antifreeze present. There is a temperature scale on the hydrometer. It shows the temperature at which the coolant will boil. It also shows the temperature at which the coolant will freeze. **Fig. 2-9 (b).**

A ball hydrometer has four or five balls in a plastic tube. Coolant is drawn into the tube. A stronger antifreeze concentration causes more balls to float.

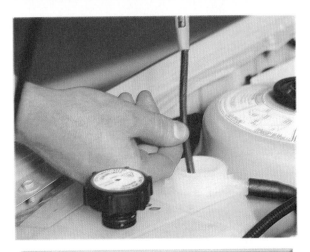

Fig. 2-9 (a) Antifreeze strength is checked by drawing the coolant into a hydrometer. *What are the two types of hydrometers?* (Terry Wild Studio)

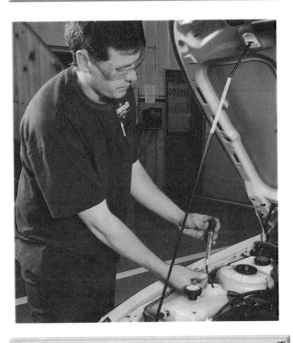

Fig. 2-9 (b) The position of the float in the coolant indicates the concentration of antifreeze present. *Why is it important to keep the tube inside the coolant container when reading the hydrometer?* (Terry Wild Studio)

SECTION 3 KNOWLEDGE CHECK

❶ What keeps the coolant from forming vapor bubbles at "hot spots" in the engine?

❷ What problem can a cooling system pressure test diagnose?

❸ What component regulates the flow of coolant through the engine?

❹ What happens when the thermostat maintains an engine temperature that is too low?

❺ What does a hydrometer measure?

CHAPTER 2 REVIEW

Key Points

Meets the following NATEF Standards for Engine Performance: diagnosing engine problems and performing engine diagnostic tests.

- A visual check of the engine and its support systems is the first step of the diagnostic process.
- A documented service history can help in planning a diagnostic strategy.
- A diagnostic strategy is a planned, step-by-step procedure.
- The color and odor of exhaust smoke can provide important diagnostic information.
- Compression, cylinder leakage, and engine vacuum tests are used to determine cylinder condition.
- Worn engine parts can cause noise, excessive oil consumption, poor performance, poor fuel economy, and high exhaust emissions.
- Pressure and temperature tests are used to determine the condition of the cooling system.
- The concentration of antifreeze determines the boiling and freezing point of engine coolant.

Review Questions

1. What are the five basic rules of trouble diagnosis?
2. What is a the difference between a "dry" and a "wet" compression test?
3. What are the steps in performing an engine vacuum test?
4. What engine problems can be diagnosed by checking the color and odor of the exhaust smoke?
5. Describe the procedures to perform the following tests: cylinder compression, cylinder leakage, and engine vacuum.
6. Name four common engine block noises.
7. What tests can be used to diagnose cooling system problems?
8. **Critical Thinking** If you know of five possible causes of a problem, which one should you check out first? (Hint: time is important.)
9. **Critical Thinking** Which is more harmful, an engine temperature that is too high, or an engine temperature that is too low?

TECHNOLOGY FORECAST

FOR AUTOMOTIVE EXCELLENCE

Diagnostics Using On-Board Computers

Getting to the root cause of engine problems has always been difficult. Service technicians must be good detectives. Modern computers and hand-held scan tools help them with their jobs. But technicians must still have a working knowledge of current engine design. They must also be able to ask car owners the right questions in order to make a good diagnosis.

Such insight will continue to be important in servicing cars and trucks of the future. Technology, though, will make the job easier.

Personal laptop computers are becoming more affordable to technicians and the repair shops where

they work. Automakers likely will make the most of that trend. New software programs allow these PCs to connect directly to the vehicle computer. Technicians can then retrieve diagnostic trouble codes (DTCs) and test various sensors and circuits. These are jobs they do today with a hand-held scan tool.

These new computers, with their greater processing power, could also run diagnostic tests and procedures. Results would then be reported to the technician, who would make the needed repairs.

Computers will continue to be important tools in the future, but they will never replace a technician's knowledge and skills.

AUTOMOTIVE EXCELLENCE
TEST PREP

Answering the following practice questions will help you prepare for the ASE certification tests.

1. A car's tailpipe blows black smoke while cruising at 45 mph [72 kph]. What is the cause?
   a Excessive oil consumption.
   b Rich air/fuel mixture.
   c A cracked cylinder head.
   d Thermostat opening too soon.

2. The compression of a four-cylinder engine is 140, 95, 135, and 138 psi, respectively. Technician A says the number two-cylinder (95) has a bad valve. Technician B says the number-two cylinder has carbon under the valve seat. Who is correct?
   a Technician A.
   b Technician B.
   c Both Technician A and Technician B.
   d Neither Technician A nor Technician B.

3. The compression of a four-cylinder engine is 140, 95, 95, and 138 psi, respectively. Technician A says there may be a crack in the cylinder block between cylinders two and three. Technician B says there may be a defective or leaking cylinder head gasket between cylinders two and three. Who is correct?
   a Technician A.
   b Technician B.
   c Both Technician A and Technician B.
   d Neither Technician A nor Technician B.

4. The cylinders of an engine test at 40 percent leakage. A large volume of air is exiting the oil-fill tube. Technician A says the engine may have worn valve guides. Technician B says the engine many have worn piston rings. Who is correct?
   a Technician A.
   b Technician B.
   c Both Technician A and Technician B.
   d Neither Technician A nor Technician B.

5. An engine has a rough idle. The vacuum gauge reading drops regularly at idle. Technician A suggests a compression test. Technician B says the engine may have worn rings. Who is correct?
   a Technician A.
   b Technician B.
   c Both Technician A and Technician B.
   d Neither Technician A nor Technician B.

6. An engine has a low vacuum reading at idle. Technician A says the ignition timing may be retarded. Technician B says the camshaft timing may be retarded. Who is correct?
   a Technician A.
   b Technician B.
   c Both Technician A and Technician B.
   d Neither Technician A nor Technician B.

7. A steady vacuum gauge reading between 15″ and 22″ of mercury (Hg) means:
   a A restricted catalytic converter, muffler, or exhaust.
   b Normal engine operation.
   c One valve is burned.
   d None of the above.

8. An engine has a light clicking noise that seems to occur at one-half crankshaft speed. Technician A says the engine may have a loose valve adjustment. Technician B says the engine may have a worn connecting rod bearing. Who is correct?
   a Technician A.
   b Technician B.
   c Both Technician A and Technician B.
   d Neither Technician A nor Technician B.

9. Blue smoke comes from the tailpipe of a vehicle during deceleration. The problem is:
   a A rich air/fuel mixture.
   b Coolant entering the cylinder through a blown head gasket.
   c Oil entering the cylinder through worn piston rings or valve guides.
   d None of the above.

10. An engine coolant system is operating at about 190°F [88°C]. The problem is:
   a Too much water in the radiator.
   b The thermostat opening too late.
   c There is no problem. This is normal.
   d The thermostat opening too soon.

CHAPTER 3

Inspecting & Testing Sensors and Actuators

You'll Be Able To:

- Identify commonly used sensors.
- Explain the difference between analog and digital signals.
- Check for normal sensor operation.
- Explain what actuators do and how they perform needed tasks.
- Explain how sensors and actuators fail.

Terms to Know:

actuator
analog signal
digital signal
scan tool
sensor

The Problem

Mr. Oversch owns a two-year-old, six-cylinder vehicle. For several weeks, the CHECK ENGINE light has been coming on. He travels about 90 miles a day to work. He tells you that the car's performance seems acceptable, but there is a ping when he accelerates. Sometimes the engine seems to misfire.

You know that the CHECK ENGINE light indicates that the PCM is seeing a fault. A scan tool shows the oxygen sensor is indicating an erratic oxygen level in the exhaust. Is the oxygen sensor at fault, or could something else be causing the problem?

Your Challenge

As the service technician, you need to find answers to these questions:

1. Could the oxygen sensor problem be related to poor fuel economy?

2. What could cause the oxygen sensor to send a false signal?

3. How can you check the oxygen sensor?

Section 1

Sensors

Electronic modules control the operation of many systems on a vehicle. The modules monitor and control engine operation to meet performance, emissions, and fuel economy goals. To do this, several operating conditions must be continuously monitored or measured. A **sensor** is a device that monitors or measures operating conditions. Each sensor generates or modifies an electrical signal based on the condition it monitors. The signals from various sensors are sent to the *powertrain control module (PCM)*. This module uses the sensor information to make decisions about actuator operation. An **actuator** is an output device controlled by the module. **Fig. 3-1.**

Sensors are also needed for automatic climate control, anti-lock braking, traction control, and other electronically operated systems. While there are many types of sensors, their operating characteristics are similar.

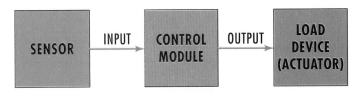

Fig. 3-1 A simplified block diagram of an automotive sensor circuit. *What does a sensor do?* (DaimlerChrysler)

Figure 3-2 lists the operating conditions in a vehicle that are monitored by the powertrain control module (PCM). They include:

- Engine speed.
- Vehicle speed.
- Engine load.
- Engine coolant temperature.
- Intake air temperature.
- Exhaust gas oxygen content.
- Throttle position.
- Crankshaft and camshaft position.
- Engine knock.

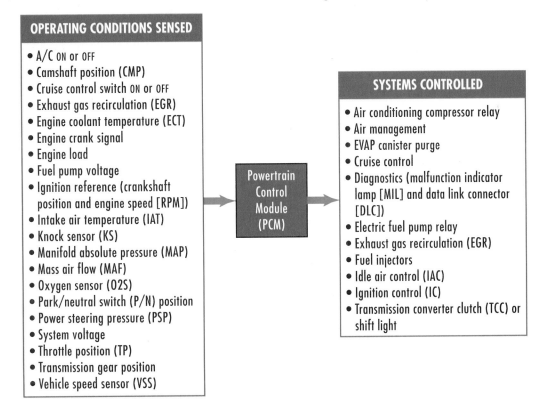

Fig. 3-2 A variety of sensors provide input to the powertrain control module (PCM). The PCM uses the inputs to control a variety of vehicle systems. *What type of device does the PCM use to control these systems?* (General Motors Corporation)

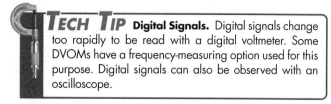

Electronic Signals

Sensors produce an output signal. The output can be either an analog signal or a digital signal. An **analog signal** is a signal that continuously changes. For example, its voltage may change from a minimum (usually zero) to a maximum (usually 5 volts). Any voltage between the minimum and maximum is possible. Many automotive sensors produce analog signals. **Fig. 3-3.**

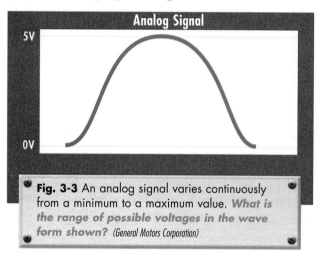

Fig. 3-3 An analog signal varies continuously from a minimum to a maximum value. *What is the range of possible voltages in the wave form shown?* (General Motors Corporation)

A **digital signal** is a signal that is either on or off. **Fig. 3-4.** Digital signals are also known as *square wave signals.* A digital signal has a rapid rise and fall time (from low to high and from high to low). The signal is either high or low, with no values in between.

Automotive computers and modules use only digital signals. Analog sensor signals must be converted to digital signals before the computer can use them. A solid state device called an analog to digital converter produces the signal conversion. Converting analog to digital signals may occur in the sensor, the ignition module, or in the computer itself.

Some digital signals transmit information by varying the number of pulses per second. This is known as a changing frequency, or *frequency modulation* (FM). The frequency of a signal is the

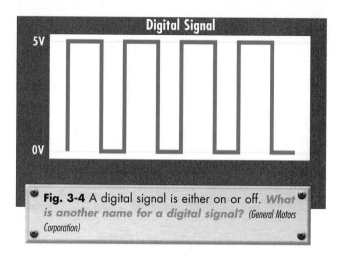

Fig. 3-4 A digital signal is either on or off. *What is another name for a digital signal?* (General Motors Corporation)

number of times it is turned on and off in a specified time, usually one second. Frequency is expressed in hertz (Hz), which are cycles per second. **Fig. 3-5(a).** Some mass air flow (MAF) sensors and manifold absolute pressure (MAP) sensors generate a frequency-modulated signal.

Digital signals may also vary in their duty cycle. This is called *pulse width modulation (PWM).* The signal stays high or low for varying lengths of time. The frequency of a PWM signal does not change. However, the amount of on and off time does change. **Fig. 3-5(b).** Fuel injectors are examples of actuators that are controlled by a PWM signal.

Reading the Data Stream

The PCM and other electronic components must be able to communicate with each other. This is done through a network of wires called a data bus or data bus network. A *data bus* is a wiring harness. It allows components connected to it to share sensor signals and other information needed for normal operation.

The information transmitted on the data bus is known as the *data stream.* Because a module can share the data stream, each sensor does not need to be connected to each module. This simplifies system wiring and reduces costs and weight.

In addition to connecting modules, sensors, and actuators, the data bus is also connected to the *data link connector (DLC).* On 1996 and newer vehicles, the DLC is located under the dash near the steering column. On earlier models the DLC may be under

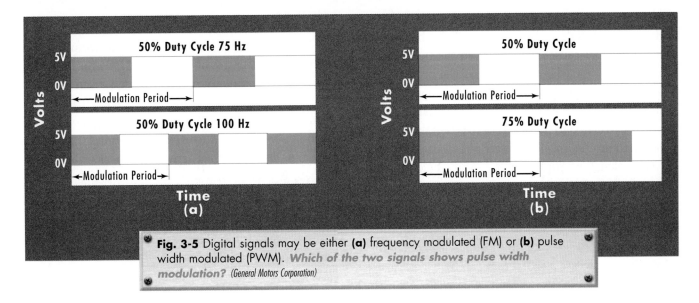

Fig. 3-5 Digital signals may be either **(a)** frequency modulated (FM) or **(b)** pulse width modulated (PWM). *Which of the two signals shows pulse width modulation?* (General Motors Corporation)

the hood or under the dash. Refer to the vehicle service manual for the location.

The DLC allows a technician to read the information on the data stream using a scan tool. A **scan tool**, or scanner, is a small, hand-held computer that communicates with the components on the data bus. A scan tool plugs into the DLC. **Fig. 3-6.** It enables the reading of *diagnostic trouble codes (DTCs)* and data from the PCM and other related modules and sensors.

On some vehicles the scan tool can be used to give operating commands to components on the data bus. By using the scan tool, a technician can check the operation of most sensors and actuators.

Speed and Position Sensors

Speed and position sensors are used to monitor the speed of rotating shafts. These sensors are found on camshafts, crankshafts, and output shafts. In camshafts and crankshafts, the sensor indicates the position of the shaft as well as its speed.

Permanent Magnet Sensors

A *permanent magnet (PM) sensor* is a small AC-voltage generator. This type of sensor is also called a *magnetic pulse generator*. It uses a wire coil, a pole piece, a permanent magnet, and a rotating trigger wheel (reluctor wheel) to produce an analog voltage signal. **Fig. 3-7.**

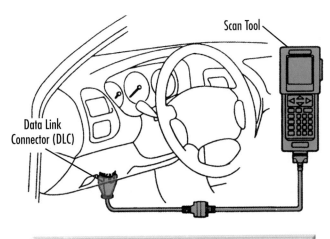

Fig. 3-6 A scan tool is connected to the DLC to read the data stream. *Where is the DLC located on 1996 and newer vehicles?*

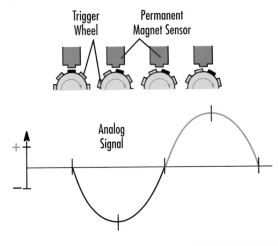

Fig. 3-7 A permanent magnet sensor produces an analog signal. *What produces the magnetic field in this type of sensor?* (DaimlerChrysler)

The trigger wheel is mounted on a rotating shaft, such as a crankshaft or distributor shaft. As the trigger wheel rotates, it causes the field from the permanent magnet to move back and forth across the pickup coil. The moving magnetic field induces an analog (AC) voltage into the coil windings. The AC voltage pulses are directly related to the speed of the shaft the trigger wheel is mounted on. Because a PM sensor generates its own voltage, no supply voltage is needed.

There are many variations of the PM sensor. Most often they are used as sensing devices for engine speed, vehicle speed, or wheel speed. The PM sensor has unique notch spacing on the trigger wheel. This allows it to generate a signal that can be used also as a position sensor. In this way, a PM crankshaft sensor can signal engine rpm. It can also signal the position of the piston in cylinder number one. **Fig. 3-8.**

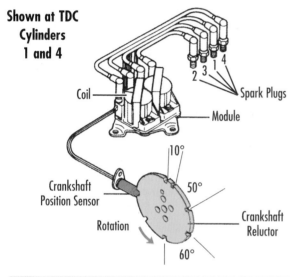

Shown at TDC Cylinders 1 and 4

Fig. 3-8 A permanent magnet sensor used as a crankshaft position sensor. *What two items of information can be obtained from a crankshaft position sensor?* (General Motors Corporation)

Testing a Permanent Magnet Sensor Test a permanent magnet sensor as follows:

1. Connect a digital volt-ohm-meter (DVOM) (use AC volt scale) to the two wires leading from the sensor.

2. Crank the engine and observe the voltage. If the output is higher than specified (usually 200–300 mV), the sensor is good.

3. As a second test, check the pickup coil windings with an ohmmeter. If the resistance is higher or lower than normal, replace the sensor.

Hall-Effect Sensors

The *Hall-effect sensor* acts like an electronic switch. It turns a voltage on or off. In the Hall-effect process, current is continually passed through a semiconductor material or chip in the sensor. When the semiconductor material is exposed to a magnetic field, a small voltage is produced. When no magnetic field is present, the Hall-effect sensor produces no voltage.

A rotating interrupter ring controls the magnetic field. The interrupter ring has alternating metal blades and windows. When used as an ignition-timing sensor in a distributor, there are as many blades and windows as there are cylinders in the engine. The ring either exposes or shields the material from the magnetic field. **Fig. 3-9.** Because the magnetic field is either exposed or shielded, the Hall-effect sensor does not rely on speed to create a signal. Unlike a PM sensor, no pickup coil or pole piece is used.

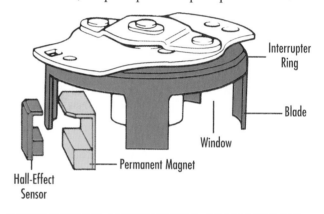

Fig. 3-9 A Hall-effect sensor used in a distributor. *What does a Hall-effect sensor do?* (Ford Motor Company)

When a metal blade is in the space between the magnet and the Hall chip, it blocks the magnetic field. When the magnetic field doesn't strike the semiconductor, the Hall switch is off. No voltage is produced.

As the interrupter ring rotates, a window moves between the magnet and chip. The window allows the magnetic field to strike the chip. This turns on the Hall circuit. A small output voltage is generated. The on-off action of the circuit creates an on-off digital (square wave) signal.

Hall-effect sensors are used for many of the same applications as PM sensors. They control fuel injection and provide a TACH signal. Some engines use dual Hall-effect crankshaft sensors. In this case a single magnet is used with two Hall-effect sensors

TECH TIP **Permanent Magnetic Sensors.** In some permanent magnetic sensors, the size of the air gap between the sensor and the trigger wheel is critical. Make sure it is properly adjusted.

and two uniquely shaped interrupter rings. One interrupter ring produces eighteen uniform pulses per revolution. The other ring produces three unevenly spaced pulses. **Fig. 3-10.**

(a) Dual Hall-Effect Crankshaft Sensor

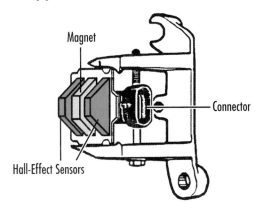

Magnet · Connector · Hall-Effect Sensors

(b) Crankshaft Damper with Interruptor Ring

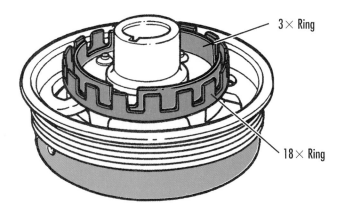

3× Ring · 18× Ring

Fig. 3-10 A dual Hall-effect sensor can produce two signals at one time. A magnet is placed between two Hall-effect sensors **(a)**. When the sensor is assembled with the interrupter ring **(b)**, the magnet is positioned between the 3× ring and the 18× ring. The Hall-effect sensors are positioned so one senses the 3× ring and one senses the 18× ring. *How many voltage toggles per revolution are produced by an interrupter ring with three blades and windows?* (General Motors Corporation)

Testing Hall-Effect Sensors Test a Hall-effect sensor as follows:

1. Use a DVOM to measure voltage on the signal line from the sensor. Refer to a wiring diagram to identify the correct wire.

2. With the ignition switch on, read the voltage.

3. Use the starter to rotate the engine a few degrees. Read the voltage again. If the sensor is good, the voltage will vary when the engine is stopped in various positions. It will vary from high (5–12 volts depending on the system) to low (near zero). It may take several attempts to get the engine to stop so that the interruptor blade is in the correct position for a reading.

Optical Sensors

Optical sensors use *light-emitting diodes (LEDs)* and *photodiodes* to produce a sensor signal. A photodiode uses the presence or absence of light to switch a reference voltage on and off. **Fig. 3-11.**

(a) Side View

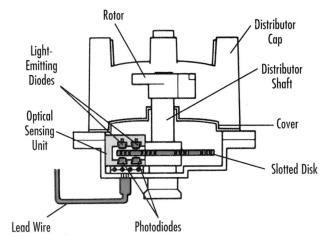

Rotor · Distributor Cap · Light-Emitting Diodes · Distributor Shaft · Optical Sensing Unit · Cover · Slotted Disk · Lead Wire · Photodiodes

(b) Top View

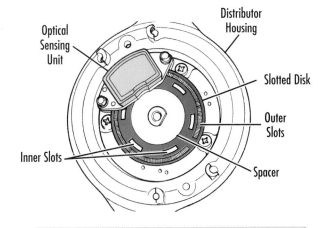

Optical Sensing Unit · Distributor Housing · Slotted Disk · Outer Slots · Inner Slots · Spacer

Fig. 3-11 An optical sensor uses light-emitting diodes (LEDs) and photodiodes to toggle a reference voltage. *What is the purpose of the slotted disk?* (DaimlerChrysler)

Like a Hall-effect sensor, the optical sensor requires a voltage feed for operation. A typical circuit has two LEDs. These are mounted over two photodiodes with a space in between. A slotted metal disk mounted on a shaft rotates between the photodiodes and the LEDs.

Testing Optical Sensors Test an optical sensor in the same manner as a Hall-effect sensor:

1. Use a wiring diagram to identify the signal wire leading from the sensor. Connect a DVOM from this wire to ground.

2. With the ignition on, check the voltage.

3. Crank the engine a few degrees and check the voltage again. As the slotted disk moves between the photodiodes and the LEDs, the reference voltage should toggle to near zero volts. If voltage is present at the sensor but the output does not toggle, the sensor is defective.

Variable Resistance Sensors

Several types of variable resistors are used as sensors for electronically controlled systems. The resistance in this type of sensor changes as the sensor monitors changing conditions.

Thermistors

A *thermistor* is a solid-state resistor. Its resistance changes with temperature. Two types are used. In a *positive temperature coefficient (PTC) sensor,* the resistance of the thermistor increases as the temperature increases. In a *negative temperature coefficient (NTC) sensor,* the resistance decreases as the temperature increases.

Thermistors are commonly used to monitor *engine coolant temperature (ECT)* and *intake air temperature (IAT)*. In either case, the thermistor is a two-wire sensor. It is connected in series between a reference voltage from the PCM and ground. The reference voltage is usually 5 volts. **Fig. 3-12**.

With the ignition on, current flow through the sensor circuit is affected by the thermistor's resistance. If the resistance is high, there is less current flow in the circuit. Less current causes less voltage drop across a resistor in the PCM. The voltage on the sensor side of the PCM resistor is monitored by the PCM as the ECT sensor input.

> **TECH TIP** **Thermistors.** You can quickly check both the ECT and IAT sensors by comparing their temperature or voltage readings on a scan tool. Do this when the engine is cool. Both sensors should read about the same.

The IAT sensor works like the ECT sensor. It is located in the engine air induction system. It provides information on the temperature of the air being drawn into the engine.

Testing Thermistors Test a thermistor as follows:

1. Using a scan tool, check the temperature readout. The temperature should be near normal for the operating conditions involved.

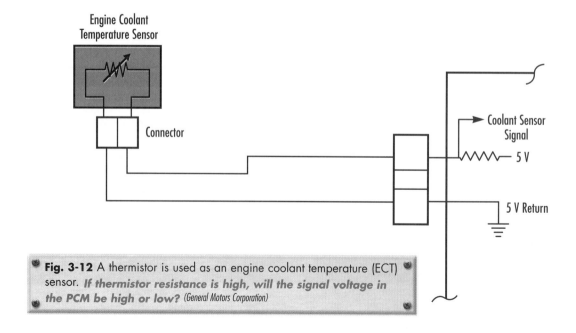

Fig. 3-12 A thermistor is used as an engine coolant temperature (ECT) sensor. *If thermistor resistance is high, will the signal voltage in the PCM be high or low?* (General Motors Corporation)

2. Using an ohmmeter, measure the resistance from one sensor terminal to the other. Compare this reading with specifications for the sensor at the test temperature.

3. Replace the sensor if the resistance is out of the specified range.

Potentiometers

A *potentiometer* is a three-wire variable resistor. It is used to monitor movement. The resistance of this sensor changes with the position of the shaft to which it is connected. One use of a potentiometer is the *throttle position (TP) sensor* on the throttle body assembly. **Fig. 3-13.**

The PCM sends a 5-volt reference signal to one terminal of the TP sensor. A second terminal is grounded at the PCM. The voltage on the third TP sensor signal wire will change as the throttle is opened or closed. In most cases, a TP sensor voltage of 0.5 volts indicates a closed throttle. At *wide-open throttle (WOT)*, the voltage will be almost 5 volts. **Fig. 3-14.**

A potentiometer is also used to indicate the position of the *exhaust gas recirculation (EGR) valve* and the vane position in a vane-type airflow meter.

Testing Potentiometer-Type Sensors Test potentiometer-type sensors as follows:

1. Connect a scan tool to the DLC (or use a DVOM) to measure voltage on the sensor signal wire.

2. Turn on the ignition switch. At closed position the reading should be less than 1 volt. At open position the reading should be almost 5 volts.

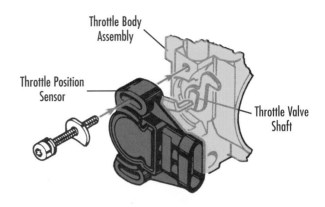

Fig. 3-13 The throttle position (TP) sensor is mounted on the throttle body assembly. *What type of variable resistor is used as a TP sensor?* (Ford Motor Company)

3. If no voltage is present, check to see if there is voltage on the reference wire from the PCM. If the voltage is 0, the reference voltage circuit is open.

Load Sensors

The PCM needs to know the amount of load on the engine. Load sensors provide this information to the PCM. This allows the PCM to calculate the correct spark advance and air/fuel ratios needed for various operating conditions.

On some engines load is calculated by using the inputs from the IAT sensor, TP sensor, and MAP sensor. This is the *speed-density* method of determining load. Other applications use a *airflow* sensor to determine the load. *airflow* sensors measure the amount of air being used by the engine.

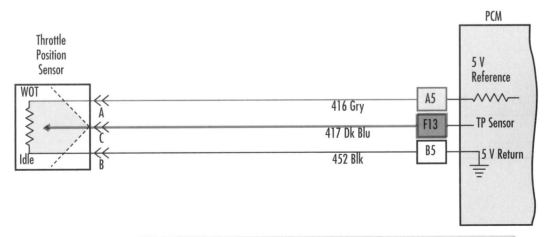

Fig. 3-14 The throttle position (TP) sensor is a potentiometer-type sensor. It receives a 5-volt reference signal from the PCM. *Identify the circuit that carries the TP sensor voltage to the PCM.* (General Motors Corporation)

Manifold Absolute Pressure Sensors

The *manifold absolute pressure (MAP) sensor* is mounted to a source of manifold vacuum. Some MAP sensors are mounted directly on the intake manifold. Others are remotely mounted. They are connected to manifold by a vacuum hose. **Fig. 3-15.**

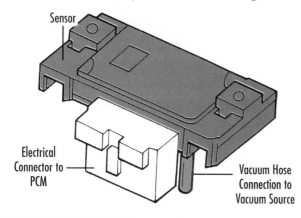

Fig. 3-15 The manifold absolute pressure (MAP) sensor is a load-sensing device. *What method of calculating engine load uses a MAP sensor?* (General Motors Corporation)

Engine vacuum is the opposite of manifold absolute pressure. When the engine is not running, there is no manifold vacuum. The pressure in the intake manifold is equal to atmospheric pressure. When engine load is low, as during closed throttle operation, vacuum is high and manifold pressure is low.

When there is more load on the engine, the throttle is opened to produce more power. Opening the throttle causes vacuum to decrease and manifold pressure to increase. Manifold vacuum and pressure change with engine load. Because of this, the MAP sensor is important in the speed-density method of calculating load.

Most MAP sensors are similar in appearance. However, several different internal designs are used. In one type of MAP sensor, the PCM monitors the voltage drop across the sensor. In another type, changes in manifold pressure cause the sensor to vary the frequency of the output signal.

On many vehicles the MAP sensor is also used to monitor atmospheric (barometric) pressure. When the key is turned on (engine off), the pressure in the manifold is atmospheric pressure. This value is sent to the PCM. It is used to calculate the amount of fuel needed by the engine. **Fig. 3-16.**

Checking Manifold Absolute Pressure Sensors

Check the MAP sensor as follows:

1. Connect a scan tool to the DLC. With the key on and engine off, read the scan display. Vacuum should be 0 or voltage should be about 4.5 volts.

2. Start the engine. Manifold vacuum should be in the range of 16–22″ [41–56 cm] of mercury. The voltage reading should be about 1–2 volts. If the readings are not as specified, make sure the vacuum line to the MAP sensor is not damaged, plugged, or disconnected.

Airflow Sensors

Several types of airflow sensors are in use. All determine engine load by directly measuring the amount of air entering the engine. Most vehicles that use airflow sensors do not use MAP sensors. However, some vehicles use both.

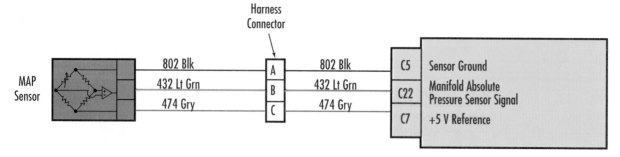

Fig. 3-16 The manifold absolute pressure (MAP) sensor is a three-wire sensor. *Which terminal of the harness connector can be used to check for signal voltage?* (General Motors Corporation)

Finding Resistance

The Wheatstone bridge is a useful circuit. It can be used to calculate the value of an unknown resistance from known resistances.

A sample circuit is shown here. The input terminals for the bridge are points C and D. The output is across terminals A and B. A voltage source (V_S) is connected across the input terminals (C and D) to energize the bridge.

If R_1 and R_2 have the same ratio as R_3 and R_4, the bridge is balanced and V_{AB}, across the bridge, is zero. Similarly, if R_1 and R_2 are equal, then R_3 and R_4 must be equal to balance the bridge and cause V_{AB} to be zero. In either case, with a balanced bridge, the formula for R_4 is:

$$R_4 = R_3 \times \frac{R_2}{R_1}$$

This equation makes the balanced bridge very useful for measuring any unknown resistance (R_4). The output voltage of the unbalanced bridge

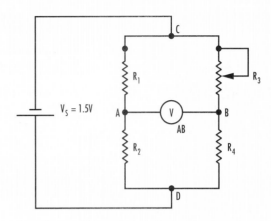

is very sensitive to the difference between the values of R_3 and R_4. The unbalanced output voltage is used to measure many variables. Temperature, pressure, and light level are some measurements made using sensors in place of R_3 and R_4.

Apply it!

Constructing a Wheatstone Bridge

Meets NATEF Science Standards for measuring electrical resistance.

Materials and Equipment
- 1.5-volt power supply (V_S)
- Galvanometer or digital multimeter
- Three 500-ohm resistors (R_1, R_2, R_4)
- 1.0 kilo-ohm potentiometer (R_3)

Let's investigate the balanced and unbalanced Wheatstone bridge.

❶ Use the digital multimeter to measure the resistance of each of the fixed resistors (R_1, R_2, and R_4). Record these values.

❷ Make sure the power supply is turned off.

❸ Construct the circuit shown. Use a length of wire as a circuit between one end of the potentiometer and its center lead. If a galvanometer is not available, use a digital multimeter to measure V_{AB}.

❹ Position the potentiometer at its center value. Then adjust its setting several times. Unbalance the bridge so you can observe both positive and negative output voltage (V_{AB}).

❺ Adjust the potentiometer to balance the bridge with a zero output voltage for V_{AB}.

❻ Disconnect the potentiometer, R_3, from the circuit without changing its setting. Use the multimeter to measure the value of its resistance.

❼ Calculate the value of R_4 using the formula:

$$R_4 = R_3 \times \frac{R_2}{R_1}$$

Use your measured values for R_1, R_2, and R_3. Compare your measured value for R_4 to the value you just calculated.

Mass Air Flow Sensors The *mass air flow (MAF) sensor* is mounted in the air induction system. This enables air entering the engine to pass through the sensor. **Fig. 3-17.**

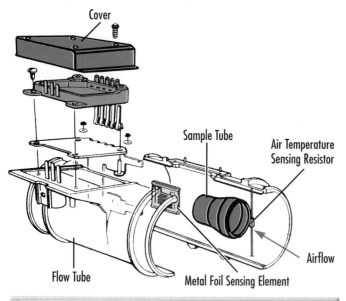

Fig. 3-17 The mass air flow (MAF) sensor directly measures the amount of air entering the engine. *What two types of signals do MAF sensors produce?* (General Motors Corporation)

Most MAF sensors measure the amount of electrical energy required to keep a heated element at a specified temperature. Incoming airflow cools the heated element. By measuring the amount of current needed to maintain the specified temperature, the PCM determines the mass of airflow into the engine.

Depending on the application, MAF sensor output may be a varying voltage or a varying frequency.

Vane Airflow Sensors Some airflow meters use a potentiometer to measure the position of a spring-loaded vane (door). The spring-loaded vane moves with the flow of air into the engine. An IAT sensor is mounted in the vane airflow housing to measure air temperature. Air entering the engine causes the air vane to move against the tension of a spring. The air vane shaft is connected to a potentiometer. The resistance of the potentiometer changes with the position of the vane. As the resistance changes, it changes the signal voltage to the PCM. **Fig. 3-18.**

Testing Airflow Sensors There are several types of airflow sensors and signals. As a result, there is no one method to test these sensors. If airflow information is displayed on the data stream, use a scan tool. A scan tool can determine if the airflow at idle is within specifications. If a vehicle with an identical powertrain is available, compare the airflow data between the two vehicles.

Voltage Generating Sensors

Some automotive sensors generate a small voltage in response to changing operating conditions. One example is the permanent magnet (PM) sensor discussed earlier. Other sensors use a variety of methods to produce an output voltage signal.

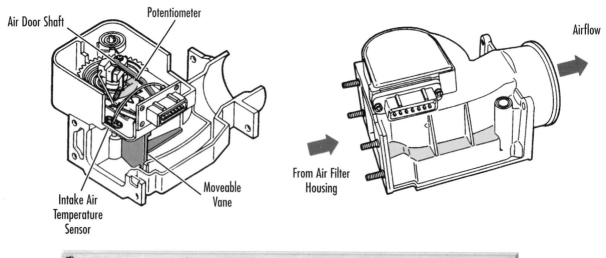

Fig. 3-18 A vane airflow sensor uses a spring-loaded vane (door) to measure intake airflow. *What type of sensor is connected to the air door shaft?* (Ford Motor Company)

Oxygen Sensors

Oxygen sensors (O2S) detect the presence of oxygen in the exhaust gases. They are similar to spark plugs in appearance. **Fig. 3-19.**

Two types of oxygen sensors are used. Both must reach a specific temperature before they produce a useful signal. In some cases the hot exhaust gases are the only source of heat. These sensors require a warm-up period before they become effective.

Other sensors have a built-in heating element so that they reach normal operating temperature more quickly. They are called *heated oxygen sensors* (*HO2S*). Heated sensors also maintain their operating temperature during periods of extended idle when a cool down might occur.

The number of oxygen sensors used and the number of wires connected to each sensor varies. It depends on the type of sensor and the application.

Several oxygen sensors may be used on the same vehicle. The sensor closest to the engine is primarily responsible for helping the PCM maintain the correct air/fuel ratio. A sensor near the inlet to the *catalytic converter* is often referred to as the "pre-cat" sensor. The "post-cat" sensor is located at the outlet from the catalytic converter.

Oxygen sensors located before and after the catalytic converter are used only on vehicles with

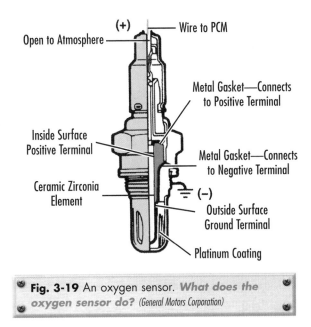

Fig. 3-19 An oxygen sensor. *What does the oxygen sensor do?* (General Motors Corporation)

on-board diagnostics-II (OBD-II). These sensors monitor catalytic converter efficiency. **Fig. 3-20.**

Zirconia Oxygen Sensors The *zirconia oxygen sensor* produces a voltage signal. This signal is based on the oxygen content in the exhaust gases.

During normal operation the output from the zirconia O2S varies from zero to about 1 volt. Low readings (below 300 mV) indicate that the exhaust gas oxygen content is high. This means the air/fuel ratio is lean (excess air). High voltage readings (over 600 mV) mean that the oxygen content of the exhaust is low. This is caused by a rich mixture (excess fuel). **Fig. 3-21.**

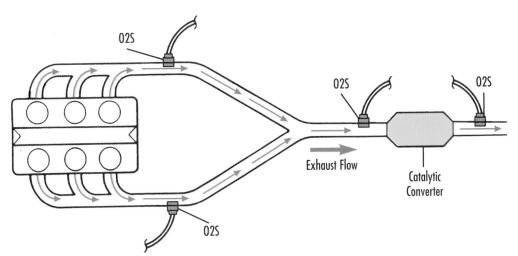

Fig. 3-20 Oxygen sensors may be used at several locations in the exhaust system. *What system uses an oxygen sensor before and after the catalytic converter?* (General Motors Corporation)

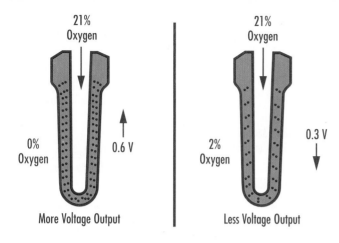

21%
Oxygen

21%
Oxygen

0%
Oxygen

0.6 V

More Voltage Output

2%
Oxygen

0.3 V

Less Voltage Output

Fig. 3-21 The oxygen sensor signal indicates the amount of oxygen in the exhaust gases. *What will the sensor voltage read if there is no oxygen in the exhaust?* *(General Motors Corporation)*

The O2S reads exhaust gas oxygen content, not the actual air/fuel ratio. Extra oxygen in the exhaust due to a cylinder misfire or exhaust manifold leak will result in a low oxygen sensor voltage reading. This will occur even if the actual air/fuel ratio is normal or slightly rich.

Titania Oxygen Sensors *Titania oxygen sensors* are similar in appearance to other O2Ss. This type of sensor changes resistance in response to the amount of oxygen in the exhaust.

In a typical application, a titania sensor is used as the post-cat O2S. The PCM sends a 2-volt signal to the sensor and monitors the resulting return voltage. A voltage near 0 volts indicates a lean exhaust. A voltage near 2 volts indicates a rich exhaust.

Oxygen sensors become less sensitive with age. When this occurs, the time it takes to respond to a change in oxygen content slows. This makes it difficult for the PCM to maintain the desired air/fuel ratio.

Oxygen sensors may become contaminated with carbon and oily deposits. These contaminants may be caused by:

• Rich fuel mixtures.
• The use of leaded fuels in vehicles requiring non-leaded fuels.
• Internal coolant leaks
• Excessive oil consumption.

Correct the condition(s) causing any contamination problems before replacing the sensor.

Checking Oxygen Sensors Check oxygen sensors as follows:

1. Use a scan tool or DVOM to observe the output voltage from the O2S. The actual signal will depend on the type of sensor and where it is located. Refer to a service manual for information on the testing of specific O2S.

2. Do not attempt to test a zirconia sensor with an ohmmeter. (It is acceptable to check the heating element for resistance.) **Figure 3-22** shows graphs of typical voltages from an O2S closest to the engine. Voltages from the post-cat sensor will be considerably different.

Knock Sensors

A *knock sensor (KS)* is used on some engines. It allows the PCM to retard ignition timing to control *detonation,* or spark knock. Detonation occurs when part of the air/fuel mixture self-ignites and "explodes." This uncontrolled burning causes engine parts to vibrate. The resulting sound waves create "spark knock" or "ping." Severe spark knock damages engine components. The KS is mounted on the engine. Here it is exposed to mechanical vibrations, including those caused by spark knock. **Fig. 3-23.**

A typical KS contains a *piezoelectric crystal.* When vibrations occur the piezoelectric crystal generates a small AC voltage. This voltage is proportional to the intensity of the knock. This signal is sent to the PCM. The PCM monitors the KS signal. It ignores most vibrations not caused by detonation. If detonation is occurring, the PCM retards the ignition timing until the knock-related vibrations stop. Normal timing is gradually restored as the conditions causing the spark knock change.

Testing Knock Sensors Test knock sensors as follows:

1. Use a scan tool to observe either ignition timing or the KS signal.

2. With the engine running at about 1,500 rpm, rap lightly with a small hammer on the engine near the sensor. Ignition timing should retard several degrees. Or, you should see a signal from the knock sensor.

3. For specific test procedures, refer to the vehicle service manual.

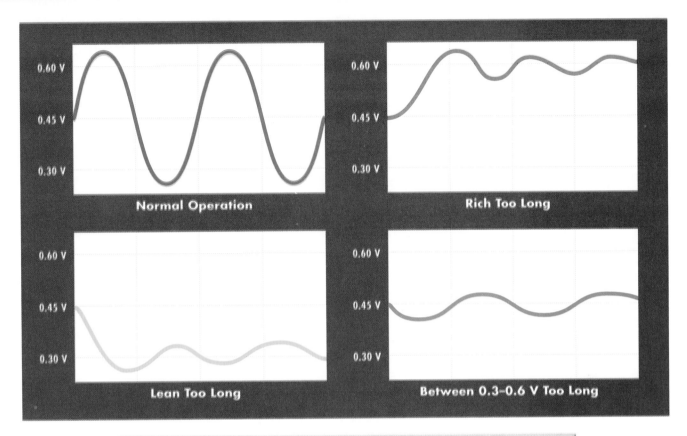

Fig. 3-22 Normal and abnormal oxygen sensor signals from the sensor(s) closest to the engine. *What would cause the oxygen voltage signal to stay near 0.30 volts for too long?* (General Motors Corporation)

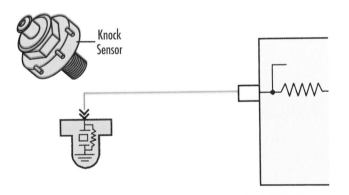

Fig. 3-23 Some engines have a knock sensor (KS) to help control spark knock (detonation). *What does the KS do when spark knock occurs?* (General Motors Corporation)

Knock sensors are electronic devices. Their operation can be affected by heat and mechanical damage. When handling these sensors always use the proper tools. Be careful not to round off the corners of the nut when removing or installing a sensor. If the knock sensor is seized, use a penetrating oil to remove any corrosion before attempting to remove it.

Switches

Sensors normally produce variable input signals. Other input signals are simple on or off signals from mechanically or pressure-operated switches. Switch input to a module is either an on or off voltage. This depends on whether the switch is open or closed. Two types of switch circuits are used.

Power Side Switches

Power side switches are located between a power supply and a module. The voltage source is usually the battery (12 volts). The module provides a ground circuit and monitors the signal. When the power side switch is open, the voltage signal to the module is zero. When the switch is closed, the signal will be the voltage supplied. The switch is also called a "pull-up" switch. When the switch closes, the voltage is "pulled-up" to its highest possible level. **Fig. 3-24.**

Ground Side Switches

Ground side switches have one terminal connected directly to ground. Power is supplied by the module, which also monitors the circuit. When the switch is open, signal voltage is at its maximum value. When the switch closes, the circuit is connected to ground. The voltage drops to near zero. Because the voltage drops, the switch is often referred to as a "pull-down" switch. **Fig. 3-25.**

Testing Switches   Test switches as follows:

1. Using a DVOM or scan tool, check for voltage on both sides of the switch. When using a DVOM, refer to the vehicle service manual to determine the proper wiring connections for the switch.

2. The voltage should change from low to high or high to low as the switch is operated. The operation of many switches can be monitored on a scan tool. The switch signal should change when the switch is operated.

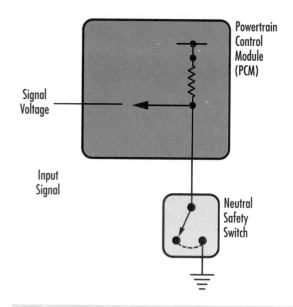

Fig. 3-25 A ground side switch is connected directly to ground. *What is another name for a ground side switch?* (DaimlerChrysler)

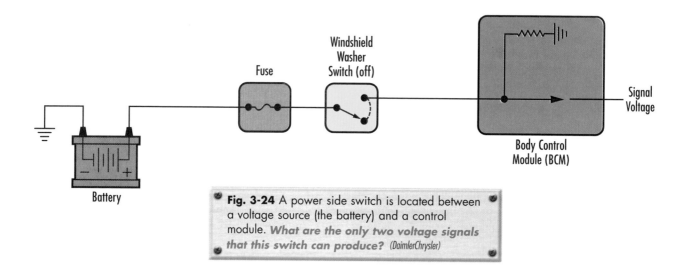

Fig. 3-24 A power side switch is located between a voltage source (the battery) and a control module. *What are the only two voltage signals that this switch can produce?* (DaimlerChrysler)

SECTION 1 KNOWLEDGE CHECK

❶ Explain the major difference between a square wave (digital) and an analog signal.

❷ Explain how to read information from the data stream.

❸ What type of signal does a Hall-effect sensor produce?

❹ Describe a potentiometer and how it is used as an automotive sensor.

❺ What two sensors are commonly used to determine engine load?

Section 2

Actuators

After electronic modules have evaluated sensor inputs, they operate output devices. These devices, called actuators, cause the desired actions to take place. Actuators are output devices that are operated by the PCM or other modules to create motion and perform other tasks. A variety of actuators are used to control automotive circuits and components. Tasks performed by actuators include controlling:

- Air/fuel ratio.
- Idle speed.
- Emission control device operation.

Solenoids

Solenoids use electromagnetism to produce motion. This motion is used to open and close air or hydraulic passages, apply a magnetic clutch, and lock and unlock doors. Solenoids can be used to control the flow of vacuum to a device. Solenoids may be either normally open (N.O.) or normally closed (N.C.).

Voltage is applied to most solenoids by the ignition switch. When the solenoid winding is grounded by the PCM, the plunger in the solenoid moves. It either opens or closes a vacuum passage. In this way the PCM controls the operation of vacuum-operated components, such as canister purge valves and exhaust gas recirculation (EGR) valves. **Fig. 3-26.**

A fuel injector is a special type of solenoid. Battery voltage is supplied to one of the injector terminals. The PCM grounds the other terminal to operate (open) the injector. **Fig. 3-27.**

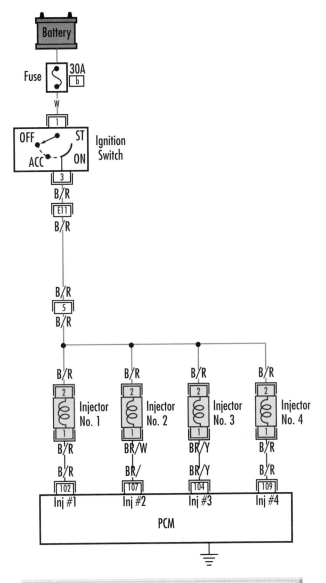

Fig. 3-27 The PCM provides the ground circuit for the fuel injectors. *When a fuel injector circuit is grounded by the PCM, does the fuel injector open or close?* (Nissan Motor Corporation)

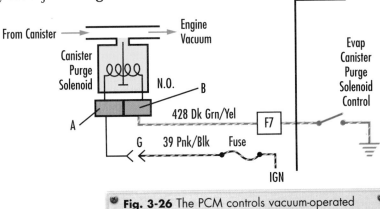

Fig. 3-26 The PCM controls vacuum-operated devices by using solenoids. *Where does this solenoid get its voltage?* (General Motors Corporation)

By controlling the on and off time (pulse width) of a fuel injector, the PCM controls the amount of fuel injected. **Fig. 3-28.**

Relays

A relay is similar to a solenoid except that the motion is used to close a set of electrical contacts. When the relay is not energized, the contacts are open. No current flows in the circuit. When the relay is energized, the contacts close and the circuit is completed. **Fig. 3-29.**

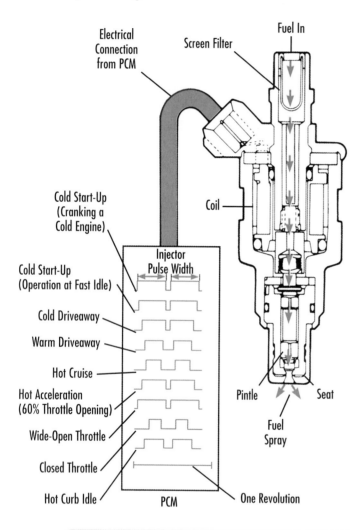

Fig. 3-28 The PCM controls the on and off time of the fuel injector to control the amount of fuel delivered to the engine. *What controls injector on and off time?* (Ford Motor Company)

The PCM energizes the fuel pump relay circuit to operate the electric fuel pump. On many vehicles relays controlled by the PCM are also used

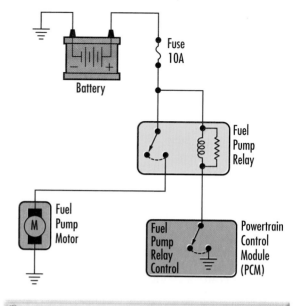

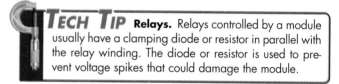

Fig. 3-29 The electric fuel pump is controlled by a PCM-operated relay. *Where is the fuel pump relay winding grounded?* (DaimlerChrysler)

TECH TIP **Relays.** Relays controlled by a module usually have a clamping diode or resistor in parallel with the relay winding. The diode or resistor is used to prevent voltage spikes that could damage the module.

to operate the electric engine cooling fan and air conditioning compressor clutch.

Testing Solenoids and Relays   Test solenoids and relays as follows:

1. Listen or feel for plunger movement. If no movement is observed, use a DVOM.

2. Use the DVOM to test for voltage to and from the windings.

3. Use an ohmmeter to check the resistance of the windings.

4. If the electrical checks are acceptable but the relay or solenoid doesn't work, the relay or solenoid is defective. Some relays and solenoids can be energized with a scan tool to test for normal operation. Refer to the vehicle service manual or the scan tool manual for specific procedures.

5. Special test procedures and equipment may be required to test some fuel injectors. Refer to the vehicle service manual for details.

Determining Rate of Change

The graph below shows the characteristics of an EFI MAP sensor. The graphed line appears to be a straight line (linear).

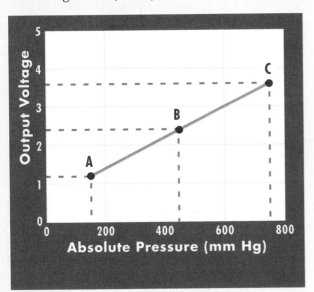

You can show mathematically that the line *is* straight. The data shown below were used to plot the line on the graph. If the rate of change between points A and B is the same as the rate of change between points B and C, the line is a straight line.

Look at the data as three sets of numbers in the form (X,Y). They are: A (150, 1.2), B (450,

	Absolute Pressure (in mm Hg)	Output Voltage (in Volts)
A	150	1.2
B	450	2.4
C	750	3.6

2.4), and C (750, 3.6). Each pair of numbers identifies a point on the graph.

You can find the rate of change of the output voltage per mm Hg from point B to point A using the formula:

$$m = \frac{Y_1 - Y_2}{X_1 - X_2}$$

where point B (450, 2.4) is (X_1, Y_1) and point A (150, 1.2) is (X_2, Y_2).

$$\frac{2.4\ \text{V} - 1.2\ \text{V}}{450\ \text{mm Hg} - 150\ \text{mm Hg}} = 0.004\ \text{V/mm Hg}$$

You can find the rate of change between points B (450, 2.4) and C (750, 3.6):

$$\frac{3.6\ \text{V} - 2.4\ \text{V}}{750\ \text{mm Hg} - 450\ \text{mm Hg}} = 0.004\ \text{V/mm Hg}$$

If the change in the vertical quantity divided by the change in the horizontal quantity is constant, then a linear relationship exists between the quantities. The above two calculations show that the rate of change of the output voltage per mm Hg is 0.004.

Apply it!

Meets NATEF Mathematics Standards for interpreting graphs and solving problems.

1. If the relationship is linear, what should be the rate of change of the output voltage per mm Hg using the sets of numbers for Point A and Point C.

2. Calculate the rate of change of the output voltage per mm Hg between A and C.

Stepper Motors

Stepper motors are similar to other small DC motors. However, the movement of the output shaft is more precisely controlled. Many fuel-injected engines use a stepper motor as an *idle air control (IAC) valve.*

The IAC valve is a reversible DC motor in the throttle body assembly. A small pintle valve controls airflow to the engine during closed throttle

conditions. **Fig. 3-30.** To control the position of the pintle valve, the PCM sends a pulsing (on and off) voltage through two sets of motor windings. By pulsing voltage to the correct winding, the PCM can position the pintle in the exact location needed for a specified idle speed.

Testing Stepper Motors Test a stepper motor by observing the pintle for normal operation. In some cases a test wiring harness is available to operate

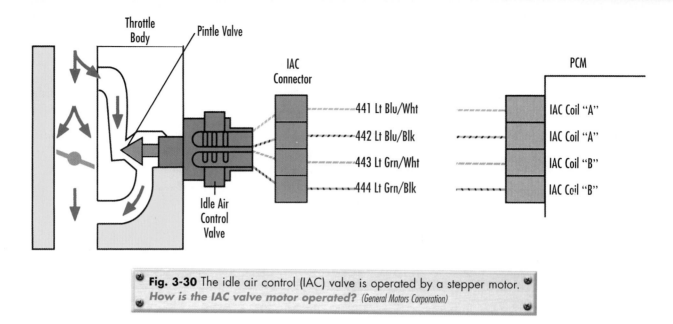

Fig. 3-30 The idle air control (IAC) valve is operated by a stepper motor.
How is the IAC valve motor operated? (General Motors Corporation)

the motor using battery voltage. Some stepper motors can be tested using commands given by a scan tool.

Other PCM-Controlled Outputs

The PCM is responsible for turning on some instrument panel warning lamps. The PCM controls the *malfunction indicator lamp (MIL)*, turning it on when certain malfunctions occur. The PCM

may also control the coolant over-temperature lamp, charging system warning lamp, and upshift lamp. Warning lamps are turned on when the PCM provides a ground for the lamp circuit. **Fig. 3-31.**

Testing Warning Lamps Most warning lamps are designed to come on as a "bulb check" when the ignition switch is turned to ON or START. If a warning lamp does not come on when it should, check the fuse and the bulb before checking the PCM circuit.

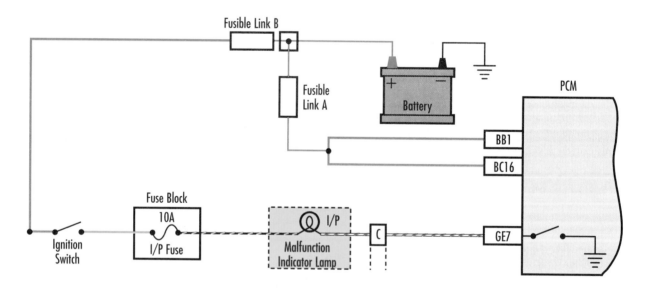

Fig. 3-31 The PCM controls the operation of several instrument panel warning lamps.
Name three warning lamps that may be controlled by the PCM. (General Motors Corporation)

SECTION 2 KNOWLEDGE CHECK

❶ Name five tasks that can be performed by an actuator.

❷ How are most engine system solenoids controlled?

❸ What is a typical use of a relay?

❹ What is the primary difference between a regular DC motor and a stepper motor?

❺ How are PCM-controlled instrument panel warning lamps turned on?

Section 3

Replacing Sensors and Actuators

Before replacing a sensor or actuator, make sure that the device is defective. Check to see if the device is properly positioned and connected. Check for defective wiring, poor connections, and leaking or missing vacuum hoses. Use a DVOM to measure the voltage supply. In some cases you may need to test a circuit at the PCM or other module to see if a signal is being received or if voltage is being supplied.

When troubleshooting an actuator, answer the following:

• Is the PCM trying to activate the device in question?

• Is voltage available?

• Is the actuator capable of producing the desired result?

Verifying the Problem

To operate correctly, a sensor or actuator must have the correct voltage supplied. Most sensors operate at 5 volts. Actuators usually operate at 12 volts. Both sensors and actuators require a good ground circuit, as well.

Blown fuses or open switches can prevent operation. Defective wiring, broken connectors, or deformed or corroded terminals can also cause problems in a circuit.

Use a DVOM to perform voltage drop or resistance tests. This will ensure that the wiring to the module is in good condition. Vehicle service manual wiring diagrams show how components are connected, the color codes for the wires, and the

location of splices, connectors, fuses, and switches. **Fig. 3-32.**

Some actuators, such as motors and solenoids, require a significant amount of current to operate. Make sure the voltage supply to the actuator is present under load. Check the wiring and the connections to the power source. Make sure the device is not shorted or grounded internally.

Safety First

Personal The exhaust system stays hot for a period of time after the engine is shut down. Use caution to avoid burns when working with oxygen sensors.

Materials You may need a special oxygen sensor wrench. Use it to prevent damage to the sensor during removal and installation.

Data Bus Problems

Sensors and actuators communicate on a data bus that connects most major electronic circuits and components. Poor connections may produce voltage drops that distort the information being transmitted. Loose wires or connections and bent or distorted terminals may result in intermittent connections. Replace damaged wires or terminals. Spray connector blocks and terminals with electronic terminal cleaner.

Mechanical Factors

Sensors and actuators may not work due to mechanical interference. Solenoid plungers may stick or bind. Motor shafts may be too tight or damaged to rotate. Vacuum lines may be plugged or kinked. Make sure the sensor or actuator has the freedom of movement required for correct operation.

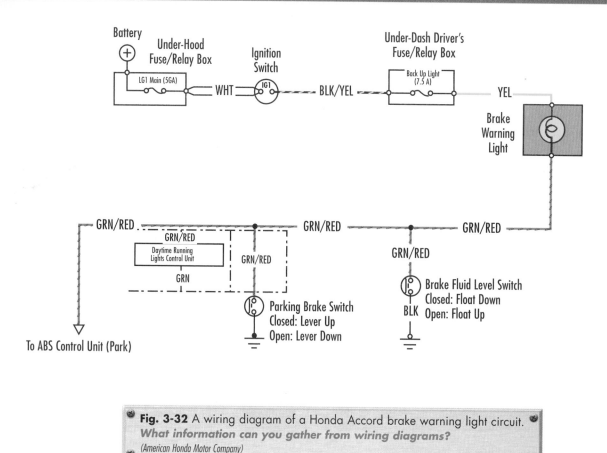

Fig. 3-32 A wiring diagram of a Honda Accord brake warning light circuit. *What information can you gather from wiring diagrams?* (American Honda Motor Company)

Proper sensor operation depends on the correct positioning of the sensor. This is needed to make an accurate measurement. The output of some PM sensors depends upon the correct air gap. The ECT sensor cannot provide an accurate reading if the coolant level is too low. Solenoid actuators cannot produce the desired range of motion if they are not properly mounted and adjusted.

Sensor and Actuator Failures

Early failures of sensors and actuators can occur for several reasons:
- Oxygen sensors become less sensitive with age. Replace the sensors if the response or cycling time is too slow.
- Potentiometer sensors may become dirty or damaged. This can cause false or intermittent signals to be sent to the PCM.
- High temperature exhaust gases may damage the exhaust gas recirculation (EGR) valve.

- Carbon deposits may interfere with normal EGR valve operation.
- Accidental voltage spikes and reversed battery polarity can damage the electronic circuits in sensors, actuators, and modules.
- Severe vibration or impact and extreme temperature changes can damage connections inside a sensor or actuator.
- Impact failure is more likely to occur when the vehicle has been in an accident.

Precautions

In replacing a sensor or actuator, take the following precautions:
- Read and follow the installation instructions.
- Make sure the ignition is off before removing any connectors.
- Remove the connectors carefully. Connectors are often held in place with plastic retainers.
- Pull only on the connector body, not the wires.

- Some connectors are filled with white silicone grease. Replace this grease as needed.
- Avoid damaging weatherproof seals when removing or installing a connector.

- Apply antiseize compound to oxygen sensor threads before installation.
- Make sure cables and harnesses are routed to prevent heat damage.

Decoding Words

Words are the basis for exchanging ideas accurately. If you do not know the meaning of a word, you could misinterpret information. This could prove to be costly to you or your customer. In your work, you must understand what you read.

There are several methods you can use to learn the meaning of a word you do not know. You may look up a definition in a dictionary or glossary. Another very useful way is to reread the material to see if the word is defined when it is first used.

In the section "Reading the Data Stream," the term *data bus* is explained in the sentence after the term is mentioned. Another example is in the section titled "Sensors". The term **sensor** is defined in the same sentence where it is first mentioned. These words are usually printed in *italic* or **bold** type.

Note how the definition is organized. It moves from the general to the specific. For example, a sensor is defined as a device that monitors or measures operating conditions. The first part of the definition states that the sensor is a device.

The second part states that its purpose is "to monitor or measure operating conditions".

Apply it!

Meets NATEF Communications Standards for writing paragraphs and organizing and comprehending written information.

In reading this chapter, look for definitions as they are presented in the text. (Hint: The words "is" or "is called" are keys to look for.)

1. Look for the definitions of the words listed as "Terms to Know" at the beginning of the chapter. Write the terms and their definitions on a sheet of paper.

2. Find at least five other terms that are defined in this chapter. Add these, with their definitions, to your list.

3. Locate each term on your list in a dictionary. Compare the definitions from this source with your definitions. If there are differences, how do the sources differ?

SECTION 3 KNOWLEDGE CHECK

1. Give three reasons why a good sensor might supply an abnormal signal to the PCM.

2. How much reference voltage is usually supplied to a sensor?

3. What mechanical problems could cause a solenoid not to work properly?

4. What vehicle maintenance problem could cause a false ECT sensor signal?

5. Name two factors that could cause electrical damage to sensors, actuators, and modules.

CHAPTER 3 REVIEW

Key Points

Meets the following NATEF Standards for Engine Performance: inspecting, testing, and replacing sensors and actuators.

- Speed and position sensors are electromagnetic, optical, and Hall-effect.
- Sensors produce either analog or digital signals.
- Actuators control specific engine management systems to provide the best performance with the lowest emissions.
- When sensors or actuators fail, driveability, engine emissions, and fuel economy are affected.
- Analog signal voltage changes continuously over a range of values.
- Digital signal voltage is either high or low.
- Electronic components communicate on a special wiring harness called a data bus.
- Sensors produce output signals based on the operating condition they are monitoring.
- The output signal of a switch is either on or off.
- Solenoids use electromagnetism to produce motion.
- A relay is used to open and close a set of electrical contacts.

Review Questions

1. What are three types of sensors commonly used to indicate rpm and crankshaft position?
2. What is the difference between an analog signal and a digital signal?
3. Name two devices that can be used to check sensor output signals.
4. What do the sensors do and how do they perform needed tasks?
5. How do sensors and actuators fail?
6. How can a data bus problem affect sensor and actuator operation?
7. When replacing a sensor or actuator, why is it important to make sure cables and harnesses are correctly routed?
8. (Critical Thinking) Explain why misfire in one or more cylinders causes the oxygen sensor to produce a low-voltage signal.
9. (Critical Thinking) How is engine operation affected by an intake air leak between the MAF and the throttle body?

TECHNOLOGY FORECAST
FOR AUTOMOTIVE EXCELLENCE

Sensors and Efficient Power

The sensor is the key to modern engine management. As the computer's "eyes" into the engine, sensors help to improve fuel economy, performance, and emissions. Engineers are designing sensors to provide even more useful and specific information.

One method being studied adds a knock sensor and an oxygen sensor to each cylinder. These sensors will work with the spark plugs. They will tell the computer the status of the combustion process in each cylinder. The computer can then retard or advance timing of the ignition as needed. It can also change the air/fuel mixture if necessary. These changes will allow an engine to produce maximum power with the least emissions.

Also under development is a compression sensor. First used on race cars, it could soon be available in passenger cars and trucks. In one version of this device, a pressure sensor monitors compression in each cylinder. It sends its data to the PCM. Again, performance can be improved while emissions are reduced.

Another sensor that is becoming popular in cars and trucks monitors the engine oil. It indicates when the oil needs to be changed.

AUTOMOTIVE EXCELLENCE
TEST PREP

Answering the following practice questions will help you prepare for the ASE certification tests.

1. Technician A says sensors are used to monitor vehicle operating conditions. Technician B says sensors may produce either analog or digital signals. Who is correct?
   - ⓐ Technician A.
   - ⓑ Technician B.
   - ⓒ Both Technician A and Technician B.
   - ⓓ Neither Technician A nor Technician B.

2. Technician A says many electronic components communicate over a data bus. Technician B says the data bus is another name for a special wiring harness that sends data signals to the PCM. Who is correct?
   - ⓐ Technician A.
   - ⓑ Technician B.
   - ⓒ Both Technician A and Technician B.
   - ⓓ Neither Technician A nor Technician B.

3. Which term identifies a pulse width modulated signal:
   - ⓐ Analog signal.
   - ⓑ Digital signal.
   - ⓒ Frequency modulated.
   - ⓓ Duty cycle.

4. Technician A says a Hall-effect sensor produces a digital signal. Technician B says it produces an analog signal. Who is correct?
   - ⓐ Technician A.
   - ⓑ Technician B.
   - ⓒ Both Technician A and Technician B.
   - ⓓ Neither Technician A nor Technician B.

5. The type of engine coolant sensor commonly used to measure temperature is called a:
   - ⓐ Resistor.
   - ⓑ Reluctor.
   - ⓒ Thermistor.
   - ⓓ Potentiometer.

6. Technician A says an engine with a MAF sensor usually does not use a MAP sensor. Technician B says both sensors are used to determine engine load. Who is correct?
   - ⓐ Technician A.
   - ⓑ Technician B.
   - ⓒ Both Technician A and Technician B.
   - ⓓ Neither Technician A nor Technician B.

7. Normal voltage output from a zirconia oxygen sensor would be:
   - ⓐ 0 to 100 millivolts.
   - ⓑ 0 to 1 volt.
   - ⓒ 0 to 2 volts.
   - ⓓ 0 to 5 volts.

8. Technician A says oxygen sensor output voltage can be checked with a test light. Technician B says it should be checked with a scan tool or DVOM. Who is correct?
   - ⓐ Technician A.
   - ⓑ Technician B.
   - ⓒ Both Technician A and Technician B.
   - ⓓ Neither Technician A nor Technician B.

9. A fuel injector is an actuator controlled by the PCM. It is a type of:
   - ⓐ Solenoid.
   - ⓑ Relay.
   - ⓒ Stepper motor.
   - ⓓ Thermistor.

10. When checking high current-draw actuators such as motors and solenoids, it may be necessary to use a digital volt-ohm-meter to check:
   - ⓐ Circuit resistance.
   - ⓑ Voltage drop.
   - ⓒ Current flow.
   - ⓓ Voltage under load.

CHAPTER 4

Diagnosing & Repairing Air Induction Systems

You'll Be Able To:

- ⊗ Inspect air induction systems.
- ⊗ Identify the components of the air induction system that require scheduled service.
- ⊗ Identify air induction throttle body problems.
- ⊗ Diagnose air metering failures.

Terms to Know:

air filter

air induction system

idle air control (IAC) valve

manifold absolute pressure (MAP) sensor

powertrain control module (PCM)

throttle position sensor (TPS)

The Problem ⟩⟩⟩⟩ Your Challenge

Jeff decides to do some routine maintenance on his car. He goes to the local parts store and purchases an air filter. After replacing the air filter, Jeff discovers that his engine is idling too fast. The current idle speed of 1,500 rpm is more than twice the specified speed.

Mike and Jerry took a look at Jeff's car in the automotive shop at school. Mike says the idle air control valve is stuck open. Jerry believes the engine needs a simple idle speed adjustment.

As a service technician, you need to find answers to these questions:

❶ Why would changing the air filter cause the idle speed to increase?

❷ Do most engines have an idle speed adjustment?

❸ What controls the idle speed in Jeff's engine?

Section 1

Air Induction Systems

The **air induction system** is the system that supplies clean air to the engine and controls the flow of air through the engine. **Fig. 4-1.** Air enters the air induction system through the air filter. The **air filter** is a ring, cylinder, or panel of filter paper that removes dirt particles and debris from the air. The air may also pass through an electronic device called an *airflow sensor.* This sensor measures the amount of air entering the engine.

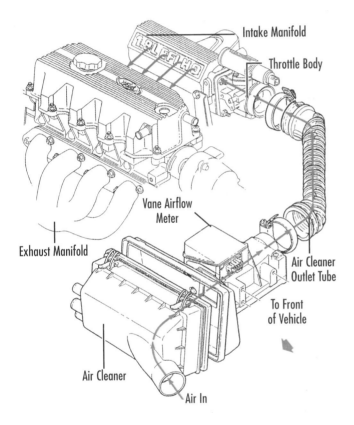

Fig. 4-1 Airflow path from a remote-mounted air cleaner to the intake manifold. *What is the purpose of an airflow sensor?* (Ford Motor Company)

The air then flows around the throttle plate in the *throttle body.* The throttle plate controls engine speed by varying the amount of air entering the engine. After the air leaves the throttle body, it enters the intake manifold. The air passes through the intake manifold and enters the cylinder head. The air mixes with fuel, passes around the intake valve, and enters the cylinder.

A problem in the air induction system will affect exhaust emissions. An air induction problem can also cause engine performance problems.

The major components of the air induction system include the:
• Air filter and housing.
• Throttle body.
• Intake manifold.

The Air Filter and Housing

The air filter removes dirt, contaminants, and small abrasive particles from the air before it enters the engine. As much as 100,000 cubic feet of air may pass through the air filter every 1,000 miles [1609 km] of driving.

Some air filters are mounted on the throttle body. Others are remotely mounted. For example, some mount on the inner fender and are connected to the throttle body by a tube.

The air filter paper, also called the *air filter element* or *filtering media,* must be strong enough to resist tearing. The filter must be flame-resistant in case the engine backfires through the air induction system. To prevent dirt from traveling around the filter, the filter must seal tightly against the *air filter housing.*

The air passing from the filter housing, around the throttle plate, and into the engine creates noise. The air filter housing muffles some of this noise. To further reduce noise, some systems have a *tuning venturi* inside the housing. **Fig. 4-2.** The shape and size of the tuning venturi compensates for noise created by air turbulence.

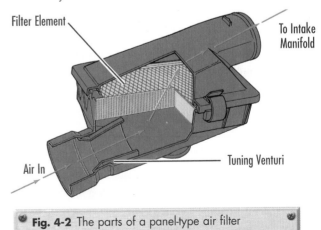

Fig. 4-2 The parts of a panel-type air filter assembly. *What is the job of the tuning venturi?* (DaimlerChrysler)

Heated Air Induction Cold air can cause throttle plate icing conditions and fuel vaporization problems. Heated air induction systems are used to prevent these problems.

Older systems passed air over the exhaust manifold to heat the air. **Fig.** 4-3. A thermostatically controlled blending door in the system regulates the temperature of the air entering the throttle body. The warm air is drawn through the throttle body and into the intake manifold.

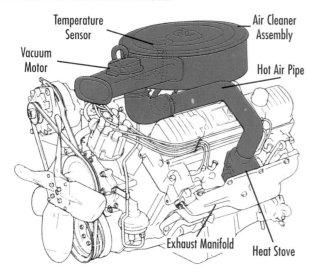

Fig. 4-3 Air is passed across the exhaust manifold to heat the air before entering the throttle body. *What problems are prevented by heated air induction?* (General Motors Corporation)

Another method of preventing the entry of cold air is to heat the throttle body. The throttle body contains passages for engine coolant. Engine coolant flowing through the passages warms the throttle body. The warm throttle body warms the air.

Changing the Air Filter Change air filters at specified intervals or sooner if they become dirty. To change an air filter, follow these steps:

1. Carefully remove the air filter housing lid or cover. Make sure nothing falls into the housing. Allowing debris to enter the air inlet or throttle body can damage the engine.

2. Inspect the air filter and housing to make sure that the filter seals correctly against the housing. A poorly fitting filter will show signs of dirty air flowing around its sealing surfaces. Make sure that any seal or gasket used in the air filter housing is intact and in place.

3. Clean the air filter housing with a shop vacuum or a clean shop towel. If required, remove the housing by removing the fastening screws. **Fig.** 4-4.

4. Make sure that the new air filter fits correctly. Tighten all air inlet connections. Start the engine and make sure it idles correctly.

Fig. 4-4 The air filter housing must be clean, with the seal firmly attached. *What might happen if dirt and debris are not removed from the air filter housing?* (Terry Wild Studio)

The Throttle Body

The *throttle body* contains the throttle plate, the throttle position sensor, and the idle air control valve.

By moving the accelerator pedal, the driver moves the throttle plate, which controls the amount of air entering the engine. The throttle plate is moved mechanically or electronically.

Throttle shaft bushings may wear with use, causing rough idle problems. Varnish can build up around the throttle plate and throttle bore. This may cause a rough idle, sluggish response, or stalling condition.

Vehicle manufacturers apply throttle bore coatings that resist dirt buildup and prevent sticking. Harsh solvents, such as carburetor cleaners, may damage these coatings. Before using any cleaner, follow the vehicle manufacturer's recommendations and procedures.

Most throttle bodies do not require routine adjustment. However, those on some vehicles may require cleaning and adjustment. Some throttle bodies found on cars may have an idle speed and emissions trim adjustment. To avoid damage to

Calculating Airflow

Good engine performance requires an adequate supply of clean air. When an engine operates at high rpms a large volume of air is needed. As much as 100,000 cubic feet of air can pass through the air filter every 1,000 miles [1609 km] of driving.

Let's find out how those numbers relate to everyday engine operation. You can calculate the volume of air in cubic feet that flows through an air filter in a minute. This calculation will help you visualize how much air must pass through an air filter every minute the engine is running.

To calculate the volume of air flowing through an air filter you must first know the engine's rpm and displacement. The engine is a four-stroke, 3.0-liter engine running at 2,500 rpm at 100-percent throttle opening.

A four-stroke engine has two crankshaft revolutions for each complete combustion cycle. At an engine speed of 2,500 rpm, each cylinder will fire 1,250 times per minute.

The engine will move its displacement volume of 3.0 liters of air 1,250 times each minute.

3 liters × 1,250 = 3,750 liters each minute

To find the volume in cubic feet, use this conversion formula:

$$1 \text{ liter} = 0.0353 \text{ ft}^3$$

Thus, 3,750 liters/min. = 3,750 × 0.0353 = 132.4 ft³/min.

This volume of air would more than fill a cube 5 feet on a side (125 ft³).

The actual airflow volume will be less than the calculated volume because no filter and intake system is 100% efficient.

Apply it!

Meets NATEF Mathematics Standards for converting measurements between the metric and English systems.

❶ Determine the volume of air flowing through an air filter at 2,800 rpm at a 100-percent throttle opening for a 2.8-liter, four-stroke engine. Show your answer in liters/min. and ft³/min.

❷ What would happen to the volume of airflow if the engine had twice the displacement volume?

components, consult the vehicle service manual before servicing any throttle body.

Mechanically Operated Throttle On a mechanical throttle, the throttle plate shaft is connected to the accelerator pedal by a cable. Pressing down on the pedal opens the throttle plate. If equipped with *throttle body fuel injection (TBI)*, the throttle plate controls the flow of the air/fuel mixture entering the engine. **Fig. 4-5.** If equipped with *multiport fuel injection (MFI)*, the throttle plate controls only the air entering the engine.

Throttle cables may stretch with use, preventing the throttle plate from opening or closing properly. When necessary, adjust throttle cables according to the procedures specified in the vehicle service manual.

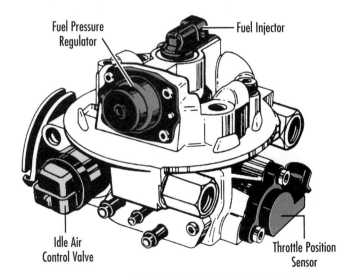

Fig. 4-5 A throttle body unit with fuel injector. *What engine problems can result if debris collects in the throttle body?* (General Motors Corporation)

Electronically Operated Throttle Instead of a mechanical throttle plate linkage, some vehicles use an *electronic throttle control* or *drive-by-wire system.*

A sensor on the accelerator pedal monitors the position of the pedal. The sensor sends a position return signal to the powertrain control module (PCM). The **powertrain control module (PCM)** is a computer that monitors and controls ignition timing, emission control devices, and other engine related systems.

The PCM computes the proper opening for the throttle plate. A small motor mounted on the throttle body opens and closes the throttle plate.

These electronic controls may be part of a *traction control system.* The system provides input to the PCM. The PCM adjusts the throttle-plate position, the air/fuel mixture, the spark advance, and the shift points of automatic transmissions and transaxles.

Throttle Position Sensor The **throttle position sensor (TPS)** is a variable resistance sensor that sends throttle plate position information to the PCM. **Fig. 4-6.** As the throttle plate position changes, the return-signal voltage from the TPS changes.

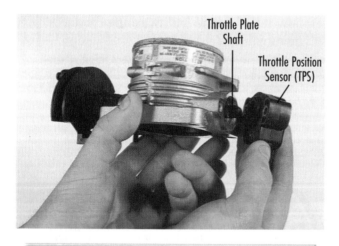

Fig. 4-6 The TPS attaches to the throttle plate shaft. *What is the purpose of the TPS?* (Terry Wild Studio)

Idle Air Control Valve The **idle air control (IAC)** valve is a valve that controls engine idle in response to signals from the PCM. As engine load at idle changes, the idle air control valve regulates the amount of air passing by the throttle plate. **Fig. 4-7.**

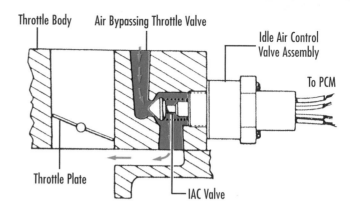

Fig. 4-7 Operation of the IAC valve. A "stepper" motor changes the position of the conical valve to control idle speed. *When would the IAC valve increase idle speed?* (Chevrolet Division of General Motors Corporation)

An example of engine-idle control is the reaction of the IAC to an increase in engine load from the power steering system. **Fig. 4-8.** A pressure switch in the power-steering system sends a return signal to the PCM indicating a pressure increase. To maintain a correct idle speed under the increased load, the engine speed must increase.

The PCM moves the IAC valve so it passes more air around the throttle plate. At the same time, the PCM signals the injectors to deliver more fuel to match the added airflow.

TECH TIP Debris in IAC valve. Debris from changing an air filter can collect in the IAC valve. The IAC valve may stick and not respond properly. This will cause a rough or erratic idle and engine stalling. You may need to remove the IAC valve for cleaning.

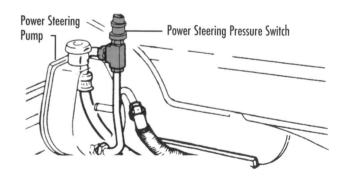

Fig. 4-8 When a load is placed on the power steering system, the power steering pressure switch signals the PCM. The PCM opens the IAC valve to increase idle speed. *What might happen if the IAC valve failed to react?* (Pontiac Division of General Motors Corporation)

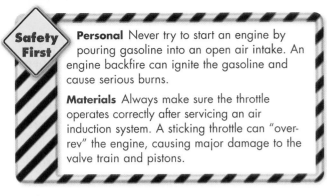

Safety First

Personal Never try to start an engine by pouring gasoline into an open air intake. An engine backfire can ignite the gasoline and cause serious burns.

Materials Always make sure the throttle operates correctly after servicing an air induction system. A sticking throttle can "over-rev" the engine, causing major damage to the valve train and pistons.

The Intake Manifold

The intake manifold connects the throttle body to the intake ports in the cylinder head. The manifold has a set of passages, or runners, through which air or an air/fuel mixture flows.

With multiport fuel injection (MFI), fuel is injected into the air as it enters the intake ports. Because MFI injects fuel directly into the intake port, each cylinder receives the same amount of fuel. MFI provides more accurate control of the air/fuel mixture than does throttle body fuel injection (TBI).

With TBI, fuel mixes with the air as it enters the intake manifold. TBI uses fewer injectors and less fuel line hardware, but it is not as efficient in balancing the air/fuel mixture. Each cylinder should receive the same air/fuel ratio mixture. But with TBI, fuel distribution between cylinders is not always uniform.

Fuel must vaporize completely to mix and flow evenly with air. Cold air does not allow for complete vaporization. Unvaporized fuel contains heavy liquid droplets. Inertia prevents these droplets from turning sharp bends and corners as the air/fuel mixture flows through the intake manifold. These droplets tend to travel in a straight line. They collect and puddle at bends in the manifold. Heating or tuning the manifold can eliminate or reduce these problems.

Tuned Intake Manifold Tuning the intake manifold improves the volumetric efficiency of normally aspirated engines. The runners of the intake manifold are designed to specific sizes and lengths. This design helps provide the highest usable pressure to each cylinder. Long runners improve low-speed performance. Short runners improve high-speed performance.

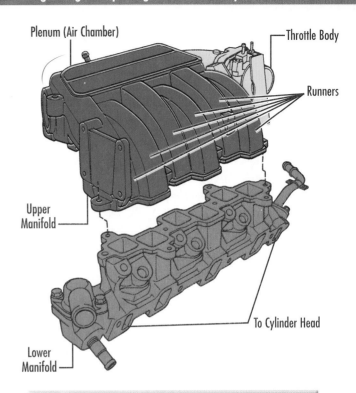

Fig. 4-9 An example of a tuned intake manifold. The long runners improve low-speed performance. *What length of runner will improve high-speed performance?* (DaimlerChrysler)

A tuned intake manifold takes advantage of the opening and closing of the intake valves to produce a "ram pressure" effect. **Fig. 4-9.**

When an intake valve opens, the air/fuel mixture flows into the cylinder. When the valve closes, this flow stops quickly. The inertia of the moving air/fuel mixture tends to keep it moving. This movement creates pressure as the air/fuel mixture piles up, or "rams," against the closed intake valves. When the valve opens, the pressure forces more air/fuel mixture into the usable volume of the cylinder. This is referred to as increasing the *volumetric efficiency.*

A tuned intake manifold is most effective at high engine revolutions per minute (rpm). To improve low- and high-rpm performance, some engines have a *variable induction system.* **Figure 4-10** shows this system on an in-line four-cylinder engine.

The variable induction system has two runners for each cylinder. The long runner, or *primary runner,* is tuned for low speed. The short runner, or *secondary runner,* is tuned for high speed.

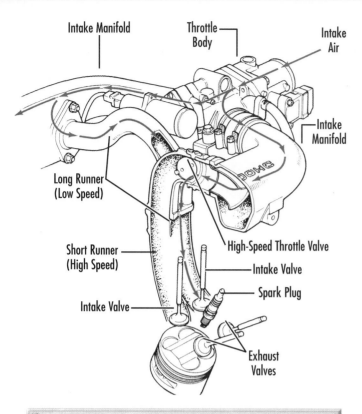

Intake Manifold
Throttle Body
Intake Air
Intake Manifold
Long Runner (Low Speed)
Short Runner (High Speed)
High-Speed Throttle Valve
Intake Valve
Spark Plug
Intake Valve
Exhaust Valves

Fig. 4-10 An electronically controlled variable-induction system has two runners for each cylinder. One runner is for low-speed operation; the other opens for high-speed operation. *What advantage is provided by a variable induction system?* (Ford Motor Company)

The primary runners are open. The secondary runners contain a PCM-controlled valve. This valve is closed at low engine rpm. The valve opens at high engine rpm.

At low engine rpm, only the primary runners are open. At high engine rpm, both runners supply cylinders with air or an air/fuel mixture. The tuned induction system improves power and throttle response.

Superchargers and Turbochargers An engine can produce more power when the air/fuel mixture enters the cylinders under pressure. Pressurizing the air/fuel mixture increases its density. A higher density air/fuel mixture creates more power during the power stroke. This is called *forced induction*. A forced induction engine provides more power than a normally aspirated engine.

Superchargers and *turbochargers* are two devices that provide forced induction. Both devices increase the pressure in the intake manifold. A supercharger is a mechanically driven air pump. The drive is provided by a belt driven off the crankshaft pulley. A turbocharger is a turbine driven air pump. Exhaust gases routed from the exhaust manifold power the turbine.

AUTOMOTIVE COMMUNICATIONS EXCELLENCE

Decoding Acronyms

WYSIWYG, pronounced "wizzy wig" by many computer users, is an acronym for the phrase "What you see is what you get." It is used when talking about displays on a computer screen. In 1941 the term "radio detecting and ranging" was shortened to the acronym RADAR. This acronym is now used as a word. Few people know it is an acronym.

An *acronym* is a word formed from the first letter or first few letters of the words in a phrase. Every area of work or interest uses acronyms. The automotive industry uses many acronyms for parts, tests, equipment, company names, and products.

As a technician, you need to know what the letters in an acronym stand for. For example, you should recognize TPS as meaning "throttle position sensor." This will save time on the job.

Apply it!

Meets NATEF Communications Standards for interpreting acronyms.

❶ As you read through this chapter, note the acronyms that are used.

❷ List the acronyms. Next to them write the words they stand for.

❸ Using a technical dictionary, look up the acronyms you have listed. Does the entry refer you to another entry? What does this tell you about the way acronyms are listed in some dictionaries?

SECTION 1 KNOWLEDGE CHECK

❶ Which part of the air induction system removes dirt from incoming air?

❷ What are two general types of fuel injection systems?

❸ Which part senses throttle plate position?

❹ Which part controls the idle speed of the engine?

❺ What is the purpose of a tuned intake manifold?

Section 2

Sensing Induction Airflow

A fuel injection system requires an accurate determination of the amount of air flowing into the engine. The PCM uses airflow information to determine the amount of fuel to be injected and the amount of spark advance needed. Airflow can be calculated or measured directly.

Calculating Airflow

On some engines the PCM calculates the airflow using the *speed density* method. This is a mathematical relationship using the following factors:

• Engine rpm.
• Throttle position.
• Intake air temperature.
• Manifold absolute pressure (MAP).
• Coolant temperature.
• Exhaust gas recirculation.

These variables are measured by a network of sensors that send signals to the PCM. Using this information, the PCM determines the amount of fuel and spark advance needed.

The manifold absolute pressure sensor plays a key role in speed density calculations. The **manifold absolute pressure (MAP)** sensor is a sensor that is mounted on or connected to the intake manifold. With the key on and engine off, the MAP sensor reads barometric pressure. With the engine running, the sensor reads the (low) pressure in the intake manifold caused by piston movement. **Fig. 4-11.**

Manifold pressure readings are the opposite of manifold vacuum readings. When manifold vacuum is high (at closed throttle, for example), manifold pressure is low. When manifold vacuum is low (at wide-open throttle), the pressure is high.

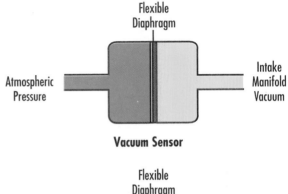

Vacuum Sensor

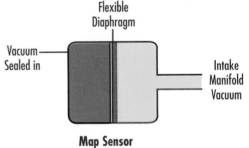

Map Sensor

Fig. 4-11 An ordinary vacuum gauge measures vacuum against atmospheric pressure. The MAP sensor measures intake-manifold vacuum against a constant sealed-in vacuum source. *Which method do you think is more accurate, and why?*

Manifold pressure is an accurate indication of engine load, and indirectly, airflow. The MAP sensor sends a varying voltage or frequency signal to the PCM as engine load and manifold pressure change.

Inputs from other sensors are also used in the airflow calculation. The *intake air temperature (IAT) sensor* measures the temperature of the air. Cool air is more dense than warm air and requires more fuel to maintain a stoichiometric air/fuel ratio, 14.7:1.

The *engine coolant temperature (ECT) sensor* measures engine temperature by monitoring coolant temperature. A cold engine requires more fuel to maintain a stoichiometric air/fuel ratio. When the engine is at operating temperature, the air is less dense and less fuel is needed.

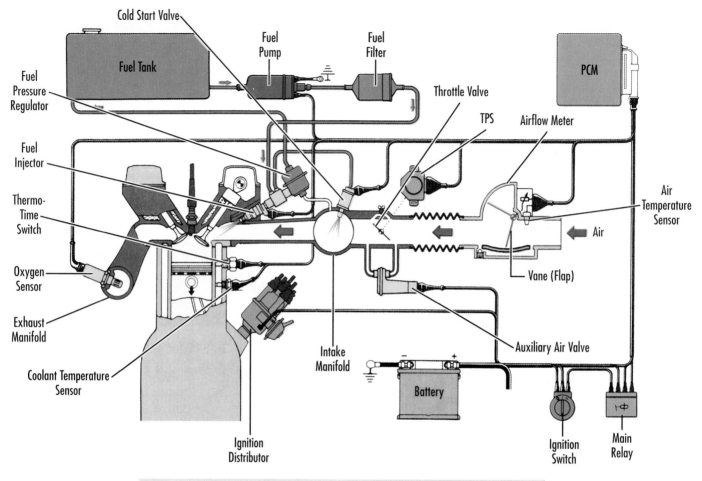

Fig. 4-12 The Bosch L-Jetronic fuel-injection system has a pivoted vane in the airflow meter. This signals the PCM as to how much air is entering the engine. *What determines the amount of vane swing?* (Robert Bosch GmbH)

Measuring Airflow Directly

A *mass airflow* (MAF) *sensor* measures airflow directly. The MAF is an electronic sensor that measures the amount of air entering the engine. The sensor sends a return signal in the form of voltage to the PCM. The MAF signal is used as a measure of airflow. The five sensors used to measure airflow directly are:

• Vane type.
• Hot-wire induction.
• Heated film.
• Airflow sensor plate.
• Karman-Vortex Path.

Each method continuously measures the amount of air flowing into the engine.

Vane-Type Sensor Some fuel injection systems, such as the Bosch L system, use the vane-type airflow sensor. **Fig 4-12.** A calibrated, spring-loaded vane is mounted in the air intake passage of the airflow meter. Air flowing past the vane forces the vane to move. The greater the airflow, the greater the movement of the vane.

A sensor attached to the vane measures its position. A vane position sensor works like a throttle position sensor. The return signal from the sensor tells the PCM how much air is flowing into the engine.

Hot-Wire Induction Sensor A hot-wire induction sensor uses a heated wire. The wire is in the path of the air passing through the airflow meter. An electric current flowing through the wire keeps the wire hot.

The wire is kept at a specific temperature by adjusting the current flow through it. Air passing by the wire cools it down. The more air that passes through the airflow meter, the more heat is lost.

The greater the heat loss, the greater the current flow needed to maintain the temperature. The PCM reads the amount of current as a measurement of airflow.

Heated-Film Sensor A heated-film sensor operates like a hot-wire induction sensor. It consists of metal foil or a grid coated with a current conducting material. **Fig 4-13.** Current flowing through the

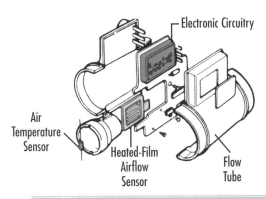

Fig. 4-13 Mass airflow sensor using a heated film and air-temperature sensor. *What heats the coating?* (Buick Division of General Motors Corporation)

material heats it. Air flowing past or through the film cools it. Like the hot-wire induction sensor, the film is maintained at a specific temperature. The more air that passes through the airflow meter, the more heat is lost. The greater the heat loss, the greater the current flow. The PCM reads the amount of current as a measurement of airflow.

Airflow Sensor Plate Mechanical fuel injection systems use an *airflow sensor plate*. **Fig. 4-14.** This is a movable plate in the intake air passage of the airflow sensor. As airflow increases, the plate moves higher. The plate moves a control plunger in the fuel distributor. The plunger controls the pressure of the fuel supplied to the fuel injectors. The greater the airflow, the greater the fuel pressure.

Karman-Vortex Path Sensor The Karman-Vortex Path sensor measures airflow by measuring air turbulence. When air passes over an obstruction, it separates into whirls and eddies. The frequency of the whirls and eddies changes in direct proportion to the amount of air passing over the obstruction. A high level of airflow produces a higher frequency of

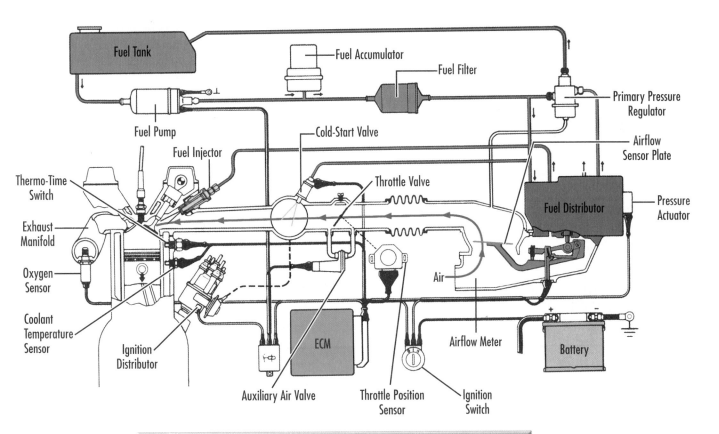

Fig. 4-14 Schematic layout of the Bosch KE continuous-injection system with an airflow sensor plate. *What causes the upward movement of the airflow sensor plate?* (Robert Bosch GmbH)

Measuring Pressure and Vacuum

You have learned that pressure and vacuum are important during engine operation. Pressure and vacuum are usually measured in inches of water or in inches or millimeters of mercury. Mercury is often used because it is about thirteen times heavier (denser) than water. Higher pressure and vacuum can be measured with a smaller column of mercury than with a column of water.

Atmospheric pressure at sea level is 14.7 psi. The pressure of the atmosphere at sea level is normally equal to the weight of a 29.92 in [760 mm] column of mercury. A barometer measures atmospheric pressure in inches or millimeters of mercury. A barometer that uses water instead of mercury would need to be about 34 feet high.

Pressures higher than 3 psi [21 kPa] are usually measured with a mechanical pressure gauge. These gauges include engine compression pressures, fuel pump pressures, and tire pressures.

Safety First

Personal Mercury is poisonous! If a manometer using mercury breaks or spills, do not pick up the mercury or handle it in any way. Call your instructor and get expert help in cleaning up a mercury spill.

Apply it!

Making a Manometer

Meets NATEF Science Standards for measuring pressure and relating the concepts of force and pressure.

Materials and Equipment
- Yardstick
- About 7 feet [2.1 m] of clear plastic tubing, with a diameter of about ½″ [13 mm]
- Clear plastic tape
- Small funnel or meat-basting bulb
- Food coloring (any color)

In this activity, you will make a simple manometer. A manometer is an instrument used to measure pressure and vacuum.

Mano comes from the Greek word *manos*, meaning "thin or rare." This suggests a vacuum. *Meter* means "to measure," as in thermometer, speedometer, and altimeter. Thus, the word *manometer* refers to a device used to measure something "thin or rare" (such as a vacuum).

❶ Fold the tubing in half to form a U-shape.

❷ Tape the center of the U to the bottom of the yardstick. Tape the tubing to the edges of the yardstick. Leave about 6″ of loose tubing on both sides at the top.

❸ Hang the yardstick so that the ends of the loose tubing are about level with your shoulders.

❹ Use a small funnel or a meat-basting bulb to fill the tube with colored water until the water level is near the 18″ mark on the yardstick.

The water is exposed to atmospheric pressure in both ends of the tube.

❺ Blow gently into one end of the tubing while watching the water level in each end of the tubing. Try to hold the level constant while someone records the measurement on both sides of the stick.

❻ Subtract the smaller number from the larger number. This will indicate the amount of pressure, in inches of water, that you created by blowing into the tube. The water in the open end was still exposed to atmospheric pressure. The water in the other end was now under more pressure. The water was pushed down by the pressure.

❼ Suck gently on the tube to create the same difference in water levels. This will indicate the amount of vacuum, in inches of water, that you have created. The water in the open is still exposed to atmospheric pressure. Now the water on the other end is under less pressure. The level rises in this end of the tube.

whirls and eddies than a lower level of airflow. The change in frequency is a measurement of airflow. The higher the frequency, the greater the airflow.

Diagnosing Airflow Sensor Failures

Measuring the airflow into an engine is important. Incorrect airflow measurements will produce incorrect air/fuel mixtures. A defective airflow sensor will send the wrong return signal to the PCM. This will cause the PCM to incorrectly adjust the air/fuel mixture. An incorrect air/fuel mixture can cause starting, fuel economy, drivability, and exhaust emission problems. Several conditions can cause a failure in the airflow sensor system.

Indirect measurement (speed density) systems may have problems with vacuum leaks that occur after the air leaves the throttle body. Air leaking into the intake manifold reduces manifold vacuum. It may also cause the idle speed to increase. Excess air leaking into the intake manifold will also cause a lean air/fuel mixture.

The cause of many vacuum and air leaks is usually a defective vacuum hose or gasket. Rubber and plastic vacuum hoses may develop leaks and cracks due to age or contamination. Manifold and throttle body gaskets may dry out, crack, and deteriorate, allowing air to enter through them.

Many air and vacuum leaks can be detected by using a probe that can inject a small amount of propane gas into the suspected area. **Fig. 4-15.** Propane gas is flammable. The gas will be drawn through the leak instead of the air. If a leak is present, the propane gas will be drawn into the cylinders and produce a noticeable change in engine operation. The gas may cause a change in engine idle speed. An engine stalling or rough running condition may change or improve.

Direct airflow measurement systems may have problems with air entering the engine after the airflow is measured. With this type of leak, the airflow sensor is not measuring all of the air entering the engine. The measurement signal sent to the PCM is not correct. An incorrect airflow measurement will cause the PCM to supply an incorrect air/fuel mixture. It is sensing too little air. It compensates by reducing the amount of fuel the fuel injectors are supplying. The resulting lean mixture will produce engine performance problems.

Fig. 4-15 A network of vacuum hoses carries vacuum to actuators. *How can a leak in the vacuum hose network affect engine performance?* (Jack Holtel)

> **TECH TIP** **Free Air.** Air leaking through the tube connecting the airflow sensor to the throttle body can cause engine stalling. This air is called "free," or unmetered, air. After servicing any air induction system, make sure all connections are tight. Check for leaks or cracks in the air inlet tube.

SECTION 2 KNOWLEDGE CHECK

❶ What are the two basic methods of measuring airflow?

❷ What is the key sensor used for speed density metering?

❸ What are five methods of measuring airflow directly?

❹ Which systems use a spring-loaded device to measure airflow?

❺ Which airflow measurement systems use current flow to measure airflow?

CHAPTER 4 REVIEW

Key Points

Meets the following NATEF Standards for Engine Performance: inspecting the air induction and filtration system and the intake manifold.

- Maximum engine life requires air induction system maintenance.
- Air induction system parts may require periodic cleaning and adjustment.
- A faulty IAC valve in the throttle body will cause idle and stalling problems.
- Airflow sensor operation is critical so the PCM receives correct air metering information.
- A tuned intake manifold provides more power output from the engine.
- Superchargers and turbochargers are two devices that provide forced induction.
- The five types of sensors used to measure airflow directly are: vane type, hot-wire induction, heated film, airflow sensor plate, and Karman-Vortex path.
- Incorrect airflow measurements will produce incorrect air/fuel mixtures.

Review Questions

1. Why is inspection of the air induction system directly related to controlling exhaust emission?
2. What components of the air induction system require periodic scheduled service and inspection?
3. How would a sticking throttle plate affect engine performance?
4. How can an air metering failure affect fuel economy?
5. What three factors affect the airflow through an engine?
6. What are some common air induction throttle body problems?
7. How are air metering failures diagnosed?
8. **Critical Thinking** How would a faulty airflow sensor cause an emissions or drivability problem?
9. **Critical Thinking** You are working on a vehicle with driveability problems. At cruising speed, the engine performs properly. However, when slowing down or stopping, the engine stalls. What could be causing the problem?

TECHNOLOGY FORECAST
FOR AUTOMOTIVE EXCELLENCE

More Efficient Combustion

Thanks to a device called a magnetic actuator, engine intake and exhaust valves may soon be operated electronically. Today's mechanical camshafts or push rods will no longer be needed.

How does it work? The actuator sits on top of each valve. It moves back and forth when electricity is sent to it by a computer. When the actuator is turned on, it pulls the valve open. When the actuator is turned off, a spring attached to the valve pushes it closed. This design will let automakers build engines with lower emissions and higher fuel economy. Performance will also be improved.

Electronic valves make these benefits possible thanks to variable valve timing. By being able to control when a valve opens and closes, the engine can operate more efficiently. The computer orders these changes based on how the vehicle is being driven.

New ways to get more oxygen into the engine are also being studied. Special filters are being developed to collect oxygen from the air coming into the engine. The oxygen will then be sent to the combustion chamber. What are the benefits of adding oxygen to the engine? Extra oxygen will help fuel burn cleaner and more efficiently.

AUTOMOTIVE EXCELLENCE
TEST PREP

Answering the following practice questions will help you prepare for the ASE certification tests.

1. The throttle plate is coated with dirt and is sticking. What should be done?
   a Spray the throttle plate with solvent.
   b Replace the throttle plate.
   c Consult the vehicle service manual for the proper procedure.
   d Do nothing. This is normal.

2. A nonfunctioning IAC valve might cause what condition?
   a The engine would stall at idle.
   b The engine would not accelerate.
   c The engine would not start.
   d None of the above.

3. An engine hesitates upon acceleration. Technician A says the TPS may not be showing the right throttle position. Technician B says the air/fuel mixture may be too lean. Who is correct?
   a Technician A.
   b Technician B.
   c Both Technician A and Technician B.
   d Neither Technician A nor Technician B.

4. Cold engine driveability problems could be caused by what condition?
   a A clogged air filter.
   b A stretched throttle cable.
   c A dirty throttle plate.
   d A lack of warm air to improve fuel vaporization.

5. Technician A says MFI delivers fuel more evenly than does TBI. Technician B says TBI is cheaper to manufacture. Who is correct?
   a Technician A.
   b Technician B.
   c Both Technician A and Technician B.
   d Neither Technician A nor Technician B.

6. An engine with a variable induction system lacks power and throttle response at high rpm. What could be causing the problem?
   a The secondary runners are too short.
   b The PCM controlled valves in the secondary runners are not operating properly.
   c The PCM controlled valves in the primary runners are defective.
   d None of the above.

7. An engine equipped with an indirect airflow measuring system is running rich. Technician A says the MAP sensor may be defective. Technician B says the throttle plate is not opening all the way. Who is correct?
   a Technician A.
   b Technician B.
   c Both Technician A and Technician B.
   d Neither Technician A nor Technician B.

8. Speed-density metering is a form of what type of airflow measurement?
   a Airflow sensor plate measurement.
   b PCM controlled airflow measurement.
   c Hot-wire induction measurement.
   d Indirect airflow measurement.

9. Technician A says electric current heats a heated-film sensor. Technician B says the temperature of the intake air heats the film. Who is correct?
   a Technician A.
   b Technician B.
   c Both Technician A and Technician B.
   d Neither Technician A nor Technician B.

10. What type of air induction sensor uses air turbulence to measure airflow?
   a A hot-wire induction system.
   b A Karman-Vortex Path system.
   c A vane-type system.
   d A heated-film system.

CHAPTER 5

Diagnosing & Repairing Fuel Systems

You'll Be Able To:

- ⊗ Test electric fuel pumps.
- ⊗ Test fuel pressure regulators.
- ⊗ Check fuel for contaminants and quality.
- ⊗ Inspect fuel filters.
- ⊗ Test fuel injectors.

Terms to Know:

air/fuel ratio
antioxidant
closed-loop operation
fuel pressure regulator
octane number
pulse width
stoichiometric ratio
volatility

The Problem 〉〉 〉〉 〉〉 Your Challenge

Clouds of black smoke come out of the tailpipe of Jack's vehicle after the engine warms up. The engine also lacks power and has poor fuel economy. Jack takes the vehicle to Jerry's Garage and tells Jerry that several other shops could not diagnose the problem. Jerry assigns the vehicle to you.

The engine is getting too much fuel. Where is it coming from? Is there a leak allowing excess fuel to enter the cylinders? Is some other factor causing the fuel system to supply too much fuel? How can you isolate the problem?

As the service technician, you need to find answers to these questions:

❶ What is the service history of the vehicle?

❷ How could this excess fuel get into the intake manifold?

❸ What is your first diagnostic step?

Section 1

Characteristics of Automotive Fuels

Automotive fuels are hydrocarbon compounds. They contain two basic elements, hydrogen and carbon. The chemical symbol for hydrogen is H. The chemical symbol for carbon is C.

Gasoline

Oil companies refine gasoline from crude oil (petroleum). **Fig. 5-1.** During the refining process, chemical additives are added to the gasoline. These additives improve the combustion properties of gasoline. They also help protect the engine and fuel system.

Fig. 5-1 A refinery breaks down and refines petroleum into many useful products. Refineries produce motor oil, gasoline, kerosene fuel oil, and gaseous products. *Automotive fuels are what type of compounds?*
(Mark Green/FPG International)

A good-quality gasoline should have:
- Proper volatility, which determines how easily the gasoline vaporizes (turns to vapor).
- The correct octane rating, which minimizes detonation (spark knock). **Fig. 5-2.**
- **Antioxidants** which are chemical compounds that prevent formation of varnish in the fuel system.
- Corrosion inhibitors, which prevent rusting of metal parts in the fuel system.
- Anti-icers, added seasonally or by geographical region, that minimize icing in the throttle body and fuel line.

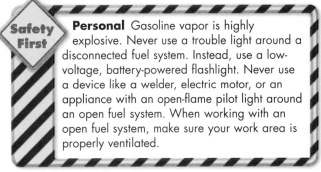

Safety First

Personal Gasoline vapor is highly explosive. Never use a trouble light around a disconnected fuel system. Instead, use a low-voltage, battery-powered flashlight. Never use a device like a welder, electric motor, or an appliance with an open-flame pilot light around an open fuel system. When working with an open fuel system, make sure your work area is properly ventilated.

- Detergents, which keep the fuel injectors clean.
- Deposit control agents, which prevent or remove fuel system deposits.

Volatility One measure of how easily a liquid (fuel) vaporizes is known as **volatility.** If volatility is too low, the fuel will not vaporize and mix properly with air. Gasoline that does not vaporize causes incomplete combustion. It leaves the cylinder as unburned hydrocarbons.

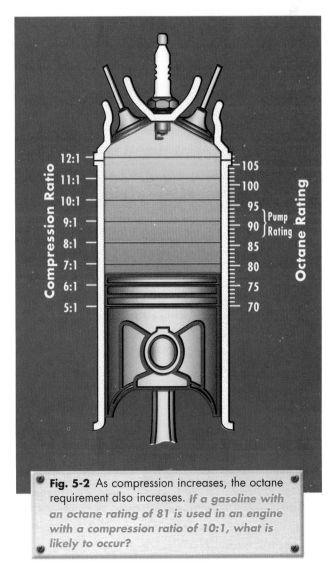

Fig. 5-2 As compression increases, the octane requirement also increases. *If a gasoline with an octane rating of 81 is used in an engine with a compression ratio of 10:1, what is likely to occur?*

If volatility is too high, the fuel may turn to vapor in the fuel system. This condition is *vapor lock*. Vapor lock prevents normal fuel flow through the fuel system.

To suit weather conditions and to meet local evaporative emissions standards, refiners adjust volatility. They increase volatility for cold weather and reduce it for hot weather. **Fig. 5-3.**

Poor-quality gasoline can cause hard starting, hesitation, stalling, and detonation. Changing brands or grades of gasoline may eliminate these problems.

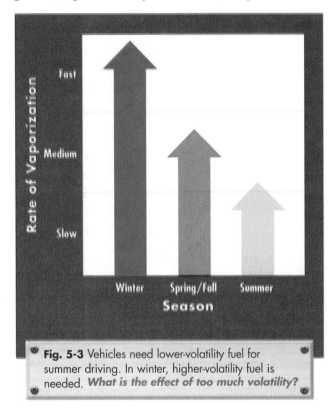

Fig. 5-3 Vehicles need lower-volatility fuel for summer driving. In winter, higher-volatility fuel is needed. *What is the effect of too much volatility?*

Octane Number The antiknock quality of a gasoline is indicated by its **octane number.** This is also known as the octane rating or octane level. The higher the octane number, the more resistant the gasoline is to detonation (spark knock). A 91-octane gasoline is more knock-resistant than an 87-octane gasoline. A gasoline that detonates easily is a *low-octane* gasoline.

The octane number is determined by testing gasoline in laboratory engines. Two common laboratory tests determine research (R) and motor (M) octane numbers. The average of these two values [(R+M)/2] is the *antiknock index (AKI)*. This number is posted on the pumps at service stations. **Fig. 5-4.**

89
Minimum Octane Rating
(R + M) / 2 Method

Fig. 5-4 Minimum octane ratings are posted on gasoline pumps. *Why is it important to select a fuel with the proper octane rating?* (Ford Motor Company)

Octane Requirements The two most important factors that determine the octane requirement of an engine are *combustion chamber design* and *compression ratio.* However, the octane requirement changes with operating conditions such as weather, altitude, and driving conditions.

Octane requirements can also be affected by changes in the mechanical condition of the engine. These include:
- Reduced cooling system efficiency.
- Lean air/fuel mixtures.
- Over-advanced ignition timing.
- Increased compression ratio.
- Failure of the *exhaust gas recirculation (EGR)* valve.

Using gasoline with a higher octane rating than is needed wastes money and resources. A high-octane or premium gasoline should be used only when recommended by the manufacturer or when detonation is an ongoing problem.

For example, engines may have a "stack up" of production tolerances as a result of machining or rebuilding. The change in tolerances reduces combustion chamber volume. The compression ratio is increased, causing detonation.

Gaseous Fuels

Some engines are designed to run on gaseous fuels, such as *liquefied petroleum gas (LPG)* or *compressed natural gas (CNG).* **Fig. 5-5.** Under pressure, petroleum gas turns into a liquid. Upon release of pressure, the liquid turns back to a gas. CNG is not a liquid. It is natural gas in a compressed state.

An advantage of LPG or CNG is that both have an octane rating over 100. LPG and CNG also burn

Increasing Oxygen in Fuel

Automotive gasoline is a mixture of many different hydrocarbons. When hydrocarbons burn in the combustion chamber of an engine, energy is released. This energy moves the piston to create mechanical energy. To burn a hydrocarbon, oxygen atoms are added to the fuel in the combustion chamber. A spark then ignites the mixture.

In a 100-percent efficient gasoline engine, all hydrocarbons would burn. Only carbon dioxide (CO_2), water (H_2O), and nitrogen (N) would be left to come out the tailpipe. Automotive engines are not 100-percent efficient. Unburned fuel, carbon monoxide (CO), and nitrogen oxides are also produced.

Increasing the amount of oxygen in the combustion chamber will burn more of the fuel. Less CO and unburned fuel will go out the tailpipe. More energy will be released.

Unlike pure hydrocarbons, oxygenated fuels contain hydrogen, carbon, and oxygen atoms.

Blending oxygenated fuels with gasoline will get more oxygen into the chamber.

Ethanol, an alcohol made from corn, is an oxygenated fuel. As ethanol burns, its oxygen atoms are freed to help convert more hydrocarbons. As a result, more fuel burns. Oxygenated fuels lower carbon monoxide levels in the exhaust.

Too much ethanol in a fuel system can harm engine components. Most manufacturers recommend a maximum of 10 percent of ethanol in any fuel.

Gasoline and ethanol will mix. However, gasoline and water will not mix. The gasoline will float on water, forming two layers.

Add water to a gasoline sample that contains ethanol. Now the ethanol will separate from the gasoline and mix with the water. Two layers will again form. The gasoline without the ethanol will float on the mixture of water and ethanol.

Apply it!

Measuring the Ethanol Content of Gasoline

Meets NATEF Science Standards for safety and proper waste disposal.

Materials and Equipment
- 10 ml gasoline
- Eyedropper
- Graduated cylinder with a stopper
- Eye protection

You will use a very small sample (10 ml) of gasoline for this test. Make sure you use gasoline containing ethanol. Before you start, review all of the *Safety First* alerts in this chapter.

Safety First

Personal Use eye protection. Gasoline is very flammable. Keep your sample away from sparks or flames. Be careful not to spill any of your sample.

1. Place the gasoline in the graduated cylinder.

2. Using the eyedropper, add 2 ml of water to the sample.

3. Shake the sample for one minute. Gently remove the stopper several times during this procedure to release pressure.

4. Two layers should now be visible in the cylinder. The bottom layer is ethanol and water, while the upper layer is the gasoline.

5. You added 2 ml of water to the cylinder. Now subtract 2 ml from the measured depth of the bottom layer of water and ethanol. The result is the amount of ethanol. Multiply this number by 10. This is the percentage of ethanol in the original sample.

6. Properly dispose of your sample according to OSHA and EPA guidelines.

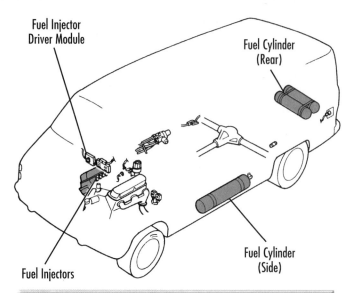

Fuel Injector
Driver Module

Fuel Cylinder
(Rear)

Fuel Cylinder
(Side)

Fuel Injectors

Fig. 5-5 LPG or CNG vehicles have special equipment that allows them to run on gaseous fuels. *What are the advantages of gaseous fuels?* (DaimlerChrysler)

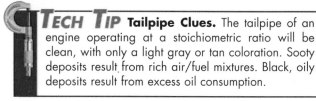

***TECH TIP* Tailpipe Clues.** The tailpipe of an engine operating at a stoichiometric ratio will be clean, with only a light gray or tan coloration. Sooty deposits result from rich air/fuel mixtures. Black, oily deposits result from excess oil consumption.

correct mixture of air to fuel. If the ratio is lower (14:1, for example), there is too much fuel for the available air. If the ratio is higher (16:1, for example), there is excess air (or a shortage of fuel).

Most vehicles manufactured since 1981 use a computer to maintain the stoichiometric ratio. This allows the catalytic converter to be most efficient in lowering exhaust pollutants.

cleanly. Clean combustion results in reduced exhaust emissions. Gaseous fuels also have disadvantages. They provide less fuel economy. They require a heavy tank capable of holding a high pressure. They also need special injection equipment to meter the fuel into the engine.

Air/Fuel Ratio

As engine demands and loads change, the fuel system must maintain a proper air/fuel ratio. **Fig. 5-6.** The **air/fuel ratio** is the proportion of air and fuel, by weight, supplied to the engines cylinders for combustion. Fifteen pounds of air to one pound of fuel is an example of the ratio of air to fuel by weight. This is written as 15:1.

The ideal air/fuel ratio is the stoichiometric ratio of 14.7:1. The **stoichiometric ratio** provides the most efficient combustion, giving the chemically

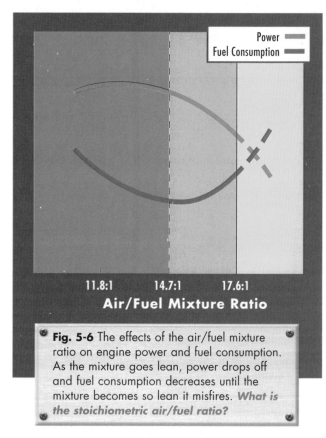

Power
Fuel Consumption

11.8:1   14.7:1   17.6:1

Air/Fuel Mixture Ratio

Fig. 5-6 The effects of the air/fuel mixture ratio on engine power and fuel consumption. As the mixture goes lean, power drops off and fuel consumption decreases until the mixture becomes so lean it misfires. *What is the stoichiometric air/fuel ratio?*

SECTION 1 KNOWLEDGE CHECK

❶ Why is gasoline called a hydrocarbon fuel?

❷ What is another name for the antiknock properties of gasoline?

❸ What are the two most important design factors that determine the octane requirement of an engine?

❹ Name four operating conditions that can change the octane requirement of an engine.

❺ What is the stoichiometric air/fuel ratio for gasoline?

Section 2

The Combustion Process

A hydrocarbon fuel and oxygen combine and burn during the combustion process. The burning gases get as hot as several thousand degrees. This high temperature causes the gases in the cylinder to expand very rapidly. This creates a high pressure that forces the piston down in the cylinder.

Factors Affecting Combustion

The combustion of the air/fuel mixture is affected by the following factors:

• Atmospheric air pressure.
• Manifold absolute pressure.
• Humidity.
• Air temperature.

Atmospheric Air Pressure At sea level and average temperature, a cubic foot of air weighs about 1.25 ounces. The total weight of thousands of cubic feet of air pressing down on the earth is about 14.7 psi [101 kPa] at sea level.

Atmospheric pressure is called *barometric pressure.* Barometric pressure is affected by altitude and weather conditions. As the altitude increases, the barometric pressure decreases. For example, the barometric pressure can be lower in Denver, Colorado, than it is in Los Angeles, California.

High- and low-pressure weather fronts affect barometric pressure. When a low-pressure storm front moves into an area, the barometric pressure usually drops below 29.92″ [101 kPa]. As the low-pressure front moves out, a high-pressure front replaces it. A high-pressure front can increase barometric pressure above 29.92″ [101 kPa].

On any given day, the barometric pressure is constantly changing. The engine's fuel management system must continually adjust to these changing conditions.

Using a Dictionary

You will often read technical material that is new to you. As you read, you will find words that you do not know. Many of these unfamiliar words will not be explained in either the text or the glossary.

In your trade, you deal with flammable materials as well as parts that ensure the safety of your customer. Your skills and understanding are critical. Therefore, you need to find the meaning of an unfamiliar word. You also need to check words that have meanings you are not sure of.

When you look up a word in a dictionary, you may find a definition directly related to automobiles. "Spark" is one example. Among the definitions, you will find one related to vehicles. For example "spark plug" is defined. If there is no auto-related definition, as with "residual," you will find one you can adapt for your use. "The leftovers from a process" is a definition that can be applied to the word "residual" in the section, "Residual Fuel Pressure."

Apply it!

Meets NATEF Communications Standards for using a dictionary.

❶ Continue to read this chapter.

❷ List words you cannot define.

❸ Look up these words in the dictionary. Write out their definitions. Save this list to help you as you study this chapter.

The lower the barometric pressure, the lower the air density. Lower air density needs less fuel for complete combustion. Higher air density needs more fuel for complete combustion. Computer-controlled engines have barometric sensors to change the air/fuel mixture as the barometric pressure changes.

Manifold Absolute Pressure A *vacuum* is a measurement of air pressure that is less than atmospheric pressure. When a piston moves down in a cylinder, the piston creates a partial vacuum, drawing air and fuel into the cylinder. An engine that uses vacuum in this manner is a normally aspirated engine.

The difference between atmospheric pressure and a partial vacuum is *absolute pressure*. Engines measure this difference with a manifold absolute pressure (MAP) sensor.

Throttle opening and engine load affect manifold pressure. A closed-throttle idle may produce a MAP reading as low as 5 psi. A wide-open throttle condition may produce a MAP reading of about 14 psi.

Barometric pressure also affects the MAP reading. For each 1,000' [305 m] above sea level, the MAP reading decreases about 1″ [25.4 cm] of mercury.

Humidity Humidity is measured as a percentage of relative water vapor. Air holding no water vapor has a relative humidity of 0 percent. Air that is completely saturated with water vapor has a relative humidity of 100 percent. A relative humidity reading of 50 percent means the air is holding half as much water as it can hold. An engine running in humid air requires slightly less fuel than an engine running in dry air.

Air Temperature An increase in air temperature decreases the density of the air. A decrease in air temperature increases the density of the air. The density of the air affects the air/fuel ratio. Dense air requires more fuel.

Abnormal Combustion

During normal combustion, the flame front travels gradually across the combustion chamber. A gasoline with good antiknock quality produces a smooth pressure rise in the cylinder. **Fig.** 5-7(**a**).

(a) Normal Combustion

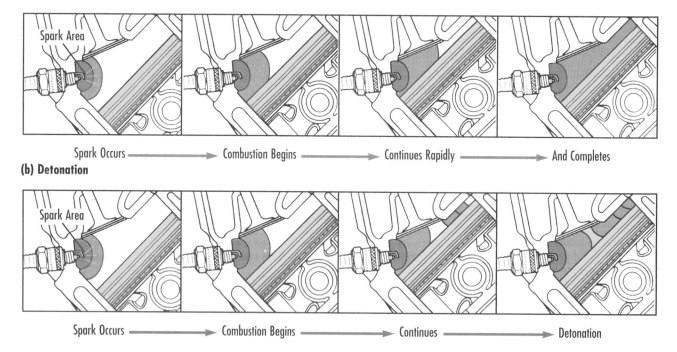

Spark Occurs ⟶ Combustion Begins ⟶ Continues Rapidly ⟶ And Completes

(b) Detonation

Spark Occurs ⟶ Combustion Begins ⟶ Continues ⟶ Detonation

Fig. 5-7 (a) Normal combustion without detonation. The fuel charge burns smoothly from beginning to end. This provides an even, powerful force on the piston. **(b)** Detonation. The last of the fuel mixture explodes or burns almost instantly to produce detonation (or spark knock). *What can detonation do to an engine?* (Champion Spark Plug Company)

Safety First

Personal Carbon monoxide is a poisonous gas. It is invisible, it has no odor, and it has no taste. Avoid working on vehicles in enclosed, unventilated spaces while a vehicle's engine is running. Make sure all engine exhaust is properly and completely vented from the workspace. To further reduce the possibility of carbon monoxide poisoning, be sure that your workspace is properly ventilated.

Symptoms of carbon monoxide poisoning include drowsiness, dizziness, headache, and nausea.

Figure 5-7(b) also shows a form of abnormal combustion known as *detonation* or *spark knock*. The flame starts across the combustion chamber. However, before the flame reaches the far side, a portion of the mixture explodes rather than burns. The two flame fronts meet, producing a very rapid pressure rise. The result is a high-pitched metallic pinging noise called detonation. A low-octane fuel or a high compression ratio can cause detonation.

The sudden power shocks of detonation can damage an engine. Damaged pistons, rings, and spark plugs are the most frequent result of prolonged (or frequent) detonation.

Composition of Exhaust Gases

When complete combustion occurs, all the hydrogen and carbon in the fuel combines with oxygen. The resulting exhaust contains only water (H_2O), carbon dioxide (CO_2), and nitrogen (N).

TECH TIP **Exhaust Fumes.** Make sure all engine exhaust is removed from any enclosed workspace. Attach a length of heat-resistant flexible tubing over the opening of the exhaust pipe(s) of the vehicle. Route the tubing to the outside through a window or suitable opening. Make sure the tubing is attached whenever the engine is operating.

These are basically harmless compounds. But fuel combustion is not always complete, so products of incomplete combustion occur.

Fuel that does not burn completely produces unburned hydrocarbons (HC) and carbon monoxide (CO). In addition, the high temperature of combustion (over 3,000°F [1,650°C]), forms nitrogen oxides (NO_X). **Fig. 5-8.**

Carbon monoxide is a poisonous gas. Unburned hydrocarbons and nitrogen oxides combine chemically in the presence of sunlight to form smog. Various federal, state, and local agencies regulate the emission of all three of these gases.

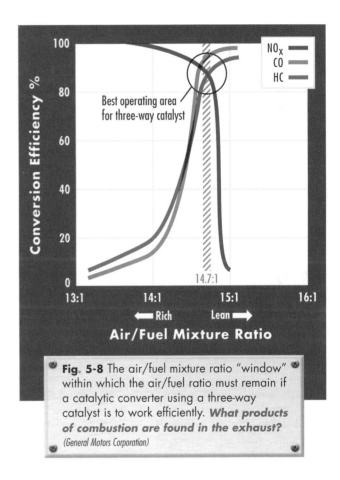

Fig. 5-8 The air/fuel mixture ratio "window" within which the air/fuel ratio must remain if a catalytic converter using a three-way catalyst is to work efficiently. *What products of combustion are found in the exhaust?*
(General Motors Corporation)

SECTION 2 KNOWLEDGE CHECK

① What element in the air does fuel combine with to complete the combustion process?

② What is a normally aspirated engine?

③ What can cause fuel detonation?

④ Complete combustion produces what three harmless compounds?

⑤ What two compounds are produced by fuel that does not burn completely?

Section 3

The Fuel Management System

The fuel management system consists of two subsystems. **Fig. 5-9.** These are the fuel supply and fuel metering systems. The fuel supply system delivers fuel from the fuel tank to the fuel metering system. The fuel metering system mixes air and fuel, and delivers this mixture to the cylinders.

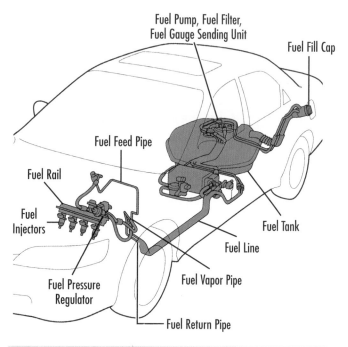

Fig. 5-9 The fuel management system. *What are the two fuel management subsystems?* (American Honda Motor Company)

The Fuel Supply System

The fuel supply system consists of several basic parts: the fuel tank, fuel pump, fuel filter, fuel pressure regulator, fuel lines, fuel rails, and vapor recovery system.

Fuel Tank

The fuel tank is a dome-shaped tank that holds the fuel. It is made of metal or plastic and contains the *fuel pump,* the *fuel level sensor,* and a portion of the *vapor recovery system.*

The fuel tank cap has a valve that prevents excessive buildup of pressure or vacuum in the fuel tank. Warm temperatures increase fuel vapor pressure. Cool temperatures decrease pressure, creating a vacuum.

Fuel Pump

The fuel pump and fuel sensor are combined as a unit, or module. This module is located inside the fuel tank. The pump in the tank allows it to pressurize the fuel in the fuel lines. **Fig. 5-10.** The pressurized fuel cannot vaporize in the fuel lines. This reduces the occurrence of vapor lock.

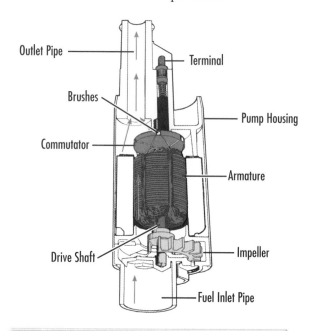

Fig. 5-10 Cutaway of a typical in-tank electric fuel pump. *Why are most fuel pumps located in the fuel tank?* (DaimlerChrysler)

When the ignition switch is turned on, the powertrain control module (PCM) activates the fuel pump relay for a few seconds. The relay supplies power to the fuel pump, which runs briefly to build pressure in the fuel system.

When the engine cranks and starts, the PCM receives an ignition reference, or "tach," signal. This signal tells the PCM that the engine is running. The PCM keeps the fuel pump relay closed and the fuel pump running. On some vehicles, the oil pressure switch serves as a backup to the fuel pump relay. If the relay does not close as it should, the oil pressure switch supplies voltage to the fuel pump when the oil pressure becomes about 2–4 psi [14–28 kPa].

TECH TIP **Fuel Pump.** Most electric fuel pumps are located in the fuel tank. If the fuel pump is receiving voltage, you should hear a low humming noise from the fuel tank.

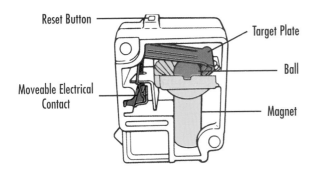

Reset Button

Moveable Electrical Contact

Target Plate

Ball

Magnet

Fig. 5-11 Construction of an inertia switch. During a collision, the ball dislodges from its seat and opens the fuel pump circuit. *Why are inertia switches used?* (Ford Motor Company)

Some vehicles have an *inertia switch* in the fuel pump circuit. **Fig. 5-11.** The inertia switch disables the fuel pump in the event of an accident. This action prevents fuel from spraying from a broken fuel line.

If accidentally bumped, the inertia switch can disable the fuel pump circuit. The switch must be reset before the fuel pump will operate. The vehicle service manual describes the location and reset procedure for the inertia switch.

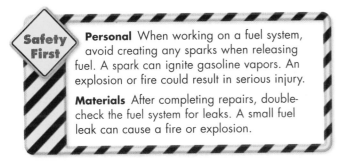

Safety First

Personal When working on a fuel system, avoid creating any sparks when releasing fuel. A spark can ignite gasoline vapors. An explosion or fire could result in serious injury.

Materials After completing repairs, double-check the fuel system for leaks. A small fuel leak can cause a fire or explosion.

Residual Fuel Pressure When a vehicle is shut off, the fuel pump maintains residual pressure in the fuel system. This pressure must be safely released before performing any work on the fuel system. These are the recommended safety practices to follow when removing fuel filters or other fuel system parts.

- Before disconnecting any fuel line, assume that it is pressurized. To prevent personal injury and property damage, follow the manufacturer's recommended procedures.
- Remove the negative battery cable to eliminate the potential for sparks.
- Remove the fuel tank cap to release pressure in the tank. Fuel expansion causes pressure in the fuel tank.

- Some systems are equipped with a *Schrader valve*. Attach a fuel pressure tester to the Schrader valve, following the manufacturer's procedures. Use the volume release valve on the tester to relieve pressure in the fuel system.
- Many systems do not have a Schrader valve. In these cases, manufacturers recommend removing the fuel tank cap and loosening a fuel line fitting. Cover the fitting with a shop towel to absorb any released fuel.
- You may need special tools to remove fuel filters or fuel lines. Consult the vehicle service manual before attempting any removal or repair.

Testing the Fuel Pump Fuel pumps must supply fuel at a specified pressure and volume. Follow the manufacturer's procedures when performing fuel pressure and fuel volume tests.

- Pressure and volume can be measured by connecting a fuel pressure gauge to the fuel system. **Fig. 5-12.**
- System design will determine where the fuel gauge should be connected. If a Schrader valve is present, connect the gauge to it.
- Connecting a pressure gauge to a system without a Schrader valve will require special adapters. It is best to connect the gauge at the inlet on the fuel rail.
- Vehicle service manuals provide modulated and unmodulated fuel pressure specifications. Modulated fuel pressure is checked with the fuel pressure regulator enabled. Unmodulated fuel pressure is checked with the fuel pressure regulator disabled. Modulated pressure should be lower than unmodulated pressure. A pinched fuel line, clogged fuel filter, or worn fuel pump can cause low fuel pressure.

Gauge

Fig. 5-12 Fuel pump pressure-testing tools. *If the gauge shows low fuel pressure, what could be the cause?* (DaimlerChrysler)

TECH TIP Leaks. A leaking fuel pressure regulator diaphragm can cause a rich fuel condition. If a rich fuel condition exists, always check the fuel pressure regulator for proper operation.

It may be possible to use a test valve on the pressure gauge to test fuel volume. Depending upon the system configuration, you may have to connect a jumper wire to the fuel pump lead, or cycle the ignition switch to run the fuel pump. With the fuel pump operating, open the test valve. Discharge the fuel into a graduated container. **Fig. 5-13.** Remove the fuel pump jumper wire when finished.

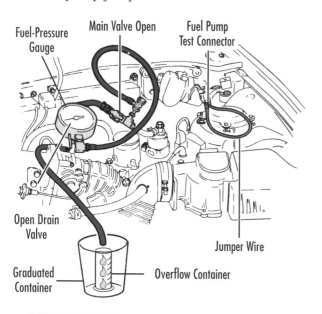

Fuel-Pressure Gauge · Main Valve Open · Fuel Pump Test Connector · Open Drain Valve · Jumper Wire · Graduated Container · Overflow Container

Fig. 5-13 Testing fuel pressure and volume. *What is unmodulated fuel pressure?* (Ford Motor Company)

Fuel volume is measured by observing the amount of fuel discharged over a specified period of time. The vehicle's service manual will provide procedures and specifications.

Fuel Filters

The fuel filters prevent contaminants from entering the fuel system and clogging the fuel injectors. Fuel

Safety First

Personal Permanently bypassing, disconnecting, or overriding any safety device is a violation of the law and can result in severe consequences for you and your employer.

Materials Bypassing safety devices can damage fuel system electronics.

contamination can cause erratic injector operation, poor engine performance and driveability problems.

The fuel system contains several filters. A paper-type fuel filter, located in the fuel line between the fuel pump and fuel injectors, prevents dirt and other contaminants from entering the fuel system. **Fig. 5-14.** These filters can become clogged and must be serviced.

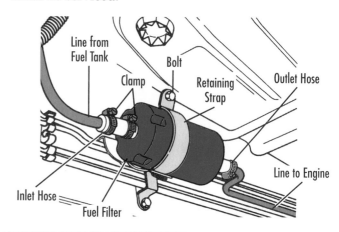

Line from Fuel Tank · Bolt · Clamp · Retaining Strap · Outlet Hose · Inlet Hose · Fuel Filter · Line to Engine

Fig. 5-14 A fuel filter mounted in the fuel line between the fuel tank and the fuel injectors. *Why do fuel-injected vehicles have several fuel filters?* (DaimlerChrysler)

Some filters are located in the fuel tank as part of the fuel pump and fuel-level sensor module. **Fig. 5-15.** These filters protect the fuel pump. They prevent debris, which may collect in the fuel tank, from getting to the pump. These filters seldom require service.

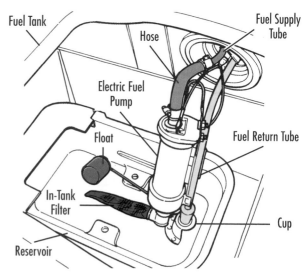

Fuel Tank · Hose · Fuel Supply Tube · Electric Fuel Pump · Float · Fuel Return Tube · In-Tank Filter · Cup · Reservoir

Fig. 5-15 A fuel filter located inside the fuel tank. *What might happen if debris in the tank collected in the fuel pump?* (Ford Motor Company)

Dirt or debris in the fuel can damage injectors. To diagnose fuel contamination, examine the contents of the fuel filter. Rusty residue indicates the presence of water. Dark, grainy residue shows that the fuel contains dirt. If the filter continues to clog, you may need to clean or replace the fuel tank.

Fuel Pressure Regulator

The **fuel pressure regulator** is a spring-loaded valve built into the fuel pump or the throttle body. It maintains a constant pressure drop across the injectors. Multiport fuel injection (MFI) systems locate the regulator on the *fuel rail.* **Fig. 5-16.**

The pressure regulator maintains the correct fuel pressure by allowing unneeded fuel to return to the fuel tank through a *fuel return line.* The return line must be unrestricted to prevent excess fuel system pressure.

If the pressure regulator is part of the fuel pump assembly, excess fuel returns directly to the fuel tank. Because it has no external fuel return line, this system is called a *single-line fuel system.*

The best fuel economy and lowest exhaust emissions require very accurate regulation of the fuel pressure. **Fig. 5-17.** Too much fuel pressure makes the air/fuel mixture too rich. An overly rich condition causes loss of power, poor fuel economy, and high exhaust emissions. Too little fuel pressure makes the air/fuel mixture too lean. This will cause hard starting, hesitation, stalling, and loss of power.

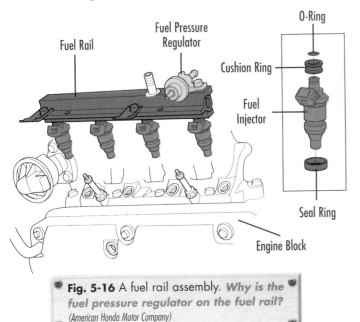

Fig. 5-16 A fuel rail assembly. *Why is the fuel pressure regulator on the fuel rail?*
(American Honda Motor Company)

Safety First

Personal When changing a fuel filter, follow the manufacturer's recommended procedures for bleeding pressure from the fuel line. Opening pressurized fuel lines will cause gasoline to spray from the fuel line fitting.

Materials You will need special tools to remove some fuel filters. Attempting to remove fuel filters without the correct tools may damage the fuel line connectors, cause a potential fuel leak, and create the risk of fire.

Causes of low fuel pressure include a low fuel level in the tank, a clogged fuel filter, a faulty fuel pressure regulator, a restricted fuel line, or a defective fuel pump.

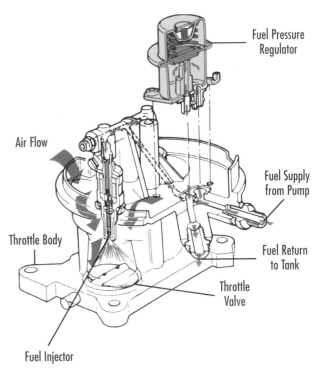

Fig. 5-17 The fuel pressure regulator in a throttle body injection system. *Why is fuel pressure regulated?*
(Ford Motor Company)

Fuel Lines

The *fuel lines* transfer fuel from the fuel tank to the throttle body or the fuel rails. Fuel lines are made of plastic, nylon, rubber, or metal. They are also called hoses, tubing, or pipes. A hose is flexible. Tubing may be flexible or rigid. Pipe is usually rigid.

Long fuel lines are usually rigid and made of nylon or steel. The fuel lines must be well supported and must not rub against sharp corners. They should not be kinked or bent unnecessarily. Replace a damaged fuel line. Use a tube bender if you need to shape metal tubing. Nylon fuel lines can be damaged by heat. Protect nylon lines or remove them from the vehicle if grinding, heating, or welding will be done nearby.

Check all fuel hoses for cracks, leaks, and hardness. Hoses deteriorate with time. They should be checked on a regular basis.

Fuel lines have many different types of couplers or connectors. These connectors are found wherever fuel lines are connected, such as in-line fuel filters and fuel rails. **Fig. 5-18.** Servicing, cutting, shaping, and flaring fuel lines and connectors may require special tools.

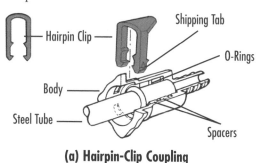

(a) Hairpin-Clip Coupling

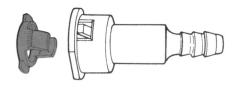

(b) Duckbill-Clip Coupling

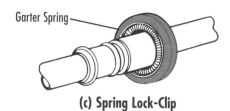

(c) Spring Lock-Clip

Fig. 5-18 Types of fuel line connectors or couplings. **(a)** Hairpin-clip **(b)** Duckbill-clip **(c)** Spring lock-clip. *Before disconnecting any fuel line, what safety precaution should you take?* (Champion Spark Plug Company)

Fuel Rails

Fuel rails are found in multiport fuel injection systems. They are an extension of the fuel lines. The fuel rails provide fuel to each injector.

Vapor Recovery System

The vapor recovery system transfers fuel vapors *(evaporative emissions)* from the fuel tank to an *evaporative emissions (EVAP) canister*. The EVAP canister stores the vapors until they are burned in the engine. This system is part of a vehicle's emission control system.

Fuel Metering System

The fuel metering system is the second part of the fuel management system. The metering system consists of the *powertrain control module (PCM)*, a network of sensors, and the fuel injectors. **Fig. 5-19.**

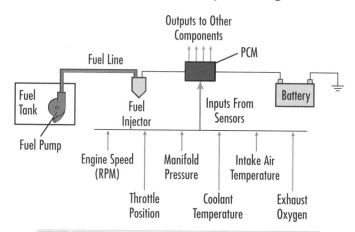

Fig. 5-19 A simplified schematic of an electronic fuel injection system. Sensors (bottom) provide information, or input, to the PCM. The PCM then determines the amount of fuel needed and opens the fuel injector to produce the required air/fuel ratio. *What components are part of the fuel metering system?*

Electronic Fuel Injection

With electronic fuel injection, the PCM controls fuel delivery by controlling injector pulse width. **Pulse width** is the duration in milliseconds that the injector is open. Most vehicles use one of two types of electronic fuel injection (EFI) systems, throttle body or multi-port.

Throttle Body Fuel Injection The throttle body fuel injection (TBI) system consists of two subassemblies, the fuel meter body and the throttle body.

The fuel meter body may have a single or double bore and one or two fuel injectors. In the example shown, the fuel pressure regulator is part of the fuel meter body.

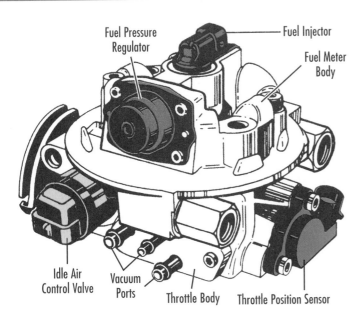

Fig. 5-20 A TBI unit using one fuel injector. *What are the two subassemblies of a throttle body?* (General Motors Corporation)

The throttle body contains the idle-air-control (IAC) valve and the throttle position (TP) sensor. It also provides vacuum ports for the manifold absolute pressure (MAP) sensor, the exhaust gas recirculation (EGR) valve, and evaporative emissions (EVAP) canister purge line. **Fig. 5-20.**

Multiport Fuel Injection The multiport fuel injection system (MFI) has one fuel injector for each cylinder. By injecting fuel directly into the intake port, the MFI system provides a metered amount of fuel to each cylinder.

Multiport systems provide more control of the air/fuel mixture than do throttle body systems. As a result, multiport systems produce lower emissions, while increasing the power output of the engine. Multiport fuel injection systems may be pulsed in banked or sequential order. **Fig. 5-21.**

Banked Multiport Fuel Injection A banked multiport fuel injection (MFI) system pulses the injectors in separate sets. Banked MFIs injectors on a four-cylinder engine pulse in sets of two. A six-cylinder engine pulses in sets of three. An eight-cylinder engine pulses in sets of four. Banked injection pulses one set of injectors for each revolution of the engine.

With banked injection, each injector delivers only one-half the required amount of fuel per pulse.

Because the injector pulses twice per cycle, it delivers the correct amount of fuel.

Sequential Multiport Fuel Injection In a sequential multiport fuel injection (SFI) system, the fuel injectors are pulsed individually in the firing order of the engine's cylinders. Each injector pulses once every two crankshaft revolutions. All the fuel needed is delivered during that one pulse. Since each injector is individually pulsed, fuel metering is more accurate than with banked injection.

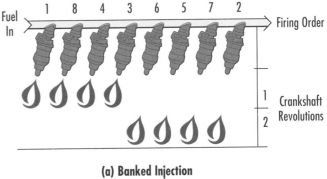

(a) Banked Injection

(b) Sequential Injection

Fig. 5-21 Fuel delivery from banked injection **(a)** compared with sequential injection **(b)**. *Which system pulses one injector at a time?* (Bendix Corporation)

When engine sensors indicate that the throttle is being opened, the PCM may add *asynchronous pulses*. The asynchronous (not synchronized) pulses do not follow the normal timing sequence. The PCM adds the extra pulses to richen the air/fuel mixture during acceleration.

While the engine is cranking at wide-open throttle, a *clear flood* mode sharply reduces pulse width.

This creates a very lean air/fuel mixture. The clear flood mode clears engine cylinders "flooded" with gasoline vapors.

On some engines, a *fuel cut-off mode* stops fuel delivery momentarily during deceleration. Data from sensors, such as the TPS and MAP, signal the PCM to go into the fuel cut-off mode. This reduces exhaust emissions and fuel consumption when the engine is not producing power.

To richen the air/fuel mixture for cold starts, some MFI systems use a *cold-start valve* to spray fuel into the intake manifold. A *thermo-time switch* limits the operating time of the cold-start valve. To help the PCM determine when to operate the cold-start valve, some systems use information supplied by the engine coolant temperature (ECT) sensor.

The Powertrain Control Module

The powertrain control module (PCM) controls fuel injector pulse width. The pulses are timed by a triggering signal from the ignition system. The triggering signal occurs so that injection begins at the proper time. Because the fuel sprays through the injector under high pressure, fuel injectors are pulsed for only a few thousandths of a second. One-thousandth (0.001 sec) of a second is a millisecond. A typical fuel injector may have a pulse width of only 2 milliseconds at idle. The pulse width at full throttle may be more than 7 milliseconds. **Fig. 5-22.**

By calculating a specific pulse width for the fuel injector, the PCM controls the air/fuel mixture. If there is a need for more fuel, the PCM increases the pulse width. If the need for fuel is less, the PCM reduces the pulse width.

The injector pulse width can be measured with some professional grade digital volt ohmmeters (DVOM). To measure these values, follow the instructions included with the DVOM. In many cases, you can also read pulse width with a scan tool.

Closed-loop operation is the operation that occurs when the PCM processes electrical inputs from a network of sensors to control fuel delivery and other actuators. The inputs are variable voltage or variable frequency signals.

The signals supply the data needed for the PCM to calculate the air/fuel mixture ratio. The PCM provides ground to control the fuel injectors supplying fuel to the engine.

The Fuel Injector

The fuel injector is a solenoid-operated valve. It is controlled by the PCM. Because the PCM controls the fuel injector, it is considered an *actuator.*

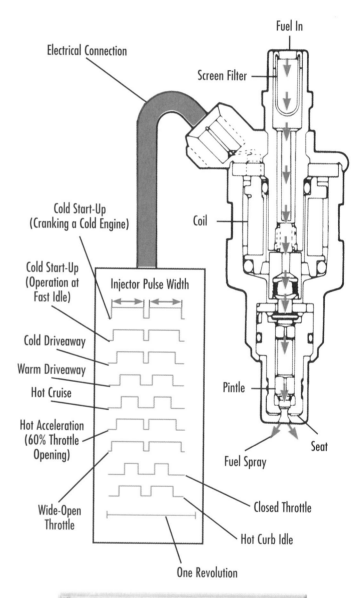

Fig. 5-22 Fuel injector pulse widths vary over a range of engine operation. The longer the pulse width, the greater the amount of fuel injected into the cylinder. *What controls the injector pulse width?*

(Ford Motor Company)

Calculating Miles Per Gallon

Karl brings his vehicle to your service station. He says he doesn't think his vehicle is getting as many miles per gallons (mpg) as it used to. Karl wants to find out if he is right.

Checkpoint 1. Karl has started to keep a log book. In it he lists the odometer reading and number of gallons he buys each time he fills his gas tank. For the first six fill-ups, he calculated the mpg by counting the number of miles between fill-ups.

	Odometer reading	Miles between fill-ups	Gallons purchased	MPG
	46,415			
1	46,830	415	15.6	26.6
2	47,232	402	15.3	26.3
3	47,620	388	15.1	25.7
4	48,043	423	15.9	26.6
5	48,482	439	16.4	26.8
6	48,706	224	9.0	24.9
7	49,098		15.9	

- Calculate the number of miles driven between his sixth and seventh fill-ups.

  49,098 miles − 48,706 miles = 392 miles

- How many mpg did his car get between his sixth and seventh fill-ups?
- Calculate the mpg. He drove 392 miles and used 15.9 gallons of gas.

$$\frac{392 \text{ miles}}{15.9 \text{ gal}} = 24.7 \text{ mpg}$$

Checkpoint 2. Karl wants to know his average mpg for the time he kept his log book. You add the seven mpg figures and divide the total by the number of items you have added (seven) to find the average.

- The sum of the seven numbers is 181.6. The average mpg is:

$$\frac{181.6}{7} = 25.9 \text{ mpg}$$

Note that the average value of a set of numbers is also called the *arithmetic mean*. The middle value of a set of numbers is called the *median*.

- Rank the seven mpg values from the largest to the smallest. The median, or middle, value is 26.3 mpg. Karl's vehicle performed better than the median three times. It had lower mileage than the median three times.

You notice that the mpg calculations for the sixth and seventh fill-ups seem low. You ask Karl to keep the log book for two more fill-ups. You will be able see what mileage he gets then.

Apply it!

Meets NATEF Mathematics Standards for finding the mean and median and for using addition, subtraction, and division.

Karl returns with figures from his eighth and ninth fill-ups. He bought 15.8 gal. of gas at 49,484 miles and 15.6 gal. of gas at 49,862 miles.

❶ How many mpg did the vehicle get for these two fill-ups?

❷ Calculate the overall average mpg for the nine fill-ups. Find the average of all nine mpg figures.

❸ You can also find the overall average by dividing the total number of miles driven by the total number of gallons of gas used. Try this method. Compare your answers. Your answer should be very close to the first number you calculated. Rounding will account for any difference.

❹ Recalculate the median mpg value for the nine fill-ups. You should calculate 25.7 mpg for the median. This value is lower than the median was for the first seven fill-ups. What does this indicate about the performance of Karl's vehicle?

❺ Why might calculating the overall mpg the second way cause you to miss a possibly important clue about the vehicle's performance?

The PCM provides a ground connection for the injector circuit. Pulse width is controlled by turning the ground circuit on and off.

When the PCM driver circuit grounds the injector circuit, current flows through the fuel injector coil. A magnetic field pulls the *injector armature* against the closing spring. As the armature moves, the pintle valve pulls away from the *pintle seat.* **Fig. 5-23.** As the injector pintle valve opens, the fuel passes through the *nozzle* in a fine, cone-shaped *spray pattern.*

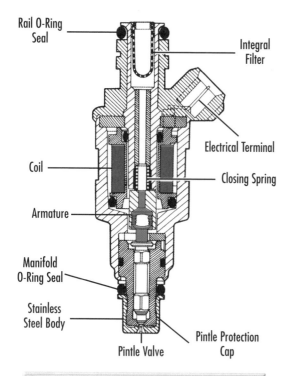

Rail O-Ring Seal
Integral Filter
Coil
Electrical Terminal
Closing Spring
Armature
Manifold O-Ring Seal
Stainless Steel Body
Pintle Valve
Pintle Protection Cap

Fig. 5-23 A cutaway view of an electronic fuel injector. *What part of a fuel injector creates a magnetic field?* (Ford Motor Company)

When the injection cycle is completed, the PCM driver opens the ground connection to the circuit. With the ground circuit open, current stops flowing through the injector coil. The pintle valve closes against the pintle seat, and fuel stops flowing through the injector.

The shape of the spray pattern and amount of fuel injected can be affected in many ways. Dirt, carbon, or varnish can clog the nozzle, making the spray pattern uneven. An uneven fuel spray pattern cannot mix properly with the air. An improper mixture causes poor fuel economy, a loss of power, and an increase in exhaust emissions.

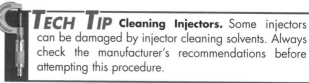

TECH TIP Cleaning Injectors. Some injectors can be damaged by injector cleaning solvents. Always check the manufacturer's recommendations before attempting this procedure.

Cleaning Fuel Injectors Deposits on the tip of a fuel injector prevent proper fuel delivery and distort the spray pattern. Rough idle, loss of power, and poor fuel economy may result.

Injectors can be cleaned by a flow of injector-cleaning solvent through them. The solvent removes gum, varnish, and other deposits. When cleaning fuel injectors, follow these general guidelines:

1. Make sure the engine is at normal operating temperature.

2. Relieve the fuel pressure and disable the fuel pump.

3. Remove and plug the vacuum hose at the pressure-regulator valve.

4. Follow the instructions for the injector-cleaning equipment you are using. Connect a container of injector cleaner to the fuel rail.

5. Start the engine and let it idle until the container is empty and the engine stalls.

6. Remove the container, restore the connections, and start the engine.

Testing Fuel Injectors Fuel injector types vary from vehicle to vehicle. However, these general steps can be used to perform a preliminary diagnosis.

1. With the engine running, use a technician's stethoscope to check for a clicking sound in each injector. A bad injector will not have a pronounced clicking noise.

2. Use a flashlight to observe the fuel spraying from injectors on TBI systems. The spray should be in an even, cone-shaped pattern.

3. With the engine off, remove the electrical connector from the fuel injector. Attach DVOM leads to each of the fuel injector terminals. Compare the resistance reading to the manufacturer's specifications. Remember that high temperatures increase the resistance reading.

4. Some manufacturer's require the use of test procedures that evaluate the injector's ability to

deliver a uniform amount of fuel (injector balance test). This requires specialized test equipment. See the service manual for details.

Replacing Fuel Injectors When replacing a fuel injector, follow procedures provided in the manufacturer's service manual.

Here are general guidelines for replacing a fuel injector in a TBI system:

1. Disconnect the negative battery cable.
2. Bleed pressure from the system.
3. Remove the fuel line and electrical connections to the injector.
4. Remove the fuel metering body to gain access to the injector.
5. Remove any hardware holding the injector in place.
6. Carefully remove the injector from the throttle body.
7. As needed, install new O-rings and washers.
8. Carefully align the new injector in the throttle body.
9. Push into position by hand, using thumb pressure to seat the injector.
10. Reconnect the fuel and electrical connections.

11. Replace all attaching hardware.
12. Reinstall the metering body.
13. Check for leaks.

Here are general guidelines for replacing a fuel injector in an MFI system:

1. Bleed pressure from the system.
2. Disconnect the negative battery cable.
3. Remove the fuel line and electrical connections to the injector.
4. Remove any hardware holding the injector in place.
5. Carefully remove the injector from the intake manifold.
6. As needed, install new O-rings and washers.
7. Carefully align the new injector in the manifold.
8. Push into position by hand, using thumb pressure to seat the injector.
9. Reconnect the fuel and electrical connections.
10. Replace all attaching hardware.
11. Check for leaks.

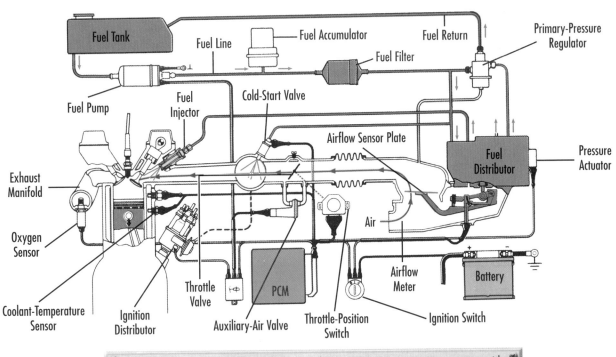

Fig. 5-24 Schematic layout of the Bosch KE continuous-injection system with an oxygen sensor. *What is the main difference between electronic and mechanical fuel injection systems?* (Robert Bosch GmbH)

Mechanical Fuel Injectors *Continuous fuel injection systems* (CIS) use mechanical fuel injectors. A CIS fuel injector continuously sprays fuel into the intake port. Instead of varying pulse widths, changing the fuel pressure to the injectors changes the air/fuel mixture.

An airflow sensor plate controls the fuel pressure. **Fig. 5-24.** The sensor plate pivots on an arm, which connects to a *fuel distributor*. The fuel distributor meters fuel to each fuel injector. As more air moves past the sensor plate, the fuel distributor increases the fuel pressure to each fuel injector.

Figure 5-25 shows the open and closed positions of the needle valve and seat in the end of the mechanical injector. Fuel flowing past the needle causes it to open and close rapidly. This rapid motion effect helps atomize the fuel.

When the engine is shut off, the fuel pump stops, and fuel pressure drops to a residual pressure. The spring in the injector forces the needle valve closed. This prevents fuel from leaking into the intake ports.

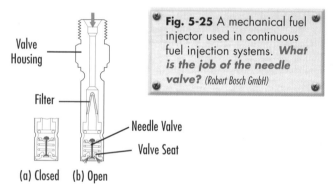

Valve Housing

Filter

Needle Valve

Valve Seat

(a) Closed (b) Open

Fig. 5-25 A mechanical fuel injector used in continuous fuel injection systems. *What is the job of the needle valve?* (Robert Bosch GmbH)

Fuel Management Sensors

The PCM uses *electronic inputs* from a network of sensors on the vehicle to calculate fuel injector pulse width.

Most PCMs supply 5 volts to power the sensor circuits. Technicians refer to this voltage as the *reference voltage*. The sensors alter the reference voltage and supply it to the PCM as a *return-signal*. The return-signal is an electrical *data stream*. The data stream may be in the form of a changing voltage or a changing frequency.

Throttle Position Sensor The throttle position sensor (TPS) sends throttle plate position information to the PCM. **Fig. 5-26.** As the throttle opens, the return signal voltage from the TPS increases. When the throttle closes, the return-signal voltage decreases.

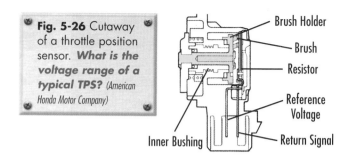

Fig. 5-26 Cutaway of a throttle position sensor. *What is the voltage range of a typical TPS?* (American Honda Motor Company)

Brush Holder
Brush
Resistor
Reference Voltage
Return Signal
Inner Bushing

The PCM sees the increase in voltage as a need for more fuel. The PCM increases the pulse width signal to the fuel injectors. This provides more fuel.

A decrease in the TPS return signal is seen as a need for less fuel. The PCM decreases the pulse width signal to the fuel injectors. This provides less fuel.

Engine Coolant Temperature Sensor The engine coolant temperature (ECT) is a *thermistor* type sensor that sends engine coolant temperature information to the PCM. **Fig. 5-27.** Thermistors change their electrical resistance as the temperature changes. As temperature decreases, the resistance of most thermistors increases. When the thermistor's resistance increases, the return signal voltage decreases, signaling the PCM that the engine is cold.

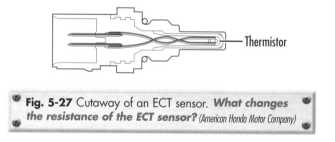

Thermistor

Fig. 5-27 Cutaway of an ECT sensor. *What changes the resistance of the ECT sensor?* (American Honda Motor Company)

Gasoline does not vaporize well when the engine is cold. A cold engine requires more fuel to start and warm up. When the ECT signals a cold engine, the PCM delivers more fuel by increasing the injector pulse width. As the coolant temperature increases, the PCM reduces injector pulse width, maintaining an appropriate air/fuel ratio.

Intake Air Temperature Sensor The intake air temperature (IAT) is a thermistor type sensor that provides intake air temperature information to the PCM. Cold air is denser and contains more oxygen than warm air. More oxygen requires more fuel to maintain a stoichiometric air/fuel ratio.

As intake air temperature decreases, the PCM increases injector pulse width and supplies more fuel. When air temperature increases, less oxygen is present. The PCM reduces the injector pulse width, decreasing the supply of fuel.

Oxygen Sensor   The oxygen sensor (O2S) provides data showing the amount of oxygen in the exhaust. The PCM uses this data to adjust the air/fuel mixture. Fuel management systems use two types of oxygen sensors. **Fig. 5-28.**

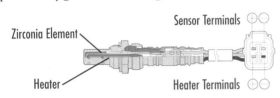

Zirconia Element
Sensor Terminals
Heater
Heater Terminals

Fig. 5-28 A zirconia and a titania oxygen sensor are similar in appearance. *How is the zirconia sensor different from the titania sensor?* (American Honda Motor Company)

The *zirconia O₂ sensor* is the most common design. This design uses a heater to quickly warm it to operating temperature. The zirconia sensor produces its own voltage. At operating temperature, about 600°F (315°C) the sensor produces an output signal of approximately zero to one-volt. A 0.2-volt signal tells the PCM that the exhaust is lean. A 0.8-volt signal tells the PCM that the exhaust is rich.

The *titania sensor* operates like a resistor. The amount of oxygen in the exhaust changes the sensor's resistance. A high signal from the sensor tells the PCM the exhaust is rich. A low signal from the sensor tells the PCM the exhaust is lean.

Manifold Absolute Pressure Sensor   The manifold absolute pressure (MAP) sensor provides intake manifold absolute-pressure information. When a vacuum exists in the manifold, absolute pressure is less than atmospheric pressure. Two types of sensors are in use.

The first type produces a voltage signal that changes with the manifold pressure. The high or low signal tells the PCM to decrease or increase the injector pulse width.

The second type produces a frequency signal that changes with manifold pressure. Frequency is measured in cycles per second or *hertz*. One hertz equals one cycle per second. The signal tells the PCM to decrease or increase the injector pulse width.

Barometric Pressure Sensor   The barometric pressure (BARO) sensor measures atmospheric pressure. As atmospheric pressure increases, the oxygen content of the air increases. As oxygen increases, the PCM increases injector pulse width to supply more fuel. When atmospheric pressure decreases, oxygen content decreases. The PCM reduces the injector pulse width to supply less fuel. On some engines barometric pressure is read by the MAP sensor when the ignition switch is first turned ON.

Mass Airflow or Volume Airflow Sensors   The mass airflow (MAF) sensor, also called the volume airflow (VAF) sensor, measures the amount of air flowing into the engine. As airflow increases, the PCM increases injector pulse width to supply more fuel. When airflow decreases, the PCM reduces the injector pulse width to supply less fuel.

The MAF sensor measures the volume and density of the air. The VAF sensor measures volume and temperature.

In some closed-loop systems, MAF or VAF sensors replace the MAP sensor. MAF and VAF sensors are direct methods of measuring airflow. MAP systems measure airflow indirectly.

Testing Sensors   While you cannot test some sensors directly, a DVOM or Lab Scope can be used to check for normal operation. All sensor circuits can be checked for the presence of a reference voltage. If a reference voltage is not found, follow the procedures that the manufacturer recommends to find the cause. Voltage changes on the return signal wire can be read using a DVOM or scan tool.

SECTION 3 KNOWLEDGE CHECK

❶ What is residual fuel pressure?

❷ What is banked fuel injection?

❸ What is pulse width?

❹ What is sequential fuel injection?

❺ The data stream from the fuel management sensors is supplied in what two forms?

CHAPTER 5 REVIEW

Key Points

Meets the following NATEF Standards for Engine Performance: checking fuel for contaminants and quality; fuel filter service.

- Fuel pumps are electrically operated and supply fuel under pressure to the fuel metering system.
- Correct fuel pressure and fuel volume are vital to proper engine operation.
- Fuel contamination can be determined by examining the residue of fuel filters.
- Multiple fuel filters are found on fuel injected vehicles. Some are located in the tank; some are located in the fuel line.
- Fuel injectors can be tested by checking the resistance of the injector coil or by performing an injector balance test.
- A properly operating fuel injector produces a cone-shaped spray pattern.
- An oxygen sensor indicates the amount of oxygen in the exhaust.
- MAF and VAF sensors provide a direct method of measuring airflow.

Review Questions

1. When testing an electric fuel pump, what two aspects of fuel pump operation are measured?

2. When inspecting a fuel filter, you find it is clogged. How would a clogged fuel filter affect fuel pressure?

3. Rusty residue in a fuel filter indicates what type of fuel contamination?

4. Dark, grainy residue in a fuel filter indicates what type of fuel contamination?

5. When making fuel pressure checks, the fuel pressure is below specified levels. What can cause low fuel pressure?

6. What type of oxygen sensor produces its own voltage?

7. When testing fuel injectors, a properly operating fuel injector shows what type of spray pattern?

8. **Critical Thinking** How would a high oxygen sensor voltage affect the air/fuel mixture?

9. **Critical Thinking** How would an injector that fails to open affect engine operation in an SFI system?

TECHNOLOGY FORECAST
FOR AUTOMOTIVE EXCELLENCE

More Bang for the Buck

Better fuel economy? More power? Both? The use of direct fuel injection technology will make improvements in each area possible. Some engineers think drivers will see gains of at least 25 percent in mpg and engine torque.

Direct fuel injection differs in important ways from today's fuel injection systems. Today's systems simply spray fuel into the intake manifold. When the intake valve opens, air is drawn into the cylinder. The air then carries the fuel along with it at normal pressure.

Some efficiency and energy are lost with today's fuel injection systems because some of the vapor-ized fuel condenses. As the fuel turns back into a liquid, it drops out of the flow. It then passes unburned through the exhaust system.

With direct fuel injection, the fuel is sent—under high pressure—directly to each cylinder. The fuel arrives as the piston comes up on the compression stroke. Air then fully surrounds the vaporized fuel so combustion can occur. More power and fuel economy result because all of the fuel is ignited by the spark plug.

Direct fuel injection is already in use by some automakers. Because it is expensive, it is usually used on higher-priced models.

AUTOMOTIVE EXCELLENCE
TEST PREP

Answering the following practice questions will help you prepare for the ASE certification tests.

1. Technician A says you should remove the fuel cap before disconnecting the fuel line. Technician B says on some engines you can release residual fuel pressure by removing the fuel pump fuse and starting the engine. Who is correct?
   - ⓐ Technician A.
   - ⓑ Technician B.
   - ⓒ Both Technician A and Technician B.
   - ⓓ Neither Technician A nor Technician B.

2. An electric fuel pump runs continuously. Technician A says there may be a short in the ECT sensor. Technician B says the fuel pump relay contacts may be stuck closed. Who is correct?
   - ⓐ Technician A.
   - ⓑ Technician B.
   - ⓒ Both Technician A and Technician B.
   - ⓓ Neither Technician A nor Technician B.

3. The fuel pump is not working. Technician A says to check the airflow sensor. Technician B says to check the oil pressure switch. Who is correct?
   - ⓐ Technician A.
   - ⓑ Technician B.
   - ⓒ Both Technician A and Technician B.
   - ⓓ Neither Technician A nor Technician B.

4. An engine hesitates and stalls. Technician A says the fuel pressure may be too low. Technician B says the fuel injector pulse width may be too wide. Who is correct?
   - ⓐ Technician A.
   - ⓑ Technician B.
   - ⓒ Both Technician A and Technician B.
   - ⓓ Neither Technician A nor Technician B.

5. An engine is running very rich. The fuel pressure is too high. What is the most likely cause?
   - ⓐ A worn fuel pump.
   - ⓑ A pinched fuel line.
   - ⓒ A clogged fuel injector.
   - ⓓ A leaking fuel pressure regulator.

6. A fuel system will not hold residual pressure. Technician A says a fuel injector may be leaking. Technician B says the fuel pump may be bad. Who is correct?
   - ⓐ Technician A.
   - ⓑ Technician B.
   - ⓒ Both Technician A and Technician B.
   - ⓓ Neither Technician A nor Technician B.

7. Technician A says that pulse width measures in thousandths of an inch. Technician B says that pulse width measures in thousandths of a second. Who is correct?
   - ⓐ Technician A.
   - ⓑ Technician B.
   - ⓒ Both Technician A and Technician B.
   - ⓓ Neither Technician A nor Technician B.

8. A fuel injection system that pulses each fuel injector in a specific firing order is:
   - ⓐ A banked system.
   - ⓑ A mechanical system.
   - ⓒ Both a and b.
   - ⓓ Neither a nor b.

9. An engine runs well before the coolant warms up to operating temperature. It begins to run rich after it warms up. What is the most likely cause?
   - ⓐ A stuck fuel injector.
   - ⓑ A clogged fuel filter.
   - ⓒ A weak fuel pressure regulator.
   - ⓓ A disconnected ECT sensor.

10. Technician A says the ECT controls engine coolant temperature. Technician B says the ECT senses engine coolant temperature. Who is correct?
    - ⓐ Technician A.
    - ⓑ Technician B.
    - ⓒ Both Technician A and Technician B.
    - ⓓ Neither Technician A nor Technician B.

CHAPTER 6

Diagnosing & Repairing Ignition Systems

You'll Be Able To:

- ⊗ Inspect and check ignition coils.
- ⊗ Inspect and test the distributor.
- ⊗ Inspect and test ignition system secondary wiring and components.
- ⊗ Check and adjust ignition timing.
- ⊗ Inspect and test ignition system triggering devices.

Terms to Know:

distributor

distributorless ignition system

duty cycle

dwell angle

ignition coil

indexing mark

The Problem

Harold Jordan is on his daily commute in the middle of rush-hour traffic. He notices that his car's engine starts to miss or cut out. Suddenly his car loses all power. Harold is able to pull his car over to the side of the expressway. He calls a tow truck.

Harold's car is towed to your service center. You suspect an ignition system problem. Harold's car has a distributorless ignition system. You understand distributor ignition systems but do not have experience with distributorless types. What should you do?

Your Challenge

As the service technician, you need to find answers to these questions:

① Do all ignition systems share the same operating principles?

② How can you apply these principles to diagnosing electronic ignitions?

③ What other information do you need?

Section 1

The Ignition System

The *ignition system* provides an ignition spark to the proper cylinder at the correct time. Delivering the ignition spark at the correct time produces the maximum power and fuel economy with the lowest exhaust emissions.

All ignition systems share the same operating principles. Ignition systems also share some or all of these basic components. **Fig. 6-1.**

• Ignition switch.
• Ignition coil.
• Distributor.
• Powertrain control module (PCM).
• Ignition module.
• Ignition wires.
• Spark plugs.

The Ignition Switch

The *ignition switch* is the master switch for the entire electrical system. When placed in the ON position, the ignition switch connects the ignition system to the battery.

The Ignition Coil

The ignition system consists of two separate but closely related electrical circuits. These circuits are a low-voltage primary circuit and a high-voltage secondary circuit. The **ignition coil** is a device that transforms low voltage from the battery into a high voltage capable of producing an ignition spark. An ignition coil has two windings, a primary winding and a secondary winding. **Fig. 6-2.**

The ignition coil has two primary winding terminals, B+ and negative. The ignition switch connects the battery to B+. The negative terminal lead is connected to the ignition module.

Depending on the application, ignition coils may have one or two secondary winding terminals. Coils used in distributorless (waste spark) systems have two secondary winding terminals.

Fig. 6-2 Cutaway view of an ignition coil. *What determines the voltage output of an ignition coil?* (General Motors Corporation)

The primary coil consists of a few hundred turns of heavy copper wire. The secondary coil consists of many thousand turns of fine copper wire. The ratio between the number of primary and secondary windings determines the voltage output of the coil.

The two coils surround an iron core. The iron core concentrates the magnetic field created when current passes through the primary coil. The ignition module controls current in the primary coil. The current flow produces a magnetic field around the primary coil. When the ignition module stops the current flow, this magnetic field collapses into the secondary coil. This coil produces a voltage as high as 50,000 volts.

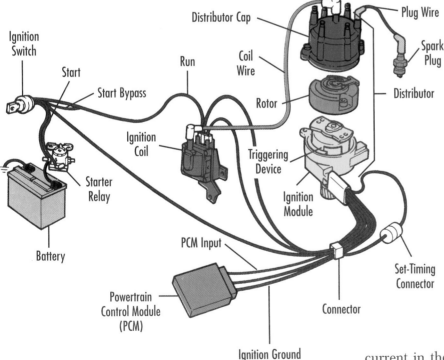

Fig. 6-1 The ignition system components in a distributor ignition. *Where is the ignition module in this system?* (Ford Motor Company)

How Does a Coil Work?

To understand how battery voltage (12 volts) can be increased to as much as 50,000 volts [50 kV] in the secondary coil, you must remember some basic rules of electricity. Electrical current flow through a conductor creates a magnetic field around the conductor. When a magnetic field moves past a conductor, a voltage is induced in the conductor.

An ignition coil consists of two coils of wire. The primary coil is connected to battery voltage and ground. One end of the secondary coil is connected to the output terminal. The type of coil determines where the other end is connected. The primary and secondary coils are close together but not touching.

As current flows through the primary coil, a magnetic field expands outside the wires for a short distance. When current flow stops, the magnetic field moves back into the coil.

When the primary circuit is closed, the moving magnetic field induces a voltage in the secondary coil. This happens when the magnetic field moves out of the primary winding.

When the primary circuit is opened (switched off), a voltage is induced a second time. This happens as the magnetic field again moves across the secondary coil. This process of producing a voltage within a wire or coil is called *magnetic induction.*

Apply it!

Inducing a Voltage in a Secondary Coil

Meets NATEF Science Standards for ignition coils.

Materials and Equipment
- 2 pieces of plastic pipe. The diameter of the pieces should be about 5 in. The diameters should differ by about ½ in. to ¼ in. You must be able to fit one inside the other.
- 2 pieces of copper wire about 70 ft. of 16–18 gauge wire about 210 ft. of 32–38 gauge wire
- 12-volt power supply
- 2 analog VOMs
- Electrical tape

To see how current flows through a coil, you can build your own coil.

❶ Wrap 50 turns of the larger diameter wire around the larger diameter pipe. This is the primary coil. Leave the two ends of each piece of wire near each other to use as terminals.

❷ Wrap 150 turns of the smaller diameter wire around the smaller diameter pipe. This is the secondary coil.

❸ Use tape to keep the wires from slipping off of the plastic pipe pieces.

❹ Place the smaller coil inside the larger coil.

❺ Complete a circuit by connecting the larger coil (primary coil) to the power supply.

❻ Connect one VOM to this coil. Set it to measure DC volts. Connect the other VOM to the smaller secondary coil to measure DC volts.

❼ Close the switch on the power supply. Read the voltage on the primary coil. What is the voltage on the secondary coil?

❽ Open the switch. Then close it again while you watch the needle on the VOM that is attached to the secondary coil. What do you see happening to the needle? You must look very carefully.

❾ Which voltage is larger, the power supply voltage or the induced voltage?

This is called *available voltage* or *maximum coil output*. On some ignition systems, it can exceed 50,000 volts [50 Kilovolts or 50 kV]. The secondary voltage builds only to the point at which current begins to flow in the secondary circuit. Normally, this is the voltage required to cause a spark to jump across the spark plug gap. This is called the "required" or "firing" voltage.

If the required voltage is 10,000 volts [10 kV], the plug will fire at this voltage. The secondary voltage will not build up higher. Think of the difference between the available and required voltage as *ignition reserve voltage*.

Firing voltage will be higher than normal when any of the following conditions exist:
• A wider than specified spark plug gap or worn plug electrodes.
• Late ignition timing.
• Higher than specified compression ratio.
• Hard acceleration or wide-open throttle operation.
• Lean air/fuel mixtures.
• Large gap in secondary wiring.

Some of these conditions may occur in only one cylinder at any given time. This is true for wide plug gaps, worn electrodes, or wire problems.

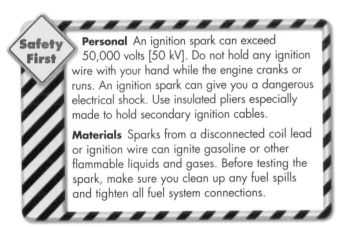

Safety First

Personal An ignition spark can exceed 50,000 volts [50 kV]. Do not hold any ignition wire with your hand while the engine cranks or runs. An ignition spark can give you a dangerous electrical shock. Use insulated pliers especially made to hold secondary ignition cables.

Materials Sparks from a disconnected coil lead or ignition wire can ignite gasoline or other flammable liquids and gases. Before testing the spark, make sure you clean up any fuel spills and tighten all fuel system connections.

When the spark jumps the gap at a voltage lower than the coil is capable of producing, the coil reserve voltage makes the spark last longer. This is called *spark duration*. The longer the spark duration, the more likely that all of the air/fuel mixture in the cylinder will ignite. Complete combustion results in better fuel economy and lower exhaust emissions. If the spark plug gap is too wide, the available voltage may not be enough to create a spark. This will result in a cylinder misfire.

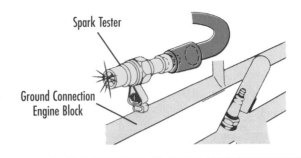

Spark Tester
Ground Connection Engine Block

Fig. 6-3 A common spark tester. *Why wouldn't you ground the spark tester to the fuel rail?*

Diagnosing the Ignition Coil Check the ignition coil output using a spark tester. **Fig. 6-3.** This is a modified spark plug that is connected to ground by a clip attached to its shell.

1. Remove a spark plug wire from its plug.
2. Connect the wire to the electrode tip of the spark tester.
3. Connect the ground clip from the spark tester to a proper ground connection, for example the engine block. DO NOT connect the ground clip to a fuel rail or any connection to the fuel system.
4. Crank the engine.

A spark should jump from the spark tester electrode to the shell. If there is a faint spark or no spark, check the distributor cap and rotor (if used) for cracks and burn marks.

The resistance of an ignition coil can be checked with an ohmmeter. **Fig. 6-4.** Connecting the ohmmeter between the terminals of the primary coil shows the resistance in the primary winding.

Ohmmeter

Primary Coil

Fig. 6-4 Testing the resistance of the primary coil with an ohmmeter. *If a coil were open, what reading would you expect to see?* (Terry Wild Studio)

On a distributorless ignition system (DIS), attaching the ohmmeter to the towers of the coil pack measures the secondary coil resistance. Removing the spark plug wires allows access to the coil terminals. **Fig. 6-5.** The vehicle service manual includes the resistance specifications for the type of coil being checked.

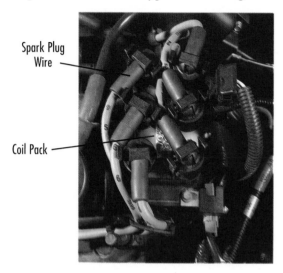

Spark Plug Wire

Coil Pack

Fig. 6-5 Spark plug wire connections to a six-cylinder DIS (waste spark) coil assembly. *How many plug wires are connected to each individual coil?* (Terry Wild Studio)

The Distributor

The **distributor** is a device designed to establish base timing and distribute the secondary ignition voltage. It is sometimes referred to as an electronic switch. Depending on the type of distributor ignition system, the distributor design may be a:

- Contact–point type–an obsolete mechanical design used on older vehicles.
- Magnetic–pickup coil type–a design using a magnetic coil and pickup in the distributor to create the timing signal.
- Hall-effect type–a design using a solid-state magnetic sensing device to create the timing signal.
- Optical-sensor type–a design using a light-emitting diode and photodiode sensor to create the timing signal.

The Powertrain Control Module

The *powertrain control module (PCM)* is a computer that manages engine and powertrain systems. Within engine management, the PCM is responsible for fuel delivery, ignition timing and some emission control devices.

Ignition timing determines when the spark occurs during the combustion cycle. When the engine is idling, the spark occurs during the compression stroke, a few degrees before the piston reaches top dead center (TDC). As the engine speeds up, the ignition system must make the spark occur earlier (advance the spark). Advanced timing of the spark gives the air/fuel mixture more time to burn completely.

The engine needs correct spark advance to deliver maximum power and fuel economy and to control exhaust emissions. If the spark occurs too early, (detonation engine knock) of the air/fuel mixture may occur. Detonation occurs when the air/fuel mixture does not burn evenly. The mixture explodes. This causes a metallic knock or pinging sound. Early ignition causes too much pressure to build in the cylinder before the piston reaches TDC. This causes rough idle, loss of power, and detonation.

If the spark occurs too late, combustion occurs late in the compression stroke. This condition is called retarded timing. With retarded timing, an engine cannot develop maximum power and fuel economy. Late spark timing may cause the engine to overheat.

The PCM calculates and electronically adjusts the spark timing by processing data from electronic sensors. A sensor mounted at the crankshaft or in the distributor senses engine speed. Input from the mass airflow (MAF) sensor, or a combination of inputs from the throttle position sensor (TPS) and manifold absolute pressure (MAP) sensor, indicates engine load. Inputs from the engine coolant temperature (ECT) sensor and intake air temperature (IAT) sensor indicate engine temperature and intake air temperature.

Some engines also use a *knock sensor (KS)*. The knock sensor is mounted on the engine block. The *knock sensor* detects the vibrations caused by detonation. The PCM uses a signal from this sensor to retard timing to this point where detonation stops.

Before the PCM can calculate ignition timing, it must receive a reference signal indicating when number-one piston is near TDC on the compression stroke. The reference signal is produced by a sensor or triggering device mounted at the crankshaft vibration damper, the crankshaft, the flywheel, or in the distributor.

Ignition triggering devices are sensors that indicate the position of the crankshaft and engine rpm. The crankshaft-position triggering device is called a *crankshaft position (CKP) sensor*. The function of the CKP is to indicate when the number-one piston nears (TDC) on the compression stroke.

There are three basic designs of ignition triggering devices: magnetic pulse, Hall-effect, and optical (photodiode). These three basic designs are manufactured in many variations. The testing procedure for each variation is specific to each vehicle model. Therefore, it can be described only in general terms.

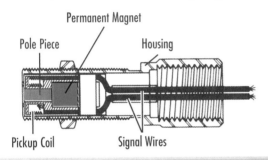

Fig. 6-6 Construction of a typical magnetic pulse sensor. *What type of signal does a magnetic pulse sensor produce?* (Air Pax Corporation)

Magnetic Pulse Sensors
Magnetic pulse sensors consist of a *pole piece* that extends from a permanent magnet and a *pickup coil*. **Fig. 6-6.**

A *reluctor,* or timing disc, mounts on the crankshaft, distributor shaft, or flywheel. **Fig. 6-7.** The slots or teeth in the reluctor indicate the position of each piston. As the reluctor rotates, a pulse is generated each time a slot or tooth passes the sensor. This signal pulse is sent to the PCM. The PCM signals the ignition module to trigger the *ignition sequence.*

When checking individual ignition components, test to see that the device does what it is supposed to do. Magnetic pulse sensors are usually tested by checking the circuit resistance. The specification for circuit resistance and the connector terminal location is found in the vehicle service manual. Check the resistance by disconnecting the magnetic sensor. Use a digital volt-ohm-meter (DVOM) to measure the resistance in ohms.

While the engine is cranking, a magnetic pulse trigger circuit should produce a small (about 300 mV) AC, or analog, signal. Therefore, another way to check the sensor is to connect a DVOM set on the AC scale to the two wires leading from the sensor pickup coil. Crank the engine and measure the resulting voltage. If it is equal to or higher than the minimum specified for the vehicle, the sensor is working normally.

The wavelike pulse produced by a magnetic pulse sensor is called an *analog signal.* This signal is converted to a digital signal by the ignition module or the PCM.

Hall-Effect Sensors Hall-effect sensors use a Hall-effect switch to supply the triggering signal to the ignition module or PCM.

An *interrupter ring* or *shutter* is mounted behind the vibration damper, on the camshaft, or on the distributor shaft. Equally spaced windows in

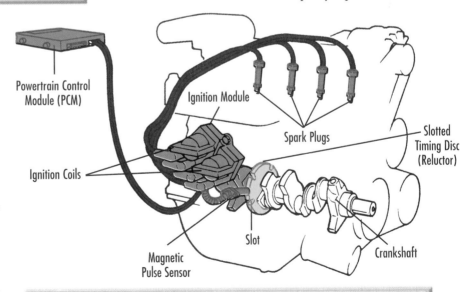

Fig. 6-7 Typical placement of a magnetic pulse sensor. Notice the timing slots cut into the crankshaft. *Why would a magnetic pulse sensor be more accurate when placed on the crankshaft?* (General Motors Corporation)

Calculating MAF Values

A mass air flow (MAF) sensor measures, in grams per second (g/sec), the mass of air that flows through the sensor. When the throttle is opened, more air flows through the sensor. The PCM uses the MAF sensor signal to determine the amount of fuel needed for the airflow. The following data were collected in a lab from a mass air flow (MAF) sensor.

RPM	Airflow (g/sec)
800	10
1000	**21**
1200	28
1400	**36**
1600	43
1800	**52**
2000	59

These data values can be plotted on the graph. Note that airflow increases as rpm increases.

The rate of increase is called the *slope*. It is usually shown by the letter m. To calculate the slope use the formula:

$$m = \frac{(y_2 - y_1)}{(x_2 - x_1)}$$

To find the rate of increase in the MAF per rpm you need to find the slope of the line between two points, (x_1, y_1) and (x_2, y_2). Let's use 1,000 rpm and 1,400 rpm as the x values of the two points. The points on the graph are $(1,000, 21)$ and $(1,400, 36)$. The slope between these points is:

$$m = \frac{y_2 - y_1}{x_2 - x_1} = \frac{36 - 21}{1,400 - 1,000} = \frac{15}{400}$$

$$= 0.0375$$

On the graph, the slope of the line (m) is 0.04 when rounded to the nearest hundredth.

You can now write a formula to calculate the expected MAF for another rpm value. You can rewrite the slope formula as:

$$(y - y_1) = m(x - x_1)$$

Here (x_1, y_1) is one of the points you already know.

Use the point $(1,000, 21)$ as the point you know to write a formula.

$$y - y_1 = m(x - x_1)$$
$$y - 21 = 0.04(x - 1,000)$$
$$y - 21 = 0.04x - 40$$
$$y = 0.04x - 19$$

If you know the value of x, you can calculate the value of y using this formula. You can find the expected airflow at 1,100 rpm, $(x = 1,100)$. You solve the equation for y, the expected airflow for 1,100 rpm.

$$y = 0.04(1,100) - 19 = 44 - 19 = 25 \text{ g/sec}$$

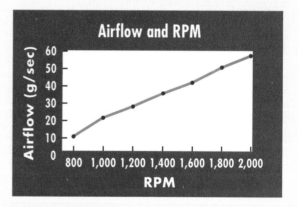

Apply it!

Meets NATEF Mathematics Standards for using tables and graphs, and for determining expected values.

❶ Refer to the data point for 1,400 rpm on the graph. What is the x value? What is the y value?

❷ Find the rate of increase in airflow per rpm between 1,400 rpm and 1,600 rpm.

❸ Was this the same slope as in the example? Why do you think this is so?

❹ Write a formula for the line between these two points.

❺ Use this formula to calculate the expected airflow at 1,500 rpm.

the shutter rotate through an air gap between a permanent magnet and the Hall-effect switch. The position and spacing of the shutter windows indicate the TDC position of each piston. **Fig. 6-8.**

When a shutter window passes through the gap between the magnet and the Hall-effect sensor, the magnetic field reaches the sensor. **Fig. 6-9.** The sensor produces a small voltage. This action creates the triggering signal sent to the PCM or ignition module.

When the window moves out of the air gap, the magnetic field cannot act on the switch. The sensor produces no voltage. This action drops the triggering signal sent to the PCM or ignition module.

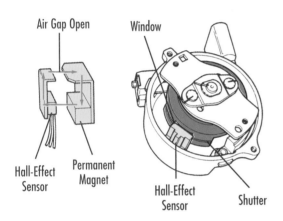

(a) No Shutter In Air Gap—Voltage Signal

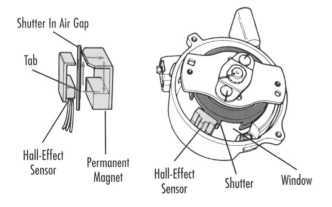

(b) Shutter In Air Gap—No Voltage Signal

Fig. 6-8 Operation of a Hall-effect sensor. **(a)** The window is passing through an air gap. The magnetic field from the permanent magnet is exposed to the Hall-effect sensor. **(b)** The shutter prevents the magnetic field from acting on the Hall-effect sensor. *What is the main difference between a magnetic pulse sensor and a Hall-effect sensor?* (Ford Motor Company)

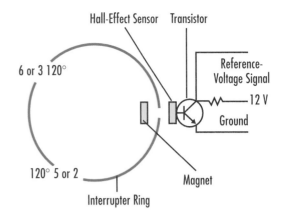

(a) Hall-Effect Crankshaft Sensor

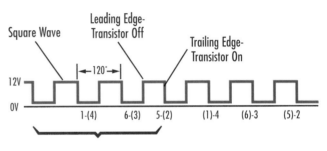

(b) Reference-Voltage Signal

Fig. 6-9 A Hall-effect sensor **(a)** generates a square-wave voltage signal **(b)**. *What is the main advantage of a square-wave signal?* (General Motors Corporation)

This Hall-effect sensor generates a "square-wave" or *digital* on/off reference signal. The reference signal indicates the position of each piston. The square-wave signal produced by this type of sensor has very sharp turn-on and turn-off times. The signal is extremely accurate.

When checking Hall-effect sensors, first test for the specified voltage at the power terminal. If the voltage is normal, check whether the voltage on the signal wire "toggles" between the signal voltage and near zero (<300 mV) while cranking the engine. You may also use a *light-emitting diode (LED) tester*. When connected to the signal wire, the LED should flash as you crank the engine. To determine connector pin connections, refer to the vehicle service manual.

Optical (Photodiode) Sensors A photodiode is an electronic device that uses the presence or absence of light to switch on and off. An LED provides the light source. Photodiode sensors are usually found in distributor ignition (DI) systems.

The timing device is a thin disc with two sets of slots around its outer edge. **Fig. 6-10.** Each set of slots controls a voltage signal in one of the photodiodes. The outer set, or high-data-rate slots, occur at 2-degree intervals of crankshaft rotation. This timing signal provides input to the PCM for engine crankshaft position and spark advance timing at engine speeds up to 1,200 rpm.

The inner, or low-data-rate, set of slots has the same number of slots as the engine has cylinders. This timing signal shows the TDC position of each piston. This signal triggers fuel injection and times spark advance at engine speeds above 1,200 rpm.

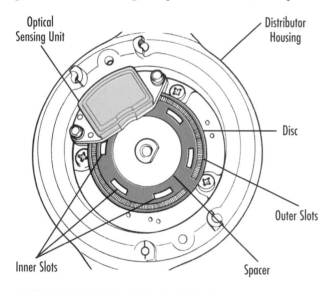

Fig. 6-10 Top view of the optical sensor distributor showing the high- and low-data-rate slots. *What type of component supplies the light source?* (DaimlerChrysler)

Two light-emitting diodes and two photodiodes are mounted on opposite sides of the slotted disc. **Fig. 6-11.** The slotted disc rotates between the LEDs and the photodiodes. As the slots move past an LED, the light beam is interrupted and the photodiode turns on and off. This creates an alternating voltage signal. An *integrated circuit* converts the voltage into digital on-and-off pulses. These pulses provide engine speed and crankshaft position signals directly to the PCM.

This type of sensor may not use a separate ignition module. Instead, the PCM uses the sensor input to control ignition timing.

Photodiode sensors can be tested by connecting a DVOM to the sensor's signal wire. The voltage

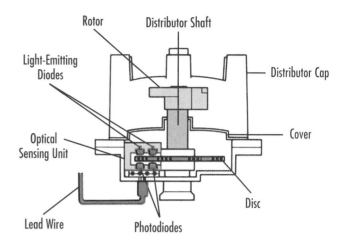

Fig. 6-11 An optical sensor uses the action of a light beam. *What is the difference between an optical sensor, and a Hall-effect sensor?* (DaimlerChrysler)

signal should change from high to low while the engine is cranking. The vehicle service manual should be used to identify the signal wire.

In most trigger circuit designs, one reluctor tooth, shutter window, or timing slot is made larger than the others. This is known as an indexing mark. The **indexing mark** is an indicator that produces a unique signal that tells the PCM when the number-one piston is nearing TDC. This signal is known as a *synchronizing pulse,* or *signature pulse.*

The Ignition Module

The ignition module is an electronic switch. It times the ignition spark. When the module switches on, current flows through the primary winding of the ignition coil. When the module switches off, current flow stops. The magnetic field in the primary winding collapses into the secondary winding. The voltage created in the secondary winding creates the spark at the spark plug.

When the engine begins cranking, the *ignition module* switches the ignition coil on and off. The module also controls the duration of current flow through the *primary winding* of the ignition coil.

This amount of time is called *dwell angle* or *duty cycle.* **Dwell angle** is the number of degrees of distributor or camshaft rotation during which current flows through the primary circuit of the ignition coil. **Duty cycle** is the percentage of time the primary circuit stays switched on.

Locating Information

Change is constant in the automotive industry. New models, new systems, new regulations, and modified or new procedures are frequently announced. Each change means new or revised information you will need to use. You also will need to know how to work on older models still in use.

To use this information you will need to find it. Fortunately much of this information has been organized to help you find it quickly. Textbooks, service manuals, and service bulletins usually have tables of contents and indexes. You can refer to these.

This textbook is organized to help you develop your ability to locate material quickly. The table of contents lists the chapters. It gives an overall view of the information in each chapter. Note that every chapter begins with a list of objectives. This list states important skills you should be able to develop as you study the chapter. The index lists subjects, components, and procedures covered in this book. It gives page numbers where you can find the information.

Most service manuals also use a table of contents to organize their information. An overall listing refers you to a specific section. There you will find a detailed listing of the topics covered in the section.

Apply it!

Meets NATEF Communication Standards for using text resources and service manuals to identify information.

1. Refer to the objectives at the beginning of this chapter.

2. Look through the chapter and identify the section headings that relate to each item on the list.

3. Choose one of the objectives.

4. Use a service manual's table of contents to find information about the objective that you selected in Step 3.

5. What similarities and differences do you find between the way these two sources of information are organized?

Dwell is reduced at idle to prevent the coil and module from overheating. Dwell is increased at high speed to produce the highest possible secondary voltage.

Diagnosing the Ignition Module Specialized diagnostic equipment is required to test ignition modules. However, a DVOM or a 12-volt test light can be used to see if the ignition module switches the primary ignition circuit on and off. Follow this procedure:

1. Use a wiring diagram to identify the correct terminals.

2. With the ignition ON, check for voltage to the module and the positive terminal of the ignition coil.

3. If voltage is present, connect a test light to the negative (ground) side of the ignition coil.

4. As the engine is cranking, the test light should flicker on and off.

5. If the light does not flicker, check for an open primary coil winding or a defective trigger circuit.

6. If the coil winding and trigger circuit are working, a defective module is indicated.

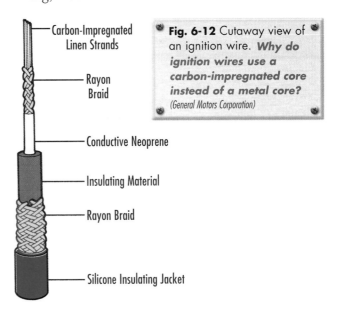

Carbon-Impregnated Linen Strands

Rayon Braid

Conductive Neoprene

Insulating Material

Rayon Braid

Silicone Insulating Jacket

Fig. 6-12 Cutaway view of an ignition wire. *Why do ignition wires use a carbon-impregnated core instead of a metal core?* (General Motors Corporation)

The Ignition Wires

The ignition wires connect the secondary winding of the ignition coil to the spark plug. The conductive core of the wire is a carbon-impregnated material. **Fig. 6-12.** An electrical resistance (about 1,000 ohms per inch) is built into the conductive core to reduce electrical interference and radio static. This problem is called electro-magnetic interference (EMI). This interference may cause erratic electrical inputs to the PCM, resulting in driveability complaints.

Diagnosing Ignition Wire Problems The core in an ignition wire may deteriorate or break. This will cause higher than normal resistance or an open circuit. Test this resistance by connecting an ohmmeter to each end of the wire. The resistance should not exceed the manufacturer's specifications. Replace any wire that is open or has higher than normal resistance.

Open ignition wires can also be detected by using an *ignition oscilloscope.* If the wire is good, the oscilloscope will display a normal waveform pattern. **Fig. 6-13.** A large break in the ignition wire will cause the waveform to show high firing voltage or other pattern variations.

Ignition wires and spark plug boots may develop cracks that can cause the spark to jump to ground. If you suspect a voltage leak, check wires for damaged insulation.

Fig. 6-13 A good ignition secondary circuit scope pattern (generic). *(Fluke Corporation)*

The Spark Plugs

The spark plug has a metal shell enclosing a ceramic insulator. **Fig. 6-14.** The *center electrode* is in the center of the insulator. The center electrode carries high voltage from the ignition coil to the plug air gap. The *spark plug* has two or more conductors called *electrodes.* The electrodes form an *air gap* between the insulated center electrode and one or more *ground electrodes.* By jumping the gap between the electrodes, the spark ignites the compressed air/fuel mixture.

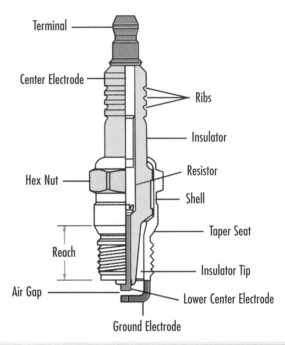

Fig. 6-14 Cutaway view of a resistor-type spark plug. The center electrode is insulated from the metal shell. The ground electrode grounds the spark plug through the engine. *What are two types of spark plug seats?* *(General Motors Corporation)*

One or more *ground electrodes* attach to the metal shell. To form an air gap at the center electrode, the ground electrode bends in. Specified gaps may range from 0.035″ [0.90 mm] to 0.080″ [2.03 mm].

Spark plugs are made in different hex sizes. The most common sizes are five-eighths and thirteen-sixteenths inches. To seal the shell to the cylinder head, some plugs use gaskets. Other plugs use tapered seats on the plugs and in the cylinder head openings. Tapered seat plugs do not use a gasket. Most automotive spark plugs have a resistance built into the center electrode to help suppress EMI. These plugs are called *resistor spark plugs.*

Spark Plug Threads Spark plugs are made with different thread diameters and thread lengths. The most common thread diameter is 14 mm, but some plugs have diameters of 10 mm or 18 mm. The thread length is also called the thread reach. *Thread reach* is the distance from the gasket seat (or top of a tapered seat) to the end of the threads.

Safety First

Materials When installing a spark plug, begin turning it by hand. Do not use a ratchet to start spark plug threads. Doing so could strip the threads and damage the plug.

If the reach is too long, the plug will protrude into the combustion chamber. The piston could strike the plug and cause engine damage. Carbon may also form on the exposed threads. This will make removal difficult. If the reach is too short, the plug may fail to ignite the air/fuel mixture properly. This will cause

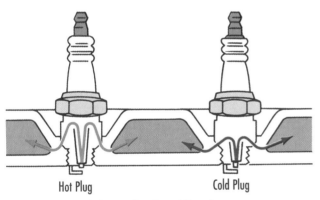

Hot Plug   Cold Plug

(a) Correct Reach and Heat Range

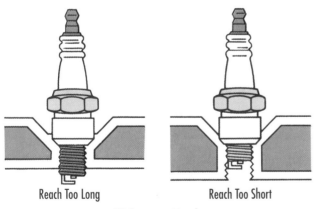

Reach Too Long   Reach Too Short

(b) Incorrect Reach

Fig. 6-15 Correct reach and heat range **(a)** and incorrect reach **(b)**. The longer the heat path, as the arrows show, the hotter the plug firing tip. *If the reach is too long, what may happen?* (General Motors Corporation)

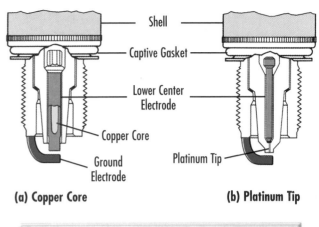

Shell

Captive Gasket

Lower Center Electrode

Copper Core

Ground Electrode

Platinum Tip

(a) Copper Core   **(b) Platinum Tip**

Fig. 6-16 Spark plugs with **(a)** a copper core and **(b)** a platinum tip. *What is one advantage of a platinum tip?* (Robert Bosch GmbH)

misfiring. The spark plugs that the engine manufacturer recommends have the correct reach. **Fig. 6-15.**

Spark Plug Heat Range The *heat range* is determined by how fast the plug transfers heat from the firing tip to the cylinder head. The length of the lower insulator and the type of metal used in the center electrode determine heat range.

The longer the length of the lower insulator, the higher the operating temperature of the plug. A short insulator transfers heat faster, so the plug has a lower operating temperature. The insulator temperature must be at least 700°F [389°C] to burn off carbon. To prevent preignition, the insulator temperature must not exceed 1,500°F [834°C].

To help conduct heat away from the lower insulator, some spark plugs have copper cores. Many plugs have electrodes made of nickel and chrome alloys that resist corrosion. To extend plug life, some plug electrodes are tipped with precious metals, such as platinum. **Fig. 6-16.**

When installing plugs, use the correct torque specification. Overtightening can damage the threads in the cylinder head. Undertightening can cause the spark plugs to overheat. An overheated plug may cause preignition damage to the engine.

Diagnosing Spark Plug Problems Spark plugs usually fail because of electrode wear or fouling deposits. Plugs may also show conditions resulting from overadvanced ignition timing, rich or lean air/fuel mixtures, and excessive oil consumption. **Fig. 6-17.**

COMMON SPARK PLUG CONDITIONS

NORMAL
Symptoms: Brown to grayish-tan color and slight electrode wear. Correct heat range for engine and operating conditions.
Recommendations: When new spark plugs are installed, replace with plugs of the same heat range.

PRE-IGNITION
Symptoms: Melted electrodes. Insulators are white, but may be dirty due to misfiring or debris in the combustion chamber. Can lead to engine damage.
Recommendations: Check for correct plug heat range, overadvanced ignition timing, lean fuel mixture, insufficient engine cooling, and lack of lubrication.

CARBON DEPOSITS
Symptoms: Dry, sooty deposits indicate a rich mixture or weak ignition. Causes misfiring, hard starting, and hesitation.
Recommendations: Make sure the plug has the correct heat range. Check for a clogged air filter or problem in the fuel system or engine management system. Also check for ignition system problems.

HIGH SPEED GLAZING
Symptoms: Insulator has yellowish, glazed appearance. Indicates that combustion temperatures have risen suddenly during hard acceleration. Normal deposits melt to form a conductive coating.
Recommendations: Install new plugs. Consider using a colder plug if driving habits permit.

ASH DEPOSITS
Symptoms: Light brown deposits on the electrodes caused by oil and/or fuel additives. Excessive deposits may foul the spark, causing misfiring and hesitation during acceleration.
Recommendations: Correct the mechanical condition with necessary repairs. Install new plugs. Also try changing gasoline brands.

DETONATION
Symptoms: Insulators may be cracked or chipped. Can lead to piston damage.
Recommendations: Make sure the fuel anti-knock values meet engine requirements. Use care when setting the gaps on plugs. Avoid lugging the engine.

OIL DEPOSITS
Symptoms: Oily coating caused by poor oil control. Oil is leaking past worn valve guides or piston rings into the combustion chamber. Causes hard starting, misfiring, and hesitation.
Recommendations: Correct the mechanical condition with necessary repairs. Install new plugs.

HEAT SHOCK FAILURE
Symptoms: Broken and cracked insulator tip. Caused by overadvanced timing, low-grade fuel, and severe operating conditions.
Recommendations: Check ignition timing. Use the proper grade of fuel. Check for proper spark plug heat range.

DEPOSIT FOULING
Symptoms: Heavy ash deposits. Ash buildup on portion of plug projecting into combustion chamber, and closest to the intake valve. Deposits are result of combustion of fuel and lubricating oil.
Recommendations: Replace plug. Check intake valve stem clearances and valve seals. Suspect defective seals if condition is found in only one or two cylinders.

INSUFFICIENT INSTALLATION TORQUE
Symptoms: Overheating of the spark plug and severe damage to spark plug elements. Caused by poor heat transfer between spark plug and engine seat.
Recommendations: Replace plug and torque to manufacturer's specifications.

Fig. 6-17 Causes of common spark plug failures. ***What are the two most common causes of spark plug failure?*** (AC Delco)

SECTION 1 KNOWLEDGE CHECK

❶ What is the purpose of the ignition system?

❷ What are the basic components of the ignition system?

❸ Which component of the ignition system has both a primary and a secondary circuit?

❹ What happens if the spark occurs too early in the cycle?

❺ What is the main purpose of the ignition module?

Section 2

Spark Distribution Systems

The *spark distribution system* delivers the spark to the correct cylinder in the correct *firing order*. Spark distribution systems are made in distributor, electronic, and direct-ignition types.

Distributor Ignition Systems

The ignition distributor usually performs two tasks. First, most distributors contain a device that triggers the primary circuit on and off. As noted earlier, this device may be a permanent magnet, Hall-effect, or optical type trigger. The second task performed by the distributor is to conduct the high ignition voltage from the coil wire to the spark plug wires. **Fig. 6-18.**

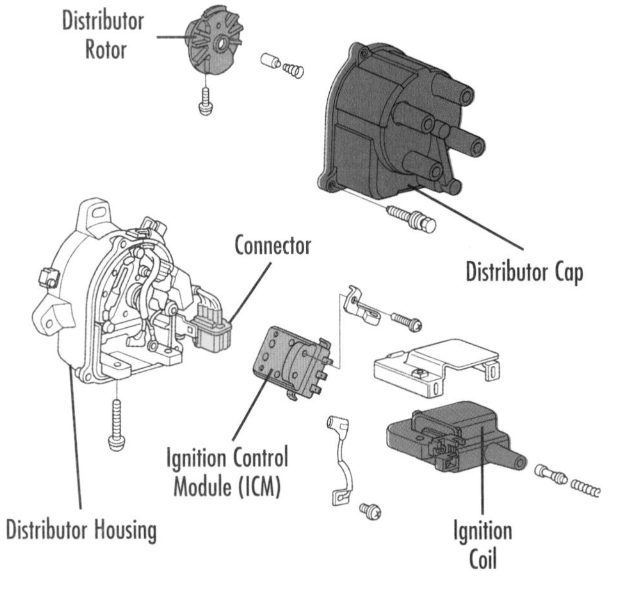

Depending on the application, the ignition module may be mounted on the inside or outside of the distributor housing. On other vehicles the module function is built into the PCM.

The distributor assembly includes a cap and rotor. The rotor mounts on the end of the distributor shaft. As the rotor turns, it delivers ignition coil output to the correct ignition wire terminal. The secondary ignition wires are connected to the distributor cap in a specific order.

Distributor ignition systems have provisions for setting the base ignition timing. Follow the manufacturer's procedures when installing a distributor or adjusting base ignition timing.

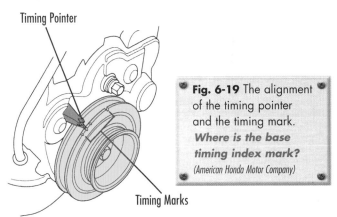

Timing Pointer

Timing Marks

Fig. 6-19 The alignment of the timing pointer and the timing mark. *Where is the base timing index mark?* (American Honda Motor Company)

Base timing is adjusted by using a *timing light*. The timing light is a bright stroboscopic (strobe) light that is usually connected to the number-one spark plug wire. Current flow through the wire causes the timing light to flash. When the light flashes, it shows the position of a timing, or index, mark on the vibration damper in relation to a timing pointer mounted on the front of the engine. **Fig. 6-19.** Some engines have timing marks located on the flywheel and bell housing.

Distributor Rotor

Connector

Distributor Cap

Ignition Control Module (ICM)

Distributor Housing

Ignition Coil

Fig. 6-18 Exploded view of a Honda distributor. *What two tasks are performed by a distributor?* (American Honda Motor Company)

With the engine running at the specified rpm, adjust the base timing by loosening the distributor hold-down bolts. **Fig. 6-20.** Turning the distributor slowly changes the timing. When the timing light shows the timing mark aligned with the timing pointer, you have correctly adjusted the base timing. To prevent the distributor from moving, you must tighten the distributor hold-down bolt(s). Restore the PCM to its normal mode.

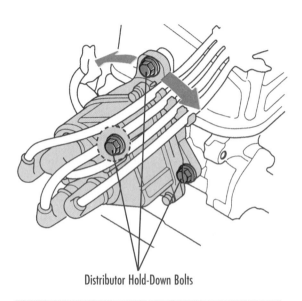

Distributor Hold-Down Bolts

Fig. 6-20 Distributor hold-down bolts. *What type of ignition timing does the position of the distributor control?* (American Honda Motor Company)

Diagnosing the Distributor Ignition System In addition to the previous tests, check the distributor cap for cracks and *carbon tracking*. Carbon tracking occurs when the spark jumps from a distributor cap terminal to another terminal or to ground. Carbon tracks appear as thin black lines on the inside of the cap. Check distributor rotors for burn-through from the center contact to the distributor shaft. Light-colored pinholes in the rotor material indicate burn-through.

TECH TIP No-Spark Condition. An electronic distributor ignition system relies on the distributor to time and distribute the secondary ignition voltage. The camshaft usually drives the distributor. The camshaft is timed to the engine through a series of gears and a timing belt or chain. If the timing belt or chain breaks, ignition timing is affected. If a no-spark condition exists, remove the distributor cap, crank the engine, and look at the rotor or triggering devices. If no motion is seen, the camshaft is not driving the distributor.

Distributorless Ignition Systems

A **distributorless ignition system** is an ignition system that does not use a distributor to establish base ignition timing or distribute the secondary ignition voltage. It uses the same type of triggering devices and modules as a distributor type system. Distributorless ignition uses one ignition coil for every two cylinders. Each coil has its own module circuit.

The cylinders are connected in pairs. When one cylinder is on its compression stroke, the *companion cylinder* is on its exhaust stroke. When the ignition system is triggered, the spark in the cylinder on compression ignites the mixture. The spark in the companion cylinder is "wasted" because the exhaust gases cannot be ignited. The next time the same pair of cylinders fire, the companion cylinder will be on compression and the other cylinder will be on exhaust. These systems are often referred to as "waste spark" systems. **Fig. 6-21.**

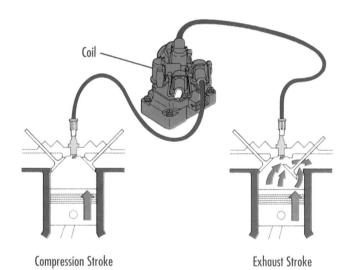

Coil

Compression Stroke   Exhaust Stroke

Fig. 6-21 A waste spark system fires two spark plugs on different cylinders at the same time. The pistons are on different strokes. One spark ignites the air/fuel mixture during the compression stroke. The other spark is wasted during the exhaust stroke. *What is the advantage of a waste spark system?* (Ford Motor Company)

Figure 6-22 shows a crankshaft position (CKP) sensor. The CKP sensor provides the ignition module and PCM with engine speed and crankshaft position information. Some distributorless ignition systems also use a camshaft position (CMP) sensor.

Crankshaft Position Sensor

Fig. 6-22 A crankshaft position sensor on a Ford 3.0L engine. *What information does the CKP provide?* (Terry Wild Studio)

Camshaft Position Sensor

Fig. 6-23 A camshaft position sensor mounted on a Ford 3.0L engine. *What information does the CMP provide?* (Terry Wild Studio)

The CMP sensor is used to synchronize the firing of the coil packs and to trigger sequential fuel injection. The CMP sensor is usually a Hall effect sensor. **Fig. 6-23.**

Diagnosing Distributorless Ignition Systems
With no coil distributor cap, rotor, or coil wire, distributorless ignition systems have fewer components to check than distributor type systems. Since one coil fires a pair of spark plugs, a coil or ignition module problem usually affects both cylinders. If one spark plug wire is open or disconnected the other plug in the pairing may fire at idle. Under load, both cylinders usually fail to fire. To diagnose a problem in distributorless ignition systems:

- Check the trigger circuit, primary circuit, and module in the same way as for other systems.
- Check for a spark from each coil assembly using the spark tester.
- Test the spark plug wires for continuity using an ohmmeter.
- Test each coil secondary winding for continuity using an ohmmeter.
- Refer to the service manual for specific test procedures and specifications.

Direct Ignition Systems

Direct ignition systems use one ignition coil and module circuit for each cylinder. They are also referred to as "coil on plug" (COP) ignition systems. A coil is mounted directly over each spark plug. This eliminates the cap and rotor and all secondary wiring.

One variation of this systems mounts an ignition coil near the spark plug and uses a short plug wire to connect the two. In either case, systems operation is the same as in the other ignition types discussed.

Diagnosing Direct Ignition Systems To diagnose a direct ignition system:

- Check the trigger circuit, primary circuit, and module in the same way as for other systems.
- Check for a spark from each coil assembly using a spark tester. For COP systems connect the tester to the coil output terminal. It may be necessary to use a secondary (coil or plug) wire to make the required connections.
- Test each coil secondary winding for continuity using an ohmmeter.
- Refer to the service manual for specific test procedures and specifications.

SECTION 2 KNOWLEDGE CHECK

❶ Which type of ignition system is driven by the camshaft?

❷ What test device is used to check base timing?

❸ Which type of ignition system has no coil lead or spark plug cables?

❹ In a waste spark ignition system, how many coils does it take to fire paired spark plugs?

❺ Where is an ignition coil located on a direct ignition system?

CHAPTER 6 REVIEW

Key Points

Meets the following NATEF Standards for Engine Performance: inspecting and testing ignition system components.

- An ignition coil has a primary and secondary winding.
- Secondary ignition voltage only builds up to the point where current begins to flow in the circuit.
- A distributor is a device designed to establish base timing and distribute the ignition spark.
- Ignition triggering devices include magnetic pulse, Hall-effect, and optical types.
- Ignition triggering devices provide crankshaft position and engine rpm information.
- The ignition module switches primary current on and off.
- Spark plug and coil wires have high resistance to control electro-magnetic interference.
- Distributorless (waste spark) ignition systems use one coil for two paired cylinders.
- Direct ignition systems use one coil for each cylinder.

Review Questions

1. Describe the procedure for inspecting and checking ignition coils.
2. How much voltage can be produced in the secondary ignition coil winding?
3. When testing a distributor, what three types of trigger circuits might you encounter?
4. What test equipment can be used to check the output signal of all ignition sensors?
5. What test device can be used to test spark plug wires for continuity?
6. How is base timing adjusted on a distributor ignition system?
7. What is the difference between distributorless (waste spark) and direct ignition systems?
8. (Critical Thinking) What can happen if spark plugs with too low of a heat range are used in an engine?
9. (Critical Thinking) What might result from a damaged slotted disc in an optical sensor?

TECHNOLOGY FORECAST FOR AUTOMOTIVE EXCELLENCE

COP Ignitions

An old idea for an ignition system will start the cars of today–and tomorrow. The "coil-on-plug" (COP) ignition system places an ignition coil at each cylinder. It was originally used on the earliest cars and trucks. It is different from today's "waste spark-type ignition," which uses one coil to fire two cylinders.

The coil-on-plug ignition is more expensive, but it is more efficient. Because it can vary the ignition timing for each cylinder, the engine can produce more power and fewer emissions. Another advantage is simplified wiring, which reduces the chance of electromagnetic interference.

Diagnosis of ignition problems is easier with coil-on-plug ignitions. With the help of a hand-held scan tool, a technician can determine which cylinder needs repair. The vehicle's computer can also warn of possible engine trouble. When a cylinder misfire is detected, the computer can turn on the "check engine" (MIL) light and set a diagnostic trouble code.

Meanwhile, the construction of the spark plug itself could change. Instead of a single-ground electrode design, it may have a four-ground electrode in the future. These multiple ground paths will allow the spark to take the most efficient path from the spark plug's platinum center electrode. The longer duration, more efficient spark should provide better combustion and yield more power.

AUTOMOTIVE EXCELLENCE
TEST PREP

Answering the following practice questions will help you prepare for the ASE certification tests.

1. An engine has detonation and high HC emissions. The most likely cause is:
   - **a** Retarded base timing.
   - **b** Overadvanced base timing.
   - **c** Both a and b.
   - **d** Neither a nor b.

2. An engine has poor power and fuel economy. Technician A says the base timing may be retarded. Technician B says the base timing may be overadvanced. Who is correct?
   - **a** Technician A.
   - **b** Technician B.
   - **c** Both Technician A and Technician B.
   - **d** Neither Technician A nor Technician B.

3. An engine with distributorless ignition has a no-spark condition. The ignition module and coil pack assembly has been replaced. What is the most likely cause?
   - **a** Defective ignition cables.
   - **b** A faulty triggering device.
   - **c** Both a and b.
   - **d** Neither a nor b.

4. Technician A says a Hall-effect sensor does not need an outside voltage supply. Technician B says a Hall-effect sensor needs an outside light source. Who is correct?
   - **a** Technician A.
   - **b** Technician B.
   - **c** Both Technician A and Technician B.
   - **d** Neither Technician A nor Technician B.

5. Two technicians are diagnosing a no-spark condition. The distributor has a slotted disk-type reluctor. Technician A says the ignition is a photodiode system. Technician B says the ignition is a magnetic pulse system. Who is correct?
   - **a** Technician A.
   - **b** Technician B.
   - **c** Both Technician A and Technician B.
   - **d** Neither Technician A nor Technician B.

6. Technician A says the ignition module may send an engine speed signal to the PCM. Technician B says the module acts as an electric switch for the primary ignition system. Who is correct?
   - **a** Technician A.
   - **b** Technician B.
   - **c** Both Technician A and Technician B.
   - **d** Neither Technician A nor Technician B.

7. An ignition coil may be defective. Technician A says the primary circuit should have high resistance. Technician B says the secondary circuit should have low resistance. Who is correct?
   - **a** Technician A.
   - **b** Technician B.
   - **c** Both Technician A and Technician B.
   - **d** Neither Technician A nor Technician B.

8. An engine equipped with a distributor ignition system backfires when cranked. A possible cause is:
   - **a** A distributor not correctly timed to the crankshaft.
   - **b** Ignition wires not installed in the correct firing order.
   - **c** Both a and b.
   - **d** Neither a nor b.

9. A six-cylinder engine with an electronic ignition system has a severe engine miss. Two cylinders have no spark. Technician A says the ignition module is not getting a good reference signal. Technician B says one of the three ignition coils is bad. Who is correct?
   - **a** Technician A.
   - **b** Technician B.
   - **c** Both Technician A and Technician B.
   - **d** Neither Technician A nor Technician B.

10. A spark plugs electrodes are burned and worn. What is the most likely cause?
    - **a** The plug heat range is too hot.
    - **b** Undertightened spark plugs.
    - **c** Both a and b.
    - **d** Neither a nor b.

CHAPTER 7

Using Computer Diagnostics: OBD-I

You'll Be Able To:

- ⊗ Use recommended precautions when handling static sensitive devices.
- ⊗ Inspect and test power and ground circuits.
- ⊗ Diagnose driveability problems with stored diagnostic trouble codes.
- ⊗ Retrieve diagnostic trouble codes.
- ⊗ Diagnose driveability problems with no stored diagnostic trouble codes.

Terms to Know:

data link connector (DLC)

diagnostic trouble code (DTC)

malfunction indicator lamp (MIL)

on-board diagnostic systems

random access memory (RAM)

serial data stream

vehicle identification number (VIN)

The Problem ⟩⟩⟩⟩⟩

Serifina Tyler was driving home from work when the malfunction indicator lamp (MIL) in her one-year-old vehicle started to blink. The engine began to run roughly. She pulled to the curb, turned off the engine, and reached for the owner's manual.

She knew that the blinking MIL indicated a problem. She turned on her hazard flashers and started reading the owner's manual. It stated that the vehicle should be serviced as soon as possible.

She restarted her car. As she continued towards her home, the MIL turned on and stayed on.

Your Challenge

As the service technician, you need to find answers to these questions:

1. Why does an engine misfire activate the MIL on an OBD-I system?
2. What are the enabling criteria for a misfire DTC?
3. When should the DTC be cleared?

Section 1

On-Board Diagnostics

In the early 1980s, vehicle manufacturers began using on-board diagnostic systems. **On-board diagnostic systems** are computer-controlled test routines that monitor engine control systems. Most on-board diagnostic systems made before 1996 are known as on-board diagnostic I (OBD-I) systems. OBD-I technology was used during the 1980s and early 1990s. The main computer in these systems was known by the following names:

• Engine control module (ECM).
• Electronic engine controller (EEC).
• Single board engine controller (SBEC).

The OBD-I system's diagnostic capabilities has the following limitations:

• It could not identify deterioration of a system.
• It could not monitor all engine-related systems.
• It used nonstandard diagnostic trouble codes, terminology, and diagnostic procedures.

Powertrain Control Module

The engine control module is now referred to as the *powertrain control module (PCM)*. The PCM is sometimes called a processor or microprocessor. It is the heart of the OBD system. Using input from a sensor network, it operates a network of actuators that regulate engine control systems. **Fig. 7-1.**

The instructions for the PCM are contained in a programmable read-only memory (PROM). The PROM may also be called a MEMory CALibration (MEMCAL) unit. The PROM can be hard-wired into the PCM or it can be removable.

The PROM contains engine *calibration* data in a read-only memory (ROM) chip. The calibration data includes *operating parameters* for functions such as idle speed, air/fuel mixture, and spark advance. Using this data, the PCM controls these engine management functions to provide the best driveability, fuel economy, and lowest emissions.

The PROM also contains operating instructions for the PCM. Whenever possible, the PCM operates in a closed-loop mode. In this mode the PCM controls the ignition timing and air/fuel mixture.

Before the PCM can go into closed-loop operation, several operating conditions must be met. These include:

• Output from the engine coolant temperature sensor must be within a specified range.
• Output from the oxygen sensor must be varying within a specified voltage range.
• A specified period of time must have passed since the last system start-up.

If certain failures occur, such as an out-of-expected-range signal from the oxygen sensor (O2S), the PCM goes into a limited operating strategy (LOS), or limp-in mode. The driver is alerted to the LOS condition by the malfunction indicator light. The **malfunction indicator lamp (MIL)** is a light that warns the driver of a problem in the systems monitored by the PCM. The MIL is sometimes referred to as the check engine light or service engine soon light.

Fig. 7-1 The PCM is the heart of the on-board diagnostics system. *What information does the PROM contain?* (Terry Wild Studio)

TECH TIP If ignition timing doesn't respond to changes in engine load, the PCM may be in LOS mode. The MIL is your only visible indicator of a problem. Before making any diagnosis, be sure that the MIL is operational.

In LOS mode, factors such as air/fuel mixture and ignition timing are controlled by preset values stored in the PCM. Driveability, fuel economy, and emissions are adversely affected.

Manufacturers may change the calibration settings of the engine to correct or improve a driveability or emissions control problem. When the calibrations are updated, OBD-I system PCMs with hard-wired PROMs must be replaced as a unit. PCMs with a removable PROM use a replaceable PROM chip.

Electronic parts such as PROM chips can be seriously damaged by static electricity. High voltage "spikes," known as electrostatic discharge, can ruin sensitive electronic circuits. When handling a PCM, never touch its connector pins or its circuit board. Also, never touch the pins of a replaceable PROM chip. **Fig. 7-2.**

The PCM also has a random access memory. **Random access memory (RAM)** is a volatile, or erasable, memory that temporarily stores information such as diagnostic trouble codes. A **diagnostic trouble code (DTC)** is a code that identifies a

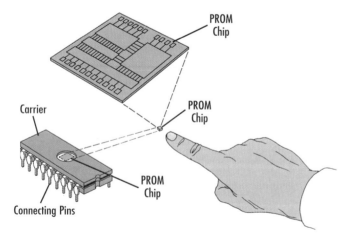

PROM Chip

Carrier

PROM Chip

PROM Chip

Connecting Pins

Fig. 7-2 The PROM chip and its carrier. *What type of memory does the PROM have?* (Ford Motor Company)

system or component malfunction. In an OBD-I system, the DTC is a two- or three-digit value. In most OBD-I systems, disconnecting the battery or removing the fuse for the PCM erases all codes stored in the RAM.

Monitored and Nonmonitored Circuits

The PCM in OBD-I systems monitors the operation of a network of sensor and actuator circuits. **Fig. 7-3.** The return signal from each sensor and actuator operation indicates how each circuit is performing. The PCM will store a DTC if the performance of a circuit is not within a range of expected values. Circuits that provide a direct input to the PCM are *monitored circuits*.

Nonmonitored circuits are circuits that are not directly monitored by the PCM. For example, OBD-I systems do not directly monitor fuel pressure or spark plug firing. OBD-I systems cannot directly detect conditions such as:
- Dirty or shorted fuel injectors.
- Worn or defective ignition parts.
- Vacuum leaks.
- Worn engine parts.

However, many of these conditions can indirectly affect a monitored circuit. A dirty fuel injector, for example, may set a DTC that shows a lean air/fuel mixture. A defective spark plug may set a DTC that shows a rich exhaust condition.

Sensors Sensor circuits contain a power or reference signal circuit, a return signal circuit, and a ground circuit. The power signal supplies the reference voltage, which is usually 5 volts. The return signal from the sensor is a changing voltage or frequency sent to the PCM.

There must be current flow in a circuit to create a voltage. To measure voltage signals, the sensor must be connected and the circuit must be "on."

The reference signal, return signal, and ground circuit voltages are measured with a digital volt-ohm-meter (DVOM). Follow the manufacturer's instructions when connecting a DVOM to any circuit. Sensitive electronic circuits can be damaged if tests are not made correctly.

The reference voltage on most systems is 5 volts. The return signal voltage should measure between 0 and 5 volts, depending upon the application.

Converting Sensor Signals

As you have learned, the PCM depends on sensors to provide the information it needs. Since the PCM is a computer, it needs such information as a voltage or frequency signal. The measured voltage or frequency signal tells the PCM about the engine condition being measured.

Most sensors use a simple circuit called a voltage divider. This circuit converts position, temperature, pressure, and other engine-related inputs into a voltage. A simple voltage divider circuit can act as a sensor to convert nonelectrical quantities into a voltage between 0 and 5 volts.

Apply it!

Wiring Voltage Divider Circuits

Meets NATEF Science Standards for electrical measurements.

Materials and Equipment
- 5-volt power source
- DVOM (an analog voltmeter may be used)
- 1000 Ω linear variable resistor or potentiometer
- Strips of sheet metal that can be drilled and bent to make brackets and a lever arm on the resistor shaft
- Insulated hookup wire
- 12″ [30 cm] piece of stiff wire, such as coat hanger wire
- 6″ × 12″ [15 cm × 31 cm] board to use as a base
- Assortment of rubber bands
- Nuts, bolts, and screws

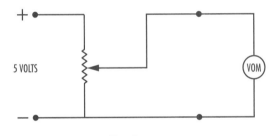

Fig. A

❶ Assemble the circuit following the circuit diagram in **Fig. A** and the drawing in **Fig. B.**

❷ Test the basic sensor. Rotate the shaft on the resistor while observing the DVOM readings. These readings illustrate the basic sensor behavior that is converting angular position into a voltage.

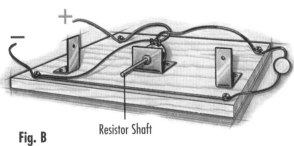

Fig. B   Resistor Shaft

❸ Change the sensor circuit so that it converts linear position into a voltage.

   a. To do this, bend a strip of sheet metal around the resistor shaft. Drill and bolt the metal to make a lever arm with a hole about level with the holes in the left and right brackets.

   b. Hook one end of the stiff wire through the hole in the lever arm. Put the other end of the wire through the hole in the left-hand bracket.

   c. Move the resistor shaft left and right, using the left end of the wire. Observe how the output voltage changes.

❹ Now convert the position sensor into a force sensor.

   a. Connect a rubber band between the lever arm and the right-hand bracket. Pull on the wire that runs through the left bracket. Observe how a force can be converted into a voltage.

   b. Repeat this step using several strengths of rubber bands.

❺ Explain how the strength of the rubber band influences the outcome of the experiment.

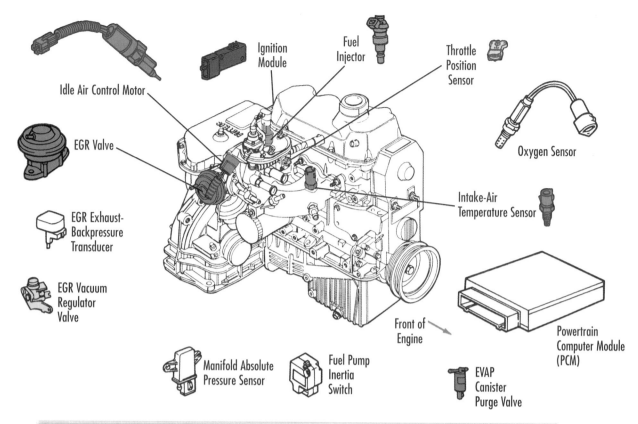

Fig. 7-3 An electronic engine management system, with its sensors and actuators. Some of these components, like the EGR valve, are mounted directly on the engine. Others, like the fuel pump inertia switch, are mounted elsewhere in the vehicle. *Which parts are actuators?* (Ford Motor Company)

The ground-circuit voltage drop should be less than 100 millivolts. High circuit resistance or poor electrical connections will increase voltage drop. High ground-circuit voltage may indicate a poor PCM ground connection.

Actuators Actuators are the solenoids and motors operated by electrical outputs from the PCM. The actuators perform a task in response to the PCM output. Tasks performed by actuators include:
• Controlling air/fuel mixture ratios.
• Adjusting idle speed.
• Opening and closing the EGR valve.
• Controlling evaporative emission (EVAP) system operation.

Fuel injectors and idle air control (IAC) motors are examples of actuators that are operated and monitored by the PCM. If the PCM senses a problem in a monitored actuator circuit, it stores a DTC.

Many diagnostic scanners can activate fuel injectors, cooling fans, fuel pump relays, and other types of actuators. This is called *interactive testing*. Dedicated scanners normally have more interactive testing features than generic scanners.

An actuator that fails to respond to an interactive test may not be defective. Before replacing a component, check for a blown fuse, a defective relay, or an open power or ground circuit.

SECTION 1 KNOWLEDGE CHECK

❶ Why is a PCM sometimes called a processor?

❷ Which part of the PCM contains its operating instructions?

❸ What is a monitored circuit?

❹ What will increase ground-circuit voltage drop?

❺ What are two examples of monitored actuators?

Section 2

Diagnostic Trouble Codes

Diagnostic trouble codes (DTCs) are set by the PCM. **Table 7-A.** The codes indicate a monitored sensor circuit that is not providing a return signal within a range of expected values. Driveability and other problems can be diagnosed by reading DTCs. They indicate the circuit or device affected. These codes are explained in the vehicle service manual.

Displaying Diagnostic Trouble Codes

Most OBD-I systems display the DTCs stored in the diagnostic memory of the PCM by:
- Flashing the code on the MIL.
- Flashing the code on LEDs in the PCM.
- Displaying the code on a scan tool or scanner.

Most OBD-I systems store two-digit codes. More complex OBD-I systems store three-digit codes. OBD-I codes are not standardized. Refer to the vehicle service manual for a complete listing of all applicable codes. Many OBD-I systems automatically erase DTCs after 50 key-on cycles if the problem has been corrected.

Table 7-A	OBD-1
Code	**Circuit or Device Affected**
12	No reference signal
13	Oxygen sensor circuit
14	Engine coolant temperature (ECT) sensor. High temperature indicated
15	Engine coolant temperature (ECT) sensor. Low temperature indicated
21	Throttle position sensor (TPS). Signal voltage high
22	Throttle position sensor (TPS). Signal voltage low
23	Intake air temperature (IAT). Low temperature indicated
25	Intake air temperature (IAT). High temperature indicated
32	Exhaust gas recirculation (EGR) system failure
33	Manifold absolute pressure (MAP) sensor. Signal voltage high
34	Manifold absolute pressure (MAP) sensor. Signal voltage low
35	Idle air control (IAC)
42	Electronic spark timing (IC)
44	Oxygen sensor circuit. Lean exhaust indicated
45	Oxygen sensor circuit. Rich exhaust indicated

Using Communication Strategies

Your customers will vary in their ability to describe a vehicle's problems. They also may vary in their ability to understand what you are telling them. For example, when you talk with a hearing-impaired customer, there are some things you can do to make effective communication easier.
- Allow enough time. Listen carefully to what your customer is telling you.
- Make sure that your customer can see your face and mouth clearly when you speak. Try to have the light fall on your face.
- Speak clearly at your normal speed.
- Avoid speaking over distracting background noises. If possible, move to a quieter area.

You may want to ask your customer, "Do you have any questions?" and, "Would you like me to put my answers in writing?"

You will encounter many other situations in which you need to adapt your communication strategies.

Apply it!

Meets NATEF Communications Standards for adapting listening and speaking strategies to the needs of the customer.

You are standing outside the open doors of your work area. There is heavy traffic on the busy street. Your customer asks you a question about her vehicle.

❶ Write a paragraph describing what you should do to make communication clearer.

❷ If there were no background traffic noise, would you change your communication strategy? Why or why not?

Using the Malfunction Indicator Lamp When you use the malfunction indicator lamp (MIL) to display a DTC, the MIL will illuminate briefly when the ignition is turned on. This is normal.

1. Place the PCM in diagnostic mode by grounding the proper terminal on the data link connector. The **data link connector (DLC)** is a plug-type connector used by a scan tool to access vehicle diagnostic information. **Fig. 7-4.** Refer to the vehicle service manual for the proper procedure.

2. Read the stored DTC on the MIL as a series of flashes representing numbers. For example, a DTC of 12 is represented as one long flash and two short flashes.

3. Write down or record DTCs in the order in which they are read.

4. Refer to the vehicle service manual for an explanation of the DTC and for accepted diagnostic and repair procedures.

Fig. 7-4 A 12-pin DLC found on some OBD-I systems. *What is the first step in reading a DTC?*

Using the Powertrain Control Module Some OBD-I systems do not have a DLC. Instead, red and green light-emitting diodes (LEDs) are built into the powertrain control module (PCM).

1. Place the PCM in the diagnostic mode. Refer to the vehicle service manual for the proper procedure.

2. Read the stored DTC as a series of flashes on the LEDs. For example, a DTC of 12 is represented as one flash on the red LED and two flashes on the green LED.

Using a Scanner The most common method of retrieving DTCs is to connect a scanner (sometimes called a scan tool) to the DLC. Two types of scanners are used—dedicated and generic. **Fig. 7-5.**

Fig. 7-5 A scanner is used to retrieve DTCs stored in the PCM. *Why would a dedicated scanner provide more information than a generic scanner?*
(Actron Manufacturing)

Dedicated scanners are designed for use on specific vehicles. These scanners normally have more diagnostic capacity and features than do generic scanners. Some of these scanners have a built-in *diagnostic database* that includes electronic specifications and troubleshooting tips.

Generic scanners are made for use on many different vehicles from different manufacturers. Most of these scanners are programmed with removable cartridges. These cartridges program the scanner for specific applications or system updates.

Scanners display DTCs and other diagnostic data in different ways. Always consult an instruction manual before connecting a scanner.

After the scanner is connected, the scanner menu may request information found in the vehicle identification number (VIN). The **vehicle identification number (VIN)** is a serial number unique to each vehicle that indicates:
• When the vehicle was made.
• The country in which it was made.
• The vehicle make and type.
• The passenger safety system.
• The type of engine.
• The line, series, and body style.
• The assembly plant where the vehicle was produced.

TECH TIP **Entering Scanner Data.** Entering the wrong VIN information or using incorrect scanner software, or a bad DLC connection may cause a NO DATA: message on the scanner. Generic scanners must be individually programmed and configured.

The VIN is inscribed on a plate mounted at the lower corner of the driver's-side windshield. **Fig. 7-6.** The VIN may also be displayed on a tag mounted on the driver's-side doorjamb.

The scanner may request the tenth character of the VIN, which is the vehicle's model year. The scanner may also request the eighth character in the VIN, which is the engine code. The engine code indicates the type and size of engine installed in the vehicle. Most scanners will also request the third character of the VIN, which indicates the type of vehicle being diagnosed.

The screen on the scanner shows a menu that may include engine, transmission, air bag, or anti-lock brake options. To retrieve engine DTCs, choose the engine menu.

The scanner may also show a diagnostic menu that includes special diagnostic test modes used only by the vehicle's manufacturer. The scanner's instruction manual will describe each diagnostic mode and the type of information it provides. To choose the correct diagnostic mode for the specific vehicle application, follow the instructions included with the scanner.

The diagnostic memory in the PCM may store a "soft" or "hard" DTC. A soft (history) DTC is not currently occurring. It may be an intermittent problem. The soft code illuminates the MIL only when the failure occurs. A hard (current) DTC is present in the system when the MIL is on.

TECH TIP **Checking the PCM.** Running the engine with a sensor disconnected should store a hard code in the diagnostic memory of the PCM. If no code is stored, check all cables and harnesses connected to the PCM. Check the voltage and ground connections at the sensor. To set some codes, the vehicle must be driven.

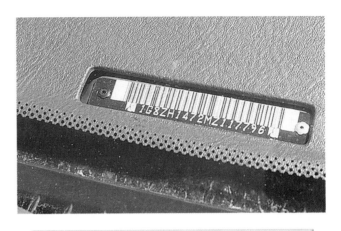

Fig. 7-6 The VIN plate is usually located at the base of the windshield on the driver's side. *In what other location is VIN information displayed?* (Jack Holtel)

Interpreting Diagnostic Trouble Codes

A DTC may not be directly related to a specific complaint. Always use an effective strategy to diagnose the condition that sets a DTC. An effective strategy might include:

1. Listing all displayed DTCs before doing any repair work. Multiple DTCs must be corrected in a specified order.

2. Checking the manufacturer's technical service bulletins for factory service information.

3. Determining when and how a specific problem occurs.

4. Accompanying the customer as they drive their car to verify the problem.

5. Identifying the problems the customer is most concerned about.

6. Inspecting the vehicle's maintenance records. Make sure that all scheduled maintenance is up to date. Find out if other service facilities have tried to repair the problem.

7. Inspecting the related sensor and its connections. For example, a cracked vacuum hose to a MAP sensor may cause a DTC in the MAP sensor circuit. A burned-through wire on the oxygen sensor may set a DTC for the oxygen sensor circuit.

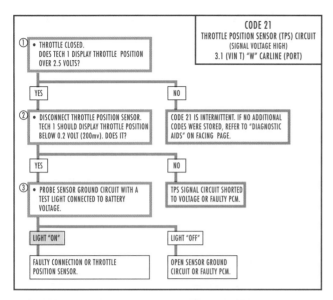

CODE 21
THROTTLE POSITION SENSOR (TPS) CIRCUIT
(SIGNAL VOLTAGE HIGH)
3.1 (VIN T) "W" CARLINE (PORT)

① • THROTTLE CLOSED.
DOES TECH 1 DISPLAY THROTTLE POSITION OVER 2.5 VOLTS?

YES → NO

② • DISCONNECT THROTTLE POSITION SENSOR. TECH 1 SHOULD DISPLAY THROTTLE POSITION BELOW 0.2 VOLT (200mv). DOES IT?

CODE 21 IS INTERMITTENT. IF NO ADDITIONAL CODES WERE STORED, REFER TO "DIAGNOSTIC AIDS" ON FACING PAGE.

YES → NO

③ • PROBE SENSOR GROUND CIRCUIT WITH A TEST LIGHT CONNECTED TO BATTERY VOLTAGE.

TPS SIGNAL CIRCUIT SHORTED TO VOLTAGE OR FAULTY PCM.

LIGHT "ON" → LIGHT "OFF"

FAULTY CONNECTION OR THROTTLE POSITION SENSOR.

OPEN SENSOR GROUND CIRCUIT OR FAULTY PCM.

Fig. 7-7 A diagnostic flowchart for a TPS circuit. *What conditions could cause a code 21?* (General Motors Corporation)

Diagnostic Flowcharts A *diagnostic flowchart* is used to troubleshoot the condition that sets a DTC. **Fig. 7-7.** The flowchart is also called a *diagnostic tree.* Each DTC has a corresponding flowchart in the vehicle service manual. The flowchart provides a step-by-step diagnostic procedure for each sensor or actuator circuit DTC. The flowchart may also provide voltage and resistance values for each sensor or actuator circuit. If more than one DTC is stored, the diagnostic flowchart may specify the order in which the DTCs should be corrected.

A DTC indicates which sensor or actuator circuit is not within expected values. For example, a DTC 21 shows a failure in the throttle position sensor circuit. The problem may be caused by:
• A defective throttle position sensor (TPS).
• An open, shorted, or grounded reference voltage wire.
• High resistance or an open in the TPS ground circuit.
• A defective PCM.

Interpreting Serial Data

Some OBD-I systems provide a serial data stream to indicate how each sensor and actuator is responding to changes in engine operation. The **serial data stream** is information displayed as voltage values or as actual readings such as degrees

of temperature, inches of vacuum, or percentage of throttle opening. **Fig. 7-8.** Connecting a scan tool or similar diagnostic display to the DLC usually retrieves this information.

Serial data is most useful when a condition does not set a DTC. This is called a no-code driveability condition. A no-code condition can occur even if the failure is in a monitored sensor circuit.

The PCM may need to sense a condition for a predetermined amount of time. If a failure occurs quickly and then returns to within its normal range, the PCM may not record a DTC.

In other cases, age may degrade a sensor's calibration. O2Ss are prone to *calibration drift.* The sensor may become contaminated with deposits that reduce its sensitivity. The sensor becomes less responsive to changes in the oxygen content of the exhaust gas. Sensor degradation will not set a DTC if the return signal is within an expected range of values.

You can test most sensors by observing the scanner's serial data stream. Serial data can be compared to specified readings to diagnose no-code conditions. For example, when the engine is cold, the serial data stream from the ECT and IAT can be compared. The two readings, the engine coolant

```
GM-DEM01. SNP: 1991 GM 3.8L TPI S
Sensors: 62  Frames: -19 to 19          Current Frame: -19

      COOLANT TEMP:      14 F         MASS AIR FLOW:      40 G/S
    O2 CROSS COUNTS:      40              O2 VOLTS:      0.17 V
              RPM:      127       THROT POS SENSOR:      0.78 V
  A/C HEAD PRESSURE"    NORM          BATTERY VOLTS:      4.00 V
          BLM CELL:      40        BLOCK LEARN MULT:      40
    C/C BRAKE SWITCH:     ON              HOT LIGHT:      ON
    IDLE AIR CONTROL:     40           IDLE RPM X 10:      50
        INTEGRATOR:      40           KNOCK RETARD:      7 DEG
      KNOCK SIGNAL:     YES            LOOP STATUS:      OPEN
  LV8(FILTERED LOAD):    40        MANIFOLD AIR TEMP:      18 F
      O2 FUEL STATE:    LEAN      SPARK ADVANCE PROGRM:     90 DEG
    THROTTLE ANGLE %:    15 %      TIME FROM START:    99.90 MIN
    2ND GEAR STATUS:     ON        3RD GEAR STATUS:      ON
    4TH GEAR STATUS:    OFF            BRAKE SWITCH:      OFF
      COMMAND GEAR:    NONE          CRUISE CONTROL:      ON
        CRUISE MODE:    IVLD          CRUISE RES/ACC:      OFF
  CRUISE SET SPEED:   1 MPH         CRUISE SET/CST:      ON
  CRUISE VACUUM SOL:     ON         CRUISE VENT SOL:      OFF
      DES SERVO POS:    15 %           PARK/NEUTRAL:      OFF
      PRNDL SWITCH:    IVLD           RAW SERVO POS:      40
      SHIFT WIRE A:      HI           SHIFT WIRE B:      LOW
      SHIFT WIRE C:     LOW           SHIFT WIRE P:      LOW
    TCC DUTY CYCLE:    15 %               TCC MODE:      OFF
          TCC SLIP:    -176      TORQUE CONV CLUTCH:      ON
  VEHICLE SPEED SENS:  40 MPH            A/C CLUTCH:      ON
        A/C REQUEST:    YES           ACT SERVO POS:      15 %
    AIR/FUEL RATIO:    0.10          CANISTER PURGE:      15 %
  CUR WEAK CYLINDER:     40             FAN HI (1):      OFF
        FAN LOW (2):    OFF          HEATED W/S REQ:      YES
          PROM ID:    3267          QUAD DRIVER 1 (A):      HI
  QUAD DRIVER 2 (B):     HI       VEH ANTI-THEFT SYS:      YES
```

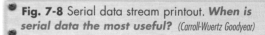

Fig. 7-8 Serial data stream printout. *When is serial data the most useful?* (Carroll-Wuertz Goodyear)

Testing a MAP Sensor

You may use a vacuum gauge and a DVOM to test most MAP sensors indirectly. The table below shows the information needed to test most MAP sensors.

Manifold Vacuum (inches of mercury (Hg))	Voltage (+/−0.04)
5	1.46
8	1.34
9	1.31
10	1.29
11	1.26
12	1.24
13	1.21
14	1.19
15	1.16
16	1.14
17	1.11
18	1.09
19	1.07
20	1.04
21	1.02
22	0.99
23	0.97

Test the MAP sensor at several vacuum values. This should be done on the vehicle. Use a scan tool or DVOM to observe signal voltage.

1. Install a vacuum/pump gauge and slowly apply vacuum.

2. Record readings at the vacuum levels shown in the table.

3. Compare each reading to the value shown in the table.

4. If the output voltages are out of specification, replace the sensor.

5. If output voltages comply with the specifications, consider other possible causes for the complaint.

Output voltage measurements for a possibly defective sensor are:

1.44 V at 5″ of Hg, 1.35 V at 10″ of Hg, 1.22 V at 15″ of Hg, and 1.10 V at 20″ of Hg.

You compare these output voltages to the values shown in the table. What conclusion do you reach?

- At 5″ of Hg, the specification requirement is between 1.42 V and 1.50 V. This is satisfied.
- At 10″ of Hg, the specification requirement is 1.25 V to 1.33 V. 1.35 V is outside of the requirements.
- At 15″ of Hg, the required interval is 1.12 V to 1.20 V and 1.22 V does not meet that specification.
- At 20″ of Hg, the acceptable measurement must be between 1.00 V and 1.08 V. 1.10 V is out of specification. The sensor needs to be replaced.

Altitude is a factor in obtaining true voltage values at different vacuum levels for your vacuum gauge. You need to consider your altitude when you use your vacuum gauge.

Apply it!

Meets NATEF Mathematics Standards for interpreting specifications and tolerance and standardizing testing equipment.

Also meets NATEF Science Standards for understanding effects of barometric pressure.

❶ Devise a plan to ensure proper diagnosis of a MAP sensor using *your* vacuum gauge. As a hint, your plan must include the concept of an "average."

❷ Implement your plan after your instructor approves it.

❸ Measure several known good sensors on four or five good running vehicles.

❹ Record your measurements. Create an accurate table of voltage readings at various vacuum levels.

❺ Compare your results with several others of your team. Did every gauge produce the same results?

temperature and the intake air temperature, should be within a few degrees of each other.

Serial data may also include return signals from other sensors or systems that affect engine operation. Many OBD-I systems use input from a detonation sensor mounted in the lower part of the engine block to control ignition timing. The detonation sensor circuit is tuned to detect vibrations caused by spark knock. If spark knock occurs, a return signal from the detonation sensor causes the PCM to retard the ignition timing. The detonation sensor should activate very slightly (if at all) under full throttle conditions.

An air leak might occur between the airflow sensor and the throttle body. The leak causes the airflow sensor to indicate that very little air is flowing through the engine at idle. This causes the PCM to reduce the fuel injector pulse width. A lean air/fuel mixture might cause a rough idle or a stall condition.

Systems displaying injector pulse width may offer diagnostic clues. A short pulse width may indicate that the PCM is trying to reduce fuel delivery to the engine because the O2S is sensing a rich exhaust mixture. The actual cause may be a ruptured fuel pressure regulator diaphragm leaking fuel into the intake manifold.

Snapshot Testing

Most scanners have a snapshot feature that can record serial data at the exact moment a condition occurs. Depending upon vehicle application, the scanner may record up to 100 frames of serial data before and after the fault occurs. Each frame represents a set of serial data that may show which sensor or actuator circuit was not within the expected range. Snapshot tests are useful for diagnosing intermittent failures.

Using a Break-Out Box

Some OBD-I systems do not display serial data. With these, a *break-out box (BOB)* can be used to test the OBD-I system. The BOB is connected between the PCM and its electrical connector. The BOB has 60 or more numbered outlets, called pins. Each numbered pin corresponds to a specific sensor or actuator circuit. When the pin is tested with a DVOM, the test is called a *pin out test*. The BOB is useful for checking the actual voltage and resistance readings of each PCM circuit.

However, a DVOM is limited in what it is able to display. The actual shape or appearance of the signal cannot be seen. A lab scope is used to observe circuit or signal activity.

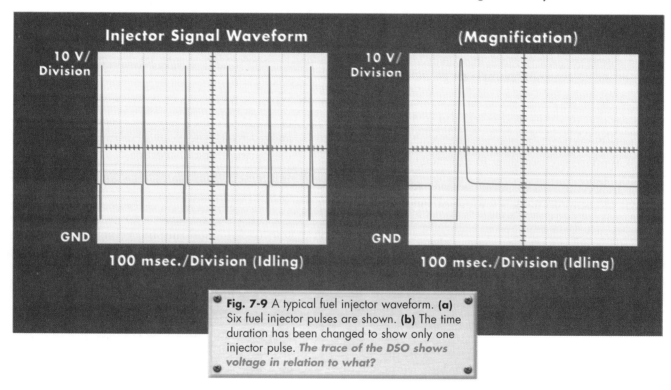

Fig. 7-9 A typical fuel injector waveform. **(a)** Six fuel injector pulses are shown. **(b)** The time duration has been changed to show only one injector pulse. *The trace of the DSO shows voltage in relation to what?*

Using a Lab Scope

A *lab scope* is the commonly used name for a *digital storage oscilloscope (DSO)*. A DSO is used to observe and measure voltage and frequency. The vertical scale displays voltage or amplitude. The horizontal scale displays time or frequency. Both scales are adjustable to provide the best pattern. Changing the scale does not change the actual signal producing the pattern. It changes only the appearance of the pattern. **Fig. 7-9.**

The pattern displayed on the screen of the DSO is called a waveform. The waveform is a representation of a signal. It displays circuit activity, showing signal amplitude and frequency. The waveform pattern provides a more accurate view of circuit activity than does a DVOM.

Using a DSO requires training and experience. Many types of waveforms can be displayed. An experienced technician must be able to identify and interpret what is shown.

The DSO can display the two basic types of waveforms produced by sensors and circuits within a vehicle.

- A sine wave is a waveform that alternates between a high and a low value. **Fig. 7-10(a).** It is also called an analog waveform. The amplitude of the signal alternates smoothly and evenly above and below the base line. Ordinary household alternating current (AC) produces a sine wave when viewed on a DSO. It alternates between 163 volts in the positive direction and 163 volts in the negative direction. The same voltage when measured with a DVOM would show the effective voltage of 110 volts to 120 volts AC.

- A digital signal has very sharp amplitude changes. It is basically an on/off signal. **Fig. 7-10(b).** The signal changes rapidly from off, or 0 volts, to the applied voltage. The CD player in your home or car produces a digital signal when played. The signal is a series of on/off pulses.

Some sensors in the on-board diagnostics system produce digital signals. The signal changes rapidly from 0 volts to the applied voltage, usually 5 volts. A square wave is a form of digital signal. However, square wave signals may vary in frequency or on/off time (pulse width).

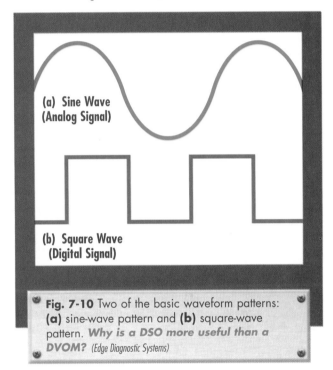

(a) Sine Wave (Analog Signal)

(b) Square Wave (Digital Signal)

Fig. 7-10 Two of the basic waveform patterns: **(a)** sine-wave pattern and **(b)** square-wave pattern. *Why is a DSO more useful than a DVOM?* (Edge Diagnostic Systems)

SECTION 2 KNOWLEDGE CHECK

❶ What are the three ways in which most OBD-I systems display DTCs stored in the PCM?

❷ What is one of the first items that a scanner menu requests?

❸ Under what conditions might the PCM fail to record a DTC?

❹ What is a break-out box (BOB) useful for checking?

❺ What is a lab scope used to observe?

CHAPTER 7 REVIEW

Key Points

Meets the following NATEF Standards for Engine Performance: retrieving and recording stored diagnostic trouble codes; obtaining and interpreting digital multimeter readings.

- Sensitive electronic components such as PROM chips can be damaged by electrostatic discharge.
- The voltage output of sensors can be read with a digital multimeter.
- Some sensors have a power feed, return signal, and ground circuit.
- DTCs are referenced and explained in the vehicle service manual.
- Three different methods can be used to retrieve DTCs depending on the vehicle type.
- The VIN identifies the engine configuration, vehicle type, and year of manufacture.
- Serial data can be used to diagnose no-code sensor faults.
- A digital storage oscilloscope (lab scope) can be used to observe both digital and analog waveforms.
- A break-out box can be used to measure voltage and resistance in some circuits.

Review Questions

1. What must you do to eliminate the possibility of damaging sensitive electronic components?
2. How would you use a digital multimeter to obtain the voltage drop in a sensor ground circuit?
3. How do you inspect and test power and ground circuits?
4. How can a driveability problem that sets a DTC be diagnosed?
5. What are three methods that can be used to retrieve diagnostic trouble codes?
6. What information is found on the vehicle's VIN plate?
7. How can a driveability problem that does not set a DTC be diagnosed?
8. **Critical Thinking** What would cause the O2S voltage to be within expected values even though a misfire is producing a rich exhaust?
9. **Critical Thinking** The knock sensor is consistently detecting engine knock. The sensor is operating properly. What could be causing the knock?

TECHNOLOGY FORECAST
FOR AUTOMOTIVE EXCELLENCE

Clean Air Laws Create Engine Improvements

Exhaust emission controls are a relatively new addition to the automobile. It wasn't until the 1960s that concern for clean air led the public and legislators to take action. Clean air laws were one factor that led to the development of on-board diagnostic systems and other engine improvements.

The Motor Vehicle Air Pollution and Control Act of 1965 ordered automakers to manufacture vehicles that created less pollution. By 1968, exhaust emission controls were standard on U.S.-built cars.

Tougher laws followed. The Clean Air Act of 1970 required tailpipe emissions to be cut by 90 percent. Automakers satisfied these laws, at first, by

making the engine more efficient. Later, additional engine improvements, the catalytic converter, on-board diagnostic systems (OBD), and the switch to unleaded gasoline were needed. Leaded gasoline was removed from the market by the mid-1980s.

Later, the Clean Air Act Amendment of 1990 ordered another seven- to eight-percent emissions reduction by 1997. This trend will continue, with automakers already producing low- and ultra low-emission vehicles (LEV and ULEV). Electric- and hybrid-powered cars and trucks likely will become common because they emit no emissions or harmless emissions like water.

AUTOMOTIVE EXCELLENCE TEST PREP

Answering the following practice questions will help you prepare for the ASE certification tests.

1. An engine has a driveability problem. Technician A says an updated PROM chip may solve the complaint. Technician B says the PCM may have to be replaced. Who is correct?
   - ⓐ Technician A.
   - ⓑ Technician B.
   - ⓒ Both Technician A and Technician B.
   - ⓓ Neither Technician A nor Technician B.

2. An engine lacks power and seems to run rich. Technician A says the fuel filter may be clogged. Technician B says the engine may be operating in limp-in mode. Who is correct?
   - ⓐ Technician A.
   - ⓑ Technician B.
   - ⓒ Both Technician A and Technician B.
   - ⓓ Neither Technician A nor Technician B.

3. An OBD-I system has stored a DTC showing a lean air/fuel mixture. Technician A says the fuel injectors may be clogged. Technician B says the fuel filter may be clogged. Who is correct?
   - ⓐ Technician A.
   - ⓑ Technician B.
   - ⓒ Both Technician A and Technician B.
   - ⓓ Neither Technician A nor Technician B.

4. The MIL stays illuminated on an OBD-I system. Which is the most likely cause?
   - ⓐ A hard DTC.
   - ⓑ A soft DTC.
   - ⓒ Both a and b.
   - ⓓ Neither a nor b.

5. A PCM has stored three DTCs in its diagnostic memory. How should the problems indicated by the codes be repaired?
   - ⓐ Replace the PCM first.
   - ⓑ Replace all the sensors indicated by the codes.
   - ⓒ Repair the problems in the order the codes are seen.
   - ⓓ Update the PROM.

6. An engine has a no-code driveability complaint. Technician A says changing the oxygen sensor usually solves this type of problem. Technician B says the serial data stream should be looked at before doing anything. Who is correct?
   - ⓐ Technician A.
   - ⓑ Technician B.
   - ⓒ Both Technician A and Technician B.
   - ⓓ Neither Technician A nor Technician B.

7. The serial data from the ECT sensor shows that the engine temperature exceeds 230°F [110°C]. The engine seems to operate normally. Technician A says to measure the engine temperature with a pyrometer. Technician B says the ECT is probably bad. Who is correct?
   - ⓐ Technician A.
   - ⓑ Technician B.
   - ⓒ Both Technician A and Technician B.
   - ⓓ Neither Technician A nor Technician B.

8. Technician A says to measure the O2S ground-circuit voltage drop with a DSO. Technician B says to use a digital voltmeter. Who is correct?
   - ⓐ Technician A.
   - ⓑ Technician B.
   - ⓒ Both Technician A and Technician B.
   - ⓓ Neither Technician A nor Technician B.

9. If a scanner (scan tool) can operate an actuator, what is this capability called?
   - ⓐ Default operation.
   - ⓑ No-code operation.
   - ⓒ Reactive operation.
   - ⓓ Interactive operation.

10. A TPS return signal reads 0 volts on a DVOM. Technician A says a fuse leading to the PCM may be blown. Technician B says the TPS circuit may be open. Who is correct?
    - ⓐ Technician A.
    - ⓑ Technician B.
    - ⓒ Both Technician A and Technician B.
    - ⓓ Neither Technician A nor Technician B.

CHAPTER 8

Using Computer Diagnostics: OBD-II

You'll Be Able To:

- ⊗ Define OBD-II terminology.
- ⊗ Understand the function of a diagnostic monitor.
- ⊗ Retrieve OBD-II DTCs.
- ⊗ Interpret PID data.

Terms to Know:

active test
diagnostic
drive cycle
enabling criteria
failure record
freeze-frame data
inspection/maintenance (I/M) ready status
monitor
parameter identification data
passive test
trip
warm-up cycle

The Problem

The engine in George Lee's car hesitates. Mr. Lee says the hesitation is most noticeable when the car is accelerating. He doesn't notice anything unusual when the engine is idling.

You take a test drive with Mr. Lee. Indeed there is a driveability problem. A glance at the instrument panel reveals that the malfunction indicator lamp (MIL) is not on. Obviously, the PCM is not sensing a problem. Or, is there more than one problem? An effective diagnostic strategy will keep you on the right track.

Your Challenge

As the service technician, you need to find answers to these questions.

1. How often does the hesitation occur?
2. Will an OBD system indicate the source of the problem?
3. If a diagnostic trouble code (DTC) is not available, what should you do next?

Section 1

OBD-II Systems

OBD-I systems were used on vehicles produced from the early 1980s through the mid-1990s. OBD-I systems were limited in the types of faults they could monitor. They were able to identify only component or system failures. OBD-I could not diagnose emission-related problems resulting from the degradation (slow failure) of components or systems.

OBD-II systems were designed to ensure the accurate monitoring and operation of all emission-related systems and components. OBD-II systems must be able to accurately detect and identify conditions that could increase vehicle emissions by one and one half times the federal standards. OBD-II systems are also able to detect emission-related problems caused by the degradation of systems and components. OBD-II systems monitor the efficiency of catalytic converters. All light-duty gasoline powered vehicles manufactured after 1995 must meet OBD-II requirements. The next generation of this system, OBD-III, is currently under development.

Purpose of OBD-II

The OBD-II system standardizes the data link connector, basic diagnostic equipment, and diagnostic procedures. Under the *Society of Automotive Engineers (SAE) J1930 standards,* most diagnostic terms, acronyms, and abbreviations are the same, regardless of the vehicle manufacturer.

Manufacturers may use slightly different components and system designs to control and reduce emissions. These designs may require procedures applicable only to the vehicle being serviced. Always refer to the specific manufacturer's service manuals.

The OBD-II system monitors the following:
- Cylinder misfire.
- Catalytic converter efficiency.
- Fuel trim adjustment.
- Exhaust gas recirculation.
- Evaporative emission control system.
- Secondary air injection.

When a component or system failure occurs with an OBD-I system, a diagnostic trouble code (DTC) is set and the malfunction indicator lamp (MIL) is

Tech Tip **Performing Interactive Actuator Tests.** Many scanners have an interactive diagnostic mode that allows the technician to manually operate many actuators. Follow manufacturer's directions during such tests.

turned on. OBD-II systems turn on the MIL when a problem causes emission levels to exceed the federal standard limits.

OBD-II Hardware

OBD-II systems are designed to detect emission-related faults and failures. The diagnostic strategy used for OBD-II is often different from that used for OBD-I. OBD-II adds many new features and technical improvements.

When the engine calibration settings are updated, OBD-I system PCMs with hard-wired *programmable read-only memory (PROM)* chips must be replaced as a unit. PCMs with a removable PROM need a PROM chip replacement. OBD-II systems use an *electrically erasable programmable read-only memory (EEPROM).* The EEPROM, also called a flash PROM, is hard-wired into the PCM. The information in the EEPROM can be updated or reprogrammed without replacing the EEPROM.

The data link connector (DLC) for OBD-II is a standardized 16-pin connector. **Fig. 8-1.** Most of the pin use complies with OBD-II standards. For example, pins 4, 5, 7, and 16 have the same assignments and the same use for all manufacturers.

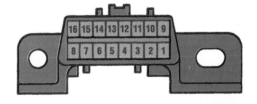

Fig. 8-1 A standard 16-pin, OBD-II DLC. *What do generic scanners require to adapt their DLC connection to different vehicle applications?*

Other pins on the DLC have different uses for specific vehicles or applications. Some pins are used only with original equipment manufacturer's (OEM), or "dedicated," scanners. Most universal, or "generic," scanners need a "test cartridge" to adapt or program their DLC connection for different vehicle applications.

Identifying the Role of a Catalyst

OBD-II systems monitor the efficiency of the catalytic converter. Oxygen sensors measure the oxygen levels going into and out of the converter. Here you will learn about the basic idea of a catalyst.

A *catalyst* is a chemical material that promotes or encourages a chemical reaction between two other materials. The catalyst itself is not consumed or changed in the reaction.

A vehicle's catalytic converter aids the combining of oxygen with potential pollutants. As a result, fewer polluting chemical compounds are emitted with the other exhaust gases.

The palladium catalyst does not generate oxygen. *Palladium* is a metallic element. It promotes the further combustion of pollutants with any excess oxygen. The palladium itself is not consumed.

Apply it!

Demonstrating Catalytic Action

Meets NATEF Science Standards for explaining chemical reactions in a catalytic converter.

Materials and Equipment
- Pint-size plastic bag with a "zipper" seal
- Packet of baking yeast
- Hydrogen peroxide, six percent strength
- Two large eyedroppers
- Wooden match

❶ Put one-half teaspoon of yeast into a pint-size plastic bag.

❷ Fill an eyedropper with hydrogen peroxide of six percent strength. Place this filled eyedropper in the bag.

❸ Flatten the bag to push out any air. Zip it closed.

❹ Hold the plastic bag by a corner. Squeeze the bulb on the eyedropper so the hydrogen peroxide drops onto the yeast. The bag will inflate as the oxygen is released. Hydrogen peroxide decomposes into water and free oxygen in the presence of a catalyst, called "catalase." Catalase is found in yeast. Although the yeast will become wet from the water, the catalase is still there, unchanged.

❺ Using the other eyedropper, squeeze the bulb to expel as much air as possible.

❻ While still squeezing the bulb, carefully insert the tip of this second eyedropper past the seal at the end of the bag. Release the bulb to draw oxygen into the eyedropper.

❼ Light the match. Let it burn for a few seconds. Blow it out so it's still glowing.

❽ Squeeze the eyedropper full of oxygen directly at the glowing match. Observe what happens. This demonstrates how oxygen promotes combustion.

Safety First

Materials Hydrogen peroxide can be corrosive at concentrations higher than six percent.

Personal Follow all fire safety precautions when using the match with its open flame.

On OBD-II vehicles, the location of the DLC is also standardized. The connector is located under the dash panel near the steering column. Be sure to consult the vehicle service manual for DLC placement. The DLC is indexed to prevent incorrect insertion of the scanner connector. Do not force the scanner connector into the DLC. Look for damage to the DLC, or to its pin connectors. Forcing a connection can result in damage to the DLC or to the scanner connector.

OBD-II Terminology

OBD-II systems use specific standardized terminology. You must learn this terminology before you can understand OBD-II diagnostics. Some of the terms used are:

- **Diagnostic** (also called a **monitor**)–a procedure used by the PCM to test relevant on-board systems. The MIL will turn on and stay on if the PCM detects a fault occurring on one or two consecutive trips.
- **Trip**–a key-on, run, key-off cycle in which all of the enabling criteria for a given diagnostic monitor are met.
- **Enabling criteria**–the sensor inputs supplied during specified driving conditions. They must be received by the PCM before a diagnostic (monitor) can be run. For example, a monitor may require input from the throttle position (TP) sensor, the vehicle speed sensor (VSS), and the mass air flow (MAF) sensor. If any of these sensor inputs are missing, the monitor cannot be run.
- **Warm-up cycle**–the period when the engine coolant temperature rises from ambient temperature to at least 160°F [70°C]. To eliminate false readings from short trips, the temperature must rise at least 40°F [22°C] during a trip.
- **Drive cycle**–a set of driving conditions that "run" all on-board diagnostics. When the monitor is run, the inspection/maintenance (I/M) readiness "flags" will set. The "flag" is an indicator that shows that the monitors have been run. Not all vehicles have I/M flags.
- **Freeze-frame data**–serial data values that are stored the instant an emission-related DTC is set and enters the diagnostic memory. Some manufacturers may refer to this as **parameter identification data (PID).** Freeze frames provide a snapshot of the conditions present when a DTC is stored. **Table 8-A.**
- **Passive test**–checks the performance of a vehicle system or component during normal operation.
- **Active test**–forces a component to operate in a specific way. A scan tool can be used to perform an active test. A valve or solenoid, for example, may intentionally be opened to cause a momentary change in the system. If the PCM doesn't "see" this change occur, it activates the MIL and stores a DTC.
- **Failure record**–may store up to five DTCs in the diagnostic memory. Unlike freeze-frame records, which store only emission-related faults, the failure records store component faults. Multiple DTCs stored in the failure record appear in the order that they occur. These DTCs must be repaired in the order that they occur.
- **Inspection/Maintenance (I/M) ready status**–shows that a vehicle's on-board diagnostics have been run. These are sometimes called I/M flags.

Table 8-A FREEZE-FRAME DATA

PID#	Acronym	Description
0005	ECT	Engine Coolant
0003	FUELSYS1	Open/Closed Loop 1
0003	FUELSYS2	Open/Closed Loop 2
0007	LONGFT1	Long Term Fuel Bank 1
0009	LONGFT2	Long Term Fuel Bank 2
0004	LOAD	Calculated Load Value
000C	RPM	Engine RPM
0006	SHRTFT1	Short Term Fuel Bank 1
0008	SHRTFT2	Short Term Fuel Bank 2
000D	VSS	Vehicle Speed

SECTION 1 KNOWLEDGE CHECK

❶ In what year did OBD-II become required on all light-duty gasoline powered vehicles?

❷ What is another name for EEPROM?

❸ What pins have standard assignments in OBD-II DLCs?

❹ What is the term given to sensor inputs supplied during specified driving conditions?

❺ When multiple DTCs are stored in a failure record, in what order do they appear?

Section 2

OBD-II Diagnostics

Diagnosing OBD-II systems requires a specific knowledge of how DTCs are created, stored, and erased. OBD-II has many classifications of DTCs. These DTCs are stored in specific sequences and by degrees of severity. When diagnosing an OBD-II system, refer to the vehicle service manual.

OBD-II Monitors

As mentioned, a monitor is a diagnostic procedure used by the PCM to test relevant on-board systems. A monitor may occur only when enabling criteria (specific sensor inputs) are received by the PCM. Many of these monitors are performed under specific operating conditions. **Fig. 8-2.** For example, a monitor may occur only when the engine reaches a certain operating temperature, speed, and load, and the transmission is in a certain gear.

OBD-II checks inputs from system sensors including:
• Manifold absolute pressure (MAP) sensor.
• Mass air flow (MAF) sensor.
• Engine coolant temperature (ECT) sensor.
• Crankshaft position (CKP) sensor.
• Camshaft position (CMP) sensor.
• Heated oxygen sensor (HO2S).
• Vehicle speed sensor (VSS).
• Intake air temperature (IAT) sensor.
• Throttle position (TP) sensor.

The *diagnostic executive* is a PCM program that controls the sequencing of tests needed to run the OBD-II monitors. This sequencing procedure organizes and prioritizes the monitor tests. The diagnostic executive includes the following processes:
• Turning the MIL on or off.
• Storing and clearing DTCs.
• Storing freeze-frame data for emissions faults.
• Displaying the status of each diagnostic.

The OBD-II system analyzes data from monitors including:
• Cylinder misfire.
• Catalytic converter efficiency.
• Fuel trim adjustment.
• Exhaust gas recirculation.
• Evaporative emission control.
• Secondary air injection.

Malfunction Indicator Lamp (MIL)

Base Engine or Any of Its Components

Camshaft Position (CMP)

Transmission or Transaxle

Ignition System

Air Conditioner (A/C) or Heater System

Fuel Level Input (FLI)

Crankshaft Position (CKP) or RPM

Mass Air Flow (MAF)

Engine Coolant Temperature (ECT)

Throttle Position (TP)

Intake Air Temperature (IAT)

Vehicle Speed

Fig. 8-2 Icons used to illustrate Ford OBD-II monitors. Icons will vary with manufacturer. *When are monitors performed?* (Ford Motor Company)

Interpreting Icons

Icons! They are everywhere. Computer screens are filled with them. Most products you buy, from bread to automobiles, have icons associated with them. Road signs around the world use a set of icons developed for international use. Icons present a simple image of a familiar idea or object. Icons have long been used in the automotive industry in dashboard lights or gauges. They may show on lights and gauges that monitor the fuel supply, the temperature of the engine, or the condition of the battery.

The automotive industry uses icons in diagnostic procedures. These picture symbols are information "short cuts." They represent items associated with the system that the diagnostic procedure tests. For example, a spark plug icon indicates the ignition system. You will see icons used in service manuals.

Apply it!

Meets NATEF Communications Standards for comprehending and using information in charts and tables.

❶ Compare the icons in Fig. 8-2 with the icons used by other manufacturers.

❷ Are the icons in foreign vehicles different from those in vehicles manufactured in the United States? If so, how?

❸ Design an icon for a vehicle that will be made 50 years from now.

OBD-II Diagnostic Trouble Codes

A DTC is displayed as an *alphanumeric designator* followed by a three-digit number. These codes are displayed on the face of a scanner connected to the DLC. **Fig. 8-3.** Code designators fall into four groups:
- Body codes: B0, B1, B2, and B3.
- Chassis codes: C0, C1, C2, and C3.
- Powertrain codes: P0, P1, P2, and P3.
- Network codes: U0, U1, U2, and U3.

Engine performance DTCs are found within the powertrain, or "P," codes. If the first number following P is 0, it indicates a generic, or SAE, code. Generic DTCs are common codes that are uniform throughout the automotive service industry.

If the first number after P is 1, it indicates the DTC is a *nonuniform,* or *manufacturer-specific, code.* Manufacturers use these codes in their own diagnostic procedures. Refer to the appropriate service manual to reference these codes. The second number indicates the system that is affected. The following are code numbers of the affected systems:
- 1–Fuel and air metering (MAP, MAF, IAT, ECT).
- 2–Fuel injector circuits.
- 3–Ignition system or misfire (KS, CKP).
- 4–Auxiliary emission controls (EGR, TWC, EVAP).
- 5–Vehicle speed control and idle system (VSS, IAC).
- 6–Computer output circuits (5-volt reference, MIL).
- 7–Transmission.
- 8–Transmission.

Fig. 8-3 Scanners are used to read DTCs. A scanner is connected to the DLC under the dash. *Where are DTCs stored?* (Actron Manufacturing; AntoXray)

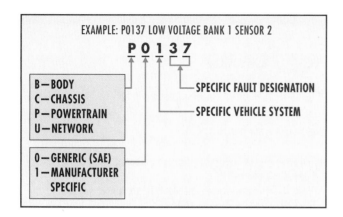

EXAMPLE: P0137 LOW VOLTAGE BANK 1 SENSOR 2

P 0 1 3 7

SPECIFIC FAULT DESIGNATION

SPECIFIC VEHICLE SYSTEM

B—BODY
C—CHASSIS
P—POWERTRAIN
U—NETWORK

0—GENERIC (SAE)
1—MANUFACTURER
SPECIFIC

Fig. 8-4 Interpreting an OBD-II DTC. *Where would you look for a list of specific fault designations?* (General Motors Corporation)

The last two numbers in an OBD-II DTC refer to the specific fault designation. A DTC of P0137 means that the number-two heated oxygen sensor on cylinder bank number one is producing a low voltage. **Fig. 8-4.** The number-one cylinder bank contains the number-one cylinder.

An OBD-II configuration may have as many as three oxygen sensors per exhaust pipe. The sensor responsible for fuel adjustment is frequently located in the exhaust manifold. It checks the air/fuel mixture. In systems that use three sensors, the second sensor is the pre-catalytic, or *pre-cat,* sensor. It is located on the input side of the catalytic converter. The final sensor is the post-catalytic, or *post-cat,* sensor. It is located on the output side of the catalytic converter. A vehicle with a dual exhaust system can have as many as six oxygen sensors.

Setting OBD-II DTCs The PCM is designed to monitor emission systems failures by comparing two or more sensor return signals. To illustrate, the throttle position (TP) sensor, engine rpm (TACH), mass air flow (MAF) sensor, and vehicle speed sensor (VSS) signals are compared. These three sensors should agree that:
• The engine is running.
• The vehicle is in gear.
• The vehicle is moving.

If one of the return signals doesn't agree with the other three, a DTC is set. The misfire monitor, for example, detects an ignition misfire in a specific cylinder by measuring small changes in crankshaft speed. The return signal from the CKP is the main signal used by the PCM to detect a cylinder misfire. **Fig. 8-5.**

If the PCM detects a misfire every 200 rpms, it is called a type-A misfire. This type of misfire produces excess hydrocarbons, causing the catalytic converter to overheat. If a type-A failure is detected and a DTC is set, the MIL blinks at a rate of once per second.

A type-B misfire occurs at least once every 1,000 rpms. Type-B misfires result in an emissions failure equal to 1.5 times the *Federal Test Procedure (FTP)* standards.

If a type-B misfire occurs, the MIL lights up and stays on and a DTC is stored. A DTC for a type-A or type-B misfire is P0300. The DTCs for individual cylinder misfires range from P0301 (number-one cylinder) to P03010 (number-ten cylinder).

Rough road conditions can cause the crankshaft speed to vary. If the varying crankshaft speed is detected, the PCM can incorrectly set a type-B misfire. Many misfire monitors use return signals from the anti-lock brake system (ABS) to help the PCM determine the roughness of the road surface. The added sensor return signal can prevent a false DTC setting.

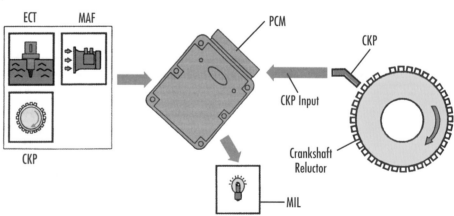

ECT   MAF

CKP

PCM

CKP

CKP Input

Crankshaft Reluctor

MIL

Fig. 8-5 Components of a misfire detection monitor. When the enabling criteria are met for a misfire, a DTC is set in the PCM. *How can driving on a rough road imitate the effect of an engine misfire?* (Ford Motor Company)

TECH TIP **Checking Catalytic Converter Temperatures.** The outlet temperature on a catalytic converter should be at least 100°F [38°C] higher than the inlet temperature. Use an infrared pyrometer to measure the temperature.

OBD-II systems monitor emissions by checking the catalytic converter for degradation. If converter efficiency decreases and emissions increase due to component degradation, a DTC is set.

To monitor catalytic converter efficiency, an oxygen sensor is placed at the inlet of the catalytic converter. The pre-cat sensor indicates the amount of oxygen entering the converter. A post-cat oxygen sensor is placed at the converter outlet. The post-cat sensor indicates the amount of oxygen leaving the converter. Under normal operation, all the oxygen in the exhaust leaving the converter is used up.

Comparing Sensor Readings The PCM compares the return signal voltages from the pre-cat and post-cat sensors. As the exhaust leaves the converter, the post-cat oxygen sensor should detect a flat output voltage. If the converter is operating efficiently, the two voltage readings will be different. **Fig. 8-6.**

If the converter is not operating efficiently, the two sensor readings will be similar. A DTC will be set and the MIL will be activated.

The catalytic converter operates most efficiently when a stoichiometric fuel ratio is maintained. To maintain efficient operation, the PCM trims the fuel injectors from slightly rich to slightly lean. The return signal voltage from the pre-cat oxygen sensor switches from about 0.8 volts (rich) to about 0.2 volts (lean). As the air/fuel mixture becomes lean, a small amount of free oxygen enters the catalytic

converter. This oxygen helps the catalytic converter oxidize the pollutants in the exhaust stream.

The PCM records short-term and long-term *fuel trim (FT)* readings. FT is displayed as a percentage. A minus 20 percent short-term FT means that the PCM is reducing fuel injector pulse width to achieve a stoichiometric ratio. A plus 20 percent short-term FT means that the PCM is increasing fuel injector pulse width. The PCM averages the FT readings. If the average reading exceeds an expected range of values, a DTC is stored.

Figure 8-7 shows examples of normal and abnormal pre-cat oxygen sensor waveforms. Waveform A shows normal operation. The signal is varying over an expected range of values. Waveform B shows low-level switching activity. The sensor isn't responding normally. A contaminated sensor normally causes "B" waveform. Waveform C shows a *low voltage bias*. The return signal is varying over a voltage range that is too low. Waveform D shows a *high voltage bias*. The return signal is varying over a voltage range that is too high. A defective oxygen sensor is indicated by waveform "B," "C," or "D."

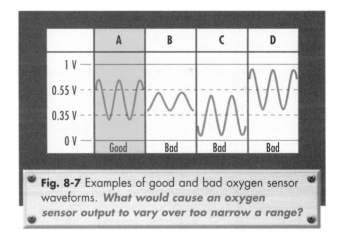

Fig. 8-7 Examples of good and bad oxygen sensor waveforms. *What would cause an oxygen sensor output to vary over too narrow a range?*

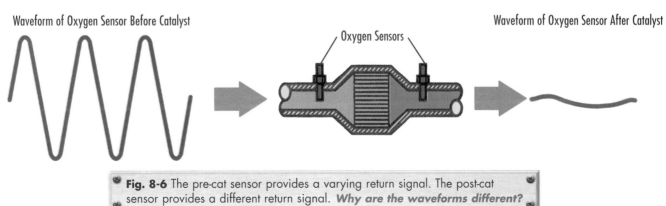

Fig. 8-6 The pre-cat sensor provides a varying return signal. The post-cat sensor provides a different return signal. *Why are the waveforms different?*

Safety First

Materials On some vehicles leaking EVAP hoses or a defective, loose, or missing fuel tank cap will activate the MIL. These problems allow fuel vapors to escape into the atmosphere, creating a source of pollution.

The evaporative emissions (EVAP) system checks the fuel tank and vapor recovery system for vapor leakage. **Fig. 8-8.** The EVAP monitor is designed to detect leaks as small as 0.040″ [0.102 cm] in diameter. The EVAP monitor does this by applying either a vacuum or pressure to the purge lines, the canister, and the fuel tank.

Vacuum should be present only when the purge valve allows intake manifold vacuum to enter the purge lines. An EVAP sensor is placed in the purge line to monitor vacuum. A DTC is set if the return signal from the sensor indicates abnormal operation.

Retrieving OBD-II DTCs OBD-II DTCs are accessed by connecting a scanner to the DLC. Unlike some OBD-I systems, "flash code" diagnosis is not possible with OBD-II.

Diagnostic procedures are shown on the scanner's menu. The menu differs with the scanner being used. Dedicated scanners may have more complicated menus.

The scanner software must be compatible with the model year of the vehicle being tested. A generic scanner may need special adapters to adapt its connection assignments to the 16-pin DLC.

Table 8-B **FREEZE-FRAME DATA**		
PID#	**Acronym**	**Description**
0005	ECT	Engine Coolant
0003	FUELSYS1	Open/Closed Loop 1
0003	FUELSYS2	Open/Closed Loop 2
0007	LONGFT1	Long Term Fuel Bank 1
0009	LONGFT2	Long Term Fuel Bank 2
0004	LOAD	Calculated Load Value
000C	RPM	Engine RPM
0006	SHRTFT1	Short Term Fuel Bank 1
0008	SHRTFT2	Short Term Fuel Bank 2
000D	VSS	Vehicle Speed

A single vehicle application has many OBD-II DTCs. For example, OBD-II systems can display over 35 generic oxygen sensor-related DTCs. A vehicle service manual gives the meaning of each DTC. The repair manual also includes specific procedures needed to diagnose a particular DTC.

A scanner may have several diagnostic modes. The PID mode, used by Ford, supplies serial data information from the PCM. When the PID mode is used, this information is continuously updated. The generic PID data stream includes calculated system values and system status reports. **Table 8-B.**

TECH TIP **Using Serial Data.** The PID mode is useful for diagnosing no-code faults. The serial data information is continuously updated. The updated information shows what values are being supplied to the PCM.

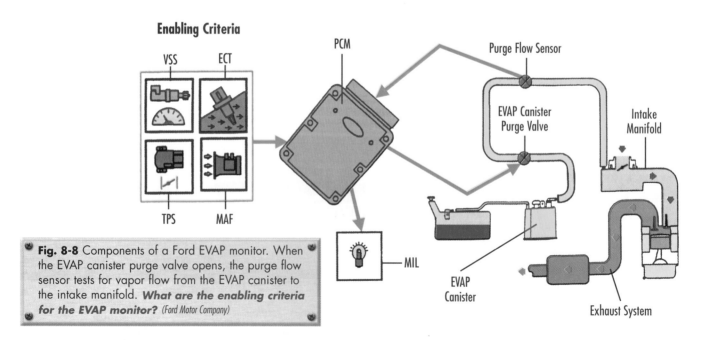

Enabling Criteria

VSS   ECT   TPS   MAF

PCM   Purge Flow Sensor   EVAP Canister Purge Valve   Intake Manifold   MIL   EVAP Canister   Exhaust System

Fig. 8-8 Components of a Ford EVAP monitor. When the EVAP canister purge valve opens, the purge flow sensor tests for vapor flow from the EVAP canister to the intake manifold. *What are the enabling criteria for the EVAP monitor?* (Ford Motor Company)

Comparing Signal Frequencies

Frequency signals are generated by MAF sensors, some MAP sensors, and some oxygen sensors. For example, Ford upstream heated and downstream heated oxygen sensors (HO2S) provide both voltage and frequency information to the PCM.

The *frequency (f)* of a signal is the number of times an entire cycle repeats within one second. Another way to state this is that frequency is the number of cycles per second (cps). One cycle per second is designated as 1 *hertz* (Hz).

The *period (p)* is the time required for one complete cycle. Usually the period of a signal is given in seconds.

The frequency of a signal with a period of 10 seconds is:

$$f = \frac{1 \text{ cycle}}{10 \text{ sec}} = 0.1 \text{ Hz}$$

Frequency is always based on seconds. Remember 100 milliseconds equals 0.1 seconds. So the frequency of a signal with a period of 600 milliseconds is:

$$f = \frac{1 \text{ cycle}}{0.6 \text{ sec}} = 1.67 \text{ Hz}$$

The frequency and the period are reciprocals. Calculate the reciprocal of f to find the period of a signal where $f = 200$ Hz:

$$p = \frac{1}{f} = \frac{1}{200 \text{ Hz}} = 0.5 \text{ sec}$$

Apply it!

Meets NATEF Mathematics Standards for using algebraic expressions and ratios.

Compare the frequency of an upstream HO2S to the frequency of the downstream HO2S in a Ford vehicle.

❶ What is the frequency if the upstream period is 0.05 sec?

❷ What is the frequency if the downstream period is 0.25 sec?

❸ What is the ratio of the upstream frequency to the downstream frequency?

❹ What is the frequency if the upstream period is 0.04 sec?

❺ What is the frequency if the downstream period is 0.5 sec?

❻ What is the ratio of the upstream frequency to the downstream frequency?

Clearing OBD-II DTCs When an OBD-II monitor detects a fault, it stores a DTC and turns on the MIL. Some faults may have to occur for two consecutive drive cycles to activate the MIL. If the MIL is turned on, three consecutive fault-free drive cycles are needed to turn off the MIL. The DTC is cleared after 40 engine warm-up cycles without the fault reoccurring.

DTCs can be manually cleared by a scanner command although this may not be the accepted practice. OBD-II DTCs shouldn't be cleared until all specified diagnostic and repair operations have been performed. It is best to allow the DTC to clear itself. This practice ensures that the fault, condition, or component causing the DTC has been repaired.

SECTION 2 KNOWLEDGE CHECK

❶ What processes are included in the diagnostic executive?

❷ How is a DTC displayed?

❸ What is a type-A misfire?

❹ How are OBD-II DTCs accessed?

❺ How many warm-up cycles are needed to clear a DTC?

CHAPTER 8 REVIEW

Key Points

Meets the following NATEF Standards for Engine Performance: supporting knowledge for diagnosing OBD-II system problems.

- OBD-II systems standardize the data link connector, basic diagnostic equipment, and diagnostic procedures. OBD-II systems also use specific standardized terminology.
- A monitor is a diagnostic procedure run to test relevant on-board systems.
- OBD-II DTCs are read with a scan tool.
- PIDs are used by some manufacturers to diagnose problems with no DTCs stored.
- OBD-II systems are designed to detect component degradation as well as failures that create emissions problems.
- A trip is a key-on, run, key-off cycle.
- A warm-up cycle means that the engine temperature reaches at least 160°F [70°C] and has increased by at least 40°F [22°C].
- A drive cycle "runs" all of the on-board diagnostics on the vehicle.
- DTCs will be cleared automatically if the fault is gone and forty warm-up cycles have occurred.

Review Questions

❶ What is an OBD-II "trip" monitor? What information does it provide to the OBD-II system?

❷ What are enabling criteria? How are the enabling criteria used in an OBD-II system?

❸ What is an OBD-II drive cycle?

❹ What is the function of a diagnostic monitor?

❺ How do you retrieve OBD-II DTCs? What do the OBD-II DTCs indicate?

❻ When reading a DTC, what differentiates a generic, or SAE, code from a nonuniform, or manufacturer-specific, code?

❼ What type of information does the PID data stream include?

❽ **Critical Thinking** Why is the OBD-II system better at detecting and diagnosing emission-related problems?

❾ **Critical Thinking** Advances in technology have led to the development of the OBD-II system, which replaces the OBD-I system. What are some of the major differences between OBD-I and OBD-II systems?

TECHNOLOGY FORECAST FOR AUTOMOTIVE EXCELLENCE

OBD-III in the 21st Century

Clean air laws require vehicles to produce as few emissions as possible. They also require owners to keep their cars and trucks in top running condition. On-board diagnostics (OBD) were developed to help consumers know when their vehicles need service. OBD systems light the dashboard's "check engine" lamp when a fault is found.

So far, two generations of OBD systems have been used. OBD-III, now under development, is expected to be phased in early in the 21st century. Many of the details remain to be decided. But some of the proposed requirements are already being discussed.

The major change from OBD-II to OBD-III centers around enforcement of clean air laws. With current OBD systems, vehicle owners can delay having emission problems fixed. OBD-III, however, would force owners to have faults repaired.

If such laws are enacted, an on-board transmitter may send out information about a vehicle's emissions system. This data will be collected by roadside readers, a local station network, or satellites. The vehicle's owner will be notified of a problem by mail and given time to have it fixed. Once a repair is made, the owner will send proof of the service to the state department of motor vehicles.

AUTOMOTIVE EXCELLENCE
TEST PREP

Answering the following practice questions will help you prepare for the ASE certification tests.

1. A vehicle experiences an intermittent no-code driveability complaint. Technician A says the enabling criteria haven't been met for a DTC. Technician B says the complaint may be caused by a vacuum leak. Who is correct?

   ⓐ Technician A.
   ⓑ Technician B.
   ⓒ Both Technician A and Technician B.
   ⓓ Neither Technician A nor Technician B.

2. Technician A says freeze-frame data record only emissions-related DTCs. Technician B says freeze frames can be used to diagnose no-code faults. Who is correct?

   ⓐ Technician A.
   ⓑ Technician B.
   ⓒ Both Technician A and Technician B.
   ⓓ Neither Technician A nor Technician B.

3. What functions can a passive diagnostic test perform?

   ⓐ Check the performance of a component.
   ⓑ Deactivate the enabling criteria.
   ⓒ Both a and b.
   ⓓ Neither a nor b.

4. What functions can an active diagnostic check perform?

   ⓐ Force the component to operate in a specific way.
   ⓑ Cause a momentary fault in the system.
   ⓒ Both a and b.
   ⓓ Neither a nor b.

5. Technician A says a monitor checks vehicle speed. Technician B says a monitor checks the EVAP system. Who is correct?

   ⓐ Technician A.
   ⓑ Technician B.
   ⓒ Both Technician A and Technician B.
   ⓓ Neither Technician A nor Technician B.

6. Technician A says a monitor is a diagnostic procedure used by the PCM to test appropriate on-board systems. Technician B says the enabling criteria must be present before the monitor can run. Who is correct?

   ⓐ Technician A.
   ⓑ Technician B.
   ⓒ Both Technician A and Technician B.
   ⓓ Neither Technician A nor Technician B.

7. A P0133 is what type of DTC?

   ⓐ Manufacturer's DTC.
   ⓑ Chassis DTC.
   ⓒ Body DTC.
   ⓓ Generic, or SAE, DTC.

8. A P1133 is what type of DTC?

   ⓐ Manufacturer's DTC.
   ⓑ Chassis DTC.
   ⓒ Body DTC.
   ⓓ Generic, or SAE, DTC.

9. Technician A says only one sensor is needed to run a monitor. Technician B says two or more sensors are needed to run a monitor. Who is correct?

   ⓐ Technician A.
   ⓑ Technician B.
   ⓒ Both Technician A and Technician B.
   ⓓ Neither Technician A nor Technician B.

10. Technician A says DTCs must be cleared as soon as they are written down. Technician B says the DTCs should be cleared only after a complete diagnosis and repairs are made. Who is correct?

    ⓐ Technician A.
    ⓑ Technician B.
    ⓒ Both Technician A and Technician B.
    ⓓ Neither Technician A nor Technician B.

Diagnosing & Repairing Emission Control Systems

You'll Be Able To:

- ⊗ Diagnose and repair positive crankcase ventilation systems.
- ⊗ Diagnose and repair exhaust gas recirculation systems.
- ⊗ Diagnose and repair air injection systems.
- ⊗ Diagnose and test catalytic converters.
- ⊗ Explain the purpose of evaporative control systems.

Terms to Know:

air injection (AIR) system

catalyst

evaporative control (EVAP) system

I/M 240 program

nitrogen oxide

The Problem

Bill Fitzpatrick's four-year-old car has failed its biannual emissions test. The car must pass the test before its registration can be renewed. Bill drives almost 20,000 miles a year. He has never had any driving related problems. Recently the MIL has blinked occasionally.

While discussing the problem with Bill, you find that the car has had normal maintenance service, as called for in his vehicle owner's manual. No emission-related problems were ever identified.

Your Challenge

As the service technician, you need to find answers to these questions:

1. Why would the car fail an emissions test?
2. What type of diagnostic strategy is needed to find the cause?
3. Would you perform an emissions test as part of the diagnostic process?

Section 1

Automotive Emissions Controls

Automotive engine emissions come from three sources: crankcase ventilation, fuel vapors, and exhaust system gases. The primary pollutant emissions are unburned hydrocarbons (HC), carbon monoxide (CO), and nitrogen oxides (NO_X). Carbon dioxide (CO_2) and oxygen (O_2) are also emitted from the exhaust. However, they are not considered pollutants. When released into the environment, unburned HC and NO_X form "smog." Smog is a cloudy formation of pollutants.

The primary subsystems of an emissions control system include the:

- Positive crankcase ventilation (PCV) system.
- Exhaust gas recirculation (EGR) system.
- Air injection (AIR) system.
- Three way catalytic converter (TWC).
- Evaporative control (EVAP) system.

Positive Crankcase Ventilation System

All engines have some blowby. Blowby is mostly unburned gasoline and combustion by-product gases that pass by the pistons into the crankcase.

The positive crankcase ventilation (PCV) system prevents blowby from escaping into the atmosphere by using vacuum to draw filtered air from the air intake into the crankcase. This air mixes with blowby gases in the crankcase. The blowby gases are then drawn through the PCV valve into the intake manifold and burned in the engine's cylinders. **Fig. 9-1.**

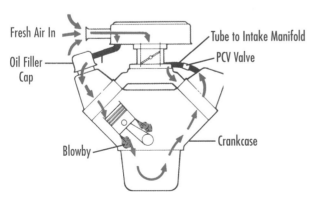

Fresh Air In
Oil Filler Cap
Blowby
Tube to Intake Manifold
PCV Valve
Crankcase

Fig. 9-1 A PCV system on a V-type engine. *Where does fresh air enter the PCV system?* (Ford Motor Company)

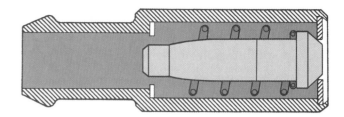

(a) Engine Off or Engine Backfire—No Vapor Flow

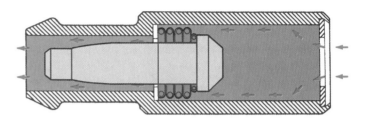

(b) High Intake Manifold Vacuum—Minimal Vapor Flow

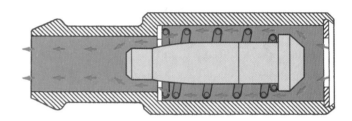

(c) Moderate Intake Manifold Vacuum—Maximum Vapor Flow

Fig. 9-2 PCV valve positions during different engine operating conditions. *What is the purpose of the PCV valve?* (DaimlerChrysler)

A PCV valve varies the flow of crankcase gases entering the intake manifold. **Fig. 9-2.** At engine idle, high intake manifold vacuum pulls the valve plunger against the tension spring. This position limits the flow of crankcase gases. Too much flow can cause a rough idle or engine stalling. Under heavier engine load, lower intake vacuum allows the valve plunger to open further. The size of the orifice increases the flow of crankcase gases. At wide-open throttle, a combination of vacuum and spring tension positions the valve plunger to allow maximum flow.

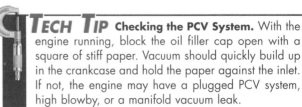

TECH TIP **Checking the PCV System.** With the engine running, block the oil filler cap open with a square of stiff paper. Vacuum should quickly build up in the crankcase and hold the paper against the inlet. If not, the engine may have a plugged PCV system, high blowby, or a manifold vacuum leak.

In the event of a backfire, the valve plunger is forced against the valve body. In this position, the backfire cannot ignite crankcase gases.

With the engine off and no vacuum present, the valve plunger is held against the valve body by spring tension.

Servicing the PCV System Proper operation of the PCV system depends upon the PCV valve being free of sludge deposits. Low-speed driving and high mileage cause deposits to accumulate in PCV passages. This causes poor crankcase ventilation. Auto manufacturers recommend replacing the PCV valve and inlet filter at scheduled intervals. The inlet filter is in the breather cap or in the air intake system. To test the PCV system, check the following:

• If the PCV vacuum side hose is made of rubber, squeeze it shut with pliers while the engine is running. The PCV valve should click as engine vacuum is applied and released. **Fig. 9-3.**

• If the PCV valve is mounted on the valve cover, remove the PCV valve from its grommet. With the engine running, check for vacuum at the PCV valve. If vacuum isn't present, check the PCV hose and intake manifold port for clogging.

• With the engine off, remove the PCV valve and shake it. A properly operating valve rattles when shaken. If no sound is heard, the valve should be replaced.

If you must repair or replace PCV system components, consult the vehicle service manual.

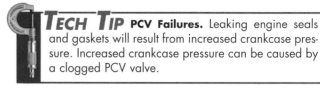

TECH TIP **PCV Failures.** Leaking engine seals and gaskets will result from increased crankcase pressure. Increased crankcase pressure can be caused by a clogged PCV valve.

Exhaust Gas Recirculation System

Nitrogen oxide (NO_X) is a chemical compound that forms when nitrogen (N) and oxygen (O_2) are bonded under high heat. Excess NO_X generally forms when peak combustion temperature exceeds 3,500°F [1,927°C].

Most engines use exhaust gas recirculation (EGR) systems to pass a small amount of exhaust gas into the intake manifold. This reduces NO_x by lowering combustion temperature. Because the exhaust gas is already burned, it is chemically inactive (inert). When the inert exhaust gas is combined with the air/fuel mixture in the cylinder, it reduces the peak combustion temperature.

The exhaust gases flow through a passage connecting the exhaust manifold to the intake manifold. The EGR valve controls the amount of exhaust flowing through this passage. **Fig. 9-4.**

Some EGR valves have diaphragms that form a vacuum chamber at the top of the valve. The chamber connects to a vacuum solenoid controlled by the powertrain control module (PCM). The PCM may use return signals from various sensors to operate or "cycle" the vacuum solenoid. These sensors include the engine coolant temperature

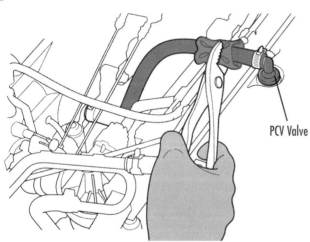

PCV Valve

Fig. 9-3 A quick check of the PCV system can be made by pinching the hose between the PCV valve and the intake manifold. *What causes the valve to make a clicking sound?* (American Honda Motor Company)

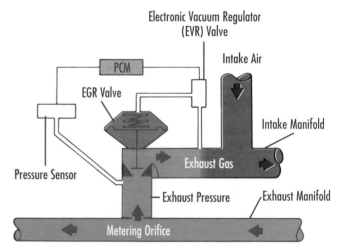

Electronic Vacuum Regulator (EVR) Valve

Intake Air

PCM

EGR Valve

Intake Manifold

Pressure Sensor

Exhaust Gas

Exhaust Pressure

Exhaust Manifold

Metering Orifice

Fig. 9-4 Basic EGR system operation. *How does the powertrain control module (PCM) control the position of the EGR valve?* (Ford Motor Company)

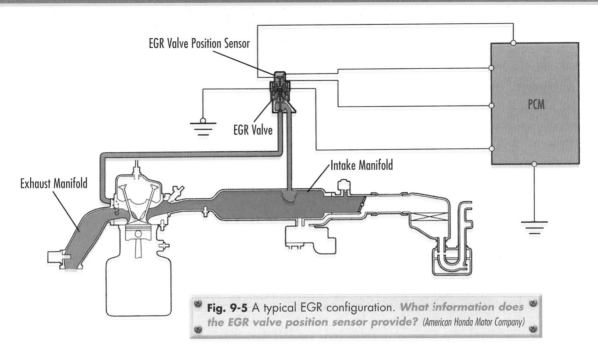

Fig. 9-5 A typical EGR configuration. *What information does the EGR valve position sensor provide?* (American Honda Motor Company)

(ECT), throttle position (TP), manifold absolute pressure (MAP)/mass air flow (MAF), intake air temperature (IAT), and tachometer (TACH).

The PCM may also use a signal from the EGR valve position (EVP) sensor to control the EGR valve. The EVP sensor provides a return signal to the PCM. This signal indicates the position of the EGR valve pintle. **Fig. 9-5.**

By changing the *duty cycle* of the vacuum solenoid, the PCM controls the amount of vacuum that reaches the EGR vacuum chamber. This changes the opening of the EGR pintle valve. This valve controls the exhaust gas flow to the intake manifold.

Some EGR systems use an exhaust backpressure sensor to control exhaust gas circulation. When the engine load is light and backpressure is low, some EGR valve vacuum is vented to the atmosphere. As engine load increases, the backpressure sensor closes the vent. This allows vacuum to open the valve further.

Many current EGR systems use an electronic EGR valve to control the flow of exhaust gas into the intake manifold. A driver circuit in the PCM electronically controls a digital EGR valve. The PCM activates the valve by grounding the valve's solenoid coil. This energizes the solenoid and raises the armature, opening the valve.

OBD-II systems monitor EGR valve operation by opening the valve during deceleration. This increases MAP pressure. The PCM averages a series of EGR tests to determine the amount of EGR flow. If the averaged values do not fall within the expected range, the EGR monitor records a failure. Failure of the EGR system may cause rough idling, detonation (spark knock), and increased emissions.

Servicing the EGR System Many different types of EGR valves are used for emissions control. Some may need special diagnostic procedures detailed in service manuals. EGR valves accumulate carbon deposits on the valve pintle and within the valve body. These deposits hold the valve off its seat. This causes rough idle or stalling because exhaust gases leak into the intake manifold.

Carbon may block the exhaust passage leading to the EGR valve. This causes excess NO_X emissions and detonation (spark knock). Detonation can damage pistons and piston rings.

Test most vacuum-operated EGR valves as follows:

• Snap the throttle open to increase engine speed to 2,000 rpm. The valve should open. The valve opening can be observed with a small mirror.

• If the valve doesn't open, refer to the vehicle service manual for diagnostic procedures. The valve may be stuck or defective. It may need to be removed for inspection. The valve pintle should move freely when opened by hand. An EGR valve that is stuck open will be very hot to the touch.

- Some EGR valves can be opened by applying vacuum with a hand vacuum pump. **Fig. 9-6.** If the valve opens, the EGR flow should stall the engine at idle speed. If the valve is cool, open it manually by lifting the vacuum diaphragm.
- If the valve opens and the engine doesn't stall, check the exhaust gas inlet passage for clogging.

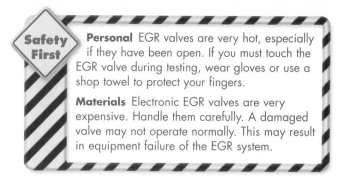

Safety First

Personal EGR valves are very hot, especially if they have been open. If you must touch the EGR valve during testing, wear gloves or use a shop towel to protect your fingers.

Materials Electronic EGR valves are very expensive. Handle them carefully. A damaged valve may not operate normally. This may result in equipment failure of the EGR system.

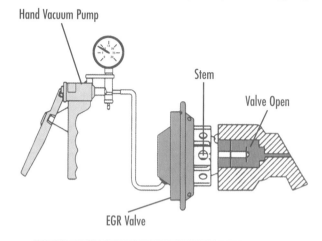

Fig. 9-6 Testing an EGR valve with a vacuum pump. *Can all EGR valves be tested with a vacuum pump?* (Ford Motor Company)

Most inoperative EGR valves will set a diagnostic trouble code (DTC) in the PCM's diagnostic memory. OBD-I systems will show an on/off serial data stream for the EGR valve. OBD-II monitors test the EGR valve during each trip, or drive cycle.

The PCM's serial data stream for electronic EGR valves will show the EGR duty cycle or the percentage of EGR opening. The EGR opening should be zero at idle speed.

If you must repair or replace the EGR system components, consult the vehicle service manual.

The Air Injection System

The **air injection (AIR) system** is an exhaust emission control system that reduces HC and CO emissions. It does this by injecting atmospheric air into the exhaust gases. Depending on the vehicle and operating conditions, the air is directed to either the exhaust manifold or the catalytic converter. In either case, the oxygen in the air helps to convert CO and unburned HC into water vapor and CO_2. This system is known as a *secondary air injection system.* The primary components of an air injection system are the air pump, diverter valve, and check valve(s). **Fig. 9-7.**

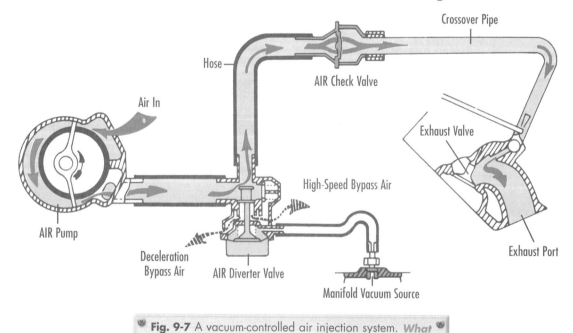

Fig. 9-7 A vacuum-controlled air injection system. *What is the role of the check valve?* (General Motors Company)

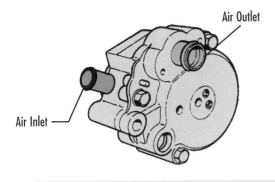

Fig. 9-8 A mechanically driven AIR pump. *What action produces water vapor and carbon dioxide?* (Ford Motor Company)

Air supplied by the pump travels through the air distribution manifold. The air manifold may be cast into the cylinder head or attached to the exhaust manifold. A belt drives the mechanical air pump from the crankshaft pulley. **Fig. 9-8.** The pump runs whenever the engine is running. On some engines the air pump is driven by an electric motor controlled by the PCM. In this case the pump only operates when turned on by the PCM. Most electric air pumps operate only during engine warmup. **Fig. 9-9.**

When the engine is cold, air is directed to the exhaust manifold. The air enters the exhaust manifold through tubes or passages provided for this purpose. In the manifold, the air mixes with the hot exhaust gases. The resulting chemical reaction oxidizes the HC and CO into water vapor and CO_2. The heat from this reaction helps both the oxygen sensor and catalytic converter reach normal operating temperature sooner.

If the injected air is allowed to enter the exhaust manifold during deceleration, the mixture in the manifold could ignite and explode. These small explosions can cause annoying popping sounds or damage to the exhaust system. To prevent this, the diverter valve diverts the air to the atmosphere during the conditions that occur during deceleration. A one-way check valve prevents exhaust gas from backing up into the diverter valve and AIR pump if the pump is not delivering air. **Fig. 9-10.**

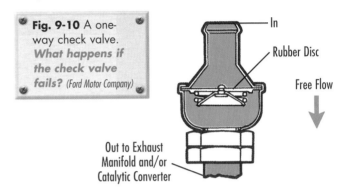

Fig. 9-10 A one-way check valve. *What happens if the check valve fails?* (Ford Motor Company)

During warmup, the engine is operating in open loop mode. The oxygen sensor signal is not being used. The extra air at the oxygen sensor has no effect on fuel delivery. When the engine goes into closed loop operation, the air must be directed away from the oxygen sensor. To do this, the PCM operates a valve that directs the air to the catalytic converter.

The air is injected at a point in the converter where it will help oxidize HC and CO without interfering with the NO_X reduction activity in the converter.

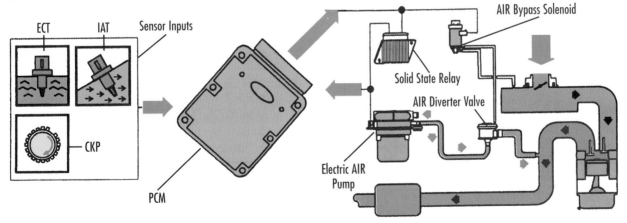

Fig. 9-9 An electric AIR pump system. The electric AIR pump is controlled by the PCM. The AIR bypass solenoid uses vacuum to operate the diverter valve. *What are the sensor inputs to the PCM?* (Ford Motor Company)

Servicing the Air Injection System The air injection system needs little scheduled maintenance. Nevertheless, the system can fail in the following ways:

- Pinched or disconnected air delivery hoses or tubes will prevent normal airflow delivery.
- Defective diverter valves may cause a backfire. To operate correctly, the diverter valve must have a manifold vacuum source.
- Defective check valves allow hot exhaust gases to burn the injection hoses, diverter valve, and AIR pump.
- A leaking check valve or distribution manifold allows oxygen or "false air" to reach the oxygen sensor. This signals a lean condition to the PCM.

The diverter valve's operating position may be shown in the PCM's serial data stream.

TECH TIP **Air Pump Drive Belt.** When adjusting belt tension, always follow manufacturer's procedures. Some vehicles use automatic tension devices to maintain belt tension. Do not pry against the AIR pump housing. Damage to the housing can result.

The Catalytic Converter

The *catalytic converter* changes harmful exhaust pollutants into harmless gases. A **catalyst** is a material that causes a chemical change without being part of the chemical reaction.

The catalytic converter is located between the engine and the muffler. Some engines may have two catalytic converters in each exhaust path. The first converter is located close to the engine so that it will heat up faster. The second is located close to the muffler. It converts the remaining pollutants into harmless gases.

The catalyst body, or substrate, is a bed of pellets or a ceramic honeycomb. The housing of the pellet-type converter is flat. The housing of the honeycomb, or monolith, converter is round or oval-shaped. The housing or shell of the converter is made of stainless steel. This helps prevent burn-through.

TECH TIP **Recycling the Converter.** A catalytic converter contains materials and elements that are very rare and expensive. Many of these elements can be recovered through a recycling process. Be sure to check for the proper method of disposing of used converters in your community.

The exhaust gas passes over a large surface area coated with a catalyst. Vehicles with catalytic converters must use unleaded gasoline. This is to prevent coating the catalyst with lead and making it ineffective. Excessive oil consumption and coolant leaks into the combustion chamber also reduce converter efficiency.

The catalyst must reach a temperature of about 480°F [250°C] before oxidation begins. When fully operational, the converter temperature ranges between 752°F [400°C] and 1,472°F [800°C]. If excess fuel enters the converter, the operating temperature can exceed 1,800°F [982°C]. At this temperature the substrate will melt and the converter will be destroyed.

The converter is equipped with heat shields to prevent unwanted heat from reaching the vehicle's floor pans.

Two-way converters use platinum or palladium to change or oxidize HC and CO into water vapor (H_2O) and carbon dioxide (CO_2). *Three-way converters* add rhodium to break up or reduce NO_X to nitrogen (N_2) and oxygen (O_2). **Fig. 9-11.**

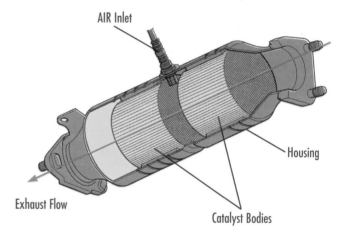

AIR Inlet

Housing

Exhaust Flow

Catalyst Bodies

Fig. 9-11 A three-way catalytic converter. *What element is added to three-way converters?* (American Honda Motor Company)

For the catalytic converter to be most effective, the air/fuel mixture must be at the stoichiometric ratio of 14.7:1. The air/fuel mixture must alternate, or "switch," from rich to lean to allow the catalyst to oxidize pollutants more efficiently. Most modern converters contain a base metal known as cerium. Cerium attracts and releases oxygen, which helps oxidize or reduce pollutants.

Vapor Pressure and Temperature

You stop at the pump to fill your fuel tank. A faint odor of gasoline fills the air. Later, you drive through the automatic car wash. When you leave the car wash, the water on your windshield dries quickly. Both events are related to the concept of evaporation and vapor pressure.

If a dish of water is left on a table for a day, the water level drops noticeably. When a liquid is exposed to air, molecules near the surface escape from it. This process is called *evaporation*. Condensation occurs when vapor returns to its normal state, such as when water vapor condenses into water droplets on a cold beverage can.

When the number of molecules evaporating equals the number of molecules condensing, an equilibrium has been reached. At a constant temperature, a solid or liquid substance evaporates and reaches an equilibrium pressure with its vapor. This is the *vapor pressure* for that substance. When the vapor pressure of a liquid equals the external pressure, the liquid boils. When the vapor pressure of water equals the atmospheric pressure at sea level (760 mm Hg), water boils at 100°C [212°F]. At higher altitudes the atmospheric pressure is lower than 760 mm of mercury. At such an altitude, water in an open container will boil at a lower temperature than 100°C.

Apply it!

Graphing Vapor Pressure

Meets NATEF Science Standards for understanding the relationship of heat and barometric pressure. Also meets NATEF Mathematics Standards for constructing and interpreting graphs.

The vapor pressure of water at various temperatures is shown. You can use this information to show the effects of pressure on the boiling point of water.

1. Use a sheet of standard graph paper. Graph the information in the table. Near the bottom of the page, label the horizontal axis "Temperature." Mark temperatures from −50°C to +120°C in increments of 10°.

2. Near the left edge of the sheet, label the vertical axis "Vapor Pressure." Mark values from 0 to 1,500 mm Hg in increments of 100 mm. Make the graph as tall as possible.

3. Plot the values given in the table as solid dots on the graph.

Results and Analysis

4. Refer to the graph. How does vapor pressure vary with the temperature?

5. Suppose that you are driving high in the mountains. The atmospheric pressure is only about 450 mm Hg. This pressure is much lower than 760 mm Hg, the atmospheric pressure at sea level. At what temperature would water boil on this mountain road?

VAPOR PRESSURE OF WATER

Temperature [°C]	Vapor Pressure [mm Hg]	Temperature [°C]	Vapor Pressure [mm Hg]
−50	0.030	50	92.5
−10	1.95	60	149
0	4.58	70	234
10	9.21	80	355
20	17.5	90	526
30	31.8	100	760
40	55.3	120	1,489

Servicing the Catalytic Converter Although catalytic converters are simple devices, they do fail. Perform the following checks:

1. If the substrate is loose in a converter, it will rattle when tapped with a rubber hammer.

2. Test the converter for clogging by using a pressure gauge attached to a special adapter. The adapter can be temporarily installed in place of the oxygen sensor. A special tap can also be temporarily threaded into the exhaust pipe. Either the adapter or tap allows a pressure gauge to be attached. In general, the backpressure shouldn't exceed 1.5 psi [10.3 kPa] at 3,000 no-load rpm.

3. Check the intake manifold vacuum using a vacuum gauge attached to the intake manifold. A 3,000-rpm, no-load condition should show a steady vacuum reading. A progressively decreasing or unusually low vacuum reading may indicate excessive backpressure.

4. Using a noncontact pyrometer, check the converter outlet temperature. When completely warmed up, the converter outlet temperature should exceed the inlet temperature by approximately 100°F [38°C].

5. Thoroughly check the ignition and fuel system for correct operation. An ignition misfire or rich fuel delivery may melt the substrate in a catalytic converter.

Safety First

Personal To prevent severe burns, let hot catalytic converters cool before servicing.

Materials Always reinstall heat shields in their original locations. Failing to do so could cause the vehicle's carpet to catch fire.

Replacing the Catalytic Converter A catalytic converter may be attached to the exhaust system, or it may be made as part of the header pipe.

If the catalytic converter is attached to the exhaust system:

1. Remove all clamps and mounting hardware that attach the converter to the exhaust system.

2. Use a hammer or similar tool to loosen the converter.

3. If necessary, use a cutter to loosen the converter from the system.

4. Remove the converter.

5. Install a new converter, using new mounting hardware.

6. Reinstall all heat shields, replacing any that are damaged.

If the catalytic converter is part of the header pipe:

1. Remove the header pipe and converter as a unit. Consult the vehicle service manual for the location of mounting hardware.

2. Install new gaskets as specified by the vehicle service manual.

3. Install a new header pipe and converter.

4. Reinstall all heat shields, replacing any that are damaged.

Legal requirements govern the replacement of catalytic converters. Many converters are covered by (OEM) warranties. A converter should be replaced only by a qualified technician. If the converter is replaced, the repair should be documented. The converter should be disposed of according to applicable regulations.

Evaporative Control System

The **evaporative control (EVAP) system** is a system that prevents gasoline vapors in the fuel system from escaping into the atmosphere. The vapors are stored in a charcoal-filled canister. When specific operating conditions are met, a purge valve is opened. Vacuum from the intake manifold draws vapors from the canister. This allows fuel vapors to flow into cylinders to be burned.

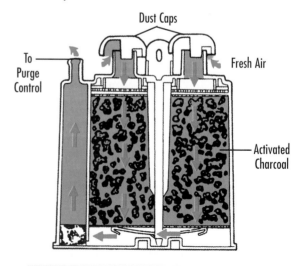

Fig. 9-12 A charcoal canister. *What is the function of this canister?* (Ford Motor Company)

Using J1930 Terminology

You have learned a great deal about automotive technology. This information includes terms, definitions, abbreviations, and acronyms. And the list continues to grow. In addition to new terms, some older terms have been modified. Different manufacturers may have used different terms for the same components. As a service technician, you know you need to learn and use these words.

The Society of Automotive Engineers (SAE) has developed a standardized list of terms and acronyms. It is referred to as *Recommended Practice J1930* or *J1930*, for short. This list was first issued in 1991 and has been updated several times. The U.S. government and the automotive industry accepted the list in 1995 as the industry standard.

J1930 lists terms, definitions, abbreviations, and acronyms. It also provides a method, or rules, to use when naming new items. Using this pattern, new terms will fit in with existing terms on the list. Some older terms have been kept because they are widely used and accepted even though their names don't follow the rules.

Here is an example. Before *J1930*, the acronym *"CPS"* could have meant "camshaft position sensor" or "crankshaft position sensor." Now, the term *"CMP sensor"* identifies the camshaft position sensor. The term CKP sensor refers to the crankshaft position sensor. Now you are less likely to be confused by these acronyms.

The definitions in *J1930* are accepted by the automotive industry. This listing can clear up confusion you may have about new terms.

Apply it!

Meets NATEF Communications Standards for using standard definitions and text resources.

❶ List the new terms and acronyms in this chapter.

❷ Write the definitions for them.

❸ Study and learn the items on your list.

❹ Form small teams. Everyone on the team should quiz each other until all are familiar with the definitions of the items on the lists.

OBD-I Evaporative Control System An OBD-I EVAP system consists of the following parts:

- A fuel tank equipped with vapor vent lines leading to a *charcoal canister*. **Fig. 9-12.** The canister stores fuel vapors when the engine is shut off.
- A fuel tank cap that prevents fuel vapors from leaking into the atmosphere.
- A vacuum-operated or electrically operated *purge valve* that controls the flow of fuel vapors into the intake manifold.

The fuel tank cap on a fuel tank in an EVAP system has a pressure release valve. This valve prevents excess pressure from building in the fuel tank. **Fig. 9-13.** Tank overfill or pinched vapor purge lines can cause a buildup of pressure. The cap also contains a vacuum release valve to prevent a vacuum from building in the tank caused by gasoline consumption or temperature changes. Most EVAP system malfunctions occur because:

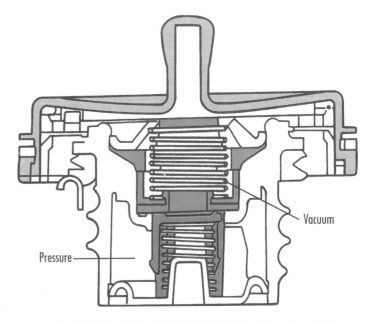

Fig. 9-13 A fuel tank cap. *How many valves does the fuel tank cap have?* (Ford Motor Company)

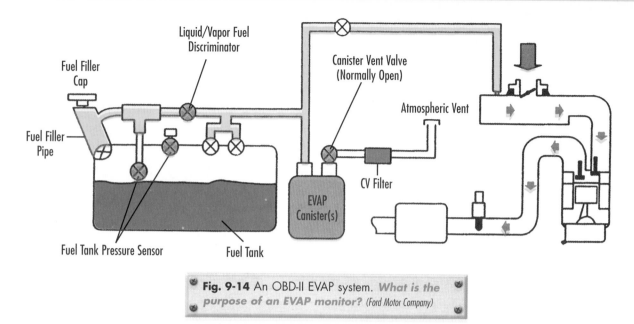

Fig. 9-14 An OBD-II EVAP system. *What is the purpose of an EVAP monitor?* (Ford Motor Company)

- The fuel tank is overfilled. Fuel station pumps have automatic fuel shut-off nozzles to prevent overfilling.
- The fuel tank cap leaks. If the fuel cap is not fully tightened, or is defective, the MIL light may be turned on in some OBD-II systems.
- The fuel is more volatile than normal. Excess pressure can develop in the fuel tank.
- The purge lines or hoses leak. Many EVAP systems perform vacuum and pressure tests on the entire system. If the tests fail, the MIL light is turned on.
- The purge valve does not open or close when it should. The purge valve must operate properly to remove fuel vapors from the system.

OBD-II Evaporative Control System OBD-II provides diagnostic monitoring of the EVAP system operation. **Fig. 9-14.** The EVAP monitor checks the operation and condition of EVAP system components. Depending upon the application, OBD-II may add these parts to the EVAP system:

- A canister vent valve.
- A leak detection pump (LDP), which pressurizes the fuel tank for the EVAP system monitor.

- A fuel tank pressure sensor.
- A domed fuel tank equipped with vent valves to prevent overfilling and allow fuel to expand without leaking.
- Rollover valves that prevent fuel from leaking from the tank inlet and outlets if the vehicle tips over.

OBD-II systems test the EVAP system for leaks by admitting a slight amount of vacuum or air pressure into the fuel tank. The operating strategies of these systems may differ. Use the vehicle service manual for diagnosis. In general, an OBD-II system uses a canister vent solenoid to close the system for testing. **Fig. 9-15.**

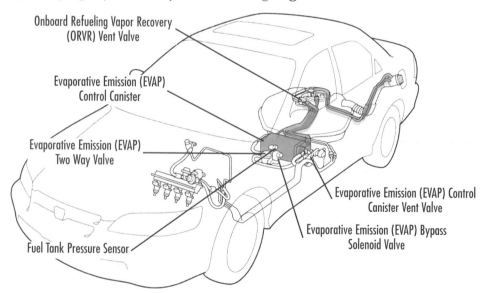

Fig. 9-15 A Honda EVAP system. *What is the purpose of the EVAP canister vent solenoid?* (American Honda Motor Company)

For vacuum-operated systems, the PCM monitor begins the test by closing the canister vent valve. The purge valve is opened to allow a small amount of manifold vacuum into the fuel tank. The PCM then closes the purge valve. A fuel tank pressure sensor measures the vacuum in the tank. If the vacuum is lost too quickly, the PCM activates the MIL and sets a DTC in its diagnostic memory.

Some EVAP systems use a leak detection pump (LDP) to pressurize the fuel tank. When the PCM activates the EVAP monitor, the canister vent valve is closed. Then the LDP begins running. If the fuel tank holds pressure, the LDP shuts off. If the LDP doesn't shut off, the PCM activates the MIL and sets a DTC in the diagnostic memory.

SECTION 1 KNOWLEDGE CHECK

❶ What are three sources of automotive emissions?

❷ What are the major subsystems of an emissions control system?

❸ What causes PCV valve failure?

❹ What is the purpose of the AIR injection system?

❺ Why do most EVAP systems fail?

Section 2

Emissions Testing

Many states have some type of vehicle emissions testing program. These programs vary according to the special needs of a state or locality. All programs require that OEM emissions hardware, including vacuum hoses, be in place and operational. **Fig. 9-16.** All programs require using an exhaust gas analyzer to test exhaust emissions.

Emissions Testing Programs

Two types of tests are used to check emissions. The *no-load test* checks exhaust emissions at either idle speed only or at idle and 2,500 no-load rpm. This test is most often used on older vehicles that are not required to meet current emissions standards.

Load tests check vehicle emissions over a range of operating conditions. The most widely used emissions testing program is the I/M 240 program.

I/M 240 programs are centralized emissions testing programs using test procedures that satisfy federal government standards. Emissions tests performed in dealerships or independent shops are decentralized test programs.

The I/M 240 test uses a *chassis dynamometer* to simulate specific load and speed conditions encountered in day-to-day driving. The test operator drives the vehicle at speeds up to 55 mph [88 kph] for 240 seconds while the emissions are measured. The emissions specifications depend upon the type of vehicle being tested.

Engine/Evaporative Family Information

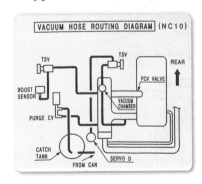

Fig. 9-16 A typical emission control information decal with vacuum hose routing. *Where is the engine displacement noted?* (Ford Motor Company)

Calculating Carbon Dioxide

The burning of fossil fuels produces large amounts of carbon dioxide (CO_2) emissions every year. *Fossil fuels* include coal, natural gas, and petroleum. Gasoline-powered vehicles produce about 20 percent of all CO_2 emissions.

The combustion of gasoline produces about 310 million metric tons of CO_2 every year. How many pounds per year is this?

310 million metric tons

= 310,000,000 metric tons

= 310,000,000,000 kg

Convert kilograms to pounds:

2.2 lb per kg x 310,000,000,000 kg

= 682,000,000,000 lb of CO_2

These numbers are very impressive. However, they are so large that we may not understand their significance.

Let's look at the numbers for one vehicle. One estimate is that burning a gallon of gasoline produces 20 pounds of CO_2. A 1999 Chevrolet Malibu with a 2.4-liter twin-cam engine could expect to get 23 mpg in city driving. The average driver drives about 12,000 miles each year. How much CO_2 will this car emit in a year?

First find how much gasoline the car will use in a year.

$$\frac{12{,}000 \text{ miles}}{23 \text{ mph}} = 522 \text{ gal}$$

Then determine how much CO_2 is produced by burning 522 gal of gasoline:

522 gal × 20 lb per gal = 10,440 lb of CO_2

Apply it!

Meets NATEF Mathematics Standards for estimating expected outcomes for a normally operating system.

❶ Assume that the Grimes family uses its car mainly for city driving. They rarely drive the car on the highway. Most of their trips are relatively short—under 5 miles. Their gas mileage is fairly poor. They average only 14 mpg. They drive 12,000 miles per year. How much CO_2 is emitted by their car in one year?

❷ Shirley Patel is a saleswoman who uses her car almost solely for highway driving. She drives 24,000 miles in a year and her car averages 27.5 mpg. How much CO_2 does her car emit in one year?

❸ What factors might account for the fact that Shirley Patel's car emits about the same amount of CO_2 as the Grimes' car emits?

The I/M 240 test may require an *EVAP system purge flow test* and an EVAP system pressure test. The purge flow test measures the amount of vapor and air flowing through the purge valve. The pressure test pressurizes the fuel tank to 0.5 psi [3 kPa]. The tank must then hold pressure for at least two minutes.

Exhaust Gas Analyzers

An *exhaust gas analyzer* is a device used to test the amount of exhaust emissions produced by a vehicle. The analyzer samples the exhaust through a probe placed in the tail pipe. The exhaust sample is drawn into the analyzer, where it is filtered. The filtered exhaust is passed through a detector. The detector measures the content and chemical composition of the exhaust gas. The exhaust gas analyzer can also be used to diagnose ignition and fuel system problems that cause driveability complaints.

Two-gas analyzers measure HC and CO. Most of these analyzers aren't sensitive enough to measure exhaust emissions on late-model vehicles. Because new vehicles produce very low exhaust emissions, the exhaust gas analyzer must be able to measure very small amounts of polluting gases.

Most exhaust gas analyzers are four- or five-gas analyzers. The four-gas analyzers are the most common. They measure levels of CO_2 and O_2 in addition to HC and CO. Though not harmful emissions, measuring CO_2 and O_2 provides data to evaluate the combustion efficiency of the engine.

The five-gas analyzer measures levels of NO_X, in addition to CO_2, O_2, HC, and CO. NO_X is a toxic pollutant, and the primary cause of emissions generated smog. High levels of NO_X are caused by:
• High combustion temperatures resulting from failure of the EGR system.
• High engine temperatures resulting from a defective cooling system or other problems.

Safety First **Personal** Carbon monoxide (CO) is an odorless, invisible gas present in all vehicle exhaust systems. Inhaling CO can cause severe illness or death. Always use exhaust ventilation equipment to remove exhaust gas.

Most analyzers require a warm-up period before they will accurately test emissions. The analyzer may also need to be zeroed while sampling clean air. Modern analyzers may use a *calibration gas* to calibrate the analyzer before it is used to test emissions.

When testing emissions, it is important to check the exhaust system for leaks. A leaking exhaust system affects emissions readings.

Exhaust gas analysis requires special training. In most states, a special license is required to perform federally mandated emissions testing.

Before testing a vehicle, make sure the engine and exhaust system are warmed to proper operating temperature. An insufficiently warmed vehicle will usually give erroneous readings. Some vehicles

may need to be driven before performing an emissions test.

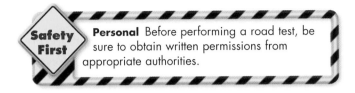

Safety First **Personal** Before performing a road test, be sure to obtain written permissions from appropriate authorities.

Actual emission values depend on many variables. These include combustion chamber design and emissions control efficiency. Automotive engines tend to produce more emissions as age and mileage increase.

An engine operating at the correct air/fuel ratio will give you the lowest emission levels. The following comments apply to no-load test results.
• HC is measured in parts per million (ppm). Misfires and lean or rich air/fuel mixtures cause excessive HC levels.
• The normal limit for CO should be less than 2 percent. Rich air/fuel mixtures cause excessive CO levels.
• NO_X is a term for various compounds of nitrogen and oxygen. High combustion temperatures cause NO_X. NO_X is not usually measured during no-load testing.
• CO_2 output should peak at 15 percent. Higher CO_2 indicates better combustion.
• O_2 content should be less than 2 percent. Higher levels show a lean air/fuel mixture. Lower levels show incomplete combustion.

TECH TIP **Contaminated Oil.** Engine oil can be contaminated with gasoline. This usually occurs during short-trip driving in cold weather. When the engine is warmed up, the gasoline vaporizes. These vapors cause high HC and CO levels (measured at idle).

SECTION 2 KNOWLEDGE CHECK

❶ For emissions testing, what requirement is placed on vehicle OEM emissions hardware?

❷ A chassis dynamometer is used for what test?

❸ What gases are measured by a four-gas analyzer?

❹ In addition to the four gases measured by a four-gas analyzer, what additional gas is measured by a five-gas analyzer?

❺ How are most exhaust gas analyzers prepared for testing?

CHAPTER 9 REVIEW

Key Points

Meets the following NATEF Standards for Engine Performance: diagnosing oil leaks and emissions problems; diagnosing emissions and driveability problems.

- Automotive engine emissions come from three sources: the crankcase ventilation, fuel, and exhaust systems.
- Proper operation of the PCV system depends upon the PCV valve being free of sludge deposits.
- Excess NO_X generally forms when the peak combustion temperature exceeds 3,500°F [1,927°C].
- The air injection (AIR) system reduces HC and CO emissions by injecting fresh air into the exhaust gases.
- The catalytic converter changes harmful exhaust pollutants into harmless gases.
- The EVAP system uses intake manifold vacuum to draw gasoline vapors from the charcoal canister into the intake manifold.
- The I/M 240 test uses a *chassis dynamometer* to simulate specific load and speed conditions encountered in day-to-day driving.

Review Questions

1. What are the primary exhaust pollutants?
2. When repairing a PCV system, when should a PCV valve be replaced?
3. When repairing an EGR system, how can a scan tool identify a problem?
4. What are the potential system failures of the air injection system?
5. How can you diagnose and test catalytic converters?
6. What is the purpose of the evaporative control (EVAP) system?
7. How can an excessive amount of backpressure be detected when checking a catalytic converter?
8. **Critical Thinking** Why might an exhaust gas analysis reveal low levels of CO_2 in the exhaust stream?
9. **Critical Thinking** Why might a cold engine and cold exhaust system provide faulty emissions readings?

TECHNOLOGY FORECAST
FOR AUTOMOTIVE EXCELLENCE

A Breath of Fresh Air

To meet future clean-air laws, cars and trucks will have to produce fewer harmful emissions. Engineers are working on many new technologies to comply with these codes.

More efficient refining of gasoline and diesel fuels is an important goal. Our future fuels must burn cleaner and without the use of sulphur, a known pollutant. Scientists are also developing fuel cells that produce power from hydrogen. The only emission is water, a harmless substance!

An advanced engine control system could become standard on future vehicles. This system would monitor the conditions in each cylinder. It would then send the right amount of air and fuel to that cylinder so the mix would burn cleanly.

New catalytic converters are also being studied. These units would work much like today's units by removing harmful compounds from the exhaust. Future catalytic converters would be electrically heated before the engine starts. Since most emissions are produced while the engine is cold, a warm converter can better remove harmful hydrocarbons (unburned fuel), oxides of nitrogen (NO_X), and carbon monoxide. Engineers are also studying the use of traps as a way to improve catalytic converters.

AUTOMOTIVE EXCELLENCE
TEST PREP

Answering the following practice questions will help you prepare for the ASE certification tests.

1. The air filter is soaked with oil. Technician A says the PCV valve may be clogged. Technician B says the piston rings may be worn. Who is correct?

   ⓐ Technician A.
   ⓑ Technician B.
   ⓒ Both Technician A and Technician B.
   ⓓ Neither Technician A nor Technician B.

2. An exhaust gas analyzer shows a high CO emission. Technician A says the air/fuel mixture may be too rich. Technician B says the air/fuel mixture may be too lean. Who is correct?

   ⓐ Technician A.
   ⓑ Technician B.
   ⓒ Both Technician A and Technician B.
   ⓓ Neither Technician A nor Technician B.

3. An engine is experiencing excessive spark knock. Changing grades of gasoline doesn't help. What could be the problem?

   ⓐ The EGR valve is stuck closed.
   ⓑ The fuel injectors are clogged.
   ⓒ The catalytic converter is clogged.
   ⓓ Too much pressure in the fuel tank.

4. An exhaust gas analyzer shows high HC and CO. Technician A says the air injection system may be pumping too much air into the exhaust. Technician B says the air injection system may not be pumping enough air into the exhaust. Who is correct?

   ⓐ Technician A.
   ⓑ Technician B.
   ⓒ Both Technician A and Technician B.
   ⓓ Neither Technician A nor Technician B.

5. Excessive pressure builds up in the fuel tank. Technician A says the fuel's volatility may be too high. Technician B says the vapor purge line may be pinched closed. Who is correct?

   ⓐ Technician A.
   ⓑ Technician B.
   ⓒ Both Technician A and Technician B.
   ⓓ Neither Technician A nor Technician B.

6. An exhaust gas analyzer shows high HC and CO. Technician A says the catalytic converter may not be hot enough. Technician B says the catalytic converter may be defective. Who is correct?

   ⓐ Technician A.
   ⓑ Technician B.
   ⓒ Both Technician A and Technician B.
   ⓓ Neither Technician A nor Technician B.

7. An engine shows low manifold vacuum at 2,500 rpm during steady throttle operation. The vacuum returns to normal at idle. What could be the cause?

   ⓐ A clogged catalytic converter.
   ⓑ A leaking intake manifold gasket.
   ⓒ Both a and b.
   ⓓ Neither a nor b.

8. The oxygen sensor return signal voltage is low. Which of the following could be the cause?

   ⓐ A stuck AIR check valve.
   ⓑ A leaking diverter valve.
   ⓒ A worn AIR pump.
   ⓓ A clogged exhaust.

9. An exhaust gas analyzer shows high emissions on a no-load test. Technician A says the thermostat may be stuck open. Technician B says the PCM may be operating in the open-loop mode. Who is correct?

   ⓐ Technician A.
   ⓑ Technician B.
   ⓒ Both Technician A and Technician B.
   ⓓ Neither Technician A nor Technician B.

10. Technician A says that low oxygen in an exhaust stream shows perfect combustion. Technician B says that low carbon dioxide in the exhaust shows perfect combustion. Who is correct?

    ⓐ Technician A.
    ⓑ Technician B.
    ⓒ Both Technician A and Technician B.
    ⓓ Neither Technician A nor Technician B.

Suspension & Steering

Tire Inspector/Technician

National Tire Company is looking for a qualified tire inspector/technician. Duties include checking tire pressure and inflating tires, inspecting tires and wheels, rotating tires, replacing tire valves, mounting tires, and installing wheels. Some automotive experience preferred but not required.

AUTOMOTIVE SWITCHBOARD OPERATOR

Part-time switchboard operator needed for automotive dealership. The right person must be outgoing, well-spoken, polite, detail-oriented, and able to handle many incoming calls. Evening and Saturday hours. We offer a competitive salary and paid vacation.

ALIGNMENT SPECIALIST

Busy auto repair shop is looking for technicians/trainees specializing in frame and wheel alignment. Ideal candidate would have an Associate's degree and 1+ years of automotive experience, but we'll train the right person. Duties include operating front end alignment equipment, straightening and aligning frames, balancing and aligning wheels, and making repairs on steering mechanisms and suspension systems.

Senior Automotive Technician

Immediate opening for an individual with 3–5 years of experience supervising maintenance technicians. We're a busy service station that takes pride in making high-quality repairs to a variety of electrical, mechanical, and pneumatic systems. We specialize in steering and suspension systems. Only hard-working candidates need apply!

Career-Focus Activity

After reading the above job ads, do the following:

- Imagine that you are applying for one of the positions. Write an online response expressing your interest.
- Focus on an automotive career that interests you. Make a chronological list of what you'll need to do to reach this career goal.
- Where could you obtain the specialized training needed to be an Alignment Specialist?

CHAPTER 1

Diagnosing & Repairing Steering Systems

You'll Be Able To:

- Identify the major components of a steering system.
- Identify the two main types of steering gears.
- Perform a power steering pressure check.
- Remove and service a rack-and-pinion steering system.
- Observe proper safety precautions when working with steering-wheel mounted air bags.

Terms to Know:

power-assisted steering

rack-and-pinion steering gear

recirculating-ball steering gear

tie rod

variable-assist power steering

The Problem

On his way to work, Mr. Ritz is happy with the steering operation of his car as he drives at higher speeds on the smooth pavement and gentle curves of the highways he uses daily. Once in downtown traffic, however, at low speeds and at stops, Mr. Ritz's vehicle makes a whining noise when he turns the steering wheel. When the vehicle is at idle, it is hard for him to steer.

Mr. Ritz brings his car to your service center. You check the power steering belt. It seems fine, but the problem remains.

Your Challenge

As a service technician, your job is to diagnose Mr. Ritz's complaint and make the necessary repairs.

1. Why is information from Mr. Ritz important?
2. What questions should you ask Mr. Ritz?
3. To isolate the complaint, what checks would you make?

Section 1

Steering System Components

The steering system gives the driver control of the vehicle's direction. The steering system may be either manual or power-assisted. A manual steering system relies solely on the driver to provide the steering force. **Power-assisted steering,** also commonly called power steering, is a system that uses hydraulic or electric power to help the driver apply steering force.

For both manual and power-assisted steering, the basic components are the same. As the driver turns the steering wheel, the *steering shaft* rotates within the steering column. The steering shaft connects to a *steering gear* (either a separate gearbox or the *pinion gear* of a *rack-and-pinion* unit). The movement of this gear causes the *steering linkage* to move left or right. The linkage connects the steering gear to the *steering arms.* The steering arms are attached to the road wheels.

A steering system has these major components:
• The steering wheel, steering column, and steering shaft are parts of the driver input portion of the steering system.
• The steering gear changes the rotary input motion of the steering column into linear output motion.
• The steering linkage connects the linear output of the steering gear to the steering arms. Because the road wheels attach rigidly to the steering arm assemblies, the wheels turn whenever the steering arms move.
• The ball sockets allow suspension action without binding the steering linkage.

Steering Wheel

The steering wheel attaches to the steering shaft by one or more threaded fasteners. In most vehicles the steering wheel secures with a single nut.

All steering wheels have a slight *interference fit* on the shaft. You will not be able to remove the steering wheel by hand. You remove the steering wheel by removing the center mounting nut. Then pull the wheel off using a steering wheel puller tool. **Fig. 1-1.**

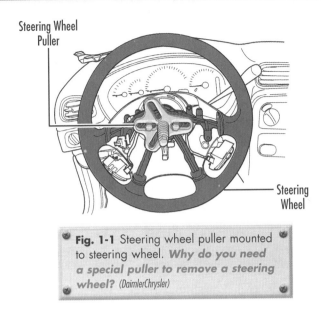

Fig. 1-1 Steering wheel puller mounted to steering wheel. *Why do you need a special puller to remove a steering wheel?* (DaimlerChrysler)

Never use a hammer or other objects to force the steering wheel off from the backside. Use only the proper puller.

Steering Column

The *steering column* is the housing that contains and supports the steering shaft. Many vehicles have a *tilt mechanism* to adjust the steering wheel or the steering column up or down. This adjustment allows more access and exit room for the driver. The design of some steering columns allows them to *telescope* in or out in addition to tilting. **Fig. 1-2.**

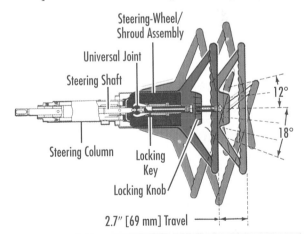

Fig. 1-2 Tilt and telescopic steering. Some systems allow the steering wheel to pivot. Others allow the entire column to raise or lower. *What are the advantages of a steering wheel with tilt and telescopic adjustments?* (DaimlerChrysler)

The steering column also houses the ignition switch and lock. The column-mounted lock allows you to lock the ignition and steering operations to inhibit theft of the vehicle. If the ignition switch or lock is damaged for any reason, replace according to manufacturer procedures.

To protect the driver, some steering columns are designed to collapse in a front-end collision. The steering column absorbs part of the impact if the driver is thrown forward into the steering wheel.

Steering Column Couplers Steering column couplers allow the steering shaft connections to pivot at various angles during steering operation. These flexible couplers serve as pivot points between the upper and lower shafts and at the steering gear connection.

Depending on the vehicle, the type of coupler will vary. However, these couplers are either *universal joints, flexible couplers, intermediate couplers,* or a combination of these couplers. The purpose of a coupler is to connect the upper and lower steering shafts together in the steering shaft assembly. **Fig. 1-3.**

Loose or worn couplers can result in unresponsive steering. A seized coupler can cause stiff or uneven operation of the steering wheel. A failed coupler can cause a complete loss of steering. You should always inspect couplers whenever the vehicle has suffered a front-end collision. Replace the damaged couplers according to manufacturer specifications.

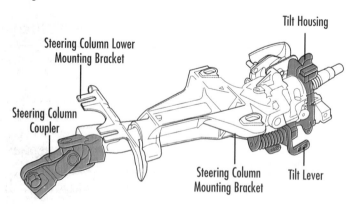

Fig. 1-3 Steering column assembly. Note steering-column coupler at lower end of column. *Why do steering assemblies need a coupler?* (DaimlerChrysler)

Labels: Tilt Housing, Steering Column Lower Mounting Bracket, Steering Column Coupler, Steering Column Mounting Bracket, Tilt Lever

Steering Column Diagnosis In the event of a front-end collision, or air bag deployment, inspect the steering column and steering wheel. Then inspect the coupler for any damage and replace if needed. Refer to the manufacturer's manual for these procedures.

If you hear a chirp, squeak, or rubbing sound, inspect the shroud, intermediate shaft, steering column, and steering wheel.

If the steering catches, binds, or sticks, check the shaft and couplers, including the tilt mechanisms, for bind. If excess play exists in the steering wheel, check the steering-shaft couplers for wear or damage and determine any needed repairs. Refer to the manufacturer's manual.

TECH TIP **Steering Alignment.** Establish an initial steering alignment position before removing the steering column. Turn the steering to the full-left or full-right lock position. Lock the steering column in position by moving the ignition to the OFF position and removing the ignition key. Be sure to maintain the position of the wheels. Reinstall the steering column with the wheels and steering column in the same full-lock position.

Steering Gear Systems

Two main types of steering gear systems are in common use. They are recirculating-ball systems and rack-and-pinion systems. A third type in limited use is the worm-and-roller steering gearbox.

Recirculating-Ball System This type of steering is used primarily on trucks, vans, some larger vehicles, and most passenger vehicles made prior to the 1980s. The **recirculating-ball steering gear** is an assembly that uses a series of recirculating balls on a worm gear to transfer steering-wheel movement to road wheel movement. **Fig. 1-4.**

The steel balls within the gear housing constantly recirculate within the guide paths. They move from one end of the ball nut through return guides to reenter the ball nut at the opposite end. The balls provide low-friction contact points between the worm gear and the internal grooves of the ball nut.

The steering-linkage system used with the recirculating-ball system is a *parallelogram linkage.* A number of separate parts connect the steering gear output to the road wheels. **Fig. 1-5.**

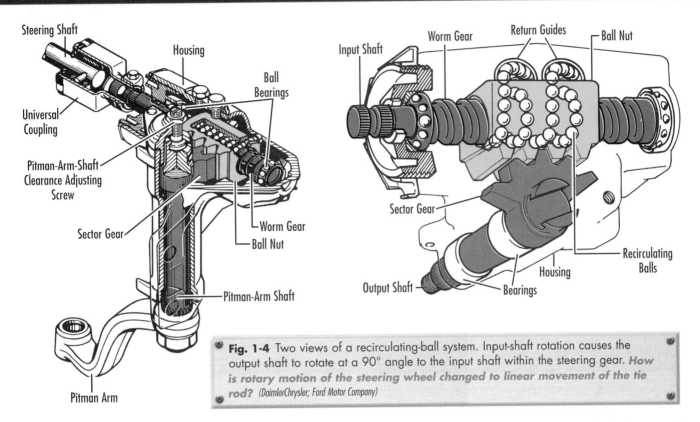

Fig. 1-4 Two views of a recirculating-ball system. Input-shaft rotation causes the output shaft to rotate at a 90° angle to the input shaft within the steering gear. *How is rotary motion of the steering wheel changed to linear movement of the tie rod?* (DaimlerChrysler; Ford Motor Company)

The steering gearbox uses an output device called a *pitman arm*. The pitman arm connects to a *center link* (also called a *drag link* or a *relay rod*). The center link runs horizontally, from left to right. The opposite end of the center link attaches to an *idler arm*. The pitman arm and the idler arm act as the center link's anchor points. The two ends of the center link pivot at their anchor points.

Two tie rods connect to the center link. A **tie rod** is an adjustable-length rod that, as the steering wheel turns, transfers the steering force and direction from the rack or linkage to the steering arm. One tie rod connects the center link to the left steering arm. The other connects to the right steering arm. To allow free motion as the road wheels turn and as the suspension compresses and rebounds, each attachment point of the tie rods uses a separate lubricated *ball socket*. **Fig. 1-6.**

Fig. 1-5 Recirculating-ball system using a parallelogram linkage. *What steering system component provides the driver input?*

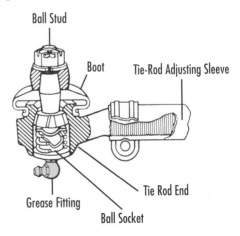

Fig. 1-6 The tie-rod ends have built-in ball sockets. *What is the function of a ball socket?* (Ford Motor Company)

Rack-and-Pinion System Most passenger vehicles have rack-and-pinion steering. This steering system uses a **rack-and-pinion steering gear** in which a pinion on the end of the steering shaft meshes with a rack of gear teeth on the major cross-member of the steering linkage. Because a rack-and-pinion system uses fewer parts, it provides space-saving and weight-saving features. **Fig. 1-7.** Because smaller cars have limited space, a rack-and-pinion design is practical for these vehicles.

The steering shaft connects to an input shaft on the rack-and-pinion unit. When the steering wheel turns, the steering shaft rotates, which rotates a pinion gear on the input shaft.

The pinion gear mates to a horizontal *toothed rack* inside the rack-and-pinion housing. As the pinion gear rotates, it forces the toothed rack to slide to the right or left inside the housing. Each end of the rack attaches to an inner and outer tie rod assembly. The tie rods are attached to the steering arms.

A rack-and-pinion system:

• Has fewer parts to inspect, maintain, and service.

• Uses fewer parts than a recirculating-ball steering system.

• Needs less space than a recirculating-ball steering system.

Another style of rack-and-pinion steering is the center-link type. Its basic function is the same as other rack-and-pinion units. However, as the name indicates, the tie rods attach at the center of the rack housing. **Fig. 1-8.**

A center-link design is more compact. Because of space limits in the chassis, some rack-and-pinion units apply the center-link style.

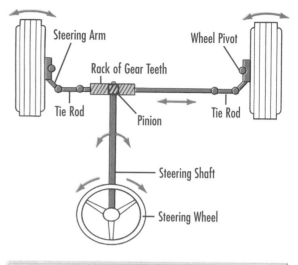

Fig. 1-7 A typical rack-and-pinion steering system. *Which parts have a function in steering output?*

Special Tools

A number of tools are available that will help you disassemble several steering system components. Some components are difficult to take apart because of either a press-fit or an interference-fit. These tools help make sure that none of the parts are damaged during disassembly.

• Pitman arm puller–Disconnects the pitman arm from the steering gear output shaft.

• Drag link socket wrench–Connects and disconnects the drag link from the pitman arm or idler arm.

• Ball joint press–Presses the ball joint tapered stud out of the steering arm (or knuckle) mounting hole.

Other special tools used with steering systems include the inner tie-rod end tool and the tie rod puller. These tools are described later in this chapter.

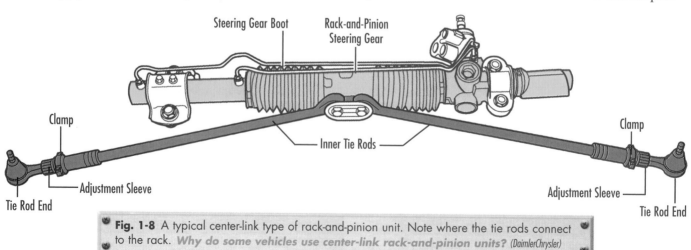

Fig. 1-8 A typical center-link type of rack-and-pinion unit. Note where the tie rods connect to the rack. *Why do some vehicles use center-link rack-and-pinion units?* (DaimlerChrysler)

Altering Force with Levers

Power steering systems are complex mechanical systems. They use gears and levers in combination. The gears and levers themselves are called simple machines. *Simple machines* are devices used to increase force. The inclined plane, wheel and axle, the wedge, and the cam are also simple machines. Simple machines may change the direction, speed, or distance another part moves.

There are three types of levers:
• First-class.
• Second-class.
• Third class.

Every lever has three parts:
• The effort point.
• The fulcrum, or pivot.
• The resistance point.

Force applied to the effort part of a lever causes the lever to pivot at its fulcrum. This transfers the force to the resistance part of the lever. The drawing below shows the differences in levers. Notice that in the first-class lever the effort and resistance move in opposite directions. In the second- and third-class levers, effort and resistance move in the same direction.

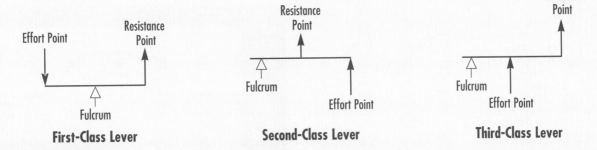

First-Class Lever   **Second-Class Lever**   **Third-Class Lever**

Apply it!

Meets NATEF Science Standards for using levers to alter force.

Identifying Levers

Look carefully at these drawings of mechanical parts. **Figs. A, B, C.** Find the effort point, fulcrum, and resistance point on each lever. Then decide which of the three types of levers each part uses.

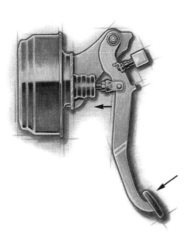

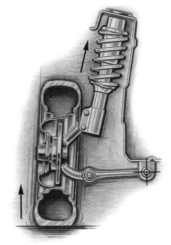

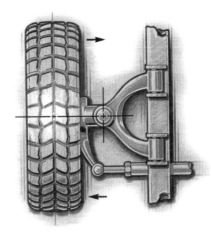

Fig. A Brake Pedal Assembly   **Fig. B MacPherson-Strut Suspension Assembly**   **Fig. C Independent Front Suspension Assembly**

TECH TIP **Hand Grease Gun.** This is not a special tool, but it is an important one! You must remember to use the hand grease gun to lubricate the grease fittings found on any ball joint or tie-rod end after your service work is completed.

Steering Ratio

Steering ratio is the number of degrees that the steering wheel must turn to turn the road wheels 1°. **Fig. 1-9.** In this figure the steering wheel turns 17.5° to turn the road wheels 1°. The steering ratio is 17.5:1.

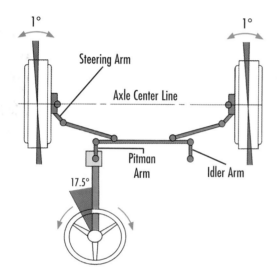

Fig. 1-9 In this example the steering ratio is 17.5:1. When the steering wheel turns 17.5°, the road wheels turn 1°. *Will a higher ratio, for example 25:1, make the road wheels easier or harder to control?*

A steering ratio can provide a mechanical advantage. It can change a small input force into a larger output force. As the driver applies a relatively small input force to the steering wheel, a much greater force is output to the road wheels. For example, a 10-pound force moves the steering wheel 17.5°. With a 17.5:1 steering ratio, a much greater force is applied to move the road wheels 1°.

Some steering gear systems provide *variable-ratio steering*. The steering ratio changes as the steering wheel turns from its straight-ahead position.

A typical change might be from a ratio of 16:1 to 13:1. The steering ratio remains constant at 16:1 for the first 40° of steering wheel movement, left or right of center. This high ratio offers better steering control for highway driving. As the steering wheel turns beyond 40°, the steering ratio decreases to 13:1. The steering wheel does not have to turn as far to turn the road wheels. This lower ratio helps the driver when cornering or parking. **Fig. 1-10.**

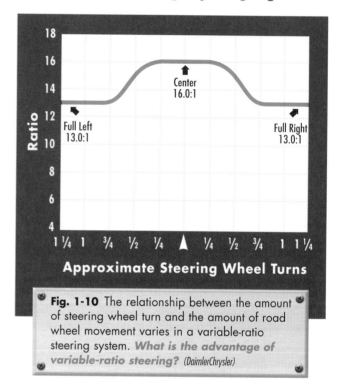

Fig. 1-10 The relationship between the amount of steering wheel turn and the amount of road wheel movement varies in a variable-ratio steering system. *What is the advantage of variable-ratio steering?* (DaimlerChrysler)

SECTION 1 KNOWLEDGE CHECK

❶ What component contains and supports the steering shaft?

❷ Why do we give a recirculating-ball steering gear that name?

❸ What are the advantages of rack-and-pinion steering?

❹ In a rack-and-pinion steering system, what moves the tie rods left or right?

❺ What effect does steering ratio have on steering effort?

Section 2

Power-Assisted Steering Systems

A power-assisted steering system uses devices to assist the driver's input force. **Fig. 1-11.** The components in a hydraulic power-assisted system include:

- A power steering pump.
- A *power steering fluid reservoir.*
- A *drive belt,* which drives the power steering pump.
- Power steering fluid.
- Hoses and tubes that connect the pump to the steering gear.
- A power-assisted steering gear assembly.

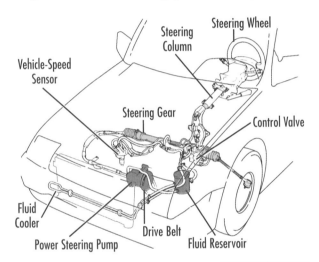

Fig. 1-11 A typical hydraulic power-assisted system using a remote-mounted fluid reservoir. *What is the purpose of the two hoses connecting the fluid reservoir?* (American Honda Motor Company)

All hydraulic power-assisted steering systems work in the same way. A hydraulic pump, the *power steering pump,* pressurizes the hydraulic fluid. Hydraulic hoses and tubes attached to the pump provide fluid to the steering gear. Pressure from the fluid is applied to a piston inside the gear housing. When the steering wheel turns, a *control valve* opens and closes fluid passages inside the gear housing. Pressurized fluid moves the piston. The piston applies force to the steering gears.

The power steering pump may have a built-in fluid reservoir, or the reservoir may mount in a remote location. A remote-mounted reservoir connects to the system with two hoses. One hose connects the reservoir to the power steering pump. This hose provides the pump with fluid. The other hose connects the reservoir to the steering gear, which allows fluid to circulate back to the reservoir.

Power Steering Pump

A power steering pump is a hydraulic pump that provides an assist to the steering system. It produces very high fluid pressures, which reduces the steering effort needed by the driver. Pressure may reach 2,000 psi [13,790 kPa]. To pressurize the fluid, the pump may use a *rotary-vane* design or a *gear-and-roller* design.

The rotary-vane type, which is very common, uses a rotor that rotates inside a *cam ring.* The rotor applies pressure to the fluid as the area between the rotor, the vane, and the cam ring becomes smaller. **Fig. 1-12.**

In most cases a belt connected to the engine's crankshaft pulley drives the power steering pump. The pump usually mounts at the front of the engine where there is belt access. **Fig. 1-13.** Some pumps, however, mount in remote locations and an electric motor drives them.

Checking the Power Steering Pump Belt Visually check the pump drive belt for excess wear. Look for cracks and missing pieces of belt material. If you find any defects, you must replace the belt.

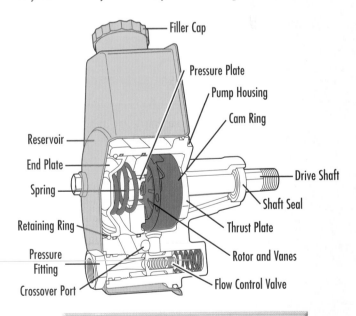

Fig. 1-12 Typical rotary-vane type power steering pump. *How does the pump pressurize fluid?* (Cadillac Division of General Motors Corporation)

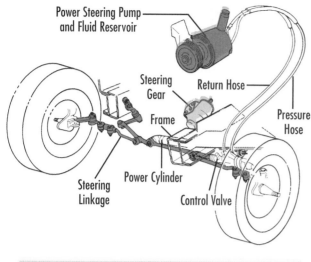

Fig. 1-13 This power steering pump has a built-in fluid reservoir. *Why do most power steering pumps mount to the front of the engine?* (Ford Motor Company)

Check that the belt fits properly on the drive pulley. Older vehicles may use a V-belt that fits into a V-groove on the pulleys. Other vehicles may use a ribbed belt that mates to a series of ribs in a drive pulley.

Check belt tension using a *belt-tension gauge*. Depending on the make and model of the vehicle, tension will vary. Newer vehicles may have an automatic belt-tensioning system.

Removing and Installing the Power Steering Pump

1. Before performing service, allow the engine to cool.

2. Remove the power steering pump drive belt. Inspect and replace if needed. Refer to the manufacturer's manual for the vehicle.

3. Disconnect the pump *suction hose.* Be careful not to spill power steering fluid. It is a good idea to place a plastic bag over the end of the hose to prevent spillage. Secure the bag with a rubber band.

4. Remove the *oil-pressure switch* from the pump, if so equipped.

5. Disconnect the pump's *pressure feed tube.* A *union bolt* or *banjo-type bolt* will likely secure this tube. A sealing washer may be under the bolt head. Be careful not to lose this part.

6. Remove the pump from the engine. Inspect the pump, pump pulley, and pump seals and gaskets. Replace if needed following any specific procedures noted in the vehicle's service manual.

Install a pump in the reverse order of the removal process.

Power Steering Fluid

Power steering systems can create very high pressures and temperatures. These systems, therefore, require a specific type of hydraulic fluid.

Some early system designs used automatic-transmission fluid. Late model vehicles need specialized power steering fluid. This is critical to the operation of power steering and to avoid part damage. Refer to the manufacturer's service manual for the fluid specification.

Checking Power Steering Fluid

1. Park the vehicle on a level surface.

2. With the engine off, inspect and check the level and condition of the fluid in the reservoir. **Fig. 1-14.**

3. If the system is hot, check that the fluid level is within the HOT level range.

4. If the system is cold, check that the fluid is within the COLD level range.

5. Start the engine and allow it to idle.

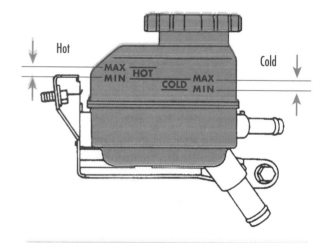

Fig. 1-14 Fluid level in reservoir will vary depending on fluid temperature. *Why should automatic-transmission fluid NOT be used in a power steering system?*

6. Turn the steering wheel lock to lock several times to boost fluid temperature.

7. Check the fluid for signs of foaming. If foaming is evident, you must bleed the power steering system of air.

Bleeding the Power Steering System

1. Raise the front wheels off the ground. With the engine off, slowly turn the steering wheel from lock to lock several times. Look for foaming in the fluid reservoir.

2. Lower the vehicle onto the ground.

3. Start the engine and allow it to idle for several minutes.

4. With the engine running, turn the steering wheel to *full lock,* either right or left, and keep it there for 2–3 seconds.

5. Turn the steering wheel to the opposite full-lock position and keep it there for 2–3 seconds.

6. Repeat this process several times.

7. Stop the engine and check the fluid in the reservoir for foaming. If foaming continues check the entire power steering system for fluid leaks. Diagnose and determine any needed repairs. Refer to the vehicle's service manual. Flush and fill power steering system and repeat the preceding steps.

8. Once you no longer see foaming, measure the fluid level in the reservoir while the engine is idling.

9. Stop the engine, wait a few minutes, and remeasure the fluid level.

Pressure-Testing the Power Steering System

Before performing a pressure test, inspect all hoses, tubes, and fittings. If any hose is soft, brittle, cracked, or shows signs of wear, it must be replaced. Any tube that is rusted, dented, cracked, or otherwise damaged must be replaced.

To determine if the pump, hoses, tubes, and steering gear are operating under proper pressures, you can perform a fluid pressure test.

1. With the engine off, install a fluid-pressure gauge on the high-pressure (feed) tube at the steering gear. Using a power rack-and-pinion system as the example, disconnect the pressure feed tube from the rack-and-pinion steering gear housing. Attach the feed tube that runs from the pump to the gear to the IN position of the gauge. **Fig. 1-15.**

2. Connect the pressure inlet port of the steering gear to the OUT position of the gauge.

3. With the engine idling and the gauge valve open, turn the steering wheel from full left to full right several times.

4. Close the valve on the gauge and note the gauge reading. You should see a pressure reading of about 925 psi [6,378 kPa], though this may vary depending on the system. Always check the manufacturer's service manual. Do not keep the pressure-gauge valve closed for more than 10 seconds. Do not allow the temperature to become too high during this test.

5. Open the valve fully.

6. Measure the fluid pressure at engine speeds of 1,000 rpm and 3,000 rpm. The difference in

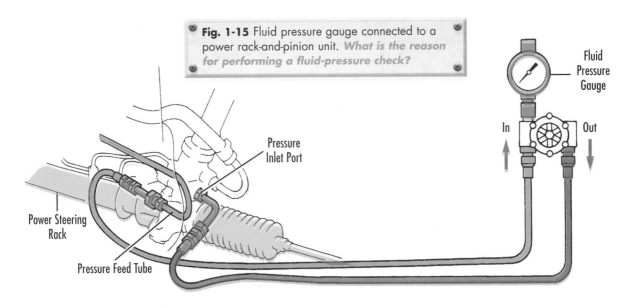

Fig. 1-15 Fluid pressure gauge connected to a power rack-and-pinion unit. *What is the reason for performing a fluid-pressure check?*

fluid pressure should be about 71 psi [490 kPa] or less.

7. With the valve fully open, turn the steering wheel to a full-lock position. Fluid pressure should be at least 925 psi [6,378 kPa]. Specific pressure specifications will vary.

8. Turn the engine off. Allow the system to cool before disconnecting the gauge. Determine any needed repairs. Refer to the vehicle's service manual.

9. Once you have reconnected the tubes to the steering gear, bleed the system and check the fluid level.

Power Rack-and-Pinion Steering Gearbox

A power rack-and-pinion steering gearbox contains an integral (built-in) power piston and cylinder. **Fig. 1-16.** It is similar in construction to a power recirculating-ball steering system, which has a power piston and cylinder built into the steering gear.

Diagnosing Problems Common power-assisted steering complaints include:
- Vibration.
- Looseness.
- Hard steering problems.

Possible causes include:
- Underinflated tires.
- Dry or worn ball joints.
- Defective or worn steering column components.
- Bad power steering pressure switch.
- Leaking steering gear assembly.
- Low power steering fluid.
- Low power steering pump pressure.
- Bent or chipped toothed rack.

Removing a Power Rack-and-Pinion Gearbox Specific procedures will vary depending on the vehicle, but the following steps provide a general outline:

1. With the vehicle raised on a hoist, disconnect the power steering hose connections at the rack-and-pinion gearbox.

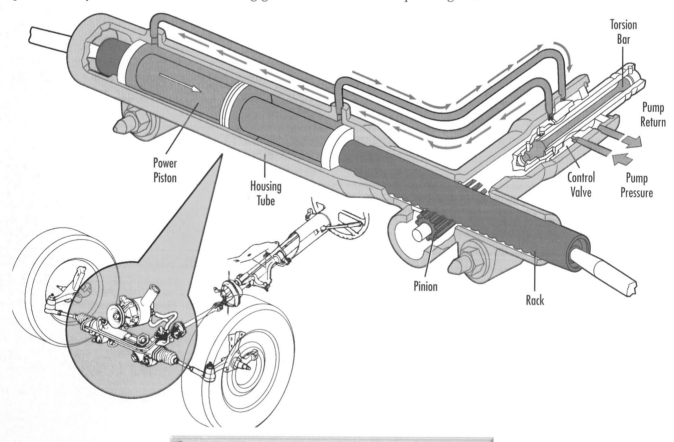

Fig. 1-16 A typical power rack-and-pinion unit. Note the integral power piston. *How is a power rack-and-pinion steering gear similar to a power recirculating-ball steering gear?* (Ford Motor Company)

Reading a Variable-Ratio Steering Graph

In a variable-ratio steering gear, the steering ratio varies as the steering wheel moves away from the straight-ahead position. Look at the graph. What does the graph indicate at Points A, B, and C in terms of variable-ratio steering?

Point A is the straight-ahead position. The steering wheel is not turned. The ratio is 16:1, which means the steering wheel must be turned 16° to pivot the front wheels 1°. At Point B, a full left turn, and at Point C, a full right turn, the ratio is 13:1, which means the steering wheel must be turned 13° to pivot the front wheels 1°. The ratio for a full right turn is equal to the ratio for a full left turn. The direction is the only difference.

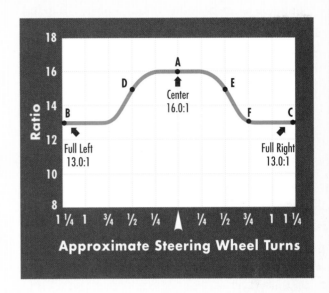

Apply it!

Meets NATEF Mathematics Standards for interpreting graphs.

❶ What is the ratio at Point F?

❷ What is the ratio at Points D and E? What do these points have in common? What is the difference?

❸ Write a brief paragraph explaining why a lower steering ratio helps in city driving during cornering or parking. Identify other situations where a low steering ratio would help.

2. Allow the power steering fluid to drain into a container.

3. Disconnect the steering sensor from the rack housing, if so equipped.

4. Disconnect the outer *tie-rod ends* from the steering arms using a tie-rod puller. This tool allows the separation of the tie-rod end from the steering arm, without damaging either component.

5. Disconnect the *steering coupler* from the pinion shaft.

6. Remove the bolts that attach the rack-and-pinion housing to the frame. **Fig. 1-17.**

7. Remove the rack unit from the vehicle. On some vehicles it may be necessary to first raise the engine and transmission or to lower the *engine cradle bracket.*

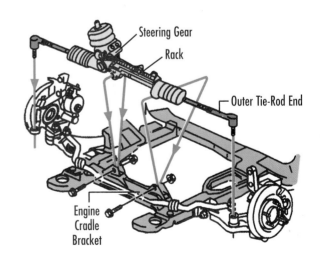

Fig. 1-17 The rack in this system mounts to the front cradle. *To remove a rack-and-pinion gearbox, what connections must you disconnect?* (Pontiac Division of General Motors Corporation)

8. Inspect the mounting bushings and mounting brackets. If cracked, oil-soaked, or otherwise damaged, replace them with new bushings and brackets.

9. Inspect tie-rod ends for faulty movement or damage to the ball-joint boot. Replace if needed.

10. Before you install the rack and pinion assembly, center the rack and install the tie rods and tie-rod ends. Adjust the tie rods to the previous length using the inner tie rod end tool. This tool allows the technician to rotate the tie rod to thread or unthread thereby adjusting the tie rods. Pre-adjusting the tie rods will provide a starting point for the *toe-in* adjustment. Front wheel alignment is necessary after any front-end steering component replacement.

Servicing a Power Rack-and-Pinion Gearbox

1. To prevent damage to the rack-and-pinion housing, mount the housing in a special rack-and-pinion vise.

2. Remove the external pressure tubes and fittings from the power rack-and-pinion.

3. Before removing the tie rods or tie-rod ends from the rack-and-pinion unit, center the rack by turning the pinion shaft left or right. Measure the length of each tie-rod assembly from a fixed point on the rack housing. Remove the outer tie-rod ends.

4. Remove the two bellows boots that protect the inner tie-rod ends from harmful contaminants. There is one bellows boot per inner tie rod. Check bellows for any damage or deterioration and replace if needed. **Fig. 1-18.**

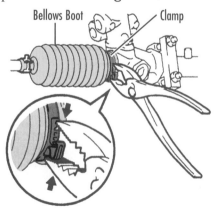

Bellows Boot   Clamp

Fig. 1-18 Remove boot from rack using pliers to disconnect the metal band clamps. *What is the function of the bellows boots?*

5. Inspect the inner tie rods for any excess movement or damage. Replace if needed following the manufacturer's procedures.

6. Remove the inner tie rods from the rack using the procedure for that make and model of vehicle.

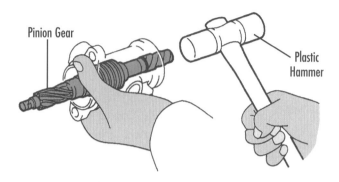

Pinion Gear   Plastic Hammer

Fig. 1-19 With the locknut removed, you can remove the pinion shaft. Inspect all bearings and oil seals and replace if needed. *Why is a plastic hammer used to remove the pinion shaft?*

7. Remove pinion bearing cap at bottom of pinion gear upper housing. Remove locknut from pinion shaft. Remove the gear from the gear housing using a plastic hammer to avoid damaging the control valve or piston shaft. **Fig. 1-19.** Remove the pinion gear upper housing.

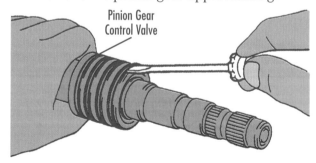

Pinion Gear Control Valve

Fig. 1-20 Inspect the Teflon® rings on the pinion-gear control valve housing and replace them if needed. *Why should you check the pinion and housing bore for galling or burrs?*

8. Remove the control valve assembly from the housing and inspect. Check the housing bore for signs of wear such as galling and burrs. Replace those components showing signs of excessive wear. **Fig. 1-20.**

9. Remove the rack from the rack housing. If the rack is placed in a vise, avoid damaging the rack. Do not close the vise too tightly. Check the rack

for straightness using a dial indicator. Carefully rotate the rack and observe the reading on the dial indicator. Maximum runout should be .0118 inch [.30 mm]. **Fig. 1-21.**

10. Inspect the rack teeth for burrs or other damage. If slight burrs are present, scrape them off the teeth. If there are gouged or chipped teeth, you must replace the rack.

11. When reassembling the rack-and-pinion assembly, you must set the total preload of the rack guide spring cap using the torque specification in the vehicle's service manual.

12. Center the rack and install the tie rods and tie-rod ends. Adjust the tie rods to the previous length. Pre-adjusting the tie rods will provide a starting point for the toe-in adjustment. Front wheel alignment is necessary after any front end steering component replacement.

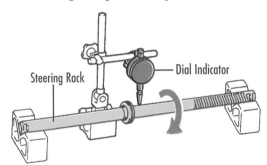

Fig. 1-21 With the rack removed, check for straightness. You must replace a bent rack. *Why is it important to secure the rack in a vise properly?*

Power Steering Pressure Switch

Most vehicles equipped with power steering have a *pressure switch* installed on the high-pressure side of the system. The switch monitors system pressure and supplies information to the engine *powertrain control module (PCM)*. When the switch senses a high pressure load, the PCM may slightly raise engine speed to increase pump pressure. **Fig. 1-22.**

Variable-Assist Power Steering

Many vehicles have speed-sensitive power steering. This is a variable-assist system. **Variable-assist power steering** is power steering that uses electronic controls to determine how much power assist the steering needs. **Fig. 1-23.** Note the electronic

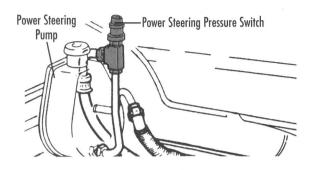

Fig. 1-22 The pressure switch mounts to the high-pressure line near the pump. *What happens when the PCM senses a higher power steering pressure load?* (Pontiac Division of General Motors)

power steering controller, the solenoid valve at the rack-and-pinion, and the steering-angle sensor at the steering column. Some vehicles may have the solenoid built into the power steering pump. Electronic variable-assist power works with either rack-and-pinion steering or recirculating-ball steering.

When the vehicle is moving at less than 20 mph [32 kph], the solenoid valve keeps the pressure orifices open, providing full power assist for low-speed and parking maneuvers. When vehicle speed increases, the solenoid valve begins to restrict fluid flow to the steering gear. The result is a slight increase in steering effort and improved road feel.

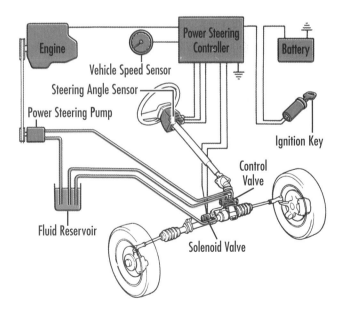

Fig. 1-23 An example of a variable-assist power steering system in a vehicle equipped with rack-and-pinion steering. *What component provides the variable rate of power assist and how?* (Ford Motor Company)

Learning About New Systems

Today's highly trained technician cannot make mistakes when servicing vehicles. A technician must not guess or use outdated information to service any system, especially a new or unfamiliar system.

In the last decade, air bags became an important part of vehicle safety. When air bags became a service item, even experienced technicians had to read and study to learn how to service them. They had learned that all new systems need careful study through reading. Successful technicians know they cannot rely on guesswork.

Your textbook has pointed out precautions you need to take when servicing air bags. However, manufacturer's service manuals will give you the specific information that you will need to take care of their air bag systems properly. Besides the basic precautions, you will learn how to install and service air bags so they will function correctly.

Apply it!

① Locate service manuals from two different manufacturers. Choose a vehicle with an air bag system from each manual.
② Read the section about air bags in each manual.
③ Take notes on specific information for each manufacturer's system.
④ Compare the information you found.

Whenever there is a need for sudden or severe steering action, the solenoid valve opens the orifices, providing full power assist.

Vehicles equipped with a variable-assist power steering system have a separate diagnostic connector. The connector allows the technician to retrieve diagnostic codes from this system.

Fig. 1-24 A typical electronic rack-and-pinion power steering gear. The electric motor consists of the hollow armature and the field coil. *What provides the input signal to the PCM?* (TRW, Inc.)

Electronic Rack-and-Pinion Power Steering

Electronic rack-and-pinion power steering uses an electric motor to provide the power assist. It does not use a hydraulic system. Instead of using a flat rack with straight teeth as in a standard *rack*, this type of steering gear uses a *helical-gear* rack driven by a fast-acting electric motor. **Fig. 1-24.**

A torque sensor mounted on the pinion shaft measures input steering torque. As torque is applied, a signal is sent to the PCM. The PCM sends a signal to the electric motor, providing the power assist.

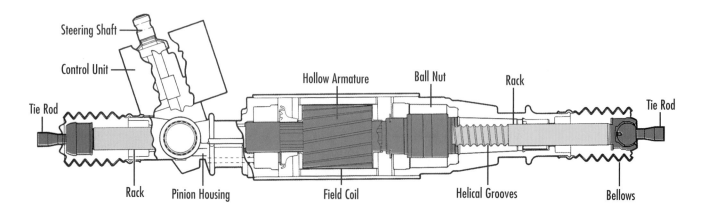

Steering-Wheel Air Bags

Many vehicles produced since the late 1980s have a steering-wheel *air bag*. The air bag is mounted in the center of the steering wheel. **Fig. 1-25.** If the vehicle hits an object with enough force, a fabric quickly inflates, using a built-in gas-charged canister. This inflated fabric protects the driver from injury caused by contact with the steering wheel, dash, or windshield.

Front *impact sensors* detect the force and direction of any impact. The sensors connect to an air-bag control module. This module is calibrated for a specific range of impact force. If the impact is strong enough, the control module triggers the air bag.

Accidental discharge of an air bag can cause injury to anyone close to the air bag. The noise a deploying air bag creates can damage hearing.

Never place an undeployed air bag face down on a solid surface, such as a worktable. If accidentally deployed, the air bag expands with great force. Never carry or handle an air bag carelessly. Always carry an air bag with the face (the interior trim side) pointing away from the body.

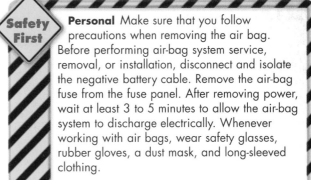

Safety First

Personal Make sure that you follow precautions when removing the air bag. Before performing air-bag system service, removal, or installation, disconnect and isolate the negative battery cable. Remove the air-bag fuse from the fuse panel. After removing power, wait at least 3 to 5 minutes to allow the air-bag system to discharge electrically. Whenever working with air bags, wear safety glasses, rubber gloves, a dust mask, and long-sleeved clothing.

When cleaning a vehicle after an air-bag deployment, always wear safety glasses, rubber gloves, a dust mask, and long-sleeved clothing. Sodium hydroxide powder is a residue of air-bag deployment. The powder can irritate skin. Flush any exposed areas with cool water. If you experience nasal or throat soreness, get some fresh air. If the soreness continues, get medical attention.

Always store air bags in a cool, dry location, away from excessive heat and static electrical activity. Store the bags facing up. Air bags are not reusable. You must dispose of them properly.

Steering Column

Clock Spring

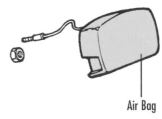

Air Bag

Fig 1-25 Placement of the air-bag module within the steering wheel. *Why are air bags mounted in the steering wheel?* (Moog Automotive, Inc.)

SECTION 2 KNOWLEDGE CHECK

❶ In a hydraulic power steering system, what produces the hydraulic pressure needed to assist the steering effort?

❷ When bleeding the system, why must you turn the steering from full left to full right several times?

❸ What procedure do you follow to check a rack for straightness?

❹ What steering systems will work with electronic variable-assist?

❺ Describe the operation of an air-bag system in the event of an impact.

CHAPTER 1 REVIEW

Key Points

Meets the following NATEF Standards for Suspension & Steering: steering systems diagnosis and repair.

- Power-assisted steering systems may be hydraulic or electronic.
- For driver comfort and convenience, many vehicles have steering columns with tilt and telescopic mechanisms. For driver safety, some steering columns are designed to collapse in a collision.
- There are two major types of steering gears.
- To connect the steering gears to the road wheels, recirculating-ball steering gear designs use a series of linkage parts. These include the pitman arm, center link, idler arm, tie rods, tie-rod ends, and steering arms.
- There are two types of rack-and-pinion link designs.
- The steering ratio is the number of degrees that the steering wheel needs to turn so that the road wheels turn 1°. The steering ratio affects steering effort.
- Power steering systems require a specific type of hydraulic fluid. Use of the proper power steering fluid allows proper system operation and prevents damage.
- The power steering pressure switch monitors system pressure and supplies information to the engine electronic control module.

Review Questions

1. Name the major components of a steering system. Which components are part of the driver input portion of the steering system?

2. Explain the differences between recirculating-ball and rack-and-pinion steering gears.

3. How does variable-ratio steering work? What advantages does variable-ratio steering offer over steering with a constant ratio?

4. Explain how to perform a pressure test on a power steering system. What is the purpose of a pressure test?

5. What is the procedure for removing a rack-and-pinion steering system?

6. How do you service a rack-and-pinion steering system?

7. What safety precautions should you observe when working with steering-wheel mounted air bags?

8. **Critical Thinking** When turning the steering from full left to full right, the steering system hesitates and binds. Where would you begin to look for problems?

9. **Critical Thinking** Foaming of power steering fluid usually means air is entering the system. What indicators are important when you are looking for air leaks?

TECHNOLOGY FORECAST
FOR AUTOMOTIVE EXCELLENCE

Electric Steering Drives the Future

Drivers may soon be able to steer their vehicles using an electric motor. This motor will be connected to the steering wheel by wires and to the car's wheels by rods. Today's steering column and hydraulic fluid will no longer be needed.

A computer will know which way the steering wheel is being turned. It will then move the wheels in the same direction. Today, only the front wheels move. The future may bring back four-wheel steering systems. These systems work by turning the back wheels slightly along with the front wheels. This makes steering easier and faster.

Benefits of electric steering include less weight, improved gas mileage, and less noise. Better comfort, feel, and stability are added benefits. Because electric steering runs without help from the engine, it could also be safer. If there is engine trouble, the vehicle will still turn easily.

AUTOMOTIVE EXCELLENCE
TEST PREP

Answering the following practice questions will help you prepare for the ASE certification tests.

1. In a steering shaft assembly, the upper shaft connects to the lower shaft by means of a:
   - **ⓐ** Union bolt.
   - **ⓑ** Banjo-type bolt.
   - **ⓒ** Intermediate coupler.
   - **ⓓ** None of the above.

2. Technician A says all recirculating-ball steering gears are manual units, while all rack-and-pinion steering gears are power assisted. Technician B says rack-and-pinion steering gears do not need power assist. Who is correct?
   - **ⓐ** Technician A.
   - **ⓑ** Technician B.
   - **ⓒ** Both Technician A and Technician B.
   - **ⓓ** Neither Technician A nor Technician B.

3. In a recirculating-ball steering system, Technician A says the pitman arm is responsible for providing the steering output from the steering gear to the steering linkage. Technician B says the pitman arm is part of the suspension system. Who is correct?
   - **ⓐ** Technician A.
   - **ⓑ** Technician B.
   - **ⓒ** Both Technician A and Technician B.
   - **ⓓ** Neither Technician A nor Technician B.

4. Technician A says a center link is part of a parallelogram steering linkage system. Technician B says a center link is responsible for controlling vehicle body sway during cornering and high crosswinds. Who is correct?
   - **ⓐ** Technician A.
   - **ⓑ** Technician B.
   - **ⓒ** Both Technician A and Technician B.
   - **ⓓ** Neither Technician A nor Technician B.

5. Variable-ratio steering depends upon:
   - **ⓐ** Vehicle speed.
   - **ⓑ** Condition of the road's surface.
   - **ⓒ** Weight of the vehicle.
   - **ⓓ** All of the above.

6. Technician A says power steering fluid operates at a very low pressure at all times. Technician B says both power steering gear hoses operate at the same pressure. Who is correct?
   - **ⓐ** Technician A.
   - **ⓑ** Technician B.
   - **ⓒ** Both Technician A and Technician B.
   - **ⓓ** Neither Technician A nor Technician B.

7. Technician A says you can use any quality hydraulic fluid in a power steering system. Technician B says you can usually only use special power steering fluid. Who is correct?
   - **ⓐ** Technician A.
   - **ⓑ** Technician B.
   - **ⓒ** Both Technician A and Technician B.
   - **ⓓ** Neither Technician A nor Technician B.

8. Technician A says the inner tie rods of a rack-and-pinion steering gear assembly are rigid extensions of the steering rack. Technician B says the inner tie rods feature a ball pivot at their inboard connection. Who is correct?
   - **ⓐ** Technician A.
   - **ⓑ** Technician B.
   - **ⓒ** Both Technician A and Technician B.
   - **ⓓ** Neither Technician A nor Technician B.

9. Technician A claims whenever you replace a rack-and-pinion tie rod, you will have to check and possibly adjust the front wheel toe-in. Technician B says the tie rod will have no effect on the wheel toe-in angle. Who is correct?
   - **ⓐ** Technician A.
   - **ⓑ** Technician B.
   - **ⓒ** Both Technician A and Technician B.
   - **ⓓ** Neither Technician A nor Technician B.

10. Technician A says the rubber bellows boots on a rack-and-pinion steering gear unit protect the inner sockets from dirt and moisture. Technician B says you should inspect the bellows boots regularly. Who is correct?
    - **ⓐ** Technician A.
    - **ⓑ** Technician B.
    - **ⓒ** Both Technician A and Technician B.
    - **ⓓ** Neither Technician A nor Technician B.

CHAPTER 2

Diagnosing & Repairing Suspension Systems

You'll Be Able To:

- Identify the different front suspension designs.
- Describe the types of springs used in front and rear suspension systems.
- Identify and service the parts of a MacPherson strut.
- Describe the three types of rear suspension designs.

Terms to Know:

anti-sway bar

ball joint

coil spring

control arm

leaf spring

torsion bar

The Problem

The Rijos have just returned from a short vacation trip using Juan Rijo's sport utility vehicle. They decided to take the scenic route on back roads rather than the direct route on expressways. During the trip, Juan believes his sport utility vehicle was leaning too much in turns. He also feels it was difficult to control on bumpy roads.

Once home, Juan brings his vehicle to your service center and describes the problems he noticed during the trip. What areas would you suspect as the cause of his problems?

Your Challenge

As the service technician, you need to find answers to these questions:

1. Did those problems exist before the trip?

2. Could either problem involve multiple areas, such as tire inflation and shock absorbers?

3. Could worn ball joints cause these problems?

Section 1

Suspension System Parts

The suspension system is located between the axles and the body or frame of a vehicle. The job of the suspension system is to:

- Support the vehicle.
- Provide a cushion effect between the body and the road.
- Allow the tires to maintain solid contact with the road.
- Maintain wheel alignment.

The suspension system provides for improved vehicle handling and for cushioning road shock. The suspension system controls the stability of a vehicle during lane changes, straight-line driving, turning, and braking.

Compression or *jounce* occurs when the suspension system moves closer to the body. During braking, the front of the vehicle noses down toward the road. When turning, the side of the body opposite the direction of a turn moves down. The result is that both sides of the body are not at the same height from the ground.

Safety First

Personal When servicing any suspension system, always be aware of the pressures and forces that exist in a compressed spring. Never attempt to remove any suspension component without taking the required steps given in the vehicle's service manual. Failure to do so could cause serious injury.

Rebound occurs as the suspension moves away from the body. After completing a stop and nosing down or after completing a turn, the body returns to a normal level position.

The basic parts of a suspension system are springs, shock absorbers, anti-sway bars, control arms, ball joints, axles, and wheel bearings. Other suspension system parts will vary, depending on the vehicle. **Fig. 2-1.** The functions of the basic parts are as follows:

- Springs support the body.
- Shock absorbers control spring action by damping the springs. This helps to control how fast the springs compress and rebound. Damping also helps to prevent the springs from "oscillating" or vibrating.

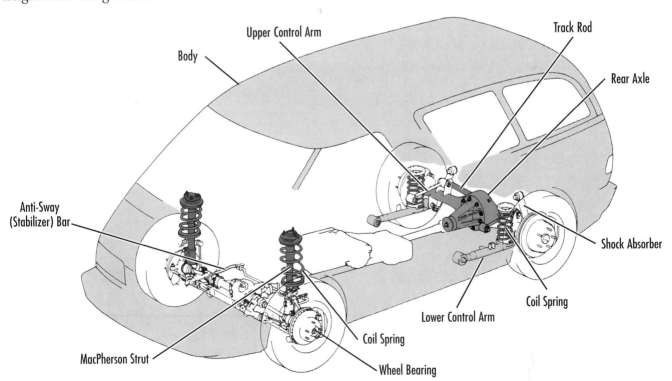

Fig. 2-1 Main parts of a vehicle suspension system. Suspension systems vary depending on the vehicle. This system has MacPherson struts in front and coil springs in the rear. *What are the functions of the suspension system?*

- Anti-sway bars are a type of torsion bar that connects the lower suspension to both sides of the frame or body. These bars prevent the body from "leaning" too far during turns.
- Control arms position the wheels on the frame or body and maintain the wheel alignment to the frame or body.
- Ball joints enable suspension parts to pivot and rotate.
- Axles provide a positioning point for the wheels. They serve as the center point for wheel position and rotation.
- Wheel bearings enable the tires and wheels at the end of the axle to rotate smoothly, freely, and safely at high speed.

Springs

Automotive suspension systems commonly use four types of springs:
- Coil springs.
- Leaf springs.
- Torsion bars.
- Air springs.

Springs are used to support *sprung weight.* This includes the weight of the vehicle body and/or frame, engine, transmission, cargo, and passengers. *Unsprung weight* refers to the weight that the springs do not support. This includes wheels, tires, brakes, lower control arms, and drive axles.

Vehicle makers try to make unsprung weight as light as possible. This is one reason why many vehicles come equipped with lighter aluminum wheels. As unsprung weight increases, so does the roughness of the ride.

Coil Springs   A length of spring-steel rod wound into a coil is known as a **coil spring**. The rod can be the same diameter from end to end, giving the spring a constant spring rate. Or, the rod may be tapered, giving the spring a variable spring rate. **Fig. 2-2.**

Spring rate and action are determined by the spring rod thickness, the number of coils, and the spacing between the coils. A variable-rate coil may compress easily initially. But, as it compresses, it becomes more difficult to squeeze together.

Leaf Springs   Springs that consist of single or multiple-spring steel bands (leaf) are called **leaf springs**. Some springs use a high-strength

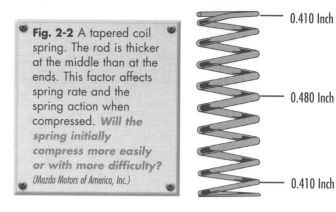

Fig. 2-2 A tapered coil spring. The rod is thicker at the middle than at the ends. This factor affects spring rate and the spring action when compressed. *Will the spring initially compress more easily or with more difficulty?* (Mazda Motors of America, Inc.)

0.410 Inch

0.480 Inch

0.410 Inch

composite material. Leaf springs are formed in an arc shape. The thickness of the band(s) and the length of the spring dictate the spring rate.

A *monoleaf spring* has one leaf. A *multi-leaf spring* has two or more bands stacked together. A leaf spring mounts to the body at each end and attaches to the axle at the center of the leaf. As vehicle weight bears down on the spring, the weight acts against the arc of the leaf.

A longitudinally mounted leaf spring positions front-to-rear. A transverse mounted leaf spring positions side-to-side parallel to the axle. If a vehicle uses longitudinal leaves, it requires one spring per side (one left and one right). If a vehicle uses a transverse leaf spring, it needs only one.

Torsion Bar   Once used only on early domestic and import vehicles, torsion bar suspension is commonly used on a variety of vehicles. A **torsion bar** is a steel rod that twists to provide spring action. **Fig. 2-3.** One end of the bar attaches to the vehicle frame, while the opposite end attaches to the lower control arm.

Torsion bars are positioned either longitudinally or transversely. Longitudinal positioning attaches the

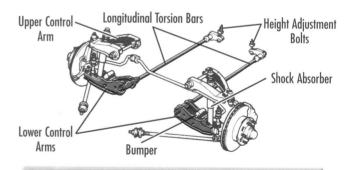

Upper Control Arm

Longitudinal Torsion Bars

Height Adjustment Bolts

Shock Absorber

Lower Control Arms

Bumper

Fig. 2-3 A front suspension system using torsion bar springs. Torsion bars mount transversely (side-to-side) or, as shown here, longitudinally (front-to-rear). *How does a torsion bar provide spring action?*

front end of the bar to the lower control arm and the rear of the bar to the frame. Transverse positioning connects the frame to the lower control arms.

Air Springs An *air spring* consists of a bag or cylinder filled with compressed air. **Fig. 2-4.** Air springs are used primarily in *automatic level control suspension systems.*

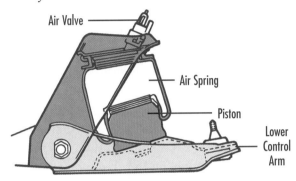

(a) Normal Ride Height

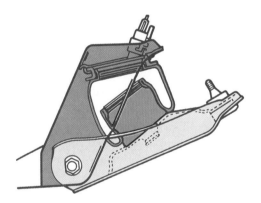

(b) Jounce (Compressed)

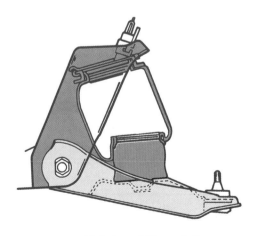

(c) Rebound (Extended)

Fig. 2-4 An air spring mounts between the frame and a suspension member. *What are the advantages of an air spring?* (Ford Motor Company)

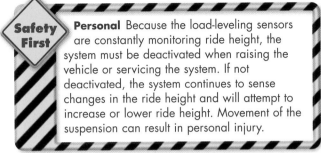

Safety First **Personal** Because the load-leveling sensors are constantly monitoring ride height, the system must be deactivated when raising the vehicle or servicing the system. If not deactivated, the system continues to sense changes in the ride height and will attempt to increase or lower ride height. Movement of the suspension can result in personal injury.

Sensors in an automatic level-control suspension system detect changes in the vehicle's ride height, or ground clearance. As a vehicle takes on weight, a height sensor detects the change in ride height. An air compressor is activated to provide additional air to the air springs. The added air raises the vehicle back to normal ride height. Upon removal of the weight, the system allows the springs to bleed-off enough air to return the vehicle to its original ride height.

Depending on the system, the air springs can also be automatically adjusted to raise and lower ride height based on the vehicle's speed. If the vehicle is stationary or traveling at slow speeds, the ride height is set to a level providing a comfortable ride and more normal ground clearance. At higher speeds, the system reduces the vehicle's ground clearance. This provides more "aerodynamic efficiency."

Shock Absorbers

Vehicles have shock absorbers to dampen spring action. This allows the spring to compress and rebound (unload) at a controlled rate. Damping prevents the suspension from continuing to load/unload uncontrollably. The two most common types of shock absorbers are "traditional" *shock absorbers* and "*struts.*"

A traditional shock absorber is a hydraulic cylinder that mounts independently of any load-carrying part. **Fig. 2-5.** The springs support the vehicle without the shock absorber affecting vehicle load. As the suspension reacts to vehicle motion, a piston moves up and down inside the shock absorber. As the shock absorber compresses and rebounds, hydraulic fluid is forced through orifices of a control valve. The valve orifices force the piston to move up and down at a controlled rate, giving the shock absorber its damping characteristics.

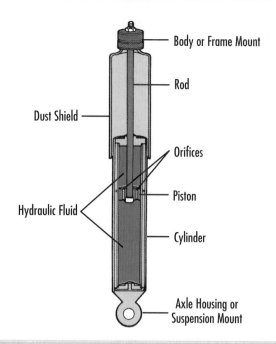

Body or Frame Mount

Rod

Dust Shield

Orifices

Piston

Hydraulic Fluid

Cylinder

Axle Housing or
Suspension Mount

Fig. 2-5 A "traditional" shock absorber. Internal valving allows the shock absorber to absorb and control spring action. *How do shock absorbers dampen spring action?* (Ford Motor Company)

Some shock absorbers also use compressed nitrogen gas. As the shock absorber cycles, the gas reduces foaming or aeration of the hydraulic fluid. Gas-charged shock absorbers have become very common.

Shock absorbers mount with the top attached to the frame or upper control arm and the lower end attached to the lower control arm or axle housing.

A strut serves as a key part of the suspension. It is located between the lower control arm and the body. Struts are also shock absorbers. A traditional shock absorber and a strut both provide damping control for the springs.

Several "special" types of shock absorbers are used. They include *spring-assisted shock absorbers, adjustable shock absorbers,* and *air-assisted shock absorbers.*

Spring-Assisted Shock Absorbers Spring-assisted shock absorbers use a coil spring installed over the shock absorber body. **Fig. 2-6(a).** For applications where added cargo load is present, the

TECH TIP **Spring-Assisted and Air-Assisted Shock Absorbers.** Spring-assisted and air-assisted shock absorbers are available to help support heavier loads. Remember, most shock absorber mounts cannot support vehicle weight. If a vehicle needs additional load carrying ability, the best solution is to install heavier or variable-rate springs.

assist spring provides additional load-carrying assistance. The shock absorber serves as a hydraulic damper for the spring action. The assist spring on the shock absorber also provides additional support for the body and frame.

Adjustable Shock Absorbers Adjustable shock absorbers provide a convenient way to tailor the handling of a vehicle. An adjustable shock absorber has a variable setting for the internal valving. **Fig. 2-6(b).** The shock absorber can be set for soft to firm damping. Adjustable shock absorbers allow adjustment for damping only and have no effect on load carrying.

Some adjustable shock absorbers have an external dial or adjustment screw. Others must be disconnected to set the adjustment. Once disconnected, they must be fully compressed and rotated in relation to the internal piston.

Air-Assisted Shock Absorbers Air-assisted shock absorbers are traditional shock absorbers with an air chamber between the inner and outer shock absorber tubes. **Fig. 2-6(c).** The chamber can be inflated with compressed air to provide additional load-carrying ability. This is helpful when the rear cargo area is heavily loaded. It is also helpful when the vehicle is used for towing. The compressed air acts as an "assist" to the normal operation of the shock absorber.

Inspecting Shock Absorbers When checking the operation and condition of shock absorbers, do the following:

• Perform a "bounce" test. Push down on the front end of the vehicle firmly and release. The front end of the vehicle should rise once and immediately settle. If the front end bounces more than once, the shock absorber is worn and should be replaced.

• Visually inspect the shock absorber for leakage. A light film of oil near the top seal is normal. If excessive leakage is seen, replace the shock absorber.

• Look for damage, such as dings and dents. If evident, replace the shock absorber.

• Inspect the shock absorber piston rod for scratches, gouges, corrosion, and bending. If there is damage, replace the shock absorber.

Servicing Shock Absorbers If the shock absorber appears faulty when you check for operation and condition, replace it. Use the procedures indicated in the specific vehicle's service manual.

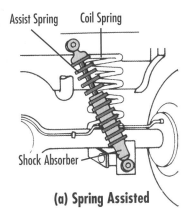

(a) Spring Assisted

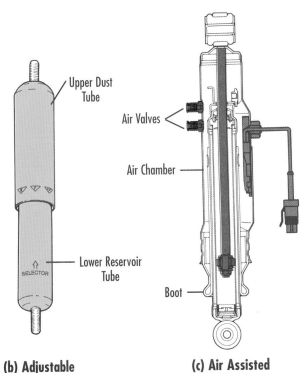

(b) Adjustable   **(c) Air Assisted**

Fig. 2-6 (a) Coil-over spring-assisted shock absorber. **(b)** Adjustable shock absorber. **(c)** Air-assisted shock absorber. *Why should spring- and air-assisted shock absorbers be used only for load assistance?*

Following is a general procedure for removing and servicing the basic shock absorber:

1. While the weight of the vehicle is on the suspension, remove the upper mount nut. This is where the top of the shock absorber attaches to a mount on the frame or body. To remove the nut, use a special hex tool to hold the shock absorber piston rod.

2. Disconnect any special air hose connectors, if applicable.

3. Raise the vehicle and support it on jack stands.

4. Remove the two bottom bolts (front) or bottom nut (rear) and washers where the shock absorber attaches to the lower control arm or axle.

5. Remove the shock absorber from the vehicle.

6. Position the new shock absorber on the vehicle.

7. Install the two bottom bolts (front) or bottom nut (rear) and washers where the shock absorber attaches to the lower control arm or axle.

8. Align the shock absorber piston rod with the upper mount. If equipped, cut the restraining wire to allow the rod to expand to the full length.

9. Install the rod in the upper mount by lowering the vehicle. Using a special hex tool to hold the rod, install the nut where the top of the shock absorber attaches to the upper mount.

10. Tighten the nuts and bolts as specified in the vehicle's service manual.

Anti-Sway Bars

The anti-sway bar is also called a stabilizer bar, sway bar, or anti-roll bar. An **anti-sway bar** is a device that helps reduce body sway. On a front suspension, the ends of an anti-sway bar connect to the two lower control arms. The center portion of the bar mounts to both sides of the vehicle frame. **Fig. 2-7.** When one front wheel moves up, the bar twists in an attempt to pick up the opposite wheel. This action helps to balance the force between the two sides of the vehicle, reducing body "roll."

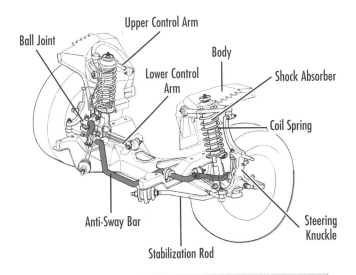

Fig. 2-7 An anti-sway bar mounts laterally and connects the suspension to the frame. *How does an anti-sway bar improve handling?* (Ford Motor Company)

How Stress Affects Springs

Applying a force to a spring creates a stress. There are three types of stress: tension, compression, and shear.

- *Tension* occurs when two forces act on opposite ends of an object and are directed away from each other. This stretches the object.
- *Compression* occurs when two forces act on opposite ends of an object but are directed toward each other. This squeezes the object.
- *Shear* occurs when parallel forces act in opposite directions on opposite sides of an object. This twists the object.

When a coil spring in the suspension system reacts to the vehicle going over a "speed bump," the coil is compressed as the wheels move over the bump (compression stress). As the wheels move down the bump, the wheels and the frame begin to move in opposite directions (tension stress). If these two stresses are controlled, the ride is smooth.

When tension stress is applied to a spring, the spring stretches. When the force is removed, the spring returns to its original shape. This property is called elasticity. As more force is applied to the spring, the spring becomes still longer.

The scientist Robert Hooke observed that the amount of stretch depended on three properties:
- The amount of force applied to the object.
- The type of material used to make the object. Metals differ in their elasticity and stretch at different rates.
- The dimensions of the object. The longer a spring is, the more it stretches. The thicker the coil is, the less it stretches.

Applying more force to any object will eventually stretch it to the point where it cannot return to its original shape.

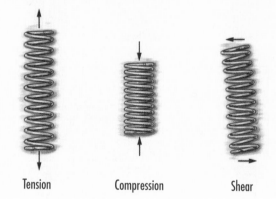

Tension   Compression   Shear

Apply it!

Testing Hooke's Idea

Meets NATEF Science Standards for explaining stress forces and the elasticity of springs.

Materials and Equipment
- Small return spring
- Spring scale
- Ruler
- Set of weights (Flat washers can be used.)

❶ Place the spring on the end of the spring scale.

❷ Measure the length of the spring.

❸ Add weight to the spring. Measure the length of the spring again.

❹ Subtract the new length of the spring from its starting length. The result is the stretch.

❺ Divide the weight on the spring by the amount of stretch. The result is the spring constant.

❻ Repeat the procedure two more times with different weights. Determine and record the spring constant. Do not exceed the elastic limit of your spring.

❼ Did the spring stretch more as you increased the amount of weight?

❽ Did the measured spring constant also increase? Explain.

The anti-sway bar mounts to the frame with rubber or urethane *isolator bushings* and retainers. **Fig. 2-8.** These allow the bar to pivot, while reducing the transfer of road noise or vibration to the body. If handling problems occur, always check these bushings. Also check the end link bushings for wear and damage.

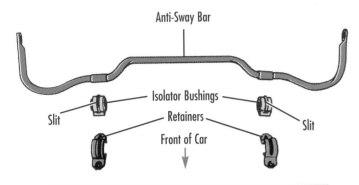

Fig. 2-8 An anti-sway bar is mounted with isolator bushings. *What function do the isolator bushings perform?* (DaimlerChrysler)

Servicing Anti-Sway Bars An anti-sway bar does not support vehicle weight. Initially, the bar may be installed with the vehicle on jack stands and the wheels hanging. Do not fully tighten the bushing mounts or end links with the suspension extended. Tighten the bushing mounts or end links only when the weight of the vehicle is on the suspension. This will prevent binding of the bar bushings.

Inspect the anti-sway bars as follows:
• Inspect the bushings that mount the bar to the frame. If these bushings appear loose, worn, damaged, or oil-soaked, they should be replaced.
• Inspect the end links and the end-link bushings. Look for signs of wear, disintegration, and missing bushings. Inspect the end link for heavy corrosion and bending. Replace any part showing signs of wear or damage.

Refer to the vehicle's service manual for specific details and procedures.

Control Arms

A **control arm** is a device that provides the connection points of the suspension for up/down pivoting movement. Because they look like the letter A when viewed from above, they are called A-type control arms. Technicians also refer to A-type control arms as *wishbones*.

The inboard end of the control arm is wider, with two pivoting connections (one at the front and one at the rear). Here the control arm attaches to the vehicle frame. The outer end tapers to a point where the control arm attaches to the spindle with a ball joint.

The wider inboard end of the control arm prevents the suspension from deflecting in or out during braking and hard cornering. The A-shape of the control arm acts as a brace. This prevents wheel alignment changes during braking or cornering.

A-type control arms may be of different lengths, where the lower arm is longer than the upper arm. This is a *short-arm/long-arm (SLA) system*. The top of the coil spring rests in a fixed spring seat at the frame. The bottom of the spring rests on the lower control arm.

Some vehicles use a lower control arm that looks like a straight beam, instead of having a shape like an A brace. **Fig. 2-9.** This single-beam style provides only one bushing attachment point at the frame.

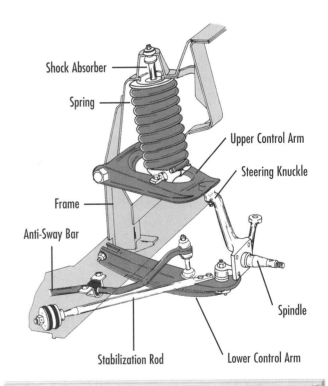

Fig. 2-9 This is an SLA design. The lower control arm is a single-beam type. It is longer than the upper control arm. Normally, the coil spring mounts between the lower control arm and the frame. When the coil spring mounts between the upper control arm and the frame as shown, technicians refer to this as an SLA-2 design. *Why do some front suspensions require the use of a stabilization rod?*

The single-beam type control arm still allows the suspension travel to take place when the wheel wants to move up or down. However, this type of control arm would deflect under braking and hard cornering. As the brakes are applied, the wheel slows, but the body and frame want to keep moving forward. This creates stress at the control arms. If the control arms move backwards in relation to the frame, tire wear and unsafe handling result.

When a single-beam-type lower control arm is used, an additional stabilization rod is added. Without proper bracing of the control arms, the wheels would move slightly forward and rearwards.

Ball Joints

Ball joints are also called *ball-and-socket joints*. A **ball joint** is a lubricated attachment that connects two suspension parts to allow pivoting movement. An automotive suspension uses ball joints at the connection points between the control arms and the wheel spindle/steering knuckle assemblies. The tapered stud provides a secure press fit at the control-arm mounting point. **Fig. 2-10.**

Ball joints allow the suspension to move up and down. They also allow the steering knuckle to pivot when the wheels turn left or right.

Ball joints must be checked for looseness and wear. Some ball joints have a built-in wear indicator. The indicator shows when the ball and internal ball socket have worn beyond the point of useable life.

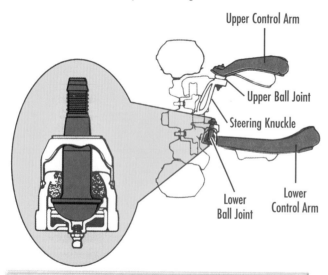

Fig. 2-10 A ball joint provides the connection between the steering knuckle and the control arm. *Why is the ball joint in the form of a ball?* (General Motors Corporation)

> **TECH TIP** **Lubricating Ball Joints.** Each ball joint has a rubber gasket or boot which holds lubricant within the joint. Make sure the ball joint is properly lubricated. A dry ball joint will wear excessively. Also, it will make a squeaking or grinding noise when the wheels are turned and during up and down movement of the suspension. Too much lubrication can rupture the gasket. A ruptured gasket can allow lubricant to escape and the ball joint to become dry.

Inspect ball joints by first checking for looseness and play. Check both the ball socket and the location where the tapered stud attaches to the steering knuckle.

- Raise the vehicle on stands and wiggle the wheel inward and outward at the top and bottom of the wheel. If excessive play or motion is found, the joint is worn. See the vehicle service manual for a specific measurement of allowable play.
- Always inspect the condition of the rubber dust cover. A damaged cover will allow dirt and moisture to enter the joint.
- Replace worn or damaged ball joints using the procedures specified in the vehicle service manual. Refer also to the Safety First feature on page SS-31. The procedures for replacing the upper and lower ball joints are similar on short-arm/long-arm suspension systems.
- When disconnecting ball joints from the spindle, be aware that ball joint studs have a taper to achieve a press fit. Disconnect the ball joint stud from the spindle using a special ball joint press tool, or puller.
- Install the new ball joint in the spindle. Tighten the nut to the torque specified in the vehicle service manual. If it is not sealed, lubricate the ball joint with a grease gun. If excessive ball joint wear was noted, check wheel alignment.

Axles

An axle is a support suspension member on which the wheels are mounted. A *live axle* supports part of the vehicle weight, and drives the wheels connected to the axle shafts. If the wheels on the live axle can pivot for steering, the axle is called a steerable live axle.

A *dead axle* is a non-drive axle. It carries a portion of the vehicle weight and the non-driven wheels. Since a dead axle is a non-drive axle, the rear axle of a front-wheel drive vehicle is a dead axle. The rear axle is a support member to which the rear wheels are attached. It is connected to the vehicle body or frame by springs and shock absorbers.

Wheel Bearings

Wheel bearings are classified as *adjustable wheel bearings* or *non-adjustable wheel bearings*. Each wheel hub has an *inner bearing* and an *outer bearing*. The bearings work together to support the wheel evenly during operation of the vehicle. **Fig. 2-11.**

Adjustable wheel bearings require periodic lubrication according to the vehicle service manual specifications. The drive wheels on front-wheel drive vehicles have two permanently lubricated, non-adjustable bearings.

Servicing Wheel Bearings The specific procedures to remove and service wheel bearings vary, depending on the vehicle make and model. Some general guidelines are as follows:

1. Raise the vehicle and support the frame on jack (axle) stands.

2. Check the hub and spindle for bearing wear and end play by grasping the tire at the bottom and top. Rock the tire in and out. Compare the movement to the specifications listed in the service manual. If the movement exceeds the specifications, the bearings are loose and must be adjusted or replaced.

3. Remove the wheel and tire assembly.

4. Remove the grease cap.

5. Remove the brake calipers, if required. Carefully suspend the calipers with safety wire. This avoids straining the flexible brake hose.

6. Remove the cotter pin, locknut, and washer.

7. Remove the outer bearing, hub, inner bearing, and wheel seal from the spindle.

8. Clean all parts and inspect for signs of excessive wear. Also look for scored or rough surfaces on the race or bearing.

9. Pack the new bearings in wheel bearing grease.

10. Install the new inner wheel bearing and seal in the hub.

11. Install the hub, outer wheel bearing, washer, and locknut on the spindle. Tighten the locknut to the specified torque value and install the cotter pin.

12. Install the grease cap.

13. Install the brake calipers, if removed earlier.

14. Install the wheel and tire assembly.

15. Remove the jack stands and lower the vehicle.

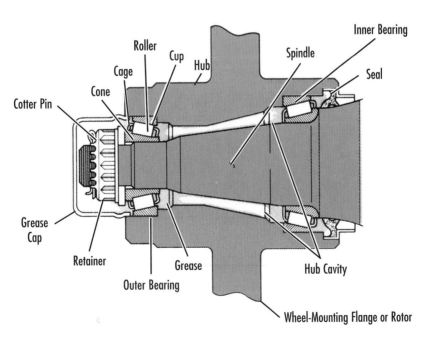

Fig. 2-11 A cutaway view of an adjustable wheel bearing in a wheel mounting flange or rotor. *Why do the drive wheels on a front-wheel drive vehicle not require lubrication servicing?* (DaimlerChrysler)

TECH TIP **Wheels and Tires.** Wheels and tires are also part of the suspension. The wheel size and offset have an effect on the overall handling and directional control of the vehicle. Tire construction and inflation pressures have an enormous effect on ride quality, directional control, and stability.

Active Suspension Systems

The purpose of an active suspension system is to maintain full suspension control during all suspension activity, including acceleration, braking, and cornering. **Fig. 2-12.**

An on-board computer receives signals from a host of sensors. These include a steering sensor,

accelerometer, and gyroscope. The sensors detect up and down movement of the wheels and angle changes of the body (forward pitch during braking, rear pitch during acceleration, or sideways pitch during cornering). The sensors send the motion information to the computer.

Instead of conventional shock absorbers and springs, an active system uses electronically controlled hydraulic actuators. The computer controls valves in the hydraulic actuators based on sensor information. The effect of the system is to maintain a proper level of ride comfort and handling.

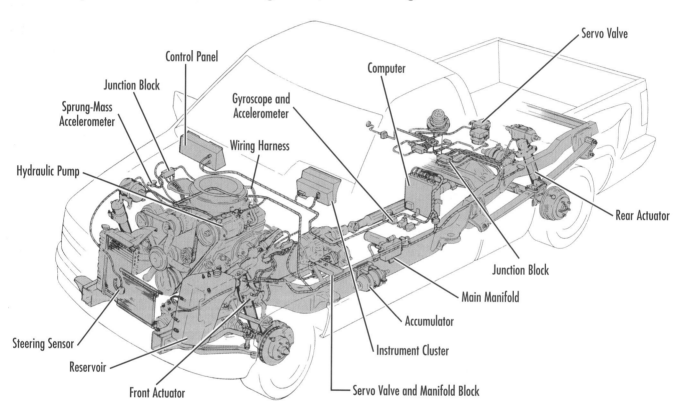

Fig. 2-12 An active suspension system uses sensors and an electronic management system. In the example shown here, hydraulic actuators change their valving and pressures according to commands sent by the on-board computer. *What components replace springs and shock absorbers in an active suspension system?* (General Motors Corporation)

SECTION 1 KNOWLEDGE CHECK

❶ What is the purpose of a shock absorber?

❷ Why should air- or spring-assisted shock absorbers not be used to provide the entire load carrying support for a vehicle?

❸ What is the purpose of the anti-sway bar?

❹ What suspension components are connected with ball joints?

❺ What purpose does the computer serve in an active suspension system?

Section 2

Front Suspension Systems

The front suspension system is responsible for:
- Supporting the weight of the vehicle front end.
- Providing wheel alignment.
- Providing steering control.
- Providing adequate handling during straight-line driving, lane changes, and turns.

When a front suspension system problem exists, before performing extensive diagnosis or beginning disassembly, do the following preliminary checks:
- Inflate the tires to the correct pressure. Make sure that all tires are the same size.
- Inspect vehicle ride height. A low ride height indicates a broken or weak spring, shock, or strut. The dimensions and points where the measurements should be made are contained in the vehicle service manual.
- Inspect the springs for collapsed or broken coils or broken leaves. Look for shiny spots where coils or leaves may have rubbed together. This is a sign that the springs are weak or were overloaded. It could also mean that the shock absorbers are worn out. Worn shocks may allow excess spring movement.

When diagnosing the cause of a front suspension system problem, relate the function of each subsystem or component to the system as a whole and determine its influence on the vehicle. Refer to the vehicle service manual for details.

If needed, replace the shock absorbers. Then move on to the more difficult replacements as required. Replace worn bushings or ball joints if they do not meet the standards in the vehicle service manual. Replace major suspension parts. Then perform a wheel alignment as specified in the vehicle service manual.

Coil-Spring Front Suspension

There are a variety of coil-spring front suspension designs. Some front suspensions have a *double A-arm* design. This design uses two control arms, one upper and one lower, and a coil spring. The shock absorber mounts between the body/frame and the lower control arm. **Fig. 2-13.**

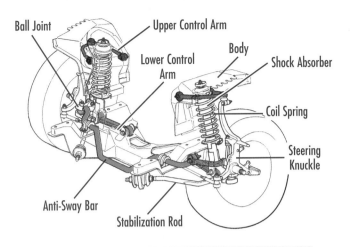

Fig. 2-13 Suspension designs vary considerably. This front suspension uses a lower control arm and an upper control arm. *(Ford Motor Company)*

As the spring compresses and rebounds during suspension travel, the control arms pivot up and down. The control arms carry the steering knuckle/spindle assembly along in an up or down direction.

The control arms attach to the frame/body with bushings. These bushings allow the control arms to pivot at this connection. Control arm bushings are generally made of rubber or a urethane material. These bushings must be checked for looseness and wear.

The outer ends of the control arms attach to the steering knuckle with lubricated ball joints. The ball joints should be checked for looseness, wear, and binding. A failed ball joint can cause the control arm to lose its connection to the steering knuckle. This results in suspension failure.

Some front suspensions use a single lower control arm. In this type, the top of the coil spring is positioned in a fixed spring seat on the frame. The bottom of the spring rests on the lower control arm.

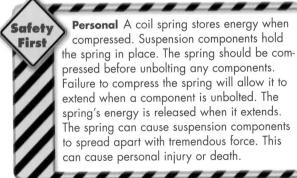

Safety First **Personal** A coil spring stores energy when compressed. Suspension components hold the spring in place. The spring should be compressed before unbolting any components. Failure to compress the spring will allow it to extend when a component is unbolted. The spring's energy is released when it extends. The spring can cause suspension components to spread apart with tremendous force. This can cause personal injury or death.

Servicing Coil-Spring Front Suspensions The specific procedures to remove a coil spring vary, depending on the vehicle make and model. However, the following are some general guidelines when removing and servicing a coil spring.

1. While the weight of the vehicle is on the suspension, install a pair of spring compressors on the coil spring.

2. Raise the vehicle and support the frame on jack stands.

3. Remove the wheel and tire assembly.

4. Remove the shock or brake components as required. Some vehicles require the removal of the brake rotor and calipers to gain access to the retaining bolts. Carefully suspend the calipers with safety wire to avoid straining the flexible brake hose.

5. Compress the coil spring using the spring compressors installed earlier. Once the coil spring is compressed enough to remove tension from the spring seats, remove the spring retainer, the spring, and insulators from the vehicle.

6. Slowly release the spring compressors and remove from the old spring.

7. Install the spring compressors on the new spring.

8. Position the new spring and insulators in the vehicle.

9. Carefully release the spring tension, making sure the ends of the coil spring properly position in both the upper and lower spring seats.

10. Install the spring retainer, if required.

11. Install the shock or brake components as required.

12. Install the wheel and tire assembly.

13. Remove the jack stands and lower the vehicle.

Strut-Type Front Suspension

A strut-type front suspension includes a shock absorber and a steering knuckle extension in a single unit. The bottom of the strut body attaches to the steering knuckle and serves as an extension of the "upright" of the knuckle.

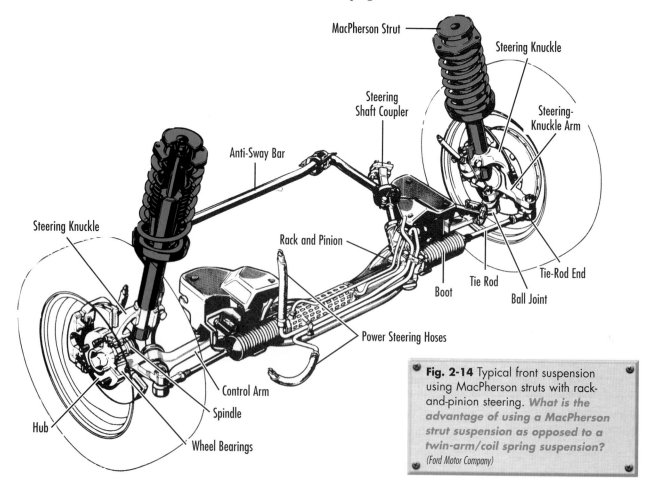

Fig. 2-14 Typical front suspension using MacPherson struts with rack-and-pinion steering. *What is the advantage of using a MacPherson strut suspension as opposed to a twin-arm/coil spring suspension?* (Ford Motor Company)

Reading Service Manuals

Twenty-five years ago, a mechanic would rarely have consulted a service manual. Today, however, vehicles are too complicated for any one technician to know everything about them. Today's technicians must consult service manuals and databases to be sure they are following the manufacturers' procedures.

Today's technician must read carefully. No longer can the tightening of nuts and bolts be left to guesswork. Specific torque values are now required to tighten fasteners correctly. *Torque* is a twisting or turning force. You must be able to read, understand, and follow directions for a specific procedure. You must also know what special tools you will need.

Most service manuals include torque specifications. They also include pictures and part numbers for special tools.

Safety First

Personal If you fail to read about special tools and specifications, you may perform incorrect repairs or maintenance. Such errors can affect the vehicle's performance, as well as your safety and the safety of your customer.

Apply it!

Meets NATEF Communications Standards for using service manuals and comprehending and applying written information.

❶ Select a service manual in your lab or school media center.

❷ Look up the procedure for wheel alignment, which requires the tightening of fasteners to a specific torque. Look up another procedure that has specific torque requirements.

❸ Look for the special tools identified by the manufacturer for the proper alignment of wheels. Look up another procedure that requires the use of special tools.

The most common type of strut is the MacPherson strut. This combines the shock absorber and coil spring in a single unit. **Fig. 2-14.**

The coil spring rests on a fixed lower spring seat that is part of the strut body. An upper spring seat holds the top of the coil spring captive on the strut. This allows installation or removal of the spring and shock absorber assembly as a unit.

The upper spring seat attaches to the vehicle frame/body. A bearing under the upper spring seat allows the strut assembly to pivot when the wheels turn. Because the top of the strut/spring attaches directly to the frame/body, there is no need for an upper control arm.

A vehicle with torsion bar springs and strut suspension may use a modified strut. A "modified MacPherson strut" uses a shock absorber mounted inside a strut tube but without a captive coil spring.

Some vehicles may also use a modified strut with a separately-mounted coil spring. The coil spring seats between the lower control arm and the frame.

Not all struts are MacPherson struts. A true MacPherson strut incorporates the strut body, coil spring, and shock absorber as an assembled unit.

Inspecting Strut-Type Front Suspensions When checking struts, do the following:

- Perform a "bounce" test. Push down on the front end of the vehicle firmly and release. The front end of the vehicle should rise once and immediately settle. If the front end bounces more than once, the strut is worn, and should be replaced.

- Visually inspect the strut for leakage. A light film of oil near the top seal is normal. If excessive leakage is seen, replace the strut.

- Look for damage such as strut body dings and dents. If evident, replace the strut.

- Inspect the strut-rod for scratches, gouges, corrosion, and bending. If damaged, replace the strut.

Refer to the vehicle service manual for any specific checks that should be performed.

Servicing Strut-Type Front Suspensions The shock absorbers used in strut-type suspensions may be serviceable. In some struts, the shock absorber attaches to the strut tube with a removable nut so that the shock absorber can be replaced. Some struts, however, use shock absorbers that are not removable. For these struts, replace the entire strut housing and shock absorber as a unit. All struts allow removal of the coil spring and upper spring seat.

The strut-type front suspension has the advantage of saving weight and space. The strut requires no upper control arm and no upper ball joint. The control arms are not needed, because the strut attaches to the vehicle body at a recessed area called the strut tower. The strut also combines the coil spring and shock absorber into one compact unit. **Fig. 2-15.**

Depending on the vehicle make and model, specific procedures for removing and servicing a MacPherson strut assembly vary. The following are some general guidelines:

1. While the weight of the vehicle is on the suspension, loosen the strut-rod nut. Loosen the nut only slightly to relieve the majority of tightening torque. Do not remove the nut at this time!

> **Safety First**
> **Personal** The strut-rod nut holds the coil spring under tension. If you remove this nut now, the coil spring will release once the bottom of the strut disconnects from the spindle. This could cause vehicle damage and personal injury.

2. Raise the vehicle and support the frame on jack stands.

3. Remove the wheel and tire assembly.

4. Remove the two bottom mounting bolts, where the bottom of the strut attaches to the spindle unit. **Fig. 2-16.**

5. A brake hose or line may attach to the strut for support. If so, disconnect the hose or line from the strut. Do not disconnect the brake line from the caliper. In some cases the brake rotor and caliper may need to be removed to gain access to the strut

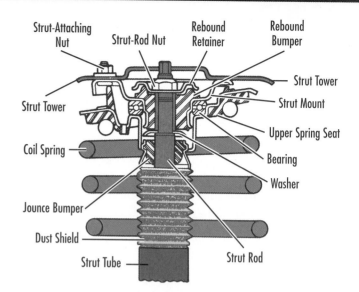

Fig. 2-15 A cutaway view of a MacPherson strut, showing the attachment to the strut tower. *Why is no upper control arm required?* (DaimlerChrysler)

bolts. Carefully suspend the caliper with safety wire to avoid straining the flexible brake hose.

6. Remove the upper strut tower nuts where the top of the strut attaches to a mount on the body. Carefully remove the strut assembly from the vehicle. Do not remove the top strut-rod nut at this time.

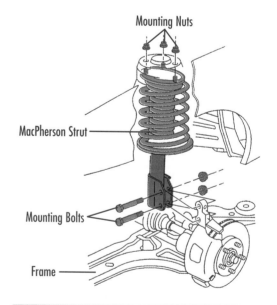

Fig. 2-16 A typical mounting for a MacPherson strut. There are three upper mounting nuts and two lower mounting bolts. *On a MacPherson strut front suspension, what allows the top of the strut to turn and pivot during steering?* (General Motors Corporation)

7. To disassemble the strut unit, place the strut in an approved strut compressor tool. **Fig. 2-17.** Some of these compressor tools are hand-held units. Some are wall or bench mounted. Always follow the specific disassembly procedure outlined in the vehicle service manual.

8. Once the coil spring is compressed enough to remove tension from the spring seats, remove the strut-rod nut.

9. Remove the upper bearing assembly, upper spring seat assembly, and dust shield (if used). Remove the coil spring from the strut. Then remove the bump stop and plate (if used). **Fig. 2-18.**

10. Store the strut spring safely. Follow the guidelines in the service manual. Inspect the strut. Some strut designs use a removable shock absorber. If this is the case, the shock absorber is attached inside the strut tube with a threaded collar nut.

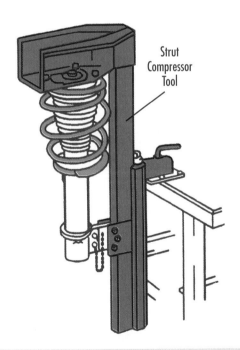

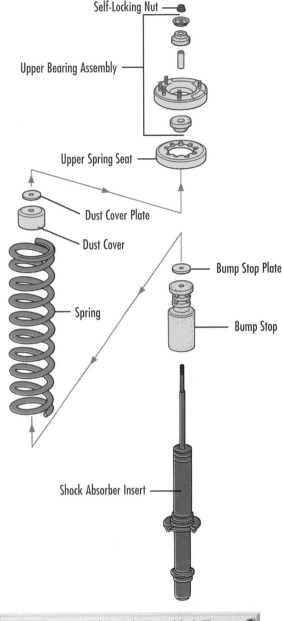

Fig. 2-18 A typical strut disassembled. The bump stop prevents the shock absorber from "bottoming out" during suspension travel. *What parts make up a typical MacPherson strut assembly?* (American Honda Motor Company)

Fig. 2-17 A strut compressor tool specified for use on Pontiac vehicles. A variety of strut compressors are available. Use only a compressor approved for the type of vehicle being serviced. In addition to being able to compress the coil spring, safety is the primary consideration. *When compressing a coil spring, what precautions must be taken?* (General Motors Corporation)

11. Using a special spanner wrench, remove the collar nut and pull the shock absorber out of the strut tube. Remove the shock absorber insert. Make sure the tube is clean inside and install a new insert. Old strut designs used "wet" shock absorber inserts. These have a supply of cooling oil inside the strut tube. More current designs do not use this separate oil bath. The unit is self-contained.

12. Place the new shock absorber properly in the strut tube. Add the correct amount of cooling oil, if required. Install a new threaded collar nut. Torque the nut to factory specifications. Install the bump stop and plate (if used).

13. Place the coil spring onto the strut. Properly position the bottom of the coil spring into the lower spring seat of the strut.

14. With the spring in a compressed state, install the dust shield (if used), followed by the upper spring seat and the upper bearing assembly.

15. Install a new strut-rod nut and torque the nut to specification.

16. Carefully release spring tension, making sure the ends of the coil spring properly position in both the upper and lower spring seats.

17. Reinstall the strut assembly in the vehicle.

18. Check the wheel alignment, as the strut position affects camber, caster, and toe angle.

Rear MacPherson struts are generally serviced using these same procedures.

TECH TIP Strut Reassembly. When reassembling a strut, never use an impact wrench to tighten the shock absorber piston rod nut. This will cause the piston rod to spin inside the shock absorber, possibly damaging the seals.

Leaf-Spring Front Suspension

Some trucks use leaf-spring front suspension and a solid or I-beam front axle. **Fig. 2-19.** Instead of using pivoting control arms, the steering knuckle connects to a solid front axle beam or a left-and-right I-beam. Also, because the leaf-spring packs mount on top of the front axle, they can help the vehicle gain additional ground clearance.

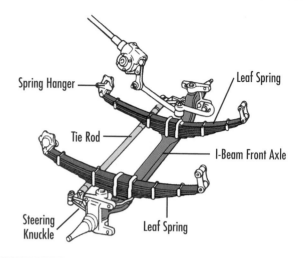

Spring Hanger — Leaf Spring — Tie Rod — I-Beam Front Axle — Steering Knuckle — Leaf Spring

Fig. 2-19 A truck front suspension using an I-beam front axle and two longitudinal leaf springs. *Where are leaf springs normally installed in relation to the axle?* (Ford Motor Company)

Servicing Leaf-Spring Front Suspensions The specific procedures to remove a leaf spring vary, depending on the vehicle make and model. However, here are some general guidelines to follow when removing and servicing a leaf spring.

1. Raise the vehicle and support the frame on jack stands.

2. Remove the wheel and tire assembly.

3. Remove the shock or brake components, as required. Carefully suspend the caliper with safety wire to avoid straining the flexible brake hose.

4. Support the axle with jack stands.

5. Remove the U-bolts and lower shock mount from the axle and leaf spring.

6. While supporting the rear of the leaf spring, remove the pins and rear shackle from the rear hanger and spring eye.

7. While supporting the front of the leaf spring, remove the pin and front spring eye from the spring hanger. Lower the spring from the axle.

8. Replace the shock absorbers or leaf springs as required. Replace worn bushings if they do not meet the standards given in the vehicle service manual.

9. While supporting the front of the leaf spring, install the pin and front spring eye in the spring hanger.

10. While supporting the rear of the leaf spring and the axle, install the pins and rear shackle in the rear hanger and spring eye.

11. Install the lower shock mount and U-bolts on the axle and leaf spring. Tighten the U-bolts as specified in the vehicle service manual.

12. Remove the jack stand support from under the axle.

13. Install the shock or brake components as required.

14. Install the wheel and tire assembly.

15. Raise the vehicle and remove the jack stand support under the frame. Lower the vehicle.

16. Perform a wheel alignment after replacing front suspension components if specified in the vehicle service manual.

 TECH TIP Leaf Spring Removal. When removing a leaf spring, always raise the vehicle to remove spring tension. Never disconnect any mounting points of a leaf spring while the vehicle weight is on the suspension.

Torsion Bar Front Suspension

Torsion bar suspension supports the weight of the vehicle by using a twisting spring action. There are a variety of torsion bar front suspension designs. Some use a *double-A arm* design. This design uses two control arms, one upper and one lower, and a torsion bar. **Fig. 2-20.** The shock absorber mounts between the body/frame and the lower control arm.

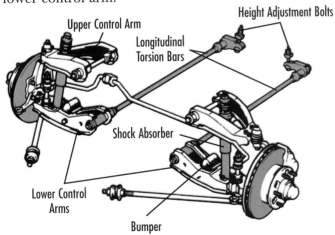

Upper Control Arm

Height Adjustment Bolts

Longitudinal Torsion Bars

Shock Absorber

Lower Control Arms

Bumper

Fig. 2-20 A front suspension system using torsion bar springs. *Where does a torsion bar mount to provide spring action?*

One end of the torsion bar is positioned in a fixed seat in the lower control arm. The other end of the torsion bar is fastened in an adjustable anchor in the frame. A screw assembly provides adjustment on the torsion bar by twisting a lever on the anchor end.

As the torsion bar twists during suspension travel, the control arms pivot up and down. The control arms carry the steering knuckle/spindle assembly along in an up or down direction.

Servicing Torsion Bar Front Suspensions The specific procedures to remove a torsion bar vary, depending on the vehicle make and model. Following are some general guidelines for removing and servicing a torsion bar.

1. Inspect the torsion bar and suspension for damaged or worn parts. Look for shiny spots where parts may have shifted and rubbed together. This is a sign that the torsion bars are weak or were overloaded or that the shock absorbers are worn out. Worn shock absorbers may allow excess torsion bar movement.

2. Raise the vehicle and support the frame on jack stands.

3. Remove the wheel and tire assembly.

4. Remove the shock or brake components, as required. Carefully suspend the caliper with safety wire to avoid straining the flexible brake hose.

5. Loosen the torsion bar locking nut on the height adjustment bolts.

6. Mark the position of the height adjustment bolts.

7. Loosen the height adjustment bolts until all tension is relieved from the control arm.

8. Remove the torsion bar, bushings, and control arm using the procedures specified in the vehicle service manual.

9. Inspect the torsion bar, bushings, and control arm for wear or damage.

10. Install the torsion bar, bushings, and control arm using the procedures specified in the vehicle service manual.

11. Tighten the height adjustment bolts until the bolts are positioned at the point marked during removal.

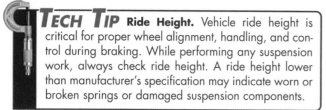

TECH TIP **Ride Height.** Vehicle ride height is critical for proper wheel alignment, handling, and control during braking. While performing any suspension work, always check ride height. A ride height lower than manufacturer's specification may indicate worn or broken springs or damaged suspension components.

12. Tighten the torsion bar locking nut on the height adjustment bolts.

13. Install the shock or brake components, as required.

14. Install the wheel and tire assembly.

15. Raise the vehicle and remove the jack stand support under the frame. Lower the vehicle.

16. Inspect the vehicle for height and level. If the vehicle is not at the correct height and level, repeat the procedures as required.

17. Perform a wheel alignment after replacing front suspension components if specified in the vehicle service manual.

Twin I-Beam Front Suspension

Some trucks use a twin I-beam independent front suspension. **Fig. 2-21.** The I-beams provide the function of lower control arms. Each wheel spindle attaches to the outer end of a separate forged I-beam

axle. The opposite end of each I-beam attaches to the frame with a flexible bushing joint. For each I-beam axle, a coil spring is located between the I-beam and an upper spring seat on the frame.

Each I-beam is braced with a radius rod. The rod attaches the I-beam to the frame, rearward of the axles. The radius rods locate the I-beam axles. They prevent the I-beams from moving forward or rearward during vehicle movement and suspension travel.

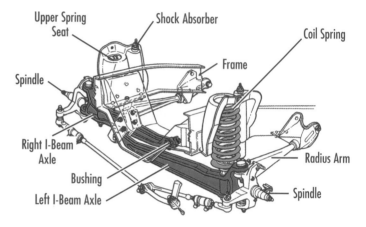

Fig. 2-21 A typical twin I-beam front suspension. The beams pivot on large rubber bushings, where they attach to the truck frame. *Why are the I-beams braced with a radius rod?* (Ford Motor Company)

SECTION 2 KNOWLEDGE CHECK

❶ Describe a double-A arm suspension design.

❷ If only a lower control arm is used, where is the top of the coil spring installed?

❸ What parts make up a MacPherson strut?

❹ What advantage is gained by mounting the leaf springs on top of the axle?

❺ What does the radius rod do on a twin I-beam front suspension?

Section 3

Rear Suspension Systems

The rear suspension system is responsible for:
• Supporting the weight of the rear of the vehicle.
• Providing wheel alignment.
• Providing adequate handling during straight-line driving, lane changes, and turns.

Before performing extensive diagnosis or beginning disassembly of any parts of the vehicle, perform the following preliminary checks:

• Inflate the tires to the correct pressure. Make sure that all tires are the same size.
• Inspect vehicle ride height. A low ride height indicates a broken or weak spring, shock, or strut. The dimensions and points where the measurements should be made are contained in the vehicle service manual.
• Inspect the springs for broken or collapsed coils or broken leaves. Look for shiny spots where coils or leaves may have rubbed together. This is

Calculating Spring Loads

A constant-rate spring, or linear-rate spring, requires a load of 600 pounds to compress the spring 1 inch. The same spring requires 1,200 pounds to compress it 2 inches. What load will be required to compress the spring 4 inches?

One way to calculate the answer is by using proportions. As a proportion:

$$\frac{1''}{600 \text{ lb}} = \frac{2''}{1,200 \text{ lb}}$$

$$\frac{1''}{600 \text{ lb}} = \frac{4''}{c}$$

where c is the load needed to compress the spring 4 inches. Solving for c, this load is 2,400 pounds.

An alternative way to find the necessary load is to find the rate of increase. Use the formula:

$$m = \frac{(y - y_1)}{(x - x_1)} \text{ or } m = \frac{(1,200 - 600)}{(2 - 1)} = 600 \text{ lb/in.}$$

Remember, the rate of increase is called the slope and is identified by the letter m. This formula shows you that for every change of 1 inch in the spring compression, a 600-pound load will be needed.

Rewrite the formula as:

$$y - y_1 = m(x - x_1)$$

Substitute for y_1, m, and x_1 in the formula and solve for y as follows:

$$y - 600 = 600(x - 1)$$

$$y = 600x - 600 + 600 = 600x$$

For an x value of 4 inches:

$$y = 600(4) = 2,400 \text{ or } 2,400 \text{ pounds.}$$

Apply it!

Meets NATEF Mathematics Standards for proportions, calculating, and evaluating algebraic expressions.

❶ What load will be required to compress the spring 2.5 inches? Arrive at a value using both methods.

❷ Consider another spring where $m = 660$ lb/in. Which of these springs would be more suitable for a heavily loaded vehicle? Explain your answer.

---

a sign that the springs are weak or were overloaded. It could also mean that the shock absorbers or shock absorber inserts are worn out. Worn shocks may allow excess spring movement.

If needed, replace the shock absorbers or shock absorber inserts. Then move on to the more difficult replacements as required. Replace worn bushings or ball joints if they do not meet the standards in the vehicle service manual. Replace major suspension parts. Then perform a wheel alignment as specified in the vehicle service manual.

Leaf-Spring Rear Suspension

Many vans, trucks, and older passenger vehicles use leaf-spring rear suspensions. Leaf springs can be designed and built to support very heavy loads.

A leaf spring attaches to the frame at both ends with hangers and a shackle. A large U-bolt attaches

the spring to the rear axle. **Fig. 2-22.** As the spring bends during suspension travel, the shackle pivots to allow the spring to lengthen and shorten. This helps soften the ride and prevent shock and vibration from reaching the body and frame. A rubber bumper on the frame cushions the axle if the suspension should "bottom out."

On a multi-leaf spring, the individual leaf plates slide on each other. "Rebound clips" wrap around the multi-leaf spring packs. These clips help to align the springs and reduce the noise caused by the spring plates sliding on each other during spring travel.

Leaf springs attach in front of and behind the axle, running parallel with the length of the frame. They cannot bend sideways. Therefore, the use of rear leaf springs helps locate the rear axle and prevent unwanted sideways movement of the axle.

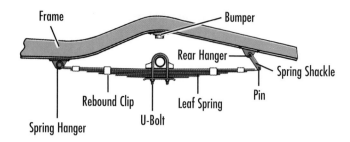

Frame Bumper Rear Hanger Spring Shackle Rebound Clip Leaf Spring Pin Spring Hanger U-Bolt

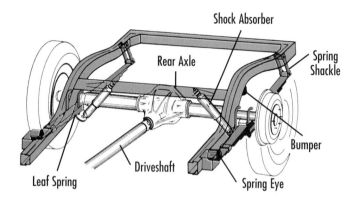

Shock Absorber Rear Axle Spring Shackle Driveshaft Bumper Leaf Spring Spring Eye

Fig. 2-22 A leaf-spring rear suspension using two longitudinal springs. *How is a multi-leaf spring pack held in alignment?*

Servicing Leaf-Spring Rear Suspensions The specific procedures to remove a leaf spring vary, depending on the vehicle make and model. However, here are some general guidelines to follow when removing and servicing a leaf spring.

1. Raise the vehicle and support the frame on jack stands.

2. Remove the wheel and tire assembly.

3. Remove the shock or brake components as required. Carefully suspend the caliper with safety wire to avoid straining the flexible brake hose.

4. Support the axle with jack stands.

5. Remove the U-bolts and lower shock mount from the axle and leaf spring.

6. While supporting the rear of the leaf spring, remove the pins and rear shackle from the rear hanger and spring eye.

7. While supporting the front of the leaf spring, remove the pin and front spring eye from the spring hanger. Lower the spring from the axle.

8. Replace the shock absorbers or leaf springs as required. Replace worn bushings if they do not meet the standards given in the vehicle service manual.

9. While supporting the front of the leaf spring, install the pin and front spring eye in the spring hanger.

10. While supporting the rear of the leaf spring and the axle, install the pins and rear shackle in the rear hanger and spring eye.

11. Install the lower shock mount and U-bolts on the axle and leaf spring. Tighten the U-bolts as specified in the vehicle service manual.

12. Remove the jack stand support from under the axle.

13. Install the shock or brake components as required.

14. Install the wheel and tire assembly.

15. Raise the vehicle and remove the jack stand support under the frame. Lower the vehicle.

16. Perform a rear wheel alignment after replacing rear suspension components if specified in the vehicle service manual.

Coil-Spring Rear Suspension

To prevent the rear axle from moving forward, rearward, or sideways, coil springs on the rear suspension require additional "locating" devices. A system of lower, or upper and lower, control arms prevents the axle from moving. The control arms attach to the frame/body with bushings. Technicians refer to these arms as *trailing arms.*

Control arm bushings allow the control arms to pivot at the frame/body. These bushings are generally made of rubber or a urethane material. They must be checked for looseness and wear.

As an added measure to prevent rear axle movement, some systems use a *track bar* or *panhard rod.* **Fig. 2-23.** This is a straight bar positioned parallel to the rear axle. One end of the bar attaches to the axle. The other end of the bar attaches to the frame or body. The bar mounts on bushings. This allows the bar to pivot when the axle moves up and down during suspension travel. The bar provides better vehicle control and handling during turns.

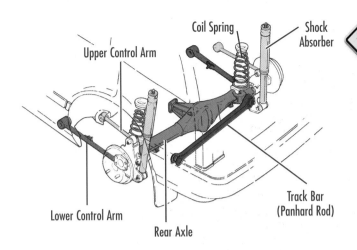

Servicing Coil-Spring Rear Suspensions The specific procedures to remove a coil spring vary, depending on the vehicle make and model. Use the procedure provided in the service manual for the vehicle you are servicing.

Strut-Type Rear Suspension

An independent rear suspension, where the rear wheels can move up and down independently of each other, is a strut-type suspension. In a rear suspension, a strut is a structural unit that locates and connects the wheel spindle to the frame. The strut also incorporates the shock absorber.

The struts may be MacPherson struts, where the shock absorber and the coil spring mount are assembled as a unit. They may instead be modified MacPherson struts. With modified MacPherson struts, the strut and coil springs are an unassembled unit, with the coil spring mounting independently between a lower control arm and the frame. **Fig. 2-24.**

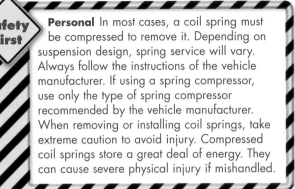

Safety First

Personal In most cases, a coil spring must be compressed to remove it. Depending on suspension design, spring service will vary. Always follow the instructions of the vehicle manufacturer. If using a spring compressor, use only the type of spring compressor recommended by the vehicle manufacturer. When removing or installing coil springs, take extreme caution to avoid injury. Compressed coil springs store a great deal of energy. They can cause severe physical injury if mishandled.

Servicing Strut-Type Rear Suspensions Specific procedures vary, depending on the vehicle make and model. The shock absorbers used in strut-type suspensions may be serviceable. If the shock absorber attaches to the strut tube with a removable nut, the shock absorber can be removed. Some struts, however, use shock absorbers that are not removable. For these struts, replace the entire strut housing and shock absorber as a unit. All struts allow removal of the coil spring and upper spring seat.

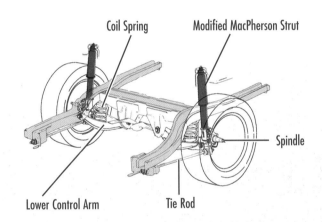

SECTION 3 KNOWLEDGE CHECK

❶ What is the function of a spring shackle?

❷ What is the purpose of a panhard rod?

❸ What should a technician look for when inspecting springs?

❹ What is a modified strut rear suspension?

❺ How can you determine if a shock absorber in a strut-type rear suspension is serviceable?

CHAPTER 2 REVIEW

Key Points

Meets the following NATEF Standards for Suspension and Steering: diagnosing and repairing leaf, coil, torsion bar, and strut suspension systems.

- The basic parts of a suspension system are springs, shock absorbers, anti-sway bars, control arms, ball joints, axles, and wheel bearings.
- Compression or jounce occurs when the suspension moves closer to the body. Rebound occurs when the suspension moves away from the body.
- Front or rear suspensions can use a variety of spring types.
- The two most common types of shock absorbers are "traditional" and struts. Shock absorbers can also be spring-assisted, adjustable, and air-assisted.
- The most common type of strut is the MacPherson strut. It combines many of the standard suspension components.
- An active suspension system uses an on-board computer and sensors to maintain full control of the suspension during all suspension activity.
- There are a variety of front suspension designs including leaf, coil, torsion bar, and strut.
- There are a variety of rear suspension designs including leaf, coil, and strut.

Review Questions

1. Describe the types of springs used in front and rear suspension systems.
2. What factors determine the spring rate of a coil spring?
3. In an air-spring suspension, how do the springs adjust for road and load conditions?
4. What are the different front suspension designs?
5. Once a MacPherson strut is removed from a vehicle, what steps do you take to service the strut?
6. Describe the three types of rear suspension designs.
7. What is the advantage of a leaf spring rear suspension?
8. **Critical Thinking** Based on their construction and effect, which rear suspension is safer–one that uses coil springs or one that uses leaf springs? Why?
9. **Critical Thinking** Based on their construction and effect, which front suspension is safer–one that uses coil springs, one that uses torsion bars, or one that uses leaf springs? Why?

TECHNOLOGY FORECAST FOR AUTOMOTIVE EXCELLENCE

Electronic Suspension Systems

Automakers want to give their cars a quiet, comfortable ride and stable handling. New electronic suspension systems should make their job easier. Ride-motion and steering wheel sensors will be able to tell if the pavement dips, curves, or has a bump. They will also sense how the driver is steering the car or truck.

Using information from the sensors, a control module will signal the suspension to be firmer or softer as needed. Electronically controlled dampers in the shock absorbers or struts will make

these changes. The system can also be used to keep the vehicle's body flat while turning and cornering. This control is possible because of actuators connected to the stabilizer bars.

Also in the future, automakers will use hydroforming. This metal-shaping process uses water pressure to shape suspension and chassis components. Benefits include parts that can be used more precisely and that are very strong for their weight. Their strength will improve ride and handling. Squeaks and rattles will be reduced.

AUTOMOTIVE EXCELLENCE
TEST PREP

Answering the following practice questions will help you prepare for the ASE certification tests.

1. Technician A says that all coil springs use the same diameter coil. Technician B says that spring rate is determined by the type of steel used, the thickness of the spring wire, and the number of coils. Who is right?
   - ⓐ Technician A.
   - ⓑ Technician B.
   - ⓒ Both Technician A and Technician B.
   - ⓓ Neither Technician A nor Technician B.

2. Which of the following statements is true?
   - ⓐ Anti-sway bars support vehicle weight when the car is parked.
   - ⓑ Anti-sway bars do not affect ride height.
   - ⓒ Anti-sway bars are only used on the front suspension.
   - ⓓ Anti-sway bars will restore weak springs.

3. Technician A says that a lower control arm ball joint is a load-carrying joint, while an upper control arm ball joint is a follower type. Technician B says that all ball joints are load-carrying type joints. Who is right?
   - ⓐ Technician A.
   - ⓑ Technician B.
   - ⓒ Both Technician A and Technician B.
   - ⓓ Neither Technician A nor Technician B.

4. Which of the following statements is not true?
   - ⓐ All ball joints are sealed and cannot be greased.
   - ⓑ Ball joints use lubricating grease to reduce friction.
   - ⓒ Upper ball joints are commonly non-load carrying joints.
   - ⓓ Lower ball joints are load carrying joints.

5. Technician A says that all strut suspensions use MacPherson-type struts. Technician B says that some vehicles use a modified MacPherson-strut-type suspension. Who is right?
   - ⓐ Technician A.
   - ⓑ Technician B.
   - ⓒ Both Technician A and Technician B.
   - ⓓ Neither Technician A nor Technician B.

6. Technician A says that ride height will only change if the vehicle is involved in an accident. Technician B says that a change in ride height can affect wheel alignment angles. Who is right?
   - ⓐ Technician A.
   - ⓑ Technician B.
   - ⓒ Both Technician A and Technician B.
   - ⓓ Neither Technician A nor Technician B.

7. Technician A says that only trucks use leaf springs. Technician B says that leaf springs can use a single leaf or multiple leaves. Who is right?
   - ⓐ Technician A.
   - ⓑ Technician B.
   - ⓒ Both Technician A and Technician B.
   - ⓓ Neither Technician A nor Technician B.

8. Technician A says that you can install rear leaf springs under or over a rear axle housing. Technician B says that passenger vehicles never use leaf springs. Who is right?
   - ⓐ Technician A.
   - ⓑ Technician B.
   - ⓒ Both Technician A and Technician B.
   - ⓓ Neither Technician A nor Technician B.

9. Technician A says that vehicles with independent rear suspensions commonly use leaf springs in the rear. Technician B says that leaf springs are only made of steel. Who is right?
   - ⓐ Technician A.
   - ⓑ Technician B.
   - ⓒ Both Technician A and Technician B.
   - ⓓ Neither Technician A nor Technician B.

10. Technician A says that an independent rear suspension on a front-wheel-drive vehicle uses control arms that allow the wheel to pivot up and down. Technician B says that only rear-wheel drive vehicles with one-piece axle housings use rear control arms. Who is right?
    - ⓐ Technician A.
    - ⓑ Technician B.
    - ⓒ Both Technician A and Technician B.
    - ⓓ Neither Technician A nor Technician B.

CHAPTER 3

Diagnosing, Adjusting, & Repairing Wheel Alignment

You'll Be Able To:

- ⊗ Diagnose vehicle driveability problems resulting from wheel alignment problems.
- ⊗ Check and adjust wheel camber.
- ⊗ Check and adjust wheel caster angle.
- ⊗ Check and adjust wheel toe angle.
- ⊗ Check steering axis inclination and included angle.

Terms to Know:

camber angle

caster angle

geometric centerline

included angle

steering axis inclination

thrust angle

toe angle

The Problem ⟩⟩⟩⟩⟩⟩⟩⟩

Mr. Rasmussen noticed that his car wanders to the left on straight roads. The car pulls hard to the left when he applies the brakes. He saw that the inside tread area of the right front tire was wearing faster than the outside tread area. He also saw that the outside tread area on the left front tire is wearing faster than the inside tread area.

You suspect that incorrect wheel alignment is causing the problems Mr. Rasmussen describes. You use special equipment to check the wheel alignment of his car.

Your Challenge

As the service technician, you need to find answers to these questions:

❶ What effect does camber angle have on steering pull and tire wear?

❷ What effect does the caster angle have on directional pull?

❸ How does a proper alignment affect driveability?

Section 1

Wheel Alignment Angles

Wheel alignment angles have an effect on vehicle driveability. Incorrect wheel alignment causes:
- Vehicle wander.
- Vehicle drift and instability.
- Vehicle pulling to the left or right.
- Hard steering effort.
- Torque steering on front-wheel-drive vehicles.
- Slow or abnormal steering return.
- Uneven tire wear.

The wheel alignment angles are:
- Camber angle.
- Caster angle.
- Toe angle.
- Scrub radius.
- Included angle.
- Turning radius.
- Setback.
- Steering axis inclination.
- Thrust angle.

Camber Angle

The angle of inward or outward tilt of a wheel is known as the **camber angle**. The camber angle is sometimes simply called *camber*. This angle is compared to true vertical when viewed from the front or rear of the vehicle. When the top of the wheel tilts in, the wheel has a *negative camber* angle. When the top of the wheel tilts out, the wheel has a *positive camber* angle. When camber is true vertical, there is "zero" camber. **Fig. 3-1.**

Camber is measured in degrees. The initial setting for camber is called the *static angle*. The intent of the static angle is to anticipate wheel travel. The static angle may initially be set positive. When the suspension compresses during normal driving, the movement pulls the wheel into a true vertical position. The camber adjustment determines what portion of the tire tread contacts the road surface.

As a vehicle's suspension goes through its loaded and unloaded movement, its wheels travel through an arc. As the suspension compresses and rebounds, camber angle changes. Vehicle manufacturers adjust camber to compensate for suspension action. They do this so that the entire tire tread is in contact with the road during normal driving.

Camber affects directional pull. A vehicle will always pull in the direction of the wheel that has the most positive camber. If camber is positive on the left-front wheel and negative on the right-front wheel, the vehicle will pull to the left. If the right-front wheel has more positive camber, the vehicle will pull to the right.

Ideally, a vehicle in motion should have the same camber angle on both the right and left wheels. However, if the right and left wheels have the same camber angle, the vehicle would pull to the right on a crowned road.

Road crown is present when the road is slightly higher in the center than at the sides. The crown allows water to run off. To compensate for road crown, vehicles have a slightly more positive camber angle on the left-front wheel than on the right-front wheel. This camber angle difference allows the vehicle to travel straight with no pull to either side.

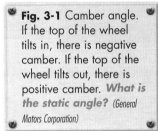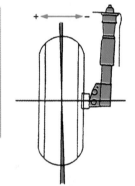

Fig. 3-1 Camber angle. If the top of the wheel tilts in, there is negative camber. If the top of the wheel tilts out, there is positive camber. *What is the static angle?* (General Motors Corporation)

Camber angle affects tire wear. Since the camber angle tilts the top of the wheel in or out, camber changes the angle at which the tire tread contacts the road. If a wheel has too much negative camber, the inside tread area of the tire wears faster than the rest of the tread. If the wheel has too much positive camber, the outside of the tread area will wear faster.

 TECH TIP Adjusting Alignment Angles. Worn, loose, or damaged suspension components will change alignment angles. Thoroughly check all components before making any alignment angle adjustments.

Caster Angle

The angle between true vertical and an imaginary line drawn through the upper and lower ball joints is known as the **caster angle**. Caster angle is sometimes referred to simply as *caster*. Caster is viewed from the side of the vehicle.

Caster is measured in degrees. When the upper ball joint is behind the lower ball joint, the wheel has *positive caster.* **Fig. 3-2.** When the upper ball joint is forward of the lower ball joint, the wheel has *negative caster.* When the upper ball joint is directly over the lower ball joint, the wheel has *zero caster.* The caster angle affects directional control of the vehicle and steering wheel return.

The caster angle on a vehicle equipped with MacPherson struts is measured differently. It is the difference between true vertical and a line drawn from the upper strut mount to the lower ball joint. Since the strut mount is a part of the vehicle body, caster is not always adjustable. **Fig. 3-3.**

A slight amount of positive caster is desirable. Positive caster helps to keep the front wheels pointed straight ahead. This prevents vehicle wander. However, excessive positive caster requires more steering effort, creates low-speed shimmy, and causes excessive steering wheel return, or "snap-back."

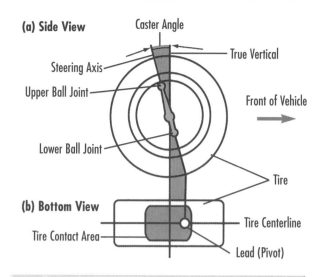

Fig. 3-2 Caster angle. **(a)** Draw a line through the upper and lower ball joints. Compare this line, called the steering axis, to true vertical. **(b)** The angle between these two lines is the caster angle. Positive caster causes the steering axis to pass through the road surface ahead of the center of the tire contact area. This allows the steering axis to lead the tire along the road surface for better control and stability. *What driveability conditions does caster affect?* (DaimlerChrysler)

Searching a Database

Experience was once the only source of information for mechanics. Vehicle systems were very simple. Anyone with mechanical knowledge could learn to fix a vehicle. Using trial-and-error methods may have wasted some time, but the job was usually done correctly.

With new developments, technical knowledge grew. Previously, one service manual may have been used for a manufacturer's whole line of automobiles. Now there are one or more service manuals per vehicle.

To keep up with the expanding field of automotive technology, computers are now used in dealership service areas. Computer databases provide the same material found in service manuals. Many of these databases can be found in CD-ROM format. Using this electronic service information (ESI) you can quickly and easily locate the information you require.

Many manufacturers also have Web sites that contain much useful information.

Apply it!

Meets NATEF Communications Standards for using databases and manuals to obtain system information.

❶ Choose a service manual for a specific vehicle model. Locate the section that deals with how to adjust the caster and camber for that vehicle.

❷ Use a computer to begin your search in one of the following ways:

a. Use a manufacturer's CD-ROM database to find the information you need. Locate the section that deals with how to adjust the caster and camber for the vehicle.

b. Perform an on-line search on the Internet. Select a search engine and look for information using key words or phrases.

❸ Compare the information in the service manual with what you located on the manufacturer's database or on the Internet.

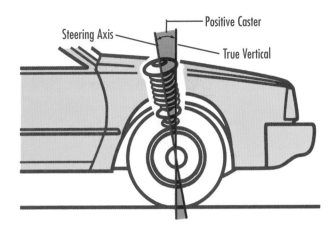

Fig. 3-3 The caster angle on a vehicle equipped with MacPherson struts is created by the difference between the true vertical line at the wheel center and the steering axis, a line drawn from the upper strut mount to the lower ball joint. *How is caster measured on a vehicle with MacPherson struts?* (DaimlerChrysler)

Rear chassis height affects the front wheel caster angle. As the load on the rear increases, the rear chassis height decreases and the front chassis height increases. The caster angle becomes more positive as the front unloads. As rear chassis height increases, front chassis height decreases. The caster angle becomes more negative as the load on the front increases. **Fig. 3-4.**

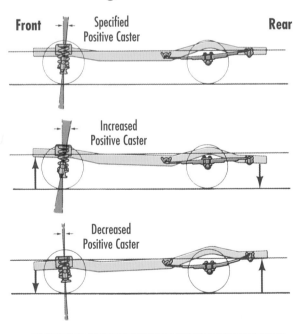

Fig. 3-4 Rear chassis height affects the front wheel caster angle. *As rear suspension loading increases, what happens to the front caster angle?* (Ford Motor Company)

A zero or negative caster adversely affects driveability. It causes the vehicle to wander, instead of tracking in a straight line. It increases steering effort by reducing steering return. The steering wheel must be manually returned to a straight-ahead position following a turn.

Caster has a great impact on vehicle directional stability. Caster angle should be equal side to side. If the right-front wheel has less positive caster than the left, the vehicle will pull to the right. If the left-front wheel has less positive caster than the right, the vehicle will pull to the left.

Technicians do not generally consider caster a factor in tire wear. Caster, however, does affect tire wear during turns.

Toe Angle

Toe angle, or toe, is the measurement by which the front of a wheel points inward or outward, compared to a true straight-ahead position. The toe angle is viewed from the top of the vehicle. Toe angle is measured in inches, millimeters, or degrees.

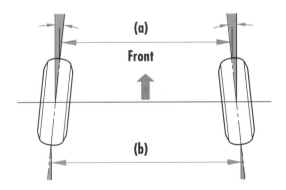

Fig. 3-5 Toe angle is the amount by which the front of the wheel points inward or outward compared to a true straight-ahead position. For wheels with toe-in, the distance between the wheels at the front **(a)** is less than the distance between the wheels at the rear **(b)**. *How does the toe angle affect tire wear?*

When the wheels are pointed in, the wheels have *toe-in*. The wheels are closer together at the front than they are at the rear of the wheels. When the wheels are pointed out, the wheels have *toe-out*. The wheels are further apart at the front than they are at the rear of the wheels. When the right and left wheels aim straight ahead and parallel, the wheels have *zero toe*. **Fig. 3-5.**

The goal is to have zero toe. The wheel with the greatest amount of toe-out will pull the vehicle in the direction of that wheel. However, the initial toe setting may be in or out to compensate for vehicle type and vehicle motion.

Typically, a slight toe-in is set on rear-wheel-drive vehicles. This setting compensates for vehicle motion and play in the steering and chassis parts. As the vehicle moves forward, the front wheels will pull out.

A slight toe-out is set on vehicles with front-wheel-drive. The torque action of front-wheel-drive pulls the wheels inboard when power is applied to move the vehicle forward. When a toe-out angle is set, the wheels will achieve a zero toe as the vehicle moves forward.

Toe is a tire-wearing angle. When a wheel has too much toe-out, the tread wears faster towards the inside of the tire. The wear may create a "feathered" edge pattern on the tread. When a wheel has too much toe-in, the tread wears faster towards the outside of the tire.

Steering Axis Inclination

The difference between a true vertical line drawn through the center of the wheel and an imaginary line drawn through the upper and lower ball joints is known as **steering axis inclination** or SAI. On a strut-equipped vehicle, consider the upper strut mount as the location of the upper ball joint. Steering axis inclination is viewed from the front of the vehicle. **Fig. 3-6.**

SAI is a negative camber angle of the steering axis. The upper ball joint is further inboard as compared to the lower ball joint. The setting of SAI affects steering stability. It also affects the wheel's ability to return to a straight-ahead position after a turn. A proper SAI setting reduces steering effort and tire wear.

SAI is not adjustable. It is a fixed angle designed into the suspension system by the manufacturer. Even though SAI is not adjustable, this angle should be checked during a wheel alignment. An incorrect SAI can be a signal of frame or suspension system damage.

Incorrect SAI could result from:
• A bent control arm.
• A bent spindle.
• Badly worn ball joints.
• Damage to vehicle frame or unibody structure.

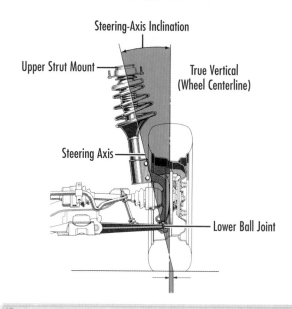

Fig. 3-6 Steering axis inclination, or SAI, is a fixed, non-adjustable angle. SAI is the angle created between a true vertical line drawn through the wheel center and the steering axis, a line drawn through the upper ball joint or upper strut mount and the lower ball joint. *What might an incorrect SAI indicate?* (Mazda Motors of America, Inc.)

Scrub Radius

Scrub radius, also called *steering offset,* is the distance between the steering axis and the centerline of the tire tread contact area. If the steering axis meets the road surface at a point outside the tire centerline, the scrub radius is negative. If the steering axis meets the road surface at a point inside the tire centerline, scrub radius is positive. **Fig. 3-7.** If the steering axis and the tire centerline intersect at the road surface, the scrub radius is zero. Scrub radius affects steering effort, steering return, and stability.

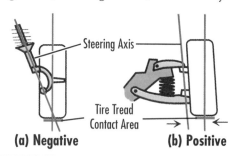

Fig. 3-7 Scrub radius is the distance between the steering axis and the centerline of the tire tread contact area. If the steering axis meets the road surface outside the centerline, the scrub radius is negative **(a)**. If the steering axis meets the road surface inside the centerline, the scrub radius is positive **(b)**. *What is a zero scrub radius?* (Ingall's Total Wheel Alignment Guide 1997)

Angles in Wheel Alignment

Caster, camber, steering axis inclination, directional pull, and toe angle are discussed throughout this chapter. Definitions and diagrams explaining them appear in different places. You can better understand these terms if you can see how they are related to each other.

In this activity, you will make your own simple model of a vehicle "front end." Then you will be able to see how factors such as caster, camber, directional pull, and toe angle affect steering and tire wear.

Apply it!

Demonstrating Caster, Camber, and Toe

Meets NATEF Science Standards for understanding the relationship between circular motion, linear motion, and friction.

Materials and Equipment
- 1.5" diameter furniture caster wheel unit, with pivot pin and plastic bearing insert.
- Masking tape, as wide as the wheel
- A flat non-slip surface (to simulate the road)

❶ Put two layers of masking tape around the wheel. The taped wheel simulates the tire. It also reduces slipping on the "road" surface.

❷ Hold the wheel unit by the plastic bearing insert so that the pivot pin is vertical. The wheel should be ahead of the pivot pin. Push the wheel unit across the surface. Notice how the wheel swings around so that its axle is behind the vertical pivot pin. The wheel has self-aligned. It now trails behind the pivot pin.

❸ Hold the unit with the wheel in front of the pivot pin. Tilt the pivot pin toward you and push. Increase the tilt angle until the wheel unit no longer rotates to put the axle behind. You have included enough positive caster to make the wheel self-aligning by moving the pivot point ahead of the steering axis.

❹ Tilt the pivot pin toward you at the same caster angle as in Step 3. Now tilt the pivot pin left (positive camber) or right (negative camber). Note what happens when you push forward. The wheel should go to the opposite side of the tilt. This illustrates how camber angle can change caster angle. Also, notice that the "tape tire" contacts the surface on only one side, leading to tire wear.

❺ Hold the pivot pin tilted toward you at the same caster angle as in Step 3. Now push the unit with the axle turned from the direction of push. You will see the wheel straighten out. This is directional pull.

❻ Hold the wheel unit by its frame instead of the pivot pin. Twist the frame slightly left or right of the direction of your push. As you push the unit, you should see and feel the "tape tire" sliding on the surface. This is what happens when the toe angle is incorrect and is why toe is a factor in tire wear.

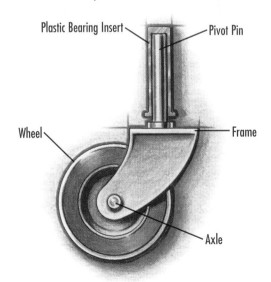

Rear-wheel-drive vehicles with unequal-length control arms have a positive scrub radius setting. Front-wheel-drive vehicles with a MacPherson strut front suspension have a negative scrub radius setting. Scrub radius, like SAI, is not adjustable.

Even though scrub radius is not adjustable, it should be checked during a wheel alignment. If incorrect, it can be a signal of suspension system or structure damage.

Included Angle

The camber angle plus the SAI angle is known as the **included angle**. Incorrect included angle may indicate a suspension or vehicle structure problem. The included angle is another reference that can be checked when diagnosing wheel alignment problems. **Fig. 3-8.**

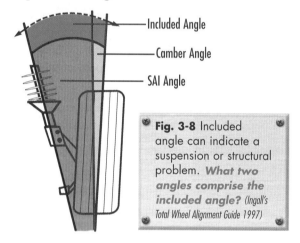

Included Angle
Camber Angle
SAI Angle

Fig. 3-8 Included angle can indicate a suspension or structural problem. *What two angles comprise the included angle?* (Ingall's Total Wheel Alignment Guide 1997)

Turning Radius

The turning radius, also called *toe-out on turns*, is the difference in the angles of the front wheels in a turn. During a turn, the two front wheels travel in concentric circles that have a common center. The outer wheel must travel a greater distance and make a wider turn than the inner wheel. Since the inner wheel travels a shorter distance, it must turn at a greater angle and follow a smaller radius than the outer wheel.

For example, when the inner wheel turns at an angle of 20°, the outer wheel may turn at an angle of 18°. The inner wheel toes out more in the turn. **Fig. 3-9.** This difference in toe angle should not vary by more than about 1.5° from the manufacturer's specifications. An incorrect turning radius angle may indicate a suspension or vehicle structure problem.

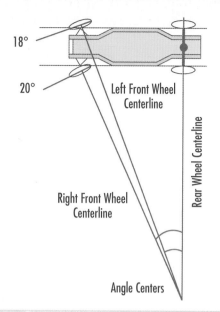

18°
20°
Left Front Wheel Centerline
Rear Wheel Centerline
Right Front Wheel Centerline
Angle Centers

Fig. 3-9 Turning radius (toe-out on turns). To avoid scrubbing, the inner wheel must turn a tighter radius. The two front wheels should turn at different angles. *Why does the inner wheel turn in a smaller radius than the outer wheel?* (Ingall's Total Wheel Alignment Guide 1997)

Setback

Wheel *setback* is the difference between the wheelbase on one side of the vehicle and the wheelbase on the other side. *Wheelbase* is the distance between the center of the front wheel and the center of the rear wheel on the same side of the vehicle. Setback, therefore, is a condition in which the distance between the left-front and left-rear wheels is different than the distance between the right-front and right-rear wheels. **Fig. 3-10.**

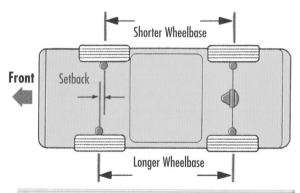

Shorter Wheelbase
Front
Setback
Longer Wheelbase

Fig. 3-10 Setback is the difference between the wheelbase on one side of the vehicle and the wheelbase on the other side. The vehicle will pull in the direction of the shorter wheelbase. *How is wheelbase measured?* (Ford Motor Company)

On front-wheel-drive vehicles, improper location of the engine cradle or subframe will create a setback condition. If the wheelbase is shorter on one side of the vehicle, the vehicle will pull in the direction of the shorter wheelbase. If a customer complains of a pull and if the steering wheel is off-center, check setback.

Thrust Angle

The **geometric centerline** is a line drawn from the midpoint of the rear of the vehicle to the midpoint of the front of the vehicle. The **thrust line** is a line drawn perpendicular to the rear axle. It bisects the total toe of the rear wheels. It points toward the front of the vehicle in the same direction as the rear wheels. If the thrust line and the geometric centerline are not parallel, a **thrust angle** exists. **Fig. 3-11.** The thrust angle is positive if it is to the right of the geometric centerline. It is negative if it is to the left of the geometric centerline. A thrust angle means that the rear wheels aim in a direction offset to the center of the front of the vehicle.

A thrust angle causes the vehicle to steer or wander in the direction opposite to the thrust line. If the rear axle aims to the left, the vehicle will pull to the right, and vice-versa. A thrust angle can cause excessive tire wear that resembles toe wear. If the vehicle has rear toe adjustments, the thrust angle can be set to zero degrees. This permits a total four-wheel alignment. If the vehicle does not have rear-wheel toe adjustments, the front-wheel toe is set to correspond with the rear-wheel toe. The wheels will then track properly with a centered steering wheel. This is called a *thrust line alignment.*

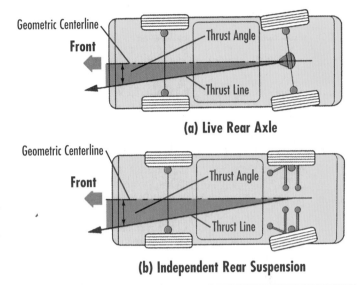

(a) Live Rear Axle

(b) Independent Rear Suspension

Fig. 3-11 Thrust angle on a vehicle with a live rear axle **(a)** and an independent rear suspension **(b)**. Thrust angle is the angle created between the geometric centerline and the thrust line. *How does the thrust angle affect the directional control of the vehicle?* (Ford Motor Company)

SECTION 1 KNOWLEDGE CHECK

❶ What is the camber angle?

❷ How does caster angle affect vehicle direction control?

❸ Why is a slight toe-out set on front-wheel-drive vehicles?

❹ What creates setback on front-wheel-drive vehicles?

❺ What condition exists when a vehicle has a thrust angle?

Section 2

Wheel Alignment Procedures

Wheel alignment requires several different checks and adjustments. Wheel alignment involves:
- Checking camber, caster, and toe angles.
- Diagnosing possible problems.
- Adjusting caster, camber, and toe angles to the specifications in the vehicle's service manual.
- Centering the steering wheel when adjustments are complete.
- Road-testing the vehicle.

Many driveability conditions may appear to be wheel alignment problems. However, broken, worn, or loose steering system or suspension components may be at fault.

Table 3-A lists those complaints, causes, and corrections relating specifically to wheel alignment.

Table 3-A	WHEEL ALIGNMENT DIAGNOSIS	
Complaint	**Possible Cause**	**Correction**
1. Hard steering	a. Low or uneven tire pressure b. Too much caster c. SAI incorrect	a. Inflate to correct pressure b. Check and adjust caster c. Check for damaged suspension or frame components
2. Wander or drift	a. Excessive toe-out b. Too little caster c. Thrust angle incorrect	a. Check and adjust toe-out b. Check and adjust caster c. Check and adjust thrust angle
3. Pulls to one side	a. Uneven caster b. Too much camber c. Incorrect setback d. Thrust angle incorrect	a. Check and adjust caster b. Check and adjust camber c. Check and adjust setback d. Check and adjust thrust angle
4. Shimmy	a. Wheels out of balance b. Too much caster c. Incorrect or unequal camber	a. Check runout. Check and rebalance wheels b. Check and adjust caster c. Check and adjust camber
5. Steering kickback	a. Tire pressure low or uneven b. Excessive positive caster	a. Inflate to correct pressure b. Check and adjust caster
6. Poor steering return	a. Excessive negative caster b. SAI incorrect	a. Check and adjust caster b. Check for damaged suspension or frame components
7. Improper tire wear	a. Wear at tread sides from under inflation b. Wear at tread center from over inflation c. Wear at one side of tread due to camber d. Feathered tread wear due to toe e. Uneven or scalloped wear f. Thrust angle incorrect	a. Inflate to correct pressure b. Inflate to correct pressure c. Check and adjust camber d. Check and adjust toe e. Rotate tires, balance tires and align wheels f. Check and adjust thrust angle
8. Excessive sway in turns	a. Caster incorrect	a. Check and adjust caster
9. Vehicle unstable	a. Low or uneven tire pressure b. SAI incorrect	a. Inflate to correct pressure b. Check for damaged suspension or frame components
10. Steering wheel off-center	a. Setback incorrect	a. Check and adjust setback

Alignment Pre-Check

Before performing any wheel alignment, always check tire pressure. Tire pressure affects vehicle pull. If the front tire is significantly underinflated, the vehicle will pull in the direction of the underinflated tire. Tire pressure also affects wandering, ride quality, tire wear, and driveability.

Check that tires on the same axle are the same size. Tires of different sizes cause pull and directional problems. Look for bent wheels and damaged hubs.

Always measure chassis height and compare measurements to the specifications in the vehicle's service manual. Incorrect height has a direct effect on camber, caster, and toe angle settings.

Inspect the suspension and steering system for damaged, missing, and worn parts. Check the steering gear, steering linkage, ball joints, wheel bearings, springs, shocks, stabilizer bar links and bushings, control arms, and control arm bushings. Replace any damaged or worn parts.

Alignment Equipment

Accurate wheel alignment requires the skillful use of specialized equipment. Specialized wheel alignment equipment provides easy-to-read diagrams and illustrations. In some instances, the equipment shows exactly how much adjustment is needed at each wheel. The display monitor shows the camber, caster, toe, thrust angle, toe-out on turns, SAI, and included angle of each wheel. The system contains alignment specifications for various vehicle makes and models. It will store information related to an individual alignment job as well as other pertinent vehicle and customer information. Equipment manufacturers provide training or documentation on the use of their equipment.

Regardless of the equipment used, the goal is to measure wheel alignment angles and to monitor changes to those angles. A skilled wheel alignment technician understands why angles change and how to achieve the required adjustments. It is critical that you understand the theory of wheel alignment, rather than simply depending on the alignment equipment.

Four-Wheel Alignment

There are three basic types of wheel alignment. A front-wheel alignment on the vehicle's geometric centerline, a thrust line alignment, and a total four-wheel alignment. The total four-wheel alignment should always be done when there are means of adjustment on all four wheels. If the rear wheels are not adjustable, then the thrust line alignment should be done. This allows the front-wheel toe to correspond to the rear-wheel toe and provides proper tracking of all four wheels. With up-to-date alignment equipment there is no need to do a front-wheel alignment using the geometric centerline of the vehicle. This will not ensure proper wheel tracking and involves the same amount of labor as the thrust line alignment. **Fig. 3-12.**

Safety First **Personal** Before performing a road test, be sure to obtain written permissions from appropriate authorities.

Fig. 3-12 A four-wheel alignment system in use. The alignment heads attach to the wheels and transmit wheel position information to the system's computer. *What is the advantage of using a computer-aided alignment machine?* (Hunter Engineering Company)

TECH TIP **Making Adjustments.** It's easier to remove shims or move components if the weight of the vehicle is not on the suspension. Raise the vehicle to unload the suspension and remove the vehicle weight from the wheels. Be sure to tighten all adjustments before placing weight on the wheels.

The location of the rear axle can dramatically affect the directional control of a vehicle. The rear wheels dictate the "thrust angle" of the vehicle. This, in turn, affects the geometry of the front wheels.

Alignment Adjustments

The general methods of making alignment adjustments are covered here. Always refer to the vehicle service manuals for the specific make and model of vehicle when making alignment adjustments. If alignment adjustments are available on the rear wheels, these adjustments must be made before adjusting the front wheels

Camber Depending on the vehicle design, there are several methods for adjusting camber. Some adjustment methods use spacer shims. Other methods use rotating eccentric control arm shafts or eccentric cam bolts. On some strut-equipped vehicles, camber is adjusted by moving the upper or lower mounting points.

If a vehicle uses multiple control arms, such as in a short-arm/long-arm (SLA) system, camber is adjusted by adding or removing shims. The shims are located where the upper control arm connects to the frame. In some designs, these shims are inboard of the upper control arm. In other designs, the shims are outboard of the upper control arm.

Adding or removing shims of equal thickness at the front and rear location affects camber angle.

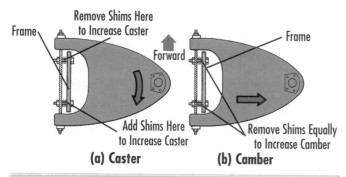

Adding or removing shims of unequal thickness at the front and rear locations affects camber and caster. **Fig. 3-13.**

When the upper control arm moves inward, the camber angle becomes more negative. When the upper control arm moves outward, the camber becomes more positive. **Fig. 3-14.**

Some vehicles have an offset upper control arm. Camber adjustment is made by rotating the control arm. The rotation moves the upper control arm inward or outward.

On other vehicles, camber is adjusted at the lower control arm. The lower control arm pivot bolt may be eccentric, or it may have eccentric washers. Camber is adjusted by rotating the eccentric adjusters. When the lower control arm moves inward, the camber angle becomes more positive. When the lower control arm moves outward, the camber angle becomes more negative.

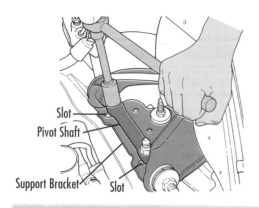

Fig. 3-14 Adjusting both caster and camber using elongated holes in the frame. *How is camber affected when the control arm is moved inward?* (DaimlerChrysler)

On some vehicles with MacPherson struts, camber is adjusted at the bottom of the strut at the steering knuckle. Loosening the two lower strut-mounting bolts creates enough movement to allow some camber adjustment. If there is not enough movement, replace the upper or lower strut mounting bolt with an adjuster bolt that uses an eccentric washer. **Fig. 3-15.**

Some vehicles equipped with MacPherson struts may provide a small amount of camber adjustment at the top of the strut mounts. The mounts have elongated mounting holes. These holes allow inward and outward movement of the strut. However, new vehicle designs do not commonly employ this method. Installing aftermarket specialty adjuster kits serves the same purpose.

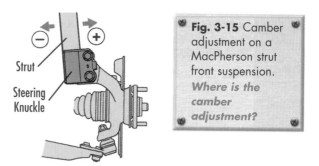

Fig. 3-15 Camber adjustment on a MacPherson strut front suspension. *Where is the camber adjustment?*

Some vehicles may allow camber and caster adjustment by sliding the upper control arm inward or outward. This type of upper control arm is attached with bolts through slotted holes in the frame. When movement at both the front and rear elongated holes is equal, only camber angle changes. If movement is unequal at these holes, both the camber angle and the caster angle change.

Caster Depending on the vehicle design, caster angle is set by a number of methods. If the suspension is a short-arm/long-arm (SLA) system, caster adjustment is similar to camber adjustment. Spacer shims are located at the upper control arm. When an unequal thickness of shims is added or removed from the front and rear shim locations, caster angle is affected.

If the vehicle has a lower strut rod, the strut rod may have a threaded adjustment. The threaded adjustment lengthens or shortens the strut rod. If you shorten the strut rod, the caster angle becomes more negative. If you lengthen the strut rod, the caster angle becomes more positive.

Measuring Camber

Manufacturer's specs give the information needed to make decisions about camber adjustment.

Camber angle tolerance is given as a range of acceptable values. This range can be written as an inequality. A *mathematical inequality* is a statement that uses the symbols $<, \leq, >, \geq$.

For example, if the camber angle tolerance is given as $1° \pm 0.5°$, the inequality is $0.5° \leq c \leq 1.5°$. This statement is read as camber angle, c, is greater than or equal to $0.5°$ and less than or equal to $1.5°$.

For a 1994 Mitsubishi Eclipse, the camber angle tolerance in degrees is $0.75° \pm 0.5°$. You measure the right rear wheel camber of your customer's Eclipse. It measures $1.4°$.

For this car, you will:

1. Determine the tolerance interval graph and indicate the optimal value.

2. Determine the adjustment for the optimal or preferred setting.

3. Determine the minimum and maximum adjustment for an acceptable measure.

4. Write a mathematical inequality for the camber, c, and tolerance interval you have determined.

The range, $0.75° \pm 0.5°$, means that the optimal, or desired, setting is $0.75°$.

The lower level of the tolerance interval is:
$$0.75° - 0.5° = 0.25°$$

The upper level of the tolerance interval is:
$$0.75° + 0.5° = 1.25°$$

Since $1.4°$ is greater than the upper level tolerance, you must adjust the camber. The necessary adjustment must be made in the negative direction. For the optimal setting, the adjustment must be $1.4° - 0.75° = 0.65°$ in the negative direction.

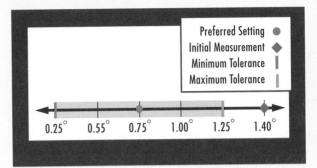

Preferred Setting	●
Initial Measurement	◆
Minimum Tolerance	|
Maximum Tolerance	|

$0.25°$ $0.55°$ $0.75°$ $1.00°$ $1.25°$ $1.40°$

The minimum adjustment would be:

$1.4° - 1.25° = 0.15°$ in the negative direction.

The maximum acceptable adjustment would be:

$1.4° - 0.25° = 1.15°$ in the negative direction.

The graph indicates the preferred setting and the initial measurement. The graph also indicates the minimum and maximum tolerance values of camber angle.

In this situation, the mathematical inequality for an acceptable camber measurement, c, is:

$$0.25° \leq c \leq 1.25°.$$

Apply it!

Meets NATEF Mathematics Standards for using angle measurements and tolerance intervals.

Assume that the camber of a front wheel of a 1994 Mitsubishi Diamante is $-0.29°$ and the tolerance is $0.31° \pm 0.50°$.

❶ Determine the tolerance interval graph and indicate the optimal value.

❷ Determine the adjustment for the optimal or preferred setting.

❸ Determine the minimum and maximum adjustment for an acceptable measure.

❹ Write a mathematical inequality for the camber, c, for the tolerance interval.

In the workplace you will probably compare the measured camber with the specs. With practice, you will learn to do this mentally without writing down all the steps you have taken.

If the upper or lower control arm has adjuster bolts, both camber and caster can be adjusted. If both front and rear bolts are adjusted an equal amount, only camber is affected. If the adjustment is unequal, the control arm will pivot slightly. Forward or rearward pivot of the control arm changes the caster angle.

If the upper control arm fastens to the frame through slotted holes, the control arm can be moved inward or outward. If the movement is unequal, the caster angle will change.

On some vehicles equipped with MacPherson struts, there may not be an adjustment for caster angle. If caster angle is incorrect, replace the control arm or strut or install aftermarket cam bolts in the lower control arm.

Toe Toe angle is adjusted at the tie rods. Rotating the tie rods effectively lengthens or shortens the rods. **Fig. 3-16.**

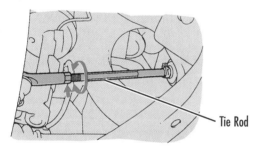

Fig. 3-16 The toe angle is adjusted at the tie rods. *How do you adjust toe?*

On some vehicles, the tie rod connects to the tie rod end using a separate "adjuster sleeve." This sleeve is simply an internally threaded tube. When rotated, the sleeve changes the length of the tie rod. **Fig. 3-17.**

When adjusting toe, any adjustment must be applied equally to both tie-rod ends. The right and left tie rods should be as equal in length as possible. Never perform the total toe adjustment on only one tie rod.

After adjusting toe, always be sure to tighten lock nuts/jam nuts fully to factory torque values.

Several specialty tools can assist in the adjustment process. An *inner tie-rod end tool* resembles pliers or a crescent wrench. It grips the tie rod, allowing you to rotate the tie rod at its adjustment connection. A *tie-rod puller* allows you to separate the tie rod end's tapered stud from the steering arm.

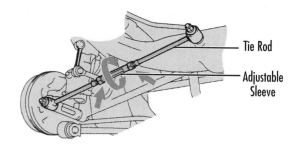

Fig. 3-17 This tie rod has a threaded adjustable sleeve. *How does the sleeve affect tie rod length?*

Checks and Adjustments

The order of tasks for wheel alignment is as follows:

1. Check tire pressure and correct as needed.
2. Check tire size for mismatch.
3. Check tires for unusual or uneven wear.
4. Measure chassis height and correct as needed. Check for unusual loads in the passenger and cargo areas.
5. Check for loose, binding, worn, or damaged parts (wheel bearings, steering linkage, steering gears, suspension pivots, ball joints, and shocks).
6. Position the vehicle on the alignment rack and connect alignment equipment.
7. Using the alignment equipment, check existing angles (camber, caster, toe, included angle, SAI, thrust angle, toe-out on turns, setback). Determine which angles require adjustment.
8. Adjust rear camber where possible.
9. Adjust/address thrust angle. Correct rear toe, if possible.
10. Adjust front caster.
11. Adjust front camber.
12. Center steering wheel.
13. Adjust front toe.
14. Road test the vehicle, checking for steering control, directional stability, drifting, and wandering. Check for pull, both while cruising and during braking.

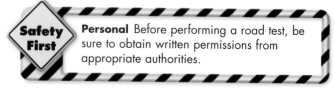

Safety First

Personal Before performing a road test, be sure to obtain written permissions from appropriate authorities.

Current alignment equipment usually requires the steering to be locked in the straight-ahead position before adjusting front toe. A special tool called a *steering wheel holder* or lock can be used to hold the steering wheel in a straight-ahead position while toe adjustment is being accomplished. Tie rods on each front wheel are usually adjusted independently to adjust the individual toe of each front wheel. This assures proper total toe and a centered steering wheel.

Rear Wheel Alignment

Rear-wheel-drive vehicles (those with a fixed-rear-axle housing assembly) offer no rear toe or camber adjustment. If the rear thrust angle is incorrect, it may be due to an improperly mounted axle housing or a damaged frame. If frame damage exists, it must be repaired before performing an alignment.

Safety First

Personal Always follow the procedures in the vehicle service manual for steering wheel removal or installation. This is especially important when working with a steering wheel assembly that has an air bag. To prevent accidental discharge of the airbag, the battery and specific fuse may require disconnection. You must always take special precautions when handling or storing any air bag assembly! Until you understand and are prepared to follow the service manual's safety instructions, *never* attempt to remove a steering wheel.

Independent rear suspensions on front- and rear-wheel-drive vehicles may provide adjustment of camber and toe. Adjustment is made by rotating the cam adjuster bolts where the rear control arms mount to the frame.

Depending on the design, cam adjusters may provide adjustment only for the toe angle. If the vehicle has rear MacPherson struts, rear camber adjustment is at the lower strut mounting point. Loosen the two mounting bolts that secure the strut to the knuckle or install adjuster bolts at these locations. For specific procedures, refer to the vehicle's service manual.

For camber adjustment, some rear hubs (on front-wheel-drive vehicles) may allow the use of shims to change the angle of the rear spindles.

Many four-wheel-drive trucks and sport utility vehicles may require the use of tapered shims to alter the spindle angle as well. Because there are so many different suspension designs, many alignment methods are available. A number of specialty tools are in common use for this purpose. **Fig. 3-18.** Always refer to the service manual.

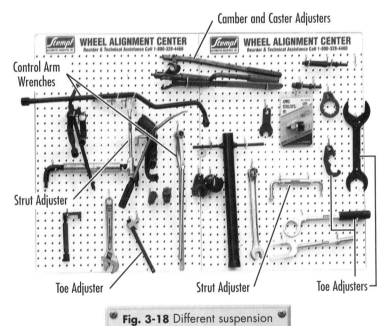

Camber and Caster Adjusters

Control Arm Wrenches

Strut Adjuster

Toe Adjuster   Strut Adjuster   Toe Adjusters

Fig. 3-18 Different suspension designs require the use of different specialty tools.

SECTION 2 KNOWLEDGE CHECK

❶ Why should a four-wheel alignment be performed instead of a two-wheel alignment?

❷ What must be checked before making any wheel alignment adjustments?

❸ What are the two commonly accepted methods of centering a steering wheel?

❹ What is the proper order of completion for the following alignment steps: center steering wheel, adjust front toe, adjust front caster, adjust front camber?

❺ What may some four-wheel-drive trucks and sport utility vehicles require you to use to adjust camber?

CHAPTER 3 REVIEW

Key Points

Meets the following NATEF Standards for Suspension & Steering: wheel alignment diagnosis; wheel alignment adjustment.

- Wheel alignment has a direct effect on vehicle driveability. Incorrect alignment causes driveability problems and uneven tire wear.
- Wheel alignment angles are camber angle, caster angle, toe angle, scrub radius, included angle turning radius, setback, steering axis inclination, and thrust angle.
- The wheel camber angle adjustment determines what portion of the tire tread contacts the road surface.
- The wheel caster angle affects directional control of the vehicle and steering wheel return.
- The ideal toe angle setting is zero. Toe-in or toe-out causes uneven tread wear.
- The steering axis inclination is a reference angle and cannot be adjusted.
- Before performing a wheel alignment, check items that can have an effect on wheel alignment, such as condition of suspension and steering parts.
- Wheel alignments should always be a total four-wheel alignment.

Review Questions

❶ What vehicle driveability problems are caused by incorrect wheel alignment?

❷ How is the camber angle of a wheel checked? What methods are used to adjust the camber angle?

❸ How is the caster angle of a wheel determined? How is the caster angle adjusted?

❹ What determines the toe angle of a wheel? What procedure is used to adjust the toe angle?

❺ How is the steering axis inclination of a vehicle determined?

❻ How is the included angle checked? Why is it important to check this angle?

❼ If the steering axis inclination angle and the included angle do not meet manufacturer's specifications, what is causing the variance?

❽ **Critical Thinking** If the tires on the front axle are a different size than the tires on the rear axle, how might alignment be affected?

❾ **Critical Thinking** If a vehicle's front tires show excessive wear on the outer edge, but no sign of pull left or right, what might cause the problem?

TECHNOLOGY FORECAST FOR AUTOMOTIVE EXCELLENCE

More Miles Between Wheel Alignments

Wheel alignments are regularly performed on vehicles today. This service helps a vehicle handle and ride smoothly. It also helps tires wear evenly. As a result, tire life is increased. Automakers are now working to increase the number of miles that can be driven between alignments. In the future, wheel alignments may be necessary only if the vehicle has been in an accident.

Engineers have developed low-friction joints that don't wear out as easily as today's components. Instead of a traditional steel-on-steel setup, these more durable joints have polished steel inside a nylon or polymer socket. These new joints are permanently sealed, causing less wear. This design ensures that the joint is properly lubricated and the proper lubricant is used. Dirt, dust, and contamination cannot enter.

To help today's technician perform a more precise wheel alignment, the use of laser-based equipment is becoming common. These devices measure exactly, allowing the technician to set accurate angles.

Newer alignment systems can guide the technician through the alignment process with step-by-step instructions. These directions, with illustrations, are viewed on a monitor.

AUTOMOTIVE EXCELLENCE TEST PREP

Answering the following practice questions will help you prepare for the ASE certification tests.

1. When does negative camber exist?
   - ⓐ When the top of the wheel leans outward in relation to a true vertical.
   - ⓑ When the top of the wheel leans inward in relation to a true vertical.
   - ⓒ When the upper ball joint is rearward of the lower ball joint when viewed from the side.
   - ⓓ When the rear axle is crooked.

2. Why is the camber angle important?
   - ⓐ Because it directly affects ride stiffness.
   - ⓑ Because it is a tire wearing angle and it affects steering control.
   - ⓒ Because it can cause vibration.
   - ⓓ Because it affects the operation of the brakes and can cause fast pad wear.

3. If the caster angle is 0° on the right front, but 3° positive on the left front, how will this affect directional control?
   - ⓐ The vehicle will handle properly and will not pull in either direction.
   - ⓑ The vehicle will always pull to the right.
   - ⓒ The vehicle will pull to the left during braking.
   - ⓓ The vehicle will pull to the right only during braking.

4. Technician A says that SAI is adjustable. Technician B says that SAI is a reference angle that you measure to check for suspension or frame damage. Who is correct?
   - ⓐ Technician A.
   - ⓑ Technician B.
   - ⓒ Both Technician A and Technician B.
   - ⓓ Neither Technician A nor Technician B.

5. Technician A says that loose/worn ball joints can prevent accurate wheel alignment. Technician B says that ball joint condition has no effect on wheel alignment. Who is correct?
   - ⓐ Technician A.
   - ⓑ Technician B.
   - ⓒ Both Technician A and Technician B.
   - ⓓ Neither Technician A nor Technician B.

6. If you move an upper control arm on a SLA type front suspension inboard, what effect will this have on camber?
   - ⓐ Camber will move in a negative direction.
   - ⓑ Camber will move in a positive direction.
   - ⓒ None.
   - ⓓ Camber will always move to a zero-camber location.

7. Technician A says that a bent strut rod that connects the lower control arm to the frame helps to prevent caster angle change during braking. Technician B says that strut rods are sometimes adjustable for caster. Who is correct?
   - ⓐ Technician A.
   - ⓑ Technician B.
   - ⓒ Both Technician A and Technician B.
   - ⓓ Neither Technician A nor Technician B.

8. How do you adjust front wheel toe on a rack-and-pinion steering system?
   - ⓐ By turning an eccentric cam bolt on the lower control arm.
   - ⓑ By adjusting tire inflation pressure.
   - ⓒ By lengthening or shortening the two tie rod assemblies via the threaded connection at the tie rod end.
   - ⓓ By bending the upper control arms.

9. Technician A says that a bent MacPherson strut can affect the camber angle. Technician B says that a bent MacPherson strut can affect suspension movement. Who is correct?
   - ⓐ Technician A.
   - ⓑ Technician B.
   - ⓒ Both Technician A and Technician B.
   - ⓓ Neither Technician A nor Technician B.

10. If the left front tire is severely underinflated, but all other tires are properly inflated, how will this affect vehicle pull?
    - ⓐ The vehicle will tend to pull to the right.
    - ⓑ The vehicle will not pull in either direction.
    - ⓒ The vehicle will pull to the left.
    - ⓓ The vehicle will only pull during braking.

CHAPTER 4

Diagnosing & Repairing Tires and Wheels

You'll Be Able To:

- ⊗ Identify different types of tire construction.
- ⊗ Identify and determine tire dimensions.
- ⊗ Identify and read tire sidewall markings.
- ⊗ Identify and determine wheel dimensions.
- ⊗ Mount and balance tires.
- ⊗ Check tire and wheel runout.

Terms to Know:

aspect ratio
bolt circle
inflation pressure
rim diameter
rim width
torque
wheel backspacing
wheel offset

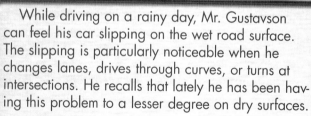

The Problem

While driving on a rainy day, Mr. Gustavson can feel his car slipping on the wet road surface. The slipping is particularly noticeable when he changes lanes, drives through curves, or turns at intersections. He recalls that lately he has been having this problem to a lesser degree on dry surfaces.

Concerned that he is putting other people and himself at increased risk, Mr. Gustavson brings his car to your service center. He says that this problem has become progressively worse in the past few months. How would you diagnose his complaint?

Your Challenge

As the service technician, you need to find answers to these questions:

1. Would worn tires create this problem?
2. Could a different tire tread design help?
3. Could worn shock absorbers add to the problem?

Tires

Tires provide the only connection between the vehicle and the road surface. For that reason, they are obviously very important. You may not plan to become a tire and wheel specialist. However, you need to understand basic tire design, tire diagnosis, and the handling of tires.

Tires serve several functions:

- Tires provide the traction required to negotiate the terrain. Tires are specially designed for use on different surfaces. These surfaces may be dry paved roads, wet roads, gravel roads, and roads covered with snow or ice.
- Tires act as part of the suspension. The air-cushioned support is part of the shock-absorbing action needed to reduce shock and vibration from the body and frame.
- The tire's construction, size, inflation pressure, and tread design enable proper handling during straight-line driving and proper operation during turns and braking.

Tire Construction

First, let us look at the difference between tube and tubeless tires. *Tube tires* have an inner tube inside the tire. This separate component holds the inflation air that supports the vehicle. Today, only some trucks and some motorcycles use tube tires. The vast majority of tires today are tubeless. **Fig. 4-1.**

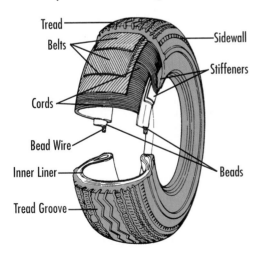

Fig. 4-1 Cutaway view showing construction of a typical tubeless radial tire. *What is the purpose of a tire inner tube?* (Ford Motor Company)

Tubeless tires contain air without the need for an inner tube. A leakproof seal is created where the tire bead seals against the wheel's flange. The tubeless tire must be nonporous. The wheel must be leak-free at all welds and seams. High-quality manufacturing standards for tires and wheels have eliminated the need for inner tubes, reducing the cost of tires. This has also reduced friction, heat buildup, and weight.

Bias-Ply, Bias-Belted, and Radial-Ply *Ply* refers to a layer of cord, fiberglass, steel, or other construction materials used to create a tire carcass. Tire makers can apply tire plies either diagonally or radially. **Fig. 4-2.** Bias plies, also called diagonal plies, run at a bias, or diagonal, angle across the tire from bead to bead. Because the plies intersect, they rub against each other, creating heat as the tire runs. Also, such tires tend to "squirm" when the vehicle is driven. Bias-belted tires include stabilizer belts placed under the tread area.

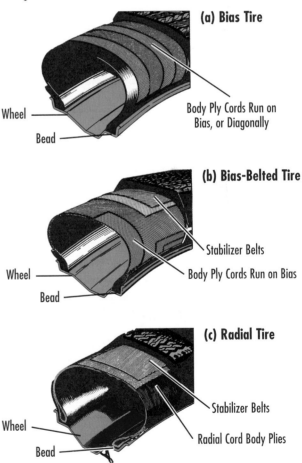

Fig. 4-2 Cross sections of **(a)** bias-ply tire, **(b)** bias-belted tire, and **(c)** radial tire. *What is the difference in construction between a radial tire and a bias tire?* (Bridgestone/Firestone, Inc.)

Radial tires feature plies that run bead-to-bead, parallel to each other, with stabilizer belts under the tread. These plies reduce heat. The tires tend to run more smoothly and quietly. A radial-ply construction is also more flexible, creating a softer sidewall. During turns more tread stays in contact with the road. **Fig. 4-3.** This results in better handling and improves high-speed stability. Radial tires also create a lower "rolling resistance." Radial tires, therefore, need less energy to roll forward. This increases fuel economy. All late-model vehicles have radial tires.

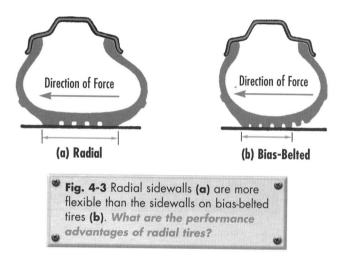

(a) Radial   **(b) Bias-Belted**

Fig. 4-3 Radial sidewalls **(a)** are more flexible than the sidewalls on bias-belted tires **(b)**. *What are the performance advantages of radial tires?*

Run-Flat Tires   Tire makers design some tires to operate with little or no inflation pressure. These tires are called "run-flat" or "zero-pressure" tires. They feature a reinforced sidewall and bead design intended to support the vehicle if a cut or puncture causes loss of inflation pressure. Typically, tire makers design these tires to operate at around 50 mph [80 kph] for 50–100 miles [80–161 km] with no internal pressure before sidewall damage takes place. This provides enough time for the driver to reach a service station or dealership.

Tire Dimensions   It is important to understand the dimensions used in tire construction and design. The most important dimensions are:

• Overall diameter, or tire diameter. This is the measurement from the top of the tire tread on one side of the tire to the top of the tire tread on the side of the tire directly opposite. This measurement runs through the center of the tire as viewed from the side.

• Section width. This refers to the tire's width measured at the tire's widest point (not the tread).

Essentially, this is the width of the tire at the sidewall area. **Fig. 4-4.**

• Tread width. This is the measurement from the outside edge of the tread to the inside edge of the tread.

• Section height. This is the height of the tire from the bead seat to the tread surface when the tire is mounted and inflated.

• Wheel size. This is the diameter of the bead area (where the tire meets the wheel).

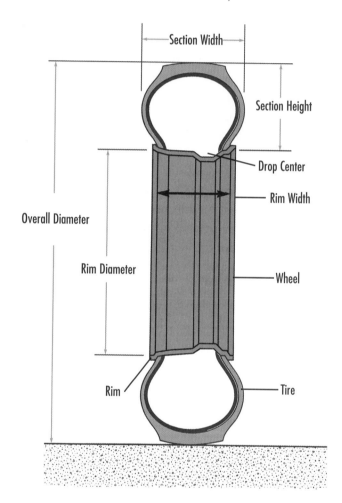

Fig. 4-4 Tire and wheel dimensions. Note tire overall diameter, section width, rim width, and section height. Rim diameter and rim width are wheel dimensions discussed later in this chapter. *Explain tire section width. What does this dimension represent?*

The size of the tires on most of today's passenger vehicles is identified using both English (customary) and metric dimensions. For example, a 245/60R15 tire has a section width of 245 mm. The second number (in this case, 60) refers to the tire's aspect ratio, explained below. The *R* means that the tire is of radial construction. The last number refers to the wheel diameter required for that tire. In this case, 15 refers to a 15-inch rim.

Aspect Ratio A tire's **aspect ratio** refers to the relationship between the tire's section height and section width. This ratio of section height to section width may also be referred to as the *profile ratio*. The section height is a percentage of the section width. For example, a 60-series tire has a section height that is 60 percent as big as the tire's section width. The lower the number of the aspect ratio, the shorter the tire, compared to its section width. **Fig. 4-5.** For example, a 235/50R15 tire is shorter than a 235/60R15 tire. The number 235 means that the section width is 235 mm. If this tire has an aspect ratio of 60, it means that the section height is 60

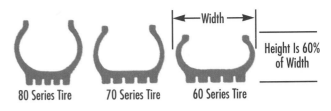

80 Series Tire   70 Series Tire   60 Series Tire

Fig. 4-5 Tire aspect ratio examples. *What are the characteristics of tires with lower aspect ratios?*

percent of the section width. If it is a 50-series tire, the section height is 50 percent of the section width.

A "lower profile" tire has a shorter section height. Because of its lower section height, it provides less sidewall height. The shorter the sidewall, the less the sidewall can flex. Lower section height results in a more responsive tire that reacts faster during turns.

Determining Tire Diameter

You can use the markings on a tire's sidewall to determine the tire's overall diameter. The sidewall information gives you the section width, the aspect ratio, and the rim diameter. For a tire with the markings 225/70R15, the rim diameter is 15″. The section width is 225 mm. The section height is 70 percent of 225 mm:

$$0.70 \times 225 \text{ mm} = 157.5 \text{ mm}.$$

You need both these dimensions to calculate the diameter of the tire. Refer to **Fig. 4-4.** The overall diameter is twice the section height plus the rim diameter. One of these numbers is given in the metric system and the other in the English (customary) system. You must decide which system you want to use.

To use the metric system, first convert 15 inches to millimeters. Remember that 1″ = 25.40 mm. So, 15 x 25.40 = 381 mm. The tire diameter in millimeters is:

$$(2 \times 157.5 \text{ mm}) + 381 \text{ mm} = 696 \text{ mm}$$

To use the English system convert 157.5 mm to inches. Use the conversion factor 1 mm = 0.039″. So, 157.5 mm × 0.039 = 6.14 in.

The tire diameter, in inches, is:

$$(2 \times 6.14 \text{ in}) + 15 \text{ in} = 27.28 \text{ in}$$

Apply it!

Meets NATEF Mathematics Standards for using and converting English (customary) and metric units.

❶ Look at the two tire diameter calculations in the example above. Does 27.28″ equal 696 mm?

❷ A customer tells you she is considering two possible replacement tires. She asks you which will be closest in diameter to her current tire, which is the tire used in the example. The two replacement tires have these markings:

Tire A: 245/60R15

Tire B: 255/60R15

Which tire's diameter is closest to her current tire's diameter?

In some cases, the tradeoff is that a shorter sidewall may mean that the tires give a stiffer ride. However, tire manufacturers can design short sidewall tires that provide a comfortable ride. Thus, this is not a hard and fast rule.

Tire Tread The part of the tire that contacts the road is called the *tire tread.* The tread serves several functions, including providing traction. Tire makers design treads to provide traction and stability in specific conditions. Although tread designs vary, in general a tread pattern with more grooves and wider tread grooves provides better traction in wet or less-than-ideal traction conditions, such as in rain, snow, gravel, and dirt.

Tire makers design most passenger vehicle tire treads to perform on both dry and wet roads. When tire makers design a tire for use on loose ground, such as snow or dirt, the tread pattern has wider tread blocks and deeper tread grooves. This gives a more aggressive grip.

In appearance and performance, tread patterns may be symmetric or asymmetric. A *symmetric tread*

pattern has the same pattern on each side of the tread (as compared to the center of the tread). An *asymmetric tread pattern* may have a different style or shape of tread on the left side of the tire's tread area as compared to the right side of the tread area.

Tires may also have *directional tread.* This means that the tire has been designed to rotate in only one direction. This is often the case with tires with asymmetric treads. The tire maker may have designed the tire for best performance if one side of the tread faces outside, or outboard, and the other side faces inside, or inboard. While you may move symmetric, *nondirectional tread* tires to any other position on the vehicle, and they may rotate in either direction, you may use a directional tread tire with only one direction of forward rotation.

An asymmetric tire moved to the other side of a vehicle will still have the outboard tread in the outboard position. Only the direction of rotation will have changed. A symmetric tire can be directional, but you can mount either sidewall outboard, depending on which side of the vehicle it is located. **Fig. 4-6.**

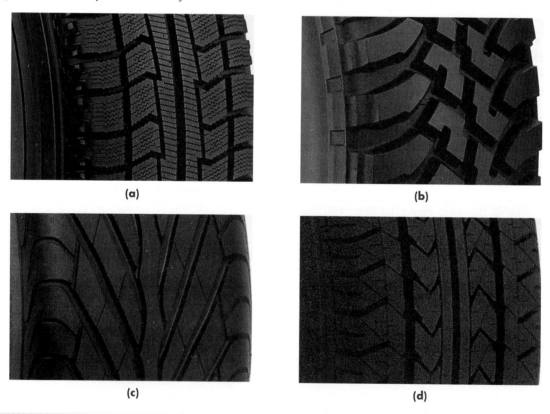

Fig. 4-6 Examples of tire tread designs. **(a)** Mud and snow (M&S) tread. Notice the great number of narrow slits (sipes). These aid traction in rain and snow. **(b)** A typical mud and snow tire for light truck use. The tread blocks help to grab and throw out snow and mud for increased traction. **(c)** A typical symmetric tread design. **(d)** A nonsymmetric or asymmetric tread. *What is the advantage of a varying tread design on a tire?* (Bridgestone/Firestone, Inc.)

To rotate the tires on a vehicle, you may move directional tires from front to rear or rear to front. However each tire must remain on the same side of the vehicle.

If you want to move a directional tire to the other side of the vehicle, you must dismount it from the wheel and remount it to reverse the position of the tire. Always pay attention to sidewall positioning marks. If the tire is directional, a sidewall marking such as an arrow will indicate the forward direction in which the tire must rotate. Also, some asymmetrical tread tires require mounting with a particular sidewall facing outboard. If you mount a

tire incorrectly, the vehicle may not handle as well. The tire may not provide directional stability.

The areas of the tread that contact the road are the tread blocks. The larger grooves that separate the tread blocks are the tread grooves. The very narrow slits that further separate the tread blocks are the tread sipes.

Tire makers design tread grooves in tires to allow water to pass through the tread on wet roads. These grooves also separate tread blocks to allow additional tread cooling. Tire makers design the smaller, narrow tread sipes to provide additional grip on wet roads. Sipes open as the tread hits the ground. They close when the tread leaves the ground.

Tire Sidewall Markings

The sidewall of a tire provides a great deal of information. Fig. 4-7. The marks may look confusing. However, they are actually easy to interpret.

Information typically provided on a sidewall includes:
- Brand name.
- Model name.
- Tire-size designation.
- Speed symbol.
- Load index.
- Maximum inflation pressure.
- Maximum load marking.
- Uniform Tire Quality Grading (UTQG) marking.
- Treadwear indication number.
- Tube requirement.
- Construction components.
- Construction type.
- Aspect ratio.
- US Department of Transportation (DOT) safety standard code.
- Manufacturer and plant code number.
- Tire size code number.
- Serial number.
- Date of manufacture.

The tire sidewall includes the brand (maker's) name and the tire

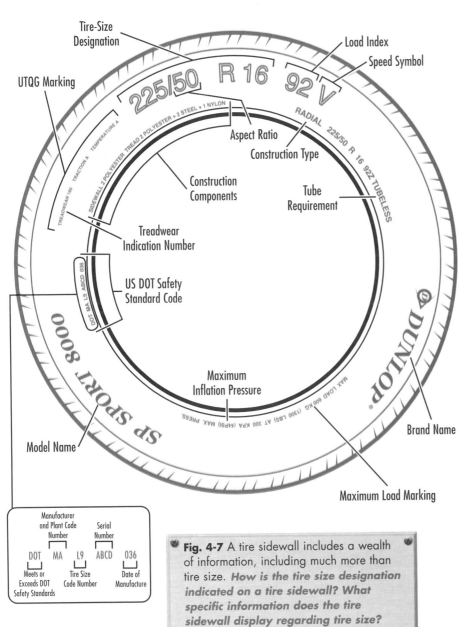

Fig. 4-7 A tire sidewall includes a wealth of information, including much more than tire size. *How is the tire size designation indicated on a tire sidewall? What specific information does the tire sidewall display regarding tire size?* (Dunlop Tire Corporation)

model. For example, a Pirelli P600 tire is the P600 model of a tire that Pirelli makes.

Tire sidewalls indicate size in either metric or alphanumeric form. A *tire-size designation* will show section width, aspect ratio, and rim diameter. For example, a designation of 215/65R14 means that the tire section width is 215 mm, the aspect ratio is 65, and the tire requires mounting on a 14-inch diameter rim. The *R* means that the tire is of radial construction.

The sidewall may include a *speed symbol*. For example, in the designation 255/50ZR16, the speed symbol *Z* indicates the speed rating. This speed rating refers to the maximum allowable operating speed of the tire.

The speed rating indicates the tire's certified top speed. It does not indicate the top speed potential of the tire. Only DOT-approved tires are speed-rated.

The speed rating system is determined by the Rubber Manufacturer's Association (RMA). This uses a letter designation from A to Z, from a low of 3 mph [5 kph] to over 168 mph [270 kph]. Passenger vehicle speed ratings are concentrated in the 100 to over-168 mph [161 to over-270 kph] range. **Table 4-A** lists currently used speed ratings related to passenger vehicle tires, and their meaning.

Note that a higher-letter does not necessarily indicate a higher speed rating. For instance, an H-rated tire carries a higher speed rating than does an S-rated tire.

Table 4-A. SPEED RATING INDEX PASSENGER VEHICLE TIRES		
Speed Rating	Speed	
	mph	kph
S	112	180
T	118	190
U	124	200
H	129	208
V	149	240
W	168	270
V*	over 130	over 210
Z	over 149	over 240
Z*	over 168	over 270

(*Unlimited rating)

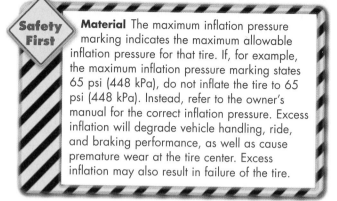

Safety First

Material The maximum inflation pressure marking indicates the maximum allowable inflation pressure for that tire. If, for example, the maximum inflation pressure marking states 65 psi (448 kPa), do not inflate the tire to 65 psi (448 kPa). Instead, refer to the owner's manual for the correct inflation pressure. Excess inflation will degrade vehicle handling, ride, and braking performance, as well as cause premature wear at the tire center. Excess inflation may also result in failure of the tire.

Tire makers use the speed rating as an industry reference number. The speed rating does not mean that you should drive at those speeds. The speed rating is simply an indication of the tire's certified top sustainable speed.

Load index, or *load range*, numbers give the maximum allowable load for that tire. For instance, a tire may have markings showing a maximum load of 2,200 lbs [999 kg]. This means that one corner of the vehicle can weigh no more than 2,200 lbs [999 kg], because that tire is capable of safely supporting only that amount of weight. In selecting any tire for any vehicle, always consider load ranges.

The *maximum inflation pressure* marking indicates the maximum pressure that the tire can handle. This is not necessarily the pressure that the tire needs to operate on a specific vehicle. The vehicle owner's manual will give the recommended inflation pressure.

The *maximum load marking* indicates the maximum amount of weight the tire can carry when the tire is inflated to its maximum pressure. Any weight above the specified maximum results in unsafe operation.

The *Uniform Tire Quality Grading (UTQG) marking* indicates subjective tire performance ratings regarding tread wear, wet braking traction, and resistance to temperature. This information is useful only in comparing different models of tires within the same manufacturer's line.

The *treadwear indication number* included in the UTQG gives a rough indication of how fast the tread will wear. The lower the number, the faster the tread will wear. A tire with a treadwear rating of 140 will probably wear faster than one with a treadwear rating of 300.

Interpreting Information

The vehicle identification number (VIN) has become an important source of data regarding the vehicle. Insurance companies and police departments consider it to be like a "fingerprint" that identifies the automobile. The 17-character VIN provides background data and production information for the vehicle. If you know how to interpret the letters and numbers, you can tell where the car was assembled. You can even tell the color of the original paint. Vehicle service manuals have charts that allow you to decode VIN information.

Another code system is used on tires. The information on the sidewalls also gives a wide range of information. Size, manufacturer, and various ratings can be found in the letters and numbers on the sidewalls. Information given in both letters and numbers is *alphanumeric*

information. Some alphanumeric information also includes other symbols such as punctuation marks and mathematical symbols. As the technician, you need to know the meanings of these letters and numbers. This will allow you to provide your customers with a safe, comfortable ride.

Apply it!

Meets NATEF Communications Standards for identifying and using information written as alphanumeric codes.

Refer to **Figure 4-7.**

❶ List the various types of information that can be found on the tire sidewall.

❷ Examine a tire. Write down the information presented by the letters and numbers on the tire.

The *tube requirement marking* on a sidewall will indicate whether the tire needs a tube. The marking will likely state either TUBE or TUBELESS.

There are two tire sidewall markings related to the tire's construction. The *construction components marking* indicates the number of plies and the materials used to make the plies for the sidewall, the radial belts, and the backup plies under the belts. The *construction type marking* indicates whether or not the tire is a radial tire.

The *US Department of Transportation (DOT) safety standard code* indicates that the tire meets accepted tire safety standards and that the US DOT has approved the tire for use on roads in the United States. If the tire is not DOT-approved, the maker does not intend it for use on public roads. For example, typical race tires are not DOT-approved. They may not offer the tread or sidewall protection needed for street use.

A tire marked M&S is designed as a "mud and snow" tire. The marking might be M&S, M+S, or M/S. They all mean the same thing. Tire makers design mud and snow tires to provide better grip and handling in less-than-ideal conditions while also pro-

viding proper performance on dry and rainy roads. You can use M&S tires year-round. They are not snow-only tires.

Checking Inflation Pressure

Inflation pressure is the measurement of the compressed air in a tire, expressed in pounds per square inch (psi) or kilopascals (kPa). Check tire inflation pressure when the tire is "cold." A tire is cold when its temperature is the same as the ambient temperature. The ambient temperature is the temperature of the surrounding air. Do not check inflation pressure when the tire is hot, after the car has been driven for a long period. Tires generate heat when the vehicle is driven. This raises inflation pressure. After a vehicle is parked, the tire begins to cool. This reduces inflation pressure. The inflation pressure listed in vehicle owner's manuals are "cold" pressures.

Some late-model vehicles use an electronic low-tire-pressure warning system. A *low-tire-pressure warning system* features a special *tire-pressure sensor* mounted inside the wheel rim. If tire pressure falls below a preset level, this tire-pressure sensor sends a radio signal to a receiver module in the vehicle. **Fig. 4-8.** A warning light then glows to indicate low tire pressure. The light gives the driver an early warning of an underinflation problem. This allows the driver to avoid excessive tire wear and handling problems.

Tire Inspection

Inspect a tire periodically for wear and damage. Check the tire tread for signs of uneven wear. Look for indications of worn suspension parts and improper inflation pressures. Look for problems that might result from poor wheel alignment or wheel balance. **Fig. 4-9.**

A thorough tire inspection involves checking tire noises. Some tire noises are caused by improper mounting, wheel defects, vibrations, or tire imbalance. To determine if there are problems, drive the vehicle on a smooth road at a variety of speeds. Listen for noise changes during acceleration and deceleration. The tire noise should remain constant, even while the engine, driveline, and exhaust noises vary.

Fig. 4-8 Tire-pressure sensor mounted on inside of wheel rim. *How does a tire-pressure sensor work?* (Smartire Systems, Inc.)

CONDITION	Rapid Wear at Shoulders	Rapid Wear at Center	Cracked Treads	Wear on One Side	Feathered Edge	Bald Spots	Scalloped Wear
EFFECT							
CAUSE	Underinflation or Lack of Rotation	Overinflation or Lack of Rotation	Under-Inflation or Excessive Speed	Excessive Camber or Improper Toe Adjustment	Incorrect Toe	Out-of-Balance and Worn Shocks or Struts	Lack of Rotation of Tires or Worn or Out-of-Alignment
CORRECTION	Adjust Pressure to Specifications When Tires Are Cool, Rotate Tires			Adjust Camber or Toe to Specifications	Adjust Toe-in to Specifications	Check Shocks and Struts. Balance Wheels	Rotate Tires and Inspect Suspension. Check Alignment

Fig. 4-9 Tire wear patterns. *What should you look for when checking the wear pattern of each tire?* (DaimlerChrysler)

Tires feature *wear indicator bars*, also called tread-wear indicators (TWI), on the tread areas. *Tread-wear indicators* are raised ribs molded into the tread. Their location is indicated by a small triangular symbol with the letters *TWI* on the tire sidewall. The height of the tread-wear indicators is such that, when they are flush with the tread surface, the tread has worn enough to require replacement of the tire. **Fig. 4-10.**

Check the tire for sidewall damage such as cuts, bulges, and scuffs. Bulges may indicate that the plies have separated or broken. Any tire with sidewall damage that appears more serious than simple cosmetic abrasions may no longer be safe. Replace such a tire as soon as possible.

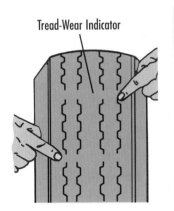

Tread-Wear Indicator

Fig. 4-10 Tread-wear indicators. When tread wears down to a predetermined level, these wear indicator bars become flush with the tread surface and easily visible. This indicates the need to replace the tires. *Why is it important to inspect a tire tread's wear indicator bars?*

You should also check for radial and lateral *runout* (even roundness). Testing for runout is discussed later in this chapter.

Tire Rotation

Rotating tires, or *tire rotation*, is the relocating of each tire and wheel assembly from one wheel position to another. Tire rotation prolongs the life of the tire. Typically, on front-wheel drive vehicle's front tires will wear faster than rear tires, because the front tires also steer the vehicle. By rotating the tires, the additional wear can be spread over all the vehicle's tires. Tires should usually be rotated every 5,000–10,000 miles [8,000–16,000 km]. The recommended rotation mileage for a specific vehicle can be found in the vehicle owner's manual.

Rotation patterns may vary, depending on the vehicle. Typically, on a rear-wheel drive vehicle, you bring the rear tires forward. Move the left-rear tire to the left-front location and the right-rear tire to the right-front

location. You also move the front tires to the rear in a crisscross pattern. Move the left-front tire to the right-rear and the right-front tire to the left-rear. **Fig. 4-11.**

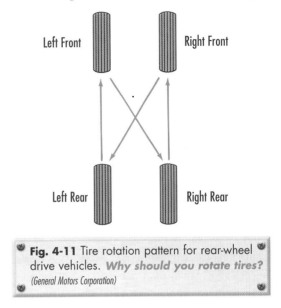

Left Front   Right Front

Left Rear   Right Rear

Fig. 4-11 Tire rotation pattern for rear-wheel drive vehicles. *Why should you rotate tires?* (General Motors Corporation)

On a front-wheel drive vehicle, move the front tires straight rearwards and move the rear tires forward in a crisscross pattern. Remember to cross the nondrive wheels. **Fig. 4-12.**

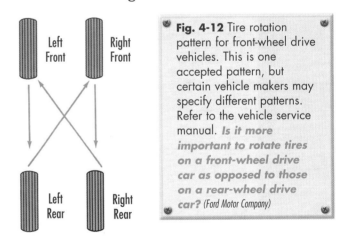

Left Front   Right Front

Left Rear   Right Rear

Fig. 4-12 Tire rotation pattern for front-wheel drive vehicles. This is one accepted pattern, but certain vehicle makers may specify different patterns. Refer to the vehicle service manual. *Is it more important to rotate tires on a front-wheel drive car as opposed to those on a rear-wheel drive car?* (Ford Motor Company)

With directional tires, the direction of each tire's forward rotation must remain constant. When moving directional tires, heed the directional arrow and any sidewall marking that indicates outboard side or inboard side. In some cases, tires may be location-specific. For example, a tire may be directional and intended for only right-front use. You cannot move this type of tire to a different location. To make sure you install the tires in the proper positions, read the sidewall information. When in doubt, refer to the vehicle service manual.

Section 2

Wheels

Wheels are made of steel or an alloy (typically an aluminum alloy). Wheel makers construct wheels in a number of ways. A one-piece wheel is usually an alloy wheel cast or machined as a one-piece part. A two-piece wheel features a rim and a center section. These two separate pieces are joined by either a series of welds or by a rivet or bolt-together connection.

A three-piece wheel involves two rim sections (inboard rim and outboard rim) and a center section. These are secured together with small bolts and nuts. Technicians often refer to three-piece wheels as "modular" wheels. This is because they allow different rim width and offset setups by using rim sections of different sizes.

Wheel makers either cast or forge alloy wheels. Forged alloy wheels are stronger. They are normally used for higher-stress applications, such as on trucks and performance vehicles.

Wheel Dimensions

Wheel dimensions include rim diameter, rim width, wheel offset, wheel backspacing, and bolt circle.

The **rim diameter** is the measurement from a point on the inside bead seat to the point on the inside bead seat directly across the diameter of the wheel. The rim diameter refers to the bead seat diameter, not the total outside diameter. For example, a 15-inch wheel features a 15-inch bead seat diameter; however, the total outside diameter may measure 17 inches.

Rim width is the measurement from the inside bead seat wall to the opposite bead seat wall. A 7-inch wide wheel will measure 7 inches between the bead seat walls.

Wheel offset is the location of the wheel's centerline as viewed from the front, relative to the location of the mounting face of the wheel hub. **Fig. 4-13.** The mounting face is the rear of the hub face that contacts the hub. If

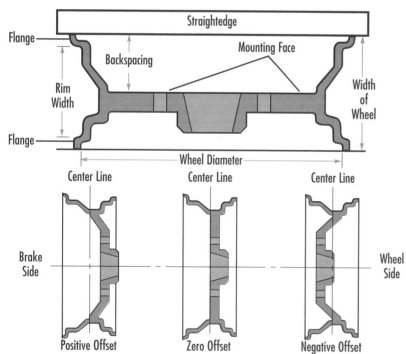

Fig. 4-13 Wheel dimensions. Rim width, backspacing, and offset are shown here. Remember that wheel diameter is measured from bead seat to bead seat, not from outer rim flanges. Using a straightedge provides a line of reference that allows easy and accurate measurement of backspacing. The straightedge is laid on top of the rim surface on the brake side of the wheel. *What is wheel backspacing?* (Tire Rack)

the mounting face is at exactly the midpoint of the rim width, the wheel has a zero offset. If the center of the rim width is outboard relative to the mounting face, the wheel has a negative offset. If the center of the rim width is inboard relative to the mounting face, the wheel has a positive offset.

The wheel offset affects track. *Track* is the distance between the center of the right tire's tread and the center of the left tire's tread. Negative offset creates a wider track, while positive offset creates a narrower track. The offset dimension is critical for clearance and handling.

Wheel backspacing is the distance from the wheel's mounting surface to the rear edge of the wheel rim. Backspacing is important only when considering inboard clearance between the wheel/tire and the wheelwell and suspension or frame.

Bolt circle refers to the size of the diameter of the mounting-fastener hole, or bolt hole, pattern in the wheel's center section. This is where the wheel bolts to the hub. Determine the bolt circle by drawing a circle passing through the center of each of the wheel's bolt holes. The bolt circle can refer also to the position of the bolt holes in the wheel's center section.

Depending on the vehicle, the number of bolt holes may vary from four to eight. The number of bolt holes affects the measurement of the bolt circle. **Fig. 4-14.** If the wheel has four, six, or eight bolt holes, measure from the center of one hole to the center of the opposite hole. If the wheel has five bolt holes, measure from the center of one hole to the outer edge of the bolt hole that is farthest away.

Wheel Fasteners

Wheel fasteners are extremely important. Never take them for granted. Using the wrong style or size of fastener or improperly tightening a fastener can damage the wheel or brake rotor or cause the wheel to fail.

Technicians often refer to a wheel fastener as a "*lug nut*," although this term is not always correct. The word "lug" is a slang term for a threaded stud used in a wheel application. When a technician refers to "lugs", this is a reference to the threaded studs to which the wheel attaches. "Lug nuts" refer to the nuts used to fasten the wheel to the studs. Some wheels use nuts, or lug nuts, while other wheels use bolts, or lug bolts. Most commonly, vehicles will feature stationary threaded studs installed onto the hub or brake rotor hub. The separate fasteners used to secure the wheels on these vehicles are nuts. However, some vehicles feature threaded bolt holes in the hubs and use separate bolts to secure the wheel.

The size and style of a wheel fastener are important. Naturally, the thread diameter and pitch of the nut or bolt must match the stud or hole at the hub. Older vehicles use standard threads on lug studs, while newer vehicles use metric threads. These threads are not interchangeable.

The "seat style" is extremely important. There are three accepted styles of fastener seats: tapered (or conical), ball (or acorn), and mag. Because seat engagement is critical to a safely secured wheel, you must *never* mix these styles.

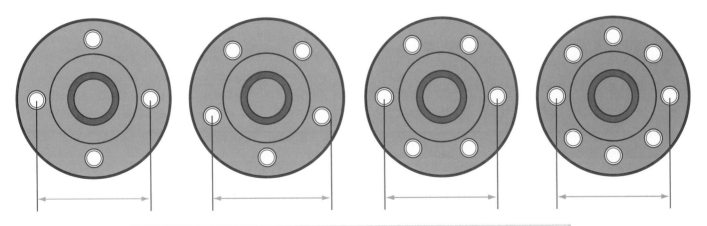

🕭 **Fig. 4-14** Bolt circle patterns and their measuring points. *How would you measure the bolt circle on a five-bolt wheel?* (Yokohama Tire Corporation)

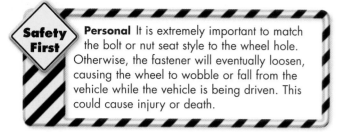

Safety First **Personal** It is extremely important to match the bolt or nut seat style to the wheel hole. Otherwise, the fastener will eventually loosen, causing the wheel to wobble or fall from the vehicle while the vehicle is being driven. This could cause injury or death.

Tapered seats are also called conical seats. They have a tapered hole in the wheel and a matching tapered shoulder under the nut or bolt head. Ball seats are also called acorn or radiused seats. They have a rounded, ball-shaped pocket at the wheel hole and a matching rounded contact area under the nut or bolt head. Mag-style holes have a countersunk, flat recess around the wheel hole. The nut or bolt for such a hole has a large flat washer that fits in a recess. **Fig. 4-15.**

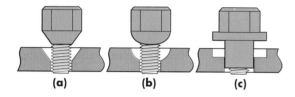

(a) (b) (c)

Fig. 4-15 Examples of three fastener seat styles: **(a)** tapered (conical), **(b)** ball (acorn), and **(c)** mag. Never mix styles of fasteners. You must match the fastener seat style with that of the wheel hole. *Why is it important to match the seat style of the fastener to the seat style in the wheel's mounting holes?* (Yokohama Tire Corporation)

Tire and Wheel Assembly

Follow the proper procedure when removing or installing a tire and wheel assembly. **Fig. 4-16.** Improper removal or installation can result in damage to the assembly or cause it to fail.

Removing a Tire and Wheel Assembly To remove a tire and wheel assembly from a vehicle:

1. Set the parking brake.
2. Position a floor jack under the lift point for the wheel to be removed.
3. Use wheel blocks or chocks to block the wheel diagonally opposite the wheel being removed.
4. Loosen the lug nuts holding the wheel to be removed ½ turn.
5. Using the floor jack, raise the vehicle until the tire is 2 inches [51 mm] above the floor.

6. Position a safety stand under the vehicle frame or lift point.
7. For vehicles using position-specific wheels, mark the wheel position before removal.
8. Remove the lug nuts and the wheel from the stationary threaded studs on the hub or brake rotor hub.

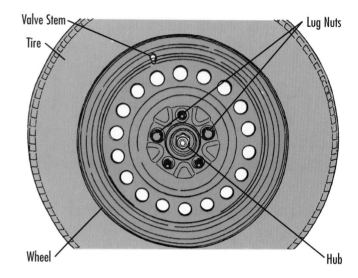

Fig. 4-16 Common wheel, tire, lug nut, and hub configuration. *Why is it important to follow the proper procedure when removing or installing a tire and wheel assembly?* (DaimlerChrysler)

Installing a Tire and Wheel Assembly To install a tire and wheel assembly on a vehicle:

1. Position the wheel on the stationary threaded studs of the hub or brake rotor hub.
2. Hand-tighten the lug nuts on the hub studs to align the tire and wheel on the hub.
3. Remove the safety stand.
4. Lower the vehicle and remove the floor jack.
5. Tighten the lug nuts using the torque pattern and tightening sequence specified. Torque is a force that gives a twisting or rotating motion. (Torque patterns are discussed in more detail in the next section.)
6. Remove the wheel chocks or blocks.
7. Release the parking brake.

Applying Torque

Torque is a turning or twisting force producing torsion and rotation around an axis. Torque is measured in pound-feet or newton-meters. The tightening

Measuring Torque

Torque is a twisting or turning force. It can cause a twisting or rotating motion. In an engine, torque is applied to the crankshaft. This happens when the piston is driven down by the combustion stroke. This force is transferred to the wheels by the transmission, transaxle, or differential. Torque is also applied to fasteners such as bolts or nuts when they are tightened by a wrench.

Nearly every bolt or nut on a vehicle has a torque specification set by the manufacturer. Undertightening may cause the fastener to loosen and create a mechanical or safety problem.

Overtightening can also cause problems. Fasteners that are overtightened may deform the part they are holding in place. Torque specifications should always be followed in tightening wheel lug nuts. Overtightening lug nuts on aluminum wheels can deform the wheel shape.

Torque is measured in pound-feet (lb-ft) or inch-pounds (in-lbs). In the metric system, the units are newton-meters (n-m). Most torque wrenches use these units.

Torque is applied by pulling the handle of the torque wrench. Some torque wrenches are made to click or beep when the torque pre-set on them is reached. Others have a dial which measures the torque when force is applied to the handle.

Apply it!

Relating Handle Length to Torque

Meets NATEF Science Standards for measuring tightening force and relating torque to force.

Materials and Equipment
- Torque wrench with a dial that measures torque produced
- Hollow metal rod to be attached to the wrench handle
- ½-in hex head bolt
- Vise
- Empty 1-gallon paint can with handle
- Sand for weight
- Scale
- Yardstick

❶ Add sand to the 1-gallon paint can to produce a 2-lb [0.9-kg] weight.

❷ Attach the bolt securely in a vise so that it will not turn.

❸ Attach the wrench to the bolt. Make sure that the handle is tilted up enough that the weight will not slip off when it is attached.

❹ Attach the metal rod to lengthen the wrench handle.

❺ Add the weight to the handle 6″ [15 cm] from the socket. Record the torque reading on the dial.

❻ Repeat the procedure with the weight 1′ [30 cm] from the socket. Record the weight.

❼ Repeat this procedure moving the weight out 6″ [15 cm] each time.

Results and Analysis

❽ Did the torque change each time you moved the weight?

❾ Did the amount of force placed on the bolt change?

❿ You may have seen someone attach a longer handle to a wrench to make it easier to tighten a nut. Is this helpful?

torque used to install a wheel on the vehicle is important. You must consider both the amount of torque and the tightening pattern.

The level of tightness must be proper, and it must be identical for all fastener locations. An undertightened fastener may eventually loosen and fail. An overtightened fastener may cause thread damage. The stud or bolt shank may be stretched too far and break, or the nut or threads in the hub may be damaged. Overtightening may also cause

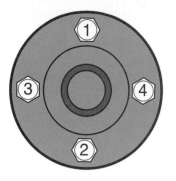   

Fig. 4-17 Torque wrench tightening patterns for 4-, 5-, 6-, and 8-bolt wheels. *What problems can occur if you tighten wheel fasteners excessively or unequally?* (Yokohama Tire Corporation)

the hub or brake rotor to warp which will adversely affect braking action.

You must tighten the fasteners on any wheel in the proper pattern. **Fig. 4-17.** Spread the torque load

Safety First

Personal It is important to tighten fasteners in the specified pattern. Not following this pattern can damage brake system components. This damage can result in loss of braking performance and increased risk of brake system failure and personal injury.

as evenly as possible across the face of the wheel's center section. Uneven tightening can distort the wheel or brake rotor. This will cause vibration when driving or brake-pedal pulsation during braking.

You can use a pneumatic wrench to remove a wheel. However, never use one to install a wheel. Without exception, to tighten any wheel fastener, use only a quality torque wrench.

Remember, when installing a wheel, *always* follow the torque values in the vehicle service manual. Tighten *only* in the recommended patterns.

SECTION 2 KNOWLEDGE CHECK

❶ What are the different ways in which wheel makers construct wheels?

❷ How would you measure the bolt circle on a 5-bolt wheel?

❸ What are the three basic seat styles of wheel fasteners?

❹ Which wheel should be blocked or chocked when changing a tire?

❺ Why is it important for you to tighten wheel fasteners in a specific pattern?

Section 3

Tire Dismounting, Mounting, and Balancing

Repairing wheels and tires requires knowledge of tire dismounting and mounting, valve stem installation, and balancing techniques.

Mount or dismount a tire with the wheel and tire off the vehicle. To remove a tire from a wheel or to install a tire on a wheel, use a tire-changing machine.

Dismounting

Before you begin to dismount a tire, determine if you want to remount the tire on the same wheel. If you do, place an *index mark* on the tire and the wheel, using a piece of chalk. This mark will provide a reference so that you can remount the tire on the wheel in the same position. Sometimes, altering the mounting position of the tire on the wheel can create a vibration or a pull. This is due to the variables of wheel and tire runout and balance.

Remove any balancing weights from the wheel using *wheel weight pliers.* This tool grips and removes clip-on style wheel weights from the wheel rim. It makes weight removal easy and reduces the chances of damaging the wheel. With the balancing weights removed, place the tire and wheel assembly on a tire changer.

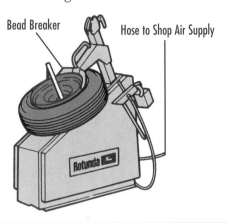

Fig. 4-18 Tire and wheel assembly on a tire changer. Note the bead breaker pressing down on the tire, immediately next to the rim edge. *How do you dislodge the tire bead from the wheel rim?* (Ford Motor Company)

A *tire changer,* or *tire mounting machine,* is a machine designed to mount or dismount a tire. **Fig. 4-18.** Some machines operate manually. However, most are pneumatic (air powered). The machine should feature a method of separating the tire bead's grip at the wheel rims and slipping the tire beads over the rim edges during tire removal or installation. Various styles of changers are available. Make sure you receive proper training in the operation of the changer you plan to use.

Once you have properly secured the tire and wheel assembly on the changer, evacuate the pressurized air from the tire. To relieve all of the internal air pressure from the tire, unthread and remove the valve core from the valve stem. To unthread a valve core, use only a quality valve core removal tool. Unthread the core slowly, allowing the pressurized air to escape before fully removing the core.

With the pressurized air released from the tire, you can release the tire bead from the wheel rim. Technicians refer to this procedure as "breaking the bead." A powerful *bead breaker* arm is used to push the tire bead free from the rim, immediately adjacent to the rim edge.

Safety First **Personal** Keep your hands and fingers away from the bead area during bead breaking. The force of the changer's bead breaker can easily cause serious injury.

Once the bead breaks loose from the rim in one spot, rotate the tire and wheel assembly to a different position to continue breaking the rest of the bead loose. You must break the bead on both the front and the rear sides of the tire.

Once you have unseated the bead from the rim, remove the tire from the wheel. Do this as follows:

1. Lubricate the tire bead and wheel rim flange with soapy water. A *bead lubricant* is available for this purpose.

2. Use a bead separator tool to lift the bead from the rim.

On some tire changers, such as a rim-clamp type changer, the bead separator tool remains stationary as the wheel and tire turn, guiding the bead up and off the wheel rim. Very wide wheels and delicate custom wheels should be handled using these changers. On other types of changers, the tool may install onto the changer's center shaft or the tool may be part of an overhead boom and rotate around the stationary wheel. Regardless of the type used, all tire changers accomplish the same task. **Fig. 4-19.**

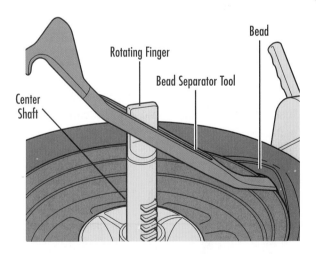

Fig. 4-19 The bead separator tool runs along the bead, lifting the bead away from the rim. On this type of tire changer, the center shaft rotates the moving bead tool around the bead. *Do all tire changers operate in the same way? What differences might exist?*

Once you have removed the top of the tire from the rim, lift the tire so that the rear tire bead contacts the upper wheel rim. Then remove the rear tire bead using the same process.

TECH TIP Valve Stems. Before mounting any tire, install a new valve stem. Installing a new valve stem assures that you will not have a valve stem leakage problem.

Mounting

Before mounting a tire, check the wheel and tire for damage and runout, make any needed tire and wheel repairs, and install a valve stem. Once these items are completed, the tire-mounting procedure can begin.

Tire and Wheel Repair Replace all damaged wheels. Bent or cracked rims and cracked center sections present dangerous failure possibilities. To assist in your detection of damage, perform lateral and radial runout checks of the wheel, hub, and axle. Depending on the damage, repair of some aluminum wheels may be possible. However, you must leave wheel repairs to shops that specialize in that type of work. Never attempt any wheel repair on your own.

It is possible for you to perform some minor tire repairs. You can usually repair a small puncture in the tread area (such as a nail might cause). Various types of plugs and patches are available to allow quick and easy repairs of small punctures. Be aware that training in the use of any plug or patch is necessary.

Insert a tire plug into the puncture to seal the air leak. **Fig. 4-20.** It is important to follow the steps recommended for specific plug kits. Improperly installed plugs can come out or can cause a tire separation.

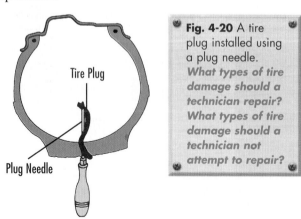

Tire Plug

Plug Needle

Fig. 4-20 A tire plug installed using a plug needle. *What types of tire damage should a technician repair? What types of tire damage should a technician not attempt to repair?*

Most technicians prefer an inside tire patch, because it is a permanent repair. Patch repairs involve cleaning and buffing the inside tire surface and installing a special vulcanized adhesive patch.

Do not attempt to repair a puncture or cut in a sidewall or bead area. Instead, replace the tire.

Remember that the average shop can repair only tread areas. Repair specialists might handle more extensive damage. If the repair involves more than a small puncture in the tread, it is best to replace the tire.

Valve Stems Inspect the valve stem. The air valve stem mounts in a hole located in the wheel rim. This allows inflation and deflation. There are two basic styles of stems: the one-piece "pull-through" style and the "modular" style. The pull-through style relies on an interference fit in the rim hole. **Fig. 4-21.** The valve stems that you will usually see will be the rubber pull-through type. You will most often find modular valve stems on custom wheels and some racing wheels.

Fig. 4-21 Pull-through rubber air valve stem. The rubber body creates a seal at the stem hole in the rim. *How often should you replace a pull-through air valve stem? Why?* (Jack Holtel)

To install a pull-through valve stem:

1. Lubricate the valve stem body.
2. Pull the valve stem through the rim hole and seat it. The rubber outer body of the stem creates an airtight seal at the rim hole.

To install a modular-style valve stem:

1. Insert the valve stem into the hole. The threaded stud portion of the stem protrudes through the outer rim surface.

2. Thread a nut over the stud and secure the valve stem in place. A rubber washer on both sides of the rim provides the airtight seal.

Because of stress during installation and removal, pull-through valve stems tend to split and deteriorate. Whenever servicing a wheel and tire assembly, install a new pull-through stem instead of reusing an old stem.

> **Safety First** ⟩ **Personal** Use caution when working with a valve core. Never stand in the path of the core. Make sure the core aims away from your body. Make sure also that no one is standing in the path of the core. The valve core can become a dangerous projectile, causing personal injury.

Mounting Procedure   The procedure for mounting a tire is as follows:

1. Remove the old valve stem. Clean the valve mounting hole in the rim.

2. Lubricate the new valve stem's outer body.

3. Remove the valve core. Pull the new stem into position using a tire valve stem installation tool. Leave the core out for now.

4. Clean and lubricate the tire beads thoroughly. Position the rear tire bead over the wheel's outer rim flange. **Fig. 4-22.**

Fig. 4-22 Before attempting to mount the tire, lubricate the rim and the tire bead. *Should you lubricate both the inner and outer beads, or do you need to lubricate only the inner (rear) bead? Why?* (Jack Holtel)

Fig. 4-23 Install the rear bead over the front of the rim first. Here the technician uses a rim-clamp style changer (also called an overhead changer). The bead tool remains stationary while the tire and wheel turn, feeding the bead over the rim. *What type of wheels should be handled with a rim-clamp style changer?* (Jack Holtel)

5. Using the changer's installation tool, hold one section of the bead down over the wheel rim.

6. As the wheel rotates on the changer, use the tool to guide the tire bead past the rim flange.

7. Drop the tire down until the outer bead contacts the outer wheel rim flange. Repeat the process until the tire is fully engaged within the rim area. **Figs. 4-23, 4-24,** and **4-25.**

Fig. 4-24 Outer bead fitted over front of wheel, using rim-clamp style changer. *What safety precautions should you take when using a tire changer?* (Jack Holtel)

8. Some tires and wheels have small colored dots. If so, orient the tire and wheel so that the dots align. This is "match-mounting." The dots indicate runout areas. On the tire the dot indicates the "high" runout peak. On the wheel the dot indicates the "low" runout peak. By matching these marks, you place the high and low runout areas together. With this orientation the high and low runout areas tend to cancel each other out. This lessens the chance of creating a severe runout problem. If there is no mark on the wheel, align the paint mark on the tire with the valve stem.

9. Install an air chuck onto the valve stem. While standing away from the changer, initially inflate the tire using 40 psi [275 kPa] to seat the tire beads. Never use more than 40 psi [275 kPa] to inflate any passenger vehicle tire initially.

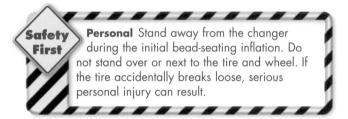

Safety First

Personal Stand away from the changer during the initial bead-seating inflation. Do not stand over or next to the tire and wheel. If the tire accidentally breaks loose, serious personal injury can result.

10. When the beads seat fully against the rim bead seats, you will hear a distinct "pop" noise. Check to see that the beads seat fully and uniformly at both the rear and front bead areas.

11. Carefully remove the air chuck from the valve stem, allowing the internal pressure to escape.

12. Install the valve core. Inflate the tire to its recommended inflation pressure.

13. Examine the bead edges for air bubbles. Air escaping at the bead areas will show in the excess lubrication as bubbles. If you see air bubbles, deflate the tire and break the beads loose. Examine the bead and rim for dirt or obstructions that may be causing the leak.

14. Mount the tire and wheel on the vehicle. Check lateral and radial runout on both the tire and wheel. A dial indicator can be used to measure runout. The dial indicator must be mounted to a rigid stand or bracket. The dial indicator's plunger is pressed against the tire or wheel surface to be measured. As the tire and wheel assembly rotates, runout is shown on the dial indicator's gauge. **Fig. 4-26.**

Fig. 4-25 When inflating the tire to seat the beads, use a clip-on air chuck. This allows you to move away from the changer during initial inflation. This helps you avoid injury. *Why is it important to follow safety precautions when inflating a tire during the mounting process?* (Jack Holtel)

Balancing

To avoid vibration or shimmy during operation, check any tire and wheel assembly for balance. *Wheel balancers* are machines designed to locate and eliminate imbalance from an assembled tire and wheel. There are two types of balancing methods: static and dynamic.

Static balancing involves placing the tire and wheel assembly on a machine called a *static balancer* (also called a *bubble balancer*). The static balancer features a pedestal with a bubble level. The assembly mounts onto the pedestal in a horizontal position. The front of the wheel faces upward.

A bubble level at the center of the balancer post indicates heavy and light areas of the assembly as the bubble moves off center. To center the bubble level, add balancing *wheel weights* to the light areas. The balancing wheel weights can be either clip-on or adhesive style.

Most technicians consider static balancing to be inaccurate and outdated. This method does not take into account the dynamic motion of the assembly as it rotates during operation.

Dynamic balancing, also called *spin balancing,* is much more accurate. **Fig. 4-27.** A *dynamic balancer*

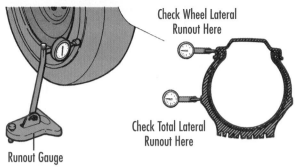

(a) Checking Lateral Runout

Check Wheel Lateral Runout Here

Check Total Lateral Runout Here

Runout Gauge

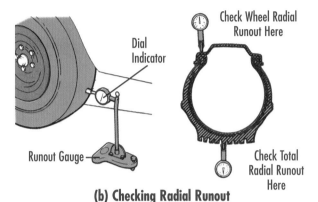

Dial Indicator

Check Wheel Radial Runout Here

Check Total Radial Runout Here

Runout Gauge

(b) Checking Radial Runout

Fig. 4-26 Using a dial indicator to measure lateral runout **(a)** and radial runout **(b)** on a tire and wheel. *What do the measurements indicate?* (Ford Motor Company)

Make sure you receive proper training for the specific balancer that you plan to use. Also, make sure you use any available safety features properly. A dynamic balancer may feature a protective hood that lowers over the tire and wheel assembly during spinning. Never remove this hood. It provides protection from flying debris, such as stones or dislodged wheel weights.

Fig. 4-27 Dynamic spin balancer. Today's computerized balancers make it easy to balance any tire and wheel assembly. Readouts show the weights you need and where to place them. *Why is the use of a dynamic balancer preferred over the use of a static "bubble" balancer?* (Hunter Engineering Company)

spins the tire and wheel assembly in an upright position. This allows dynamic balancing in more than one plane. The balancer's digital readout indicates how much weight is required and at what location you need to place weights.

Other balancers, called "on the car" balancers, are motorized units with rollers that contact the tire tread while the tire and wheel assembly is on the car. The rollers spin the entire rotating assembly on the car. This assembly includes the tire, wheel, brake rotor, and hub.

SECTION 3 KNOWLEDGE CHECK

❶ How do you break a tire bead loose from its grip at the wheel rim?

❷ To seat the tire beads, how much air pressure should you use to inflate a tire initially?

❸ Why should you check a tire and wheel assembly for balance?

❹ Which method of balancing a tire and wheel assembly is more accurate?

❺ What procedure is required to repair a small puncture in the tread area?

CHAPTER 4 REVIEW

Key Points

Meets the following NATEF Standards for Suspension & Steering: diagnosing and repairing wheels and tires; mounting tires; balancing wheels and tires; checking tire and wheel runout.

- Tires serve several important functions. These functions include providing traction, acting as part of the suspension system, and enabling proper handling and operation.

- Aspect ratio refers to the relationship between the tire's section height and section width. The lower the ratio, the shorter the sidewall in relation to the section width.

- The tire tread provides traction. A tire tread may be symmetric or asymmetric as well as directional or nondirectional.

- Tire inflation pressure should be checked when the tire's temperature is the same as the ambient temperature.

- Tires should be rotated at recommended mileage intervals. The rotation pattern depends on the vehicle and if the tires have directional tread.

- The fastener torque must be considered when installing a wheel on a vehicle. Both the amount of torque and the torquing pattern are important.

Review Questions

❶ Identify the different types of tire construction.

❷ What tire dimensions are considered to be the most important? How do you determine each of these dimensions?

❸ What information is typically provided on a tire sidewall? What do the numbers and letters on the tire size indicate?

❹ What are the dimensions of a wheel? How do you determine each of these dimensions?

❺ What is the procedure for mounting a tire onto a wheel?

❻ How do you measure tire and wheel runout? When should you check a tire and wheel assembly for runout?

❼ Identify and describe the two methods of balancing wheels.

❽ (Critical Thinking) In addition to indicating low tire pressure, what are other possible uses of tire pressure monitoring systems?

❾ (Critical Thinking) What are some benefits of run-flat tires besides those mentioned in the text?

TECHNOLOGY FORECAST

FOR AUTOMOTIVE EXCELLENCE

Run-Flat Tires Reduce Roadside Changes

Changing a flat tire can be a difficult and dangerous job. Drivers could be hit by passing vehicles. They could be hurt if the jack collapses. The road surface might not support the weight of a jack. Finally, drivers could become the victims of crime.

Those reasons will likely lead to wider use of "run-flat" or "extended mobility" tires. Run-flat tires are made by adding rubber and stiffeners to the tire sidewalls, making them stronger. They don't require special rims.

A run-flat tire can support the weight of the vehicle, even with no air. When a run-flat tire loses its air, the driver will notice little or no difference.

A sensor system alerts the driver to the loss of pressure. Engineers say these deflated tires can operate safely for 50 miles at 55 mph.

Besides safety, there is another benefit. With run-flat tires, a vehicle doesn't have to carry a spare tire or a jack. That means trunks and cargo areas will have more room.

Run-flat tires are not a new idea. Bullet-proof tires have been available for presidential limousines and armored cars since the 1960s. Run-flat tires are available today at some tire retailers and on a few luxury or high-performance vehicles.

AUTOMOTIVE EXCELLENCE
TEST PREP

Answering the following practice questions will help you prepare for the ASE certification tests.

1. Technician A says the tire bead seating and the air valve and stem create the pressure seal on a tubeless tire and wheel assembly. Technician B says only the sidewall and tread create the pressure seal. Who is correct?

   ⓐ Technician A.
   ⓑ Technician B.
   ⓒ Both Technician A and Technician B.
   ⓓ Neither Technician A nor Technician B.

2. Technician A says a tire marked 235/60R15 has an overall height of 235 mm, is 60 mm wide, and is used for racing. Technician B says the tire has a 235-mm section width, has an aspect ratio of 60, and is of radial construction. Who is correct?

   ⓐ Technician A.
   ⓑ Technician B.
   ⓒ Both Technician A and Technician B.
   ⓓ Neither Technician A nor Technician B.

3. Technician A says the lower the treadwear rating number of a tire, the longer the tread will last. Technician B says the higher the treadwear rating number, the longer the tread will last. Who is correct?

   ⓐ Technician A.
   ⓑ Technician B.
   ⓒ Both Technician A and Technician B.
   ⓓ Neither Technician A nor Technician B.

4. Technician A says the tire sidewall marking DOT indicates Department of Transportation approval for road use. Technician B says this marking indicates where the tire was made. Who is correct?

   ⓐ Technician A.
   ⓑ Technician B.
   ⓒ Both Technician A and Technician B.
   ⓓ Neither Technician A nor Technician B.

5. Technician A says the sidewall marking M+S means the tire is rated as a mud and snow tire. Technician B says this marking is a speed rating. Who is correct?

   ⓐ Technician A.
   ⓑ Technician B.
   ⓒ Both Technician A and Technician B.
   ⓓ Neither Technician A nor Technician B.

6. When measuring a wheel rim diameter, Technician A says you should consider the distance between the bead seats at opposite sides of the rim diameter. Technician B says you should consider the distance between the bead seats from the front to the back of the rim. Who is correct?

   ⓐ Technician A.
   ⓑ Technician B.
   ⓒ Both Technician A and Technician B.
   ⓓ Neither Technician A nor Technician B.

7. How do you measure the bolt circle on a 4-bolt wheel?

   ⓐ From the center of the wheel to the outside edge of a bolt hole.
   ⓑ From the center of one bolt hole to the center of the opposite bolt hole.
   ⓒ From the center of one bolt hole to the center of the adjacent bolt hole.
   ⓓ From the edge of the wheel rim to the center of one bolt hole.

8. How much inflation pressure should be used to seat tire beads initially?

   ⓐ 10 psi [69 kPa] or less.
   ⓑ 100 psi [690 kPa] or more.
   ⓒ 40 psi [275 kPa] maximum.
   ⓓ 60 psi [414 kPa].

9. Technician A says you should stand directly over the tire you are inflating during bead seating. Technician B says you should never stand over or next to the tire you are inflating during bead seating. Who is correct?

   ⓐ Technician A.
   ⓑ Technician B.
   ⓒ Both Technician A and Technician B.
   ⓓ Neither Technician A nor Technician B.

10. Technician A says the hood cover on a dynamic balancer prevents loss of inflation pressure. Technician B says the hood cover offers protection from flying debris. Who is correct?

    ⓐ Technician A.
    ⓑ Technician B.
    ⓒ Both Technician A and Technician B.
    ⓓ Neither Technician A nor Technician B.

METRIC CONVERSION FACTORS

When you know:	You can find:	If you multiply by:
Length		
inches	millimeters	25.4
feet	centimeters	30.48
yards	meters	0.9144
miles	kilometers	1.609
millimeters	inches	0.039
centimeters	inches	0.39
meters	yards	1.094
kilometers	miles	0.621
Area		
square inches	square centimeters	6.45
square feet	square meters	0.0929
square yards	square meters	0.836
square miles	square kilometers	2.59
acres	hectares	0.405
square centimeters	square inches	0.155
square meters	square yards	1.196
square kilometers	square miles	0.386
hectares	acres	2.47
Weight (mass)		
ounces	grams	28.3
pounds	kilograms	0.454
short tons	metric tons	0.907
grams	ounces	0.0353
kilograms	pounds	2.2
metric tons	short tons	1.1
Liquid volume		
ounces	milliliters	29.6
pints	liters	0.473
quarts	liters	0.946
gallons	liters	3.785
milliliters	ounces	0.0338
liters	pints	2.113
liters	quarts	1.057
liters	gallons	0.264
Temperature		
degrees Fahrenheit	degrees Celsius	0.6 (after subtracting 32)
degrees Celsius	degrees Fahrenheit	1.8 (then add 32)

GLOSSARY

A

absolute pressure The difference between atmospheric pressure and a partial vacuum.

accessory system The system that provides comfort, safety, security, and convenience devices.

accumulator In an anti-lock brake system, a device that stores fluid under pressure. Fluid is stored to provide pressure for one or two brake applications.

active suspension system The system that maintains full suspension control during all suspension activity, including acceleration, braking, and cornering.

active test A test that forces a component to operate in a specific way.

actuator An output device that is controlled by the powertrain control module.

add To increase fluid or pressure to the correct level or amount.

adjust To bring components to specified operational settings.

adjustable shock absorber A shock absorber with a variable setting for the internal valving.

adjustable wheel bearing A wheel bearing that requires periodic lubrication.

aerobic sealant Gasket material that hardens in the presence of air.

air bag A balloon-type passenger-safety device that inflates automatically on vehicle impact.

air bag control module The module that monitors system operation, controls the air bag warning light, and stores trouble codes.

air filter Any device such as porous paper or wire mesh filter that prevents airborne particles from entering air-breathing machinery.

air filter element A flame-resistant and tear-resistant material used as the filter in an air filter.

air gap Space between spark plug electrodes.

air induction system The system that supplies clean air to the engine and controls the flow of air through the engine.

air injection (AIR) system An exhaust emission control system that reduces hydrocarbon and carbon monoxide emissions.

air spring In the suspension system, a flexible bag filled with air which compresses to absorb shock.

air valve In a brake system, the second port of the control valve. The valve stays closed until the driver pushes the brake pedal, allowing air to enter the rear chamber of the vacuum power booster.

air/fuel mixture The combination of air and fuel that is used to fuel an automotive engine.

air/fuel ratio The proportion of air and fuel, by weight, supplied to the engine cylinders for combustion.

air-assisted shock absorber A shock absorber equipped with a chamber that can be inflated with compressed air to provide additional load-carrying ability.

airflow sensor Sensor that monitors the amount of air entering a vehicle's throttle body.

airflow sensor plate A movable plate in the intake air passage of the airflow sensor. In a mechanical fuel injection system, the plate moves and applies pressure to fuel supplied to the fuel injectors. The greater the airflow, the greater the fuel pressure.

alphanumeric designator A code consisting of a letter and a number, used for diagnostic trouble codes (DTCs), that represents a system or an area within a vehicle.

alternating current (AC) Electrical current that changes its direction of flow in a regular and predictable way.

anaerobic sealant Gasket material that hardens in the absence of air.

analog multimeter A device that uses a moving needle to indicate the measurement of voltage, current, or resistance.

analog signal A signal that continuously changes from positive to negative.

analyze To examine the relationship of components of an operation.

GLOSSARY

antifreeze A chemical, usually ethylene glycol, added to the engine coolant to raise its boiling temperature and lower its freezing temperature.

anti-icer A chemical compound that minimizes icing in the throttle body and fuel line.

antiknock index The average of R (research) and M (motor) octane numbers. An indicator of the anti-knock quality of a gasoline.

anti-lock brakes (ABS) system Brakes installed with the service brakes to prevent wheel lockup during braking.

antioxidant A chemical compound that prevents formation of varnish in the fuel system.

antiseize compound A lubricant that prevents bolt threads from locking or seizing.

anti-sway bar A type of torsion bar that connects the lower suspension to both sides of the frame or body. This helps reduce body sway. Also called an anti-roll bar.

apply piston The piston that applies the force on the hydraulic fluid in a brake system. Also called input piston.

armature The part of the starter that rotates. It contains many individual windings, or coils, through which electric current flows, producing magnetic fields.

aspect ratio The relationship between a tire's section height and section width. Also called profile ratio.

assemble To fit together the components of a device.

asymmetric tread pattern A tire tread that has a different style or shape of tread on either side of the tire.

asynchronous pulses Pulses that do not follow the normal timing sequence.

automatic level control suspension system A system that compensates for variations in vehicle load by maintaining the body at a predetermined height.

Automotive Service Excellence (ASE) A non-profit national institute dedicated to improving the quality of vehicle service and repair. ASE promotes testing and certification for service and repair professionals.

automotive stethoscope A diagnostic tool that helps isolate and amplify noises.

automotive system A system made up of two or more parts that work together to perform a specific task.

auxiliary parking brake A parking brake that uses a separate set of brake linings.

available voltage The amount of voltage available in the primary ignition coil. Also called maximum coil output.

axle A support suspension member on which the wheels are mounted.

B

backing plate The metal plate on which many of the drum brake components are mounted.

ball joint A lubricated attachment that connects two suspension parts to allow pivoting movement.

ball joint press A tool that presses the ball joint tapered stud out of the steering arm (or knuckle) mounting hole.

ball socket Connects to the tie rods at attachment points to allow free motion.

ballast module A device that produces the high voltage required for high-intensity discharge (HID) operation.

banked multiport fuel injection (MFI) system A system that pulses the injectors in separate sets.

barometric pressure Atmospheric pressure, measured in inches of mercury.

battery An electrochemical device that stores electric current in chemical form so that it can be released as electricity for cranking the engine and powering the electrical load.

battery jumper box A portable power pack used for jump starting vehicles.

bead breaker arm A tool used to push the tire bead free from the rim.

belt tension gauge A measurement tool used to check belt tension.

GLOSSARY

bleed To flush air or contaminated fluid out of a system; as in to bleed a brake system.

blowby The leakage of air/fuel mixture and combustion that escapes past the piston rings and into the crankcase.

body and frame system The system that supports the vehicle and provides enclosures or compartments for the engine, passengers, luggage, and cargo.

body control module (BCM) A device that uses inputs from a variety of sensors to control the operation of motors and actuators.

bolt A heavy metal pin threaded like a screw.

bolt circle The size of the diameter of the mounting-fastener hole, or bolt hole, pattern in the wheel's center section. This is where the wheel bolts to the hub.

bolt strength markings Markings that indicate the tensile strength of a bolt, which is the amount of stress a bolt can take without breaking.

bottom dead center (BDC) The piston position when the piston has reached the lower limit of its travel in the cylinder, when the cylinder volume is at its maximum.

brake An energy-conversion device used to slow, stop, or hold a vehicle or mechanism.

brake bleeding The process of flushing air or contaminated fluid from the braking system.

brake drum A metal drum mounted on a vehicle wheel to form the outer shell of the brake; the brake shoes press against the drum to slow or stop drum-and-wheel rotation for braking.

brake fluid A chemically inert hydraulic fluid used in hydraulic brake systems to transmit force and motion through a closed system of tubing or brake lines.

brake grabbing A condition in which the brakes apply more quickly than expected compared to the amount of pedal effort applied.

brake hardware The assorted small parts that attach to the brake shoes.

brake lines The tubes and hoses connecting the master cylinder to the wheel cylinders or calipers in a hydraulic brake system.

brake pads The parts of disc brake systems that apply friction to the brake rotor and convert vehicle motion into heat.

brake pedal fade A temporary reduction or fading out of brake effectiveness.

brake rotor In a disc brake system, a rotating disc against which the disc pads are clamped to provide braking action.

brake shoe A carrier to which the brake lining is attached, used to force the lining in contact with the brake drum.

brake spoon A tool used to adjust drum brakes.

braking system The system that slows and stops the vehicle.

break-out box (BOB) A device used to check the actual voltage and resistance readings of each power-train control module (PCM) circuit.

brush In an electric motor, a block of conducting substance, such as carbon.

burner The structure that produces light in a high-intensity discharge (HID) lamp.

C

calibration A precise factory setting made to produce a given output or effect.

calibration drift When a component, due to age or contamination, no longer accurately represents the factory setting.

caliper In a disc brake system, the housing that contains the pistons and the brake pads.

cam A round collar with a high spot, or lobe.

cam follower The part of a cam mechanism that rides on the contour surface of a cam.

camber angle Simply called camber. The angle of inward or outward tilt of a wheel, measured in degrees.

cam-rod parking brake actuator Also called the eccentric shaft and rod design, this parking brake system is common on Asian imports.

GLOSSARY

camshaft A shaft having a series of cams for operating the valve mechanisms.

carbon monoxide (CO) A poisonous gas that is a product of incomplete combustion of hydrocarbon fuel with air.

carbon tracking A condition that occurs when the spark jumps from a distributor cap terminal to another terminal or to ground. Carbon tracks appear as thin black lines on the inside of the cap.

caster angle Simply called caster. The angle between true vertical and an imaginary line drawn through the upper and lower ball joints, measured in degrees.

catalyst A material that causes a change without being part of the chemical reaction.

catalytic converter A muffler-like device for use in an exhaust system. It converts harmful gases into harmless gases by promoting a chemical reaction between the catalysts and the pollutants.

center electrode The center of a spark plug insulator. It carries high voltage from the ignition coil to the plug air gap.

center high-mounted stop (brake) light (CHMSL) A light located at the rear of the vehicle at approximately driver's eye level for better visibility.

center link The link in a steering system that connects the pitman arm to the steering arm. Also referred to as drag link or relay rod.

central processing unit (CPU) A microprocessor in the instrument cluster that controls digital displays.

charcoal canister A container filled with activated charcoal; used to trap gasoline vapor from the fuel tank and fuel system while the engine is off.

charge To bring to "full" state; e.g., battery or air conditioning system.

charging current test A test that measures the maximum current output of the generator at a specified voltage.

charging system The system that supplies all current needed for vehicle operation, lights, and accessories while the engine is running. It is also responsible for keeping the battery charged.

charging voltage test A test that determines whether a generator's output is within the normal range.

chassis dynamometer A machine that measures the amount of power delivered to the drive wheels.

check To verify condition by performing an operational or comparative examination.

chemical generator A generator that uses a chemical reaction to create current flow. A battery is a chemical generator.

circuit board Electronic circuits printed on a plastic board.

circuit breaker A resettable protective device that opens an electric circuit to prevent damage when the circuit is overheated by excess current flow.

clamping diode A diode that provides a path to use up unwanted voltage. Used in situations in which high voltage could damage the module that controls the circuit.

clean To rid components of extraneous matter for the purpose of reconditioning, repairing, measuring, or reassembling.

clear flood mode The mode that clears engine cylinders flooded with gasoline vapors.

clockspring A device that consists of two or more wires coiled inside a plastic housing mounted at the top of the steering column. It maintains a "hard-wired" connection between the steering column wiring and the rotating steering wheel.

closed-loop operation The operation that occurs when the powertrain control module (PCM) processes electrical inputs from a network of sensors and provides output controls to a series of actuators.

coefficient of friction The ratio of the force of friction to the normal force that presses two surfaces together.

coil spring A length of spring-steel rod wound into a coil.

cold cranking amps (CCA) The amount of current a fully charged battery can supply at 0°F (−17.8°C) for 30 seconds.

cold-start valve This valve sprays fuel into the intake manifold during cold engine starting.

GLOSSARY

combustion chamber The space between the top of the piston and the cylinder head, in which the fuel is burned.

commission A percentage of the labor costs charged to customers that is then added to an employee's salary.

commutator A series of copper segments placed side by side to form a ring around the armature shaft.

companion cylinder The second of a pair of piston cylinders in a waste spark system.

composite master cylinder A master cylinder made of cast aluminum, with an attached plastic reservoir, holding the brake fluid.

compressed natural gas (CNG) Natural gas (principally methane) stored under pressure for use as an automotive fuel.

compression In a suspension system, bump travel of the suspension. Also called jounce.

compression (force) Two forces act on opposite ends of an object but are directed toward each other, squeezing the object.

compression engine A diesel engine in which the fuel is injected into the cylinders, where the heat of compression ignites it. This engine burns diesel fuel oil instead of gasoline.

compression ratio The volume in the cylinder with the piston at bottom dead center (BDC) divided by the volume in the cylinder with the piston at top dead center (TDC).

compression ring Ring fitted onto the piston that forms a sliding seal between the piston and the cylinder wall; prevents blowby of combustion gases.

compression stroke The piston movement (from bottom dead center to top dead center) immediately following the intake stroke. During this movement, the intake and exhaust valves are closed while the air or air/fuel mixture in the cylinder is compressed.

conductor Any material that contains free electrons that allow electrical current to flow easily.

connecting rod In the engine, the rod that connects the journal on the crankshaft with the piston.

connecting rod bearing A two-part bearing between the crankshaft and the connecting rod.

connecting rod journal A housing holding the connecting rod bearing and the oil lubricating the bearing.

connector A device used to connect and disconnect multiple wires.

continuity tester A device that checks for a complete circuit path, or continuity from one point to another.

continuous fuel injection systems (CIS) A system that uses mechanical fuel injectors to continuously spray fuel into the intake port.

control arm A device that provides the connection points of the suspension for up/down pivoting movement. It positions the wheels on the frame or body and maintains the wheel alignment to the frame or body.

control valve (brake) A valve that regulates the flow of atmospheric pressure and vacuum to the two separate chambers in the vacuum power booster.

control valve (power steering) In the power steering pump, a valve which opens and closes fluid passages inside the gear housing.

coolant The liquid mixture of about 50 percent antifreeze and 50 percent water used to carry heat out of the engine.

cooling system The system that removes heat from the engine by the forced circulation of air or coolant and thereby prevents engine overheating. In a liquid-cooled engine, the system includes the coolant, water pump, water jackets, radiator, and thermostat.

cooling system pressure test A test that diagnoses cooling system leaks.

cooling system thermostat A device that regulates the flow of coolant through the engine and cooling system, keeping the engine within the correct operating temperature range.

corrosion inhibitor A chemical compound that prevents rusting of metal parts in the fuel system.

cranking vacuum test A test that measures engine vacuum while the engine is cranking.

cranking voltage/current draw test A test that measures battery voltage and current during engine cranking.

crankshaft A one-piece steel casting or forging that serves as the main rotating member, or shaft, of the engine. The crankshaft has offset journals to which the connecting rods are attached; it converts their up-and-down (reciprocating) motion into circular (rotary) motion.

crankshaft position (CKP) sensor A sensor that indicates when the number-one piston nears top dead center (TDC) on the compression stroke.

crash (impact) sensors The sensors in an air bag system that close an electrical circuit to activate the system when sufficient impact occurs.

cruise (speed) control system A system designed to maintain a constant cruising speed. It controls throttle position to maintain the driver-selected vehicle speed.

cruise control vacuum servo A device that uses a vacuum-operated diaphragm to hold the throttle linkage open.

current flow The movement of electrical energy through a conductor.

cylinder Machined opening in the engine block; this cylindrical or tubular chamber houses the piston of a reciprocating engine.

cylinder block The basic framework of the engine, in and on which the other engine parts are attached. It includes the engine cylinders and the upper part of the crankcase. Also called the engine block.

cylinder bore Machined opening in the engine block; this cylindrical or tubular chamber houses the piston of a reciprocating engine or pump.

cylinder compression The pressure developed in the cylinder as the engine cranks.

cylinder compression test A test that measures the compression in each cylinder as the engine cranks. There are two types: wet compression test and dry compression test.

cylinder head The part of the engine that covers and encloses the cylinders. It contains cooling fins or water jackets and the valves.

cylinder head gasket As part of a reciprocating engine, a flat compressible seal which seals the cylinder head and the engine block.

cylinder leakage test A test that checks a cylinder's ability to hold pressure. The percentage of leakage is measured as air pressure is added to the cylinder while the piston is at top dead center (TDC) of its compression stroke.

D

damper A device located on the front of the crankshaft, used to dissipate the energy of vibration and reduce vibration.

data bus A wiring harness that allows components connected to it to share sensor signals and other information.

data link connector (DLC) A plug-type connector used by a scan tool to access vehicle diagnostic information.

data stream The information transmitted on the data bus.

daytime running lights (DRLs) A safety feature in which the headlights are automatically illuminated when the engine is running.

dead axle A rear axle that does not provide drive. It serves only as a suspension member on which to attach wheels.

deceleration sensing Sensing that a vehicle is losing speed.

deposit control agent Chemical compound that prevents or removes fuel system deposits.

detonation Commonly referred to as spark knock or ping. In the combustion chamber of a spark-ignition engine, an uncontrolled second explosion (after the spark occurs at the spark plug), with spontaneous combustion of the remaining compressed air/fuel mixture, resulting in a pinging sound.

diagnosis Answers the question "What is wrong?" or "What caused or is causing the problem?"

GLOSSARY

diagnostic A procedure used by the powertrain control module (PCM) to test relevant on-board systems.

diagnostic database An electronic record within a dedicated scanner that contains electronic specifications and troubleshooting tips.

diagnostic executive A powertrain control module (PCM) program that controls the sequencing of tests needed to run the OBD-II monitors.

diagnostic strategy A planned step-by-step procedure used to locate a problem.

diagnostic testing Testing a vehicle or a vehicle's system in order to determine the cause or causes of a problem. Usually refers to automated or computerized testing.

diagnostic tree A diagram that resembles a tree. The diagram, or flowchart, lists testing methods and possible results in order to help diagnose automotive problems.

diagnostic trouble code (DTC) A code that identifies a system or component malfunction.

diesel engine An engine in which the fuel is injected into the cylinders, where the heat of compression ignites it. This engine burns diesel fuel oil instead of gasoline.

differential carrier Gear system that permits one drive wheel to turn faster than the other drive wheel. In a rear wheel anti-lock braking system, the place where wheel lockup is sensed.

digital multimeter (DVOM) A device that measures voltage, current, and resistance and displays readings numerically.

digital ratio adapter controller (DRAC) A device that receives signals from a speed sensor and converts the signals to digital signals for the anti-lock brake system (ABS) control module.

digital signal A signal that is either on or off. The voltage is either high or low, with no values in between. Also called square wave signal.

digital storage oscilloscope (DSO) A tool used to observe and measure voltage and frequency. Also known as a lab scope.

dimmer switch A two-position switch operated by the driver to select the high or low headlight beam.

diode A solid-state electronic device that allows the passage of an electric current in one direction only.

direct current (DC) Electrical current that flows in a single direction.

direct ignition system An ignition system that mounts the ignition coils directly to the spark plugs.

directional tread A tire that has been designed to rotate in only one direction.

disassemble To separate a component's parts as a preparation for cleaning, inspection, or service.

disc brake A brake in which brake pads, in a vise-like caliper, grip a revolving disc to stop it.

discharge To empty a storage device or system.

displacement The volume swept out when the piston moves from one end of the cylinder to the other.

distributor A device designed to establish base timing and distribute the ignition spark.

distributorless ignition system An electronic ignition system without a separate ignition distributor. Sensors signal the position of the crankshaft to the powertrain control module (PCM), which then electronically times and triggers the system and controls distribution of the secondary voltages.

double A-arm design Suspension system which uses two control arms, one upper and one lower, and a coil spring.

double flare A type of replacement brake line where the end looks like a trumpet bent back over itself.

drag link The link in a steering system that connects the pitman arm to the steering arm. Also referred to as center link or relay rod.

drag link socket wrench A tool that connects and disconnects the drag link from the pitman arm and idler arm.

drain To use gravity to empty a container.

drive belt A rubber/fabric belt that transmits engine power to accessory items such as the power steering pump, generator, and air conditioning compressor.

GLOSSARY

drive cycle A set of driving conditions that "run" all on-board diagnostics.

drive plate On a vehicle with an automatic transmission, a light plate bolted to the crankshaft, to which the torque converter attaches.

drive train Transmission system from engine output shaft to driven road wheels.

drive-by-wire system Control of engine or other vehicle functions by electronic rather than mechanical means.

drop center A wheel center that is smaller in diameter than the rim.

drum brake A brake in which curved brake shoes press against the inner circumference of a metal drum to produce the braking action.

dual-braking system A type of braking system that has a dual-piston master cylinder, two fluid reservoirs, and two separate hydraulic systems.

duo-servo drum brakes Brakes in which the action of one shoe reinforces the action of the other shoe.

duty cycle The percentage of time a circuit or device stays switched on.

dwell angle The number of degrees of distributor or camshaft rotation during which current flows through the primary circuit of the ignition coil.

dynamic balancer A balancer that spins the tire and wheel assembly in an upright position. This allows balancing in more than one plane. Digital readouts show where to place weights.

E

electrical (electronic) system The system that produces and directs electrical power needed to operate the vehicle's electrical and electronic components.

electrically erasable programmable read-only memory (EEPROM) Computer memory chips that can be reprogrammed. Also called a flash PROM.

electricity The flow of electrons through conducting materials.

electrode An insulated center rod attached to the shell of a spark plug.

electrolyte A compound that conducts an electrical current in a water solution.

electromagnet A coil of wire (usually wrapped around an iron core) that produces magnetism as electric current passes through it.

electromagnetic display A gauge that uses electromagnetism to move an indicating needle (pointer).

electromagnetic field The space around an electromagnet that is filled with invisible lines of force. The strength of the field depends on the level of current flow and the number of wires in the coil.

electromagnetic induction The characteristic of a magnetic field that causes an electric current to be created in a conductor as it passes through the field.

electromagnetic interference (EMI) Interference that may cause erratic electrical inputs to the powertrain control module (PCM) resulting in driveability complaints.

electromagnetism Magnetism achieved by an electromagnet.

electron An atomic particle with a negative electrical charge.

electronic Of or pertaining to electrons or electronics.

electronics Electrical assemblies, circuits, and systems that use electronic devices such as transistors and diodes.

electronic circuit tester A device that is used to safely test electronic circuits. It uses light-emitting diodes (LEDs) to display test results. Also called a logic probe.

electronic ignition system A system that electronically controls current flow in the primary ignition circuit.

electronic inputs Data from a network of sensors and actuators on a vehicle.

enabling criteria The sensor inputs supplied during specified driving conditions.

engine A machine that turns heat energy into mechanical energy. A device that burns fuel to produce mechanical power.

GLOSSARY

engine block The basic framework of the engine, in and on which the other engine parts are attached. It includes the engine cylinders and the upper part of the crankcase. Also called the cylinder block.

engine control module (ECM) Now known as the powertrain control module (PCM).

engine coolant temperature (ECT) sensor A sensor that measures engine temperature by monitoring coolant temperature.

engine vacuum The low-pressure condition created as the crankshaft turns, pulling the piston down in the cylinder.

engine vacuum test A test that checks vacuum readings through a range of engine operations. Also known as manifold absolute pressure test.

ergonomics The study of workplace design.

ethanol An alcohol made from corn that can be used as fuel.

EVAP system purge flow test A test that measures the ability of the EVAP system to maintain a pressure or a vacuum.

evaporative control (EVAP) system A system that prevents gasoline vapors in the fuel system from escaping into the atmosphere.

evaporative emissions Fuel vapors.

evaporative emissions (EVAP) canister A canister that stores fuel vapors until they are burned in the engine.

exhaust gas analyzer A device used to test the amount of exhaust emissions produced by a vehicle.

exhaust gas recirculation (EGR) system An emissions control device that recycles a small part of the exhaust back through the intake manifold to lower the combustion temperature.

exhaust manifold A metal casting with several passages through which exhaust gases leave the engine combustion chambers and enter the exhaust system.

exhaust stroke The piston stroke (from bottom dead center to top dead center) immediately following the power stroke, during which the exhaust valve opens so the exhaust gases can escape from the cylinder to the exhaust manifold.

exhaust system The system that collects the exhaust gases and discharges them into the air. Consists of the exhaust manifold, exhaust pipe, catalytic converter, muffler, tail pipe, and resonator (if used).

exhaust valve Valve that releases burned gases from a cylinder.

expansion tank A tank connected by a hose to the filler neck of an engine radiator; the tank provides room for heated coolant to expand and to give off any air that may be trapped in the coolant.

F

failure record A record of up to five diagnostic trouble codes (DTCs) in the diagnostic memory, relating to component faults.

fastener A device, such as a screw, nut, or bolt, that holds automotive parts together.

field coil A coil in a generator or starter motor that produces a magnetic field as current passes through it.

filament A small wire-like conductor inside an incandescent bulb. Current flowing through the filament causes it to glow white-hot and give off light and heat.

fill To bring fluid levels to a specified point or volume.

filtering media A flame-resistant and tear-resistant material used as the filter in an air filter.

find To locate a particular problem, e.g., shorts, grounds, or opens in an electrical current.

firing order The numbered sequence in which the cylinders of a multi-cylinder engine fire.

fixed caliper disc brakes Disc brakes that use a caliper that is fixed in position and cannot move.

flare nuts Used to connect replacement rigid brake lines.

flat rate A predetermined amount of time it should take to perform a specific service or repair job.

fleet A group of five or more vehicles owned and maintained by a single company.

GLOSSARY

flex plate On a vehicle with automatic transmission, a light drive plate bolted to the crankshaft, to which the torque converter attaches. Also called a drive plate.

floating caliper disc brakes Disc brakes that use a caliper that is free to move sideways on bushings and guide pins.

flush To use a fluid to clean an internal system.

flywheel A mechanical device that is used to store energy, which helps smooth out the engine power surges from the power strokes.

forced induction The process of compressing the air/fuel mixture before it enters the cylinders.

fossil fuel A fuel, such as natural gas, petroleum, or coal, that forms from the remains of plants and animals.

four-stroke cycle engine An engine in which the completed power cycle consists of four piston strokes —intake, compression, power, and exhaust.

freeze-frame data Serial data values that are stored the instant an emission-related diagnostic trouble code (DTC) is set and enters the diagnostic memory.

frequency (f) In a signal, the number of times an entire cycle repeats within one second.

frequency modulation (FM) Changing frequency in order to transmit information.

friction The resistance to motion between two objects or surfaces that touch.

fuel cut-off mode The mode which stops fuel delivery momentarily during deceleration.

fuel distributor A device that meters fuel to each fuel injector.

fuel filter A device which prevents contaminants from entering the fuel system and clogging the fuel injectors.

fuel injector A solenoid-operated valve that injects fuel into cylinder.

fuel level sensor A sensor in the fuel tank that measures fuel levels.

fuel line A line that transfers fuel from the fuel tank to the throttle body or the fuel rails.

fuel metering system The system that consists of the powertrain control module (PCM), a network of sensors, and the fuel injectors.

fuel pressure regulator A spring-loaded valve built into the fuel pump or the throttle body. It maintains a constant pressure drop across the injectors.

fuel pump The electrical or mechanical device in the fuel system that forces fuel from the fuel tank to the fuel-injection system.

fuel rail In a multiport fuel injection system, this provides fuel to each injector.

fuel return line The line that returns unneeded fuel to the fuel tank.

fuel supply system The system that delivers fuel to the engine cylinders. Consists of the fuel tank and lines, gauge, fuel pump, fuel-injection system, and intake manifold.

fuel tank A dome-shaped tank that holds the fuel and contains the fuel pump, the fuel level sensor, and a portion of the vapor recovery system.

fuel trim (FT) A measure of the air/fuel mixture from the optimal amount. Displayed as a percentage.

full lock The result of turning the steering wheel all the way in either direction.

full-wave rectification The condition in which both halves of a sine wave are used to provide a DC voltage.

fuse A device that opens an electric circuit when excessive current flows, to protect equipment in the circuit.

fusible link A short length of insulated wire connected in a circuit. It is designed to melt when current flow exceeds the rating for the circuit.

G

gasket A thin layer of soft material such as paper, cork, rubber, copper, synthetic material, or a combination of these. It is used to create a tight seal between two flat surfaces.

GLOSSARY

gasoline A liquid blend of hydrocarbons, obtained from crude oil; used as the fuel in most automobile engines.

gear ratio The number of rotations a pinion gear must turn to rotate a driven gear one time.

gear reduction Increases torque ratio by increasing the gear ratio.

generator The device that converts mechanical energy into electrical energy.

geometric centerline A line drawn from the midpoint of the rear of the vehicle to the midpoint of the front of the vehicle.

gravity bleeding Brake bleeding that uses gravity and atmospheric pressure to bleed the brake system.

ground electrode An electrode that attaches to the metal shell of a spark plug, bending in to form an air gap at the center electrode.

H

Hall-effect sensor A switch used to toggle a reference voltage applied to the sensor.

halogen lamp A bulb filled with halogen gas. Halogen is a chemically inactive gas, which protects the filament from burnout and allows it to operate at a higher temperature.

hazardous materials Materials and wastes that pose a danger to human health and the environment.

header pipe A pipe connected to the exhaust manifold that carries exhaust gases to the catalytic converter.

heat range A spark plug specification that determines how fast a spark plug transfers heat from the firing tip to the cylinder head.

heated oxygen sensor An oxygen sensor that maintains its operating temperature during periods when engine cool down might occur.

heated-film sensor A sensor that measures airflow by measuring current flow. The device consists of a metal foil coated with a conducting material. Current flowing through the material heats it. Air flowing past or through the material cools it. The film is maintained at a specific temperature; as the air cools it, more current is needed to heat it.

helical-gear rack A gear design where gear teeth are cut at an angle to shaft.

high voltage bias When return signal voltage from an oxygen sensor varies over a voltage range that is too high.

high-intensity discharge (HID) lamp A lamp in which light is produced when high voltage creates an arc between two electrodes. Mercury vapor or tight xenon gas is used in HID lamps.

high-rate discharge test A test that measures battery voltage while the battery is supplying a large current for 15 seconds. Also called a load test.

hold-pressure position In anti-lock brake systems (ABS), when the ABS control module holds the hydraulic position constant.

hone To restore or resize or bore a cylinder by using rotating cutting stones.

horn relay A relay that controls current flow from the battery to the horns.

humidity A measurement shown as a percentage of relative water vapor.

hydraulic actuator In a braking system, a unit that can increase brake pressure, decrease brake pressure, or hold brake pressure steady.

hydraulic booster A brake booster that uses hydraulic pressure supplied by the power-steering pump, or a separate electrically driven pump, to assist in applying the brakes.

hydraulic pressure The pressure applied to a liquid to create the force.

hydraulics The process of applying pressure to a liquid to transfer force or motion.

hydrocarbon (HC) Any compound composed of hydrogen and carbon. A product of incomplete combustion of hydrocarbon fuel with air.

hydrometer An instrument that determines the density of a liquid.

I

identify To establish the identity of a vehicle or component prior to service; to determine the nature or degree of a problem.

GLOSSARY

idle air control (IAC) valve A reversible DC motor in the throttle body assembly that controls airflow to the engine while the engine is at idle.

idler arm In parallelogram steering linkage, a link that supports the tie rod and transmits steering motion to both wheels through the tie-rod ends.

ignition coil A device that transforms low voltage from the battery into a high voltage capable of producing an ignition spark.

ignition module In an electronic ignition system, the electronic control unit that opens and closes the primary circuit; may be a separate unit or a function of the powertrain control module (PCM).

ignition reserve voltage The difference between the voltage required to create a spark at the spark plugs, and the voltage required to maintain the spark.

ignition switch The key-operated main power switch that opens and closes the circuit that supplies current to the ignition and other electrical systems.

ignition system In the spark-ignition engine, the system that furnishes high-voltage sparks to the cylinders, at the proper time, to fire the compressed air/fuel mixture.

impact sensor A switch that is designed to close when an impact occurs that is severe enough to warrant air bag deployment.

incandescent bulb A bulb that uses a tungsten filament placed in a vacuum inside a glass bulb. Used for side marker, license plate, and interior lights.

included angle The camber angle plus the steering-axis inclination (SAI) angle.

incremental adjuster The drum brake self-adjuster that moves the shoes outward whenever the gap is large enough to turn adjusting screw.

indexing mark An indicator that produces a unique signal that tells the powertrain control module (PCM) when the number-one piston is nearing top dead center (TDC).

inertia switch A switch that disables the fuel pump in the event of an accident. This action prevents pressurized fuel from spraying from a broken fuel line.

inflation pressure The measurement of the compressed air in a tire, expressed in pounds per square inch (psi) or kilopascals (kPa).

inflator module The module that contains the air bag, ignitor, solid propellant, and cover.

injector armature The part of the fuel injector that opens and closes the pintle valve.

inner bearing Works with the outer bearing to support the wheel evenly during operation of the vehicle.

inner tie rod end tool A tool that resembles pliers or a crescent wrench. It grips the tie rod, allowing rotation of the tie rod at its adjustment connection.

input piston A type of piston in the input cylinder. Also called apply piston.

install To place a component in its proper position in a system.

inspection/maintenance (I/M) ready status A record that shows a vehicle's on-board diagnostics have been run.

insulator A nonconducting material or shield covering an electrical conductor.

intake air temperature (IAT) sensor A sensor that measures the ambient air temperature.

intake manifold A set of tubes, or casting with several passages, through which air or air/fuel mixture flows from the throttle valves to the intake ports in the cylinder head.

intake stroke The piston stroke (from top dead center to bottom dead center) immediately following the exhaust stroke, during which the intake valve opens and the cylinder fills with air/fuel mixture from the intake manifold.

intake valve Valve that controls the admission of the air/fuel mixture into the cylinder of an engine.

integral braking system A system in which the brake booster, master cylinder, pump, accumulator, and pressure modulator are combined as a single unit.

integral master cylinder A one-piece, cast-iron component with a dual reservoir.

GLOSSARY

integral parking brake A parking brake that uses the same brake shoes or pads as the service brakes.

integrated circuit (IC) Many very small solid-state devices capable of performing as a complete electronic circuit, usually manufactured as a chip.

interactive testing Tests that involve activating the actuator being tested.

interference nut A self-locking nut that will not loosen.

internal combustion engine Engine in which energy is provided by combustion within a working chamber, causing direct mechanical displacement of a piston, rotor, turbine, or other mechanical element.

interrupter ring A rotating mechanical device consisting of blades or shutters and windows. Used for restricting light or a magnetic field, as in electronic ignitions. Also called a reluctor or trigger wheel.

ISO (International Standards Organization) flare A type of replacement brake line where the end looks like a bubble.

isolator bushing Mounts the anti-sway bar to the frame. Allows the bar to pivot, while reducing the transfer of road noise.

J

jam nut A nut that is added second and is tightened against the adjusting nut to keep it from moving.

jobber A person who buys automotive parts for resale. A jobber can be either a wholesaler who sells directly to automotive service businesses or a retailer who sells to individuals who service their own vehicles.

jounce Bump travel of a wheel suspension. Also called compression.

jump starting The process of starting the engine in one vehicle by connecting it to the battery in another vehicle.

jumper wire A short piece of wire used as a temporary connection between two points in a circuit.

K

Karman-vortex path sensor A sensor that measures airflow by measuring air turbulence.

key-off load A device that draws current even when all switches are turned off. Examples are computer and radio memory circuits. Also called parasitic drain.

kinetic energy The energy of motion.

kinetic friction The resistance between objects that are in contact and in relative motion.

knock sensor (KS) A sensor that detects the onset of detonation in an engine.

L

lab scope A tool used to observe and measure voltage and frequency. Also known as a digital storage oscilloscope (DSO).

lateral acceleration sensor In anti-lock brake systems (ABS), device that senses hard cornering during braking.

leading-trailing drum brakes Brakes in which the action of one shoe does not affect the other shoe.

leaf spring A spring that consists of single or multiple-spring steel bands (leaves).

lighting system The system that includes headlights and taillights, directional signals, brake warning lights, interior convenience and courtesy lights, and instrument control-panel information and warning lights.

liquid crystal display (LCD) A display panel that is placed in front of an incandescent or halogen lightbulb.

liquified petroleum gas (LPG) An automotive alternative fuel consisting mainly of propane.

listen To use audible clues in the diagnostic process; to hear the customer's description of a particular problem.

live axle The shaft through which power travels from drive axle gears to driving wheels.

load index Tire designation, with a letter (A, B, C, etc.), used to identify a given-size tire with its load and inflation limits. Also called load range.

load test A test that measures terminal voltage while the battery is supplying a large current for 15 seconds. Also called high-rate discharge test.

load-sensing proportioning valves Valves that adjust the hydraulic pressure to the rear brakes according to changes in the load. The valves sense vehicle load by measuring the distance between the truck bed and the axle.

logic probe A device that is used to safely test electronic circuits. It uses light-emitting diodes (LEDs) to display test results. Also called electronic circuit tester.

low voltage bias When return signal voltage from an oxygen sensor varies over a voltage range that is too low.

low-tire pressure warning system A system using a tire-pressure sensor that sends a radio signal to a receiver module when tire pressure falls below a preset level.

lubricate To employ the correct procedures and materials in performing the prescribed lubrication service.

lubricating system In the engine, the system that supplies engine parts with lubricating oil to prevent contact between any two moving metal surfaces.

lug nut A wheel fastener.

M

MacPherson strut A strut that combines a coil spring and a shock absorber into a single assembly using only a beam-type lower control arm.

magnetic field The space around a magnet that is filled by invisible lines of force.

magnetic pulse generator (See magnetic pulse sensor.)

magnetic pulse sensor A sensor that uses a permanent magnet to signal the powertrain control module (PCM) to begin the ignition sequence.

main bearings In the engine, the bearings that support the crankshaft.

malfunction indicator lamp (MIL) A light that warns the driver of a problem in the systems monitored by the powertrain control module (PCM). Also referred to as the check engine light or the service engine soon light.

manifold absolute pressure (MAP) sensor Sensor that monitors the engine's intake manifold pressure.

manifold absolute pressure test A test that checks vacuum readings through a range of engine operations. Also called the engine vacuum test.

manometer A device for measuring a vacuum.

manufacturer-specific A trait or code that is not common throughout the industry, but rather is used only by a particular manufacturer.

mass airflow (MAF) sensor A sensor that measures the amount of air entering an engine.

master cylinder The liquid-filled cylinder in the hydraulic braking system where hydraulic pressure is developed when the driver depresses a foot pedal.

master technician An automotive technician who is ASE-certified in all eight auto/light truck areas.

material safety data sheet (MSDS) An information sheet that identifies chemicals and their components. It also lists possible health and safety problems and describes safe use of the chemical.

maximum coil output The amount of voltage available in the primary ignition coil. Also called available voltage.

maximum inflation pressure Tire marking that indicates the maximum pressure that the tire can handle.

mechanical generator A generator driven by the engine which converts mechanical energy to electrical energy.

memory calibration (MEMCAL) unit Computer chips that must be replaced as a unit. Also called the PROM.

metering valve A valve that controls, or delays, the flow of brake fluid to the front brakes.

GLOSSARY

metric system A system of measurement that uses meters, liters, and grams. Also called the System of International Units (SI).

micrometer A hand-held precision measuring instrument used to make linear measurements.

monitor A procedure used by the powertrain control module (PCM) to test relevant on-board systems.

monitored circuit A circuit that provides a direct input to the powertrain control module (PCM).

monoleaf spring A spring that consists of a single steel band (leaf).

mount To attach or place a tool or component in the proper position.

muffler In the engine exhaust system, a device through which the exhaust gases must pass to reduce noise.

multi-leaf spring A spring that consists of multiple-spring steel bands (leaves).

multimeter A tester that measures voltage, current, and resistance. Also called volt-ohm-meter (VOM).

multiport fuel injection (MFI) A system which has one fuel injector for each cylinder.

N

National Automotive Technicians Education Foundation (NATEF) An organization that certifies automotive training programs.

National Institute for Occupational Safety and Health (NIOSH) An organization that tests and certifies safety equipment.

negative camber angle A wheel condition in which the top of the wheel tilts in.

negative caster angle A wheel condition in which the upper ball joint is forward of the lower ball joint.

negative temperature coefficient (NTC) sensor A thermistor in which the resistance increases as the temperature decreases.

neutron An atomic particle with no electrical charge.

nitrogen oxide A chemical compound formed when nitrogen and oxygen bond under high heat.

non-adjustable bearing A wheel bearing that is permanently lubricated.

nondirectional tread A tire that can rotate in either direction, allowing it to be placed at any position on the vehicle.

nonintegral braking system A system that uses traditional brake system components, such as the master cylinder and brake booster.

nonmonitored circuit A circuit that is not directly monitored by the powertrain control module (PCM).

nonuniform code A manufacturer-specific diagnostic trouble code (DTC), not common throughout the industry.

nonvented rotor A solid metal brake rotor with no air vents to aid in heat removal. Used mainly on smaller vehicles where brake overheating is not a serious problem.

nozzle The opening, or jet, through which fuel or air passes as it is discharged.

O

Occupational Safety and Health Administration (OSHA) The organization created to enforce safe and healthful working conditions.

octane number An indicator of the antiknock quality of a gasoline. The higher the octane number, the more resistant the gasoline is to detonation.

Ohm's law Mathematical relationship between voltage, resistance, and current in an electrical circuit.

ohmmeter An instrument used to measure resistance in ohms.

oil pan The detachable lower part of the engine that encloses the crankcase and acts as an oil reservoir.

oil pump A device that circulates oil from the oil pan, through the engine, and back to the oil pan.

oil-control ring A ring fitted onto the piston that scrapes excess oil from the cylinder wall and returns it to the crankcase.

on-board diagnostic system (OBD) Computer-controlled test routines that monitor and control vehicle systems.

GLOSSARY

on-car brake lathe A device used to perform brake rotor machining on the vehicle.

one-channel ABS An anti-lock brake system (ABS) that is only present on the rear wheels.

one-shot adjuster The drum-brake self-adjuster that makes a single adjustment once the clearance between the lining and drum reaches a predetermined gap.

open circuit An incomplete electrical circuit.

open-circuit voltage (OCV) test A test that checks the state of charge of a battery.

optical sensor A sensor that uses LEDs and photodiodes and a reluctor wheel to create a sensor signal.

oscilloscope A test device that displays voltage changes within a certain time period.

outer bearing Works with the inner bearing to support the wheel evenly during vehicle operation.

output piston In a hydraulic braking system, the piston that moves in direct relationship to the motion of the input piston.

overhead-camshaft engine An engine in which the camshaft is mounted in the cylinder head instead of in the cylinder block.

overrunning clutch A device that prevents the engine from driving the starter. This roller or sprag clutch transmits torque in only one direction.

oxygen sensor A sensor that detects the amount of oxygen in the exhaust gases.

P

palladium A metallic element used as a catalyst in a catalytic-converter to further the reduction of exhaust pollutants.

panhard rod A transverse rod pivoted at one end to the chassis or vehicle shell and at the other to a beam axle, which it constrains in lateral movement. Also called a track bar.

parallel circuit The electric circuit formed when two or more electric devices have their terminals connected together, positive to positive and negative to negative, so that each may operate independently.

parallelogram linkage A steering-linkage system used with the recirculating-ball system.

parameter identification data Serial data values that are stored the instant an emission-related diagnostic trouble code (DTC) is set and enters the diagnostic memory.

parasitic drain Current flow that is present even when all switches in a vehicle are turned off. Examples are computer and radio memory circuits. Also called key-off load.

park switch A switch that supplies power to the wiper motor after the windshield wiper switch is turned off.

park/neutral position switch A switch that prevents the starter relay or solenoid from closing when the vehicle is in gear.

parking brake equalizer A device that balances the braking forces so both rear brakes are applied evenly.

parking brakes The brakes used to keep a parked vehicle from moving. They are usually on the rear wheels and are mechanically operated.

passive test A test that checks the performance of a vehicle system or component during normal operation.

pathogen Any microorganism or virus that causes a disease.

period (p) The time required for one complete cycle or operation to complete.

perform To accomplish a procedure in accordance with established methods.

permanent magnet A ceramic magnetic material that does not require current flow through a field coil to create its magnetic field.

permanent magnet (PM) sensor A small AC-voltage generator that uses a wire coil, a pole piece, a permanent magnet, and a rotating trigger wheel (reluctor wheel) to produce an analog voltage signal. Also called a magnetic pulse generator.

personal protective equipment (PPE) Equipment worn by workers to protect against hazards in the environment.

GLOSSARY

photocell A device that uses light energy to create current flow.

pickup coil In an electronic ignition system, the coil in which the timing signal voltage is induced by the moving teeth on a reluctor or armature.

piezoelectric (pee·ay·zo·uh·LEK·trik) converter A device that uses physical pressure to create current flow.

pin out test A test using a digital volt-ohm-meter (DVOM), to check the individual pins, or outlets, of a break-out box (BOB).

pinion gear Located on the starter, it is the smaller of two meshing gears.

pintle seat Part of the electronic fuel injector against which the pintle valve seats.

pintle valve The valve that controls the fuel flow in a fuel injector.

piston A cylindrical plug that fits inside a cylinder. It receives and transmits motion as a result of pressure changes applied to it.

piston clearance The small gap between a piston and the cylinder wall.

piston pin The cylindrical or tubular piece that attaches the piston to the connecting rod.

piston rings Rings fitted into grooves in the piston. There are two types: compression rings for sealing the compression pressure in the combustion chamber, and oil rings to scrape excessive oil off the cylinder wall.

pitch The length from a point on a fastener thread to a corresponding point on the next thread. It is calculated by dividing one inch by the number of threads per inch.

pitman arm On a recirculating-ball steering gear, the arm that connects the steering gear output-shaft to the steering linkage. As the shaft turns, the pitman arm swings back and forth, transferring output-shaft movement into movement of the steering linkage.

pitman arm puller A tool that disconnects the pitman arm from the steering gear output shaft.

planetary gears A gear set used in automatic transmissions and transfer cases, which use a central sun gear surrounded by two or more planet pinions. The planet pinions are meshed with the ring gear.

plate In a battery, a flat rectangular sheet of spongy lead. Sulfuric acid in the electrolyte chemically reacts with the lead to produce an electric current.

ply A layer of cord, fiberglass, steel, or other material used to create a tire carcass.

pneumatic motors Motors that are powered by compressed air.

polarity The quality of an electronic component or circuit that determines the direction of current flow.

pole piece A soft iron piece that is assembled over the end of a field coil.

pole shoe Part of the field coil starter. A soft iron core wrapped with a coil of copper wire.

positive camber angle The wheel condition in which the top of the wheel tilts in.

positive caster angle The wheel condition in which the upper ball joint is behind the lower ball joint.

positive crankcase ventilation (PCV) system Prevents blowby from escaping into the atmosphere by using vacuum to draw blowby gases from the crankcase into the intake manifold.

positive temperature coefficient (PTC) resistor A solid state device used as a circuit breaker that opens a circuit when an over-current condition occurs.

positive temperature coefficient (PTC) sensor A thermistor in which the resistance increases as the temperature increases.

potentiometer (puh·TENT·shee·ah·muh·ter) A three-wire variable resistor used to monitor movement or control output.

power brake booster A device that uses hydraulic pressure or a vacuum to supply additional energy, boosting brake application pressures.

power brakes A service brake system that uses either a vacuum and atmospheric pressure, or hydraulic pressure, to provide most of the force required for braking.

power piston In a power brake booster, the interface between the front and rear pushrods. It is attached to or suspended from the diaphragm.

GLOSSARY

power plant The engine or power source that produces the power to move the vehicle.

power steering A system that uses hydraulic or electric power to help the driver apply steering force. Also called power-assisted steering.

power steering fluid reservoir Container that holds the reserve of power steering fluid.

power steering pump A hydraulic pump that provides an assist to the steering system.

power stroke The piston movement (from top dead center to bottom dead center) immediately following the compression stroke, during which the valves are closed and the fuel burns. The expanding compressed gas forces the piston down the cylinder, transmitting power to rotate the crankshaft.

power-assisted steering (See power steering.)

power-assist unit A device that uses hydraulic pressure or a vacuum to supply additional energy, boosting brake application pressures.

powertrain control module (PCM) An electronic module or computer that receives input from various engine and powertrain sensors, and responds by sending output signals to various engine and powertrain actuators.

powertrain system The system that includes the engine transmission, drive train, and axles that carry the power to the drive wheels.

predelivery service The preparation of a vehicle provided by a dealer before delivery to the customer.

pressure bleeding The most commonly used brake bleeding procedure. The pressure bleeder provides the pressure normally provided by the master cylinder.

pressure differential valve The valve that senses the pressure in each branch of a hydraulic circuit.

pressure test To use air or fluid pressure to determine the condition or operation of a component or system.

primary runner The intake manifold runner tuned for low speed in a variable induction system.

profile ratio The relationship between a tire's section height and section width. Also called aspect ratio.

programmable read-only memory (PROM) A computer memory chip that, once programmed, can only be read.

proportioning valve A valve that reduces the amount of braking force at the rear wheels on front disc and rear drum brake systems.

proton An atomic particle with a positive electrical charge.

pulse width In fuel injection systems, the duration during which a fuel injector supplies fuel.

pulse width modulation (PWM) A digital signal with a variable duty cycle. The duty cycle stays high or low for varying lengths of time. The frequency of PWM does not change, only the length of the duty cycle.

purge valve In evaporative-control systems, a valve used on some charcoal canisters to limit the flow of fuel vapor to the intake manifold.

pyrometer An instrument that checks temperature. A "contact" pyrometer must touch the heat source to check temperature. A "non-contact," or "infrared," pyrometer measures the heat radiated from the heat source.

R

rack A straight length of toothed gearing.

rack-and-pinion steering gear Steering gear in which a pinion on the end of the steering shaft meshes with a rack of gear teeth on the major cross member of the steering linkage.

radial tire A tire featuring plies that run bead-to-bead, parallel to each other, with stabilizer belts under the tread.

radiator In the engine cooling system, the heat exchanger that removes heat from coolant passing through it, receives hot coolant from the engine, and returns the coolant to the engine at a lower temperature.

random access memory (RAM) A volatile, or erasable, memory that temporarily stores information such as diagnostic trouble codes.

GLOSSARY

ready To prepare a system or component for service, installation, or operation.

rear-wheel anti-lock systems (RWAL) Anti-lock brake systems (ABS) that control both rear wheels at the same time.

rebound Extension of suspension travel beyond static condition.

recall When a manufacturer asks vehicle owners to return their vehicles to the dealer for inspection and possibly repair. A recall is frequently safety related.

reciprocating engine An engine in which the pistons move up and down inside the cylinders.

recirculating-ball steering gear An assembly that uses a series of recirculating balls on a worm gear to transfer steering wheel movement to road wheel movement.

rectification The process in which AC voltage is changed to DC by the diodes.

reduce-pressure position In anti-lock brake systems (ABS), when the ABS control module reduces hydraulic pressure. This occurs when the wheel(s) slow down too fast or lock up.

reference voltage Five volts from the powertrain control module (PCM) to power the sensor circuits.

relay An electrical device that opens or closes a circuit in response to a voltage signal.

relay rod The link in a steering system that connects the pitman arm to the steering arm. Also referred to as the center link or drag link.

reluctor wheel A toothed metal ring. The teeth pass through a sensor's magnetic field, causing a timing signal that is sent to the Powertrain Control Module (PCM) or ignition module.

remove To disconnect and separate a component from a system.

repair To restore a malfunctioning component or system to operating condition.

replace To exchange an unserviceable component with a new or rebuilt component; to reinstall a component.

reserve capacity The measure of how many minutes a battery can supply a load of 25 amps at 80°F (27°C).

residual pressure check valve A valve that maintains a residual pressure of about 6–18 psi in the brake lines.

resistance The opposition to a flow of current through a circuit or electrical device; measured in ohms.

resistor Any material that has limited electrical conductivity.

resistor spark plug A spark plug with resistance built into the center electrode to help suppress electromagnetic interference (EMI).

resonator An acoustic chamber with a specific resonant frequency used in induction and exhaust systems.

return-signal Altered reference voltage supplied to the powertrain control module (PCM) from sensors.

rheostat A variable resistor.

rim diameter The measurement from a point on the inside bead seat to the point on the inside bead seat directly across the diameter of the wheel. It only refers to the bead seat diameter, not the total outside diameter.

rim width The measurement from the inside bead seat wall to the opposite bead seat wall.

ring gear Mounted on the flywheel and the drive plate, it is a large gear that meshes with and is driven by a pinion gear.

rivet A metal pin used to fasten parts together.

road crown The condition in which the road is slightly higher in the center than at the sides. The crown allows water to run off.

rocker arm A pivoted lever that transfers cam or pushrod motion to the valve stem.

rod bolt Bolts that attach the connecting rod to the rod cap, which connects to the crankshaft.

GLOSSARY

rotor (brake) A disc-shaped device made from cast iron or sintered iron. They have flat friction surfaces machined onto both sides against which the brake pads are pressed. They are attached to the spindles or hubs and rotate on the wheel bearings.

rotor (generator) The part of the generator that rotates inside the generator housing. It creates the rotating magnetic field of the generator.

run-flat tire A tire designed to operate with little or no inflation pressure. It has a reinforced sidewall and bead design intended to support the vehicle if a cut of puncture causes loss of inflation pressure. Also called a zero-pressure tire.

S

safety The practice of protecting yourself and others from danger and possible injury. It is the first priority of every employee in the automotive workplace.

safety system The system that provides added passenger safety and includes devices such as air bags, anti-lock brakes, side-door panel steel rails, and shock-absorbing bumpers.

scan tester A device used to read a vehicle's data stream and trouble codes. Also called a scanner or scan tool.

scientific method A logical approach to problem solving in which researchers make an observation, collect information, form a hypothesis, perform experiments, and analyze the results.

scores Grooves or deep scratches on a smooth surface. Scoring is caused by debris or metal-to-metal contact.

screw thread A fastener that has a spiral ridge, or screw thread, on its surface.

scrub radius The distance between the steering axis and the centerline of the tire tread contact area. Also called steering offset.

secondary air injection system A system in which air is pumped to the cylinder-head exhaust ports, exhaust manifold, or catalytic converter to promote the chemical reactions that reduce exhaust-gas emissions.

secondary cells A battery that can be recharged, such as an automotive battery.

secondary runner The intake manifold runner tuned for high speed in a variable induction system.

select To choose the correct part or setting during assembly or adjustment.

self-adjuster A device on drum brakes that compensates for lining wear by automatically adjusting the shoe-to-drum clearance.

self-powered test light A device that checks for a complete circuit path, or continuity from one point to another. Also called a continuity tester.

self-tapping screw A screw that cuts its own threads when turned into drilled holes.

semiconductor Any material that acts as an insulator under some conditions and as a conductor under other conditions.

sensor A device that monitors or measures operating conditions. It generates or modifies an electrical signal based on the condition it is monitoring.

sequential multiport fuel injection (SFI) system A system in which the fuel injectors are pulsed individually in the firing order of the engine's cylinders.

serial data stream Information displayed as voltage values or as actual readings, such as degrees of temperature.

series circuit An electrical circuit in which the devices are connected end to end, positive terminal to negative terminal. The same current flows through all the devices in the circuit.

service To perform a specified procedure when called for in the owner's or service manual.

service brakes The primary braking system used to slow or to stop a vehicle.

service history A written service record for a vehicle.

Service Technicians Society (STS) An association for automotive and transportation professionals that provides a forum to exchange technical information and to share industry trends.

GLOSSARY

setback The difference in length between the wheelbase on one side of the vehicle and the wheelbase on the other side.

setscrew A fastener used to secure a collar or gear on a shaft.

shear When opposite forces act in opposite directions on opposite sides of an object, causing the object to twist.

shock absorber A device placed at each vehicle wheel to regulate suspension rebound and compression.

short circuit When two or more electrical conductors touch each other where no connection is intended.

short-arm/long-arm (SLA) system An A-type control arm in which the lower arm is longer than the upper arm.

shutter (See interrupter ring.)

signature pulse The signal given off by the indexing mark. Also called synchronizing pulse.

simple machine A device used to increase or multiply force.

sine wave A voltage pattern created by an analog signal.

single-line fuel system A system without a fuel return line.

sliding caliper disc brakes Disc brakes that use a caliper that is held in place by a retainer, a spring, and a bolt.

slip ring (See commutator.)

snap ring A fastener used to secure or locate the ends of a shaft.

snap-throttle vacuum test A test that shows the condition of the pistons and the piston rings as the engine speed decreases to idle.

solenoid An electromechanical device which, when connected to an electrical source such as a battery, produces a mechanical movement.

solid-state device A device, such as a diode, resistor, or transistor, that has no moving parts.

spark distribution system The system that delivers the spark to the correct cylinder in the correct firing order.

spark duration The length of time a spark lasts when it jumps the spark plug gap.

spark knock In the combustion chamber of a spark-ignition engine, an uncontrolled second explosion (after the spark occurs at the spark plug), with spontaneous combustion of the remaining compressed air/fuel mixture, resulting in a pinging sound. Also called detonation.

spark plug The assembly, which includes an electrode, an insulator, and a shell that provides a spark in the engine cylinder.

specific gravity The weight per unit volume of a substance as compared with the weight per unit volume of water.

speed symbol The sidewall inscription that states the speed rating, the maximum allowable operation speed of the tire.

split point The point during hard braking when the proportioning valve reduces the amount of pressure to the rear brakes.

spontaneous combustion Fire caused by chemical reactions with no spark.

spool valve A valve that shuttles back and forth to open and close ports for the pressurized power steering fluid. It resembles a spool that carries sewing thread.

spray pattern The way a liquid or fuel is released from a valve nozzle.

spring A device that changes shape under stress or force but returns to its original shape when the stress or force is removed. In the suspension system, the spring absorbs road shocks by flexing or twisting.

spring-assisted shock absorber Shock absorber that uses a coil spring installed over the shock absorber body. This provides additional load-carrying assistance and serves as a hydraulic damper for the spring action.

sprung weight Vehicle weight, including the engine, body and frame, transmission, and cargo, that is supported by the springs.

GLOSSARY

square wave signal A signal that is either on or off. The voltage is either high or low, with no values in between. Also called digital signal.

star wheel An adjusting nut with indexing arms for positive and accurate rotation, as on a brake or clutch.

starter A high-torque electric motor that cranks the engine.

starter relay An electrical device that opens or closes the high-voltage circuit to the starter.

starter solenoid An electromechanical device that moves the pinion gear into mesh with the ring gear on the engine flywheel or drive plate.

starting system The system that converts electrical energy from the battery to mechanical energy from the starter. This energy is used to crank the engine.

static angle The initial setting for camber.

static friction The resistance between objects that are in contact but at rest.

stator frame The iron core housing for the windings, or phases, of a stator.

stator winding The stationary coils or windings in a generator.

steering arm The arm attached to the steering knuckle, and to which the tie rod end attaches, that turns the knuckle and wheel in and out for steering.

steering axis inclination (SAI) The difference between a true vertical line drawn through the center of the wheel and an imaginary line drawn through the upper and lower ball joints.

steering column The housing that contains and supports the steering shaft.

steering column couplers Mechanisms that allow the steering shaft connections to pivot at various angles during steering operation. They serve as a pivot point between the upper and lower shafts and at the steering gear connection.

steering gear That part of the steering system located at the lower end of the steering shaft; changes the rotary motion of the steering wheel into linear motion of the front wheels for steering.

steering linkage The part of the steering system that connects the steering gear to the steering arms.

steering offset The distance between the steering axis and the centerline of the tire tread contact area. Also called scrub radius.

steering ratio The number of degrees that the steering wheel must turn to turn the road wheels 1°.

steering shaft The shaft extending from the steering wheel to the steering gear; usually enclosed by the steering column.

steering system The system that enables the driver to control the direction of vehicle travel.

steering wheel The wheel which the driver uses to control the direction of the vehicle.

step-bore cylinder A cylinder with two different internal diameters.

stepper motor A small DC motor. In an electronic cruise control system, the device that opens the throttle to maintain the desired cruising speed. It replaces the vacuum diaphragm.

stoichiometric (stoy•kee•oh•MEH•trik) ratio The ratio (14.7:1) that provides the most efficient combustion, giving the chemically correct mixture of air to fuel.

strut A bar that connects the lower control arm to the vehicle body or frame; used with a beam-type lower control arm which has only one point of attachment to the body or frame. Also called a brake-reaction rod.

sulfation A condition resulting when the lead sulfate within a battery remains on the plate.

supercharger A pump that precompresses the air or air/fuel mixture before it enters the engine cylinder.

supplemental restraint system (SRS) An air bag system.

suspension system The system that absorbs the shocks of the tires and wheels from bumps and holes in the road.

symmetric tread pattern A tire tread that has the same pattern on each side of the tread.

synchronizing pulse The signal given off by the indexing mark. Also called the signature pulse.

GLOSSARY

system of international units (SI) A system of measurement that uses meters, liters, and grams. Also called the metric system.

T

tail pipe The rearmost pipe of an exhaust system.

telescope Adjustment that allows the driver to extend or compress the length of the steering column.

tension Effort that elongates or stretches a material.

terminal 1) An electrical connection point. 2) A mechanical stamping made of tin-coated brass or steel.

terminal voltage The voltage measured across the positive and negative terminals of a battery.

test To verify a vehicle's condition through the use of meters, gauges, or instruments.

thermistor A small solid-state device used for temperature sensing.

thermo-time switch A switch that limits the operating time of a component, based on time and/or temperature.

thread reach On a spark plug, the distance from the gasket seat to the end of the threads.

throttle body The air control device for fuel-injected engines. The amount of air that enters is primarily controlled by the position of the throttle valve that opens or closes as the driver presses and releases the accelerator pedal.

throttle body fuel injection (TBI) A gasoline fuel-injection system that sprays fuel under pressure into the intake air passing through the throttle body on the intake manifold.

throttle position (TP) sensor A variable-resistance sensor on the throttle body that continuously sends a varying voltage signal, which is proportional to the throttle valve position, to the electronic control module.

thrust angle The relationship between the position of the rear axle and the centerline of the vehicle.

thrust line A line drawn perpendicular to the rear axle, pointing toward the front of the vehicle, and in the same direction as the rear wheels.

tie rod An adjustable-length rod that, as the steering wheel turns, transfers the steering force and direction from the steering gear or linkage to the steering arm.

tie rod puller A tool that allows you to separate the tie rod end's tapered stud from the steering arm.

tilt mechanism The device that allows the driver to adjust the steering wheel or steering column up or down.

timing light A bright stroboscopic light that is used to set ignition timing. It is usually connected to the number-one spark plug wire. Current flow causes the timing light to flash and show the position of a timing mark on the vibration damper in relation to a timing pointer mounted on the front of the engine.

tire The casing-and-tread assembly that is mounted on a vehicle wheel and usually filled with compressed air to transmit braking and motion forces to the road.

tire changer (See tire mounting machine.)

tire mounting machine A manual or pneumatic machine designed to mount or dismount a tire.

tire rotation The relocating of each tire and wheel assembly from one wheel position to another.

tire tread The part of a tire that contacts the road, providing traction.

tire-pressure sensor (See low-tire pressure warning system.)

tire-size designation The inscription on a tire's sidewall that indicates section width, aspect ratio, and rim diameter.

titania oxygen sensor A sensor that changes resistance in response to the amount of oxygen in the exhaust.

toe angle The measurement by which the front of the wheel points inward or outward, compared to a true straight-ahead position, measured in degrees. Also called toe.

toe-in The condition in which the wheels are pointed in and closer together at the front than they are at the rear.

toe-out The condition in which the wheels are pointed out and are farther apart at the front than they are at the rear.

GLOSSARY

toe-out on turns The difference in the angles of the front wheels in a turn. Also called turning radius.

tone wheel In an anti-lock brake system (ABS), a toothed ring that rotates with the wheel. The teeth pass through the wheel speed sensor's magnetic field, causing a voltage signal that is sent to the ABS system.

toothed rack A straight length of toothed gearing.

top dead center (TDC) The piston position when the piston has reached the upper limit of its travel in the cylinder and the centerline of the connecting rod is parallel to the cylinder walls.

torque Twisting or turning force, measured in pound-feet or newton-meters.

torque converter In an automatic transmission, a fluid coupling that transmits engine torque to the transmission.

torsion bar A steel rod that twists to provide spring action.

track The distance between the center of the right tire's tread and the center of the left tire's tread.

traction control system A system used with an anti-lock brake system (ABS) to prevent unwanted wheelspin during acceleration.

trailing arm Suspension linkage supporting wheel assembly aft of a transverse pivot axis.

transfer cable A cable in a parking brake system that runs between the rear wheels or from the equalizer to each rear wheel.

transient voltage An unexpected voltage spike that could damage sensitive electrical and electronic components.

transistor A semiconductor device used to control the flow of an electric current; can act as an electric switch, amplifier, or detector.

tread wear indicator (TWI) Raised ribs molded into tire tread. When the ribs are flush with the tread surface, the tread has worn enough to require replacement of the tire. Also called wear indicator bar.

treadwear indication number A number within the Uniform Tire Quality Grading (UTQG) that gives a rough indication of how fast the tread will wear. The lower the number, the faster the tread will wear.

trip A key-on, run, key-off cycle in which all of the enabling criteria for a given diagnostic monitor are met.

tube tire A tire with an inner tube inside the tire. Today only some trucks and motorcycles use tube tires.

tubeless tire A nonporous tire that contains air without the need for an inner tube.

tuning venturi A device that reduces noise inside the air filter housing.

turbocharger A centrifugal supercharger or air pump, driven by the engine exhaust gas, that forces an additional amount of air or air/fuel mixture into the intake manifold.

turning radius The difference in the angles of the front wheels in a turn. Also called toe-out on turns.

U

understeer A condition that results when the front tires lose adhesion during cornering. The vehicle wants to push, instead of turning.

Uniform Tire Quality Grading (UTQG) Tire marking that indicates subjective tire performance ratings regarding tread wear, wet braking traction, and resistance to temperature.

United States Customary (USC) system A system of measurement that uses inches, feet, miles, pints, gallons, and pounds. Also called standard.

universal joint A connecting joint that allows power to be transmitted between two rotating shafts that are operating at an angle to each other.

unsprung weight Vehicle weight, including the wheels, tires, brakes, drive axles, and lower control arms, that is not supported by the springs.

GLOSSARY

V

vacuum A measurement of air pressure that is less than atmospheric pressure.

vacuum bleeding A procedure using a hand-held vacuum pump used to draw fluid and trapped air from the brake system at the bleeder screws.

vacuum booster A device that uses vacuum to supply additional energy to boost brake application pressures.

vacuum gauge A device that measures vacuum in inches of mercury (Hg).

vacuum test A test used to determine the integrity and operation of a vacuum-operated component and/or system.

vacuum tube fluorescent (VTF) display A display that produces light in a manner similar to a television tube.

vacuum valve A value whose action is controlled by vacuum. In a power/brake booster, the first port of the control valve. The valve stays open, allowing vacuum to enter the rear chamber booster assembly.

valve overlap The number of degrees of crankshaft rotation during which the intake and exhaust valves are open at the same time.

valve seat The surface against which a valve comes to rest to provide a seal against cylinder leakage.

valve spring The coil spring attached to each valve that closes the valve after the cam lobe has rotated past the valve-open position.

valve stem Narrow cylindrical rod to which valve disc is attached.

valve train A series of parts that open and close the valves by transferring cam-lobe movement to the valves.

vane airflow sensor A sensor with a movable vane connected to a variable resistance; calibrated to measure the amount of air flowing into the engine.

vapor lock The condition in which the fuel turns to vapor in the fuel system and blocks the normal fuel flow through the fuel system.

vapor pressure The measure of pressure developed by a liquid in a closed container.

vapor recovery system The system that transfers fuel vapors from the fuel tank to an evaporative control system (EVAP) canister.

variable induction system An air induction system with two runners for each cylinder, one tuned for low speed and one tuned for high speed. The combination of runners improves power and throttle response.

variable-assist power steering Power steering that uses electronic controls to determine how much power assist the steering needs.

variable-ratio steering Steering in which the steering ratio changes as the steering wheel turns left or right from its straight-ahead position.

vehicle identification number (VIN) A serial number unique to each vehicle. The number indicates the year, assembly plant, and country in which the vehicle was made; the vehicle make and type; the passenger safety system; the engine type; and the line, series, and body style.

vehicle speed sensor (VSS) A small AC-signal generator driven by a shaft in the transmission, that generates a voltage pulse proportional to vehicle speed.

vented rotor A disc brake rotor with fins in the space between its machined surfaces. These fins allow air to pass through the rotor to cool it.

volatility A measure of how easily a liquid (fuel) vaporizes.

verify To establish that a problem exists after hearing the customer's complaint and performing a preliminary diagnosis.

voltage A measurement of the pressure that causes electrical energy (current) to flow.

voltage drop The voltage produced as current passes through a resistance.

voltage drop test A test that checks for high resistance and voltage loss across a cable, component, or connection.

voltage regulator A device used to control the generator output voltage.

GLOSSARY

volt-ohm-meter (VOM) An analog tester that measures voltage, current, and resistance. Also called a multimeter.

volumetric efficiency A measure of how completely the engine cylinder fills with air/fuel mixture during the intake stroke.

W

warm-up cycle The period when the engine coolant temperature rises from ambient temperature to at least 160°F [70°C].

warranty work Vehicle repair work paid for by the vehicle manufacturer. The warranty covers repairs only for a specified period of time or mileage.

water jackets The space between the inner and outer shells of the cylinder block or cylinder head, through which coolant circulates.

wear indicator bar (See tread wear indicator (TWI).)

wheel A disc or spokes with a hub (revolving around an axle) at the center and a rim around the outside for mounting the tire.

wheel alignment A series of tests and adjustments to ensure that the wheels and tires are properly positioned on the vehicle and running with the specified relationship with the road and with each other.

wheel backspacing The distance from the wheel's mounting surface to the rear edge of the wheel rim.

wheel balancer A machine designed to locate and eliminate imbalance from an assembled tire and wheel.

wheel bearings Bearings that enable the tires and wheels at the end of an axle to rotate smoothly, freely, and safely at high speed.

wheel brake mechanism A part of the service brake located at each of the vehicle's wheels.

wheel cylinder In a drum brake system, the mechanism that brings brake pads to bear on disc or rotor by a clamping or pinching action.

wheel offset The location of the wheel's centerline as viewed from the front, relative to the location of the mounting face of the wheel hub.

wheel speed sensor A sensor that inputs speed to the ABS control module, located on each wheel.

wheel weight Small weights added to the rim of a wheel to balance the wheel.

wheel weight pliers A tool that grips and removes clip-on style wheel weights from the wheel rim.

wheelbase The distance between the center of the front wheel and the center of the rear wheel on the same side of the vehicle.

wide-open throttle (WOT) The maximum opening of the throttle.

wiring diagram A drawing that shows the wires, connectors, and load devices in an electrical circuit.

wishbone The name given to an A-type control arm.

Y

yaw sensing Sensing that a vehicle is sliding or spinning.

Z

Zener diode A diode that can conduct current flow in a reverse direction without being damaged.

zero camber angle The wheel condition in which the camber is true vertical.

zero caster angle The wheel condition in which the upper ball joint is directly over the lower ball joint.

zero toe The condition in which the wheels aim straight ahead and are parallel.

zero-pressure tire (See run-flat tire.)

zirconia oxygen sensor A sensor that produces a voltage signal based on the oxygen content in the exhaust gases.

INDEX

Page numbers in italics are references to illustrations and tables.

INDEX

INDEX

INDEX

INDEX

INDEX

INDEX

INDEX

INDEX

CREDITS

Cover Photo: Ron Kimball Photography, Inc.

AC Delco, xxiv

Altered Images Photography
    A. Greg Alter, xix, EL-82

American Honda Motor Company, viii, xxiii, xxv

AutoXray, xxv, HB-47

Bridgestone/Firestone, Inc., xxviii

DaimlerChrysler, i, iv, ix, xii, xviii, xix, xx, EL-104, EP-2, SS-1

Fluke Corporation, HB-1

Ford Motor Company, x, xiii, xvii, xviii, xxii, xxiv, xxv, xxvi, xxvi, xxvii

Fountain Industries, HB-51

FPG International
    Mark Green, xxiii

FPPF Chemical Company, HB-50

Jerome Gantner, iv, HB-2, EL-2

General Motors Corporation, vi, xi, xv, xvii, xviii, xix, xxi, xxvii
    Chevrolet Division, xiv, BR-100, EL-62
    Delco Moraine Division, xi
    Oldsmobile Division, xi, xxviii, BR-2, SS-60
    Pontiac-GMC Division, xiii, BR-14, EP-1

Jack Holtel, iv, v, vi, vii, viii, ix, x, xi, xv, xvi, xxi, xxii, xxvii, xxviii, HB-1, HB-18, HB-34, HB-37, HB-38, HB-39, HB-40, HB-41, HB-42, HB-43, HB-44, HB-45, HB-47, HB-48, HB-49, HB-50, HB-51, HB-52, HB-53, HB-57, HB-58, BR-1, BR-32, BR-52, EL-1, EL-26, EP-18, EP-32, EP-56, EP-110, EP-124, EP-136, SS-44

Hunter Engineering Company, v

David S. Hwang, HB-50

Lisle Products/Wesley Day Advertising, HB-45, HB-49

Mazda North American Operations, SS-2

Midtronics, xvi

New England Stock
    Frank Siteman, EP-92

Petosky Plastics, HB-51

Photo Researchers
    Bill Bachman, xxvi, xxix, SS-20
    David R. Frazier, EP-70

Robert Bosch GmbH, xxii

Sears, Roebuck & Company, vii, HB-38, HB-41, HB-42, HB-43, HB-45, HB-47, HB-48, HB-49, HB-50, HB-51, HB-53, HB-57

Snap-On Tools, xviii, HB-1, HB-41, HB-47, HB-50, HB-52

The Stock Market
    Gary D. Landsman, xvii, EL-46

Superstock, xii, BR-86

Terry Wild Studio, vii, xiii, xv, xx, xxi, xxiii, xxiv, HB-18, HB-34, HB-40, HB-43, HB-44, HB-45, HB-47, HB-48, HB-49, HB-52, HB-57, BR-74

Acknowledgments: Special thanks to Delaware County Technical School, Folcroft, Pennsylvania and Miami Valley Career Technical Center, Clayton, Ohio.

Image sources have granted permission for reproduction of artwork and photos, but in no way imply endorsement of the material in the textbook, the course itself, or the techniques taught.